Transportation Electrification

IEEE Press
445 Hoes Lane
Piscataway, NJ 08854

Transportation Electrification

Breakthroughs in Electrified Vehicles, Aircraft, Rolling Stock, and Watercraft

Edited by

Ahmed A. Mohamed
City University of New York
NY, USA

Ahmad Arshan Khan
CNH Industrial
MI, USA

Ahmed T. Elsayed
Boeing Defense, Space & Security (BDS)
AL, USA

Mohamed A. Elshaer
Ford Motor Company
MI, USA

IEEE Press Series on Power and Energy Systems

IEEE PRESS

WILEY

Published by John Wiley & Sons, Inc., Hoboken, New Jersey.
Published simultaneously in Canada.

Library of Congress Cataloging-in-Publication Data
Names: Mohamed, Ahmed A., editor. | Khan, Ahmad Arshan, editor. | Elsayed, Ahmed T., editor. | Elshaer, Mohamed A., editor.
Title: Transportation electrification : breakthroughs in electrified vehicles, aircraft, rolling stock, and watercraft / edited by Ahmed A. Mohamed, Ahmad Arshan Khan, Ahmed T. Elsayed, Mohamed A. Elshaer.
Description: Hoboken, New Jersey : Wiley, [2023] | Series: IEEE press series on power and energy systems | Includes bibliographical references and index.
Identifiers: LCCN 2022046752 (print) | LCCN 2022046753 (ebook) | ISBN 9781119812326 (cloth) | ISBN 9781119812333 (adobe pdf) | ISBN 9781119812340 (epub)
Subjects: LCSH: Electric vehicles. | Electric airplanes. | Electric boats.
Classification: LCC TL220 .T66 2023 (print) | LCC TL220 (ebook) | DDC 629.22/93–dc23/eng/20221013
LC record available at https://lccn.loc.gov/2022046752
LC ebook record available at https://lccn.loc.gov/2022046753

Cover Design: Wiley
Cover Image: © RomanBabakin/Getty Images

Set in 9.5/12.5pt STIXTwoText by Straive, Pondicherry, India

Contents

About the Editors

Ahmed A. Mohamed (El-Tallawy) is an associate professor of Electrical Engineering (EE) at the City College of the City University of New York (CUNY). He is the EE PhD program advisor and the director of the CUNY Smart Grid Interdependencies Laboratory. Prof. Mohamed's research interests include critical infrastructure interdependencies, transportation electrification, and microgrids. He has numerous publications in these fields as book chapters and articles in premier journals and conference proceedings. Prof. Mohamed is the recipient of the 2019 NSF CAREER Award, among several other honors and awards. Several of Prof. Mohamed's publications received best-paper awards.

Ahmad Arshan Khan is director of Power Electronics and Electric Machines at CNH Industrial. He received his PhD in Electrical Engineering from Florida International University and M.S. in Electrical Engineering from Illinois Institute of Technology. Since 2010, he has worked for Fiat Chrysler, Ford Motor Company, AVL, and Eaton Corporation. From 2016 to 2018, he worked as an adjunct faculty member at the University of Michigan, Dearborn. He published several technical papers and holds three US patents. In 2013, he received a prize-paper award from IEEE IAS Electric Machines Committee and "Electrified Powertrain Engineering Innovation Award" from Ford Motor Company in 2017.

Ahmed T. Elsayed received his B.Sc. and M.Sc. degrees in Electrical Engineering from Shoubra Faculty of Engineering, Benha University, Egypt, in 2006 and 2010, respectively. In 2016, he received his PhD degree in Electrical and Computer Engineering from Florida International University, Miami, Florida. He is with the Boeing Company since 2017, where he worked on the development of multiple programs including the New Mid-Market Airplane (NMA) and B-52 modernization. Currently, Ahmed is a technical lead engineer (TLE) with Boeing Defense, Space and Security (BDS). He is leading the design and analysis for multiple proprietary and defense programs.

Mohamed A. Elshaer is a technical expert at Ford Motor Company. He received his PhD degree in Electrical Engineering from The Ohio State University. Stemming from his 11 years in the automotive industry, Dr. Elshaer's work on Pro Power Onboard for Ford's iconic F-150 truck was recognized by TIMES as one of the top 100 inventions of the year in 2021. He is a recognized expert with deep knowledge and intuition about global market trends, end customers' needs, and technological advancement. Dr. Elshaer is currently leading the engineering team at Ford in developing next-generation core technologies in the power conversion space.

List of Contributors

Robert Abboud
Beacon Power, LLC
Tyngsborough, MA
USA

Rohama Ahmad
City University of New York, City College
Department of Electrical Engineering
New York, NY
USA

Abir Alabani
The University of Manchester
Department of Electrical and Electronic Engineering
Manchester
UK

Luigi Alberti
University of Padova
Department of Industrial Engineering
Padova
Italy

Brian Battle
Beacon Power, LLC
Tyngsborough, MA
USA

Matteo Beligoj
University of Padova
Department of Industrial Engineering
Padova
Italy

Sahil Bhagat
Hitachi Rail STS
Traction Power and EMC
Abu Dhabi
United Arab Emirates

Giampaolo Buticchi
University of Nottingham Ningbo China
Faculty of Science and Engineering
Ningbo, Zhejiang
China

Lujia Chen
The University of Manchester
Department of Electrical and Electronic Engineering
Manchester
UK

Satish Chikkannanavar
Electrified Systems Engineering, Ford Motor Company
Vehicle Powertrain Electrification Center
Allen Park, MI
USA

Ian Cotton
The University of Manchester
Department of Electrical and Electronic Engineering
Manchester
UK

Ayman M. EL-Refaie
Marquette University
Electrical and Computer Engineering
Department
Milwaukee, WI
USA

Ahmed T. Elsayed
Boeing Defense, Space & Security (BDS)
Huntsville, AL
USA

Mohamed A. Elshaer
Ford Motor Company
Energy Storage Systems
Dearborn, MI
USA

William Franks
IPRE, Independent Power and Renewable
Energy LLC
Wells, ME
USA

Pramod Ghimire
Norwegian University of Science and
Technology (NTNU)
Department of Marine Technology
Trondheim, Trøndelag
Norway
Kongsberg Digital (KDI), Maritime Simulation
Horten, Vestfold
Norway

Qinghua Han
The University of Manchester
Department of Electrical and Electronic
Engineering
Manchester
UK

Ali Hosseinipour
Lehigh University
Electrical and Computer Engineering
Bethlehem, PA
USA

Zhen Huang
The University of Nottingham
Faculty of Engineering, University Park
Nottingham
UK

Hasan Iqbal
Aligarh Muslim University
Department of Electrical Engineering, ZHCET
Aligarh, Uttar Pradesh
India

Ahmad Arshan Khan
CNH Industrial, Electrification System
Integration
Livonia, MI
USA

Javad Khazaei
Lehigh University
Electrical and Computer Engineering
Bethlehem, PA
USA

Mahdiyeh Khodaparastan
EnBW North America
Boston
USA

Rajeev Kumar
Indian Institute of Technology Mandi
School of Mechanical and Materials
Engineering
Mandi, Himachal Pradesh
India

Gunho Kwak
Ford Motor Company
Product Development, Ford Ion Park
Romulus, MI
USA

Xiaoyu Lang
The University of Nottingham
Faculty of Engineering, University Park
Nottingham
UK

Michele Mattetti
University of Bologna
Department of Agricultural and Food Sciences
Bologna
Italy

Nabeel Mehdi
North Carolina State University
Department of Industrial and Systems Engineering
Raleigh, NC
USA

Hamid Metwally
Zagazig University
Electrical Power and Machines Department
Faculty of Engineering
Zagazig
Egypt

Ahmed A. Mohamed
City University of New York, City College
Department of Electrical Engineering
New York, NY
USA

Ahmed A. S. Mohamed
Eaton Research Laboratory, Eaton Corporation
Golden, CO
USA
Zagazig University
Electrical Power and Machines Department
Faculty of Engineering
Zagazig
Egypt

Rajesh Manjibhai Pindoriya
Thapar Institute of Engineering and Technology Patiala
Department of Electrical and Instrumentation Engineering
Patiala, Punjab
India

Bharat Singh Rajpurohit
Indian Institute of Technology Mandi
School of Computing and Electrical Engineering
Mandi, Himachal Pradesh
India

Prem Ranjan
The University of Manchester
Department of Electrical and Electronic Engineering
Manchester
UK

Amir Ranjbar
Canoo
Power Electronics Department
Torrance, CA
USA

Haroon Rehman
Aligarh Muslim University
Department of Electrical Engineering, ZHCET
Aligarh, Uttar Pradesh
India

Deepak Ronanki
Indian Institute of Technology Delhi
Department of Energy Science and Engineering
New Delhi, Delhi
India

Adil Sarwar
Aligarh Muslim University
Department of Electrical Engineering, ZHCET
Aligarh, Uttar Pradesh
India

Arif I. Sarwat
Florida International University
Department of Electrical and Computer Engineering, Energy Power Sustainability & Intelligence (EPSi) Lab
Miami, FL
USA

Elia Scolaro
University of Padova
Department of Industrial Engineering
Padova
Italy

Ahmed A. Shaier
Zagazig University
Electrical Power and Machines Department
Faculty of Engineering
Zagazig
Egypt

Jaskaran Singh
City University of New York, City College
Department of Electrical Engineering
New York, NY
USA

Constantine Spanos
Consolidated Edison Company of New York
Distribution Engineering
New York, NY
USA

Mohd Tariq
Aligarh Muslim University
Department of Electrical Engineering, ZHCET
Aligarh, Uttar Pradesh
India
Florida International University
Department of Electrical and Computer Engineering, Energy Power Sustainability & Intelligence (EPSi) Lab
Miami, FL
USA

Rishi Kant Thakur
Indian Institute of Technology Mandi
School of Mechanical and Materials Engineering
Mandi, Himachal Pradesh
India

Diego Troncon
CNH Industrial
Electrification System Integration
Modena
Italy

Behrooz Vahidi
Amirkabir University of Technology (Tehran Polytechnic)
Department of Electrical Engineering
Tehran
Iran

Jiajun Yang
University of Nottingham Ningbo China
China Beacons Institute
Ningbo, Zhejiang
China

Tao Yang
The University of Nottingham
Faculty of Engineering, University Park
Nottingham
UK

Aydin Zaboli
Amirkabir University of Technology (Tehran Polytechnic)
Department of Electrical Engineering
Tehran
Iran

Mehdi Zadeh
Norwegian University of Science and Technology (NTNU)
Department of Marine Technology
Trondheim, Trøndelag
Norway

Introduction

Ahmed A. Mohamed[1], Ahmad Arshan Khan[2], Ahmed T. Elsayed[3], and Mohamed A. Elshaer[4]

[1] *City University of New York, City College, Department of Electrical Engineering, New York, NY 10031, USA*
[2] *CNH Industrial, Electrification System Integration, Livonia, MI 48150, USA*
[3] *Boeing Defense, Space & Security (BDS), Huntsville, AL 35808, USA*
[4] *Ford Motor Company, Energy Storage Systems, Dearborn, MI 48124, USA*

Background

The transportation sector has consistently produced the highest levels of CO_2 emissions in the United States as compared with other sectors, including industrial, residential, and commercial. Electrifying the transportation sector has hence emerged as a key requisite to combating global warming. According to the Energy Information Administration, prior to 2020, the transportation sector was responsible for about 50% of CO_2 emissions in the United States. This percentage dropped by about 15% in 2020 due to the global COVID-19 pandemic; however, transportation remained the prime source of emissions. In 2021, the CO_2 emissions related to transportation rose by about 11% from their 2020 levels. It is anticipated that emissions will continue increasing as societies gradually reopen.

Electrification of the transportation sector reduces CO_2 emissions due to the direct elimination of exhaust gases. In addition, electrified transportation has significantly higher well-to-wheel energy efficiency – that is the consumed energy to input energy traced back to its original fuel source. Electrifying the transportation sector increases the demand for electricity from clean sources. Therefore, in countries and states with aggressive greenhouse gas emission reduction targets, transportation electrification and renewable energy deployment (e.g. solar, wind, and hydro) go hand in hand.

Fundamentally, electric transportation means of all kinds rely on electric machines and their drives for propulsion, and potentially energy storage to recapture regenerative braking energy, if any. While each of them has its own unique problems, their electric power systems often share common characteristics, challenges, and design targets. For instance, airplanes and shipboard power systems (SPSs) must satisfy the energy demanded for propulsion and heavily non-linear loads (e.g. starting motors), while being isolated from the main grid. The energy efficiency of both electric vehicles (EVs) and subway systems partially depends on effective recuperation of regenerative braking energy.

Overview

Technological advances and lessons learned related to one of the transportation modes can greatly benefit other modes; however, the different industry sectors and research groups have mostly been working in silos. The overarching goal of this book is to attempt to bridge these silos and inform professionals and researchers working on a mode of transportation (e.g. electrification of passenger vehicles) about challenges and solutions adopted in other modes. The book covers recent technological breakthroughs pertinent to the electrification of vehicles, aircraft, rolling stock, and watercraft. The focus is on new technologies that are poised to lead to significant advances cutting across multiple modes of transportation. Examples of these technologies include applications of energy storage systems (ESSs), wide-bandgap power electronic devices, wireless power transfer, and electric machines and controls.

Vehicles

The popularity of EVs is rapidly increasing due to their environmental benefits, energy savings, and acceleration performance. However, the adoption of EVs faces an uphill battle as it competes with fossil fuel vehicles. EVs are more expensive than internal combustion engine (ICE) vehicles, and range anxiety is still a significant consideration for pure EV users. For EVs to penetrate the market, improving charging technology and increasing the system's power density and efficiency are necessary. Solutions to electricity grid distribution issues, data integrity, and user privacy are essential for shifting to electric mobility and clean energy.

Chapter 1 provides a detailed overview of the current state-of-the-art electric machines used in traction/propulsion applications. It covers major global trends, challenges, and tradeoffs of various traction motor technologies. The impact of advanced materials and manufacturing and various approaches to reduce or eliminate rare-earth materials in electrical machines are discussed in this chapter too. The main focus of this chapter is on light-duty vehicles. However, locomotive, aerospace, ship propulsion, off-highway vehicles are included to show similarities and differences in motor technologies in each mode of transportation.

Charging technologies hold the key to accelerating EV adoption. With significant demand to reduce the charge time, the need to charge the HV battery faster emerged. Long-range EVs require a high-rate charge combined with high-rate discharge. Special consideration is needed to address degradation aspects on high-energy-density cathodes during high-rate cycling. Chapter 2 provides an overview of energy storage and charging technologies for EVs – giving the reader a comprehensive understanding of critical technological challenges and a summary of future innovations. Challenges in achieving high-energy density were discussed in detail, and key advancements in Cathode and Anode technologies were presented. Popular approaches for fast-charging Lithium-ion batteries and issues such as overcharging, thermal runaway, and cell degradation were included. The chapter also contains a comprehensive summary of the current development status for solid-state batteries.

Increasing the battery capacity drives the need to increase the onboard charging power. While the number of public DC fast charge stations is growing rapidly, the time it takes to charge the battery is still a major inconvenience for most EV users. Charging at a high rate requires passing a high current in the charging cable wire. The charging wire used in 400 VDC fast charging stations can be considerably heavy once power exceeds 150 kW. Hence, moving to an 800 V-battery architecture will enable charging the HV battery in less than 15 minutes.

Wide-bandgap materials play a key role in advancing transportation electrification. These materials enable operation at higher voltage levels, lower switching loss, higher operating junction temperature, and higher thermal conductivity. Device-level and application-level challenges associated with SiC MOSFET and GaN HEMTs are discussed in Chapter 3.

The flexibility in receiving power for charging the HV battery is essential in easing consumers' range anxiety. With the emergence of wireless charging technology, it is now possible to charge the HV battery at the same power levels as when the vehicle is plugged into the grid. A comprehensive overview of wireless power transfer technology is provided in Chapter 4, serving as a resource for readers interested in learning about this emerging technology and its different categories. The topics covered span the development of static wireless charging systems to dynamic and semi-dynamic wireless charging systems.

In Chapter 5, authors discuss the origin of acoustic noise and vibration (ANV) in electric machine drives and evaluate various control strategies to reduce ANV. Comparison between different random pulse width modulation techniques is shown in terms of efficiency, time response, current harmonics, and torque ripples.

Finally, renewable energy systems and smart grids will play an important role in supporting the EV infrastructure. Blockchain is a promising technology to build decentralized applications, which can integrate various systems in a trustless manner. Chapter 6 explores the different use cases of blockchain technologies that are relevant to e-mobility and charging infrastructure.

Aircraft

The aerospace industry, like many other industries, has witnessed giant technological leaps. It started with very absurd trials that involve jumping from towers, large kite gliding, or even covering a human body with feathers, then developed into today's overly complex and luxurious airplanes. Throughout this journey, many technologies were developed and utilized. Some are proprietary to the aviation industry and others were borrowed concepts from other industries like automotive. Nevertheless, this advancement came with a high environmental bill. According to Air Transport Action Group (ATAG), the global aviation industry produces around 2.1% of all human-induced carbon dioxide (CO_2) emissions. Another report issued by International Air Transport Association (IATA) anticipates this contribution to increase to 3% if the growth remains at its current rate without developing more environment-friendly technologies. This attracts more focus from the industry and international regulators to assist the progressive increase in transportation electrification, which gave rise to more electric aircrafts (MEAs). MEAs utilize electric solutions to replace the existing heavier and less efficient hydraulic and pneumatic counterparts. Since the aircraft technologies stretch over a wide span of vehicle sizes, from personal air vehicle (PAV) that carries 1–4 passengers to 400 passenger-commercial airplanes or 800 passengers like A380, no single solution fits all. Different solutions and technologies should be adopted to fit the significantly varying vehicle size. Each category/scale has its own challenges and merits. On the small scale, some electrification technologies developed by the automotive industry were adopted. In early 2019, Boeing tested its PAV prototype, which is an example of an electric vertical takeoff and landing (EVTOL) aircraft. Interestingly, it is often referred to as flying cars or flying taxis. Airbus tested its version of the PAV, known as Vahana, at the start of 2018. In October 2021, NASA and GE Aviation announced a partnership to mature a megawatt-class hybrid electric engine that could power a single-aisle aircraft. Boeing joined the project later in 2022. Recently, while focusing on MEA, Airbus announced an ambitious plan to develop the world's zero-emission commercial aircraft by 2035.

The aircraft section of this book aims at providing an insight into recent technological advancements. Chapters 7 and 8 provide an overview of the onboard microgrids concept and its supporting technologies. Additionally, Chapter 8 presents an assessment of the aircraft network stability. Fault-tolerant capabilities are discussed in Chapter 9. The standardization efforts for the electrified components are a crucial part and are discussed in Chapter 10.

Railway

Railway systems generally use less energy and produce less CO_2 emissions than personal vehicles and trucks. Moreover, electric rail systems are substantially more energy-efficient than their diesel-powered counterparts. Whereas diesel-powered trains are capable of transferring only about 35% of the energy produced by combustion to the wheels, supplying electricity directly from an overhead power line or a third rail transfers about 95%.

Electric railway systems receive their power from the main power grid through connection points (i.e. dedicated subway substations) that are distributed across train paths. Depending on the system, the power can be in the form of alternating current (AC), or direct current (DC) in which case rectifiers (typically uncontrolled/diode rectifiers) are placed within the subway substations. Whether it is overhead-catenary or third rail-based and whether it is AC or DC system, designers and operators have to handle issues related to earthing, bonding, stray currents, and electromagnetic compatibility to be able to reap the benefits of electrifying the railway system. Chapters 11–14 discuss these issues in detail.

Electric train cars are typically equipped with the capability to perform regenerative braking. Unlike vehicles, recycling this energy is quite more challenging. In fact, most of the produced regenerative braking energy is typically dumped in many of today's electric railway systems worldwide. Methods to efficiently recuperate regenerative braking energy exist. Finding the most economically viable solution for a given site, design optimization, control, and operation are essential elements for the success of any of those methods. Chapters 15 and 16 discuss this aspect in detail and present a case study based in New York City.

Railway substations may provide a cost-effective means to enable high penetration levels of renewable energy sources. Distributed energy resources (e.g. solar and batteries) can be integrated behind the meter or in front of the meter into railway substations. This will potentially mitigate the impact of renewable energy on the power grid while providing a collaborative value proposition that can benefit both the transportation company as well as the power distribution utility. With renewable power generation and energy storage, railway substations can be re-imagined as community resiliency hubs. Chapter 17 discusses this opportunity in depth.

Agriculture and Watercraft

Diesel engines are the most widespread power sources in the agriculture industry. The lack of standard driving cycles like automotive industry is one of the major challenges in agriculture machine electrification. In Chapter 18, an introduction to conventional powertrain used in farm tractors and various challenges for electrification of powertrain and auxiliaries are provided. Chapter 19 gives an overview of various system architectures and electrical components used in electrified farm tractors. Chapter 20 presents a life cycle cost analysis-based method to evaluate the economic feasibility of tractor powertrain electrification.

The watercraft-building industry is transitioning toward all-electric ship (AES) development. In AES, all aspects are driven by electric means including propulsion. Therefore, shipboard power systems (SPS) design is gaining more attention with arising need for more reliability and efficiency. The SPS can be regarded as an isolated microgrid. The integration of an energy storage system (ESS) in SPS is found to help achieve lower operating costs, higher fuel efficiency, and improved mission capabilities. The watercrafts can have different missions, applications, and cruising profiles, which impact their SPS design. The watercraft can be part of the military fleet (e.g. destroyers) or used as the main transportation with intermittent cruising profile in cities like Venice or between neighboring islands. This variability in mission goals requires new methods for optimizing the design of SPS. Designing and modeling of SPSs are crucial. Chapter 21 discusses this issue in detail and Chapter 22 discusses the recent effort toward electrification and decarbonization of SPSs.

Note on this Book

This book will adopt hybrid bottom-up and top-down approaches. The book will look at all modes of transportation panoramically, comparatively determines their current status of development, and identifies distinctive and common barriers, challenges, gaps in technology and knowledge, opportunities, and possible solutions. It will also include chapters that dive into details of recent breakthroughs and advances in each mode of transportation. We encourage the reader to read the book in its entirety maintaining a holistic vision of how/whether those advances can be applicable to other modes.

1

Electrical Machines for Traction and Propulsion Applications

Ayman M. EL-Refaie

Marquette University, Electrical and Computer Engineering Department, Milwaukee, WI 53233, USA

1.1 Introduction

Over the past few decades, there has been growing interest in electrification. This led to growing interest in hybrid/electrical traction applications. Many hybrid/electrical vehicles have been commercially introduced. Even though the main focus has been on areas like energy storage and power electronics, there is growing recognition of the importance of traction motors and generators. Various technologies for traction motors/generators have been developed. There is wide recognition of the need for advanced motors and generators in order to meet the aggressive targets (in terms of power density, efficiency, and cost) of the electric drive train. This recognition led several of the main Original Equipment Manufacturer (OEMs) like General Motors (GM) to decide to make large investments in developing and producing their own traction motors and generators [1]. In addition, over the past decade, hybrid/electric propulsion for aerospace applications became a big initiative as a part of the overall transportation electrification efforts. This chapter will provide a comprehensive review of the state of the art highlighting the key global trends and tradeoffs of various technologies. The chapter will also discuss future trends and potential areas of research. The chapter will cover light-duty vehicles (with more focus), medium- and heavy-duty vehicles, off-highway vehicles (OHVs), locomotives, and ship propulsion. In addition, the chapter will cover the fast-growing area of hybrid/electric propulsion for aerospace applications. Most of the material for Sections 1.2–1.6 come from [2] while most of the material for Section 1.7 comes from [3]. The goal of the chapter is to serve as a comprehensive reference for engineers working in the traction/propulsion area (Figure 1.1).

1.2 Light-Duty Vehicles

The light-duty vehicle market has been leading the way in terms of hybrid and electric vehicles. Companies like Toyota and Honda have been pioneers in commercializing hybrid vehicles. The Toyota machines have been viewed for many years as the state of the art. A lot of effort went into trying to /benchmark the motors/generators in the various vehicles. A lot of this work has been done at Oak Ridge National Lab (ORNL) and Argonne National Lab (ANL). Table 1.1 provides a summary of the various motors/generators used in the various vehicles, as well the technology used in each machine as well as the machine rating [4–7].

Transportation Electrification: Breakthroughs in Electrified Vehicles, Aircraft, Rolling Stock, and Watercraft,
First Edition. Edited by Ahmed A. Mohamed, Ahmad Arshan Khan, Ahmed T. Elsayed, and Mohamed A. Elshaer.

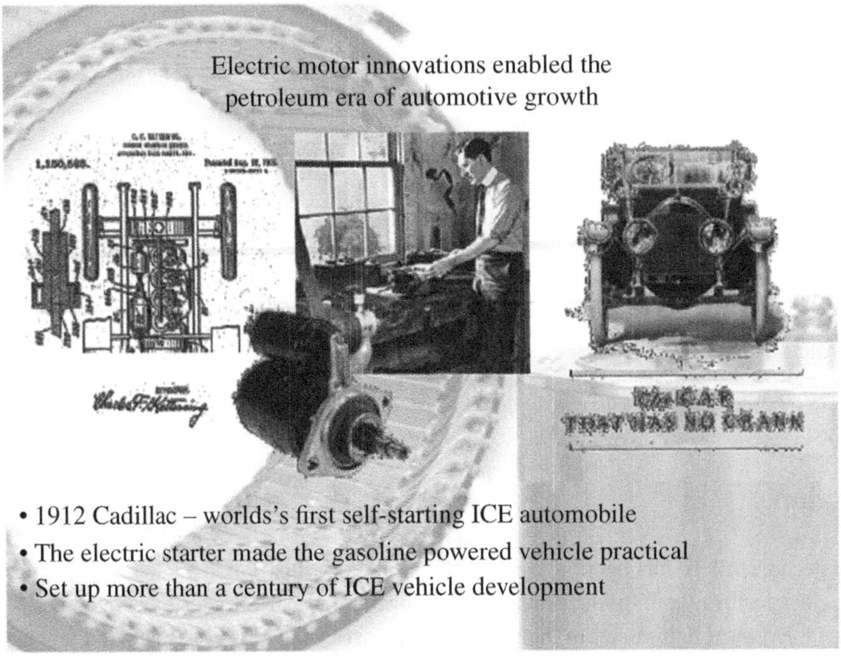
Electric motor innovations enabled the
petroleum era of automotive growth
• 1912 Cadillac – worlds's first self-starting ICE automobile
• The electric starter made the gasoline powered vehicle practical
• Set up more than a century of ICE vehicle development

Figure 1.1 Historical importance of electric traction motors. *Source:* Savagian [1]/IEEE.

Several key observations come out of looking at this table:

- Almost the entire light-duty hybrid vehicle industry has shifted to PM machines in order to meet the increasing power density and efficiency requirements. GM's EV1 was one of the very first vehicles to use induction machines in the early 1990s. Tesla was one of the few companies that used induction machines for a number of years but shifted to IPM machines in the newer models (Model 3 followed by Models S and X).[1]
- The Honda vehicles (as well as Hyundai which can be classified as mild hybrid) focused on surface PM (SPM) and Inset PM machines with segmented stator structures and fractional-slot concentrated windings. Segmented structures have the potential of increasing copper slot factor and reducing manufacturing costs. On the other hand, the segmented structures compromise the stator back iron rigidity. The Honda designs focused on 0.5 slot/pole/phase configurations, which tend to have low winding factor of 0.866. Also, it can be seen that the Honda designs utilize low DC bus voltage.
- The Toyota designs (which can be classified as full hybrid) focused on interior PM (IPM) machines with continuous laminations and regular integral-slot distributed windings. Relatively recently, Toyota is using fractional-slot concentrated windings in the Prius 2010 generator. Some of the Toyota designs have a single layer of magnets and some have multiple layers of magnets to increase reluctance torque contribution. Also, it can be seen that the more recent designs utilize higher DC bus voltage (by using a boost converter). This reduces the flux-weakening requirements, especially for higher-speed machines like the Toyota Camry and Lexus motors. Currently, there is a trend to explore higher DC bus voltages up to 800 V [8]. This will help further improve efficiency and reduce the size of electrical connections.

Table 1.1 Summary of the motors/generators in the key hybrid vehicles.

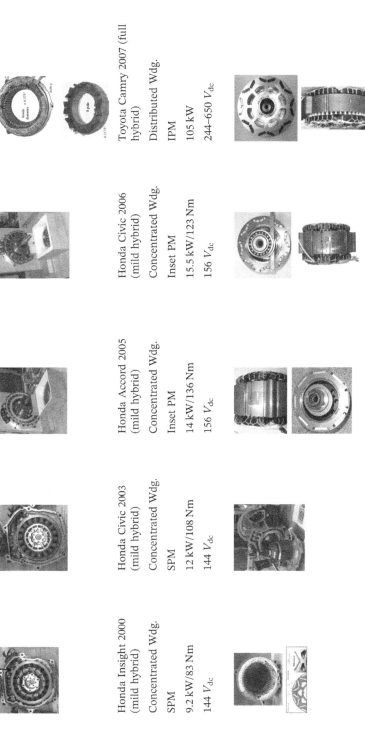

Motor/ generator		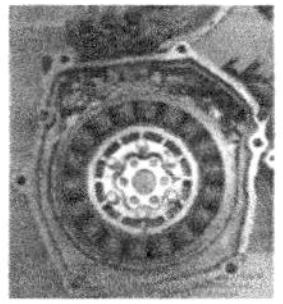			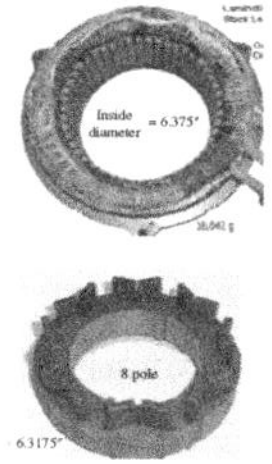
Vehicle	Honda Insight 2000 (mild hybrid)	Honda Civic 2003 (mild hybrid)	Honda Accord 2005 (mild hybrid)	Honda Civic 2006 (mild hybrid)	Toyota Camry 2007 (full hybrid)
Stator	Concentrated Wdg.	Concentrated Wdg.	Concentrated Wdg.	Concentrated Wdg.	Distributed Wdg.
Rotor	SPM	SPM	Inset PM	Inset PM	IPM
Rating	9.2 kW/83 Nm	12 kW/108 Nm	14 kW/136 Nm	15.5 kW/123 Nm	105 kW
DC bus voltage	144 V_{dc}	144 V_{dc}	156 V_{dc}	156 V_{dc}	244–650 V_{dc}
Motor/ generator	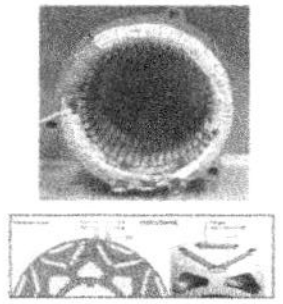				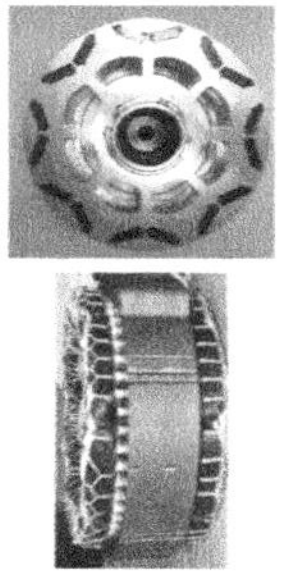
Vehicle	Lexus 2005 Rx400h (full hybrid)	Toyota Prius 1998 (full hybrid)	Toyota Prius 2002 (full hybrid)	Toyota Prius 2004 (full hybrid)	Toyota Prius 2010 (full hybrid)

(*Continued*)

Table 1.1 (Continued)

Stator	Distributed Wdg.	Distributed Wdg.	Distributed Wdg.	Distributed Wdg.	Distributed Wdg.
Rotor	IPM	IPM	IPM	IPM	IPM
Rating	123 kW	30 kW	33 kW	50 kW	60 kW
DC bus voltage	650 V_{dc}	273 V_{dc}	274 V_{dc}	500 V_{dc}	650 V_{dc}
Motor/ generator	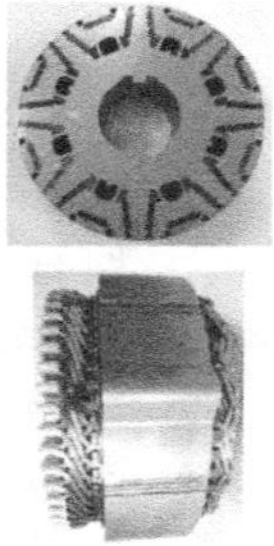			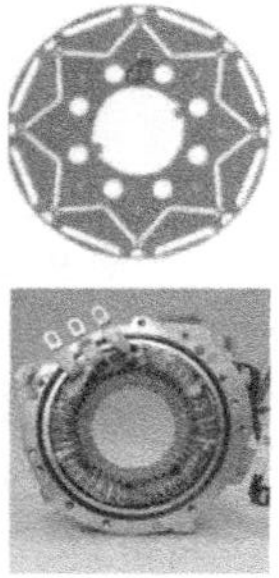	
Vehicle	Toyota Prius 2017 (full hybrid)	BMW i3 2016 (plug-in hybrid)	Hyundai Sonata 2011 (mild hybrid)	Nissan Leaf 2012 (pure EV)	Tesla Model 3 (pure EV)
Stator	Hairpin Distributed Wdg.	Distributed Wdg.	Concentrated Wdg.	Distributed Wdg.	Distributed Wdg.
Rotor	IPM	IPM	Inset IPM	IPM	Carbon-wrapped IPM
Rating	53 kW	125 kW/250 Nm	30 kW/205 Nm	80 kW	211 kW/450 Nm
DC bus voltage	600 V_{dc}	300 V_{dc}	270 V_{dc}	375 V_{dc}	350 V_{dc}

- The Honda designs as well as the Prius have a top speed of around 6000 rpm. In more recent designs like the Camry and Lexus (as well as some of the GM and Ford designs), the machines are designed for higher top speeds up to 14,000 rpm, in order to increase the machine power density.
- IPM machines are becoming the dominant type.
- 30–70 kW ratings/machine for full hybrids (cars).
- >120 kW for full-hybrid SUVs.

In more recent electrified vehicles, the machine power ratings can go as high as 200 kW and even higher in some cases.

- Liquid cooling is becoming an industry common practice in order to meet higher performance requirements. The peak air gap shear stress for all these machines ranges from 8 psi up to 17 psi depending on how aggressive the cooling is and how long is the transient power requirement.
- Higher coolant inlet temperature up to 105 °C is still targeted for future designs (to eliminate additional cooling loop).
- Electric machines tightly integrated with ICE/drive train housing.
- GM as well as several other OEMs are pursuing IPM machines with bar-wound (hairpin) windings [1]. This winding has a higher fill factor (and hence power density) but the main concern is the AC losses in the windings, especially at higher speeds. The performance of the machine will rely heavily on the vehicle duty cycle and how long the motor can thermally operate at higher speeds, and also depends on the cooling mechanism implemented (Figure 1.2).

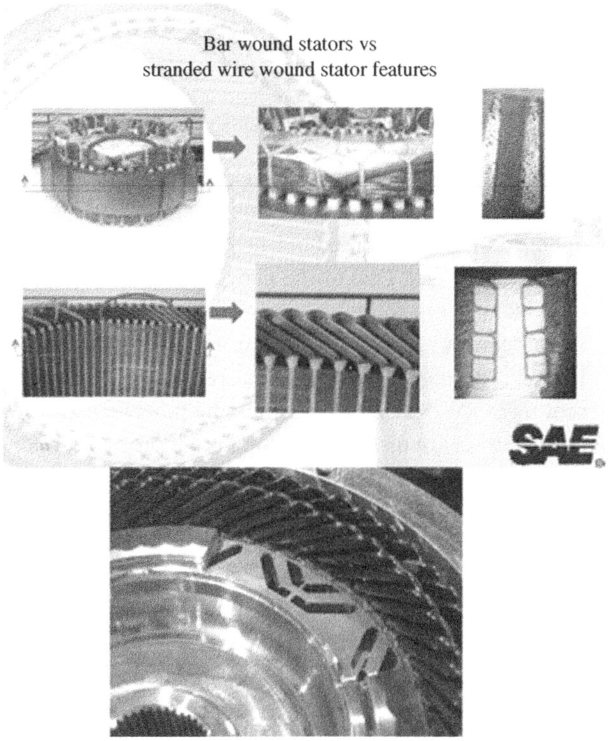

Figure 1.2 GM/Remy IPM motors with bar-wound (hairpin) windings. *Source:* Savagian [1]/IEEE.

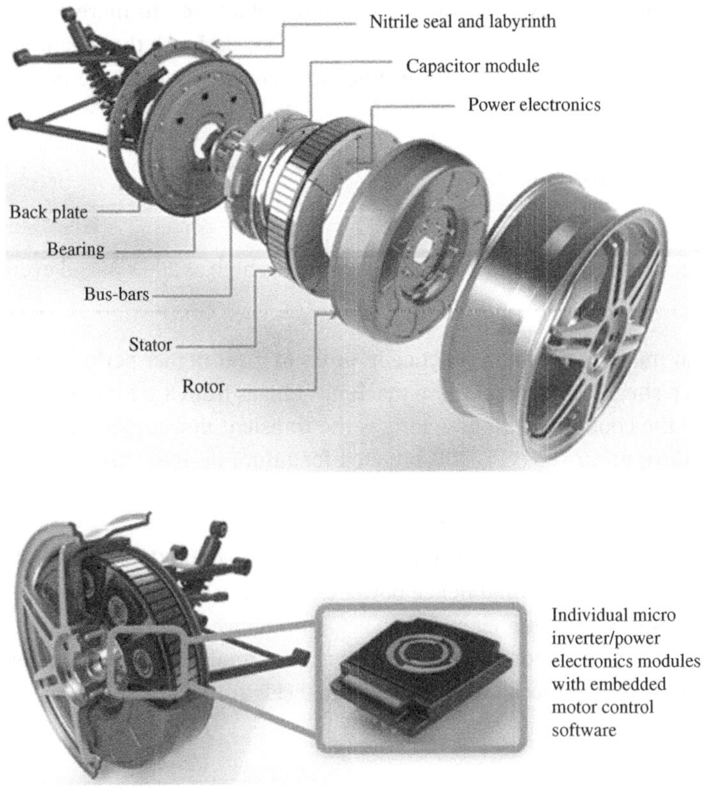

Figure 1.3 Protean electric wheel motor. *Source:* Protean electric wheel (http://www.proteanelectric.com/live/docs/PD18-3-LV1.2-Data-Sheet.pdf).

Figure 1.4 Mitsubishi wheel motor. *Source:* Mitsubishi Motors North America, Inc. (www.mitsubishimotors.com).

In addition to the regular central traction motors, numerous wheel-hub motor concepts have been developed (some were installed in a few vehicles) but no commercial autos yet. One of the main technical challenges of wheel motors is the un-sprung mass [9]. Many ideas have been developed to minimize this issue. Protean Electric[2] developed an inside-out radial-flux PM wheel motor (54 kW/83 kWpk) in which the power electronics and controls are distributed along the motor circumference as shown in Figure 1.3. Mitsubishi[3] (50 kW) and TM4[4] (80 kW) also developed inside-out radial-flux PM wheel motors as shown in Figures 1.4 and 1.5. GM developed a dual rotor axial-flux PM with an epoxy-based stator [10, 11]. The motor is shown in Figure 1.6. Despite all these concepts and the significant progress made, cost, fault tolerance, system complexity, and unsprung mass continue to be significant challenges for wheel motors.

1.3 Medium- and Heavy-Duty Vehicles

Even though the light-duty vehicle section has been dominated by liquid-cooled (as previously mentioned in Section 1.2), this is not the case with medium-, and heavy-duty vehicles for example delivery trucks, buses, and class 8 (super) trucks. In these vehicles, induction machines are still considered the main workhorse. This is probably due to the historical robustness and cost-effectiveness of induction machines. Table 1.2 summarizes some of the available traction systems for medium- and heavy-duty vehicles.[5–9] The majority are liquid-cooled induction machines. There is also growing interest in IPM machines. UQM is considered one of the leading companies in providing PM liquid-cooled traction systems for medium and heavy-duty vehicles[10] Switched reluctance (SR) machines are also used in some heavy-duty vehicles.[11] In these types of vehicles, the well-known issues of SR machines including acoustic noise as well as torque ripple and vibration might not be of great concern. SR Drives (part of Emerson) is a lead supplier of SR machines. PM machines have higher power density and efficiency at the machine torque–speed envelope. Induction machines have advantages in terms of partial load efficiency (also there are very low-drag losses if the machine is unloaded). Also, induction machines do not have the issues of fault tolerance[12]

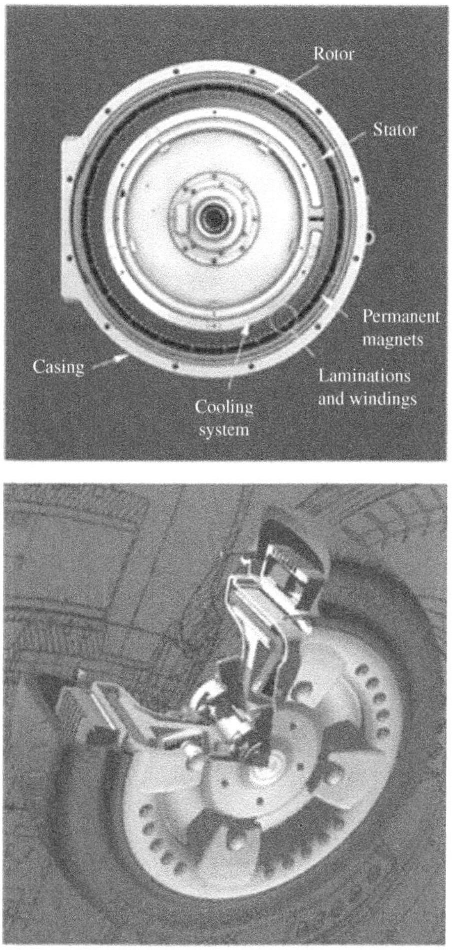

Figure 1.5 TM4 wheel motor. *Source:* DANA TM4 Inc. (http://www.tm4.com/moteurs_generateurs.aspx).

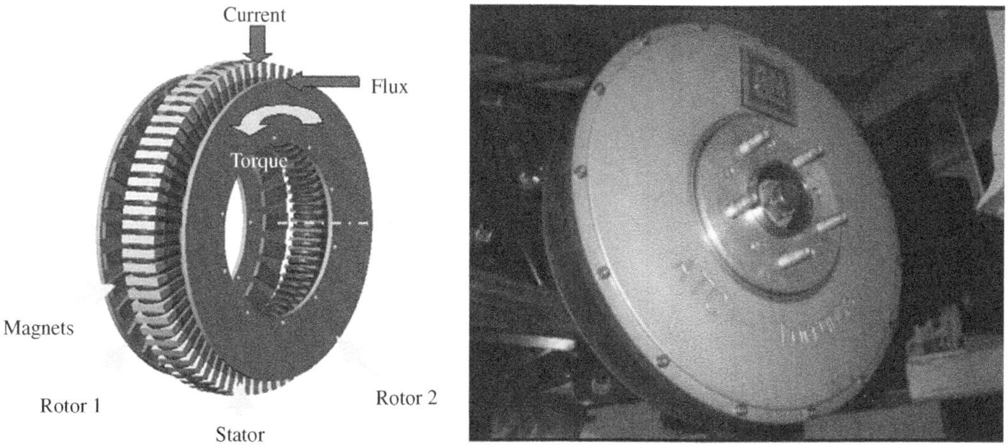

Figure 1.6 GM axial-flux PM wheel motor. *Source:* Rahman et al. [10]/IEEE.

Table 1.2 Summary of hybrid systems for medium and heavy-duty vehicles.

Allison transmission	BAE	Azure dynamics	Solectria	UQM	Oshkosh
• Allison EP-50/40 • Motor/ generator • 75 kW nominal 150 kW peak • Induction machine by Remy	• HybriDrive® propulsion systems • Induction machine traction motor 160 kW (215 hp) Continuous • 200 kW (268 hp) peak	• 50 kW Continuous shaft power at 1000–2500 rpm • 97 kW peak shaft power • Induction machine air-cooled • AC90 Motor with DMOC645 Controller	• AC55 is a single output, 78 kW 3-phase AC induction motor • Nominal speed of 2.5 krpm and a maximum speed of 8 krpm	• PM liquid cooling • 200 kW peak, 115 kW continuous motor power • 200 kW peak, 120 kW continuous generator power/regenerative braking • 150 kW peak, 100 kW continuous motor power • 150 kW peak, 100 kW continuous generator power	• The ProPulse system combines a 300 hp. diesel engine with a 225 kW induction generator and ultracapacitors (1.4 MJ) to drive two 140 hp induction traction motors

(and the braking torque produced in case of a fault) or uncontrolled generation mode (UCG) as in the case of a PM machine [12, 13].

1.4 Off-Highway Vehicles

There is growing interest in hybrid-electric drive systems in construction and OHVs to replace existing mechanical and hydraulic systems to achieve higher performance.[13] Table 1.3 provides a summary of these examples. As can be seen, Siemens Motors (as well as GE) and advanced AC control systems are very popular, especially in big mining trucks. It can also be seen that induction machines are the dominant type. Also SR machines are being used due to their robustness and as previously mentioned, acoustic noise, and vibrations are not key issues in such applications. PM machines are also being evaluated but there are concerns about the shock/vibration levels that the machine can tolerate mainly due to the brittleness of PMs [14]

1.5 Locomotives

Historically, induction machines are the dominant type of traction motors in locomotives. This can be seen in Table 1.4, which summarizes the traction motors used in various locomotives [15]. The typical maximum speeds for these motors are in the range of 3000–6000 rpm. All these motors are air-cooled. More recently, PM motors have been developed for high-speed rail,[14] for example, the AGV PM traction motor developed by Alstom and shown in Figure 1.7. This motor is closed and self-ventilated and has a maximum speed of 4500 rpm. The next step in order to meet the demand-

Table 1.3 Examples of OHV motors and generators.

Application	Company	Machine type and rating
Tractor	Caterpillar (CAT D7E with electric drive train)	2∗110 HP induction motors
Loader	Volvo (Volvo L220F hybrid wheel loader)	700 Nm and 50 kW peak induction motors
	LeTourneaue Technologies (50-Series for loaders and dozers, e.g. new hybrid wheel loader L-1150)	1050 HP switched reluctance motors and an SR generator
Mining trucks	Caterpillar 795F-AC Rear axle mounted	3400 HP wheel induction motor
	Hitachi EH5000AC	3000 HP induction motor
	Liebherr TI 274	3000 HP Siemens induction wheel motors
	Komatsu 860E	2400 HP Siemens induction motors

Source: Adapted from http://machinedesign.com/article/hybrid-drives-for-construction-equipment-0707.

Table 1.4 Summary of the motors/generators in various locomotives.

Locomotive	Motor type and rating	Locomotive	Motor type and rating
Italian Pendolino ETR 460 tilting train generation, max. Speed 250 km/h, in operation since 1992.	12∗500 kW induction motors traction motors made by Alstom	German DB 152 electrical locomotive, max. speed 170 km/h, in operation since 2001	4∗1600 kW induction machines made by Siemens
German DB ICE 3 high-speed train, max. speed 330 km/h, traction motors, power 16 × 500 kW, in operation since 1999	16∗500 kW induction motors made by Siemens max speed of 6000 rpm	German DB-Railion 189 electrical freight locomotive, max. speed 140 km/h, in operation since 2003	4∗1600 kW induction machines made by Siemens
Swiss Railways SBB FLIRT RABe 521/523, max. speed 160 km/h, in operation since 2004	4∗500 kW Induction motors made by TSA Traktionssysteme	Swiss SBB Re 460 electrical locomotive, max. speed 230 km/h, in operation since 1992	4∗1525 kW induction machines made by ABB
Spanish Talgo 350 high-speed train, max. speed 350 km/h, traction motors, in operation since 2005	8∗1000 kW Induction motors made by Siemens	Indian 3-phase electric freight locomotive WAG-9, max. speed 100 km/h, in operation since 1996	6∗850 kW induction motors made by ABB
Korean KTX high-speed train, operational speed 300 km/h, in operation since 2004	12∗1130 kW self-commutated synchronous machines (replaced by induction machines in KTX-II) made by Rotem, power	USA – Houston TX, MetroRail, Avanto light rail vehicle, max. Speed 105 km/h, in operation since 2001	4∗200 kW at 2500 rpm induction machines made by Siemens
Austrian ÖBB Taurus electrical locomotive 1016/1116, max. speed 230 km/h, in operation since 2000	4∗1600 kW induction machines made by Siemens	Czech Republic – Prague 15 T low floor tramway, max. speed 60 km/h	16∗60 kW induction motors made by ŠKODA ELECTRIC
Chinese Railways DJ4 electrical locomotive, max. speed 120 km/h, in operation since 2006	8∗1200 kW induction machines made by Siemens and Zhuzhou Electric Locomotives Works	Austria – Graz, Cityrunner, low floor tramway, max. speed 70 km/h, in operation since 2001	8∗50 kW induction motors made by ŠKODA ELECTRIC

Source: Adapted from SKF [15].

ing requirements for light-weight high-speed rail is to move toward a liquid cooling of both the power converter and traction motors. This is currently under investigation and development.

1.6 Ship Propulsion

Historically: DC and synchronous machines have been the dominant type of ship propulsion.[15] Figures 1.8 and 1.9 show examples of such machines by LDW. More recently, PM machines have been developed for ship propulsion.[16] Figure 1.10 shows an example of a

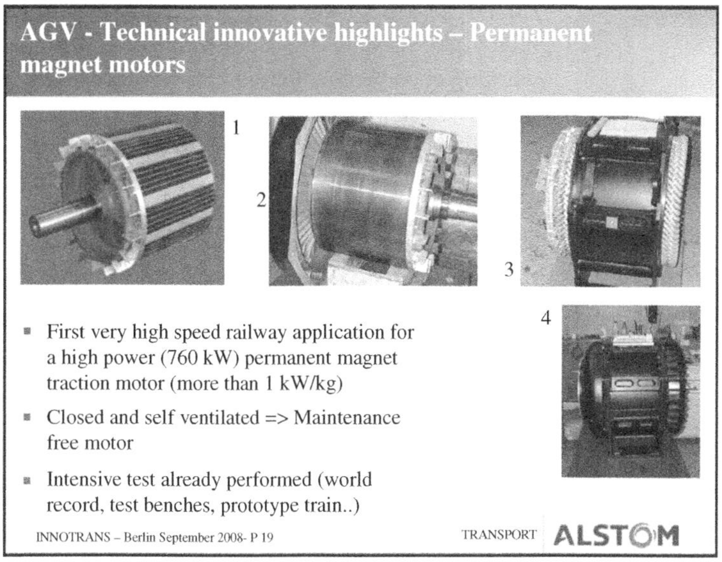
AGV - Technical innovative highlights – Permanent magnet motors
1
2
3
4
First very high speed railway application for a high power (760 kW) permanent magnet traction motor (more than 1 kW/kg)
Closed and self ventilated => Maintenance free motor
Intensive test already performed (world record, test benches, prototype train..)
INNOTRANS – Berlin September 2008- P 19
TRANSPORT
ALSTOM

Figure 1.7 AGV high-speed PM traction motor by Alstom.

Figure 1.8 21 000-kW variable speed synchronous propulsion motor for cruise liner. *Source:* http://www.ship-technology.com/contractors/propulsion/ldw.

Figure 1.9 950 kW DC motor, delivered by LDW, as main drive for the research vessel "Solea."

shipboard propulsion high-speed 36.5 MW PM motor developed by DRS. Some companies are also working on superconducting machines for marine propulsion applications.[17] Figure 1.11 shows the potentially significant increase in power density by replacing conventional machines with high-temperature superconducting machines by American superconductors. Table 1.5 summarizes some of the key motors used in podded ship propulsion. As can be seen, there are a mix of wound field synchronous machines, induction machines, and PM machines.[18,19] Also there have been several rim-driven machines proposed.[20] Few examples are shown in Figures 1.12–1.14 [16].[21,22]

Figure 1.10 36.5 MW permanent magnet motor systems (PMMS) by DRS. *Source:* http://www.drs.com/Products/PESG/PDF/PMMS.pdf.

1.7 High Specific Torque/ Power Electrical Machines

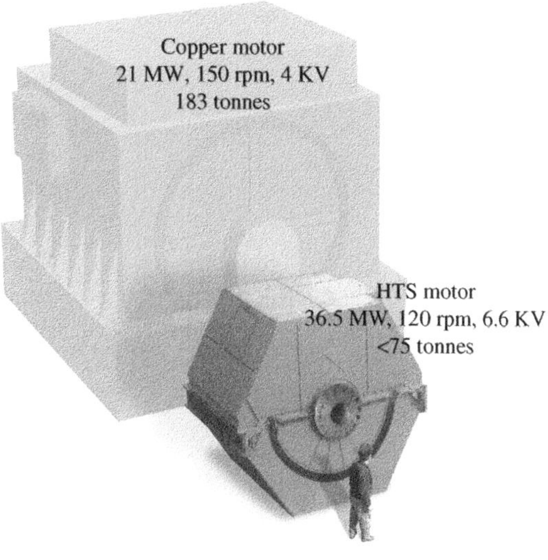

Figure 1.11 Comparison of conventional 21 MW motor and HTS 36.5 MW motor by American Superconductor. *Source:* http://www.amsc.com/products/motorsgenerators/documents/MP_DS_365_1007_A4_r3.pdf.

There has been growing and continued interest in high-speed and high specific power electrical machines. In [17], a survey of high-speed machines based on various application and machine topologies is reported as shown in Figures 1.15 and 1.16. Another survey has been presented in [18] highlighting the key technologies that go into high-speed machines, as well as some performance figure-of-merit (FoM). For example, RPM$\sqrt{\text{kW}}$ is used to identify the safe limits for tip speeds to avoid running into mechanical and rotor dynamics issues. In [19], a survey of high specific power electrical machines is disclosed. The focus in [19] is on the electric machine-specific power and a general comparison between different machine topologies is presented. This section will provide a more comprehensive survey of high specific power machines in terms of highlighting specific examples that are considered the state of the art. More performance details for these specific examples will be reported. Key "system" tradeoffs and considerations will also be discussed. The section will mainly focus on electrical machines for land vehicles, and aerospace applications.

1.7.1 Electrical Machines for Land Vehicles

In this section, high specific power electrical machines designed for hybrid/electric vehicles are covered. The focus will be on machines that stand out in terms of having significantly higher power

Table 1.5 Summary of the motors used in podded ship propulsion.

Company	Mermaid (RR)	Azipod (ABB)	Azipod compact (ABB)	SSP (Siemens)	Converteam
Machine type	Wound field synchronous	Wound field synchronous and induction	PM synchronous	PM synchronous	Induction
Power rating	5–25 MW[a]	Induction <5 MW[b] Synchronous 5–25 MW	1.3–4.5 MW[c]	5–30 MW[d]	1–12 MW[e]

[a] http://www.rollsroyce.com/Images/Product%20Range_tcm92-8658.pdf.
[b] http://www05.abb.com/global/scot/scot293.nsf/veritydisplay/9eccbc669a5fcff1c125768f004a9bc9/$file/azipod%20project%20guide%206-3.pdf.
[c] http://www05.abb.com/global/scot/scot293.nsf/veritydisplay/d77d9ac59ba74cb7c12576d3004cfe3f/$file/azipod%20co%20brochure_2010.pdf.
[d] http://www.industrysolutions.siemens.com/marine/en/.
[e] http://www.converteam.de/majic/dl/4/doc/Markets/Navy/Final_NAVAL.pdf.

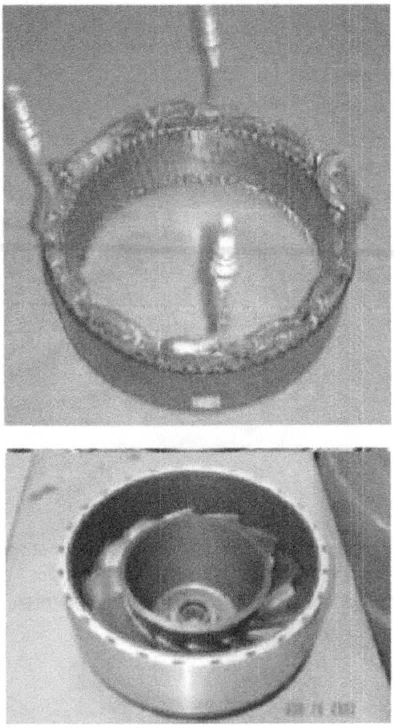

Figure 1.12 110 kW, 4100 rpm rim-driven PM motor by SatCon. *Source:* Nichols et al. [20]/ Synapticon GmbH.

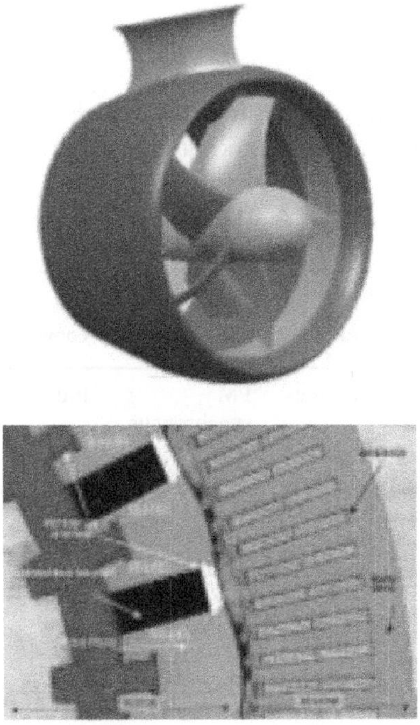

Figure 1.13 4.47 MW rim-driven IPM motor by GDEB. *Source:* Kane et al. [21]/IEEE.

density and/or torque density compared to many other machines in that crowded space. In this section, the focus will be on electrical machines used in light-duty vehicles (since they usually target the highest specific power) while a comprehensive review of electrical machines used in other types of vehicles has been presented in [2]. Also, a comprehensive summary of the teardown and test results of several mainstream central traction motors in light-duty vehicles was disclosed.[23] These include 2004 Prius, 2006 Accord, 2007 Camry, 2008 LS 600 h, 2010 Prius, 2011 Sonata, 2012 Sonata generator, 2012 LEAF, 2013 LEAF charger, 2013 Camry PCU, 2014 Accord, 2016 BMW i3, and 2017 Prius. Those machines have "peak" specific power ranging from 1.1–3 kW/kg. The DC bus voltage ranges from 270 to 650 V. Table 1.6 provides a summary of some of the salient examples in this space (where n_{peak} is the peak power speed while n_{max} is the maximum speed). These machines are mainly traction motors (some are central traction motors and others are wheel motors). They have significantly higher specific power and/or specific torque compared to other mainstream traction motors covered in previous references.[23] They target some specialized applications, for example, racing cars as in the case of McLaren motor. Also, YASA Motors have been used in some high-performance cars like Ferrari and the company has been acquired by Mercedes Benz. The following points can be observed:

- All high-performance machines in this space are PM machines. In addition to the high specific power, maximum efficiency of 94% or higher could be achieved.
- Central traction motors are mainly radial-flux inner-rotor PM machines.
- Wheel motors use other topologies including axial-flux PM (YASA) as well outer-rotor PM (Protean), which lend themselves to better integration with wheels as hub motors.
- All machines are liquid-cooled (masses listed are based on dry machines, the cooling liquid mass is not included).
- The McLaren and KIT machines have the highest corner speeds and achieve the highest specific power density. Both are targeting racing vehicles (McLaren is a mature product while KIT targets university racing and not a commercial product).

Racing applications are not cost-sensitive and hence there is room to come up with higher-performance designs.
- The wheel motors are lower-speed motors so their specific power is not as high but they have significantly higher specific torque. Usually, there is a tradeoff between specific power and specific torque.
- The UQM is a central traction motor and its performance falls within the performance of other mainstream central traction motors.
- For automotive applications, it is important to remember that available space comes at a premium. So, power density/volume is of equal if not more importance than specific power. Therefore, kg/l (which is effectively the ratio of power density and specific power) is included in Table 1.6.
- The DC bus voltage varies from 400 to 800 V, which is considered low voltage. This is suitable for the power ratings of these machines.

Figure 1.14 100 kW rim-driven PM motor by Brunvoll. *Source:* [22] and http://www.norpropeller.no.

1.7.2 Electrical Machines for Aerospace Applications

In this section, high specific power electrical machines designed for aerospace applications will be covered. Aerospace is the key area that requires substantial improvement in power density. This includes More Electric Aircraft (MEA) [23, 24], hybrid/electric propulsion[24] [25, 26] green taxiing [27], and UAVs and Vertical Takeoff and Landing (VTOL).[25] All these machines are used for relatively new non-traditional applications. These include:

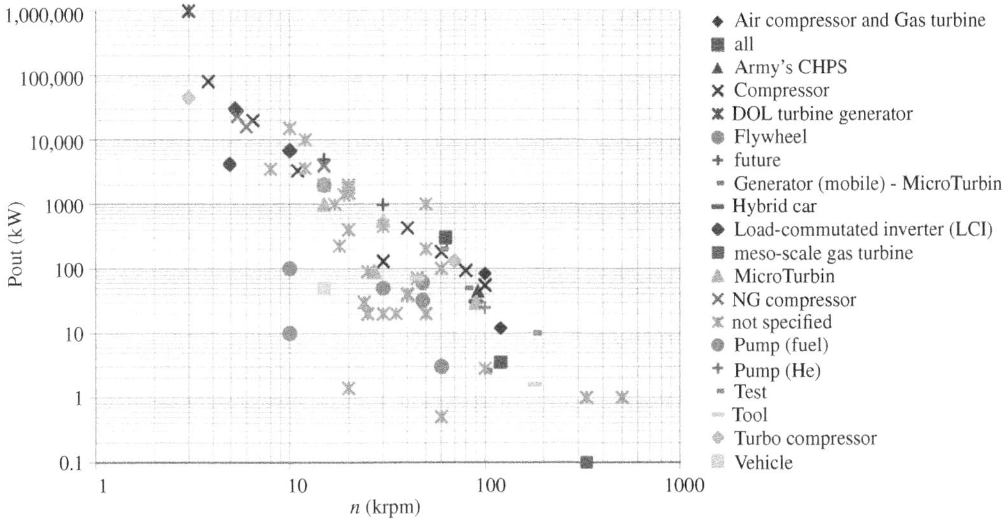

Figure 1.15 Survey of high-speed machines by application [17].

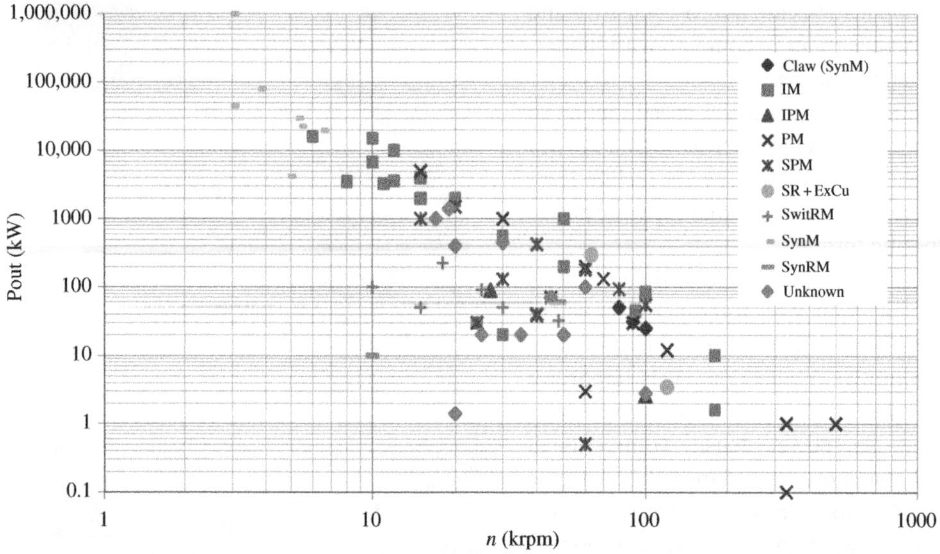

Figure 1.16 Survey of high-speed machines by topology [17].

(i) generators as well as motors that drive propellers in the case of hybrid/electric propulsion, (ii) wheel motors for green taxiing, and (3) motors that provide both vertical lift as well as propulsion in the case of UAVs and VTOL. These applications are very different from the conventional role of electrical machines in aerospace applications which mainly included generators (connected to the engine vis power takeoff (PTO) shafts, small motors driving various loads as well as electrical actuators. These new applications have significantly higher demands when it comes to specific power and efficiency (since both have a significant impact on specific fuel consumption). Tables 1.7–1.9 provide a detailed summary of various electrical machines broken down by power ratings. Since the previously mentioned applications are fairly new, the amount of information about these high specific power machines available in the public domain is still relatively limited. These tables include the main bulk of information that is available. Most of the internal dimensions of the various machines are not readily available and hence it is difficult to calculate useful quantities like air gap shear stress, torque per airgap volume and rotor tip speed. Still, the information included in these tables should be very useful for engineers and researchers working in this field in terms of understanding and establishing technology trends. The general trends include:

- Higher power (>100 kW) electrical machines cover applications ranging from small aircraft (four seats or more) all the way up to large commercial aircrafts (Honeywell machine). Electrical machines between 10 and 100 kW mainly cover smaller planes. Electrical machines <10 kW mainly cover UAVs and remote-controlled small planes.
- All machines are PM (radial inner rotor, radial outer rotor, and axial flux).
- Higher power machines ≥100 kW are largely liquid-cooled. Lower power machines <100 kW are largely forced air-cooled.
- Lower-speed machines lend themselves better to tooth windings (due to the lower frequency, which is typically 10s–100s Hz) as well as direct-conductor cooling (due to larger fewer slots).

Table 1.6 High-power density machines for land vehicles.

	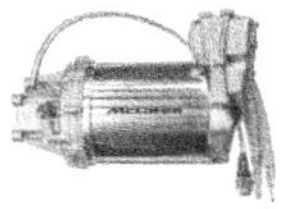	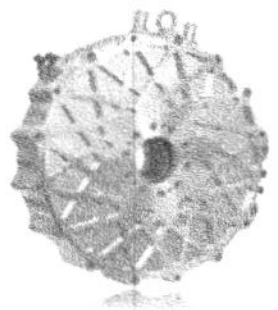		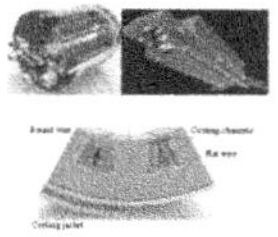	
Manufacturer	McLaren	YASA	Protean	Karlsuhe Institute of Technology (KIT)	UQM
Reference	http://www.mclaren.com/appliedtechnologies/products/item/e-motor-120kw-130nm	http://www.yasamotors.com/products/yasa-750	http://www.proteanelectric.com/en/specifications	[28]	https://www.neweagle.net/support/wiki/ProductDocumentation/EV_Software_and_Hardware/Electric_Motors/UQM/PowerPhase%20HD%20250%20web.pdf
Application	McLaren P1 Hybrid supercar and Formula-E EV	2 direct drive motors for pure EV (Regera hypercar)	2 in-wheel motors for pure EV (VW Bora compact sedan)	Electric vehicle (Audi race)	EV/HEV
Machine topology	Surface PM (SPM)	Axial flux PM	Outer rotor SPM	Interior PM (IPM)	SPM
Cooling method	50/50 water/glycol	Oil cooling	Liquid cooling jacket	50/50 water/glycol indirect slot cooling	Water jacket cooling
Mass (kg)	26	33	34	14	95
DC bus voltage (V)	545	800	400	450	360–440
Efficiency (%)	96 (@120 kW and 13 krpm)	≥95	≥93 (including inverter)	97 (max eff.)	94 (max eff.)
P_{rated} (kW)	100	75	54	70	115
P_{peak} (kW)	120	199	75		200
kW_{rated}/kg	3.8	2.3	1.6	5	1.2
kW_{peak}/kg	4.6	6	2.2		2.1
n_{rated} (rpm)	9545	1800	793	7400	2440
n_{peak} (rpm)	8815	2400	716		2122
n_{max} (rpm)	15 000	3250	1600	15,000	
T_{rated} (Nm)	100	398	650	90	450
T_{peak} (Nm)	130	792	1000		900
Nm_{rated}/kg	3.8	12	19.1	6.4	4.7
Nm_{peak}/kg	5	24	29.4		9.5
kg/l	3.81	5.2	2.13	9.93	2.97

Table 1.7 High-power density machines for aerospace applications ≥100 kW.

		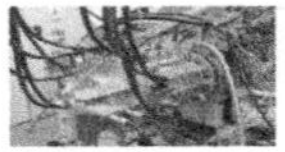			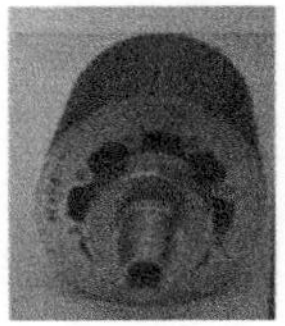
Manufacturer	Siemens	Honeywell	Siemens	ENSTROJ – Slovenia	Rolls Royce-University of Sheffield
Reference	https://nari.arc.nasa.gov/sites/default/files/attachments/Korbinian-TVFW-Aug2015.pdf	[29]	https://nari.arc.nasa.gov/sites/default/files/attachments/Korbinian-TVFW-Aug2015.pdf	http://www.enstroj.si/Electric-products/emrax-motorsgenerators.html	[30]
Application	Four or more seats' plane	Hybrid electric aircraft propulsion	Generator lab approval for series hybrid propulsion	Electric glider Apis EA2	Starter-generator for small civil turbofan
Machine topology	Halbach array SPM	Wound-field synchronous (2 sets of 3-ph windings)	SPM	Axial-flux PM	SPM
Cooling method	Directly cooled conductors	Engine oil cooling, conduction and end-winding spray	Directly cooled conductors	Combined cooling; indirect cooling: air + water	Water jacket cooling
Mass (kg)	50	126.5	24.4	20.3	22.7
DC bus voltage (V)	580	300–600 (2 3-ph diode rectifiers in series or in parallel)	580	700	540
Efficiency (%)	≥95	97	≥95	93–98	96
P_{rated} (kW)	260	1000 (only tested up to 540 kW)	170	100	100
P_{peak} (kW)				200	150
kW_{rated}/kg	5.2	7.9	7	4.9	4.4
kW_{peak}/kg				9.8	6.6
n_{rated} (rpm)	2500	19,000	6250	4000	27,000
n_{peak} (rpm)				4000	27,000
n_{max} (rpm)		20,000	6500	5000	27,000
T_{rated} (Nm)	993	110	260	250	35
T_{peak} (Nm)				500	53
Nm_{rated}/kg	19.9	4	10.7	12.3	1.6
Nm_{peak}/kg				24.6	2.4
kg/l	1.21	2.09	2.18	4	3.02

Table 1.8 High-power density machines for aerospace applications ≥10 kW and <100 kW.

Manufacturer	Rotex – Czech Republic	Siemens and EADS – Germany	ACENTISS – Germany	Yuneec – China	University of Nottingham
Reference	http://www.rotexelectric.eu/rotexen/index.php/template-info/bldcmotors/2012-01-30-22-30-21	http://www.siemens.com/press/en/feature/2013/corporate/2013-06-airshow.php?content%5b%5d=CC&content%5b%5d=I&content%5b%5d=IDT&_sm_au_=iMVts5DMjPZ7RNvM	http://www.acentiss.de/de/produkte/produkte-duplexmotor	https://en.wikipedia.org/wiki/Yuneec_Power_Drive_20 https://youtu.be/sKOJ8UzUnYs	[31]
Application	Electric Powered small aircraft	Series hybrid electric drive for Diamond Aircraft 2-seater motor glider	Electric Powered small aircraft	Ultralight aircraft (Espyder)	Green taxiing motor
Machine topology	Outer rotor SPM	Surface PM	Two electric motors on a common drive shaft of the propeller	Outer rotor SPM	Halbach array outer rotor SPM
Cooling method	Air or liquid cooling	Direct oil cooled winding	Air cooling	Air cooling	Air cooling
Mass (kg)	20	13	11	8.2	108 (active)
DC bus voltage (V)	800	545	58	67	
Efficiency (%)	≥95	95			
P_{rated} (kW)	50	65	32	20	
Ppeak (kW)	80	80	40		59.1
kW_{rated}/kg	2.5	5	2.9	2.4	
kW_{peak}/kg	4	6.2	3.6		0.5
n_{rated} (rpm)	2200	5000	2200	2400	
n_{peak} (rpm)	2200	5000	2200	2400	80.8
n_{max} (rpm)	2400	11,000	2500		1800
T_{rated} (Nm)	400	110	139	80	35
T_{peak} (Nm)	790	130	174		6979
Nm_{rated}/kg	10.9	9.5	12.6	9.7	
Nm_{peak}/kg	17.4	11.8	15.8		64.6
kg/l	3.17			1.96	4.25

Table 1.9 High-power density machines for aerospace applications <10 kW.

	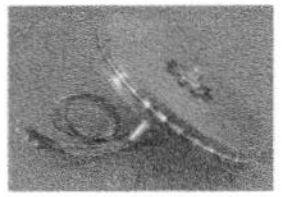		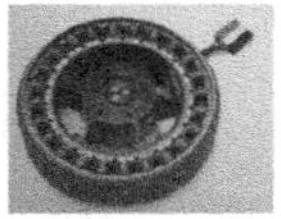	
Manufacturer	Launchpoint	KDE Direct	Joby Motors	ThinGap
Reference	http://www.launchpnt.com/portfolio/transportation/electric-vehicle-propulsion	https://www.kdedirect.com/collections/xf-brushless-motors/products/kde700xf-295-g3	http://www.jobymotors.com/public/public/views/pages/prodcuts.php	http://www.thingap.com/standard-products
Application	Unmanned Aerial Vehicle (UAV)	Remote-controlled electric helicopter series	Remote-controlled model planes	UAV
Machine topology	Axial flux ironless SPM with dual Hallbach arrays	Outer rotor SPM	SPM	Outer rotor SPM
Cooling method	Air cooling	Air cooling	Air cooling	Air cooling
Mass (kg)	0.64	0.695	1.8	1.59
DC bus voltage (V)		50.4–67.2	40–450 (depending on winding connections)	
Efficiency (%)	95	93	85–95	91
Prated (kW)		7.2	8.2	4
P_{peak} (kW)	5.22	12.9	12.6	11.3
kW_{rated}/kg		10.4	4.6	2.5
kW_{peak}/kg	8.2	18.5	7	7.1
n_{rated} (rpm)		14,900	6000	7987
n_{peak} (rpm)	8400	19,800	6000	7987
T_{rated} (Nm)		4.6	13	4.83
Tpeak (Nm)	6	6.2	20	13.55
Nm_{rated}/kg		6.6	7.3	3
Nm_{peak}/kg	9.3	8.9	11.1	8.5
kg/l	1.8	3.78	3.78	1.44

Higher-speed machines (>1 kHz frequency) lend themselves better to distributed windings as well as indirect-conductor cooling (immersed stator, spray cooling, and cooling jacket).

- As previously noted, all the masses are "dry" masses not including the cooling liquid mass. If "wet" masses are included, the gap in specific power between liquid- and air-cooling will decrease.
- All machines are considered low-voltage machines (DC bus voltage ≤ 800 V).
- Lower-speed machines (<3000 rpm) tend to have higher specific torque while higher-speed machines (8000–24,000 rpm) tend to have higher specific power.
- Most of these machines have significantly higher specific power and/or specific torque compared to machines used in the automotive sector. The emphasis is more on specific power and not power density and this is reflected in the kg/l values.

1.7.3 Key System Tradeoffs and Considerations

Beyond the electrical machine, there are several system level tradeoffs and considerations that are equally or more important than the machine specific power. Even though the various applications previously discussed might have different system requirements, they might have some system tradeoffs in common. In general, the following discussion is more geared toward hybrid/electrical propulsion applications with special focus on larger commercial planes which represent the most significant application space moving forward. These system tradeoffs include.

1.7.3.1 Specific Power vs Efficiency

Even though typically the focus especially in aerospace applications is on specific power, efficiency is another key performance metrics. Typically, there is a tradeoff between specific power and efficiency. Specific fuel consumption (SFC) is dependent on both specific power and efficiency. Depending on the overall system architecture, sometimes it is better to design an electrical machine with lower specific power and higher efficiency.

1.7.3.2 Fault Tolerance

As shown, PM machines are really the dominant type since they have the entitlement in terms of high specific power and/or efficiency. For safety-critical applications, it is important to take fault tolerance into consideration. This can lead to a significant reduction in specific power (for example, back emf might need to be limited). Some of the proposed designs that are either ironless and/or have airgap windings usually have very low inductances. This leads to very high fault-currents which would not be acceptable from a system perspective.

1.7.3.3 System Voltage

System voltage is a key parameter. Even though all the machines presented in the paper are low voltage, for MW-class systems and depending on the aircraft size, there will be a need for higher system voltages in excess of 2 kV DC bus voltage. This is mainly to reduce the cables mass which can be the most dominant factor in the overall system specific power. The higher system voltage poses a challenge in terms of the insulation systems required to withstand such voltage levels at altitude (corona effects are more severe at higher altitudes and reduced pressure). This will lead to a different and much thicker insulation build which will make the thermal management of electrical machines much more challenging (depending on the machine aspect ratio and whether spray cooling can be effective or not).

1.7.3.4 Machine Controllability

Machine parameters affecting machine controllability are key factors that have to be considered while designing high specific power machines. Similar to the comments about fault tolerance, if a machine is designed with a low inductance (for example, designs that are ironless and/or have airgap windings), this poses control challenges to keep current ripple under control (to minimize its impact on losses and torque ripple) as well as the control stability. Another design parameter is the machine fundamental frequency. The higher the machine fundamental frequency, the higher the required switching frequency to maintain high quality current waveform. The higher switching frequency can have adverse effect on insulation system and/or sizing of filters. In addition, it can lead to higher switching losses in the power converter and hence reduction in overall system efficiency. A typical switching frequency of 15–20 kHz is used depending on the machine power rating.

1.8 How Does the Future Look Like?

This section will discuss how does the future look like as well as ongoing and potential areas of research. The main focus will be on light-duty vehicles and aerospace applications.

Short-term future, PM machines will continue to dominate the light-duty vehicle market. Also, they started to extend into medium and heavy-duty markets as well as high-speed rail, off – highway vehicles and ship propulsion as previously mentioned in Sections 1.4–1.6 [32, 33].

In terms of PM machines, there are several research areas of interest. These include developing concepts for integrated motor and drive. This can have a significant impact on reducing system cost as well as EMI issues. Couple of examples is shown in Figures 1.17 and 1.18 [34].[26] Another area of research is developing traction motors with 105 °C coolant inlet temperature as previously mentioned. This can have a significant impact on eliminating a separate cooling loop for the motor and inverter, which can cause significant reduction in system cost.

Long-term, there is a lot of concern about the prices, and availability of rare-earth permanent magnets in the future as discussed in the previous section. There is increased activity to try to develop non-rare earth PM traction motors that can have comparable performance to the rare-earth PM motors. Some of the key examples on the commercial side is in [35]. Continental announced that they are at the verge of commercializing wound-field synchronous traction motors as shown in Figure 1.19. Also, BMW is pursuing wound-field synchronous traction motors for their future vehicles.

Jaguar also recently released the plug-in electrical CX75 that has two SR generators provided by SR drives.[27]

Toyota also recently made an announcement that they are working on an advanced induction motor.[28] The concern about the sustainability of rare earth permanent magnets is also obvious

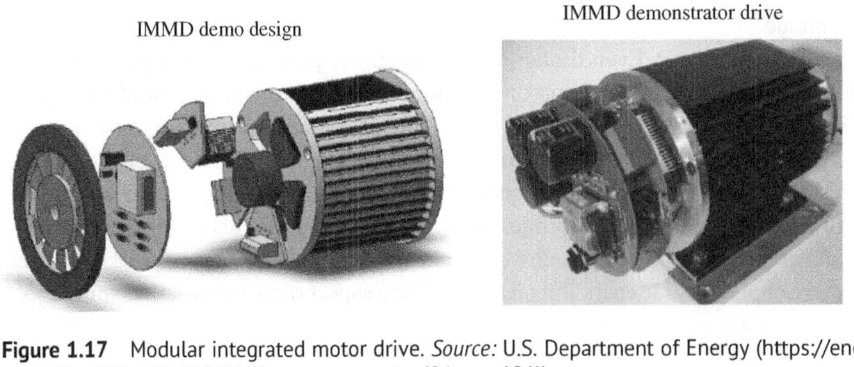

Figure 1.17 Modular integrated motor drive. *Source:* U.S. Department of Energy (https://energy.gov/sites/prod/files/2014/04/f15/2013_apeem_report.pdf (page 191)).

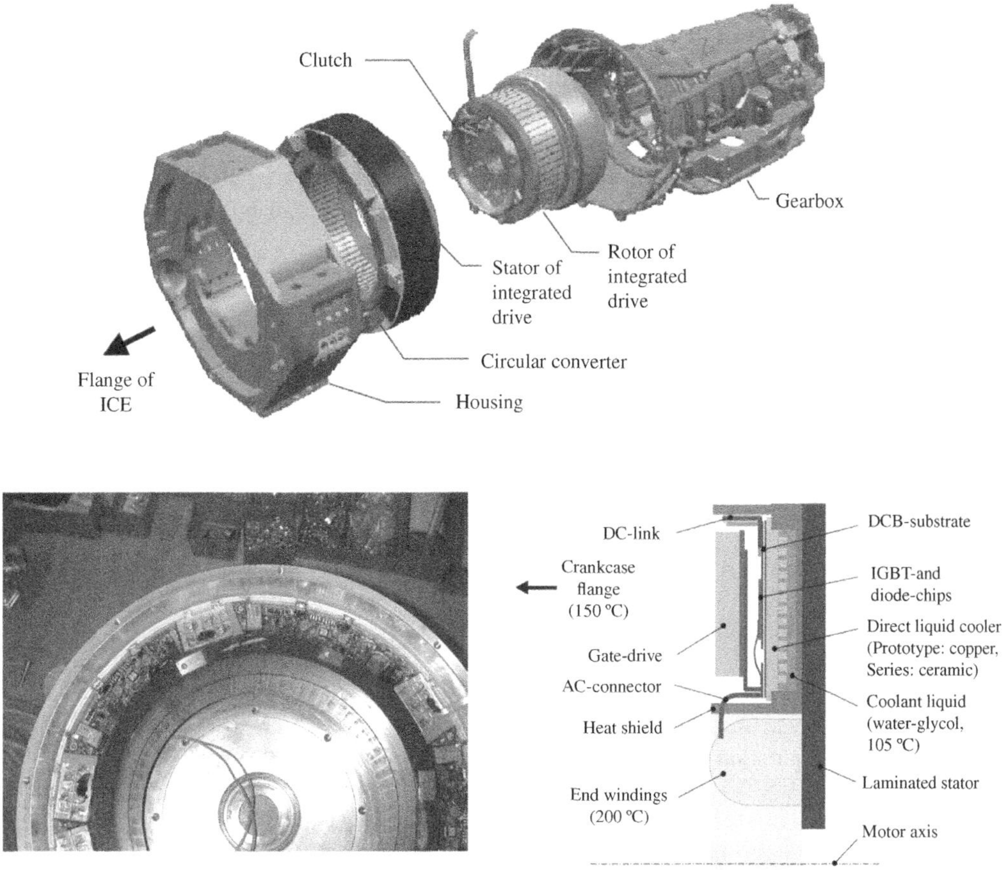

Figure 1.18 Daimler ring integrated motor drive. *Source:* Brown et al. [34].

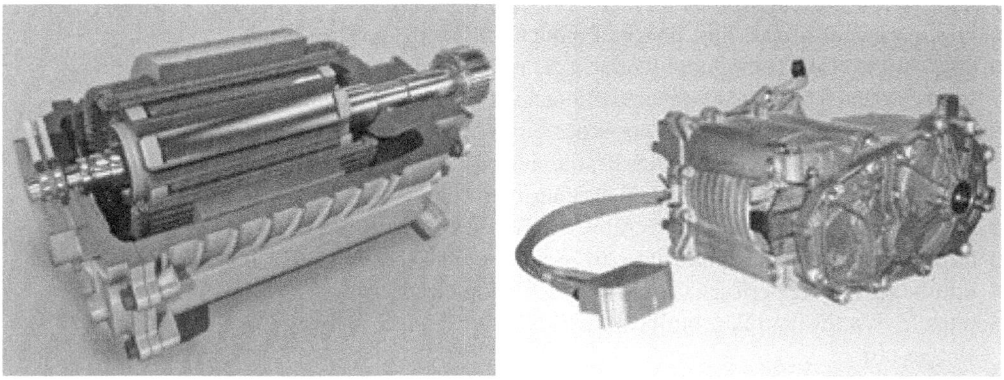

Figure 1.19 Continental wound-field synchronous traction motors 5–120 kW. *Source:* Tadros et al. [35].

in the fact that the US Department of Energy started funding research and development in the area of non-rare earth PM traction motors.[29] So far, the focus has been on light-duty vehicles, but there is a big market for medium and heavy-duty vehicles that needs to be properly addressed.

A critical area (especially for pure electrical traction applications) that needs to be properly addressed is fault tolerance and limp-home capability. Many of the concepts for safety-critical applications like aerospace applications can potentially leveraged in the traction sector [36].

In terms of the hybrid/electric propulsion for aerospace applications, an increased level of research and development to achieve very high specific powers will continue. This will include developing and optimizing novel machine topologies, develop advanced thermal management as well as developing and implementing advanced materials and manufacturing.

Notes

1 https://arstechnica.com/cars/2019/04/motor-technology-from-model-3-helps-tesla-boost-model-s-range-10/
2 www.proteanelectric.com/live/docs/PD18-3-LV1.2-Data-Sheet.pdf
3 www.mitsubishimotors.com
4 www.tm4.com/moteurs_generateurs.aspx
5 http://xoomer.virgilio.it/parimirco/Wheel%20Hub%20Motors%20-Nagashima-%20Session%20D2%20.ppt
6 www.atstransmission.com
7 www.baesystems.com/
8 www.azuredynamics.com/
9 http://metadope.com/Bus/pdf/AC55.pdf
10 www.oshkoshcorporation.com/
11 www.uqm.com/
12 www.srdrives.co.uk/
13 http://machinedesign.com/article/hybrid-drives-for-construction-equipment-0707
14 http://www.unife.org/uploads/2008/presentations%20innotrans/rs/01_Presentation%20Alstom.pdf
15 http://www.ship-technology.com/contractors/propulsion/ldw/
16 http://www.drs.com/Products/PESG/PDF/PMMS.pdf
17 http://www.amsc.com/products/motorsgenerators/documents/mp_ds_365_1007_a4_r3.pdf
18 http://www.rolls-royce.com/Images/Product%20Range_tcm92-8658.pdf
19 http://www05.abb.com/global/scot/scot293.nsf/veritydisplay/9eccbc669a5fcff1c125768f004a9bc9/$file/azipod%20project%20guide%206-3.pdf
20 http://www05.abb.com/global/scot/scot293.nsf/veritydisplay/d77d9ac59ba74cb7c12576d3004cfe3f/$file/azipod%20co%20brochure_2010.pdf
21 http://www.industrysolutions.siemens.com/marine/en/
22 http://www.converteam.de/majic/dl/4/doc/Markets/Navy/Final_NAVAL.pdf
23 https://energy.gov/sites/prod/files/2017/06/f34/edt087_burress_2017_o.pdf
24 https://aerospaceamerica.aiaa.org/features/fly-the-electric-skies/
25 https://www.theguardian.com/business/2017/apr/21/electric-flying-car-lilium-google-uber-vtol-jet-taxi
26 https://energy.gov/sites/prod/files/2014/04/f15/2013_apeem_report.pdf (page 191)
27 http://www.contionline.com/generator/www/com/en/continental/pressportal/themes/press_releases/3_automotive_group/powertrain/press_releases/pr_2010_04_29_elektromotoren_en.html
28 http://online.wsj.com/article/SB10001424052748703583404576080213245888864.html
29 http://www.cleancitieseastbay.org/DOE-Funding-opp-12-16-10.sflb.pdf

References

1 Savagian, P. (2010). Motors for automotive electrification: a GM perspective. *SAE Hybrid Vehicle Technologies 2010 Symposium*, February 2010, San Diego.

2 El-Refaie, A.M. (2013). Motors/generators for traction/propulsion applications: a review. *IEEE Veh. Technol. Mag.* 8 (1): 90–99.

3 El-Refaie, A. and Osama, M. (2019). High specific power electrical machines: a system perspective. *CES Trans. Electr. Machines Syst.* 3 (1): 88–93. https://doi.org/10.30941/CESTEMS.2019.00012.

4 Staunton, R.H., Burress, T.A., and Marlino, L.D. (2006). Evaluation of 2005 HONDA ACCORD hybrid electric drive system. ORNL/TM-2006/535.

5 Hsu, J.S., Ayers, C.W., Coomer, C.L. et al. (2007). Report on Toyota/Prius motor torque capability, torque property, no-load back EMF, and mechanical losses. ORNL/TM-2004/185.

6 Burress, T.A., Coomer, C.L., Campbell, S.L. et al. (2007). Evaluation of the 2007 Toyota Camry hybrid synergy drive system. ORNL/TM-2007/190.

7 Burress, T.A., Coomer, C.L., Campbell, S.L. et al. (2008). Evaluation of the 2008 Lexus LS 600H hybrid synergy drive system. ORNL/TM-2008/185.

8 Aghabali, I., Bauman, J., Kollmeyer, P.J. et al. (2021). 800-V electric vehicle powertrains: review and analysis of benefits, challenges, and future trends. *IEEE Trans. Transp. Electrification* 7 (3): 927–948. https://doi.org/10.1109/TTE.2020.3044938.

9 Mraz, S.J. (2010). Hub motors for all-electric vehicles still have some technological challenges to overcome. Machine http://Design.com (accessed 2021).

10 Rahman, K.M., Patel, N.R., Ward, T.G. et al. (2006). Application of direct-drive wheel motor for fuel cell electric and hybrid electric vehicle propulsion system. *IEE Trans. Industry Appl.* 42 (5): 1185–1192.

11 Nagashima, J. (2005). Wheel hub motors for automotive applications, EVS-21, Monaco.

12 Welchko, B.A., Jahns, T.M., Soong, W.L., and Nagashima, J.M. (2003). IPM synchronous machine drive response to symmetrical and asymmetrical short circuit faults. *IEE Trans. Energy Convers.* 18 (2): 291–298.

13 Han, S.-H., Jahns, T.M., Guven, M.K. et al. (2006). Impact of maximum back-EMF limits on the performance characteristics of interior permanent magnet synchronous machines. *Industry Applications Conference, 2006. 41st IAS Annual Meeting*. Conference Record of the 2006 IEEE, vol. 4, pp. 1962–1969, 8–12 October 2006.

14 Jahns, T.M., Han, S.-H., El-Refaic, A.M. et al. (2006). Design and experimental verification of a 50 kW interior permanent magnet synchronous machine. *Industry Applications Conference, 2006. 41st IAS Annual Meeting*. Conference Record of the 2006 IEEE, vol. 4, pp. 1941–1948, 8–12 October 2006.

15 SKF. Solutions for traction motors. https://www.skf.com/binaries/pub12/Images/0901d1968019754c-6815EN_tcm_12-57716.pdf.

16 Gieras, J. (2008). *Advancements in Electric Machines*. Springer.

17 Moghaddam, R.R. (2014). High speed operation of electrical machines, a review on technology, benefits and challenges. *2014 IEEE Energy Conversion Congress and Exposition (ECCE)*, Pittsburgh, PA, pp. 5539–5546.

18 Gerada, D., Mebarki, A., Brown, N.L. et al. (2014). High-speed electrical machines: technologies, trends, and developments. *IEEE Trans. Ind. Electron.* 61 (6): 2946–2959.

19 Zhang, X. and Haran, K.S. (2016). High-specific-power electric machines for electrified transportation applications-technology options. *2016 IEEE Energy Conversion Congress and Exposition (ECCE)*, Milwaukee, WI, pp. 1–8.

20 Nichols, S., Foshage, J., and Lovelace, E. (2006). Integrated motor, propulsor (IMP) and drive electronics for high power density propulsion. *Electric Machine Technology Symposium EMTS06*, Philadelphia, PA: American Society of Naval Engineers (ASNE).

21 Kane, D.M. and Warburton, M.R. (2002). Integration of permanent magnet motor technology. *Power Engineering Society Summer Meeting*. IEEE, vol. 1, pp. 275–280; vol. 1, 25–25 July 2002.

22 Brunvoll presents a rim driven thruster RDT, 2005, Brunvoll AS, Molde, Norway. www.brunvoll.no; http://www.norpropeller.no

23 Roboam, X., Sareni, B., and Andrade, A.D. (2012). More electricity in the air: toward optimized electrical networks embedded in more-electrical aircraft. *IEEE Ind. Electron. Mag.* 6 (4): 6–17.

24 Cao, W., Mecrow, B.C., Atkinson, G.J. et al. (2012). Overview of electric motor technologies used for more electric aircraft (MEA). *IEEE Trans. Ind. Electron.* 59 (9): 3523–3531.

25 Neuman, T. (2016). Fly the electric skies. *IEEE Spectr.* 53 (6): 44–48.

26 Yoon, A., Xuan Yi, J. Martin, Y.C. and Haran, K. (2016) A high-speed, high-frequency, air-core PM machine for aircraft application. *2016 IEEE Power and Energy Conference at Illinois (PECI)*, Urbana, IL, pp. 1–4.

27 Galea, M., Xu, Z., Tighe, C. et al. (2014). Development of an aircraft wheel actuator for green taxiing. *2014 International Conference on Electrical Machines (ICEM)*, Berlin, pp. 2492–2498.

28 Schiefer, M. and Doppelbauer, M. (2016). Indirect slot cooling for high-power-density machine with concentrated winding. *# Presentation by M. Doppelbauer in Airbus Symposium*, Ottobrun (11 March 2016).

29 Anghel, C. (2015). High-efficiency and high-power-density generator rated for 1MW. *Electric & Hybrid Aerospace Technology Symposium*, Bremen, Germany.

30 Orr, E. (2017). Keynote presentation by Rolls Royce at More Electric Aircraft MEA.

31 Galea, M., Xu, Z., Tighe, C. et al. (2014). Development of an aircraft wheel actuator for green taxiing. *2014 International Conference on Electrical Machines (ICEM)*, pp. 2492–2498.

32 Weeber, K.R., Shah, M.R., Sivasubramaniam, K. et al. (2010). Advanced permanent magnet machines for a wide range of industrial applications. *2010 IEEE Power and Energy Society General Meeting*, pp. 1–6, 25–29 July 2010.

33 El-Refaie, A., Nold, R., Haran, K. et al. (2012). Testing of advanced permanent magnet machines for a wide range of applications. *2012 International Conference on Electrical Machines*, Marseille, pp. 1860–1867

34 Brown, N.R., Jahns, T.M., and Lorenz, R.D. (2007). Power converter design for an integrated modular motor drive. *Industry Applications Conference. 42nd IAS Annual Meeting*, New Orleans, LA. Conference Record of the 2007 IEEE, pp. 1322–1328, 23–27 September 2007.

35 Tadros, Y., Ranneberg, J., and Schäfer, U. (2003). Ring shaped motor-integrated electric drive for hybrid electric vehicles. EPE, Toulose.

36 EL-Refaie, A.M. (2009). Fault-tolerant PM machines: a review. *Electric Machines and Drives Conference*, Miami, FL. IEMDC '09. IEEE International, pp. 1700–1709, 3–6 May 2009.

2

Advances and Developments in Batteries and Charging Technologies

Satish Chikkannanavar[1] *and Gunho Kwak*[2]

[1] *Electrified Systems Engineering, Ford Motor Company, Vehicle Powertrain Electrification Center, Allen Park, MI 48101, USA*
[2] *Ford Motor Company, Product Development, Ford Ion Park, Romulus, MI 48174, USA*

2.1 Introduction

The scope for the use of Li-ion batteries (LIBs) is expected to expand further in the future as global interest in climate change continues, to replace internal combustion engines with electric motors. LIBs are the most versatile energy storage systems found in our daily life. Besides the greater role they have played in the field of consumer electronic devices since 1990s, we have seen their new application in the electrification field for the past decade, where batteries help in propulsion in the xEVs. With suitable battery sizing, automotive companies have designed and launched full hybrid electric vehicles (FHEVs), plug-in hybrid electric vehicles (PHEVs), and battery electric vehicles (BEVs or EVs), offering customers a range of choice for their personal or mass transport. The need for higher electric vehicle (EV) range and higher fuel economy are the main drivers for exploration in this field. With the need for higher range, there comes the need to charge the high voltage battery at faster rate using a network of fast chargers. In this chapter, we would like to cover various topics on (i) advances in cathodes and anodes covering high energy density for EV applications, (ii) high power/energy cell design for PHEVs and FHEVs (iii) post LIBs: solid-state batteries, (iv) advances in charging batteries, (v) degradation considerations, and (vi) future outlook.

2.2 Advances in Cathodes/Anodes Covering Energy Density Increase for EV Applications

The need for energy storage devices that accelerate the transition from fossil-fuel-based power to electric power has motivated significant research into the development of cathode materials for rechargeable metal-ion batteries based on different standard reduction potential of Li^+, K^+, Ca^{2+}, Na^+, Mg^{2+}, Al^{3+}, Zn^{2+}, and Fe^{2+}. No one would disagree that LIBs have the best battery performance among the various secondary batteries currently in the market. The LIBs have been by far the most successful, enabling the wireless revolution of portable electronics such as cell phones, laptops, digital assistant, audio devices, watches, and power tools and emerging as the battery-of-choice for EVs and intermittent energy storage [1]. However, as a power source for EVs, LIBs

Transportation Electrification: Breakthroughs in Electrified Vehicles, Aircraft, Rolling Stock, and Watercraft,
First Edition. Edited by Ahmed A. Mohamed, Ahmad Arshan Khan, Ahmed T. Elsayed, and Mohamed A. Elshaer.

have several technical issues that need to be addressed, which relate to energy density, cost (US $/kWh), reliability, and safety. The available energy of a battery is a function of its voltage and capacity, and energy density can be maximized by minimizing the weight and volume of active material. Also, having a large potential difference between the electrodes and avoiding oxidation or reduction reactions with excessive consumption of the electrolyte are other factors [2].

A typical LIB consists of a transition metal oxide as cathode and a graphite as anode. Cathode is the main Li-ion source and the electrochemical reduction reaction occurs during discharge. The formulation for cathode is made of active material of lithium source; conductive additive (typically carbon black), which increases electrical conductivity; and a polymer binder like as polyvinylidene fluoride (PVDF), which is serves as a kind of adhesive. Meanwhile, anode stores the lithium ions and undergoes thorough oxidation when using the battery, thus generating electricity in the external circuit. Typically, anodes consist of graphite or silicon active material and binder. In this chapter, we shall survey cathode and anode materials that can increase the mileage and improve performance of EVs through high energy density.

2.2.1 Cathode Challenges for High Energy Density

In an LIB, cathode is the main component which drives the battery cost. Also, the battery capacity is dictated by the cathode. Hence, the emphasis should be on achieving high energy density. Over the past decade, the discovery of new materials and the development of a fundamental understanding of the structure–property–performance relationship has played an important role. In order for the batteries in EVs to have high energy density, power, and cycle life, the performance of the cathode is very important. Thus, the development of high-performance cathode active materials for LIBs has become a key research area. There are currently five cathode active materials that produce adequate performance as batteries, as shown in Table 2.1.

Table 2.1 Five major cathode active materials used in the LIB industry.

	LCO	LFP	LMO	NCA	NCM
Cathode elements	Cobalt	Iron	Manganese	Nickel Cobalt Aluminum	Nickel Cobalt Manganese
Chemical formula	$LiCoO_2$	$LiFePO_4$	$LiMnO_2$	$Li(NiCoAl)O_2$	$Li(NiMnCo)O_2$
Specific energy (Wh/kg)	150–200	90–120	100–150	200–260	200–270
Power density (W/L)	1200–3000	4500	2000	4000–5000	6500
Thermal runaway (°C)	150	270	250	150	240–305
Safety	–	++	+	––	Ni content dependent
Comparative advantage	Capacity Stability	Stability Cost	Power Cost	Capacity Power	Capacity Balanced performance
Applications	IT/consumer device	Non-IT EV	Non-IT ESS	Non-IT EV	EV ESS
Year introduced	1991	1996	1996	1999	2008

Cathodes lithium nickel cobalt manganese oxide (NCM) and lithium nickel cobalt aluminum oxide (NCA) have the widest application spectrum including transportation in xEVs. The interaction of three metal components (Ni, Mn, Co) with Li offers cathode characteristics which have enabled such performance. Strategies for improving LIB characteristics are divided into categories related to the system development such as electrode density, high voltage window, efficient battery module/pack design, and categories related to the development of four key components namely cathode, anode, electrolyte, and separator. Four of these key components' development strategies are methods that can improve cell capacity, operating voltage, wide temperature range, durability, energy density, and power of LIBs [3].

Doping in NCM system of cathodes increases the conductivity of electrons and ions and limits the movement of Ni^{2+} ions to the lithium layer, reducing cation mixing and maintaining structural stability. It also strengthens transfer metal oxide bonds, inhibiting oxygen release and affecting phase changes during cycling. It can minimize side reactions on active material and enhance the rate of Li diffusion with charge transfer. The most frequently used doping materials for Ni-rich active materials, which are now in the spotlight as the next generation of cathode, are Al, Ti, Mg, Si, Zr, W, Mo, Ga, Sn, Vb, Zr, B, and others [4]. Typical dopants have electrochemical inactivity, high polarity, large ion radius, and strong binding power with oxygen. Without doping, microcracks occurred in the interparticle boundaries, the primary particles separated from each other to form NiO-phase, and electrolytes penetrated inside through the interparticle boundaries. However, with doped, there are no microcracks and internal deformation due to volume change during cycling is suppressed and that helps prevent capacity reduction. High-Ni cathode materials that have increased nickel content to more than 80% are attracting attention due to their high capacity and relatively lower cost. However, as charging and discharging are repeated, not only the microcracks occur inside the material particles but also durability is rapidly reduced due to side reactions with electrolytes.

Coating on cathode active materials is unlike doping, and is associated with improving the surface characteristics rather than stabilizing the internal structure of the Ni-rich cathode. Surface coating and modification can prevent active material from direct contact with the electrolyte and prevent electrolyte decomposition or oxidation, inhibit the dissolution of metal ions, and accelerate the charge transfer. It also improves cell cycle life and capacity retention and contributes to thermal stabilization [5]. The electrochemical properties of cathode active material and the kinetics of lithium ions are greatly influenced by boundary conditions such as various coating materials with sizes, coating thickness, uniformity, coating density, and conductivity. The typical coating materials for Ni-rich cathode are metal oxides such as Al_2O_3, ZrO_2, $Zr(OH)_4$, TiO_2, $B2O_3$, MoO_3, and WO_3; fluorides such as AlF_3; phosphates such as $Ni_3(PO4)_2$, $FePO_4$, and TiP_2O_7; Li-conducting oxides such as $LiAlO_2$, Li_2TiO_3, Li_4SiO_4, Li_3ZrO_2, Li_2ZrO_3, Li_2O-$2B_2O_3$, Li_3PO_4, and Li_2WO_4; and dual-functional reformation including Li borate coating and boron doping [6, 7].

Most of the disadvantages of Ni-rich NCM cathodes are high Ni content, which leads to unstable Ni turning into an impurity; Ni–O phase, with electrolytes breaking down the surface of cathode material; and microcrackage due to anisotropic volume change. To compensate for these shortcomings, concentration gradient cathodes were synthesized by adjusting the composition of the center and surface of the particles differently. This strategy can protect the surface of the cathode material from electrolytes. By increasing the content of Ni at the center, the materials will retain high capacity, and the surface being Mn rich, has a stable structure, which helps prevent the spread of crack. It also produces crystalline tissue in the form of rods, which facilitates the spread of lithium ions. This effect not only improves thermal stability but also improves capacity retention. Figure 2.1 shows the progress over the years on the design of concentration gradient Ni-rich cathodes. We can notice that with increase in capacity, the cycle life performance is hard to improve, but with concentration gradient

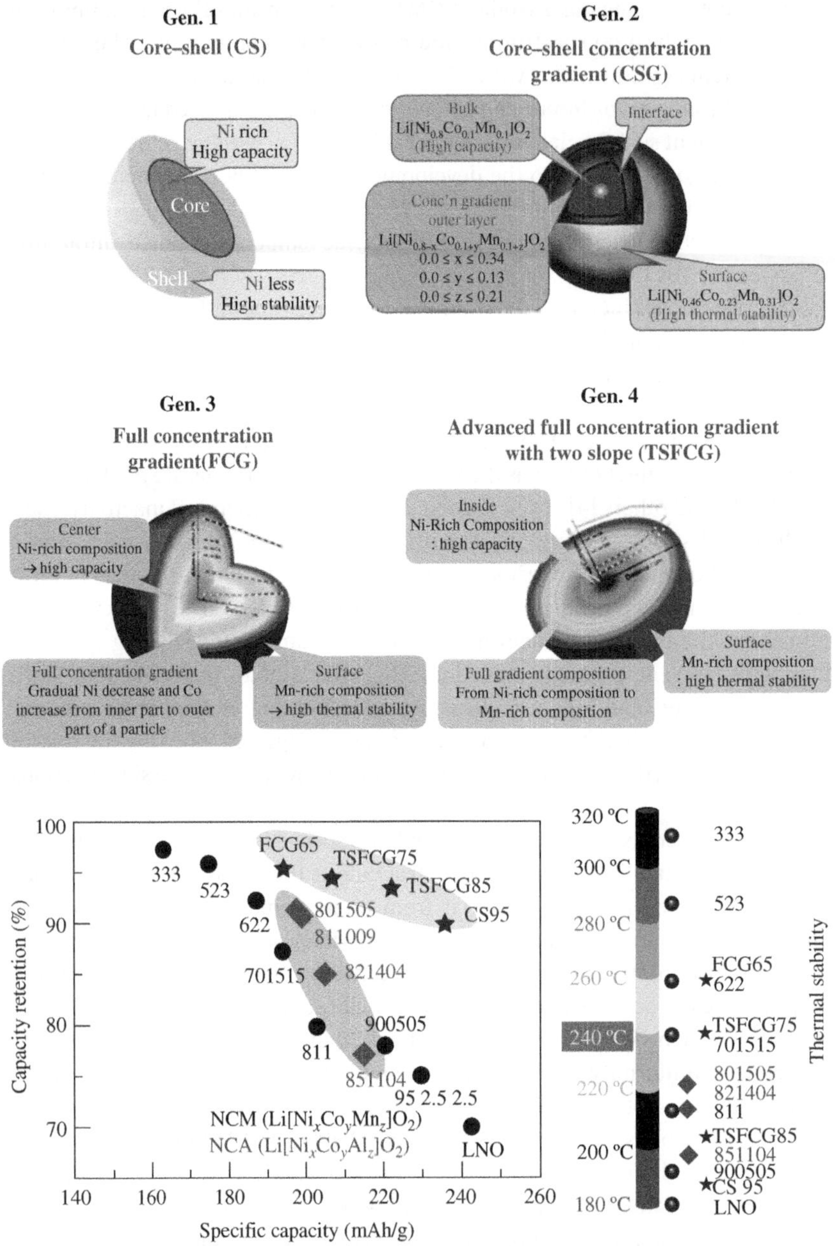

Figure 2.1 Concentration gradient cathode materials for high energy density LIBs with capacity retention, specific capacity, and thermal stability. *Source:* Yoon et al. [8]/with permission of Yang-Kook Sun.

structures. Also, next to the chart of capacity life performance, the bar chart shows the benefits of surface stabilization, with improved and higher onset for temperature for thermal event [9].

2.2.2 Anode Challenges for High Energy Density

There are many candidates for anode materials such as natural and artificial graphite, lithium metal, and silicon. Natural graphite is a low-cost active material that is capable to exhibit stable electrochemical performance and low-volume expansion during the reversible lithium intercalation and deintercalation processes. By stably hosting Li^+ into the carbon interlayers forming LiC_6, graphite yields a theoretical capacity of 372 mAh/g. Artificial graphite is a manufactured product made by high-temperature treatment of amorphous carbon materials. The production involves the calcination of the intermediate precursor material at around 800–1200 °C, followed by crushing and grinding, from which the particles are classified according to their particle size [10]. Therefore, the price is more expensive than natural graphite, but due to its high purity, structural stability, and production quality, it has a comparative advantage over cycle life and a wide range of operating temperatures. Thus, most suppliers produce anodes for LIBs with a blend of both natural and artificial graphite to balance the cost and performance. However, for graphite anodes, the maximum capacity of 372 mAh/g is 10 times lower than that for lithium metal (3800 mAh/g). Lithium metal can be the ideal anode material, but shows problems with dendrite growth, which creates internal short. Silicon has 10 times more energy density through a reversible alloying process that offers significantly higher theoretical capacity of 4200 mAh/g than graphite, and is seen as an alternative. But it was difficult to commercialize it due to the problem of poor electrode performance with silicon's low electrical conductivity and also due to large volume expansion during cycling usage [11]. Several hurdles have been reported that limit its applications, including a large crystallographic volume expansion (~300%) upon full lithiation, low lithium diffusion rate, and high reactivity at full charge, resulting in particle cracking, isolation, electrolyte decomposition, and electrode delamination defects. These chemical reactivity and volume expansion pose problems such as solid electrolyte interface (SEI) instability, coulombic efficiency, and durability for calendar and cycle life [12]. Thus, researchers in various fields have tried to create a variety of silicon-based electrodes by understanding and analyzing these physicochemical problems. Advances in material design strategies of various nanostructures such as SiO_x, core-shell, Si nanoparticles coated with carbon, Si vapor deposited on current collector, porous carbon particles filled with Si, nanowires, and nanotubes have been developed over the years to significantly improve the physicochemical property of Si-active materials [13–15]. However, improving the performance of Si anodes through structural control is limited by focusing only on silicon individual particles and/or graphite interface, which requires stabilization technology at the electrode and cell design level.

2.3 High Power/Energy Cell Designs for xEVs

For the batteries designed for xEVs, there are two categories: power batteries for applications in HEVs, and energy batteries for use in PHEVs and BEVs. There is no choice, but we have to deal with the trade-off between energy and power performance when designing and producing the batteries. Products designed to maximize energy density suffer from low power performance, and fast charging/discharging cell design is caught in a dilemma of low energy density. In the case of HEVs, power tends to be more important, but in the case of PHEVs, recent mobility electrification trends are increasingly demanding to satisfy both conflicting characteristics. In this chapter, we will discuss the correct LIB material and cell selection methods for each application range, along with the differences in cell design between power cells and energy cells.

The basic difference between an energy cell and a power cell can be divided into how deep and how fast. The power cell is designed to deliver intermittent high current loads over a short period of time, making it ideal for use at high speeds that can support high loads and produce large torque. On the other hand, energy cells are designed to carry continuous currents over a long period of time, making them ideal for cyclic, long drive applications. It means that there must be trade-off in the

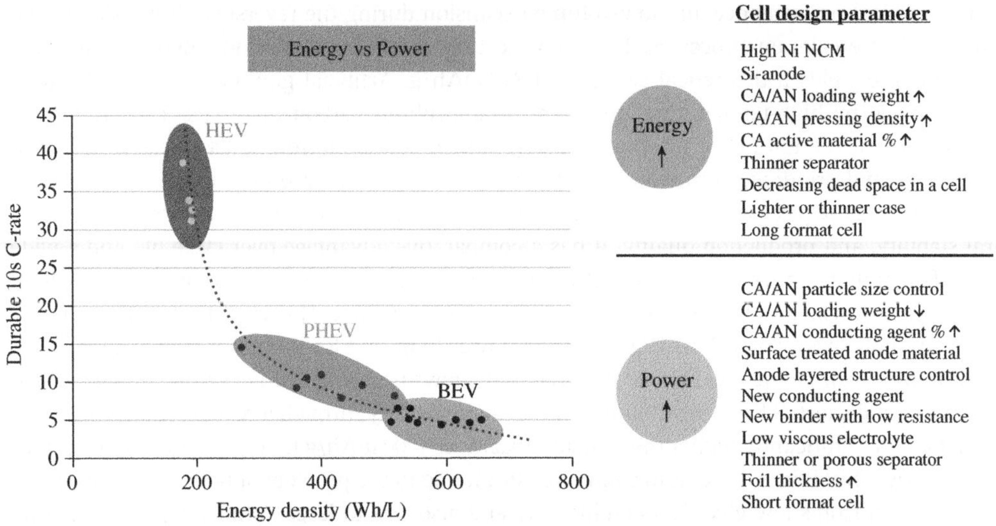

Figure 2.2 Cell design parameters of power cells and energy cells (Ford internal data).

cell design required for power and energy. Achieving this trade-off through fundamental material modification, new cell chemistry, innovative cell design with fabrication, advanced module/pack design, and improved battery management system have been attempted [16, 17]. As shown in Figure 2.2, the "Power cell (kW↑)" is typically designed with small particle sizes and low loading levels to maximize the lithium ion mass transport. It also consists of carbon additives, low-resistance binders, low-viscosity electrolytes, and with thin porous separator to increase electric and ionic conductivity. Thus, it is characterized by thin, porous electrodes coated on more current collectors, and it must be thicker in itself to handle high electron current density. This reduces energy density as more volumes/weights are used. Power cells tend to cost more on a kWh basis than energy cells, as the proportion of inactive substances that reduce energy density is high, e.g., expensive current collectors and tabs usages are high. The short-format cell for low resistance has also negative effect on cost competitiveness. In contrast, "energy cells (kWh↑)" are designed to maximize the fraction of active materials which increase the energy density with high loading weight and minimize dead space with lighter and thinner assembly parts. As the fraction of inactive substances such as carbon additives and current collectors is reduced for high energy density cells, it is characterized by a design that thickens the active material of large particles on the thinner current collector and compresses with low pores to create high electrode density. Thus, these dense electrodes degrade the diffusion rate of lithium ions, hinder the movement of electrons and ions, and increase the resistance of the electrode, resulting in poor charging and discharge power performance.

2.4 Post Li-Ion Batteries: Solid-State Batteries

Designing the batteries which are long lasting, cost competitive, fast charging, and safety is essential to the expansion of the xEV market. However, current state of the art LIBs still cannot provide the long driving range compared to the internal combustion engines. Also, LIBs are still high cost by US \$/kWh, and take longer time to charge. A solid-state battery (SSB) is a secondary battery in which all the component materials are in solid phase. In general, compared to conventional liquid bases LIBs, SSBs are classified as superior secondary batteries, with the characteristics including excellent safety due to low possibility of explosion and fire. No separator is needed to physically block the cathode and anode, which can reduce cell volume/weight and also reduces cost. Thermal management system and heat-diffusion protection components that account for a significant portion of the battery pack can be eliminated, contributing to US $/kWh reduction and increased energy density. Unlike the mono-polar structure of the conventional LIB, the multi-polar structure allows multiple electrode connections to reduce the volume of the module. In addition, it can contribute to improving life expectancy due to low level side reactions such as high temperature deterioration, gas generation, and expansion due to solid electrolytes.

However, despite these superior performance and advantages, SSBs still have many technical challenges to solve, such as reduced ion mobility of the solid electrolytes, hard to expand capacity with secure process capability, and dendrite generation problems that reduce charging efficiency [18]. In this chapter, we will discuss the current development status of SSBs and the challenges and problems for the mass production.

2.4.1 Roadmap and Collaborative Relationships

The popular discussion of SSBs, which has drawn attention as a game changer, has led to progress in both academia and industry. In Figure 2.3, automotive original equipment manufacturers (OEMs), major cell manufacturers, startups, government-funded research institutions, and stakeholders in the LIB industry are all engaged in strategic alliances to develop and mass produce SSBs that will shift the existing LIB paradigm.

2.4.2 Current Development Status and Key Challenges

In order to compete with internal combustion engines, batteries must be safe, have sufficient long driving range, and offer cost competitiveness. To this end, the automotive industry requires high energy density (~400 Wh/kg, ~1000 Wh/L) and higher power specifications of 5000 W/kg depending on application usage. With the combination of Li-metal anode and high voltage cathode active materials, solid-state technology is the most demanding technology in terms of performance and mass productivity as it approaches automotive requirements. As you can see in Figure 2.4, the key point of SSBs is to eliminate some essential components that must be installed in current state of the art LIBs. It is possible to increase energy density by filling the space with more anode and cathode active materials. The energy density for raising the driving range is achieved by using Li-metal or silicon-based anode with high voltage cathode materials at the cell level and simplified bipolar system which can produce more than 14 V in a single cell and by omitting thermal/cooling systems at the pack level. In addition, SSBs will have significantly better safety than conventional LIBs, due to the elimination of flammable liquid electrolytes. Solid electrolytes offer lower power for the batteries, due to lower Li-ionic conductivity compared to liquid electrolytes, and have poor cycle life than conventional LIBs, which is due to high interfacial resistance between solid electrolyte and electrode. Therefore, various solid electrolytes such as sulfides, oxides, and polymers are being studied to improve interfacial resistance between electrode and solid electrolytes, but there is still a difference from liquid electrolyte performance as shown in Table 2.2.

It is an essential problem that solid electrolytes have low ion conductivity, and various challenges are to be addressed. The contact interfaces between active materials and electrolyte have to be maximized, while minimizing surface resistance. There are many observations regarding the timing of full-scale introduction of SSBs, but there will not be any doubt to the premise that they will be able to replace existing LIBs. However, many challenges have not been solved through solid electrolyte materials, interfacial resistance between active materials and electrolytes, and secure manufacturing

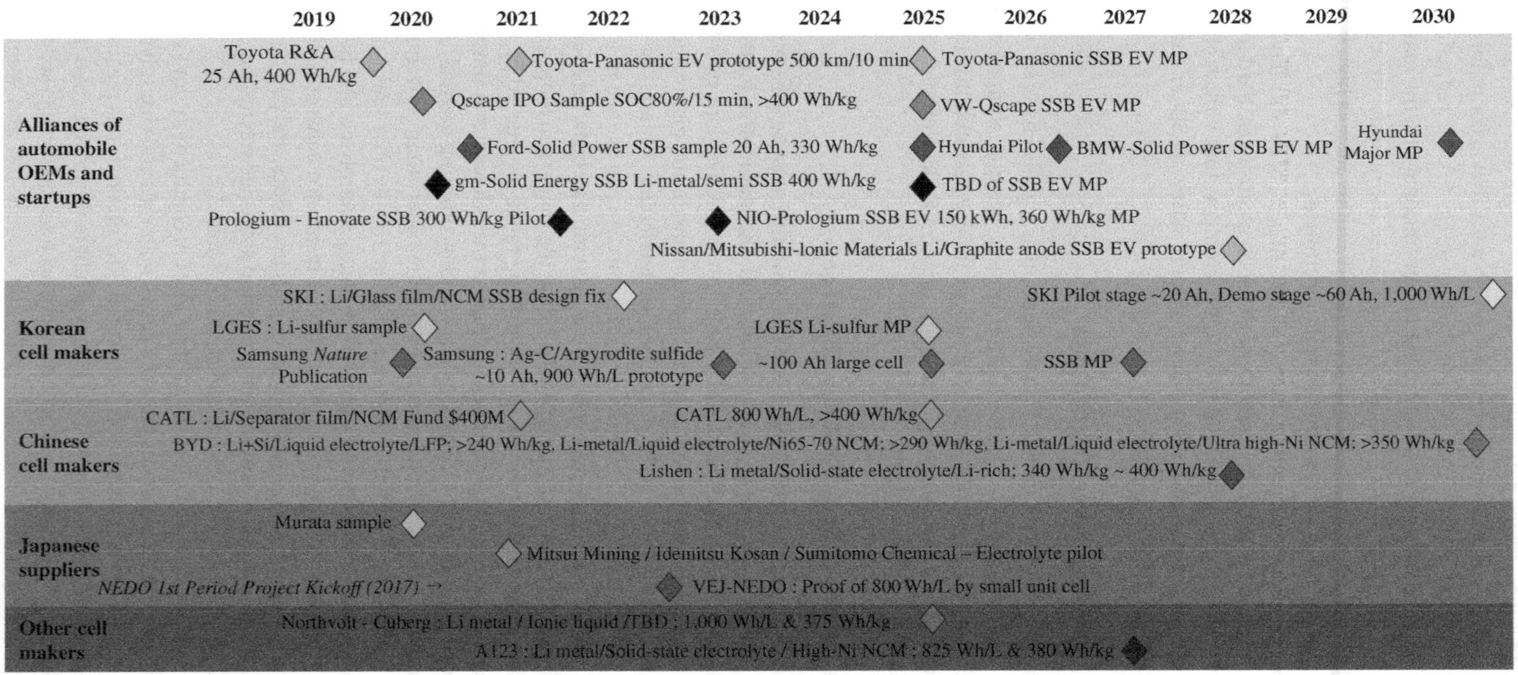

Figure 2.3 The roadmap and collaborative relationships for SSBs.

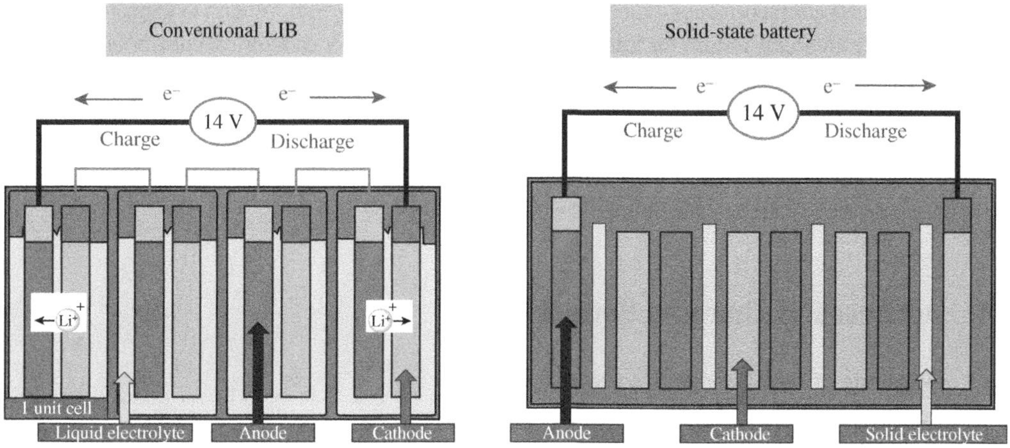

Figure 2.4 Comparison of current LIBs vs. SSBs bipolar system.

Table 2.2 Comparison of solid electrolytes.

Types	Pros	Cons	Key players with anode
Sulfides	– Good mechanical strength and flexibility – Scalable with productivity – Low grain-boundary resistance – Higher ionic conductivity (~10^{-2} S/cm)	– Unstable in air – Compatibility issues – Low oxidation stability	– Solid power/Ford/BMW: sulfidic ceramic with Li – QuantumScape/VW: sulfidic ceramic with anode-free Li – Panasonic/Toyota: sulfidic ceramic with Li – Samsung: sulfidic ceramic with Li Ag-C coated
Oxides	– High chemical and electrochemical stability – Good mechanical strength – Air/humidity stable – High oxidation voltage – Medium ionic conductivity (~10^{-3} S/cm) – Thermal stability	– Difficult to scale-up – Poor compatibility – Not flexible – High cost	– ProLogium/NIO/FAW: oxidic ceramic with Li or graphite/Si – QuantumScape/VW: oxidic ceramic with anode-free Li – Panasonic/Toyota: oxidic ceramic with Li
Polymers	– Good electrochemical properties – Air/humidity stable – Flexibility – Scalable – Low cost	– Poor mechanical strength – Low ionic conductivity (~10^{-4} S/cm) – Low oxidation voltage – Low shear modulus	– Ionic materials/Nissan/Renault/Mitsubishi/Hyundai: doped Pi-conjugated polymer with Li or graphite – Solid energy/GM: semi-solid solvent-in-salt with Li – S ion power/BASF/AIRBUS: hybrid ceramic-polymer with Li

processes with cost competitiveness. Electrolyte materials for SSBs are still not up to expected performance and there are many issues. Because manufacturing process of cell requires enormous pressure and high temperature, it is difficult to establish mass production facilities. Since it is in solid state, it is hard to avoid interfacial resistance between heterogeneous powders and interfacial resistance of electrodes and electrolytes. Like conventional LIBs, coating with slurry during SSB electrode manufacturing significantly reduces capacity or rate characteristics. Therefore, it is currently inferior to LIBs other than safety and requires continuous development.

2.5 Advances in Charging Batteries

2.5.1 Methods of Fast Charging Batteries

Since their early invention, LIBs have had charging methods which followed safety during usage in consumer electronic devices. With the proliferation of usage in xEV applications, the need for fast charging has become a popular demand. Among the various charging approaches [19], most common approach is constant current followed by constant voltage (CC-CV) method (Figure 2.5). In this approach, a current or power (CC or CP) is used at constant rate to charge the battery from low state of charge (SOC) to a certain cut off upper voltage level. Beyond this point, charge current is reduced in what is called as constant voltage (CV) mode, to charge the battery to the full level. As this step is slow and takes longer time until current tapers to a low value (e.g., C/20 or lower), CC-CV approach takes longer time. Using higher currents can cause battery performance degradation [20, 21]. In multistage constant current (MCC) approach, constant currents at decreasing levels are applied to reduce the thermal influence of high currents during the charging step. This is followed by CV mode charging to complete the charging to topmost SOC. Literature reports on charging

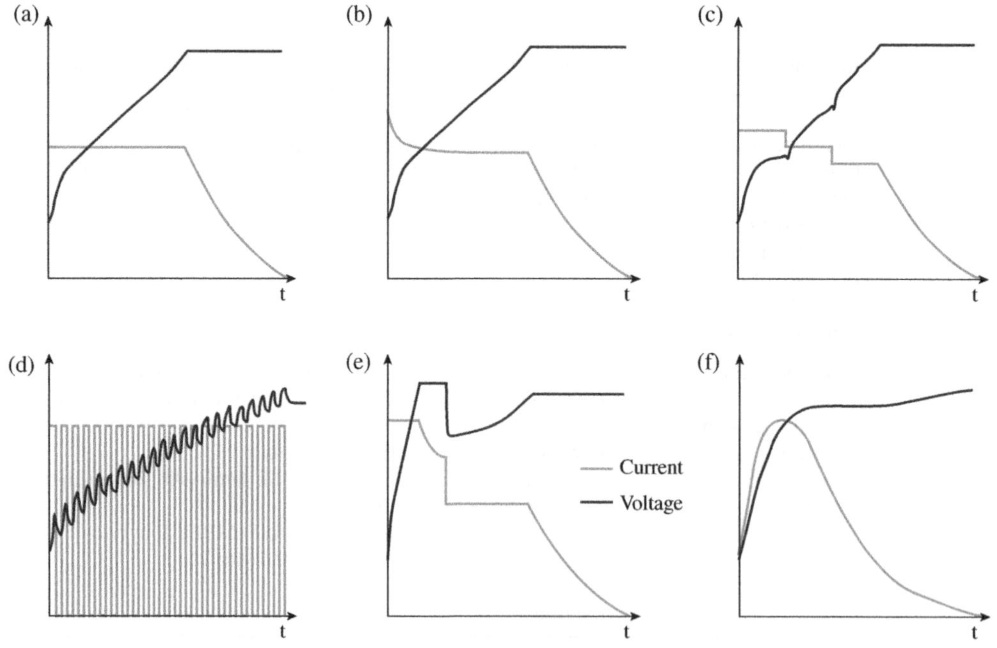

Figure 2.5 Various approaches for fast charging the Li-ion batteries: (a) CC-CV, (b) CP-CV, (c) MCC-CV, (d) pulse charging, (e) boost charging, and (f) variable current profile (VCP). *Source:* Tomaszewska et al. [19]/Elsevier/CC BY-4.0.

methods point to the effect of charge rates and impact due to Li plating [22, 23]. Note that, for designing an appropriate charge approach, studies should involve the cell designs with Li reference electrode, which can enable tracking graphite anode potential to account for influence of Li plating and related effects on the battery life.

Pulse charging is an approach where current pulses are used with either rest or discharge pulses to charge the batteries. This approach eliminates the effects of concentration polarization, which can cause overpotential or Li plating concerns. Li et al. [24] first investigated this method, and they could charge the battery in 1 hour using 1C pulse charging, while it would take 3.5 hours for CC-CV method, due to CV step. Also, it was observed that CC-CV charged batteries degraded faster relative to pulse charge method, which can be attributed to concentration gradient and resistance increase due to SEI growth factors. The charging and cycle life studies by Abdel-Monem et al. [25] using different approaches indicate that CC-CV method causes more degradation after 1700 cycles, while using discharge pulses helps to establish uniform concentration and counters resistance growth. Typically, customers charge their EVs generally using AC Level 1 and AC Level 2 chargers, and both these methods are slow rate. In direct current fast charge (DCFC) method, the battery pack can be charged at higher rates from low SOC to at least 80% SOC in the time window of 20–30 minutes. Thus, DCFC chargers enable fast charging of EVs, saving the time of charging for the customers. Accompanying the DCFC usage on routine basis by customers can cause faster deterioration of battery capacity.

2.5.2 Li Plating Effects

The fundamental limitation of current generation batteries is that they have limited rate capability for charging, especially at higher rates at high SOCs. That is mainly due to the limitation of the anode for charge acceptance at moderate to high rates. A severe limitation due to Li plating and related effects limit us from charging the LIBs which use graphite as anode. The Li plating is caused by accumulation of Li ions on the anode surface, instead of intercalation or insertion into the sites inside the anode structure (Figure 2.6). At full charge, lithiated graphite would have a composition LiC_6. During battery charging at high charge rates and certain cold and moderate temperatures, Li ions can plate on the anode surface, instead of intercalation into layers inside the graphite [22, 23]. Accompanying to the Li plating, the graphite potential can drop below 0 V, while overall cell voltage would keep rising during continued charging. With Li plating, other degradation pathways can set in, causing faster capacity degradation, affecting the battery life performance [26–28]. There have been major efforts at understanding the process of Li plating, its manifestation, correlation to anode potential, and also mitigation strategies. The Li plating and its effects are directly correlated to the charging

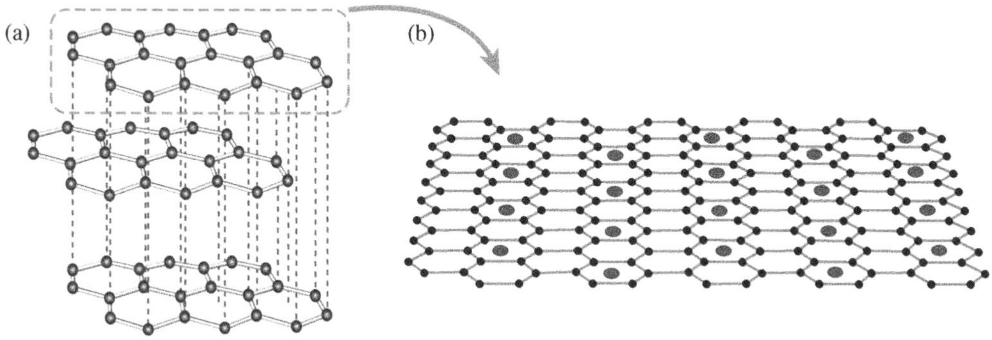

Figure 2.6 (a) 3D structure of graphite showing hexagonal lattice and (b) fully charged or lithiated graphite layer showing Li ions intercalated in, giving composition LiC_6.

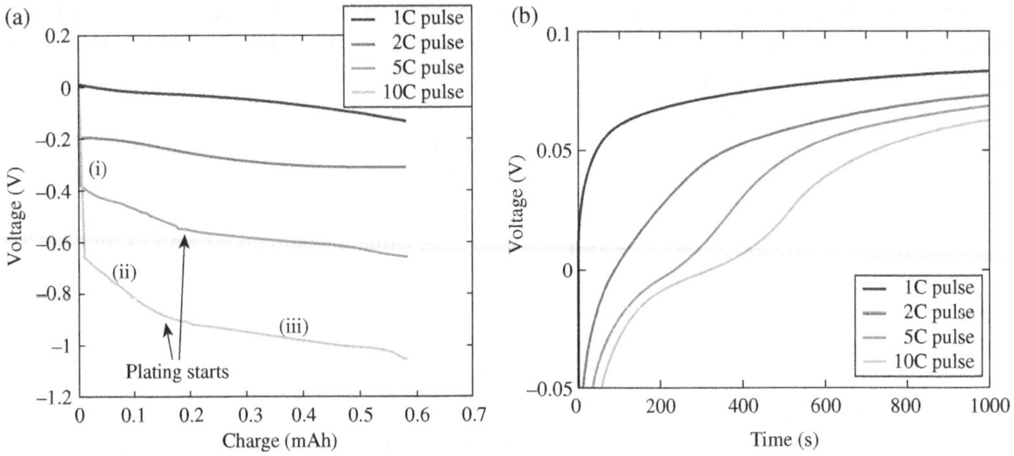

Figure 2.7 (a) Anode potential for graphite electrodes during charging tests at various current pulse rates. At high rate pulse (10 C) three regions can be seen: voltage showing initial drop (i), followed by high sloping line (ii) and lower slope line (iii). Li plating onset happens at spots pointed by arrows. (b) shows the voltage relaxation post pulse current tests, which show the effect of plated Li as a function of pulse C-rates. *Source:* Reproduced from Uhlmann et al. [22]/with permission of Elsevier.

rates used for the batteries. We expect the effects to be severe at colder temperatures; however, that does not mean plating can not occur at warmer temperatures. In fact, the high energy density of batteries plays into Li plating occurring at moderate and warm temperatures too.

When charging LIBs, changes happen at electrodes and electrolyte level. These include change in structure and/or volume due to Li insertion and extraction, SEI growth, Li plating, gassing, etc. At the cell level, the resistance can grow over the usage time and cells can swell and expand, which in turn can affect the integrity and performance at module or array level. For battery cells assembled into module or array, aspects such as thermal performance and suitable compression force have to be factored in. Otherwise, excess heat generation effects can lead to resistance increase and influencing the battery life and safety. For a battery pack being used in a vehicle for xEV applications, thermal, mechanical, and environmental influences have to be factored in (Figure 2.7). When tested at 23 °C, graphite half cells exhibited three distinct regions under increasing charge pulse C rates; a steep drop (region 1), followed by high slope (region 2) and later low slope section (region 3), as shown in Figure 2.7a. The slope change regions shown by arrows point to onset of Li plating [22]. Beyond the pulse time, the anode potential relaxes as shown in Figure 2.7b, and they carry signatures of Li plating. The type of usage such as fast charging, customer driving nature, and exposure to hot, cold, and moderate climate conditions would influence the battery performance. Figure 2.8 illustrates the scaling of things that impact the battery operation and influence battery performance during the battery usage in xEVs. All of this is related to the systems engineering and execution toward the better design of high voltage batteries for applications in xEVs [19]. Note that, battery array or module design should address the optimal thermal management and thermal system performance so as to enable the battery life requirement to meet the battery design target. The xEV manufacturers routinely provide warranty for the battery for usage time up to 100,000 mi and 8 years.

2.5.3 Overcharge Induced Thermal Runaway

During real life usage, while fast charging, LIBs can undergo accelerated degradation and perhaps thermal propagation. Such an outcome for LIBs is mainly contributed by the material level changes happening inside, which release large amounts of thermal energy, which further causes exothermal

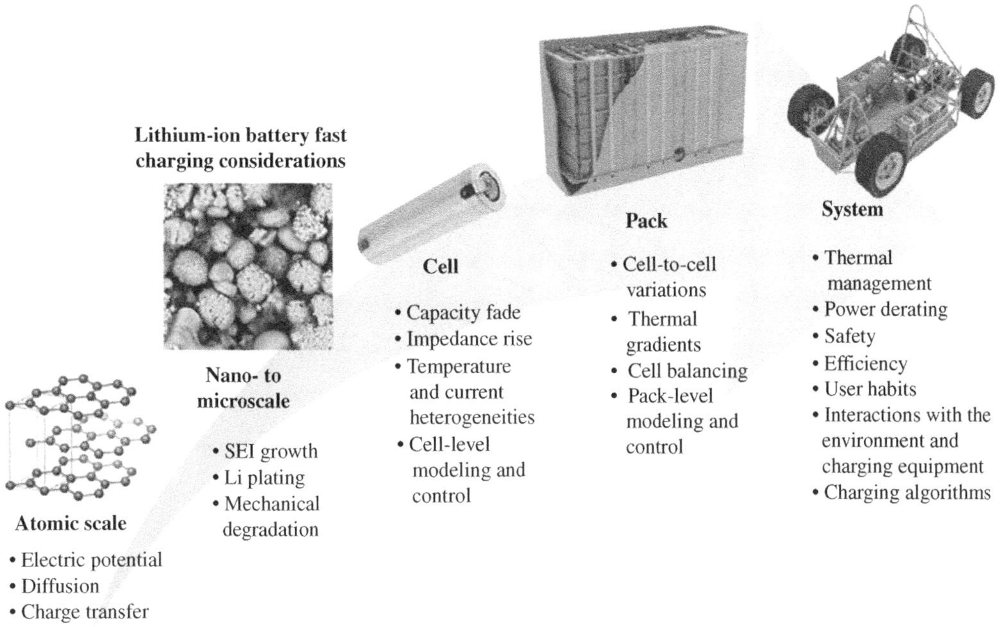

Figure 2.8 Size scaling and process parameters at electrode, battery cell, module, battery pack, and vehicle level during the fast charge (DCFC) usage of LIBs. *Source:* Reproduced with permission from Tomaszewska et al. [19]/Elsevier/CC BY-4.0.

reactions and changes at cathode and anode materials. Figure 2.9 illustrates the changes happening inside the battery during usage involving fast charging. LIB cells in general are safe when operated in certain voltage window [29, 30]. During certain usage cases like fast charging and abuse tolerance tests (see Freedomcar manual) [31], changes inside the battery can lead to thermal runaway [32, 33]. Such battery failure process consists of four stages [34].

Stage 1: The battery SOC may fall in the window 100% < SOC < 120%, and the cell voltage could exceed the cut-off voltage (V_{cell} ~ 4.2 V) and may increase slowly beyond 4.2 V. The cells are generally built with excess anode material, for safety margin, and they can withstand plated lithium during early stages of overcharging [35]. Li-plating-related side reactions can cause gas generation inside the cell, and the internal resistance (IR) of the battery may increase slowly.

Stage 2: Battery SOC could keep rising, with 120% < SOC < 140% and cell voltage keeps increasing (V_{cell} > 4.5 V). The electrolyte solvent oxidation would start, as the battery potential exceeds over the safe and stable battery operating window. A dissolution reaction of transition metal ions such as Mn^{2+} may be triggered in positive electrodes of certain chemistries, due to the excessive deintercalation of lithium [36, 37]. With V_{cell} increasing with more Li deposition, reactions at the anode can become aggressive [37, 38], and this would lead to thicker SEI film, increasing the IR of the cell [38, 39]. The joule heating effect of the overcharge current causes a significant increase in the battery temperature (>200 °C).

Stage 3: With battery at 140% < SOC < 160%, V_{cell} reaching higher values (but ≤6 V), the exothermic reactions inside the battery begin to dominate, due to the further removal of lithium from the positive electrode. Oxidative decomposition of the electrolyte solvents produces a greater amount of heat [40, 41], accompanied by gas generation and causes the battery to expand excessively. Also, with cathodes getting stripped of Li, structure becomes unstable and this may cause

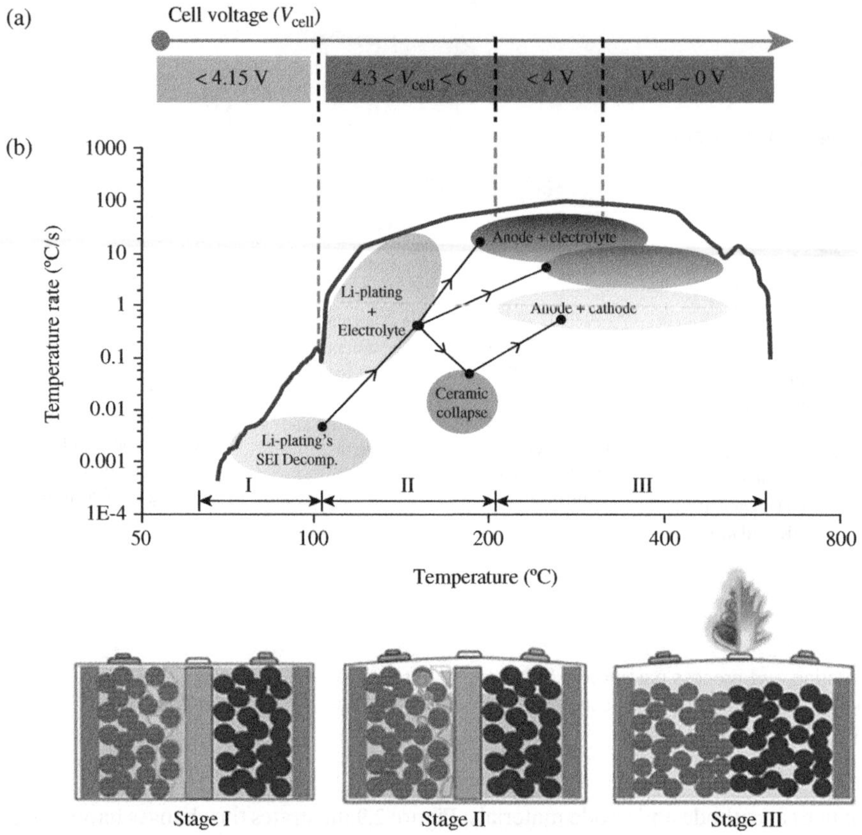

Figure 2.9 The top panel (a) shows the correlation of cell voltage correlation during overcharge and other changes occurring inside the battery. *Source:* Tomaszewska et al. [19]/Elsevier/CC BY-4.0. Schematic (b) shows the various reactions during fast charging, leading to thermal propagation in LIBs, where Li plating may cause cell to rupture with or without fire through stages I, II, and III. *Source:* Tomaszewska et al. [19]/Elsevier/CC BY-4.0.

decomposition into oxides with release of large amount of heat. By now the cell voltage would start dropping fast [42–44].

Stage 4: Battery SOC reaching 150–160% would cause oxidative decomposition of electrolyte, causing the generation of large volume gases and the battery ruptures. With separator being displaced, shorting between cathode and anode causes thermal runaway [45], followed by a quick drop in cell voltage to 0 V.

2.6 Degradation Considerations

With the increasing demand for high rate charging combined with high rate discharging for current LIBs for long range EV applications, this calls for better design of Ni-rich NCM and NCA cathodes along with advanced anode materials. In this section we shall survey the aspects of degradation on high energy density cathodes during high rate cycling. In recent years, advances in doping and surface coating on Ni-rich NCM and NCA cathodes have yielded robust improvement in cycle life performance when the cathodes are paired with graphite anodes [46]. When the Ni-rich NCM and NCA cathodes were cycled at 25 and 45 °C, the cathodes which were doped with Al performed

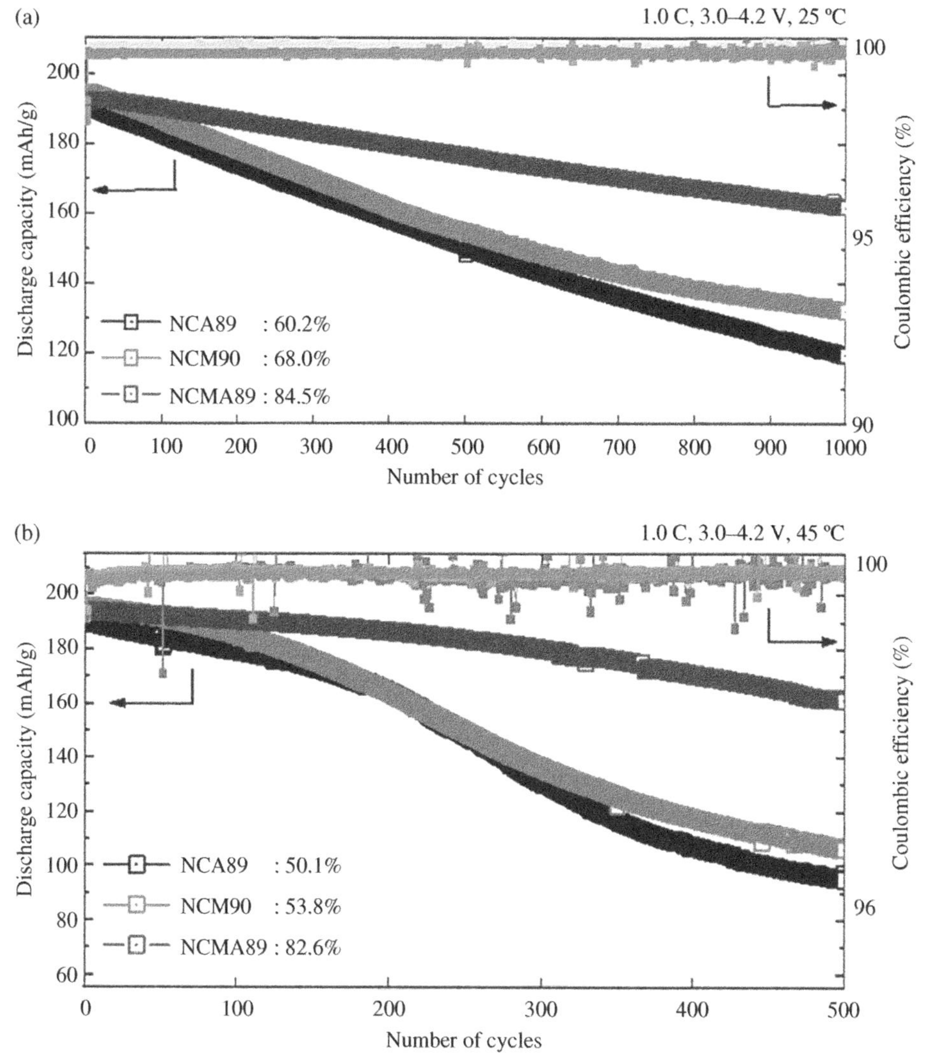

Figure 2.10 Cycling performance of Ni-rich NCM and NCA cathodes along with NCMA cathodes in pouch cells at (a) 25 and (b) 45 °C during 1C rate cycling. *Source:* Reproduced from Kim et al. [46]/with permission of American Chemical Society.

really well during 1C rate cycling (Figure 2.10). Without the Al doping, NCA89 and NCM90 both suffer from various degradations such as gas generation and particle cracking. However, with doping, NCMA89 cathodes perform well and capacity life degradation is slowed due to containment of gasing and particle cracking effects.

Enabling DCFC on EV batteries is associated with faster degradation of battery life, which is due to factors such as excess heat generation, resistance increase, and gas generation. So, attention must be paid to limit repeat DCFC cycles on EV batteries, to meet battery design life. Also, there is safety aspect too, wherein repeat fast charging can also cause Li plating, in turn causing thermal runaway and fire hazard [28]. The authors reported that using commercial DCFC cycles caused faster degradation and mechanical distortion of NCA 18650 cells than IR-based charge protocol (Figure 2.11). Excess heat generation and side reactions can lead to excess gas generation and swelling of cells, which eventually leads to the rupture of the cylindrical metal enclosure of cells. Thus, fast rate

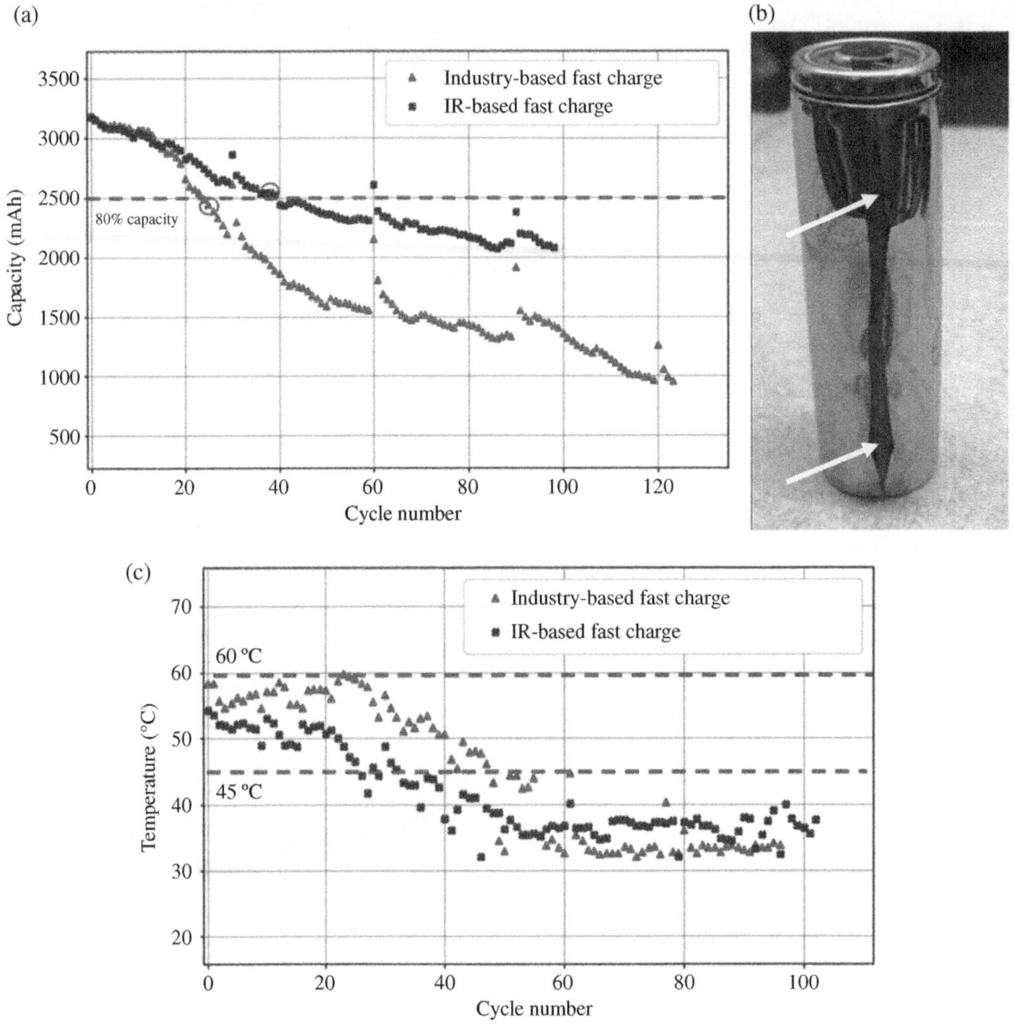

Figure 2.11 (a) NCA battery capacity comparison test; (b) NCA battery mechanical distortion and (c) NCA peak ambient temperature comparison for two charging methods. Note that temperature rise can accelerate cell capacity degradation. *Source:* Reproduced with permission from Sebastian et al. [28]/John Wiley & Sons, Inc.

charging and fast discharging can accelerate the capacity degradation and excess gas generation, in turn affecting cell integrity, causing rupture or vent.

2.7 Future Outlook

With the proliferation of LIBs in EV applications, a greater interest has triggered in both academic and industrial research on advancing capability and performance for cathodes, anodes, electrolyte design, etc. This in turn has led to higher energy density for the xEV batteries. Also, improvements such as fast charging capability (DCFC) are being improved by careful choice of cathode, anode, and electrolyte systems. The current metric of 30 minutes for fast charging (from 5% SOC to

80% SOC) is being upgraded by some battery manufacturers to 15 minutes, while still maintaining performance and safe battery operation. Also, battery cost is being incrementally reduced over the past decades, and at present cost is in the range of ~$120–130/kWh and in the coming years (2023–2025), it is supposed to hit numbers well below $100/kWh. With electrification of sports utility vehicles (SUVs) and trucks, the architectures with 800 V or higher voltage battery, inverter, and charger systems are gaining more popularity. Having a 800 V battery would also enable faster and efficient charging capability, which would enable higher EV range and reduce range anxiety issue for the EV customers. This would spur innovations in power electronics, wireless charging, and also battery integration for 800 V systems and beyond. The biggest challenge for industry with the state-of-the-art LIBs is the manufacturing quality concerns, causing battery fire and other failures, which result in costly warranty returns [47]. Robust improvements are expected in the coming years and that would help the battery and xEV industries. Also, this would help customers with transition toward larger adaptation of EVs, instead of ICE vehicles. This in turn could help the efforts of countering the global climate change concerns.

Acronyms

LIB	lithium-ion battery
EV	electric vehicle
PHEV	plug-in hybrid electric vehicle
HEV	hybrid electric vehicle
xEV	HEV, PHEV, and EV
SSB	solid-state battery
LCO	lithium cobalt oxide ($LiCoO_2$)
LFP	lithium iron phosphate ($LiFePO_4/C$)
LMO	lithium manganese oxide ($LiMn_2O_4$)
NCA	lithium nickel cobalt aluminum oxide ($LiNiCoAlO_2$)
NCM	lithium nickel cobalt manganese oxide ($LiNiCoMnO_2$)
NMP	*N*-methylpyrrolidone
PVDF	polyvinylidene fluoride
ESS	energy storage system
SEI	solid electrolyte interface
CC-CV	constant current followed by constant voltage
MCC	multistage constant current
DCFC	direct current fast charge

References

1 Goodenough, J.B. and Park, K.S. (2013). The Li-ion rechargeable battery: a perspective. *J. Am. Chem. Soc.* 135: 1167–1176. https://doi.org/10.1021/ja3091438.

2 Armand, M. and Tarascon, J.M. (2008). Building better batteries. *Nature* 451: 652–657. https://doi.org/10.1038/451652a.

3 Kwak, G.H., Park, J.H., Lee, J.U. et al. (2007). Effects of anode active materials to the storage-capacity fading on commercial lithium-ion batteries. *J. Power Sources* 174: 484–492. https://doi.org/10.1016/j.jpowsour.2007.06.169.

4 Schipper F.; Bouzaglo, H.; Dixit, M.; Erickson, E. M.; Weigel, T.; Talianker, M.; Grinblat, J.; Burstein, L.; Schmidt, M.; Lampert, J.; et al. From surface ZrO2 coating to bulk Zr doping by high temperature annealing of nickel-rich lithiated oxides and their enhanced electrochemical performance in lithium ion batteries. *Adv. Energy Mater.* 2017, 1701682, https://doi.org/10.1002/aenm.201701682

5 Z. Xie, Y. Zhang, X. Min, A. Yuan, J. Xu, One-step bulk and surface co-modification of $LiNi_{0.8}Co_{0.15}Al_{0.05}O_2$ cathode material towards excellent long-term cyclability. *Electrochim. Acta* 2021, 379, 138124, https://doi.org/10.1016/j.electacta.2021.138124

6 A. Manthiram, B. Song, W. Li, A perspective on nickel-rich layered oxide cathodes for lithium-ion batteries. *Energy Storage Mater.* 2017, V6, 125–139, https://doi.org/10.1016/j.ensm.2016.10.007

7 Han, B.H., Psulauskas, T., Key, B. et al. (2017). Understanding the role of temperature and cathode composition on interface and bulk: optimizing aluminum oxide coatings for Li-ion cathodes. *ACS Appl. Mater. Interfaces* 9 (17): 14769–14778. https://doi.org/10.1021/acsami.7b00595.

8 Yoon, C.S., Park, K.J., Kim, U.H. et al. (2017). High-energy Ni-rich $Li[Ni_xCo_yMn_{1-x-y}]O_2$ cathodes via compositional partitioning for next-generation electric vehicles. *Chem. Mater.* 29, 24: 10 436–10 445. https://doi.org/10.1021/acs.chemmater.7b04047/.

9 Sun, Y.K., Myung, S.T., Park, B.C. et al. (2009). High-energy cathode material for long-life and safe lithium batteries. *Nat. Mater.* 8: 320–324. https://doi.org/10.1038/nmat2418.

10 Zeng, X., Li, M., El-Hady, D.A. et al. (2019). Commercialization of lithium battery technologies for electric vehicles. *Adv. Energy Mater.* 9: 1900161. https://doi.org/10.1002/aenm.201900161.

11 Jo, Y. and Lee, J. (2018). Electrochemical performance of graphite/silicon/carbon composites as anode materials for lithium-ion batteries. *Korean Chem. Eng. Res.* 56 (3): 320–326. https://doi.org/10.9713/kcer.2018.56.3.320.

12 Zhang, X., Wang, D., Qiu, X. et al. (2020). Stable high-capacity and high-rate silicon-based lithium battery anodes upon two-dimensional covalent encapsulation. *Nat. Commun.* 11: 3826. https://doi.org/10.1038/s41467-020-17686-4.

13 Luo, W., Chen, X.Q., Xia, Y. et al. (2017). Surface and interface engineering of silicon-based anode materials for lithium-ion batteries. *Adv. Energy Mater.* 7: 1701083. https://doi.org/10.1002/aenm.201701083.

14 Zhu, G., Zhang, F., Li, X. et al. (2019). Engineering the distribution of carbon in silicon oxide nanospheres at the atomic level for highly stable anodes. *Angew. Chem. Int. Ed.* 58: 6669–6673. https://doi.org/10.1002/anie.201902083.

15 Mao, Y., Karan, N.K., Song, M. et al. (2017). Investigation of solid electrolyte interphase formed on Si nanoparticle composite electrodes using hyperpolarized 129 Xe nuclear. *Energy Fuels* 31 (5): 5622–5628. https://doi.org/10.1021/acs.energyfuels.7b00250.

16 Etacheri, V., Marom, R., Elazari, R. et al. (2011). Challenges in the development of advanced Li-ion batteries: a review. *Energy Environ. Sci.* 4 (9): 3243–3262. https://doi.org/10.1039/c1ee01598b.

17 Scrosati, B. and Garche, J. (2010). Lithium batteries: status, prospects and future. *J. Power Sources* 195 (9): 2419–2430. https://doi.org/10.1016/j.jpowsour.2009.11.

18 Kato, Y., Hori, S., Saito, T. et al. (2016). High-power all-solid-state batteries using sulfide superionic conductors. *Nat. Energy* 1: 16030. https://doi.org/10.1038/nenergy.2016.30.

19 Tomaszewska, A., Chu, Z., Feng, X. et al. (2019). Lithium-ion battery fast charging: A review. *eTransportation* 1: 100011. https://doi.org/10.1016/j.etran.2019.100011.

20 Zhang, S., Xu, K., and Jow, T. (2006). Study of the charging process of a LiCoO2-based Li-ion battery. *J. Power Sources* 160: 1349. https://doi.org/10.1016/J.JPOWSOUR.2006.02.087.

21 Ouyang, M., Chu, Z., Lu, L. et al. (2015). Low temperature aging mechanism identification and lithium deposition in a large format lithium iron phosphate battery for different charge profiles. *J. Power Sources* 286: 309. https://doi.org/10.1016/J.JPOWSOUR.2015.03.178.

22 Uhlmann, C., Illig, J., Ender, M. et al. (2015). In situ detection of lithium metal plating on graphite in experimental cells. *J. Power Sources* 279: 428. https://doi.org/10.1016/j.jpowsour.2015.01.046.

23 Lain, M.J. and Kendrick, E. (2021). Understanding the limitations of lithium ion batteries at high rates. *J. Power Sources* 493: 229690. https://doi.org/10.1016/j.jpowsour.2021.229690.

24 Li, J., Murphy, E., Winnick, J., and Kohl, P.A. (2001). The effects of pulse charging on cycling characteristics of commercial lithium-ion batteries. *J. Power Sources* 102: 302. https://doi.org/10.1016/S0378-7753(01)00820-5.

25 Abdel-Monem, M., Trad, K., Omar, N. et al. (2017). Influence analysis of static and dynamic fast-charging current profiles on ageing performance of commercial lithium-ion batteries. *Energy* 120: 179. https://doi.org/10.1016/J.ENERGY.2016.12.110.

26 Petzl, M., Kasper, M., and Danzer, M.A. (2015). Lithium plating in a commercial lithium-ion battery – a low-temperature aging study. *J. Power Sources* 275: 799. https://doi.org/10.1016/j.jpowsour.2014.11.065.

27 Waldmann, T., Hogg, B., and Wohlfahrt-Mehrens, M. (2018). Li plating as unwanted side reaction in commercial Li-ion cells – a review. *J. Power Sources* 384: 107. https://doi.org/10.1016/j.jpowsour.2018.02.063.

28 Sebastian, S.S., Dong, B., Zerrin, T. et al. (2020). Adaptive fast charging methodology for commercial Li-ion batteries based on the internal resistance spectrum. *Energy Storage* 2: e141. https://doi.org/10.1002/est2.141.

29 Zhang, C., Jiang, Y., Jiang, J. et al. (2017). Study on battery pack consistency evolutions and equilibrium diagnosis for serial-connected lithium-ion batteries. *Appl. Energy* 207: 510. https://doi.org/10.1016/j.apenergy.2017.05.176.

30 Li, Z., Lu, L., Ouyang, M., and Xiao, Y. (2011). Modeling the capacity degradation of LiFePO4/graphite batteries based on stress coupling analysis. *J. Power Sources* 196: 9757. https://doi.org/10.1016/j.jpowsour.2011.07.080.

31 Orendorff, C.J., Lamb, J., Steele, L.A.M. (2019). Recommended practices for abuse testing rechargeable energy storage systems (RESSs) https://www.sandia.gov/ess-ssl/wp-content/uploads/2019/05/SNL-Abuse-Testing-Manual_July2017_FINAL.pdf (accessed 15 November 2021).

32 Hofmann, A., Uhlmann, N., Ziebert, C. et al. (2017). Preventing Li-ion cell explosion during thermal runaway with reduced pressure. *Appl. Therm. Eng.* 124: 539. https://doi.org/10.1016/j.applthermaleng.2017.06.056.

33 Zhu, X., Wang, Z., Wang, Y. et al. (2019). Overcharge investigation of large format lithium-ion pouch cells with $Li(Ni_{0.6}Co_{0.2}Mn_{0.2})O_2$ cathode for electric vehicles: thermal runaway features and safety management method. *Energy* 169: 868. https://doi.org/10.1016/j.energy.2018.12.041.

34 Ren, D., Feng, X., Lu, L. et al. (2017). An electrochemicalthermal coupled overcharge-to-thermal-runaway model for lithium ion battery. *J. Power Sources* 364: 328. https://doi.org/10.1016/j.jpowsour.2017.08.035.

35 Krueger, S., Kloepsch, R., Li, J. et al. (2013). How do reactions at the anode/electrolyte interface determine the cathode performance in lithium-ion batteries? *J. Electrochem. Soc.* 160: A542. https://doi.org/10.1149/2.022304jes.

36 Zheng, H., Sun, Q., Liu, G. et al. (2012). Correlation between dissolution behavior and electrochemical cycling performance for $LiNi_{1/3}Co_{1/3}Mn_{1/3}O_2$-based cells. *J. Power Sources* 207: 134. https://doi.org/10.1016/j.jpowsour.2012.01.122.

37 Arora, P. (1998). Capacity fade mechanisms and side reactions in lithium-ion batteries. *J. Electrochem. Soc.* 145: 3647. https://doi.org/10.1149/1.1838857.

38 Sharma, N. and Peterson, V.K. (2013). Overcharging a lithium-ion battery: effect on the Li_xC_6 negative electrode determined by in situ neutron diffraction. *J. Power Sources* 244: 695. https://doi.org/10.1016/j.jpowsour.2012.12.019.

39 Arora, P., Doyle, M., and White, R. (1999). Mathematical modeling of the lithium deposition overcharge reaction in lithium-ion batteries using carbon-based negative electrodes. *J. Electrochem. Soc.* 146, 10: 3543. https://doi.org/10.1149/1.1392512.

40 Kumai, K., Miyashiro, H., Kobayashi, Y. et al. (1999). Gas generation mechanism due to electrolyte decomposition in commercial lithium-ion cell. *J. Power Sources* 81: 715. https://doi.org/10.1016/S0378-7753(98)00234-1.

41 Ohsaki, T., Kishi, T., Kuboki, T. et al. (2005). Overcharge reaction of lithium-ion batteries. *J. Power Sources* 146: 97. https://doi.org/10.1016/j.jpowsour.2005.03.105.

42 Wu, L., Nam, K.W., Wang, X. et al. (2011). Structural origin of overcharge-induced thermal instability of Ni-containing layered-cathodes for high-energy-density lithium batteries. *Chem. Mater.* 23: 3953. https://doi.org/10.1021/cm201452q.

43 He, Y.B., Ning, F., Yang, Q.H. et al. (2011). Structural and thermal stabilities of layered $Li(Ni_{1/3}Co_{1/3}Mn_{1/3})O_2$ materials in 18650 high power batteries. *J. Power Sources* 196: 10322. https://doi.org/10.1016/j.jpowsour.2011.08.042.

44 Lin, C.K., Ren, Y., Amine, K. et al. (2013). In situ high-energy X-ray diffraction to study overcharge abuse of 18650-size lithium-ion battery. *J. Power Sources* 230: 32. https://doi.org/10.1016/j.jpowsour.2012.12.032.

45 Zheng, S., Wang, L., Feng, X., and He, X. (2018). Probing the heat sources during thermal runaway process by thermal analysis of different battery chemistries. *J. Power Sources* 378: 527. https://doi.org/10.1016/j.jpowsour.2017.12.050.

46 Kim, U.H., Kuo, L.Y., Kaghazchi, P. et al. (2019). Quaternary layered Ni-rich NCMA cathode for lithium-ion batteries. *ACS Energy Lett.* 4: 576. https://doi.org/10.1021/acsenergylett.8b02499.

47 LaReau, J.L. (2021). LG will reimburse GM $1.9 billion for cost of Chevy Bolt battery recall. https://www.freep.com/story/money/cars/general-motors/2021/10/12/gm-lg-chevy-bolt-recall-battery/6101025001 (accessed 15 December 2021).

3

Applications of Wide Bandgap (WBG) Devices in the Transportation Sector. Recent Advances in (WBG) Semiconductor Material (e.g. Silicon Carbide and Gallium Nitride) and Circuit Topologies

Amir Ranjbar

Canoo, Power Electronics Department, Torrance, CA 90503, USA

3.1 History of Semiconductor Technology Evolution

Efficient conversion of power from DC/DC, AC/DC, and DC/AC has been the main motivation toward the development and evolution of semiconductor technologies. Figure 3.1 shows the basic schematic of a linear regulator, also known as low drop out (LDO), that is used to convert an input DC voltage to an output DC voltage that is smaller than the input. In this architecture, an NPN pass transistor provides a variable resistance by operating in the linear region. The output voltage is then sensed through a set of feedback resistors and is fed to an error amplifier that is used to compare the output voltage with a reference voltage and drive the base of the pass transistor through a current amplifier circuit. This simple architecture was used by several multibillion-dollar power supply companies until early 1960s [1]. As shown here, the same architecture can also be used in AC/DC power conversion.

The major limitation of LDOs is the fact that they can only perform a step-down voltage conversion where the output voltage is lower than the input voltage. In addition to that, since the pass transistor operates in the linear region, that results in high conduction loss. This power loss will increase as the output power is increased and/or a higher step-down ratio is required. Finally, since both input and output share the same ground, this architecture cannot provide a DC isolation.

The introduction of the metal-oxide silicon field effect transistor (MOSFET) by Bell Labs in 1959 revolutionized the power supply industry. The high switching speed of MOSFETs allowed for the introduction of various switched-mode power supply (SMPS) topologies that provided an alternative and more efficient way of performing both step-down as well as step-up voltage conversions. The simple concept of a Buck regulator, as the most basic switched-mode power supply, is shown in Figure 3.2. In this architecture, the input voltage is switched and by varying the duty cycle, the average DC voltage delivered to output is controlled. The proper selection of components for the LC filter converts the square wave modulated voltage, V_1, into a low ripple DC voltage on the output. Based on this concept, various other switching regulators including Boost, fly-back, forward, push-pull, half-bridge, full-bridge, and Cuk converters were developed during early 1960s to late 1970s [1–3]. These topologies enabled both step-down and step-up voltage conversions. They also provided isolation between input and output. Figure 3.3 shows some of these topologies [1]. Note that over years, these topologies have evolved into many additional topologies with improved performances.

Transportation Electrification: Breakthroughs in Electrified Vehicles, Aircraft, Rolling Stock, and Watercraft, First Edition. Edited by Ahmed A. Mohamed, Ahmad Arshan Khan, Ahmed T. Elsayed, and Mohamed A. Elshaer.

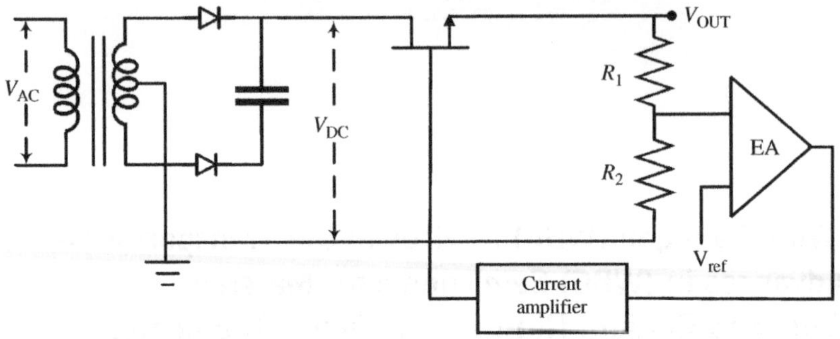

Figure 3.1 Basic architecture of a linear regulator used for DC/DC and AC/DC power conversion until early 1960s.

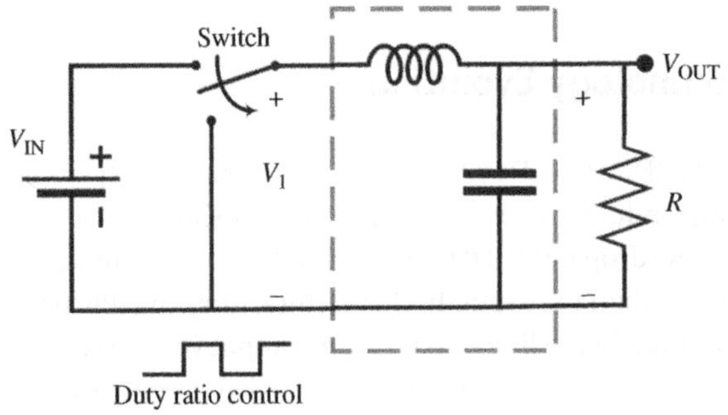

Figure 3.2 Basic concept of a buck-switching regulator.

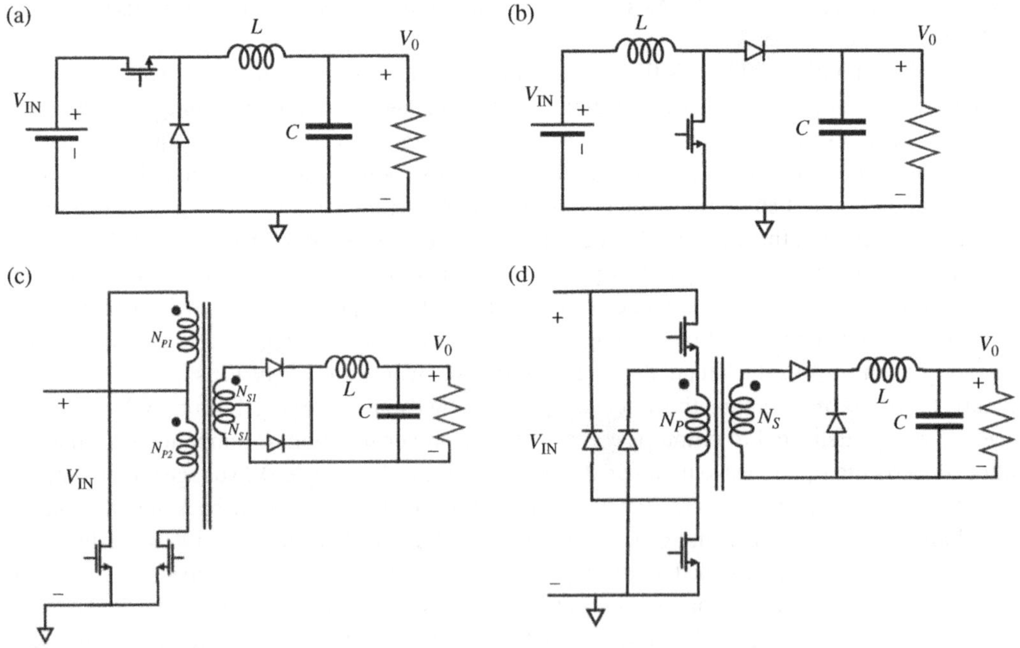

Figure 3.3 Examples of switching regulators developed by introduction of MOSFETs. Schematic of a (a) buck regulator, (b) boost regulator, (c) push-pull regulator, and (d) forward regulator.

3.2 Semiconductor Technologies for Transportation Electrification

In transportation electrification, Si-MOSFETs are still being widely used in DC/DC converters, onboard charger (OBC) modules, as well as traction inverters. In terms of power, however, the Si-MOSFETs are limited to converters with up to 10 s of kilowatts of power [4]. For higher power applications, such as high voltage (HV) traction inverters that are used to drive >100 kW motors, the high conduction loss of Si-MOSFETs is a limiting factor. Therefore, Si-IGBTs were introduced. Compared with Si-MOSFETs that represent resistance when they are in on state, Si-IGBTs are represented by a voltage source during on state. Therefore, although it is true that when IGBTs are on, the voltage drop across the IGBTs is at least one diode drop; however, compared with Si-MOSFETs with the same rating, Si-IGBTs exhibit a significantly lower on-state conduction loss.

The reason for a lower conduction loss in Si-IGBTs, as compared with Si-MOSFETs, is the fact that unlike Si-MOSFETs that are majority carrier devices where only electrons flow, in Si-IGBTs, holes are injected into their drift region. Therefore, the electric current flow in an IGBT is composed of both electrons as well as holes. The injection of holes as minority carriers significantly reduces the effective resistance of the IGBTs against current flow. In other words, conductivity is improved or equivalently the conduction loss is reduced.

As expected, nothing comes for free and the price paid for the higher conductivity of the IGBTs during on state and therefore lower conduction loss is their longer turn-off time. The reason for a longer turn-off time in Si-IGBTs is that there is no effective way of stopping the holes as the device is turned off. The electron mobility is controlled by the voltage applied to the gate. Therefore, as the gate voltage is reduced to zero, the flow of electron is stopped. However, reducing the gate voltage does not remove the holes that are accumulated in the drift region. There is essentially no way to remove the holes from the drift region, unless by voltage gradient and recombination. This is the reason why Si-IGBTs exhibit tail current during turn-off events until all the holes are swept away or recombined. The IGBT tailing current is shown in Figure 3.4. The big tail of current increases the turn-off loss in IGBTs. The higher switching loss, in turn, results in a limitation on the application switching frequency to limit the excessive heat.

3.2.1 Trends in Transportation Electrification

As the electrification of transportation becomes mainstream in the automotive industry, efforts are made to improve efficiency and power density in high-power electrified powertrains. Figure 3.5 shows a simplified schematic of a typical electrified powertrain.

In the schematic, shown in Figure 3.5, to increase efficiency, the HV battery voltage is increased. For the same power, a higher bus voltage means lower electric current, which then results in a lower copper loss and, therefore, higher efficiency. On the other hand, to effectively increase the power density, the switching frequency must be increased. The increase in the switching frequency will result in a smaller DC-link capacitor in traction inverters. The need for higher voltage and higher frequency capabilities in electrified powertrains resulted in the introduction of wide-bandgap (WBG) devices such as SiC-MOSFET and GaN devices. Table 3.1 compares the

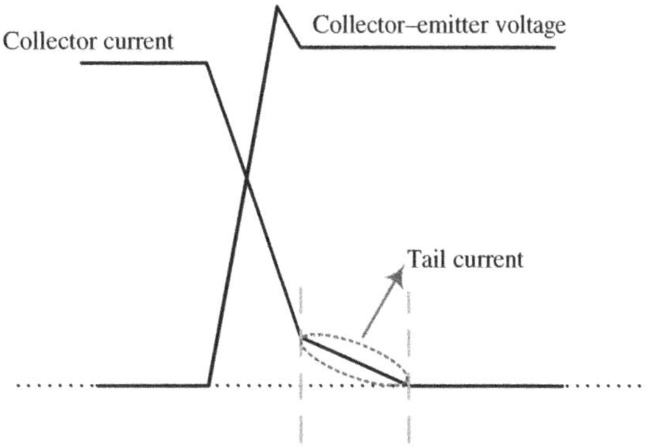

Figure 3.4 IGBT tail current during turn-off results in high switching loss.

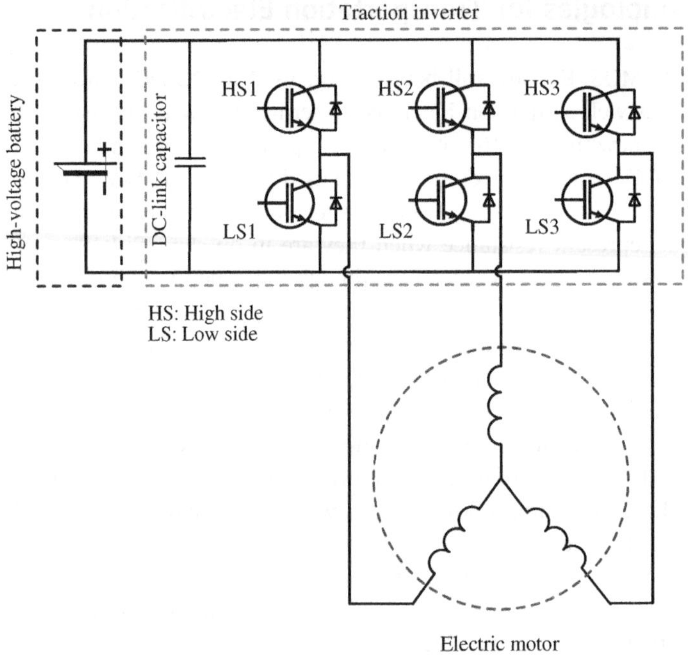

Figure 3.5 Simplified schematic of a typical electrified powertrain.

Table 3.1 Comparing properties of Si vs SiC vs GaN.

Properties	Si	SiC	GaN
Bandgap energy (eV)	1.12	3.26	3.50
Electron mobility (cm^2/Vs)	1400	900	1250
Hole mobility (cm^2/Vs)	600	100	200
Breakdown field (MV/cm)	0.3	3.0	3.0
Thermal conductivity (W/cm °C)	1.5	4.9	1.3
Coefficient of thermal expansion (CTE) $\times 10^{-6}$ (°C^{-1})	2.6	4.3	3.2
Maximum junction temperature (°C)	150	600	400

main characteristics of silicon as a conventional semiconductor technology compared with SiC-MOSFET and GaN as WBG devices [5].

As seen in Table 3.1, the wider (three times higher) bandgap energy associated with SiC-MOSFET and GaN devices means the energy to move electrons from their valence band to the conduction band is three times higher. Therefore, WBG devices behave more like an insulator and less like a conductor.

The higher bandgap energy in SiC-MOSFETs and GaN devices allows them to withstand a much higher breakdown voltage (10 times higher) as compared with Si devices. The higher withstand voltage capability in WBG devices as compared with silicon enables a reduction in device thickness, which, by itself, translates into a lower on-state resistance as well as higher power density at the die level.

The benefits associated with higher withstand voltage capability in WBG devices come at the expense of lower electron mobility. As seen in Table 3.1, the electron mobility in SiC is 64% of the electron mobility in Si. On the other hand, the hole mobility in SiC is only 17% of hole mobility in Si. Similar characteristics can be observed in the case of GaN devices. Therefore, although WBG devices have higher withstand voltage capabilities and therefore are better insulators, they are not great conductors. What this means from application's standpoint is that the gain of WBG devices is low as compared with silicon devices.

Comparing GaN vs SiC characteristics in Table 3.1, GaNs not only have slightly higher bandgap energy as compared with SiC but they also have higher electron mobility and higher hole mobility (both still lower than silicon). However, one of the key reasons why GaNs have not been able to penetrate high-power applications such as electrified powertrains is the fact that they have a relatively low thermal conductivity. Based on Table 3.1, the thermal conductivity of GaN is ~15% lower compared with Si and ~75% lower compared with SiC.

A unique characteristic associated with GaN devices is that unlike Silicon and SiC-MOSFET counterparts that are in enhancement mode, GaNs are in depletion mode in nature. Enhancement mode devices are normally off devices that require a positive gate voltage to turn on. Depletion mode devices, on the other hand, are normally on and require a negative gate voltage to turn off. Figure 3.6 shows the typical architecture of a depletion mode GaN. Because of the piezoelectric nature of GaN, two-dimensional electron gas (2DEG) forms between GaN and AlGaN interface [6]. The 2DEG layer provides a conductive path between the source and drain terminals and shorts them together. Therefore, the device is normally on at zero gate bias and requires a negative voltage to turn off.

In power electronics applications, depletion mode devices are not preferred choices. Figure 3.7 shows a typical half-bridge architecture, commonly used in power electronics applications including switched-mode power supplies as well as traction inverters. Since depletion mode GaN devices are normally on, that means they will cause a battery short circuit when no bias is present at their gate. Therefore, depletion mode GaNs cannot be used in such applications.

Therefore, alternative approaches have been proposed by various GaN manufacturers to convert the depletion mode GaNs into enhancement mode GaNs. Figure 3.8 shows a GaN device in cascode configuration. In a cascode configuration, the high voltage (HV) GaN is placed in series with a low voltage (LV) silicon MOSFET. The GaN gate is grounded and the MOSFET gate is being driven to control the GaN device. Therefore, in addition to providing an enhancement mode functionality, this architecture also makes it possible to use typical silicon MOSFET gate drivers to drive the GaN device.

The cascode architecture, despite its benefits, has practical challenges that make it an unfavorable solution. First of all, cascode GaN configuration exhibits higher parasitic capacitance since the

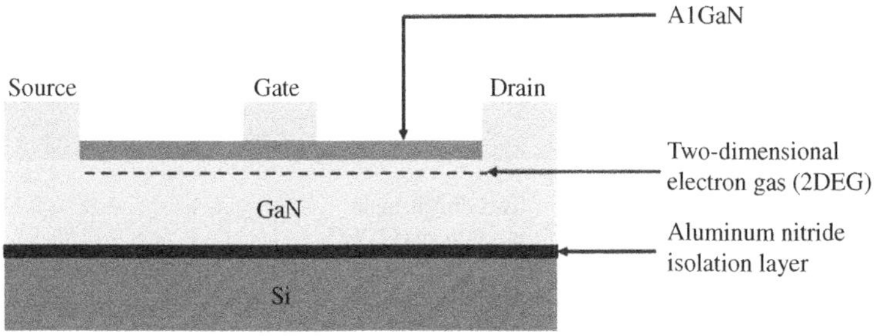

Figure 3.6 Typical depletion mode GaN/AlGaN architecture. *Source:* Adapted from Ref. [6].

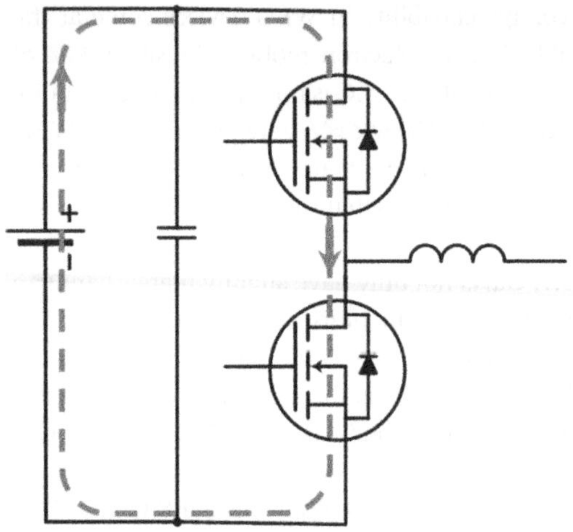

Figure 3.7 Typical half-bridge architecture in power electronics applications.

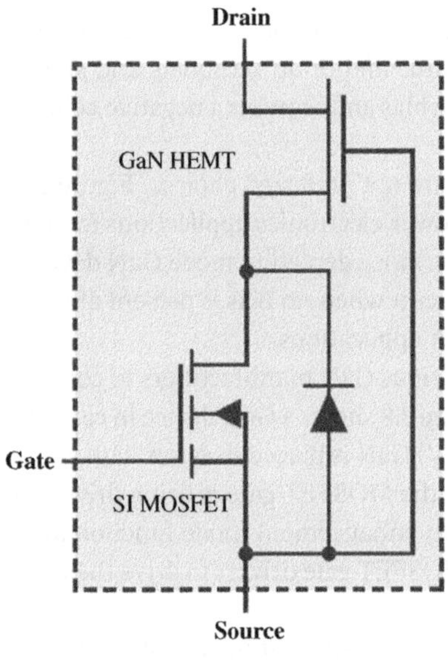

Figure 3.8 GaN device in cascode-configuration. *Source:* Ref. [7]/Texas Instruments Incorporated.

GaN gate-to-source (C_{GS}) capacitance and the Si-MOSFET output capacitance (C_{OSS}) are placed in parallel. This jeopardizes one of the key advantages of the GaN devices, which is smaller parasitic capacitance and, therefore, faster switching capabilities. Second, in cascode configuration, the voltage distribution between the GaN and MOSFET in off mode can cause the MOSFET to avalanche. This is due to the lower drain-to-source capacitance (C_{DS}) in the LV Si-MOSFET as compared with the HV GaN device. One solution would be to add an external capacitor in parallel with the MOSFET's drain and source. However, this results in higher switching losses by slowing down the switching transitions.

Figure 3.9 shows an alternative enhancement mode GaN architecture, developed by efficient power conversion (EPC) [8]. In this architecture, there is a depletion region under the gate terminal that blocks the conductive path between the source and drain terminals. The eGaNs require a positive gate voltage above its threshold point to enhance the channel between the drain and source and conduct current.

Comparing the enhancement mode GaN (also known as eGaN) in Figure 3.9 vs depletion mode GaN (also known as dGaN) in Figure 3.6, one can notice that the fabrication process of eGaN is more complicated. However, the operation and control of eGaNs are much easier as compared with dGaNs. Therefore, eGaNs are preferred choices in power electronics applications.

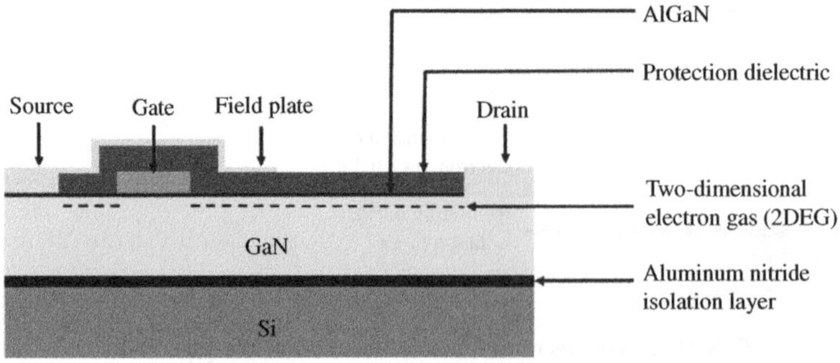

Figure 3.9 Enhancement mode GaN developed by EPC. *Source:* Ref. [8]/IEEE.

3.3 Challenges Associated with GaNs in Practical Applications

Looking at Table 3.1, GaN has several advantages over SiC technology including higher bandgap energy as well as higher electron mobility. GaN's application, however, has been limited as compared with SiC-MOSFETs. One of the key reasons for that, as briefly mentioned earlier, is GaN's lower thermal conductivity. Therefore, GaN has limited capability in higher power applications where more heat needs to be dissipated. Figure 3.10 provides an insight into the region in terms of power and frequency where GaNs can be most effective [4]. As shown in Figure 3.10, GaNs are best fitted into applications with up to 10 kW power and >100 kHz switching frequency. In transportation electrification, this specification mainly involves DC–DC converters and on-board chargers (OBCs). These are the primary application targets for GaNs in transportation electrification.

While GaNs can provide system-level benefits in applications mentioned earlier, there are certain challenges associated with GaNs that have limited their further penetration into the transportation electrification market. These challenges can be classified into two main categories of device physics level and application level [9]. The following section will try to briefly address these challenges.

3.3.1 Device Physics Level Challenges with GaNs

Device physics level challenges associated with GaNs are the ones that originate from the device architecture and its specific characteristics. These issues include electron trapping, punch-through current, and substrate lattice and CTE mismatch.

3.3.1.1 Electron Trapping

The known inverse piezoelectric effect in GaNs device architecture is the main root cause for electron trapping [10]. The trapped charges normally accumulate close to gate, source, and drain edges when the device is exposed to an external electric field [9]. The trapped charges adversely impact device performance by introducing excessive leakage currents. One of the major issues caused by

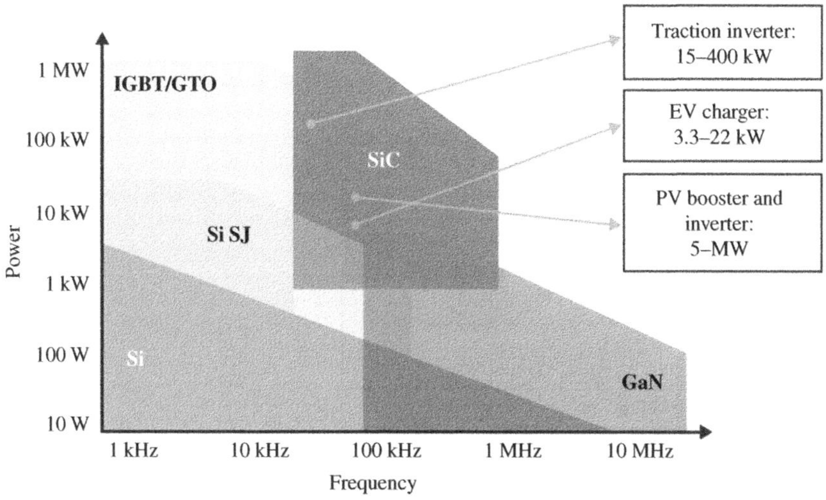

Figure 3.10 Semiconductor device application based on power and frequency range. *Source:* Ref. [4]/Texas Instruments Incorporated.

electron trapping is the dynamic increase in channel resistance, Ron, as a result of trapped charge leakage proceeding from the gate edge to the drain edge. This phenomenon is also known as current collapse [10].

3.3.1.2 Gate Edge Degradation

The structural defects such as inverse piezoelectric effect, time-dependent trap formation, percolative conductive path formation, electrochemical reactions, and gate metal diffusion result in failure mechanisms in GaN/AlGaN layer [11–13]. These defects occur next to the gate edge of the GaN device where the electric field strength is maximum. That, in turn, results in electrons tunneling from the gate edge into the AlGaN barrier layer and, therefore, increasing the gate leakage current. This not only adversely affects the output power and efficiency but also reduces the device's reliability.

3.3.1.3 Punch Through Current

The source to drain leakage is also called punch-through current and is primarily caused by defects in GaN and buffer layers [14]. As shown in Figure 3.11, the charge from the source leaks to the drain by taking an alternate path through the GaN layer. Such behavior can only be observed when the device is in its semi-OFF state.

3.3.1.4 Substrate Choice

Although the bulk GaN layer as a substrate improves the device performance, the fabrication process of bulk GaN multilayers is a complex task and commercially not a feasible choice [9]. Therefore, as shown in Figure 3.11, an alternative technique is used for device fabrication where GaN is grown over substrates such as Si, sapphire, and SiC. The challenge associated with this approach is the lattice and coefficient of thermal expansion (CTE) mismatch between GaN and the substrate material. Table 3.2 compares the lattice and CTE values for GaN vs various potential substrate materials. As seen later, SiC offers better characteristics with a 3.5% lattice mismatch. But the 33% of CTE mismatch between GaN and SiC is still significant and can cause problems as the devices go through power cycling in the application. Not to forget that SiC is more expensive compared with Si.

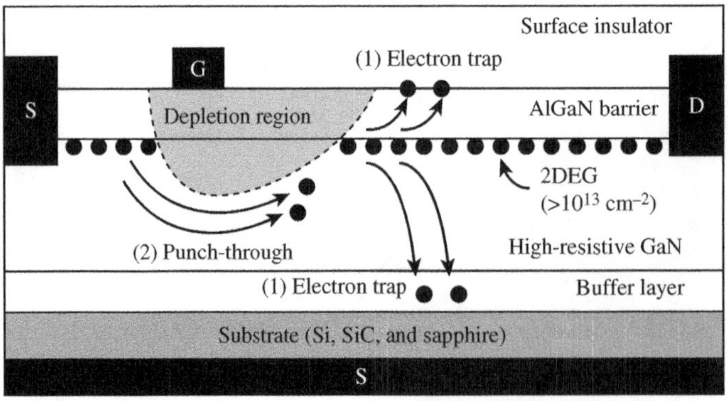

Figure 3.11 Device-level issues of GaN: (1) electron trapping and (2) punch-through current. *Source:* Ref. [14]/IEEE.

Table 3.2 Lattice constant and CTE of semiconductor material.

Substrate	GaN	Si <111>	Sapphire (crystal of Al_2O_3)	SiC 6H	Ge <111>
Lattice constant	3.19	3.84	2.75	3.08	4.0
Coefficient of thermal expansion (CTE)	5.6	2.6	7.5	4–4.2	5.9

Source: Ref. [15].

3.3.2 Application Level Challenges with GaNs

Other than the device physics level challenges associated with GaNs, there are also application-/circuit-level challenges for GaN devices. Below is a summary of challenges associated with GaNs in the application level.

3.3.2.1 GaN's Narrow Gate Voltage Margin

Figure 3.12 shows the output characteristics of EPC2001C GaN device. As seen here, the device is fully enhanced at a gate voltage of 5 V. Table 3.3 shows the EPC2001C absolute maximum ratings. Based on Table 3.3, the absolute maximum permissible gate voltage is only 6 V, which means there will be only a 1 V safety margin with respect to the nominal gate voltage of 5 V. Therefore, the gate driver design must include special considerations to limit the voltage overshoot across the gate-to-source terminals. Other than the device parasitic elements, this overshoot also depends significantly on gate loop stray inductance. Therefore, an optimized circuit layout becomes significantly more important in GaN applications as compared with silicon applications where the safety margin is much wider.

There are various other ways to mitigate the V_{GS} overshoot problem including adding an external capacitor in parallel with C_{GS} and/or adding a gate resistor to the gate loop. Note, however, that both of these two techniques will increase the device-switching loss and will limit the maximum

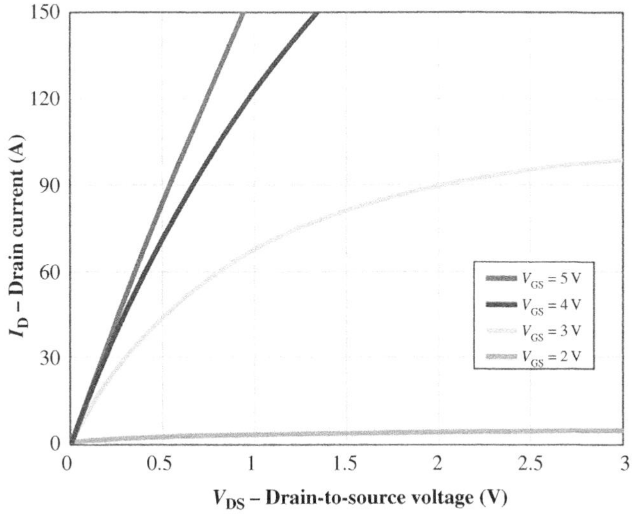

Figure 3.12 EPC2001C output characteristic curves. *Source:* Ref. [16].

Table 3.3 EPC2001C absolute maximum ratings.

Maximum ratings			
V_{DS}	Drain-to-source voltage (continuous)	100	V
	Drain-to-source voltage (up to 10,000 5 ms pulses at 125 °C)	120	V
I_D	Continuous ($T_A = 25C$, $\theta_{jA} = 13$)	36	A
	Pulsed (25 °C, $T_{Pulse} = 300$ μs)	150	A
V_{GS}	Gate-to-source voltage	6	V
	Gate-to-source voltage	–4	V
T_j	Operating temperature	–40 to 150	C
	Storage temperature	–40 to 150	C

Source: Ref. [16].

operating switching frequency [17, 18]. Placing a Zener diode-based clamp circuit in parallel to the gate is an alternative technique that can be utilized to limit the gate voltage overshoot. The nonlinearity of the Zener diodes, however, adds complexity to the design and, therefore, this technique is less preferred [19]. Despite all these techniques, GaN devices still suffer from overshoots during turn-on events that can result in catastrophic failures.

The other characteristic associated with a narrow gate voltage margin in GaN devices is the Rdson change by a small variation in gate voltage. Figure 3.13 shows the typical Rdson change by gate voltage in EPC2001C. As seen in Figure 3.13, with gate voltages of <4 V, a small change in gate voltage results in a significant change in Rdson. For example, looking at the $I_D = 80$ A curve, reducing the gate voltage from 4 to 3.5 V, the Rdson increases by ~28%. Therefore, special care must be taken to the design of the gate driver board such that not only the V_{GS} overshoot is tightly controlled but a tight voltage regulation is also achieved.

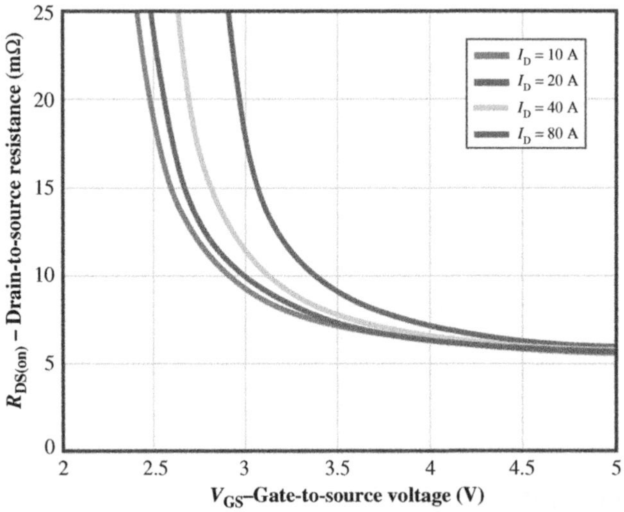

Figure 3.13 Rdson vs VGS for various drain currents for EPC2001C. *Source:* Ref. [16].

3.3.2.2 dv/dt Immunity and False Turn-On in GaN Devices

A typical GaN device including its parasitic capacitance is shown in Figure 3.14. In this figure, as the device turns off, it faces a high dv/dt slew rate across its drain to the source. This high dv/dt, in turn, rapidly charges the device's parasitic capacitances as illustrated in Figure 3.14. The charge current passing through the gate will raise the gate voltage. If the gate voltage is raised beyond its threshold limit, it can falsely turn the device on. This phenomenon is also called the Miller effect and can cause shoot-through in half-bridge architectures. Note that the same phenomenon exists in silicon devices as well as SiC-MOSFETs. However, the issue is worse in the case of GaN devices as the turn-on threshold voltage is lower in the case of GaN devices. For example, the minimum threshold voltage in the case of EPC2001C is only 800 mV [16].

3.3.2.3 di/dt Immunity in GaNs

Figure 3.15 shows how the step voltage induced across the common source inductance (CSI) as a result of high di/dt can induce an opposing voltage across the gate-to-source of a GaN device in the off state. The induced voltage across the gate-to-source is a negative voltage, which, if not damped properly, can cause ringing due to the RLC tank created in the gate loop. If the positive voltage created as a result of ringing at the gate goes beyond the device threshold, it can falsely turn it on and cause a shoot-through.

Note that this phenomenon is also present in all other switching devices. However, the outcome is more severe in the case of GaNs as GaN's turn-on threshold voltage is lower as compared with other switching devices such as silicon and silicon carbide. In addition to that, as seen in Table 3.3, the absolute maximum permissible voltage is limited to a positive voltage of 6 V and a negative voltage of −4 V. Therefore, not only the positive ringing across the gate-to-source can cause false turn-on and shoot-through but the negative ringing across the gate-to-source can also exceed device limitations and cause catastrophic failure.

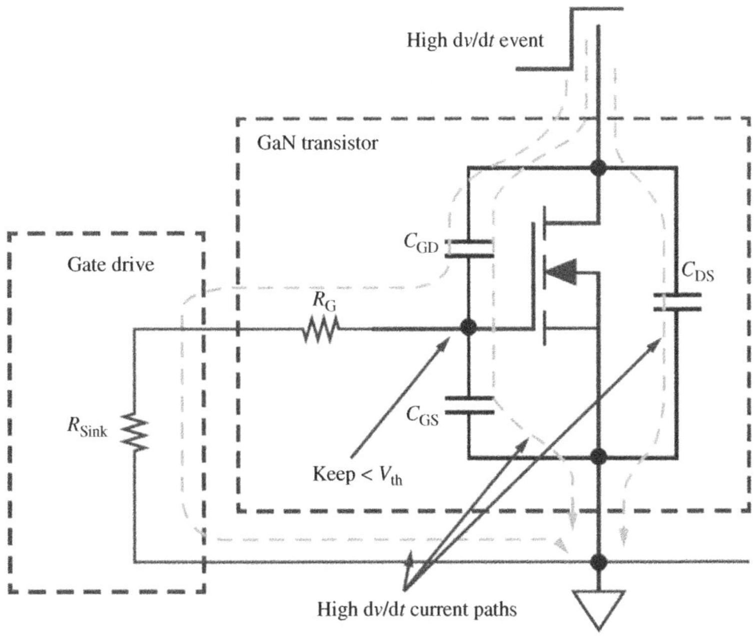

Figure 3.14 Effect of dv/dt on a GaN device in the off-state. *Source:* Strydom et al. [20].

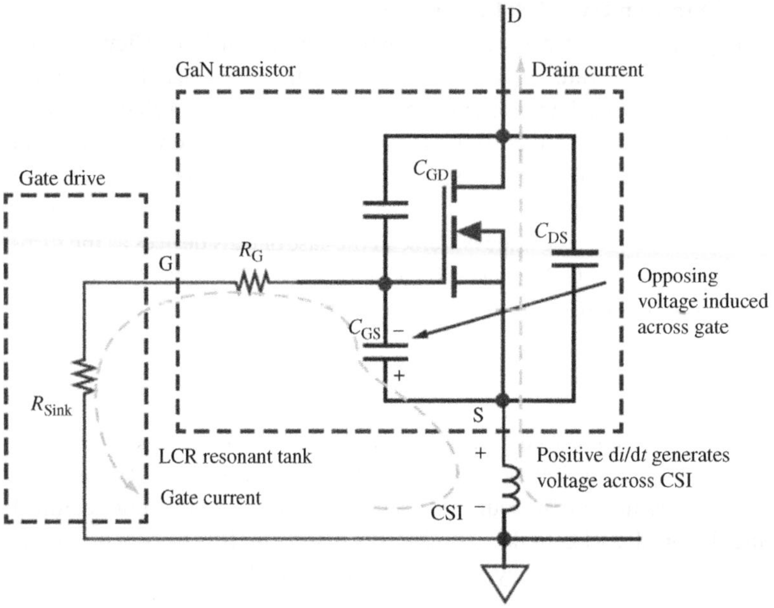

Figure 3.15 Impact of a positive di/dt of an off-state device with common-source inductance. *Source:* Strydom et al. [20].

One potential solution to this issue is to increase the damping in the gate loop in off-state by increasing the sink resistor, R_{Sink}, in Figure 3.15. However, increasing the gate loop sink resistor would negatively impact the dv/dt immunity, which can also cause false turn-on [20]. Therefore, the di/dt immunity still remains a challenge with GaNs in practical applications.

While these are some of the key challenges associated with GaNs in both the device level and application level, efforts are being made to address these issues so the benefits associated with GaNs can be utilized in transportation electrification.

3.4 SiC-MOSFET Challenges in Transportation Electrification

After discussing the challenges with GaNs, it may be instructive to discuss SiC-MOSFETs and their suitability for transportation electrification. Unlike GaNs, SiC-MOSFETs have been more successful in penetrating transportation electrification. The device-level challenges associated with GaNs, discussed earlier, do not exist in SiC-MOSFETs. In addition to that, SiC-MOSFETs have a thermal conductivity, which is >3 times higher than both silicon and GaN devices. Therefore, SiC-MOSFETs have been considered for high-power applications such as traction inverters. Despite all the advantages associated with SiC-MOSFETs, there are also challenges associated with them that need to be well understood in order to utilize them effectively in transportation electrification.

3.4.1 Low Gain of SiC-MOSFETs

To investigate the impact of low gain in SiC-MOSFETs, it may be instructive to look at the transconductance (also known as output characteristics) of a typical SiC-MOSFET as compared with an IGBT of the same rating. Figure 3.16 compares the output characteristics of a 1200 V/300 A IGBT (MBM300GS12AW) by Hitachi [21] vs a 1200 V/300 A SiC-MOSFET (BSM300D12P2E001) by

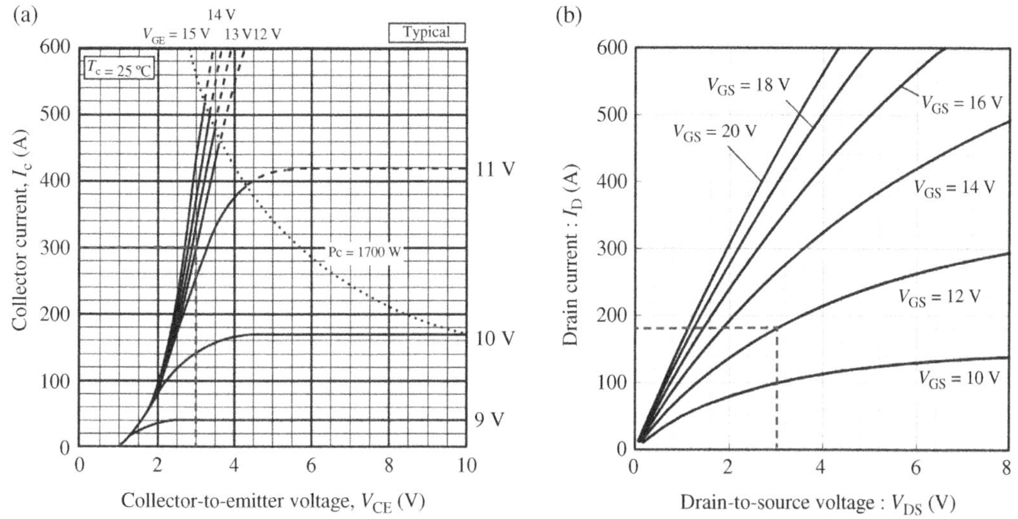

Figure 3.16 Comparing output characteristics of Si-IGBT vs SiC-MOSFET. (a) MBM300GS12AW, 1200 V/300 A Si-IGBT module by Hitachi and (b) BSM300D12P2E001, 1200 V/300 A SiC-MOSFET module by Rohm.

Rohm [22]. Looking at this picture, one can see that achieving 300 A at a V_{CE} of 3 V requires 12 V at the gate in the case of the Si-IGBT. With 12 V at the gate of the SiC-MOSFET, assuming the same V_{DS} drop of 3 V, only 180 A is achievable. This is a direct impact of lower gain in SiC-MOSFETs. It can also be seen that the output current capability of the IGBT module is characterized by a maximum gate voltage of 15 V. Looking at the IGBT curves in Figure 3.16a, from 12 and 15 V, the curves are closely overlapped. Therefore, increasing the gate voltage further does not have much impact on current capability. But in the case of SiC-MOSFET output characteristic, shown in Figure 3.16b, the output is characterized up to V_{GS} of 20 V. Looking at Figure 3.16b, one can observe that there is relatively a big gap between the curves at 18 V vs 20 V. The lower gain of SiC-MOSFETs will also impact the gate driver board design as with SiC-MOSFETs, a higher gate drive supply voltage as well as a higher source/sink current capability is required to enable a faster transition from desaturation (high loss) region to ohmic (low loss) region.

3.4.2 Fault Detection in SiC-MOSFETs

In transportation electrification, one of the key aspects associated with the reliable operation of power inverters is fast detection and isolation of desaturation fault, also known as DESAT. Figure 3.17 illustrates the ohmic region vs desaturation region in a typical output characteristic curve of a power module [23]. In Figure 3.17, the knee point represents the threshold voltage above which the device needs to shut down to prevent damage due to excessive conduction loss.

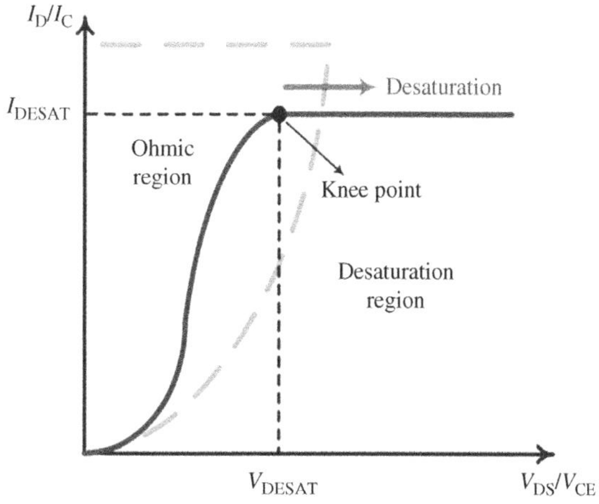

Figure 3.17 Ohmic region vs desaturation region in a power module's output characteristic.

Figure 3.18 shows a typical DESAT fault detection circuitry. This circuit is employed

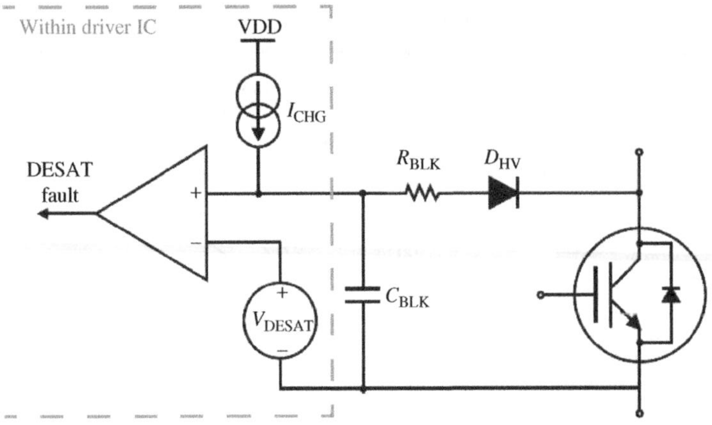

Figure 3.18 Typical DESAT detection circuitry. *Source:* Ref. [4]/Texas Instruments Incorporated.

to detect desaturation in semiconductor devices. In this circuit, when the MOSFET is on, the I_{chg} is applied to the anode of the diode D_{HV} and it conducts. Therefore, the voltage across the drain-to-source of the MOSFET is measured and fed into the internal comparator to be compared with the V_{DESAT}. The V_{DESAT} is the voltage level at which the DESAT fault is triggered and the MOSFET is disabled. The V_{DESAT} level is normally adjustable in gate driver ICs and is determined based on the V_{DS} value at the knee point in Figure 3.17.

Looking at Figure 3.16, it can be noted that, unlike Si-IGBTs, SiC-MOSFET output characteristic does not have a clear knee point. Therefore, proper selection of V_{DESAT} in Figure 3.18 and, therefore, detecting DESAT fault in SiC-MOSFETs is a major challenge.

Additionally, due to smaller die size for the same voltage, SiC-MOSFETs have lower short circuit capability as compared with Si-IGBTs. Therefore, DESAT fault needs to be detected much faster in the case of SiC-MOSFETs as compared with IGBTs to prevent damage [4]. This poses a significant challenge for SiC-MOSFETs in electrified powertrain applications where having proper short circuit capability is a must to have reliable performance and prevent catastrophic failures.

3.4.3 Driving SiC-MOSFETs

In traction inverter applications, where the inverter drives an electric motor, the highly inductive load of the motor acts as a big current source. Therefore, as the inverter is switching, the load current does not change, but rather is transferred to the body diode of the opposite side switch. Once the opposite switch turns on, the total load current needs to be supplied by the SiC-MOSFET. Now, as shown in Figure 3.16, the SiC-MOSFET gain is low. Therefore, the transition from off-state to fully on-state needs to happen faster as compared with IGBTs. Otherwise, the SiC-MOSFET will operate in the lossy desaturation region, causing excessive loss that can damage the device.

To enable a faster switching transition, also known as slew rate, the gate driver needs to be able to supply a higher source current during turn-on events. This sets a hard requirement for gate driver IC. The minimum source current requirement in the case of SiC-MOSFETs is >10 A. This compares with 3–4 A of peak source current requirement in the case of Si-IGBTs [24–26].

3.4.4 Maximum Gate Voltage Swing in SiC-MOSFETs

The turn-on threshold voltage in SiC-MOSFETs is normally lower as compared with Si-IGBTs. That, in addition to the higher turn-on slew rate, results in the need for a high negative voltage at the gate of a SiC-MOSFET during turn-off and while in the off state. Si-IGBTs, however, can successfully turn off at 0 V. Adding a small negative supply of up to −5 V can prevent any potential hazard associated with false turn-on in Si-IGBTs. Therefore, in Si-IGBTs, as the switch transitions from off state to on state, the gate voltage transitions from 0/−5 to +15 V.

In SiC-MOSFETs, however, due to the reasons mentioned earlier, a turn-off gate voltage of at least −8 V is required to safely turn the device off and prevent any false turn-on while the device is in off state. Additionally, as discussed earlier, due to the lower gain of the SiC-MOSFETs, they require a higher positive voltage of typically ~20 V at the gate for turn-on [27]. Therefore, during each off-to-on transition, the gate of a SiC-MOSFET is exposed to a voltage swing of >28 V. So, the voltage swing at the gate of a SiC-MOSFET is at least 40% higher than Si-IGBTs. The higher voltage swing at the gate of a SiC-MOSFET not only requires a higher source current from the gate driver board but it also adversely impacts device reliability.

3.4.5 Layout Considerations

Based on the earlier discussion, it is important to note that not only the SiC-MOSFETs can switch faster but they must also switch faster to compensate for their lower gain. As the slew rate and/or switching frequency is increased, there are more EMI concerns to face.

Referring to Figure 3.19, the higher slew rate in a SiC-MOSFET causes higher di/dt in both power loops as well as gate loops. The higher di/dt in the power loop imposes a higher voltage overshoot across the device because of the high $L_{Parasitic}di/dt$ when the device turns off. The higher di/dt in the gate loop causes excessive oscillations at the gate, which, in turn, can cause a false turn-on during the off state. Both excessive overshoot across the device during turn-off and high oscillations in the gate loop can cause catastrophic failures. Therefore, to prevent these issues, having an optimized gate loop layout as well as a power loop layout where the parasitic inductance is minimized becomes extremely critical. An optimized layout can also prevent the high slew rate signals from interfering with sensitive signals such as input commands from the controller.

3.5 Advanced Power Module Packaging to Accommodate WBG Devices

Figure 3.20 shows the internal architecture of a conventional IGBT-based power module. The semiconductor dies (IGBTs and diodes) are soldered onto a direct bonded copper (DBC) substrate. The basic DBC technology was developed by General Electric in the 1970s [28]. It involves two layers of copper on top and bottom with a ceramic layer in the middle.

Therefore, the DBC serves multiple functionalities such as a patterned Cu area for die interconnection on top, heat-spreading on the bottom Cu, and the ceramic in the middle to provide not only a dielectric capability for isolation but also to provide a high thermally conductive layer to spread the heat out of the dies and transfer that to the bottom Cu area and eventually to the baseplate. The DBC material used in most conventional power modules is Al_2O_3, also known as Alumina.

The baseplate acts as a heat-spreader to spread the heat generated by dies on top Cu and transfer that to the heat exchanger/heatsink on the bottom side. The baseplate also provides solid mechanical support for the entire module to be mounted onto the heatsink.

The bottom side of the semiconductor dies (also known as power devices) are normally soldered onto the top-side Cu pattern on the DBC. Aluminum wirebonds are conventionally used to provide top-side interconnection between multiple dies in series and/or parallel.

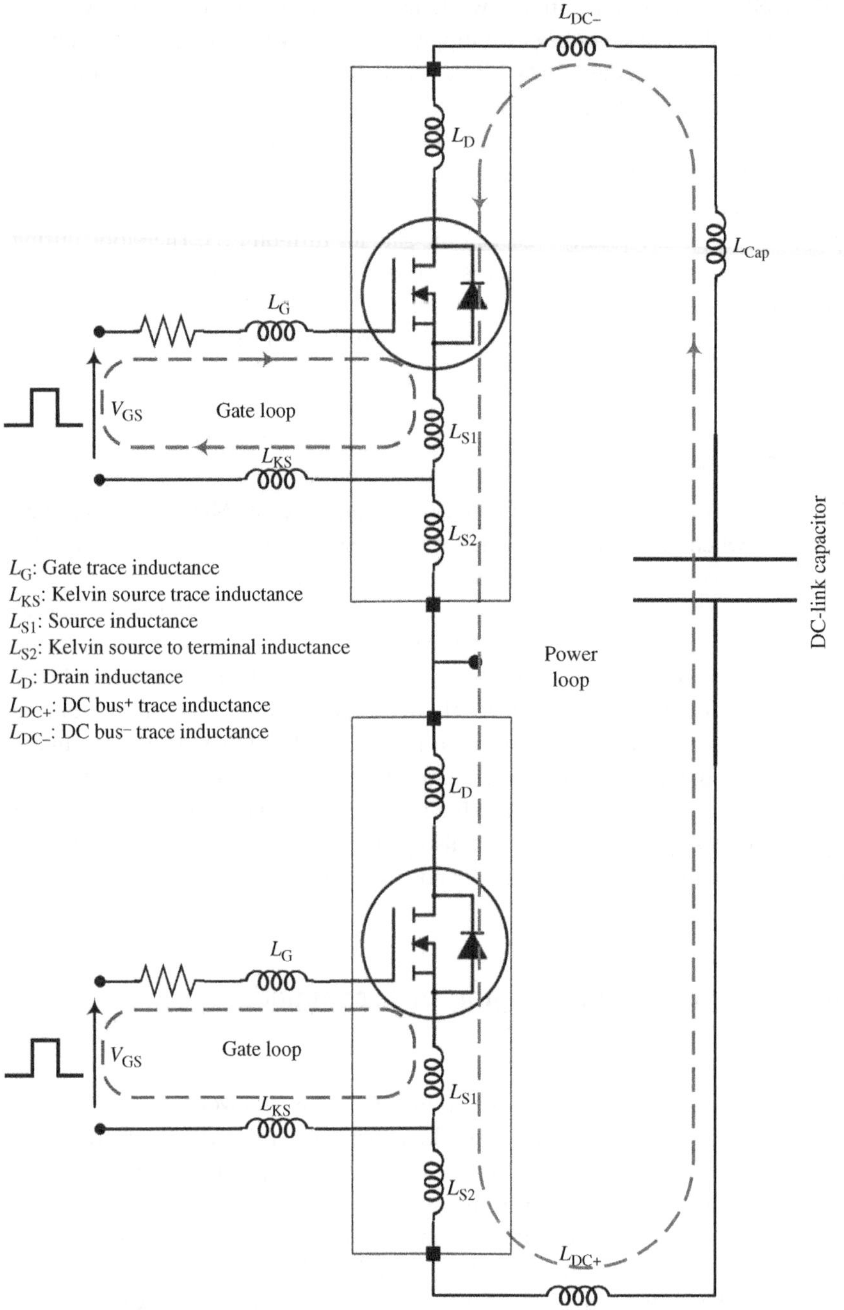

L_G: Gate trace inductance
L_{KS}: Kelvin source trace inductance
L_{S1}: Source inductance
L_{S2}: Kelvin source to terminal inductance
L_D: Drain inductance
L_{DC+}: DC bus+ trace inductance
L_{DC-}: DC bus− trace inductance

Figure 3.19 Importance of gate loop and power loop in HV inverters with WBG devices.

The area above the dies where the interconnection happens is normally filled with silicone gel encapsulant. This encapsulation provides two main functionalities. First, it provides mechanical support and integrity to the module and second, it prevents contamination from entering the sensitive environment involving dies and wirebonds. In addition to these two main functionalities, the encapsulation also needs to be able to provide voltage isolation capability up to the breakdown

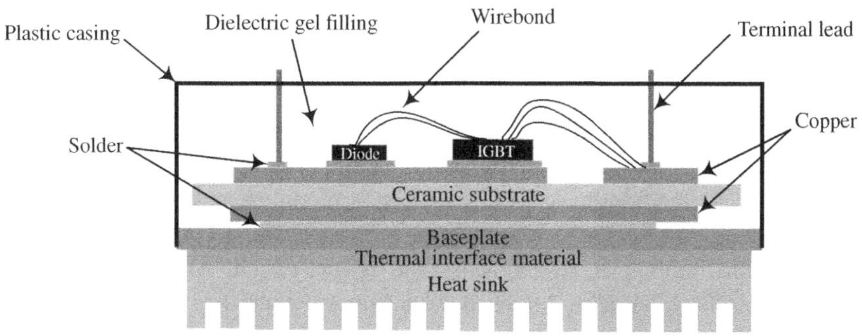

Figure 3.20 Internal architecture of a conventional power module.

capability of the dies. In the absence of encapsulation, the air has a much lower voltage breakdown capability.

As the semiconductor technology evolves from Si-MOSFET to Si-IGBT and recently to WBG devices, the power module packaging needs to be improved to enable such transition. There are features associated with WBG devices that call for advanced power module packaging. Looking at Table 3.1, SiC-MOSFET dies have much higher junction temperature capabilities as compared with Si-IGBTs. Commercially available SiC-MOSFET dies can operate up to 200 °C [29]. That means all the material used inside a power module package must also be able to operate up to 200 °C at a minimum.

3.5.1 Advanced Substrate Materials

As the operating junction temperature increases to up to 200 °C in SiC-MOSFET applications, a DBC material with a higher thermal conductivity is required to enable more efficient heat dissipation. Additionally, as the operating junction temperature increases, the thermal gradient between the junction and coolant increases. As a result of that, higher thermal stress is applied to the DBC, which can cause immature failures such as delamination. To reduce the thermal stress applied to the DBC under a higher operating junction temperature, a substrate material with a lower CTE is required. Table 3.4 shows alternative substrate materials that have higher thermal conductivity and lower CTE as compared with Al_2O_3 [30]. Based on this table, as compared with Al_2O_3, the CTE of AlN and Si_3N_4 are lower and closer to the CTE of SiC-MOSFET die. This means under the same operating temperature, a lower thermal stress is applied to the AlN and Si_3N_4 substrates. Therefore, it is expected that a higher power cycle capability and better lifetime are achieved when using AlN or Si_3N_4 with SiC-MOSFET dies [31].

Table 3.4 Advanced substrate materials.

Material	Thermal conductivity (W/mk)	CTE ppm (°C)
SiC	120	4
Al_2O_3	20–35	7.2
AlN	150–180	4.6
Si_3N_4	60–70	3

3.5.2 Advanced Die Attach Methods

In conventional power modules, the dies are soldered onto the DBC. The melting point of the most commercially available solders such as SnAg is in the range of 220–225 °C. To prevent remelting of the solder during normal operation, the maximum operating junction temperature must be limited to <175 °C to provide a 20% margin with respect to the melting point of the solder material. Therefore, advanced bonding materials/methods have been developed by module manufacturers that can allow a maximum junction temperature of up to 200 °C. One of these techniques that is still under development for further optimization is nano-silver sintering. Compared with solder paste, nano-silver has a melting point of 961 °C that is well beyond the maximum operating junction temperature of 200 °C in SiC-MOSFETs. The other benefit of nano-silver sintering as compared with soldering is in terms of CTE. As the operating junction temperature increases, the thermal gradient between the die temperature and the coolant temperature increases. Therefore, considering the low CTE of the semiconductor dies, it is desirable to reduce the CTE of the bond layer to prevent/reduce excessive thermal stress on the dies that can result in immature failures. Table 3.5 compares the major characteristics of SnAg solder paste as compared with nano-silver sinter paste [32]. The lower CTE of sinter paste in addition to its four times higher thermal conductivity helps reduce the thermal stress on the dies at higher junction temperatures and, therefore, the module power cycle capability and lifetime are improved.

3.5.3 Interconnection

Aluminum wirebonds, used for interconnection between series and/or parallel dies, account for one of the major weak points in power module architecture when it comes to power cycling capability. Bond lift-off and heel crack are common failure modes in aluminum wirebonds because of high thermal stress during power cycling. Increasing the operating junction temperature in WBG devices further increases the thermal stress the wirebonds are exposed to. Therefore, alternative ways of die interconnection are desirable to improve electrical conductivity as well as thermal conductivity and eventually improve the module lifetime.

The other major shortcoming associated with wirebonds is their high parasitic inductance. The high parasitic inductance of conventional aluminum wirebonds not only causes EMI issues but also limits the maximum slew rate as well as application-switching frequency.

Now, as discussed earlier, the low gain of WBG devices calls for a higher slew rate to limit their switching loss. Additionally, one of the major advantages of using SiC-MOSFETs in transportation electrification is to increase the switching frequency to reduce the size of passive components such as DC-link capacitors.

Table 3.5 Comparing characteristics of SnAg solder paste vs Ag sinter paste.

Property	Unit	Ag sinter layer	SnAg solder layer
Liquidus	°C	961	221
Electrical conductivity	MS/m	41	7.8
Thermal conductivity	W/mK	250	70
CTE	μm/mK	19	28
Tensile strength	MPa	55	30

Table 3.6 Comparing characteristics of copper vs aluminum for wirebond applications.

Feature	Unit	Copper	Aluminum
Electrical resistivity	μΩ cm	1.7	2.7
Thermal conductivity	W/m K	400	220

Considering the points mentioned earlier, extensive research and development have been done by major module manufacturers to find alternative solutions to conventional aluminum wirebonds. The goal of these studies has been focused on the reduction of stray inductance while increasing the electrical conductivity and thermal conductivity, as well as improving module lifetime.

Several power module manufacturers have started using copper wirebonds instead of aluminum wirebonds. Compared with aluminum, copper has a lower electrical resistance, which leads to higher electrical conductivity. Copper also has higher thermal conductivity as compared with aluminum. Table 3.6 compares the electrical resistivity and thermal conductivity of aluminum vs copper [33]. Copper wirebonds can result in a 40% improvement in overall electrical conductivity and up to twice better thermal conductivity.

One of the main challenges with replacing aluminum wirebonds with copper wirebonds is the less flexibility of copper, as compared with aluminum, which makes them more difficult to bend and form for die interconnection. An example of a copper wirebond used in the Danfoss DCM1000 power module is shown in Figure 3.21. Danfoss developed a technology called Danfoss Bond Buffer (DBB) that enables the application of copper wirebonds [34].

On-semi has proposed an application of flip-chip technology to replace wirebonds in their VE-Trac dual power modules. The flip-chip technology in power modules uses spacers for die interconnection. The spacers are thick, short copper pillars that, compared with conventional aluminum wirebonds, significantly reduce both electrical resistance as well as parasitic inductance and improve thermal conductivity. Figure 3.22 shows both internal architecture as well as outside packaging of the On-Semi VE-Trac dual IGBT module that uses spacers for die interconnection [35]. As seen in this figure, there are two DBCs in this module that not only enable using spacers but also provide double-sided cooling (DSC) capability for better thermal performance. The major challenge with flip-chip technology in high-power applications is the CTE mismatch between the power device and the copper pillars when the module goes through power cycling.

Power Overlay (POL) is another technology developed by General Electric Aviation Systems (GEAS) that uses planar interconnect technology, using polyimide and copper layers to eliminate conventional wirebonds in power modules. This technology is briefly shown in Figure 3.23 [36]. The POL technology not only significantly reduces the stray inductance, enabling higher voltage operation but it also helps with the integration of passive components inside the power module. Latter helps with improved dynamic current sharing between parallel dies that will eventually improve power module lifetime.

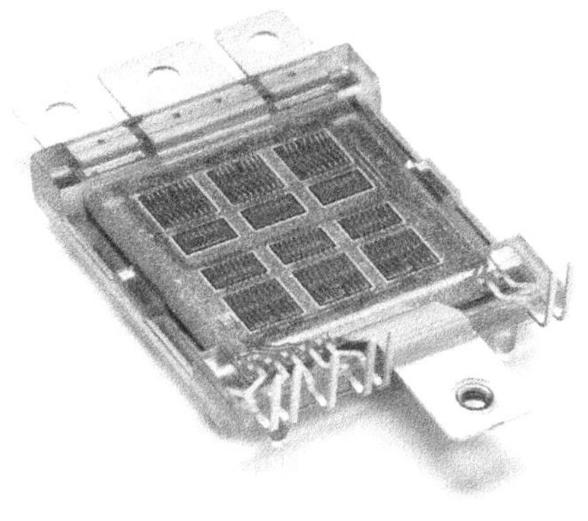

Figure 3.21 Danfoss DCM1000 module that uses DBB technology with copper wirebonds.

Semikron has developed skin technology where the interconnection between semiconductor dies is provided by using a flexible foil (PCB) that is silver

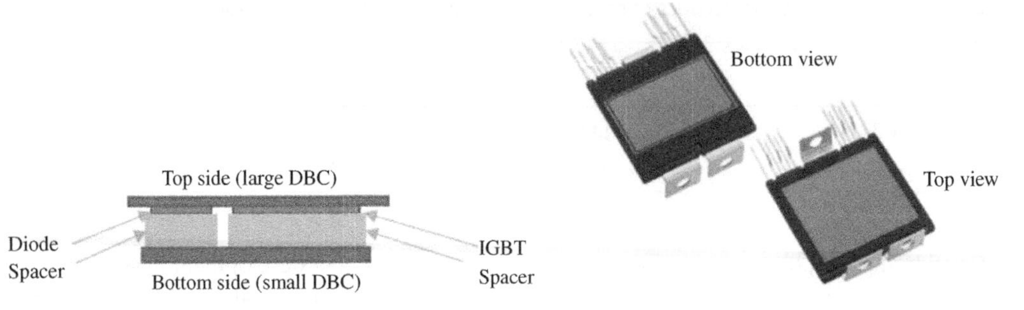

Figure 3.22 On-Semi VE-Trac dual module with spacers for die interconnection.

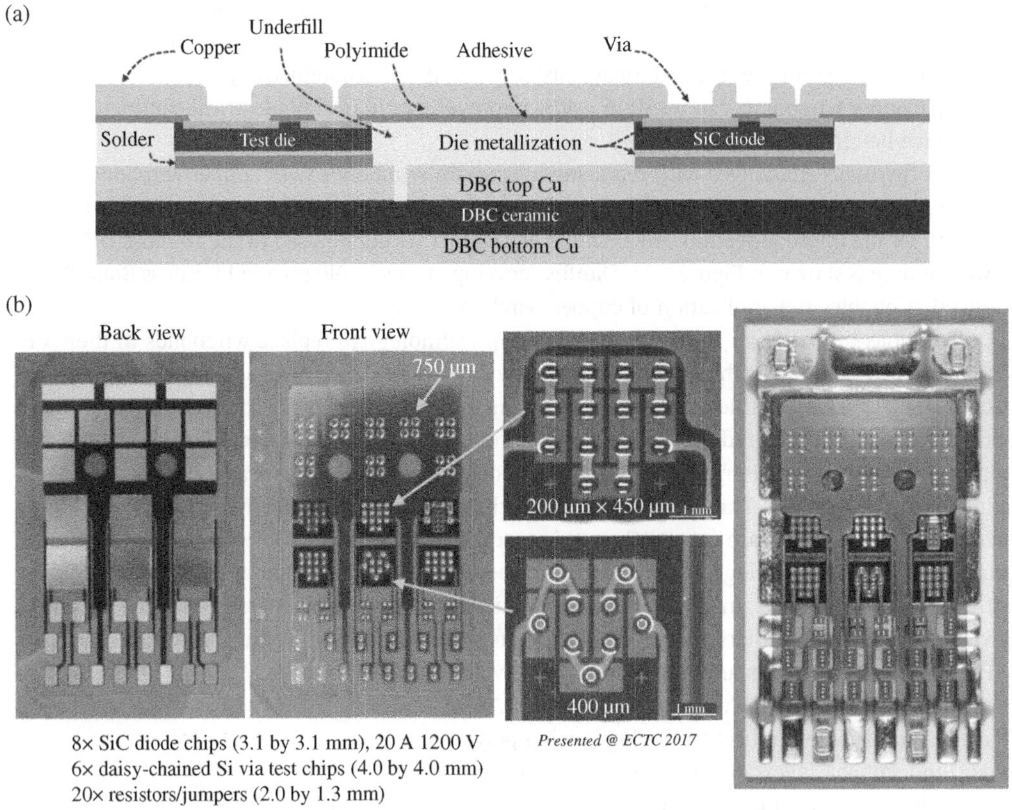

Figure 3.23 GE POL power module internal architecture. (a) POL technology architecture and (b) inside on actual POL-based power module.

sintered on top of the power devices and the DBC substrate. The flexible foil helps reduce the module stray inductance significantly. Compared with conventional wirebond technology, skin technology helps reduce stray inductance by up to 80%. The skin interconnection method is the basis for Semikron's latest power module technology, known as direct pressed die (DPD). In this technology, not only the flexible PCB provides a low stray inductance interconnection but the bonding of dies onto the CBC is also achieved by using nano-silver sintering. Since both the top side as well as the bottom side of the dies are silver sintered, this technology is also known as

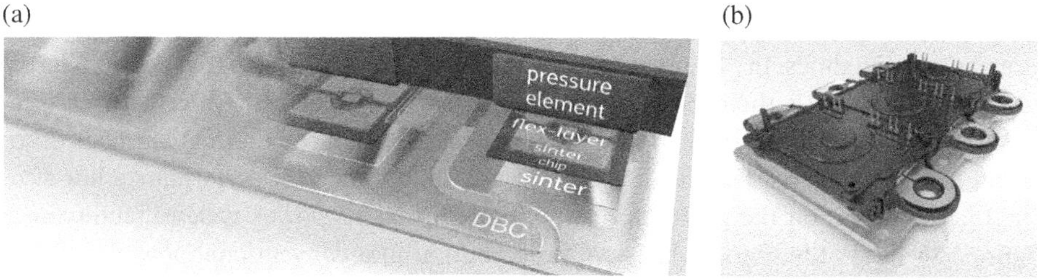

Figure 3.24 Semikron's (a) DPD technology and (b) eMPack module. *Source:* https://www.semikron.com/innovation-technology/packaging-technology/direct-pressed-die-dpd.html.

double-sided sintering (DSS). The DDS technology results in significant improvement in power density, thermal conductivity, and module lifetime. Details of the DPD technology and eMPack and Semikron's power module developed based on DPD technology[1] are shown in Figure 3.24. Semikron claims that with this technology, the module power cycle capability is improved by 300%.[2]

3.5.4 Advanced Encapsulation Materials

As discussed earlier, the encapsulation material, widely used in conventional power modules, is a dielectric gel, also known as silicone gel. One of the main challenges associated with the application of Silicone gel is its maximum operating temperature capability. The maximum operating temperature of dielectric material is determined based on its glass transition temperature range. Most of the commercially available silicone gels have a glass temperature transition of 165–175 °C. Therefore, in WBG devices, where the max-operating junction temperature can reach up to 200 °C, silicone gels will not be able to operate.

This is the reason why silicone gels have been replaced by Epoxy resin molding in advanced power module packages. Compared with silicone gels, epoxy resin moldings not only have a higher operating temperature of >200 °C, but they also have higher mechanical strength. Figure 3.25 shows a few examples of advanced power modules that use epoxy molding (also known as transfer molding) for encapsulation.[3–5]

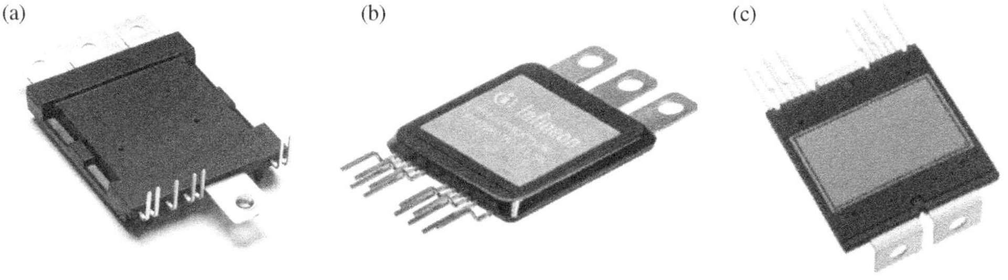

Figure 3.25 Examples of advanced power modules that use Epoxy resin for encapsulation: (a) Denfoss DCM1000, (b) Infineon FF400R07A01E3_S6, and (c) On-Semi NVG800A75L4DSC.

3.5.5 Advanced Cooling Methods

In conventional power modules, the DBC substrate is soldered onto a baseplate that acts as a large heat spreader. The baseplate is then mounted onto the heatsink/cooler using thermal interface material (TIM). Due to the low thermal conductivity of TIM material, the thermal gradient from device junction to coolant increases. This not only results in low thermal performance but also increases the thermal stress on the power devices, which impacts its power cycle capability and lifetime. Figure 3.26 shows the thermal resistance distribution in a conventional power module. As seen here, TIM accounts for 22% of total thermal resistance in a conventional power module architecture [37].

Therefore, direct cooling methods are becoming popular in advanced power module packages where the flat baseplate is replaced by either a wave baseplate or a pinfin baseplate. The wave or pinfins are in direct contact with the coolant. In a study done by Infineon, it was noted that pinfin baseplates can help reduce thermal resistance by up to 35%. Figure 3.27 compares the physical appearance of flat baseplates as compared with wave baseplates and pinfin baseplates by Infineon [38]. According to Infineon, the flat baseplate version is cost-optimized for 100kW inverters while the wave baseplate can be used in inverters up to 150 kW. The pinfin baseplate architecture is used for performance drivetrains.

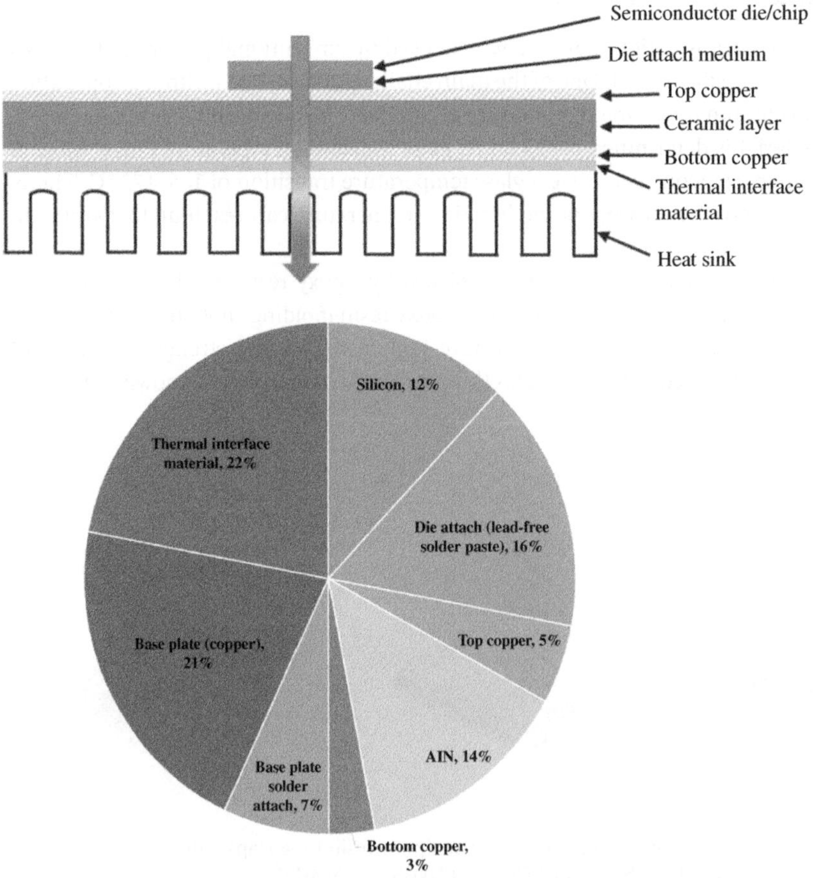

Figure 3.26 Thermal distribution within a conventional power module.

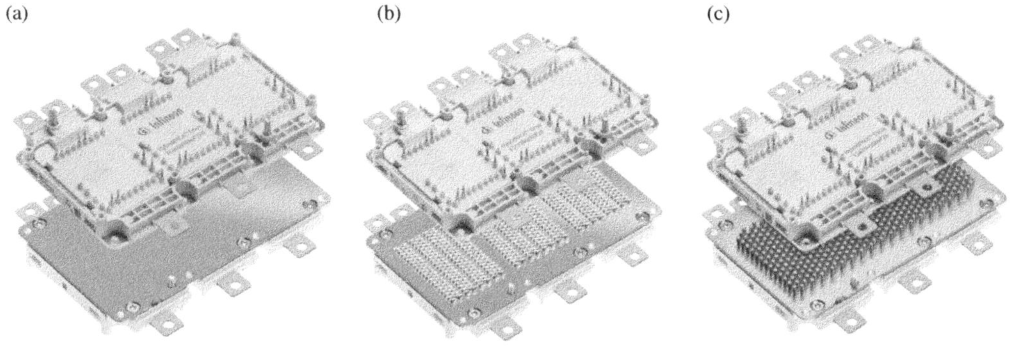

Figure 3.27 Flat baseplate vs wave baseplate vs pinfin baseplate in advanced power modules. (a) HybridPACK™ Drive Flat, (b) HybridPACK™ Drive Wave, and (c) HybridPACK™ Drive Pinfin.

On the other hand, as shown in Figure 3.26, the baseplates account for 21% of overall power module thermal resistance. Therefore, efforts are made to eliminate baseplates from the module architecture and provide direct cooling of the DBCs. The Infineon DSC module and On-Semi VE-Trac module, shown in Figure 3.25, are examples of power modules where the baseplate is eliminated to reduce the overall power module thermal resistance.

3.6 Summary

The introduction of WBG devices into transportation electrification can help improve overall system efficiency and power density. This is achieved by higher voltage capability, lower switching loss, higher operating junction temperature, and higher thermal conductivity of WBG devices. Proper integration of WBG devices into transportation electrification, however, requires full characterization and understanding of challenges associated with them in this application.

Advanced power module packaging technologies such as nano-silver sintering to improve maximum operating junction temperature as well as increasing module lifetime, wirebond-less architectures to reduce power module stray inductance and enable higher voltage operation, and advanced cooling architectures that can help reduce thermal resistance between junction and coolant are major improvements made to the power module packaging to accommodate WBG devices.

In addition to the need for advanced module packaging, the system also needs to go through an optimization that can provide a proper platform to accommodate WBG devices. Some of these optimizations include advanced gate driver board design that can address the low gain of WBG devices and enable higher slew rate required by WBG devices, advanced safety features and fault diagnostics that can address the low short circuit capability of WBG devices, and optimized layouts to prevent excessive electromagnetic interference (EMI) as a result of higher slew rate and higher switching frequencies to name a few.

Notes

1 https://www.semikron.com/innovation-technology/packaging-technology/direct-pressed-die-dpd.html.
2 https://www.semikron.com/about-semikron/technical-articles/advanced-power-modules.html.
3 https://www.danfoss.com/en-us/about-danfoss/our-businesses/silicon-power/danfoss-dcm-1000-power-module-technology-platform/.

4 https://www.infineon.com/cms/en/product/power/igbt/automotive-qualified-igbts/automotive-igbt-modules/ff400r07a01e3_s6/.
5 https://www.onsemi.com/pdf/datasheet/nvg800a75l4dsc-d.pdf.

References

1 Pressman, A.I. (1998). *Switching Power Supply Design*, 2e. McGraw-Hill. https://d1.amobbs.com/bbs_upload782111/files_41/ourdev_652687FB25XY.pdf (accessed November 2021).
2 Cuk, S. (2015). *Power Electronics: Topologies, Magnetics and Control*, vol. 1. TESLAco.
3 Mohan, N. (2003). *Power Electronics: Converters, Applications, and Design*. John Wiley & Sons, Inc.
4 Texas Instrument. (2021). IGBT & SiC Gate Driver Fundamentals, slyy169. https://www.ti.com/lit/eb/slyy169/slyy169.pdf?ts=1637101824100&ref_url=https%253A%252F%252Fwww.google.com%252F (accessed November 2021).
5 Choi, H. (2016). *Overview of Silicon Carbide Power Devices*. Fairchild Semiconductor.
6 Lidow, A., Strydom, J., De Rooij, M., and Reusch, D. (2014). *GaN Transistors for Efficient Power Conversion*. Wiley.
7 Texas Instrument. (2017). Design considerations of GaN devices for improving power converter efficiency and density, slyy124 https://www.ti.com/lit/wp/slyy124/slyy124.pdf?ts=1637173263881&ref_url= https%253A%252F%252Fwww.google.com%252F (accessed November 2021).
8 Lidow, A. (2010). Is it the end of the road for silicon in power conversion? *2010 6th International Conference on Integrated Power Electronics Systems (CIPS)*, 17 November 2011, Atlanta, GA, pp. 1–8. IEEE.
9 Dhakal, S. (2018). Circuit level reliability considerations in wide bandgap semiconductor devices. PhD Thesis. The University of Toledo.
10 Binari, S.C., Klein, P., and Kazior, T.E. (2002). Trapping effects in GaN and SiC microwave FETs. *Proc. IEEE* 90 (6): 1048–1058.
11 Zanoni, E., Meneghini, M., Chini, A. et al. (2013). AlGaN/GaN-based HEMTs failure physics and reliability: mechanisms affecting gate edge and Schottky junction. *IEEE Trans. Electr. Dev.* 60 (10): 3119–3131.
12 Brunel, L., Lambert, B., Mezengea, P. et al. (2013). Analysis of Schottky gate degradation evolution in AlGaN/GaN HEMTs during HTRB stress. *Microelectron. Reliab.* 53 (9-11): 1450–1455.
13 Sun, H., Bajo, M.M., Uren, M.J., and Kuball, M. (2014). Implications of gate-edge electric field in AlGaN/GaN high electron mobility transistors during OFF-state degradation. *Microelectron. Reliab.* 54 (12): 2650–2655.
14 Nakajima, A., Takao, K., and Ohashi, H. (2013). GaN power transistor modeling for high-speed converter circuit design. *IEEE Trans. Electr. Dev.* 60 (2): 646–652.
15 Digikey. (2014). Gallium nitride (GaN) versus silicon carbide (SiC) in the high frequency (RF) and power switching applications. https://www.richardsonrfpd.com/docs/rfpd/Microsemi-A-Comparison-of-Gallium-Nitride-Versus-Silicon-Carbide.pdf (accessed November 2021).
16 EPC2001C datasheet. (2021). https://epc-co.com/epc/Portals/0/epc/documents/datasheets/EPC2001C_datasheet.pdf (accessed November 2021).
17 Jones, E.A., Wang, F., and Ozpineci, B. (2014). Application-based review of GaN HFETs. *2014 IEEE Workshop on Wide Bandgap Power Devices and Applications (WiPDA)*, 13–15 October 2014, Knoxville, TN, pp. 24–29. IEEE.

18 GaN Systems Inc. (2016). How to drive GaN enhancement mode power switching transistors, GaN Systems Application 2014 Notehttps://www.mouser.com/catalog/additional/GaN%20Systems_GN001%20App%20Note%202014-10-21.pdf (accessed November 2021).

19 Barchowsky, A., Kozak, J.P., Hontz, M.R. et al. (2017). Analytical and experimental optimization of external gate resistance for safe rapid turn on of normally off GaN HFETs. *Applied Power Electronics Conference and Exposition (APEC)*, 26–30 March 2017, Tampa, FL, pp. 1958–1963. IEEE.

20 EPC. (2017). Using enhancement mode GaN-on-silicon power FETs (eGaN® FETs), EPC application note AN003. https://epc-co.com/epc/Portals/0/epc/documents/product-training/Using_GaN_r4.pdf (accessed November 2021).

21 Hitachi. MBM300GS12AW datasheet https://datasheet.octopart.com/MBM300GS12AW-Hitachi-datasheet-10733748.pdf (accessed November 2021).

22 Rohm. (2018). BSM300D12P2E001 datasheet. https://fscdn.rohm.com/en/products/databook/datasheet/discrete/sic/power_module/bsm300d12p2e001-e.pdf (accessed November 2021).

23 Ranjbar, A. (2020). SiC-MOSFETs: challenges in transportation electrification. ElectronicDesign.com, https://www.electronicdesign.com/power-management/whitepaper/21144910/sic-mosfets-challenges-in-transportation-electrification (accessed November 2021).

24 Infineon. (2021). EiceDRIVER™ 1ED31xxMC12H Compact datasheet. https://www.infineon.com/dgdl/Infineon-1ED31xxMC12H-_1ED-X3_Compact-DataSheet-v02_00-EN.pdf?fileId=5546d46277fc7439017802de09e5671d (accessed November 2021).

25 NXP. (2020). MC33GD3100 datasheet. https://www.nxp.com/docs/en/data-sheet/MC33GD3100_SDS.pdf (accessed November 2021).

26 Infineon. (2022). Infineon-2EDF8275F-DataSheet-v02_07-EN. https://www.infineon.com/dgdl/Infineon-2EDF7275F-DataSheet-v02_07-EN.pdf?fileId=5546d462636cc8fb0163b08fd9203057 (accessed October 2022).

27 OnSemi. (2022). SiC MOSFETs: gate drive optimization. https://www.onsemi.com/pub/Collateral/TND6237-D.PDF (accessed October 2022).

28 Oakridge National Laboratory. (2006). High-temperature high-power packaging techniques for HEV traction applications. https://digital.library.unt.edu/ark:/67531/metadc932216/m2/1/high_res_d/974605.pdf (accessed November 2021).

29 OnSemi. (2022). Onsemi Gen 1 1200 V SiC MOSFETs & modules: characteristics and driving recommendations. https://www.onsemi.com/pub/collateral/and90103-d.pdf (accessed October 2022).

30 Guillemet, T. (2013). Diamond-based heat spreaders for power electronic packaging applications. Material chemistry. Université Sciences et Technologies - Bordeaux I, 2013. English. NNT: 2013BOR14841. tel-01062919.

31 Berry, D.W. (2014). Design, analysis, and experimental verification of a mechanically compliant interface for fabricating reliable, double-side cooled, high-temperature, sintered silver interconnected power modules. https://vtechworks.lib.vt.edu/bitstream/handle/10919/64898/Berry_DW_D_2014.pdf?sequence=1 (accessed November 2021).

32 Gobl, C. and Faltenbacher, J. (2010). *Low Temperature Sinter Technology Die Attachment for Power Electronic Applications*. Nuremberg, Germany: CIPS.

33 Yannou, J.-M. and Avron, A. (2012). Analysis of innovation trends in packaging for power modules. *7th European Advanced Technology Workshop on Micro-packaging and Thermal Management*, 1–2 February. IMAPS.

34 Danfoss. (2019). Danfoss Silicon Power | DCM™1000, Ahead of the curve – DCM™1000 customized power modules for advanced power transmission. https://files.danfoss.com/download/Drives/DCM-brochure.pdf (accessed November 2021).

35 OnSemi. (2019). AND9988/D VE-Trac^{TM} Dual technical guide. https://www.onsemi.com/pub/Collateral/AND9988-D.PDF (accessed November 2021).

36 Yang, L., Eddins, R., George, R. et al. (2019). *Reliability of Silicon Carbide Power Modules Using POL-kW Packaging Technology*. Anaheim, CA: APEC.

37 Balakrishnan, M. (2016). Evaluation of Thermal Management Solutions for Power Semiconductors. A Thesis submitted for the degree of Doctor of Philosophy in the Department of Electronic and Electrical Engineering, The University of Sheffield.

38 Infineon. (2019). New Infineon HybridPACK^{TM} power modules enable fast and flexible electrification of vehicles. https://www.infineon.com/cms/en/about-infineon/press/market-news/2019/INFATV201905-062.html (accessed November 2021).

4

An Overview of Inductive Power Transfer Technology for Static and Dynamic EV Battery Charging

Ahmed A. S. Mohamed[1,2], *Ahmed A. Shaier*[2], *and Hamid Metwally*[2]

[1] *Eaton Research Laboratory, Eaton Corporation, Golden, CO 80401, USA*
[2] *Zagazig University, Electrical Power and Machines Department, Faculty of Engineering, Zagazig 44519, Egypt*

4.1 Introduction

The public transportation sector consumes about 70% of fossil fuels in the world, which makes it a major source of harmful emissions of greenhouse gases (GHGs) and air pollution [1]. Immaculate transportation technologies are crucial to minimizing the dependency on fossil fuels and the emission of GHGs. Electric vehicles (EVs) are one of the main players in this domain due to their merits that are clearly demonstrated in emission, performance, and safety. The charging infrastructure of EVs is one of the prime obstacles that hinder the penetration of EVs in the global market. Wireless charging technology presents an exemplary solution for EV charging due to being convenient, automatic, durable against vandalism, reliable in harsh weather conditions, and can be implemented on the road, in public parking, in private parking, etc. [2]. Wireless power transfer (WPT) technologies can be classified into four prime categories [3]: near-field [4], far-field transfers [5], acoustic [6], and mechanical force [7], as depicted in Figure 4.1. In near-field technology, power transfers through electromagnetic fields (EMFs), which stay within a short distance around the transmitter. Within this distance, the fields can be separated to transmit energy over either magnetic fields by coils (inductive) [3], or electric fields by capacitors (capacitive) [8]. Among the WPT charging technologies, inductive charging provides promising merits for EV because: (i) it is free from noise, (ii) its components are electrically isolated, (iii) its capability of transferring high power through a relatively large air gap (10–40 cm) [9], which covers the ground clearance for most of the vehicles, and (iv) it does not include any moving parts, therefore, it requires minimal maintenance [2, 10].

In an inductive power transfer (IPT) system, the power transfers from a transmitter to a receiver through the air by magnetic induction. This technology presents a new revolution in the EV-charging industry. Unlike conventional plug-in tethered to a charger, no power connection is needed; instead, an EV can charge its battery automatically and remotely without physical connections or human intervention. There are three visions for deploying IPT technology for EV charging: during long-term parking (static or stationary charging), during high-speed travel in freeways (dynamic or in-motion charging), and during transient stops like bus stops and traffic signals (quasi-dynamic or opportunistic charging). This chapter discusses these three visions in terms of component structure, challenges, and opportunities, and state-of-the-art activities.

Transportation Electrification: Breakthroughs in Electrified Vehicles, Aircraft, Rolling Stock, and Watercraft, First Edition. Edited by Ahmed A. Mohamed, Ahmad Arshan Khan, Ahmed T. Elsayed, and Mohamed A. Elshaer.

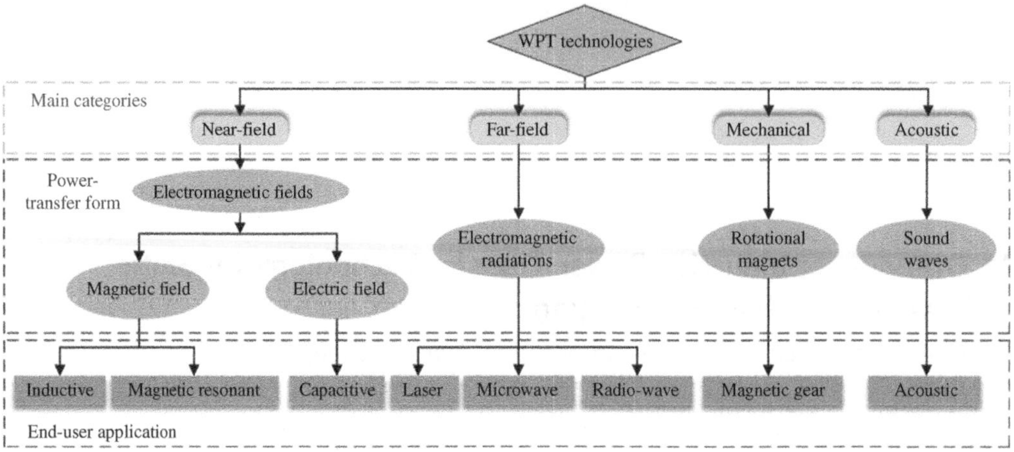

Figure 4.1 Categories of WPT technologies.

4.2 IPT System Components

Inductive charging systems fall into two basic categories, closely coupled and loosely coupled. The former has magnetizing inductance much larger than the leakage inductance as in a conventional transformer. Therefore, the coupling coefficient approaches 0.95 and it is called a strongly coupled circuit. The latter has a much less magnetizing inductance when compared to the leakage inductance and this appears in most applications such as guide rail systems and battery charging for EV. In a loosely coupled system, the power is transmitted through a relatively large air gap, so the coupling coefficient is small, ranging from 0.01 to 0.5 [11]. Figure 4.2 shows a simplified block diagram of an exemplary loosely coupled IPT system.

Due to the large leakage inductance because of the large air gap and to supply the system with the reactive power required to magnetize this air gap, resonant circuits are used on both the transmitter and receiver sides. The transmitter inductance is tuned to reduce the power source rating. On the other side, the receiver inductance is tuned to enlarge the output power. A high-frequency (HF) inverter (10–100 kHz) is used to drive the transmitter side in order to increase the transmission efficiency and reduce the system footprint [12]. Often, the transmitter side is fixed with several moving receiving pads. So, the transmitter inductance (L_1) consists of either one long coil called a track to transmit the power continuously, and in this case, the system is called a distributed IPT or it consists of separate pads for transmitting the power at discontinuous regions and the system here is called

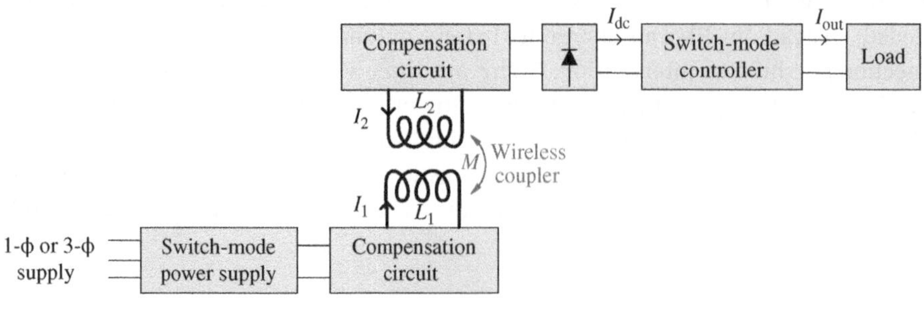

Figure 4.2 Simplified block diagram of an exemplary loosely coupled IPT system.

lumped IPT. The receiver side assembly contains one or many coils, depending on the desired application. These coils are compensated to operate at the angular frequency (ω) of the transmitter side to obtain the maximum transmitted power [13].

In the IPT systems, the maximum power that can be transferred from a transmitter to a receiver coil regardless of the compensation network is defined as uncompensated power (P_{su}), which is represented by the multiplication of the short circuit current (I_{sc}) of the receiving coil and open circuit voltage (V_{oc}), as given in (4.1) [14]. The uncompensated power is a function of the transmitter coil current (I_1), angular operating frequency (ω), mutual inductance (M), and self-inductance of receiver coil (L_2). The actual output power (P_{out}) depends on the receiver circuit parameters and type of compensation network, which are defined by the receiver circuit quality factor (Q_2), as indicated in (4.2) [14].

$$P_{su} = V_{oc}\, I_{sc} = \omega \cdot {I_1}^2 . \frac{M^2}{L_2}; \quad V_{oc} = j\omega \cdot M \cdot I_1; \text{and} \quad I_{sc} = \frac{MI_1}{L_2} \tag{4.1}$$

$$P_{out} = \omega \cdot {I_1}^2 \cdot \frac{M^2}{L_2} \cdot Q_2 \tag{4.2}$$

The output power can also be expressed in terms of Q_2, coupling coefficient (k), and primary circuit Volt-Ampere (VA_1), as given in (4.3) [14]. Increasing the coupling factor significantly increases the power transfer capability. The coupling factor depends on the self-inductances of primary and secondary coils (L_1 and L_2) and the mutual inductance between them, as indicated in (4.4).

$$P_{out} = V_1 I_1\, k^2 Q_2 = \text{VA}_1\, k^2 Q_2 \tag{4.3}$$

$$k = \frac{M}{\sqrt{L_1\, L_2}} \tag{4.4}$$

4.3 Static IPT System

The conventional configuration of an EV inductive charger for static inductive power transfer (SIPT) is depicted in Figure 4.3. The system includes two galvanically isolated sides: primary (transmitter or grid) and secondary (receiver or vehicle). The former includes a transmitter coil, which is

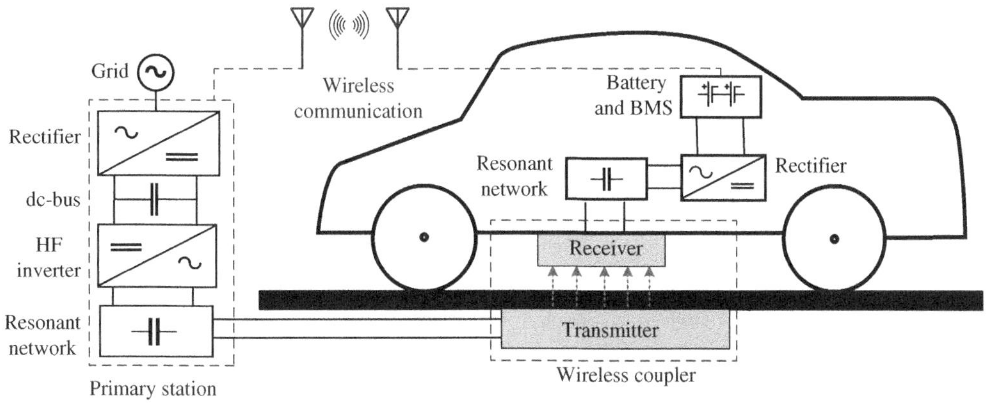

Figure 4.3 Components of SIPT system for EV charging.

connected to the power source through a grid-rectifier, HF inverter, and compensation network. The latter consists of a receiver coil, which is coupled to the EV battery through a compensation network and rectifier. Sometimes, a DC–DC converter is used between the rectifier and vehicle battery for voltage regulation and to manage the charging process. The transmitter is embedded in the road and is responsible for generating HF EMFs that are coupled to the receiver when the vehicle exists above the system. The linked EMFs induce power in the secondary circuit, which is rectified and stored in the storage element (battery). The system operates at a high frequency ranging from 79 to 90 kHz [15], which helps reduce the system components' size (transmitter, receiver, power converters, etc.), and improves the power transfer efficiency [16]. For operation with high power and high efficiency, resonance capacitors are used and connected on both transmitter and receiver sides. These capacitors help compensate for the large leakage inductance due to the large air gap and supply the required reactive power in order to magnetize this air gap. These capacitors can be connected in series, or in parallel, and can be a combination of LC circuits [17, 18].

4.3.1 Coupler Components

The inductive pad is the most sensitive part of the IPT system as it is responsible for the power transmission from the source to the vehicle. An IPT system includes two pads: ground and vehicle, and each pad incorporates three main parts: conductive wires, flux concentrator, and EMF shield. Special types of wires are used in IPT systems to reduce skin and proximity effects due to the HF operation. These wires have small AC resistance that leads to high system quality factor and efficiency [19]. Several classes of wires have been tested, demonstrated, and reported in the literature for IPT system, such as Litz wire [20, 21], magneto-plate wire (LMPW) [22, 23], magneto-coated wire (LMCW) [24, 25], tubular conductor [26, 27], REBCO wire [28], and Cu-clad-Al wire (CCA) [20, 29]. The performance of these different wires is summarized and compared in Table 4.1, which shows a trade-off between performance and cost.

Due to the loosely coupling nature of EMFs in an IPT system, flux concentrators are used to direct EMFs from the transmitter to receiver coils, which enhances coupling performance and coefficient as well as helps reduce leakage EMFs around the system. Typical flux concentrators used in IPT systems are made of magnetic material that is highly conductive to magnetic fields. Ferrite cores

Table 4.1 Characteristics of different types of wire.

Characteristic	Litz wire	LMPW	LMCW	Tubular copper	REBCO	CCA
R_{skin}	Medium	Very low	Very low	Low	Low	Medium
R_{prox}	Medium	Very low	Very low	Low	Low	Medium
R_{hys}	–	Low	Low	–	–	–
R_{ac}	Medium	Very low	Very low	Low	Medium	Slightly high
Cost	High	Very high	Slightly high	Medium	High	Low
Complexity	Medium	High	Medium	Low	High	Low
Density (m^3/kW)	Medium	High	High	Slightly high	Medium	small
Flexibility	High	Low	Low	Medium	Slightly high	High
References	[20, 21]	[22, 23]	[24, 25]	[26, 30]	[28, 31]	[20, 29]

Table 4.2 Characteristics of magnetic materials.

Material	Cost	Flexibility	Weight	References
Ferrite	High	Hard and fragile	High	[35, 43]
Magnetizable concrete	Very low	High	Very high	[41, 42]
Flexible magnetic	Low	Very high	Low	[38, 39]
Nanoparticles	Medium	Medium	Low	[35, 36]

are the most commonly used in IPT systems due to their high magnetic permeability, and low electrical conductivity [9]. These ferrite cores can be shaped as either one plate [32], multiple bars [33], or discrete tiles [15, 34]. In [35, 36], a ferromagnetic nanoparticle material was proposed for the IPT system, which offers higher power transfer capability, improves the shielding performance, and reduces the system weight [37]. The utilization of flexible magnetic cores for the IPT system was discussed in [38, 39], which leads to less core losses and better-quality factors, and system performance. Ferrite, magnetic nanoparticles and flexible core are more convenient for vehicle pads; however, they are not ideal for transmitter pads as they are not compatible with the road. These materials are fragile and might not withstand mechanical stress, vibrations, and high temperature on the road. Therefore, a magnetizable concrete was developed and proposed for a transmitter design in [40, 41]. It is very cheap since the magnetic particles can be made of recycled materials [42]. A comparison among different materials of flux concentrator is presented in Table 4.2.

An IPT system produces strong leakage EMFs due to the large air gap, which might exceed the safety limits identified by the different international standard organizations, such as International Commission on Non-Ionizing Radiation Protection (ICNIRP). The ICNIRP 2010 is being considered for IPT systems, which recommends limits for external magnetic field density (B) of 27 μT for humans, and 15 μT for pacemakers [44, 45]. To comply with these limits, a conductive plate is usually attached to the magnetic flux concentrators in both transmitter and receiver pads. This plate is made of a conductive material (usually aluminum) and acts as another coil in which eddy currents are generated due to the HF EMFs [46–48]. These eddy currents produce EMFs that oppose the original fields and reduce them. This helps minimize the leakage flux around the system but adversely affects the system performance by reducing the coupling factor and efficiency [49, 50]. This type of shield is called passive shielding, as indicated in Figure 4.4a.

For high-power IPT systems (>100 kW), it is a challenge to manage the leakage EMFs around the vehicle and keep them within safe limits using passive shielding [51]. Therefore, active shielding, as a more effective technique, was proposed for the IPT system [52]. In this case, extra turns are added to each coil and wounded with reverse polarities, as depicted in Figure 4.4b. When a current flows through the coils, the shielding turns will generate intentionally opposite EMFs that cancel part of the main fields and minimize the leakage EMFs [50, 53]. Active shields are more effective compared to passive; however, the extra turns increase the coil's cost, weight, and losses, and impact the system efficiency. To reduce the negative impact of the active shield on the system, the resonant reactive shield was proposed in [50, 53]. It depends on the use of a passive compensation loop coil with a resonant capacitor, as indicated in Figure 4.4c. This type does not need to be powered externally and has less impact on the system efficiency than the active shield [53]. The performance of different shielding techniques is summarized in Table 4.3.

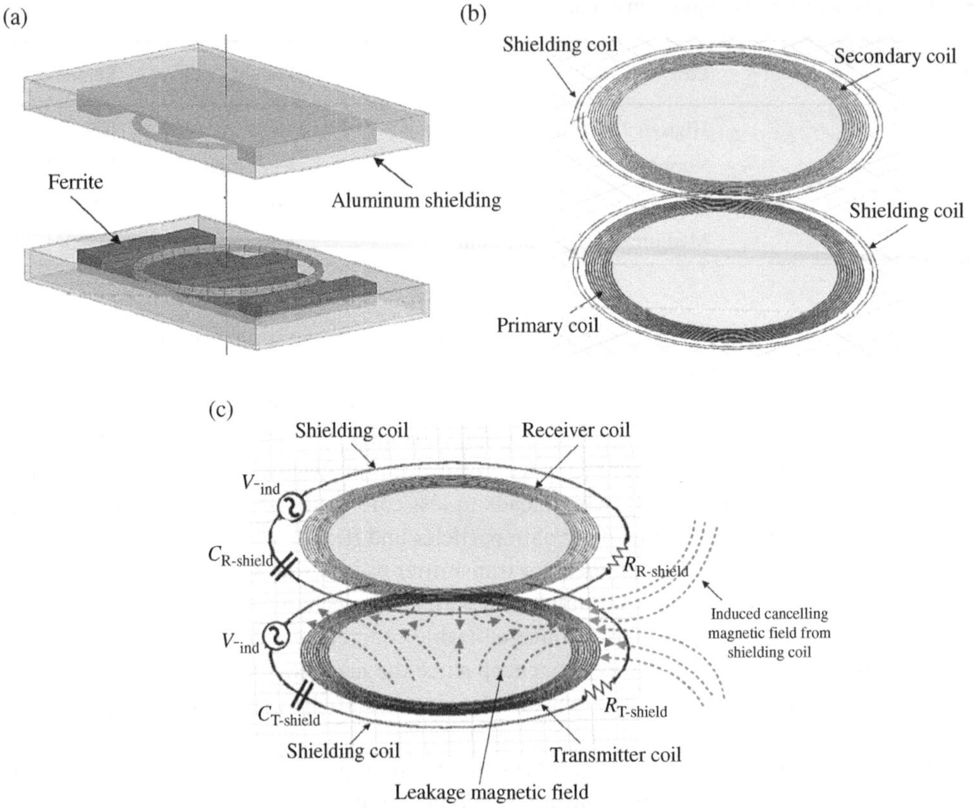

Figure 4.4 EMF shielding in the IPT system: (a) passive, (b) active, and (c) reactive.

Table 4.3 Comparison of various shielding types.

Shielding method	Passive	Active	Reactive
Effect on efficiency	Low	High	Low
Active source required	No	Yes	No
Eddy current losses	High	Low	Low
Effectiveness	High shielding effect suitable for low power	High shielding effect suitable for high power	High shielding effect suitable for medium power

Source: Adapted from Refs. [50, 52–56].

4.3.2 Structures of Inductive Pad

Wires, flux concentrators, and shields are stacked to form planar pads, which are classified into two types based on the coupled flux components: non-polarized (NPPs) and polarized (PPs) pads. The former consists of a single coil that generates vertical flux components to be coupled with the receiver, such as a circular (CP) and a rectangular (RP) pad. PPs generate vertical and horizontal flux components, and both are coupled with the receiver coil and are responsible for power transfer. Typically, PP consists of more than one coil, such as double-D (DD), double-D quadrature (DDQ), and bipolar (BP) pad [57]. Most pad structures can operate as transmitter and receiver pads, such as CP, RP, DD, etc. Examples of different pad structures are summarized and compared in Tables 4.4 and 4.5, based on shape, coupling performance, tolerance to misalignment, shielding, polarization, interoperability, etc.

4.3.3 Research and Development (R&D) and Standardization Activities

Plentiful R&D and demonstration activities are happening in the SIPT system for EV charging with different power levels and air gaps to serve the requirements of different EVs. For light-duty (LD) EV, the charging power ranges from 3 to 22 kW. However, for medium- and heavy-duty (MD/HD) EVs, the system power should be more than 22 kW, according to international standards and guidelines [79, 80]. Also, there is a wide range of operating frequencies combined with different maximum power and air gap of IPT systems for EV applications. Although most of IPT systems operate at frequency ranges from 80 to 90 kHz, few studies considered higher frequencies with air-core coils [81, 7]. Table 4.6 summarizes and compares the specifications of some available products of IPT systems for EV applications. The efficiency numbers are characterized as: grid-to-battery (G2B), which shows the overall system efficiency or dc-to-dc converter efficiency, which includes all the components from the grid dc-bus to the battery dc-bus excluding the front-end rectifier.

As an emerging technology, a stationary inductive charging system necessitates standards that provide clear specifications and guidelines for technology development, testing, installation, and operation. EV supply equipment (EVSE) companies and automotive companies, which are looking to bring this technology to the market will rely on these standards to streamline development, reduce costs, accelerate adoption, and comply with interoperability objectives, safety measures, and efficiency. In addition, standardization will help facilitate the interoperable operation of the system with different vehicles. Several national and international entities are developing standards, guidelines, specifications, and recommended practices for a stationary inductive charger for different vehicles and operation environments, such as the Society of Automotive Engineering (SAE), International Electrotechnical Commission (IEC), the Japan Automobile Research Institute (JARI), International Organization for Standardization (ISO), Underwriters Laboratories (UL), the National Technical Committee of Auto Standardization (NTCAS), etc. Some of these standards describe the system configurations for different power levels and air gaps, such as SAE J2954 and IEC 61980-1, while others provide guidelines for data communication and interface, such as IEC/TS 61980-2. A summary of most of the released standards for EV inductive stationary charging technology is presented in Table 4.7. The table shows the developing organization/group, a brief description, and the status, which is expressed as published, revised, stabilized, or active. These standards cover several topics related to technology. However, there are still some challenges facing the standardization process as summarized hereunder:

- Standards considered incompatible shapes of coils such as circular/rectangular and double-D (DD), which do not work with each other efficiently [15]. This raises interoperability concerns and complicates the design to make the system interoperable and increases the system cost.
- Many details and information are missing in the standard related to compensation topologies, power converters, control, and system operation.
- These standards are developed for LD EVs with slow charging (up to 20 kW). However, standards that cover fast wireless charging for light-, medium-, and heavy-duty EVs do not exist.

Table 4.4 Comparison of different single-coil structures of an inductive pad.

Pad structure	Circular	Rectangular	Flux pipe	Solenoid
Shape	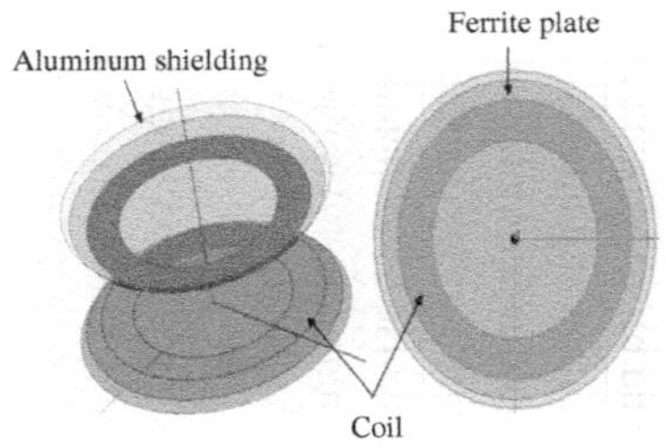	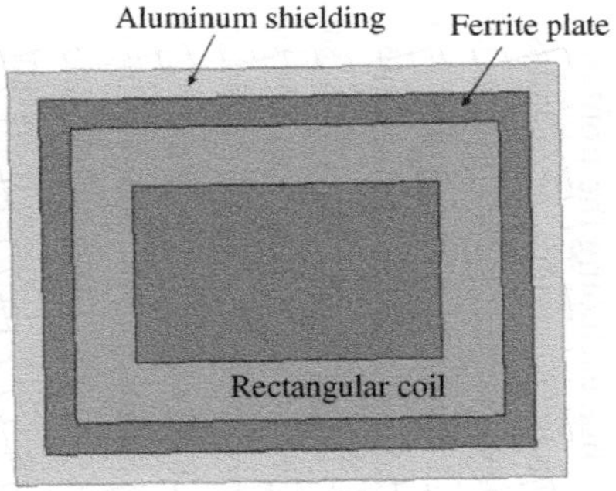	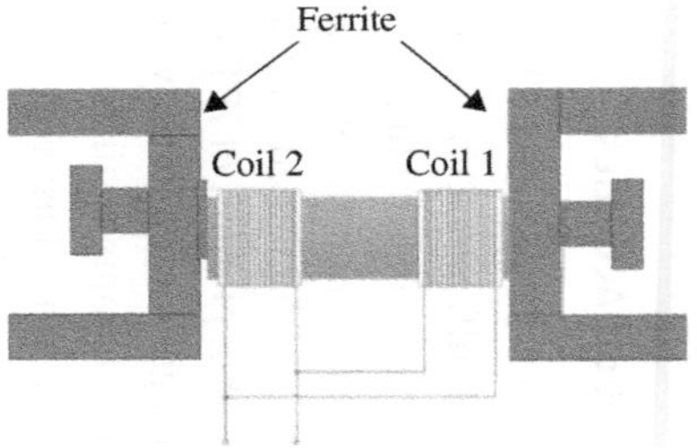	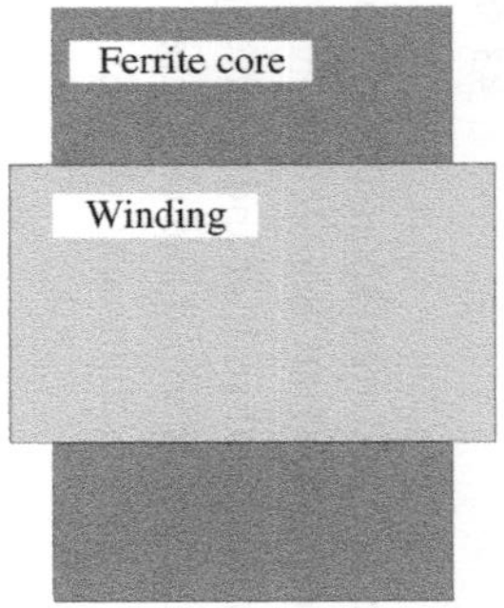
Misalignment	Poor	Medium	Poor	Good
Magnetic flux	Single-sided	Single-sided	Double-sided	Double-sided
Shielding impact	Low	Medium	High	Very high
Charging zone	Small	Small	Medium	Medium
Distance	Low	Low	Medium	Medium
Polarization	Non-polarized	Non-polarized	Polarized	Polarized
Interoperability	Very low	Very low	Low	Slightly low
Common use	Transmitter	Transmitter and receiver	Transmitter and receiver	Transmitter and receiver
Leakage flux	High	Medium	Medium	Medium
# Coils	1	1	1	1
References	[2, 57–60]	[61–65]	[2, 33, 59, 60, 66, 67]	[62, 68–70]

Table 4.5 Comparison of different multiple-coil structures of an inductive pad.

Pad structure	Double-D	Bipolar pad	Quadrupole	Double-D quadrature (DDQ)
Shape	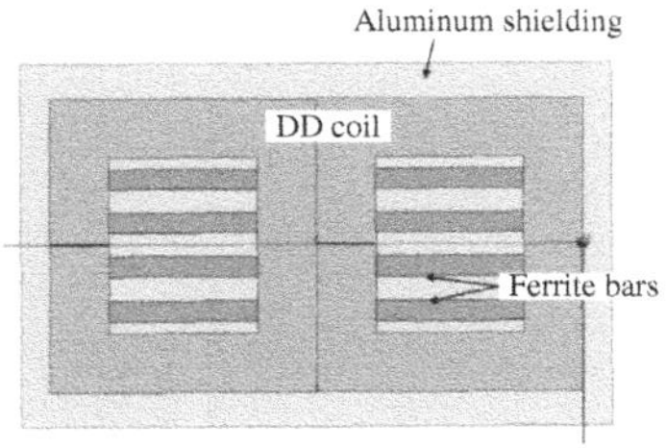	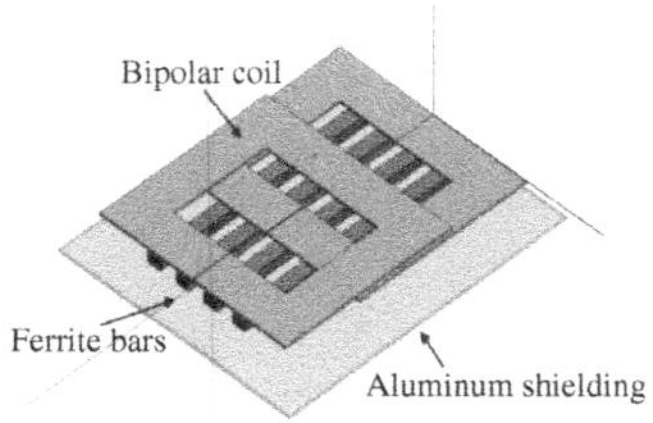	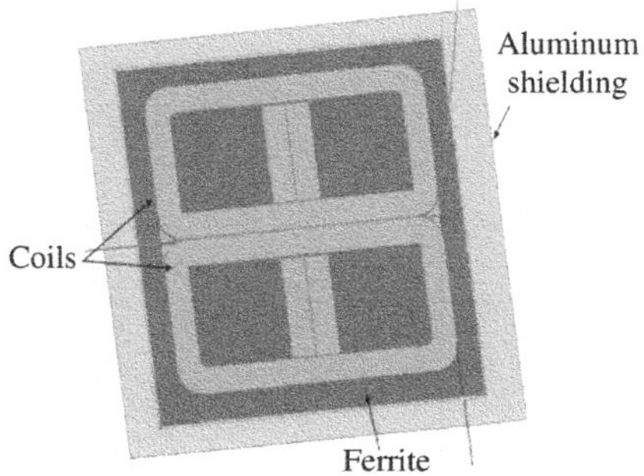	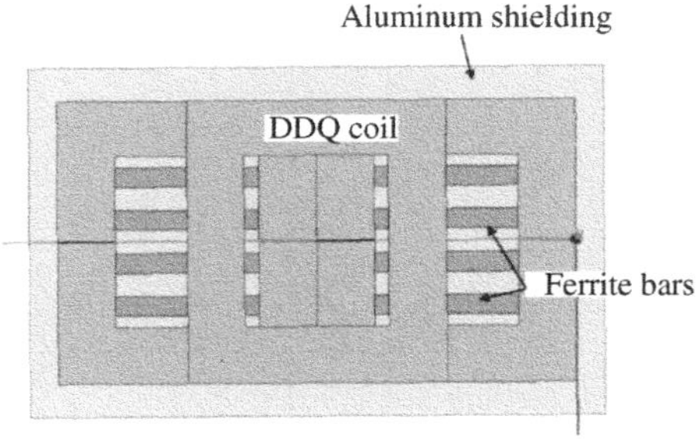
Misalignment	Medium	Medium	High	High
Magnetic flux	Single-sided	Double-sided	Double-sided	Double-sided
Shielding impact	High	High	High	High
Charging zone	Medium	Large	Large	Large
Distance	Medium	High	High	High
Polarization	Polarized	Polarized	Polarized	Polarized
Interoperability	Non-interoperable with NPPs	High	High	High
Common use	Transmitter	Receiver	Transmitter and receiver	Receiver
Leakage flux	Extremely low	Extremely low	Low	Extremely low
# Coils	2	2	4	3
References	[33, 58, 61, 62, 71, 72]	[59, 60, 62, 71, 73, 74],	[75–78]	[60–62, 71, 72, 74]

Table 4.6 Current commercial EV SIPT.

Power (kW)	Supply voltage	Frequency (kHz)	Air gap (mm)	Efficiency	References
3.6	240 VAC	85	100–250	G2B: >90%	[82]
3.6–7.2	208–240 VAC	20	102	G2B: 90%	[83]
2–10	208–240 VAC	85	≤304.8	G2B ≥85%	[84]
3.6	240 VAC	85	160–220	G2B: >90%	[85]
50	277 VAC	23.4	165–190	dc-to-dc: 89–92%	[86]
60–180	460 VAC	20	40	G2B: >90%	[87]
10	240 VAC	20	200	G2B: >92%	[88]
200	NA	20	203.2	G2B: 94%	[89]
25	240 VAC	20	241.3	G2B: 90%	[11]

Table 4.7 Summary for the different standards of EV-inductive static charging.

Developer	Name	Description	Status	Date	References
SAE	J2954	Introduces guidelines which determine acceptable criteria for interoperability principle, EMF compatibility, minimum performance, safety degree, and testing for wireless charging of light-duty and plug-in EVs. This version considers unidirectional charging operation, from grid to vehicle (G2V).	Published	2020	[79, 90]
	J1773	Defines the minimum acceptable limits of interface requirements for EVs, besides inductive charging in areas of the same geographical nature. It is recommended to transmit power at higher frequencies than power line frequencies.	Stabilized	2014	[91]
	J2847/ 6_202009	Defines the accepted limits for communication between an EV and an inductive battery-charging system for WPT.	Revised	2020	[92]
	J2836/ 6_201305	Defines the communication cases that are used for EVs and the wireless EVSE for wireless power transmission as defined in SAE J2954. It provides communication requirements between the onboard charging system and wireless EVSE to support WEVSE detection, charging operation, and charging operation monitoring.	Revised	2021	[93]
IEC	IEC 61980-1	Universal requirements. It introduces recommendations that apply to equipment to transmit power wirelessly using the inductive charging concept, to supply power to storage devices such as insulating batteries or supply the power to grid when needed. The areas presented in this issue are the characteristics for the desired safety limit of a supply device, communications between EV and transmitter device to enable and control the WPT system, and specific EMF compatibility requirements for a supply device.	Active, most current	2020	[80, 94]

Table 4.7 (Continued)

Developer	Name	Description	Status	Date	References
	IEC/TS 61980-2	Presents particular conditions for communication between EVs and infrastructure in order to facilitate charging in the WPT system. Work has also been done to reach the communication requirements for charging vehicles with two and three wheels, and communication requirements during bidirectional charging. This release does not address safety requirements during maintenance or trolley buses, and trucks designed for off-road use.	Active, most current	2019	[95]
	IEC/TS 61980-3	Specifies special requirements for the magnetic field WPT (MF-WPT) generated in the wireless charging system. Work has also been done to reach the communication requirements for required safety by an MF-WPT system, the requirements to assure efficient and safe MF-WPT power transfer, and specific EMF compatibility requirements for MF-WPT systems.	Active, most current	2019	[96]
JARI	G106:2000	Provides the general requirements to provide inductive charging for EVs.	Published	2000	[97]
	G108:2001	Provides the software interface for the inductive charging system of EVs.		2021	
	G109:2001	Describes the use of the IPT system to transfer power wirelessly. It also defines universal requirements for the wireless charging process.		2001	[90, 98]
UL	UL9741	Provides universal requirements for the interchange charging operation considering the bidirectional charging process to supply the power to grid and feed traditional loads.	Active	2017	[62, 99, 100]
	UL-SUBJECT 9741	Defines the requirements for each unidirectional and bidirectional operation for EVs. Unidirectional operation supplies power from the utility grid to charge the EV battery. Bidirectional operation serves the same function but additionally provides power to the utility grid from the EV.	Active	2021	[101]

4.4 Dynamic IPT System

In a dynamic IPT (DIPT) system, EVs can be charged while driving without the need to stop or wait for the charging process to complete. This technology is intended to support long-trip travel on freeways to overcome range anxiety, which is one of the main obstacles to vehicle electrification. In addition, the DIPT technology has the potential to significantly increase driving range and enable using smaller onboard batteries [102]. Therefore, it is a promising solution for self-driving vehicles as well as heavy-duty vehicle electrification. Dynamic charging is achieved by burying the transmitter coil into the ground and attaching the receiving coil at the bottom of the vehicle. These coils are supplied by high voltage and high frequency from an AC source. They are coupled with each other by the magnetic field when the vehicle passes over the transmitter coil to transmit the nominal power with maximum efficiency. There are some challenges that hinder the dynamic charging penetration, such as the high overall cost, the necessity to have a separate lane for charging, which is a challenge in crowded cities, and a perfect alignment must be achieved to avoid the occurrence of loss of the transported power [5]. In [103], different coil and ferrite configurations that show good characteristics with dynamic charging are reviewed taking into consideration the safety limits of EMFs. The study shows that dynamic inductive charging has the potential to achieve efficiency up to 94% by performing an appropriate coil design, including shape and ferrite. From the interoperability perspective, it is expected that the vehicle pad that supports dynamic charging is the same as the one used for static charging, which can be one of the configurations presented in Tables 4.4 and 4.5. However, on the transmission side, two main types of transmitters are considered; a single long coil track (stretched) and a segmented coil array [104].

4.4.1 DIPT with a Single Long Coil Track

It is a single coil with a length larger than the vehicle length. It ranges around 10 to 100 m long inside the ground track, enabling it to charge more than one EV at the same time [105]. It consists of a long transmitting wire track and a primary power station connected to the grid and contains a resonant network, HF inverter, and rectifier with power factor correction, as shown in Figure 4.5.

This system has several merits such as easy to control and operate, simple in configuration, and less use of compensation elements. Additionally, this configuration enables fixed mutual inductance between the track and receiver coil with movement, which results in a smooth power transfer profile. The long track configuration is commercially available in Korea and is called online EV (OLEV) technology [105–108]. On the other hand, this configuration system has several disadvantages including its high losses, cost, and maintenance [102]. Also, it produces redundant EMFs when the vehicle does not cover the transmitter track, and the coupling coefficient is small. So,

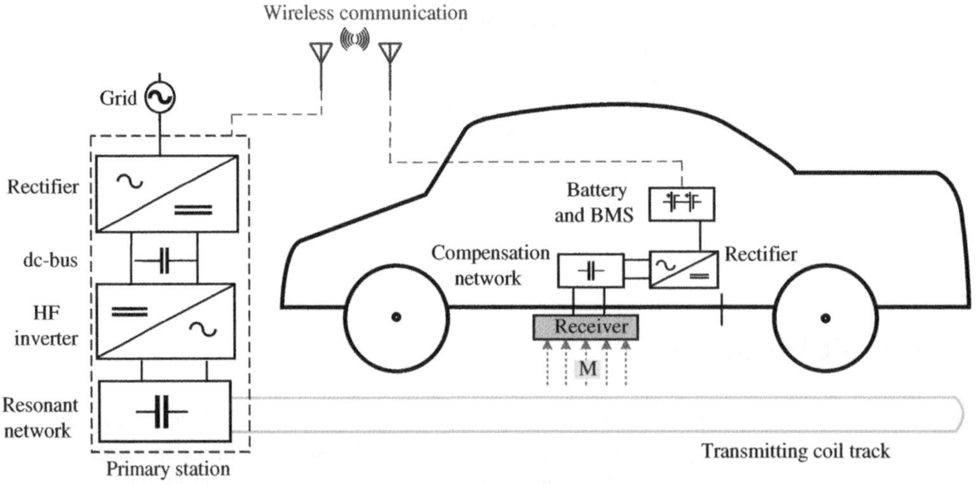

Figure 4.5 Components of the DIPT system with single long coil track.

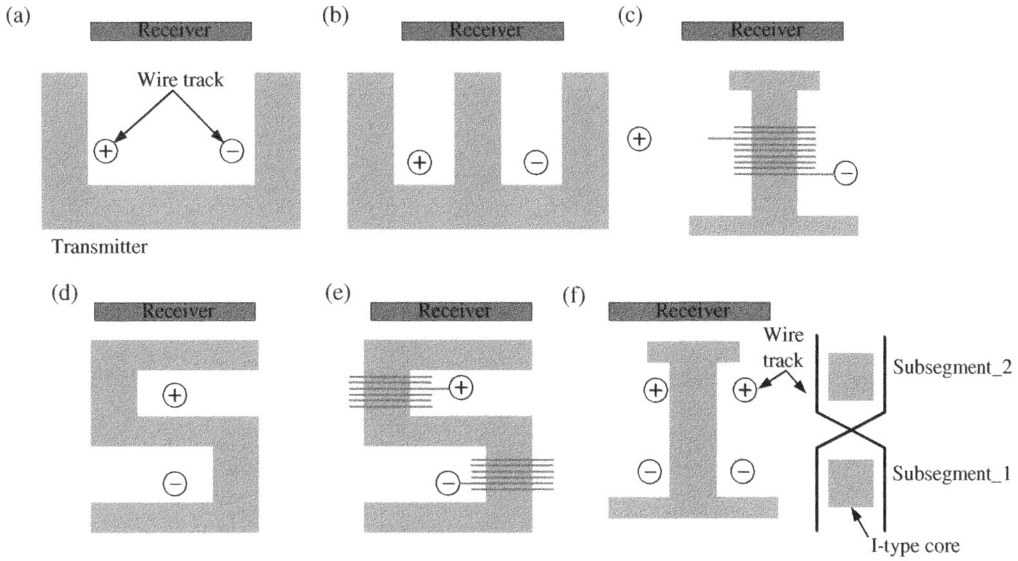

Figure 4.6 Single transmitting wire track used in the DIPT system: (a) U-type, (b) E-type, (c) I-type, (d) S-type, (e) ultra slim S-type, and (f) X-track (cross segmented).

the overall transmission efficiency is low [109]. There are different configurations of long transmitter track, which are classified based on the shape of magnetic core into U-type [55, 56], E-type [55, 56], I-type [110], S-type [55, 111], ultra slim S-type [111], and X-track (cross segmented) [112]. These types are indicated in Figure 4.6 and their performances are summarized and compared in Table 4.8.

In the DIPT system, it is necessary to preserve power from loss as well as to protect living organisms from harmful EMFs. Therefore, when using a long transmitter track on the ground side, it must be activated only when an EV passes over it. To meet these requirements, the long transmitter track must be divided into several smaller sub-tracks. Each sub-track can be activated by supplying it with an HF current through a switch box fed by an inverter, as shown in Figure 4.7 [55, 112]. The centralized switching track, shown in Figure 4.7a, is the first type of long sectionalized track consisting of a few sub-tracks, a bundle of supply cables, and a central switching box. One of several pairs of supply cables is connected to the inverter through the switch box at a time. One of the disadvantages of this topology is that the inverter can be used to activate only one sub-track. It also

Table 4.8 Performance comparison of different types of single transmitting wire tracks used in the DIPT system.

Parameters	U-type	E-type	I-type	S-type	Ultra slim S-type	X-track
Leakage EMF	High	Low	Medium	Small	Very low	Very low
Air gap	Medium	Low	Large	Large	Large	Large
Track width	Very large	Medium	Small	Small	Small	Small
Efficiency	Low	High	High	Low	Low	High
Output power	Small	Small	High	High	High	High
Lateral misalignment	Large	Small	Large	Large	Very large	Large
Studies	[55, 56]	[55, 56]	[110, 112]	[55, 111]	[56, 111]	[110, 112]

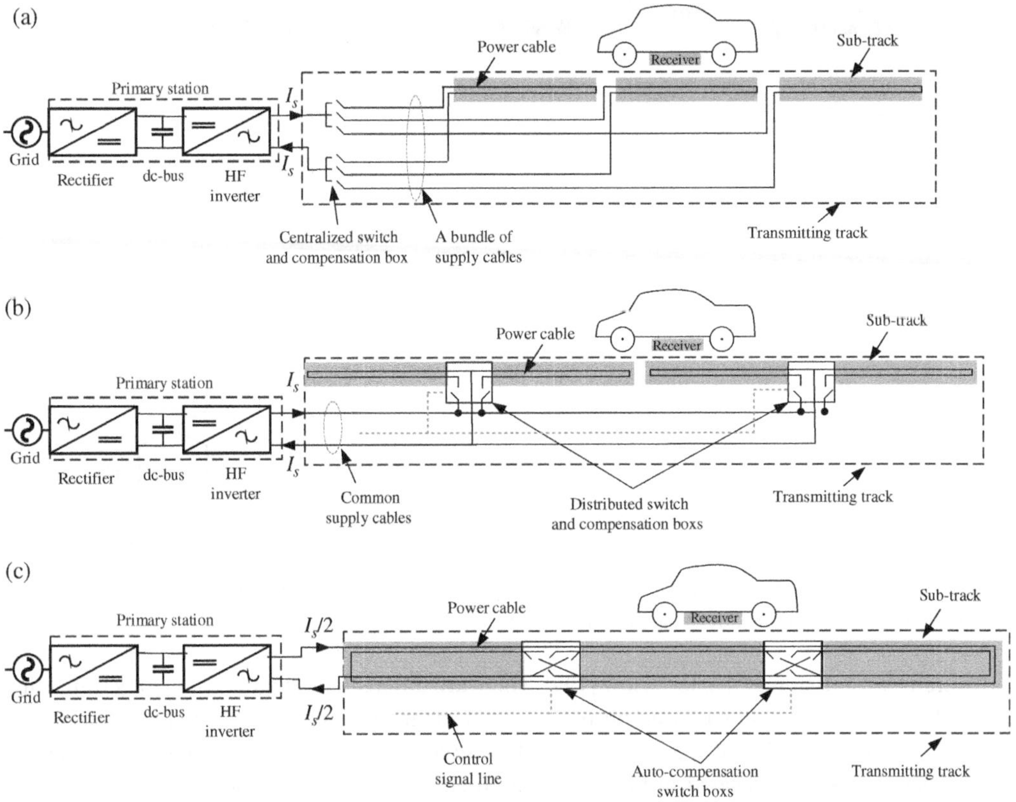

Figure 4.7 Types of sectionalized long transmitting track: (a) centralized switching, (b) distributed switching, and (c) cross segmented switching.

requires a very large number of bundles of power cables. Therefore, the second type of long partitioned track was proposed, which is the distributed switching track shown in Figure 4.7b. It consists of a few sub-tracks, a pair of common power supply cables, and several switch boxes that are connected between two sub-tracks. In this type, the length of the cables is less than that in the centralized type, especially when using many cables. Because of common power supply cables, the cost of development and conductive losses of the distributed switching track are high. Therefore, the third type was resorted, which is the cross segmented track (X-track) shown in Figure 4.7c [55, 112]. The cross segmented track contains auto-compensation switch boxes, segmented sub-tracks, roadway harnesses, and control signal lines. This type has the shortest length of cables, and several sub-tracks can be operated by the same inverter, which leads to a much lower cable cost (less than half the cost of the other two types) [112]. The sub-track is made of a twisted power cable, core, and copper nets so that the EMFs under each sub-track are small and within the ICNIRP2010 limits [45].

4.4.2 DIPT with Segmented Coil Array

The segmented coil array consists of multiple coils connected to each other in series or in parallel, and buried inside the ground forming the charging track, as depicted in Figure 4.8. Similar to static charging, each segmented transmitter coil is roughly the same size as the receiving coil (usually within 100 cm^2) [113], and has its own compensation circuit [102, 114]. In this type, the receiver

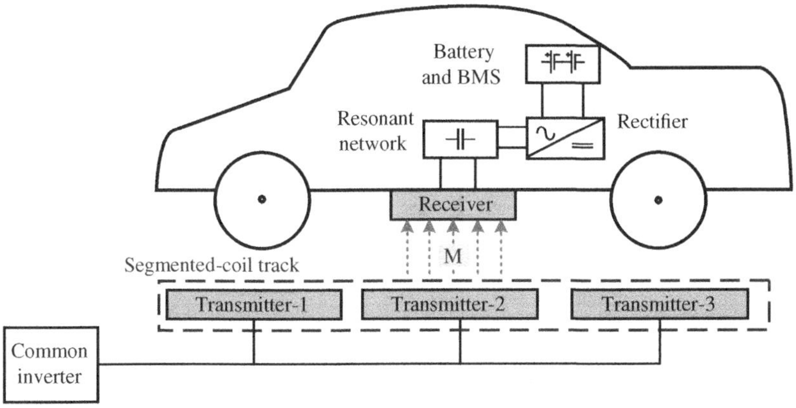

Figure 4.8 Simple configuration of segmented transmitter coil track for the DIPT system.

coil is coupled with one coil from a segmented transmitter array. So, only the closest transmitter to the receiver will be energized. When the vehicle moves away from the energized coil, its power is turned off and the next coil is energized. This leads to a higher coupling coefficient, less leakage EMFs, and higher system efficiency. Also, having a compensation circuit for each segmented coil makes the design much more simple and flexible [105, 115]. However, this configuration is expected to be more costly because of a large number of transmitter coils and the associated compensation components and HF inverters. A single inverter can be used to supply several coils, which helps reduce the overall cost but requires more transmission lines from the inverter to the coils. Another challenge for this structure is the horizontal distance between the transmitter coils within an array. This distance should be large enough to eliminate self-coupling between adjacent coils. When the vehicle travels over this distance, coupled magnetic fields between transmitter and receiver coils are very low, which results in a very low power transfer (near zero when the receiver coil is at mid-distance between two transmitters). Therefore, a pulsation power transfer profile is realized when the vehicle is in motion. Reducing the distance between coils helps reduce the ripple in the power profiles, but special attention should be given to the design of the resonant circuits to compensate for the self-coupling [105, 109]. In [113], the value of power pulsation remains about 50% of the maximum power when the distance between the transmitter coils is about 30% of the transmitter length. A comparison between several configurations of the segmented coil array is presented in [116].

There are two main types of feeding arrangements for a dynamic charging system when using ground-based segmented coils. In the first arrangement, each segmented coil is connected to a separate HF inverter, as shown in Figure 4.9a [117, 118]. This inverter activates and deactivates the system when the vehicle is aligned or not aligned with the segmented coil, respectively. This arrangement is simple in construction and allows the use of low-power inverters, as well as connecting each segmented coil to an independent compensation circuit, which improves the system performance. However, this configuration requires a large number of inverters and sensors to operate.

The other arrangement is achieved by feeding a set of segmented coils from a single HF inverter and sharing a common HF AC distribution feeder. A switching device and an independent compensating circuit are connected to each segmented coil, as shown in Figure 4.9b [113, 119]. When the vehicle is aligned with a segmented coil, the switching device activates the coil and then deactivates it after the vehicle leaves. This system requires less HF inverters, but longer track coils and more switching and sensing devices [120, 56]. In addition, extending a high-power HF distribution feeder raises safety concerns due to EMFs.

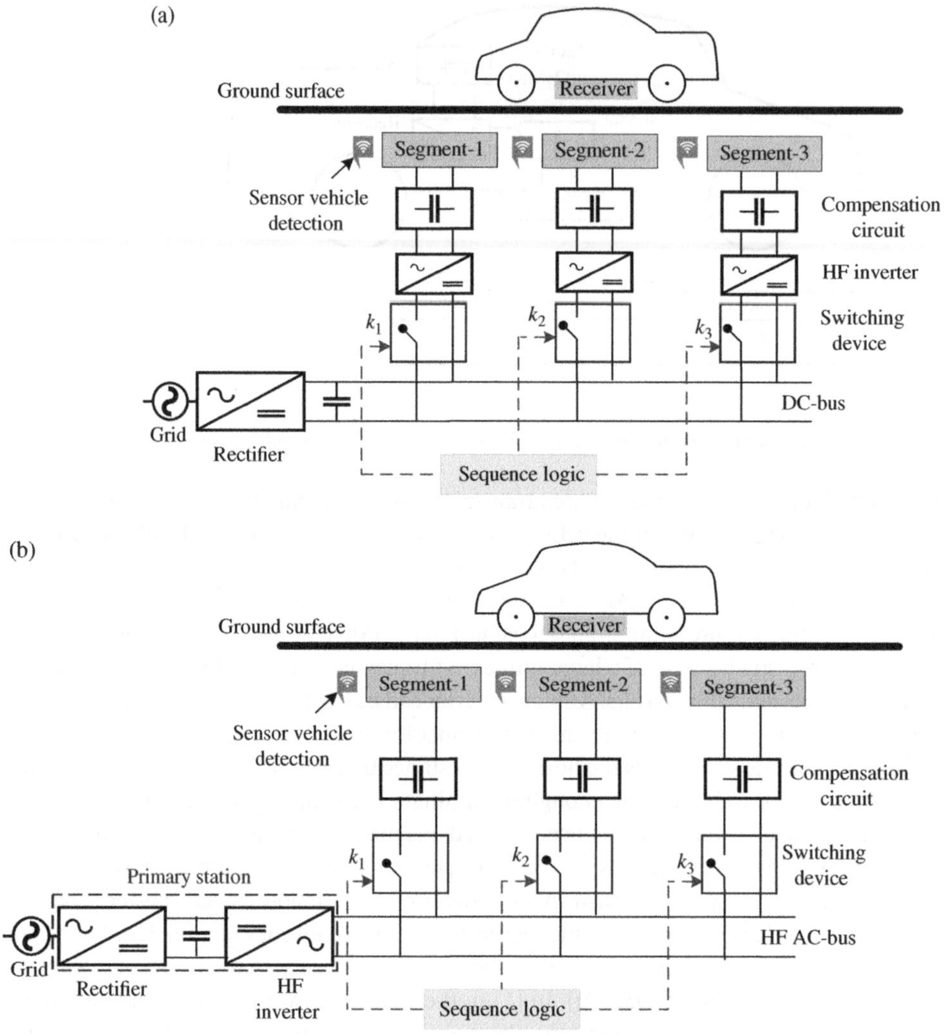

Figure 4.9 Supply arrangements for the segmented transmitter of the DIPT system using a common (a) DC-bus and (b) HF AC-bus.

There are other types of the DIPT system that are fed with a segmented transmitter, and each uses a different method to control the charge of the individual segment inside the transmitter track. One of these types is shown in Figure 4.10a, where the power is transferred from the primary power station through a magnetic coupler to a specific charging segment, instead of the direct connection in the aforementioned arrangements. The activation or deactivation of each charging segment is controlled by a simple bidirectional AC switch [121, 122]. Another arrangement was introduced in [112, 117] and is shown in Figure 4.10b. It uses a two-turn transmitter path configuration, where switching boxes can be used to change the direction of the current when it passes in one of the turns in order to activate or deactivate the magnetic fields in the area around the specified transmitter segment in the path. Thus, each segment can be controlled individually and independently. The number of plugins required is small, but all the charging segments are connected in series and all switching boxes need to be rated at the full current rating of the transmitter track.

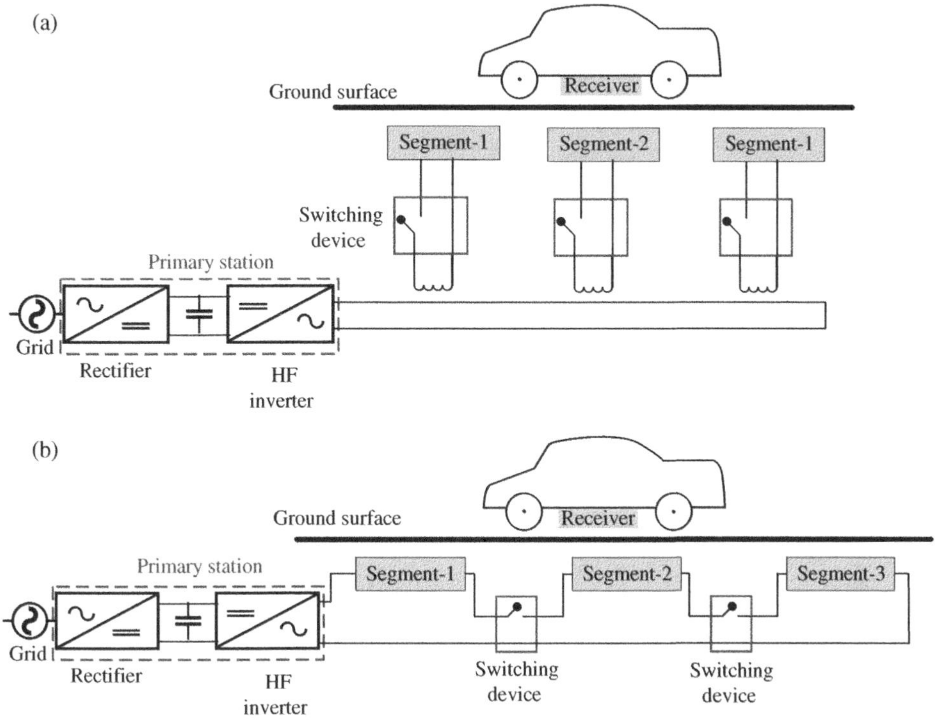

Figure 4.10 Additional supply arrangements for the segmented transmitter of DIPT (a) using a common HF inverter with a long magnetic path and (b) using a common HF inverter with a series connection of transmitter segments.

There are three major obstacles to the aforementioned systems, which make them not exemplary for dynamic charging applications. (i) Because of the mechanical vibrations and pressures caused by the movement of trucks, buses, and vehicles, the roadways are unfavorable places for the presence of electronic circuits, as there is difficulty in the process of maintaining or replacing any component, and these operations are very expensive. It is necessary for the system to continue to work even if one or more transmitter segments are found to be down. Regardless of the system shown in Figure 4.10a, the remaining systems do not have any means of isolating the power supply and the charging segments, which increases the possibility that the entire transmission track will fail if any failure occurs in a single charging segment. (ii) Most charging systems are designed to operate at a nominal frequency of 20 kHz, which is contrary to the recommendations of SAE J2954 to operate at a frequency of 85 kHz [123]. Therefore, dynamic charging models will face several hurdles to be able to operate at a frequency of 85 kHz such that the power source is required to be rated at hundreds of kVA due to the high transfer power demanded for electric cars. It is also difficult to find switching devices that operate at high frequency and power. (iii) All the aforementioned systems do not contain any kind of communication systems between the EV and infrastructure, which may lead to an increase in the load on the electrical network during peak times or traffic congestion due to charging a large number of cars at the same time.

In [118], a double-couple was proposed to feed the DIPT system. This system consists of an HF power track buried underground and feeding the power to an "intermediate coupler circuit" (ICC) at the specified segment, as shown in Figure 4.11a. The output of the intermediate coupling circuit is connected to a controlled rectifier to convert HF AC voltage to DC voltage. The HF inverter is fed

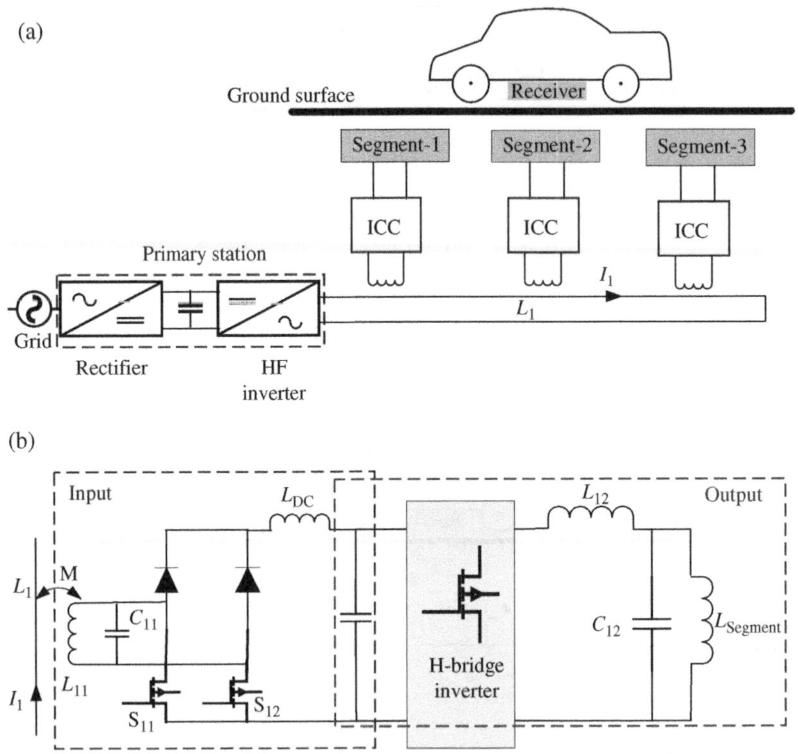

Figure 4.11 Schematics of (a) double-coupled system and (b) intermediate coupler circuit (ICC).

by the rectified dc voltage, which controls the transmitter segments as shown in Figure 4.11b. The ICC provides an independent control for each transmitter segment. This arrangement solves the isolation problem facing other systems and helps reduce the impact of dynamic charging on an electrical network at peak times. Additionally, it has the potential to reduce the losses in the transmission track by operating it at a low frequency, while the transmitter segments operate at a high frequency, which improves the system efficiency. On the other side, the costs of the overall system are expected to be higher because of the excess power electronics involved in addition to low reliability due to the use of a central power supply unit for the entire system [117].

4.4.3 R&D and Standardization Activities

This section presents historical information on using the DIPT system for EV charging, as well as R&D activities that have been conducted to develop the system, whether the transmitter is a single long track or several separate coils that are extended through the ground. A brief summary of the projects that contribute to standardization for the DIPT system is presented.

4.4.3.1 Historical Background

The "Transformer System for Electric Railways" was first patented in 1894, which marked the beginning of the concept of EV being charged during movement on highways (dynamic EV) [124]. The basic configuration of the system at that time was completely similar to the IPT system, in addition to some other features and improvements that were claimed in the patent, including the need to transmit high power over a large air gap distance, reduce conductive and eddy current losses, and raise the system efficiency, and it is still under research and development until now [111]. In the United States, the interest in EVs increased after the oil crisis in the 1970s, when several research teams began working on the use of vehicles that would be able to travel on highways without the need to stop to charge their batteries or fill up fuel in order to overcome the shortage of oil [124–129]. The first development of highway cars began in 1976 to confirm the technical feasibility and then create the first model of the IPT system that transmits power of 8 kW, but this model was not fully functional [130]. In 1979, the Santa Barbara electric bus project began, where another prototype was developed [128, 129]. In 1992, the Partners for Advanced Transit and Highways (PATH) project was kicked off, in which laboratory and field tests were carried out, and designs for the IPT system were made and installed in a bus. Then, the project constructed roads and supplied power transmission systems to them and studied the potential impacts on the environment. This project was able to transmit a power of 60 kW over an air gap distance of 76 mm with an efficiency of 60% [131, 132]. The prototype of PATH did not gain commercial popularity due to the large weight of the coils forming the power transmission system, high acoustic noise, high cost, and high current due to the low operating frequency of 400 Hz. Also, the distance over which the power is transmitted is small and not commensurate with the ground clearance of EVs. Despite the obstacles that led to the failure to non-spread of this system, it was a good start and opened the way for researchers to develop and improve the technology [131, 132].

4.4.3.2 R&D on DIPT

Most of the research has been conducted on the segmented coil array as a transmitter with a goal of achieving the optimal operating conditions in terms of transmitting the maximum power at the highest possible efficiency commensurate with the ground clearance of EVs. In this arrangement, the transmitter consists of multiple pads with different configurations similar to those used in static charging, such as circular, rectangular, Double-D (DD), DDQ, DQD, bipolar, etc. In [133], a method was proposed to choose the length of the transmission coil taking into account the vehicle speed, power consumption per kilometer, power loss, and charging efficiency of the system. A long transmitter coil, a rectangular receiver coil, and an LCC-S compensation topology were used. The transmission power and efficiency were calculated. Then the minimum value of the transmission coil current was inferred from the maximum value of the charging power. Power transmission has been achieved with an efficiency higher than 85%. In [134], a segment transmitter array is proposed. Each coil is in DD format and the mutual inductance is evaluated using the finite element analysis. An analysis and study of the proposed system were done in order to obtain the optimal dimensions and the best horizontal separation between the transmitter pads to improve the system properties. A double-spiral repeater was used in [135], to improve the power transmission efficiency, increase the transmission distance, and increase the variation in the linear misalignment tolerance. When optimizing the system parameters, represented by the horizontal distance between the segmented coils, transmission efficiency, number of segmented coils, and load conditions, an efficiency of about 60% is obtained at a transmission distance of 35 cm and 81% at a distance of 10 cm. In [136], the performance of segmented coils was improved and a stable charging method was identified for high-power applications. A mixture of coils was used in the track where a rectangular coil was combined with a DD coil and then a rectangular coil side by side. These coils are excited to improve the number of active transmitters and reduce the variation of the output voltage. 2.5-kW power transmission with 85% dc–dc efficiency was achieved. In [137], the relationship between several parameters such as path length, operating efficiency, and vehicle speed has been studied and analyzed. The optimum path length at which the highest possible transfer efficiency is achieved has been identified. In [118], a double-coupled system was used for the segmented transmitter dynamic charging system. An ICC is used to control the transmitter segments, the system operates at a frequency of 20 kHz, and transmits 5-kW power at a maximum efficiency of 92.5%.

In [105], six side-by-side coils are connected in parallel with a very small distance between them and used to reduce the occurrence of power fluctuations. Therefore, the coil self-coupling coefficient is taken into account in the circuit design. The six coils are fed from a common inverter, in order to simplify the power electronic circuit. Each segmented coil at the transmitter side has an independent compensation circuit. This design helps reduce the voltage pressure on the compensation capacitors and to adjust the path length. A power transfer of 1.4 kW with a dc–dc transmission efficiency of 89.7% was demonstrated, showing a maximum power fluctuation of ±2.9% of the average power. A two-segmented LCC compensation circuit was developed in [138]. The effect of using the LCC compensation and inter-coupling on the neighbors' coils in the transmission path was investigated. A 2.34-kW power transmission with a dc–dc efficiency of 91.3% was achieved. In [139], a new transmission coil structure was proposed, whereby bipolar coils are used symmetrically on adjacent unipolar transmitter coils. This leads to the natural separation between the bipolar and unipolar coils. This configuration has the potential to reduce the self-coupling between adjacent unipolar coils, thus facilitating the design of the compensation circuit. Also, this design shows better stability for mutual coupling between the transmitter and receiver during vehicle movement, which makes it suitable for dynamic charging. The power transmission was achieved with an efficiency of up to 90%, with a fluctuation of power within ±22.5%.

Researchers are working day and night to develop a dynamic charging system so that it can be practically applied within cities. This application makes the wireless charging process easier, safer, and helps the spread of environmentally friendly electric cars, as it will facilitate the idea of creating self-driving cars. A summary of the latest DIPT system prototypes showing power level, air gap, and efficiency is presented in Table 4.9.

Even though there are some ongoing discussions and thoughts within the research communities to develop standards for the DIPT system for EV charging, to the authors' knowledge, there are no current standard projects. However, it is crucial to start developing standards and recommended practices for this technology to facilitate system development, demonstration, and operation. Such

Table 4.9 Latest Prototypes of the DIPT System.

Power (kW)	Frequency (kHz)	Air gap (mm)	Efficiency (%)	Studies
0.3	100	170	77–90	[115, 140]
1	90	100	>90	[141]
1.5	23	100	75	[142]
3	20	10	88	[55, 143]
6	20	170	72	[55, 56]
15	60	120	74	[55]
20	22	162	93	[144]
20–30	12.9	500	85	[14, 117]
25	20	–	86	[145]
50	20	152–254	90	[117, 146]
80	20	100	88–90	[102]
120	15–20	40	90	[117]
200	20	60	90	[117, 147]

a standardization effort requires strong collaboration among diverse stakeholders, including road authorities, governments, EVSE companies, and original equipment manufacturers (OEMs). Additionally, demonstration and pilot projects will be very helpful to explore challenges in this technology, identify solutions, and also help develop standard practice. Examples of these projects are FABRIC and UNPLUGGED.

FABRIC project started in 2014 and ended at the end of 2017 [148]. It has demonstrated the feasibility of using the DIPT in various use cases, which include high power (20 kW), and movement from stationary to speed (100 km/h). It also explored interoperability, electromagnetic compatibility, and the impact of EMFs on human health in the countries participating in this project [149]. This project concluded that the ease of access to the power electronics components underground leads to the possibility of repairing and improving the work of the wireless charging system. It also concluded that the overall efficiency of the DIPT system is a key factor in decision-making for the implementation of the system, as it is expected that efficiency from 80% to 90% is achievable when the transmitter coil matches the vehicle pad. Road construction plays an important role in the efficiency and reliability of the system where setting a coil on the ground side requires careful study to avoid overlapping fields.

UNPLUGGED is the second project that ran from 2012 to 2015. This project tested the effect of the IPT system for EVs on clients in urban regions and the feasibility of the technology to extend the driving range [150]. To achieve these goals, UNPLUGGED examined the interoperability, practical issues, technical feasibility, and social and economic effects of inductive charging. The economical design and feasibility study for in-route dynamic charging were included. The production, development, and implementation of two inductive chargers: 3.7 kW (tested for efficiency) and 50 kW (considering two different vehicles, with different restrictions and conditions; the charger is able to give full power to both, improving flexibility) were achieved in UNPLUGGED project [151]. Such projects will be helpful to inform decision-makers about the feasibility of dynamic inductive charging for EVs and contribute to the standardization effort. Here are the main topics that need to be included within standards to help tackle challenges associated with the DIPT system:

- Standards must incorporate appropriate transmitter design including coil configuration, compensation topology, power converters, coil detection, control, and data communication.
- Design considerations must be presented to ensure the interoperability between static and dynamic charging systems. So, an EV with a wireless pad must be able to charge using a static and dynamic system. Additionally, since the system will be installed and operate on - public roadways, it must be applicable for all types of EVs regardless of model, size, manufacture, etc.
- Safety topics related to EMFs, object detection, and access control must be addressed showing the requirements and testing procedures to ensure the compatibility of the system to these requirements.
- It should include methods to integrate the transmission stations (transmitter pad, compensation circuit, and converter) with the road and the impact of different road materials, such as concrete and asphalt on the system performance.
- It must present details for system packaging, thermal management, and cooling process for outdoor installation and operation.
- The standard should incorporate how to integrate the system into the grid, including grid interface, impacts, and mitigation techniques.
- Standards should include information about system level design and characteristics in terms of power level and roadway coverage, considering vehicle speed, vehicle efficiency, road condition, grid availability, etc.

4.5 Quasi-Dynamic IPT System

Quasi-dynamic charging is a concept that allows an EV to charge during transient stops, such as traffic signals, intersections, and bus stops and while traveling at low speeds [152]. In [152], static, dynamic, and quasi-dynamic charging are compared in terms of initial cost in the public transportation system. If there is a possibility of dynamic charging, the power transmission can start wirelessly when the car is on the dynamic charging path, but when dynamic charging is not available, the system automatically switches to the ability to receive power from static charging systems. This system will take advantage of dynamic charging in addition to simplifying the complexity of the control process and greatly reducing the cost of the infrastructure, as well as helping fully align the transmitter and receiver sides. If the quasi-dynamic charging system is implemented at traffic lights, it will be very beneficial, as a transmitter pad can be placed in each stopping lane, and thus each transmitter pad can be controlled separately through an independent converter, or by using one common converter to control several transmitter pads. The excitation of the transmitter can be controlled or canceled via the control device located at the traffic lights [5]. In [153], an analysis of the principle of bi-directional wireless charging and discharging of EVs was made to demonstrate the possibility of using it at traffic lights. Three scenarios were implemented using laboratory investigations. First, electric cars were charged at constant power levels at the intersection points, which led to increased efficiency and driving range. Second, the variable charging scenario independently adjusted the charging level based on the battery's charge status. Third, in which the operations of vehicle to grid (V2G) and grid to vehicle (G2V) were tested. For each scenario, a comparison was made between the EV-driving range and the range gain per kWh consumed was quantified considering the influence of misalignments. Through this study, it was concluded that quasi-dynamic charging at traffic lights represents an appropriate and promising solution to significantly expand the driving range and operating time in the city, especially at high power levels.

4.6 Technology Challenges and Opportunities

In addition to what has been investigated and explored in the literature, there remains a need for further research and development to improve the performance of the wireless charging system and to find radical solutions to the challenges that hinder the penetration of technology into the global market. Some of these challenges are presented along with some of the recommended guidelines as follows:

- *Road infrastructure:* Road infrastructure is one of the most important factors that help EVs penetrate the global market and make it more widespread. Good road infrastructure allows easy installation, repair, maintenance, and less installation cost of wireless charging systems, which encourages governments and organizations to popularize the use of the technology.
- *Interoperability:* Extra effort is required to explore the interoperable operation among different pad designs, dynamic and static charging, and different EV models and sizes. How to test and evaluate interoperability? Which designs are more interoperable with others? Investigate the impact of different installation conditions for the primary pad: above-, flush-, and under-ground on the interoperability of the IPT system.
- *Safety:* There are quite concerns related to the safe operation of the system. The first safety concern is related to EMFs, which requires additional effort to explore novel shield designs, especially active and reactive shielding. Another safety concern is associated with the heat generated in metal objects near the system, which requires an effective foreign metal object

detection method. In addition, a living object detection mean is necessary to prevent access of pets and animals to the hot region during operation. All the current foreign object detection techniques are exploring a way to detect the object and shut down the system. However, it is important to explore what needs to be done after shutting the system down.
- *Durability:* This system will be installed outdoors in for public use. So, it needs to be robust enough to handle the harsh environmental and operating conditions. In addition, the way to integrate the pads with both the vehicle and road is an open question that requires extra research effort.
- *Cost:* More effort is required to bring the system's cost down by using less expensive materials (wires, magnetic cores, and shielding materials), manufacturing, installation, and maintenance process.
- *Fast Charging:* There is a current need to design chargers that are able to bring the charging time down to less than 15 minutes. Therefore, exploring high-power wireless chargers (>200 kW) is a gap that needs to be filled. Novel pad designs with new magnetic materials and wires are crucial so that the system can transfer high-power efficiently at a reasonable cost.

4.7 Conclusion

This chapter presents an inclusive review of the current state-of-the-art static, dynamic, and quasi-dynamic inductive charging systems for EVs. Different types of wires (Litz, LMPW, LMCW, etc.), magnetic materials (ferrite, magnetizable concrete, etc.), and shielding (passive, active, and reactive) are presented and compared. Structures of transmitters used in static and dynamic charging are discussed and compared. The challenges associated with each type along with the research and development activities to tackle these challenges are presented and discussed including standardization efforts. In conclusion, inductive charging offers great advantages for EV charging in terms of flexibility, automation, interoperability, and safety. Dynamic and quasi-dynamic wireless charging has the potential to significantly extend the driving range for EVs, reduce idle time, enable the use of a smaller battery, and eliminate the range anxiety problem. Considering the diverse charging modes is expected to increase the spread of EVs enabling more efficient and environmentally friendly transportation. Additionally, wireless charging aligns very well with automated vehicles so that the vehicles will be able to self-drive and self-charge.

References

1 Morrow, W.R., Lee, H., Gallagher, K.S., and Collantes, G. (2010). *Reducing the US Transportation Sector's Oil Consumption and Greenhouse Gas Emissions*. Cambridge: Harvard Kennedy School.

2 Budhia, M., Covic, G.A., and Boys, J.T. (2011). Design and optimization of circular magnetic structures for lumped inductive power transfer systems. *IEEE Trans. Power Electron.* 26 (11): 3096–3108.

3 Mohamed, A.A.S., Shaier, A.A., Metwally, H., and Selem, S.I. (2020). A comprehensive overview of inductive pad in electric vehicles stationary charging. *Appl. Energy* 262: 114584. https://doi.org/10.1016/j.apenergy.2020.114584.

4 Qiu, C., KT, C., Ching, T.W., and Liu, C. (2014). Overview of wireless charging technologies for electric vehicles. *J. Asian Electr. Veh.* 12 (1): 1679–1685.

5 Ahmad, A., Alam, M.S., and Chabaan, R. (2018). A comprehensive review of wireless charging technologies for electric vehicles. *IEEE Trans. Transport. Electrification* 4 (1): 1. https://doi.org/10.1109/TTE.2017.2771619.

6 Roes, M.G., Duarte, J.L., Hendrix, M.A., and Lomonova, E.A. (2012). Acoustic energy transfer: a review. *IEEE Trans. Ind. Electron.* 60 (1): 242–248.

7 Thakur, R. and Natale, A. (2009). High efficiency wireless power transmission at low frequency using permanent magnet coupling. *Cardiol. Clin.* 27 (1).

8 Hanazawa, M. and Ohira, T. (2011). Power transfer for a running automobile. *2011 IEEE MTT-S International Microwave Workshop Series on Innovative Wireless Power Transmission: Technologies, Systems, and Applications*, pp. 77–80. doi: 10.1109/IMWS.2011.5877095.

9 Mohamed, A.A.S. and Mohammed, O. (2018). Physics-based co-simulation platform with analytical and experimental verification for bidirectional IPT system in EV applications. *IEEE Trans. Veh. Technol.* 67 (1): 275–284. https://doi.org/10.1109/TVT.2017.2763422.

10 Lu, X., Wang, P., Niyato, D. et al. (2015). Wireless charging technologies: fundamentals, standards, and network applications. *IEEE Commun. Surv. Tutorials* 18 (2): 1413–1452.

11 Mohamed, A.A.S., Meintz, A., Schrafel, P., and Calabro, A. (2018). In-vehicle assessment of human exposure to EMFs from 25-kW WPT system based on near-field analysis. *2018 IEEE Vehicle Power and Propulsion Conference (VPPC)*, pp. 1–6. https://doi.org/10.1109/VPPC.2018.8605011.

12 Ongayo, D. and Hanif, M. (2015). An overview of single-sided and double-sided winding inductive coupling transformers for wireless electric vehicle charging. *2015 IEEE 2nd International Future Energy Electronics Conference (IFEEC)*, pp. 1–6. https://doi.org/10.1109/IFEEC.2015.7361593.

13 Zaheer, A., Covic, G.A., and Kacprzak, D. (2014). A bipolar pad in a 10-kHz 300-W distributed IPT system for AGV applications. *IEEE Trans. Ind. Electron.* 61 (7): 7. https://doi.org/10.1109/TIE.2013.2281167.

14 Covic, G.A. and Boys, J.T. (2013). Modern trends in inductive power transfer for transportation applications. *IEEE J. Emerg. Sel. Top. Power Electron.* 1 (1): 28–41. https://doi.org/10.1109/JESTPE.2013.2264473.

15 Mohamed, A.A.S., Shaier, A.A., Metwally, H., and Selem, S.I. (2020). Interoperability of the universal WPT3 transmitter with different receivers for electric vehicle inductive charger. *eTransportation* 100084. doi: 10.1016/j.etran.2020.100084.

16 Mohamed, A.A.S., Allen, D., Youssef, T., and Mohammed, O. (2016). Optimal design of high frequency H-bridge inverter for wireless power transfer systems in EV applications. *2016 IEEE 16th International Conference on Environment and Electrical Engineering (EEEIC)*, pp. 1–6. 10.1109/EEEIC.2016.7555646.

17 Mohamed, A.A.S., Berzoy, A., de Almeida, F.G.N., and Mohammed, O. (2017). Modeling and assessment analysis of various compensation topologies in bidirectional IWPT system for EV applications. *IEEE Trans. Ind. Appl.* 53 (5): 4973–4984. https://doi.org/10.1109/TIA.2017.2700281.

18 Zhang, W. and Mi, C.C. (2016). Compensation topologies of high-power wireless power transfer systems. *IEEE Trans. Veh. Technol.* 65 (6): 4768–4778. https://doi.org/10.1109/TVT.2015.2454292.

19 Takanashi, H., Sato, Y., Kaneko, Y., et al. (2012). A large air gap 3 kW wireless power transfer system for electric vehicles. *2012 IEEE Energy Conversion Congress and Exposition (ECCE)*, pp. 269–274. 10.1109/ECCE.2012.6342813.

20 Barth, D., Klaus, B., and Leibfried, T. (2017). Litz wire design for wireless power transfer in electric vehicles. *2017 IEEE Wireless Power Transfer Conference (WPTC)*, May 2017, pp. 1–4. 10.1109/WPT.2017.7953819.

21 Rossmanith, H., Doebroenti, M., Albach, M., and Exner, D. (2011). Measurement and characterization of high frequency losses in nonideal Litz wires. *IEEE Trans. Power Electron.* 26 (11): 3386–3394. https://doi.org/10.1109/TPEL.2011.2143729.

22 Mizuno, T., Ueda, T., Yachi, S. et al. (2014). Dependence of efficiency on wire type and number of strands of Litz wire for wireless power transfer of magnetic resonant coupling. *IEEJ J. Ind. Appl.* 3 (1): 35–40. https://doi.org/10.1541/ieejjia.3.35.

23 Shinagawa, H., Suzuki, T., Noda, M. et al. (2009). Theoretical analysis of AC resistance in coil using magnetoplated wire. *IEEE Trans. Magn.* 45 (9): 3251–3259. https://doi.org/10.1109/TMAG.2009.2021948.

24 Konno, Y., Yamamoto, T., Chai, Y. et al. (2017). Basic characterization of magnetocoated wire fabricated using spray method. *IEEE Trans. Magn.* 53 (11): 1–7. https://doi.org/10.1109/TMAG.2017.2719047.

25 Yamamoto, T., Konno, Y., Sugimura, K. et al. (2019). Loss reduction of LLC resonant converter using magnetocoated wire. *IEEE J. Ind. Appl.* 8 (1): 51–56. https://doi.org/10.1541/ieejjia.8.51.

26 Jawad, A.M., Nordin, R., Gharghan, S.K. et al. (2018). Single-tube and multi-turn coil near-field wireless power transfer for low-power home appliances. *Energies* 11 (8): 1969. https://doi.org/10.3390/en11081969.

27 Pantic, Z. and Lukic, S. (2013). Computationally-efficient, generalized expressions for the proximity-effect in multi-layer, multi-turn tubular coils for wireless power transfer systems. *IEEE Trans. Magn.* 49 (11): 5404–5416. https://doi.org/10.1109/TMAG.2013.2264486.

28 Sekiya, N. and Monjugawa, Y. (2017). A novel REBCO wire structure that improves coil quality factor in MHz range and its effect on wireless power transfer systems. *IEEE Trans. Appl. Supercond.* 27 (4): 1–5. https://doi.org/10.1109/TASC.2017.2660058.

29 Sullivan, C.R. (2008). Aluminum windings and other strategies forHigh-frequency magnetics design in anEra of high copper and energy costs. *IEEE Trans. Power Electron.* 23 (4): 2044–2051. https://doi.org/10.1109/TPEL.2008.925434.

30 Filipović, D. and Dlabač, T. (2010). A closed form solution for the proximity effect in a thin tubular conductor influenced by a parallel filament. *Serbian J. Electr. Eng.* 7 (1): 13–20.

31 Shi, Y., Dennis, A.R., Huang, K. et al. (2018). Advantages of multi-seeded (RE)–Ba–Cu–O superconductors for magnetic levitation applications. *Supercond. Sci. Technol.* 31 (9): 095008.

32 Mohamed, A.A.S., An, S., and Mohammed, O. (2018). Coil design optimization of power pad in IPT system for electric vehicle applications. *IEEE Trans. Magn.* 54 (4): 1–5. https://doi.org/10.1109/TMAG.2017.2784381.

33 Budhia, M., Boys, J.T., Covic, G.A., and Huang, C.-Y. (2013). Development of a single-sided flux magnetic coupler for electric vehicle IPT charging systems. *IEEE Trans. Ind. Electron.* 60 (1): 318–328. https://doi.org/10.1109/TIE.2011.2179274.

34 Kim, H., Song, C., Kim, J., et al. (2012). Shielded coil structure suppressing leakage magnetic field from 100W-class wireless power transfer system with higher efficiency. *2012 IEEE MTT-S International Microwave Workshop Series on Innovative Wireless Power Transmission: Technologies, Systems, and Applications*, pp. 83–86. 10.1109/IMWS.2012.6215825.

35 Delgado, A., Oliver, J.A., Cobos, J.A., et al. (2019). Optimized design for wireless coil for electric vehicles based on the use of magnetic nano-articles. *2019 IEEE Applied Power Electronics Conference and Exposition (APEC)*, pp. 1515–1520. 10.1109/APEC.2019.8721998.

36 Delgado, A., Salinas, G., Rodríguez, J., et al. (2018). Finite element modelling of Litz wire conductors and compound magnetic materials based on magnetic nano-particles by means of equivalent homogeneous materials for wireless power transfer system. *2018 IEEE 19th Workshop on Control and Modeling for Power Electronics (COMPEL)*, pp. 1–5. 10.1109/COMPEL.2018.8460012.

37 Yoon, T.-J., Lee, W., Oh, Y.-S., and Lee, J.-K. (2003). Magnetic nanoparticles as a catalyst vehicle for simple and easy recycling. *New J. Chem.* 27 (2): 227–229. https://doi.org/10.1039/B209391J.

38 Sun, X., Zheng, Y., Peng, X. et al. (2014). Parylene-based 3D high performance folded multilayer inductors for wireless power transmission in implanted applications. *Sens. Actuators Phys.* 208: 141–151. https://doi.org/10.1016/j.sna.2013.12.038.

39 Sun, X., Zheng, Y., Li, Z., et al. (2013). Stacked flexible parylene-based 3D inductors with Ni 80 Fe 20 core for wireless power transmission system. *2013 IEEE 26th International Conference on Micro Electro Mechanical Systems (MEMS)*. IEEE, pp. 849–852.

40 Esguerra, M. and Lucke, R. (2004). Application and production of a magnetic product. US6696638B2. https://patents.google.com/patent/US6696638B2/en (accessed 24 October 2019).

41 R. Tavakoli, Echols, A., Pratik, U. et al. (2017). Magnetizable concrete composite materials for road-embedded wireless power transfer pads. *2017 IEEE Energy Conversion Congress and Exposition (ECCE*, pp. 4041–4048. 10.1109/ECCE.2017.8096705.

42 Carretero, C., Lope, I., and Acero, J. (2019). Magnetizable concrete flux concentrators for wireless inductive power transfer applications. *IEEE J. Emerg. Sel. Top. Power Electron.* 1–1. https://doi.org/10.1109/JESTPE.2019.2935226.

43 Ibrahim, M. (2014). Wireless inductive charging for electrical vehicules : electromagnetic modelling and interoperability analysis. Thesis, Paris 11. http://www.theses.fr/2014PA112369 (accessed 23 October 2019).

44 International Commission on Non-Ionizing Radiation Protection (2010). Guidelines for limiting exposure to time-varying electric and magnetic fields (1 Hz to 100 kHz). *Health Phys.* 99 (6): 818–836. https://doi.org/10.1097/HP.0b013e3181f06c86.

45 ICNIRP (1998). International commission on non-ionizing radiation protection. Guidelines for limiting exposure to time-varying electric, magnetic, and electromagnetic fields (up to 300 GHz). *Health Phys.* 74 (4): 494–522.

46 Kitano, Y., Omori, H., Morizane, T., et al. (2014). A new shielding method for magnetic fields of a wireless EV charger with regard to human exposure by eddy current and magnetic path. *2014 International Power Electronics and Application Conference and Exposition*, pp. 778–781. 10.1109/PEAC.2014.7037956.

47 J. Zhou, Y. Gao, C. Zhou, et al. (2017). Optimal power transfer with aluminum shielding for wireless power transfer systems. *2017 20th International Conference on Electrical Machines and Systems (ICEMS)*, pp. 1–4. 10.1109/ICEMS.2017.8056143.

48 Tan, L., Li, J., Chen, C. et al. (2016). Analysis and performance improvement of WPT Systems in the environment of single non-ferromagnetic metal plates. *Energies* 9 (8): 576. https://doi.org/10.3390/en9080576.

49 Feliziani, M. and Cruciani, S. (2013). Mitigation of the magnetic field generated by a wireless power transfer (WPT) system without reducing the WPT efficiency. *2013 International Symposium on Electromagnetic Compatibility*, pp. 610–615.

50 Kim, S., Park, H.-H., Kim, J. et al. (2014). Design and analysis of a resonant reactive shield for a wireless power electric vehicle. *IEEE Trans. Microwave Theory Tech.* 62 (4): 1057–1066. https://doi.org/10.1109/TMTT.2014.2305404.

51 Mohammad, M., Pries, J., Onar, O. *et al.* (2019). Design of an EMF suppressing magnetic shield for a 100-kW DD-coil wireless charging system for electric vehicles. *2019 IEEE Applied Power Electronics Conference and Exposition (APEC)*, pp. 1521–1527. https://doi.org/10.1109/APEC.2019.8722084.

52 Choi, S.Y., Gu, B.W., Lee, S.W. et al. (2014). Generalized active EMF cancel methods for wireless electric vehicles. *IEEE Trans. Power Electron.* 29 (11): 5770–5783. https://doi.org/10.1109/TPEL.2013.2295094.

53 Kim, J., Kim, J., Kong, S. et al. (2013). Coil design and shielding methods for a magnetic resonant wireless power transfer system. *Proc. IEEE* 101 (6): 1332–1342. https://doi.org/10.1109/JPROC.2013.2247551.

54 Mohamed, A.A. and Shaier, A.A. Shielding techniques of IPT system for electric vehicles' stationary charging. In: *Electric Vehicle Integration in a Smart Microgrid Environment* (ed. M.S. Alam and M. Krishnamurthy), 279–293. CRC Press.

55 Choi, S.Y., Gu, B.W., Jeong, S.Y., and Rim, C.T. (2015). Advances in wireless power transfer systems for roadway-powered electric vehicles. *IEEE J. Emerg. Sel. Top. Power Electron.* 3 (1): 18–36. https://doi.org/10.1109/JESTPE.2014.2343674.

56 Mi, C.C., Buja, G., Choi, S.Y., and Rim, C.T. (2016). Modern advances in wireless power transfer systems for roadway powered electric vehicles. *IEEE Trans. Ind. Electron.* 63 (10): 6533–6545. https://doi.org/10.1109/TIE.2016.2574993.

57 Biswas, M.M. (2018). Comparative study of inductive wireless power transfer pad topologies for electric vehicle charging. University of Akron. http://rave.ohiolink.edu/etdc/view?acc_num=akron1536943828810247.

58 Patil, D., Balsara, P.T., Fahimi, B. et al. (2019). Dynamic wireless power transfer for electric vehicle. PhD thesis, University of Texas, Dallas.

59 Patil, D., McDonough, M.K., Miller, J.M. et al. (2017). Wireless power transfer for vehicular applications: overview and challenges. *IEEE Trans. Transport. Electrification* 4 (1): 3–37.

60 Aditya, K. (2016). Design and implementation of an inductive power transfer system for wireless charging of future electric transportation. http://ir.library.dc-uoit.ca/handle/10155/712 (accessed 25 October 2019).

61 Ahmad, A., Alam, M.S., Chabaan, R., and Mohamed, A. (2019). Comparative analysis of power pad for wireless charging of electric vehicles. *SAE Technical Paper, 0148–7191.*

62 Ahmad, A., Alam, M.S., and Chabaan, R. (2018). A comprehensive review of wireless charging technologies for electric vehicles. *IEEE Trans. Transp. Electrification* 4 (1): 38–63. https://doi.org/10.1109/TTE.2017.2771619.

63 Villa, J.L., Sallan, J., Sanz Osorio, J.F., and Llombart, A. (2012). High-misalignment tolerant compensation topology for ICPT systems. *IEEE Trans. Ind. Electron.* 59 (2): 945–951. https://doi.org/10.1109/TIE.2011.2161055.

64 Chen, W., Liu, C., Lee, C.H.T., and Shan, Z. (2016). Cost-effectiveness comparison of coupler designs of wireless power transfer for electric vehicle dynamic charging. *Energies* 9 (11): 906. https://doi.org/10.3390/en9110906.

65 Choi, S.Y., Gu, B.W., Jeong, S.Y., and Rim, C.T. (2015). Advances in wireless power transfer systems for roadway-powered electric vehicles. *IEEE J. Emerg. Sel. Top. Power Electron.* 3 (1): 18–36. https://doi.org/10.1109/JESTPE.2014.2343674.

66 Bosshard, R. (2015). Multi-objective optimization of inductive power transfer systems for EV charging. *ETH Zurich.* https://doi.org/10.3929/ethz-a-010664107.

67 Kim, S., Zaheer, A., Covic, G., and Boys, J. (2014). Tripolar pad for inductive power transfer systems. *IECON 2014 - 40th Annual Conference of the IEEE Industrial Electronics Society*, pp. 3066–3072. 10.1109/IECON.2014.7048947.

68 Nagatsuka, Y., Ehara, N., Kaneko, Y., et al. (2010). Compact contactless power transfer system for electric vehicles. *The 2010 International Power Electronics Conference - ECCE ASIA*, pp. 807–813. 10.1109/IPEC.2010.5543313.

69 Nagatsuka, Y., Noguchi, S., Kaneko, Y. et al. (2010). Contactless power transfer system for electric vehicle battery charger. *Presented at the The 25th World Battery, Hybrid and Fuel Cell Electric Vehicle Symposium & Exhibition, China.*

70 Chigira, M., Nagatsuka, Y., Kaneko, Y. (2011), et al. Small-size light-weight transformer with new core structure for contactless electric vehicle power transfer system. *2011 IEEE Energy Conversion Congress and Exposition*, pp. 260–266. 10.1109/ECCE.2011.6063778.

71 Liu, C., Jiang, C., and Qiu, C. (2017). Overview of coil designs for wireless charging of electric vehicle. *2017 IEEE PELS Workshop on Emerging Technologies: Wireless Power Transfer (WoW).* IEEE, pp. 1–6.

72 Budhia, M., Covic, G.A., Boys, J.T., and Huang, C.-Y. (2011). Development and evaluation of single sided flux couplers for contactless electric vehicle charging. *2011 IEEE Energy Conversion Congress and Exposition*, pp. 614–621. 10.1109/ECCE.2011.6063826.

73 Zaheer, A., Kacprzak, D., and Covic, G.A. (2012). A bipolar receiver pad in a lumped IPT system for electric vehicle charging applications. *2012 IEEE Energy Conversion Congress and Exposition (ECCE)*, pp. 283–290. 10.1109/ECCE.2012.6342811.

74 Covic, G.A., Kissin, M.L.G., Kacprzak, D., et al. (2011). A bipolar primary pad topology for EV stationary charging and highway power by inductive coupling. *2011 IEEE Energy Conversion Congress and Exposition*, pp. 1832–1838. 10.1109/ECCE.2011.6064008.

75 Ahmad, A., Alam, M.S., and Mohamed, A.A.S. (2019). Design and interoperability analysis of quadruple pad structure for electric vehicle wireless charging application. *IEEE Trans. Transport. Electrification* 1–1. https://doi.org/10.1109/TTE.2019.2929443.

76 Lin, F., Covic, G.A., and Boys, J.T. (2018), A comparison of multi-coil pads in IPT systems for EV charging. *2018 IEEE Energy Conversion Congress and Exposition (ECCE)*, pp. 105–112. 10.1109/ECCE.2018.8557369.

77 Turki, F., Detweiler, M., and Reising, V. (2016). Performance of wireless charging system based on quadrupole coil geometry with different resonance topology approaches. *2016 IEEE PELS Workshop on Emerging Technologies: Wireless Power Transfer (WoW)*, pp. 104–109. 10.1109/WoW.2016.7772074.

78 Jeong, S.Y., Choi, S.Y., Sonapreetha, M.R., and Rim, C.T. (2015). DQ-quadrature power supply coil sets with large tolerances for wireless stationary EV chargers. *2015 IEEE PELS Workshop on Emerging Technologies: Wireless Power (2015 WoW)*, pp. 1–6. 10.1109/WoW.2015.7132800.

79 J2954: Wireless Power Transfer for Light-Duty Plug-In/ Electric Vehicles and Alignment Methodology - SAE International. https://www.sae.org/standards/content/j2954_201605 (accessed 24 September 2019).

80 IEC 61980-1 : Electric vehicle wireless power transfer (WPT) systems – Part 1: General requirements. https://global.ihs.com/doc_detail.cfm?document_name=IEC%2061980%2D1&item_s_key=00656450 (accessed 6 December 2021).

81 F. Lu, H. Zhang, H. Hofmann, and C. Mi (2015). A high efficiency 3.3 kW loosely-coupled wireless power transfer system without magnetic material. *2015 IEEE Energy Conversion Congress and Exposition (ECCE)*, pp. 2282–2286. 10.1109/ECCE.2015.7309981.

82 WiTricity. The next wireless revolution: electric vehicle wireless charging, power and efficiency. https://witricity.com/wp-content/uploads/2018/03/WIT_White_Paper_PE_20180321.pdf (accessed 1 October 2019).

83 Evatran. Buy plugless wireless EV charging. *Plugless Power.* https://www.pluglesspower.com/shop (accessed 1 October 2019).

84 HEVO: Wireless Charging for Electric Vehicles. https://hevo.com/ (accessed 22 October 2022).

85 Qualcomm Halo Wireless Electric Vehicle Charging. https://www.qualcomm.com/solutions/automotive?id=41#media-centre (accessed 3 October 2019).

86 Wu, H.H. and Masquelier, M.P. (2015). An overview of a 50kW inductive charging system for electric buses. *2015 IEEE Transportation Electrification Conference and Expo (ITEC)*, pp. 1–4. 10.1109/ITEC.2015.7165747.

87 Conductix-Wampfler (2013). Charging electric buses quickly and efficiently: bus stops fitted with modular components make Charge & Go simple to implement. http://www.conductix.us/en/news/2013-05-29/charging-electric-buses-quickly-and-efficiently-busstops-fitted-modular-components-make-charge-go (accessed 10 October 2019).

88 Miller, J.M. and Daga, A. (2015). Elements of wireless power transfer essential to high power charging of heavy duty vehicles. *IEEE Trans. Transport. Electrification* 1 (1): 26–39. https://doi.org/10.1109/TTE.2015.2426500.

89 Calabro, A., Cohen, B., Daga, A., et al. (2019). Performance of 200-kW inductive charging system for range extension of electric transit buses. *2019 IEEE Transportation Electrification Conference and Expo (ITEC)*, pp. 1–5. https://doi.org/10.1109/ITEC.2019.8790490.

90 Alam, M.S., Ahmad, A., Khan, Z.A. et al. (2018). A bibliographical review of electrical vehicles (xEVs) standards. *SAE Int. J. Altern. Powertrains* 7 (1): 6398.

91 J1773A: SAE Electric Vehicle Inductively Coupled Charging - SAE International. https://www.sae.org/standards/content/j1773_201406 (accessed 25 October 2019).

92 J2847/6A: Communication for Wireless Power Transfer Between Light-Duty Plug-in Electric Vehicles and Wireless EV Charging Stations - SAE International https://www.sae.org/standards/content/j2847/6_202009 (accessed 6 December 2021).

93 J2836/6A: Use Cases for Wireless Charging Communication for Plug-in Electric Vehicles - SAE International. https://www.sae.org/standards/content/j2836/6_202104 (accessed 6 December 2021).

94 IEC 61980-1:2015 I IEC Webstore. https://webstore.iec.ch/publication/22951. (accessed 25 October 2019).

95 IEC/TS 61980-2: Electric vehicle wireless power transfer (WPT) systems – Part 2: Specific requirements for communication between electric road vehicle (EV) and infrastructure. https://global.ihs.com/doc_detail.cfm?document_name=IEC%2FTS%2061980%2D2&item_s_key=00786994#product-details-list (accessed 6 December 2021).

96 IEC/TS 61980-3 : Electric vehicle wireless power transfer (WPT) systems – Part 3: Specific requirements for the magnetic field wireless power transfer systems. https://global.ihs.com/doc_detail.cfm?&document_name=IEC%2FTS%2061980%2D3&item_s_key=00786995&item_key_date=800631& (accessed 6 December 2021).

97 Alam, M.S., Ahmad, A., Khan, Z.A. et al. (2018). A bibliographical review of electrical vehicles (xEVs) standards. *SAE Int. J. Altern. Powertrains* 7 (1): 63–98.

98 Leskarac, D., Panchal, C., Stegen, S., and Lu, J. (2015). PEV charging technologies and V2G on distributed system and utility interfaces. *Veh. Grid Link. Electr. Veh. Smart Grid* 79: 157–209.

99 Ahmad, A., Alam, M.S., and Chabaan, R. (2017). A comprehensive review of wireless charging technologies for electric vehicles. *IEEE Trans. Transp. Electrification* 4 (1): 38–63.

100 UL Outline I UL 9741. https://standardscatalog.ul.com/ProductDetail.aspx?productId=UL9741 (accessed 6 December 2021).

101 UL - SUBJECT 9741 - UL Outline of Investigation for Electric Vehicle Power Export Equipment (EVPE) I Engineering360. https://standards.globalspec.com/std/14386115/subject-9741 (accessed 6 December 2021).

102 Panchal, C., Stegen, S., and Lu, J. (2018). Review of static and dynamic wireless electric vehicle charging system. *Eng. Sci. Technol. Int. J.* 21 (5): 5.

103 Kumar, K., Gupta, S., and Nema, S. (2021). A review of dynamic charging of electric vehicles. *2021 7th International Conference on Electrical Energy Systems (ICEES)*, pp. 162–165. 10.1109/ICEES51510.2021.9383634.

104 Bi, Z., Kan, T., Mi, C.C. et al. (2016). A review of wireless power transfer for electric vehicles: prospects to enhance sustainable mobility. *Appl. Energy* 179: 413–425. https://doi.org/10.1016/j.apenergy.2016.07.003.

105 Lu, F., Zhang, H., Hofmann, H., and Mi, C.C. (2016). A dynamic charging system with reduced output power pulsation for electric vehicles. *IEEE Trans. Ind. Electron.* 63 (10): 6580–6590. https://doi.org/10.1109/TIE.2016.2563380.

106 Suh, N.P., Cho, D.H., and Rim, C.T. (2011). Design of On-Line Electric Vehicle (OLEV). In *Global Product Development* (ed. A. Bernard), 3–8. Berlin, Heidelberg: Springer.

107 Sun, L., Ma, D., and Tang, H. (2018). A review of recent trends in wireless power transfer technology and its applications in electric vehicle wireless charging. *Renew. Sustain. Energy Rev.* 91: 490–503. https://doi.org/10.1016/j.rser.2018.04.016.

108 Ko, Y.D. and Jang, Y.J. (2013). The optimal system design of the online electric vehicle utilizing wireless power transmission technology. *IEEE Trans. Intell. Transp. Syst.* 14 (3): 1255–1265. https://doi.org/10.1109/TITS.2013.2259159.

109 García-Vázquez, C.A., Llorens-Iborra, F., Fernández-Ramírez, L.M. et al. (2017). Comparative study of dynamic wireless charging of electric vehicles in motorway, highway and urban stretches. *Energy* 137: 42–57. https://doi.org/10.1016/j.energy.2017.07.016.

110 Huh, J., Lee, S.W., Lee, W.Y. et al. (2011). Narrow-width inductive power transfer system for online electrical vehicles. *IEEE Trans. Power Electron.* 26 (12): 3666–3679. https://doi.org/10.1109/TPEL.2011.2160972.

111 Choi, S.Y. and Rim, C.T. (2015). Recent progress in developments of on-line electric vehicles. *2015 6th International Conference on Power Electronics Systems and Applications (PESA)*, pp. 1–8. 10.1109/PESA.2015.7398964.

112 Choi, S., Huh, J., Lee, W.Y. et al. (2013). New cross-segmented power supply rails for roadway-powered electric vehicles. *IEEE Trans. Power Electron.* 28 (12): 5832–5841. https://doi.org/10.1109/TPEL.2013.2247634.

113 Miller, J.M., Jones, P.T., Li, J.-M., and Onar, O.C. (2015). ORNL experience and challenges facing dynamic wireless power charging of EV's. *IEEE Circuits Syst. Mag.* 15 (2): 40–53. https://doi.org/10.1109/MCAS.2015.2419012.

114 Gil, A. and Taiber, J. (2014). A literature review in dynamic wireless power transfer for electric vehicles: technology and infrastructure integration challenges. *Sustainable Automotive Technologies 2013*, Cham, pp. 289–298. https://doi.org/10.1007/978-3-319-01884-3_30.

115 Lee, K., Pantic, Z., and Lukic, S.M. (2014). Reflexive field containment in dynamic inductive power transfer systems. *IEEE Trans. Power Electron.* 29 (9): 4592–4602. https://doi.org/10.1109/TPEL.2013.2287262.

116 Song, K., Koh, K.E., Zhu, C. et al. (2016). A review of dynamic wireless power transfer for in motion electric vehicles. *Wireless Power Trans. Fundam. Technol.* 109–128.

117 Patil, D., McDonough, M.K., Miller, J.M. et al. (2018). Wireless power transfer for vehicular applications: overview and challenges. *IEEE Trans. Transport. Electrification* 4 (1): 1. https://doi.org/10.1109/TTE.2017.2780627.

118 Chen, L., Nagendra, G.R., Boys, J.T., and Covic, G.A. (2015). Double-coupled systems for IPT roadway applications. *IEEE J. Emerg. Sel. Top. Power Electron.* 3 (1): 37–49. https://doi.org/10.1109/JESTPE.2014.2325943.

119 Miller, J.M., Onar, O.C., White, C. et al. (2014). Demonstrating dynamic wireless charging of an electric vehicle: the benefit of electrochemical capacitor smoothing. *IEEE Power Electron. Mag.* 1 (1): 12–24. https://doi.org/10.1109/MPEL.2014.2300978.

120 Shin, S., Shin, J., Kim, Y., et al. (2012). Hybrid inverter segmentation control for online electric vehicle. *2012 IEEE International Electric Vehicle Conference*, pp. 1–6. https://doi.org/10.1109/IEVC.2012.6183257.

121 Beh, H.Z.Z., Covic, G.A., and Boys, J.T. (2015). Wireless fleet charging system for electric bicycles. *IEEE J. Emerg. Sel. Top. Power Electron.* 3 (1): 75–86. https://doi.org/10.1109/JESTPE.2014.2319104.

122 Nagendra, G.R., Boys, J.T., Covic, G.A., et al. (2013). Design of a double coupled IPT EV highway. *IECON 2013 - 39th Annual Conference of the IEEE Industrial Electronics Society*,, pp. 4606–4611. https://doi.org/10.1109/IECON.2013.6699878.

123 S. Standard (2017). Wireless power transfer for light-duty plug-in/electric vehicles and alignment methodology.

124 Hutin, M. and Leblanc, M. (1894). Transformer system for electric railways. US527857A. https://patents.google.com/patent/US527857A/en (accessed 14 January 2021).

125 Bolger, J.G. (1975). Supplying power to vehicles. *Google Pat.*

126 Bolger, J.G., Kirsten, F.A., and Ng, L.S. (1978). Inductive power coupling for an electric highway system. *28th IEEE Vehicular Technology Conference*, vol. 28, pp. 137–144. https://doi.org/10.1109/VTC.1978.1622522.

127 Zell, C.E. and Bolger, J.G. (1982). Development of an engineering prototype of a roadway powered electric transit vehicle system: A public/private sector program. *32nd IEEE Vehicular Technology Conference*, vol. 32, pp. 435–438. https://doi.org/10.1109/VTC.1982.1623054.

128 Lashkari, K., Shladover, S.E., and Lechner, E.H. (1986). Inductive power transfer to an electric vehicle. *Presented at the International Electric Vehicle Symposium (8th: 1986: Washington DC).*

129 Lechner, E. and Shladover, S.E. (1986). Roadway powered electric vehicle: an all-electric hybrid system, no. CONF-8610122.

130 Shladover, S.E. (1988). Systems engineering of the roadway powered electric vehicle technology. *Presented at the international electric vehicle symposium (9th).*

131 Bolger, J.G. (1994). Urban electric transportation systems: the role of magnetic power transfer. *Proceedings of WESCON '94*, pp. 41–45. 10.1109/WESCON.1994.403598.

132 D. Empey, I. Systems Control Technology, P. Transit, Highways & University of California, and BIOTS (1994). Roadway powered electric vehicle project: track construction and testing program. Phase 3d.

133 Tan, L., Zhao, W., Liu, H. et al. (2020). Design and optimization of ground-side power transmitting coil parameters for EV dynamic wireless charging system. *IEEE Access* 8: 74595–74604. https://doi.org/10.1109/ACCESS.2020.2988622.

134 Buja, G., Bertoluzzo, M., and Dashora, H.K. (2016). Lumped track layout design for dynamic wireless charging of electric vehicles. *IEEE Trans. Ind. Electron.* 63 (10): 6631–6640. https://doi.org/10.1109/TIE.2016.2538738.

135 Sampath, J.P.K., Vilathgamuwa, D.M., and Alphones, A. (2016). Efficiency enhancement for dynamic wireless power transfer system with segmented transmitter array. *IEEE Trans. Transport. Electrification* 2 (1): 76–85. https://doi.org/10.1109/TTE.2015.2508721.

136 Wang, H. and Cheng, K.W. (2021). An improved and integrated design of segmented dynamic wireless power transfer for electric vehicles. *Energies* 14 (7): https://doi.org/10.3390/en14071975.

137 Zhang, W., Wong, S., Tse, C.K., and Chen, Q. (2014). An optimized track length in roadway inductive power transfer systems. *IEEE J. Emerg. Sel. Top. Power Electron.* 2 (3): 598–608. https://doi.org/10.1109/JESTPE.2014.2301460.

138 Zhu, Q., Wang, L., Guo, Y. et al. (2016). Applying LCC compensation network to dynamic wireless EV charging system. *IEEE Trans. Ind. Electron.* 63 (10): 6557–6567. https://doi.org/10.1109/TIE.2016.2529561.

139 Li, X., Hu, J., Wang, H. et al. (2020). A new coupling structure and position detection method for segmented control dynamic wireless power transfer systems. *IEEE Trans. Power Electron.* 35 (7): 6741–6745. https://doi.org/10.1109/TPEL.2019.2963438.

140 Lukic, S. and Pantic, Z. (2013). Cutting the cord: static and dynamic inductive wireless charging of electric vehicles. *IEEE Electrification Mag.* 1 (1): 57–64. https://doi.org/10.1109/MELE.2013.2273228.

141 Throngnumchai, K., Hanamura, A., Naruse, Y., and Takeda, K. (2013). Design and evaluation of a wireless power transfer system with road embedded transmitter coils for dynamic charging of electric vehicles. *World Electr. Veh. J.* 6 (4): https://doi.org/10.3390/wevj6040848.

142 Onar, O.C., Miller, J.M., Campbell, S.L., et al. (2013). A novel wireless power transfer for in-motion EV/PHEV charging. *2013 Twenty-Eighth Annual IEEE Applied Power Electronics Conference and Exposition (APEC)*, pp. 3073–3080. 10.1109/APEC.2013.6520738.

143 D. Empey (1994). Roadway powered electric vehicle project track construction and testing program phase 3D," *Syst. Control Technol. Inc Tech Rep.*

144 Miller, J.M., Scudiere, M.B., McKeever, J.W., and White, C. (2011). Wireless power transfer. *Presented at the Oak ridge National Laboratery's Power Electronics Symposium.*

145 Tavakoli, R. and Pantic, Z. (2018). Analysis, design, and demonstration of a 25-kW dynamic wireless charging system for roadway electric vehicles. *IEEE J. Emerg. Sel. Top. Power Electron.* 6 (3): 1378–1393. https://doi.org/10.1109/JESTPE.2017.2761763.

146 Wave I Wireless Advanced Vehicle Electrification I Inductive Charging. *Wave.* https://waveipt.com (accessed 28 January 2021).

147 Primove. *Bombardier Transportation.* https://localhost:4503/content/bbd-transport/en.html (accessed 28 January 2021).

148 Wireless power transfer–task 26 final report.pdf. http://www.ieahev.org/assets/1/7/Task_26_Final_Report_v1.7_(FINAL2).pdf (accessed 20 July 2020).

149 Benders, B., Vermaat, P., Bludszuweit, H.B., and Theodoropoulos, T. Interoperability considerations. http://www.fabric-project.eu/www.fabric-project.eu/images/Deliverables/FABRIC_D33.3_Interoperability_considerations_2017_update.pdf (accessed 20 July 2020).

150 Wireless charging for Electic Vehicles I UNPLUGGED Project I FP7 I CORDIS I European Commission. https://cordis.europa.eu/project/id/314126 (accessed 22 Febraury 2021).

151 Final Report Summary - UNPLUGGED (Wireless charging for Electic Vehicles) I Report Summary I UNPLUGGED I FP7 I CORDIS I European Commission. https://cordis.europa.eu/project/id/314126/reporting (accessed 22 Febraury 2021).

152 Jang, Y.J., Jeong, S., and Lee, M.S. (2016). Initial energy logistics cost analysis for stationary, quasi-dynamic, and dynamic wireless charging public transportation systems. *Energies* 9 (7): https://doi.org/10.3390/en9070483.

153 Mohamed, A.A.S., Lashway, C.R., and Mohammed, O. (2017). Modeling and feasibility analysis of quasi-dynamic WPT system for EV applications. *IEEE Trans. Transp. Electrification* 3 (2): 343–353. https://doi.org/10.1109/TTE.2017.2682111.

5

Effectiveness Analysis of Control Strategies in Acoustic Noise and Vibration Reduction of PMSM-Driven Coupled System for EV and HEV Applications

Rishi Kant Thakur[1], Rajesh Manjibhai Pindoriya[2], Rajeev Kumar[1], and Bharat Singh Rajpurohit[3]

[1] *Indian Institute of Technology Mandi, School of Mechanical and Materials Engineering, Mandi, Himachal Pradesh 175075, India*
[2] *Thapar Institute of Engineering and Technology Patiala, Department of Electrical and Instrumentation Engineering, Patiala, Punjab 147004, India*
[3] *Indian Institute of Technology Mandi, School of Computing and Electrical Engineering, Mandi, Himachal Pradesh 175075, India*

5.1 Chapter Organization

The organization of this chapter is as follows: Origin of ANV including mechanical, electromagnetic, and aerodynamic sources and its consequences in the PMSM-based coupled system is given in Section 5.2. A state-of-art related to control aspects for ANV reduction including vibration reduction strategy at site (mechanical) and vibration reduction at source (electrical) is given in Section 5.3. The complete detail of the experimental setup used in this chapter is given in Section 5.4. Further, the methodology of various control strategies including PTPWM and RPPM and their implementation for ANV reduction are given in Section 5.5. In Section 5.6, the torsional vibration at resonance is analyzed using lumped parametric modeling and MPF and further experimental validation is also studied. In Section 5.7, the relative analysis of various control strategies over vibration reduction is presented. In Section 5.8, the result and discussion and in Section 5.9, the conclusion of the chapter are presented.

5.2 Origin of ANV and its Consequences in the PMSM-Based Coupled System

In electrified vehicles, vibrations are induced due to power supply fluctuations like a ripple in torque or because of poor design like the eccentric center of mass, misaligned shafts, improper placement of bearing, support, etc. Misalignment causes impact loads over the adjoining components like bearings, couplers, etc., and produces noise. For rotating electrical machines, ANV has been classified into the following three categories: electromagnetic noise, mechanical noise, and aerodynamic noise as given in Figure 5.1 [1].

The schematic layout of the energy conversion process and generation of ANV in the electric machine drive is shown in Figure 5.2 [1]. The controllable high-frequency three-phase stator currents (i_a, i_b, and i_c) are the input supply of the PMSM drive coupled with the DC generator. The magnetic forces are produced in the electric machines by the interaction of a three-phase stator current with the material of the machines. The main sources of mechanical vibration are magnetic

Transportation Electrification: Breakthroughs in Electrified Vehicles, Aircraft, Rolling Stock, and Watercraft, First Edition. Edited by Ahmed A. Mohamed, Ahmad Arshan Khan, Ahmed T. Elsayed, and Mohamed A. Elshaer.

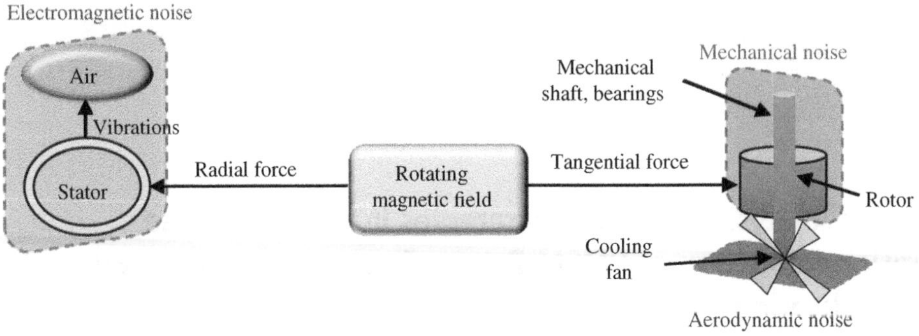

Figure 5.1 Generation of vibration and noise of different origins in an electrical machine.

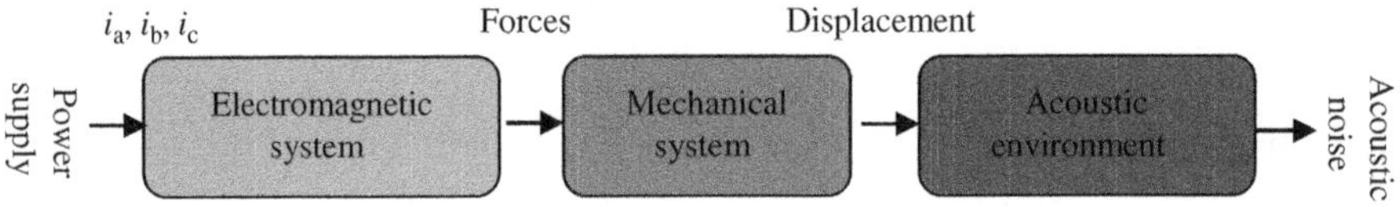

Figure 5.2 Conversion of electrical energy into vibro-acoustic energy.

forces, which are to excite the machine's frame and stator core in the corresponding frequency range. The stator and frame vibrations cause the surrounding medium of air to excite and vibrate and finally generate acoustic pressure variations (and thereby noise).

5.2.1 Mechanical Noise

Mechanical components in working conditions are under irregular cyclic loading, which leads to fatigue failure of components. Various influencing factors include stress concentration, surface irregularities, geometrical discontinuity, and load distribution and its nature like static or dynamic, which reduces the functional life of the shaft [2]. Under fatigue load, endurance limit is the criteria of failure, which incorporates factors like stress concentration factor, size factor, notch sensitivity factor, surface roughness factor, loading condition factor, thermal condition factor etc., overyielding, and ultimate stress and various criteria like Gerber parabola, Soderberg line, and Goodman line are used in S-N (endurance limit vs millions of revolution) curve as safety criteria to calculate shaft life [3]. In various practical applications, the root causes of vibration are mechanical sources like in diesel engines, especially in marine propulsion engines, the blast of diesel inside the combustion chamber produces driving fluctuation torque on the propulsion shaft, leading to torsional vibration [3]. In PMSM, magnet breakage, bearing the inner or outer race failure, is the main cause of vibration [4].

5.2.2 Electromagnetic Noise

Electrical machines work on the principle of electromagnetic energy conversion process between electrical and mechanical energy. The electromagnetic noise is generated due to two sources, which are Maxwell force and magnetostriction force. Maxwell forces are of two types: radial forces and tangential forces. The radial component of air gap stress (σ_{RR}) due to the radial component of Maxwell force can be defined by (5.1) [5].

$$(\sigma_{RR}) = \frac{b_R^2 - b_\theta^2}{2\mu_0} \tag{5.1}$$

The magnitude of the tangential force component ($\sigma_{R\theta}$) of the Maxwell force is very small and, therefore, negligible in comparison to the radial component. Therefore, tangential components acting tangentially on stator teeth provide the working electromagnetic toque.

$$(\sigma_{R\theta}) = \frac{2b_R b_\theta}{2\mu_0} \tag{5.2}$$

where b_R is the radial component of flux density, b_θ is the tangential component of flux density, and μ_0 is the permeability of free space.

If the stator produces $B_{m1}\cos(\omega_1 t + k\alpha + Ø_1)$ magnetic flux density wave and the rotor produces $B_{m2}\cos(\omega_2 t + k\alpha + Ø_2)$ magnetic flux density wave, then their product is [6]

$$\begin{aligned} &0.5B_{m1}\,B_{m2}\cos\left[(\omega_1 + \omega_2)\,t + (k + l)\,\alpha + (Ø_1 + Ø_2)\right] \\ &+ 0.5B_{m1}\,B_{m2}\cos\left[(\omega_1 - \omega_2)\,t + (k - l)\,\alpha + (Ø_1 - Ø_2)\right] \end{aligned} \tag{5.3}$$

where B_{m1} and B_{m2} are the magnitude of the electrical machine stator and rotor magnetic flux density waves, respectively. ω_1 and ω_2 are the angular frequencies of the stator and rotor magnetic fields, respectively. $Ø_1$ and $Ø_2$ are phases of the stator and rotor magnetic flux density waves, respectively. $k = 1, 2, 3, \ldots$ and $l = 1, 2, 3, \ldots$ The product expressed by (5.3) is proportional to the magnetic stress wave in the air gap with amplitude $P_{mr} = 0.5B_{m1}B_{m2}$, angular frequency $\omega_r = \omega_1 \pm \omega_2$, order $r = k \pm l$, and phase $(Ø_1 + Ø_2)$.

In three-phase inverter-fed motors, the parasitic torques are produced due to higher time harmonics in the stator winding currents [7]. The order of time harmonics (h) is $h = 2Nk \pm 1$ where N is number of stator phases, $k = 0,1,2,3, \ldots$. The electrical machines phase current in the Nth phase is given by (5.4) [8]:

$$i_N = \hat{I}\left[\sin\left(\omega t + (N-1)\frac{2\pi}{N}\right) + K_{I3}\sin 3\left(\omega t + (N-1)\frac{2\pi}{N}\right) + \cdots\right] \tag{5.4}$$

where $\hat{I}$ is peak stator phase current, ω is the angular frequency of supply, i is instantaneous current and K_{Ip} is peak value relative to fundamental of pth harmonic in the current waveform.

The expression of torque for three-phase PMSM is given by (5.5) [8, 9]:

$$T = 1.5\hat{I}\hat{B}mDLPK_{w1} \times \left[\begin{array}{l} 1 + \dfrac{K_3K_{w3}K_{I3}}{K_{w1}} + \dfrac{K_5K_{w5}K_{I5}}{K_{w1}} + \cdots + \\ \left(\begin{array}{l} K_{I7} - K_{I5} + \dfrac{K_3K_{w3}K_{I9}}{K_{w1}} - \dfrac{K_3K_{w3}K_{I3}}{K_{w1}} + \\ \dfrac{K_5K_{w5}}{K_{w1}} + \dfrac{K_7K_{w7}}{K_{w1}} + \dfrac{K_9K_{w9}K_{I3}}{K_{w1}} + \cdots \end{array}\right)\cos 6\omega t + \\ \left(\begin{array}{l} K_{I13} - K_{I11} + \dfrac{K_3K_{w3}K_{I9}}{K_{w1}} - \dfrac{K_5K_{w5}K_{I7}}{K_{w1}} - \\ \dfrac{K_7K_{w7}K_{I5}}{K_{w1}} - \dfrac{K_9K_{w9}K_{I3}}{K_{w1}} + \cdots \end{array}\right)\cos 12\omega t \end{array}\right] \tag{5.5}$$

where m is turns in series per pole per phase, D is stator bore diameter, K_{w1} is the first harmonic winding factor, P is the number of pole pairs, and L is the active axial length of machines. Torque ripple frequency will be 6, 12, 18, ... cycles per pole pair of rotor position.

As mentioned earlier and from (5.5), for a three-phase PMSM, the torque ripple frequency is equal to six times the fundamental supply frequency [9, 10]. Then the three-phase PMSM torque expression of (5.5) can be given in a very simplified manner which consists of an average component and harmonics of order multiple of six as given by (5.6):

$$T = T_0 + \sum_{n=1}^{\infty} T_{6n} \cos\left(n6\omega t + \varnothing_{6n}\right) \tag{5.6}$$

where T_0 is average torque and T_{6n} is the sixth order of harmonics.

Since the torque harmonics/ripples are a function of stator current time-harmonics/ripples which, in turn, contributes to the production of ANV in PMSM coupled with DC generator.

5.2.3 Aerodynamic Sources

Aerodynamic source of vibration founds in aerospace applications such as fluttering of wings of an aeroplane, rain wind-induced vibration in long-span electric cables, tall buildings, automobiles, etc. [11] Aerodynamic forces cause vibration in high-speed trains in pantograph, bogie, inter-coach spacings, etc., leads to noise generation, which limits the maximum speed of the train as well [12]. Also, around wind turbine blades, vortex-induced vibration and formation of galloping lead to damage of blades, especially when vibration frequency matches with the natural frequency of blades, hence acting as a criterion of design [13]. In the present chapter, electrical fluctuations are considered as the source of ANV, which is analyzed numerically and further validated with experimental results.

5.3 Recent Trends of Control Strategies for ANV Reduction

Transportation electrification has incrementally gained its potential market compared to conventional IC engines because of its performance and low expense and toxic emission but various sources of ANV like electromagnets, power fluctuations, motor drives, misaligned shafts, bearings, etc., reduce the overall functional life. Generally, mechanical and electrical control strategies are implemented for ANV reduction, which is a severe transportation electrification anomaly. In the mechanical control strategy, usually the reduction of vibration is done by implementing control at the site of vibration like shafts, turbine blades, gear trains, automobile suspension system, etc., and in the electrical control strategy, the control algorithms are implemented over a source of vibration like reduction in power supply fluctuation.

5.3.1 Control Aspects at the Site of Vibration (Mechanical)

Mechanical control strategies are basically classified into active and passive control [14]. In passive vibration control, no external source of power is required for vibration reduction, but in active control, external source of power is essential [15]. Recently, various ways of passive vibration control are implemented like tuned vibration absorbers, elastic support isolators, and tuned mass dampers [16, 17]. A tuned mass damper is used to shift the natural frequency of the system to below or above the prespecified value to avoid resonance. But because of their limitations like low-frequency band applicability and also, they lead to make the overall system bulky. Hence, they are replaced by active vibration control or feedback control strategies, which have very large bandwidth and application without affecting the basic structure, geometry, and dynamics of the system. The most commonly used active vibration control method is the use of a smart material like piezoelectric material that can act as a sensor as well as an actuator [18]. When sensing and actuation are done automatically in a close loop, the structure over which piezo patches are placed is called a smart structure [15]. Active vibration control founds its application in the marine propulsion system, wind turbines, earthquake secure buildings, large turbine and generator shaft, tuned dampers in vehicles, etc. [19]. Some recent advancements include semi-active control strategies like shape memory alloy, magnetic, and electric rheological dampers are also used in some specific applications including electrified vehicles [15].

5.3.2 Control Aspects at the Source of Vibration (Electrical)

In this section, the state of art related to various pulse width modulation (PWM) techniques and their effect on the ANV of electrified vehicles is illustrated. PWM-based voltage source inverters are the driving force in industrial and commercial applications. In the PWM technique, high-frequency acoustic noise is unavoidable. The proposed PTPWM and RPPM techniques are able to reduce this anomaly to a significant extent compared to conventional PWM with no requirement for additional circuits [20]. The DC link voltage and current harmonics are generated in SVPWM during the intrinsic switching process. In some real-world applications like medical, electrified vehicles, elevators etc., guidelines like IEC 60034-9 norm of low ANVs are to be strictly obeyed [21, 22]. Hence, ANV is to be addressed properly. Nowadays PWM-based three-phase voltage source inverters are widely used in PMSM drives, and advanced control methods based on PWM are also applied to improve the overall performance of PMSM drives [23]. However, the PWM in the PMSM drive coupled with the DC generator will generate the phase current harmonics/ripples nearby the carrier frequency and its multiples, which can result in high-frequency ANV [24]. For the reduction in the ANV of PMSM drives, new random pulse width modulation (RPWM) techniques have been developed, in which carrier periods, pulse position, and pulse width vary from one switching cycle to another [23–29]. Compared with the fixed switching-frequency PWM techniques, RPWM-based methods spread the power of noise over a wide range of frequency domains and the peaky PWM noise in phase voltage is reduced by about 9–10 dB [10, 30].

Random modulation techniques include a variation of carrier periods, pulse width, and position compared with the fixed switching-frequency PWM technique, which disperses noise over the varied frequency range and the peaky PWM noise in phase voltage is decreased [31]. ANV produced is influenced by the motor structure and the control strategy. Radial force in PMSM is analyzed in [32], and the structure of the permanent magnet, airgap length, and geometry of the stator core is optimized. Random PWM (RPWM) methods, including random pulse position (RPP) [33], random cantered distribution (RCD) [14], variable delay random (VDR) [31], random zero voltage vector distribution (RZD), and random switching frequency (RSF) PWM [33–34], have been discussed in detail and suitable for distribution of harmonics on the wide frequency spectrum. Among these conventional RPWM strategies, RSFPWM has a dominating effect on high-frequency harmonic diffusion as well as on vibration suppression [33–34]. The RSFPWM method for propulsion drive in a naval application is evaluated in [34]. In [33], five RPWM strategies, including RZD, RCD, RPP, RSF, and VDR, are compared, which shows that RSFPWM has a good performance and lower current total harmonic distortion (THD).

The random pulse width is the most prominent method to modulate the switching signal. The fundamental idea behind random PWM is to vary in a randomized manner one or more of the gating signal parameters. To usefully introduce them, the switching signal is modulated by varying one or more of its parameters; the switching frequency (T_k), the pulse width (α_k), or the pulse position (ε_k) are shown in Figure 5.3.

The comparison of different random PWM techniques is given in Figure 5.4 [35].

A schematic diagram of vector control of the PMSM drive system is shown in Figure 5.5. Summary of the possibilities of switching signals with respect to various parameters is given in Table 5.1.

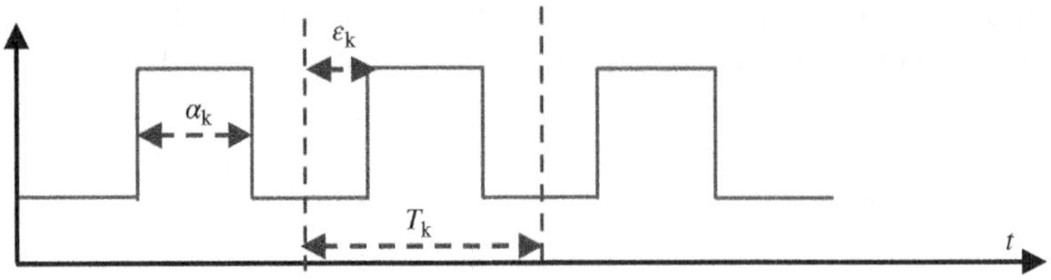

Figure 5.3 Gating pulses controlling the power-processing unit.

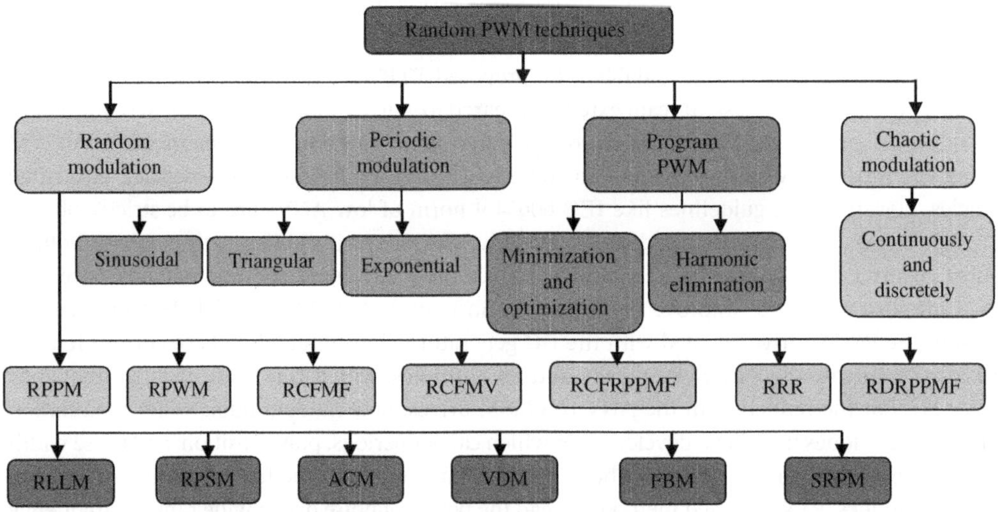

Figure 5.4 Random PWM techniques. *Source:* Adapted from Gamoudi et al. [35].

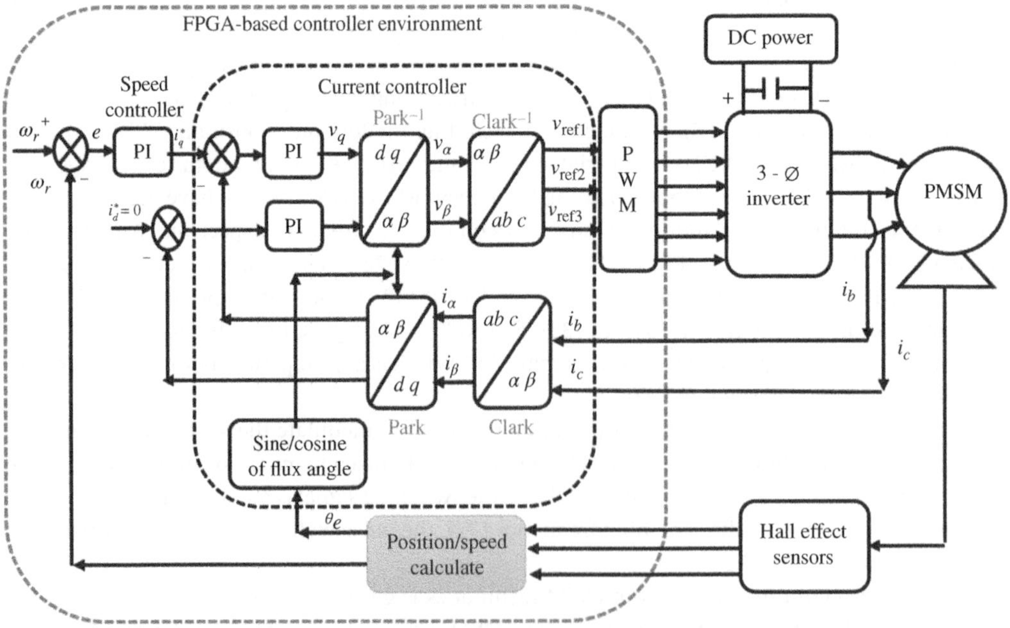

Figure 5.5 A schematic diagram of vector control of PMSM drive.

Table 5.1 Summary of the possibilities of switching signal parameters.

S. no.	Scheme	T_k	α_k	ε_k	$d_k = \frac{\alpha_k}{T_k}$	Sampling frequency	Average inductor current
1	PWM	Constant	Constant	Constant	Constant	Constant	Constant
2	RPPM	Constant	Constant	Random	Constant	Randomized	Randomized
3	PPWM	Constant	Random	Constant	Random	Constant	Constant
4	RCFMFD	Random	Random	Constant	Constant	Randomized	Randomized
5	RCFMVD	Random	Random	Constant	Random	Randomized	Randomized
6	RCFRPPMFD	Random	Random	Random	Constant	Constant	Randomized
7	RDRPPMFCF	Constant	Random	Random	Random	Randomized	Randomized
8	RRRM	Random	Random	Random	Random	Constant	Randomized

Source: Adapted from Lai et al. [36].

Table 5.2 Comparison of various RPWM techniques [37].

	Deterministic PWM		PTPWM		RPWM	
PWM techniques	Range	Average	Range	Average	Range	Average
Largest spectral component of current noise (μAdB) (10 kHz–150 kHz)	118–126	120.3	97–107	102.0	107–116	109.3
The largest spectral component of current noise (μAdB) (150 kHz–3 MHz)	108–111	108.6	91–94	92.3	91–98	93.8
Efficiency (%)	86–92	90	85–92	90	87–92	90
Torque ripple coefficient	0.01–0.15	0.051	0.04–0.18	0.065	0.04–0.19	0.065
Step response to current command (f_{sw} = 12 kHz)	Setting time = 3 ms (apx.), Overshoot = 20–30%					

Generation of random PWM signal depends on switching frequency, duty cycle, and sampling frequency.

A comparison of various RPWM techniques with respect to efficiency, time response, current harmonics, and torque ripple of machines is given in Table 5.2. From Table 5.2, it is observed that efficiency is more and torque ripple is less in the case of RPWM techniques compared to conventional PWM techniques.

5.4 Detailing of PMSM-Driven Experimental Setup

For experimental verification of modeling for vibration prediction corresponding to various control strategies mentioned in this chapter, a small-scale experimental setup is established inside an acoustic chamber. The detailed schematic representation of this experimental setup is shown in Figure 5.6.

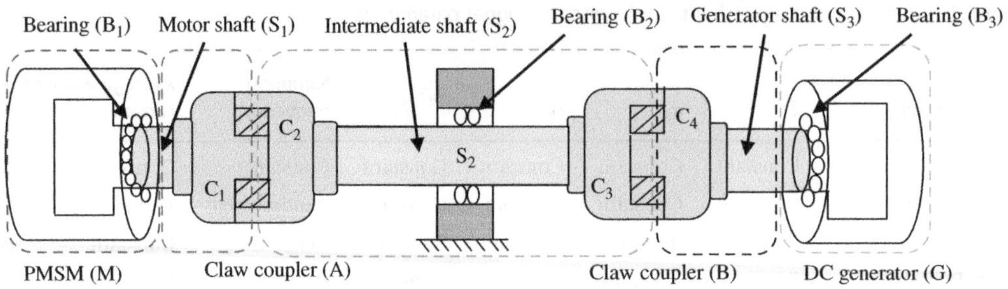

Figure 5.6 Detailed schematic representation of PMSM-driven coupled system.

As shown in Figure 5.6, the power is transmitted from PMSM drive (M) to a load DC generator (G) through a coaxially coupled transmission shaft from motor shaft (S_1) to generator shaft (S_3) through intermediate shaft (S_2). The C_1 disc of claw coupler (A) is pinned with the motor shaft (S_1) and the C_2 disc of claw coupler (A) is pinned with the intermediate shaft (S_2) to transmit power from S_1 to S_2. Similarly, the C_3 disc of the claw coupler (B) is pinned with the intermediate shaft (S_2) and the C_4 disc of the same coupler is pinned with generator shaft S_3 to transmit power from S_2 to S_3. B_1 bearing provides roller contact between motor hub (M) and motor shaft (S_1) and B_3 bearing provides roller contact between generator hub (G) and generator shaft (S_3), respectively.

The experimental tests are done on a 1.07-kW, 3000-rpm, and 4-poles PMSM drive to validate the effectiveness of the proposed control techniques. The system parameters of the PMSM drive coupled with DC generator are given in Table 5.3. A field programmable gate array (FPGA) is used to improve the system's performance. The clock frequency of FPGA is 20 MHz. The pictorial view of an experimental setup is shown in Figure 5.7 [10]. A low-cost acoustic chamber has been developed in the laboratory to investigate the ANV of a PMSM-driven system. For simulation studies, MATLAB/Simulink tool is employed. In the laboratory setup, PMSM is coupled with a DC

Table 5.3 Experimental specification of the PMSM drive.

Specification	Value	Unit
Rated power	1.07	kW
Rated speed	3000	rpm
No. of poles	04	
Rated torque	3.6	N-m
Rated current	6.29	A
Rated DC bus voltage	300	V
Stator resistance	3.07	Ω
Stator inductance	6.57	mH
Rotor inertia	1.4–1.8	kg-m^2
Spartan 3AN FPGA kit	20	MHz clock frequency
IGBT-based inverter stack	600, 30	V, A
Accelerometer sensitivity (PCB Piezotronics 352C03)	10	mV/g
Microphone sensitivity (1/2″ free-field) (NI USB 4432)	10	mV/Pa

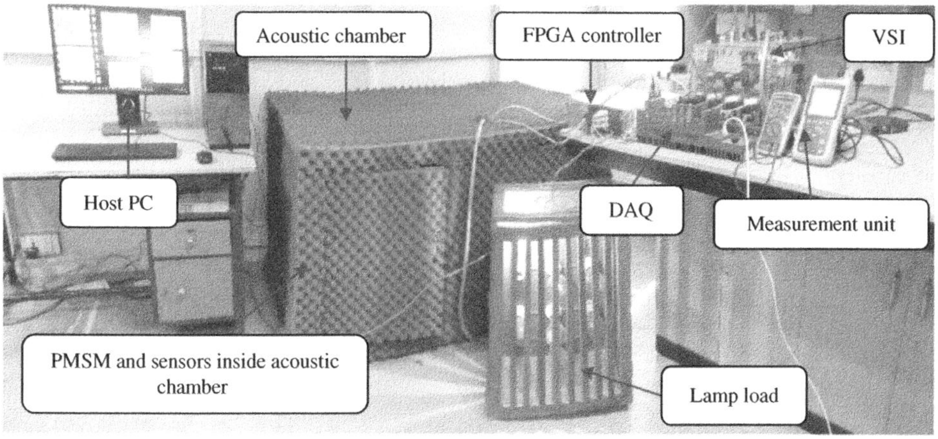

Figure 5.7 An experimental setup for investigation of acoustic noise and vibration of PMSM coupled with DC generator.

generator and fed from a three-phase Semikron converter. The data acquisition (DAQ) system for ANV measurement is from National Instruments (NI).

5.5 Methodology of Various Control Strategies and Their Implementation for ANV Reduction

5.5.1 Pseudorandom Triangular Pulse Width Modulation Technique (PTPWM)

The PTPWM technique is implemented by varying the switching frequency in a cycle. It is achieved by randomly changing the slope of a triangular carrier wave. The PTPWM technique replaces the reference triangular signal with a random signal (*triangular*) to generate a random-frequency PWM. With reference to any randomized PWM technique, power carried (Ψ_k) by the kth harmonic of voltage is given as the following function of a random variable (ε) [37],

$$\Psi_k \propto \left| E\left\{ \sum\nolimits_{\delta=1}^{\varepsilon} \overline{a}_\delta \frac{\sin\left(\pi \frac{k}{\delta} \overline{a}_\delta\right)}{\left(\pi \frac{k}{\varepsilon} \overline{a}_\delta\right)} \cdot \mathrm{e}^{-j2\pi k/\delta\varepsilon} \right\} \right|^2 \tag{5.7}$$

where $E\{\cdot\}$ is statistical expectation and $\overline{a}_\delta$ is the duty ratio of the switching cycle δth switching interval. For random variable ε, if one of H equally possible integer ε continued within the ε_1 to ε_k range, then (5.7) can be written as

$$\Psi_k \propto \left| \frac{1}{H} \sum\nolimits_{i=1}^{H} \sum\nolimits_{\delta_i}^{\varepsilon_i} \overline{a}_{\delta i} \frac{\sin\left(\pi \frac{k}{\varepsilon} \overline{a}_{\delta i}\right)}{\left(\pi \frac{k}{\varepsilon} \overline{a}_{\delta i}\right)} \cdot \mathrm{e}^{-j2\pi \frac{k}{\varepsilon_i} \delta_i} \right|^2 \tag{5.8}$$

In the case of sinusoidal PWM (SPWM), statistical expectation can be removed from (5.7) and, hence, power spectrum is discrete with a separate sine-function envelope for each distinct pulse width. Suppose ε is considered a random variable, then from (5.8), when ε_i increases, the harmonic power in the signal is shifted toward a higher-frequency range. The power of each harmonic for any δ_ith periodic pulse is inversely proportional to ε_i due to the narrowing of pulses in increasingly smaller intervals.

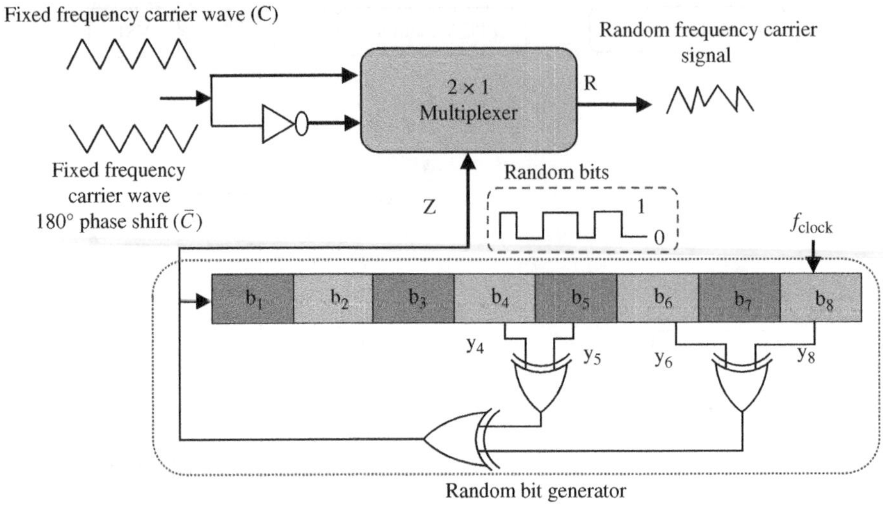

Figure 5.8 Block diagram of the proposed PTPWM technique.

The block diagram of the PTPWM technique is shown in Figure 5.8. Random bit generator as shown in Figure 5.8 consists of EXCLUSIVE OR (XOR) gates and linear shift registers that help in generating pseudorandom bit signals. A clock is used to give a switching signal to shift register with f_{clock} switching frequency. Eight shift registers are connected in cascaded form where the output of one shift register is an input to the second. The output of 4-, 5-, 6-, and 8-bit shift registers combine to be fed as an input to the XOR gates. The respective outputs of the XOR gates are further given as an input to the 2×1 multiplexer given in (5.9) [38]:

$$y_4 \oplus y_5 \oplus y_6 \oplus y_8 \tag{5.9}$$

where $\oplus$ is XOR operator.

As shown in Figure 5.8, the triangular carriers with fixed frequency *(C)* and the triangular carriers with the fixed frequency with opposite phase $(\overline{C})$ are input to the 2 × 1 multiplexer. Then *C* and $(\overline{C})$ are randomly selected by the output random bits (0) or (1) of the random bit generator. Choice of *(C)* and $(\overline{C})$ is dependent on the output *(Z)* of the pseudorandom binary sequence random bits generator. In case *(Z)* is (1), then *(R)* is selected as *(C),* and if *(Z)* is (0), then *(Z)* is selected as $(\overline{C})$.

Figure 5.9 shows the waveforms of the proposed method as given in Figure 5.8. The output *(Z)* of the PRBS random bits generator is similar to the conventional random lead–lag PWM. The random lead–lag PWM is the early version among random PWM, and its pulses are randomly placed between the first position and the last position in the modulation interval.

An intermediate waveform for the PTPWM technique is shown in Figure 5.9. Two carrier waves with 180° phase shift to each other as given in Figure 5.9a,b are processed in the multiplexer with the random bits produced by the random bit-generator as given in Figure 5.9c. For a time period of (0.1210–0.1225 seconds), the carrier wave is selected according to the random bit and accordingly the multiplexer gives a pseudorandom triangular carrier frequency signal as shown in Figure 5.9d.

5.5.2 Random Pulse Position Pulse Width Modulation Technique (RPPM)

The block diagram of the RPPM technique is shown in Figure 5.10. In the RPPM technique, the pulses of switching signals are randomly placed in individual switching intervals. The simplest

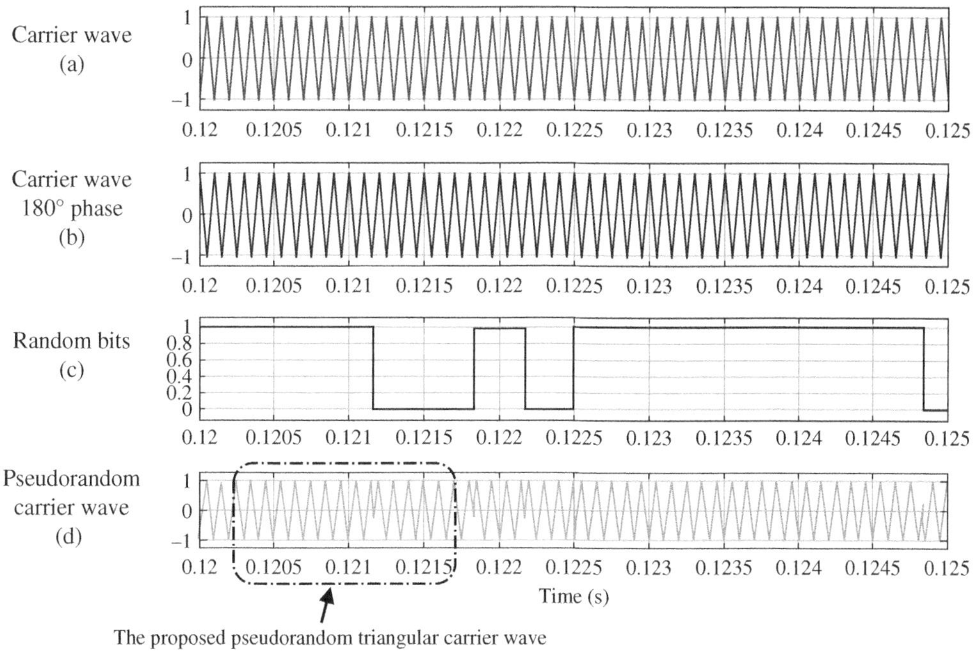

Figure 5.9 Intermediate waveforms for the PTPWM technique.

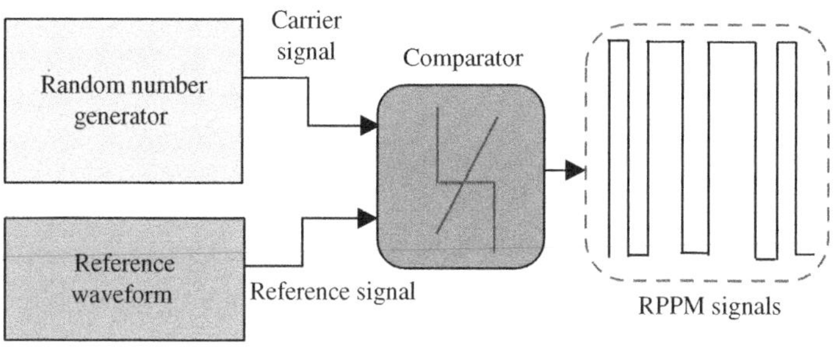

Figure 5.10 Block diagram of the generation of RPPM signals in MATLAB/Simulink tool.

approach is generally used to place a random pulse position either at the beginning or at the end of the switching interval. The randomly generated fractional numbers, n, having uniform probability distribution, are compared with the desired duty ratios of the switching signals for individual phases of the inverter.

A schematic layout of a generation of FPGA-based RPPM signal for a three-phase inverter is shown in Figure 5.11. A generated RPWM, by varying the value of the output frequency (f^*) (maximum frequency range 20 kHz) and modulation index (m^*) (the maximum modulation index) is 0.8. Intermediate waveforms for a generation of RPPM signals for a three-phase inverter are shown in Figure 5.12. A random bit generator generates a random number as shown in Figure 5.12a and those random numbers are compared with a carrier wave as shown in Figure 5.12b. Finally, generated RPPM signals based on a comparison between random number and carrier signals are shown in Figure 5.12c.

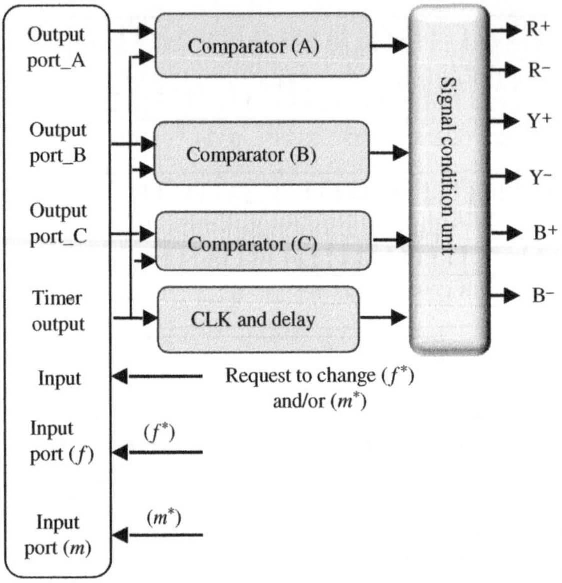

Figure 5.11 Schematic layout of an FPGA-based generated RPPM signals for a three-phase inverter.

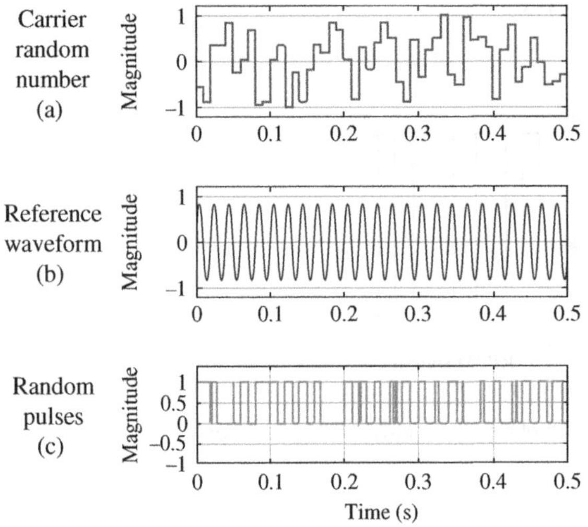

Figure 5.12 Intermediate waveforms for a generation of RPPM signals for a three-phase inverter.

5.6 Analysis of Torsional Vibration Response at Resonance

The torque output at the PMSM shaft or motor shaft (S_1) corresponding to various control strategies as mentioned in Section 5.5 is shown in Figure 5.13.

All the torque waveforms corresponding to each control strategy can be written in the form as shown in (5.10).

$$T_i = \left(T_{\text{avg}}\right)_i + (T_a)_i \sin(314 \times t) \tag{5.10}$$

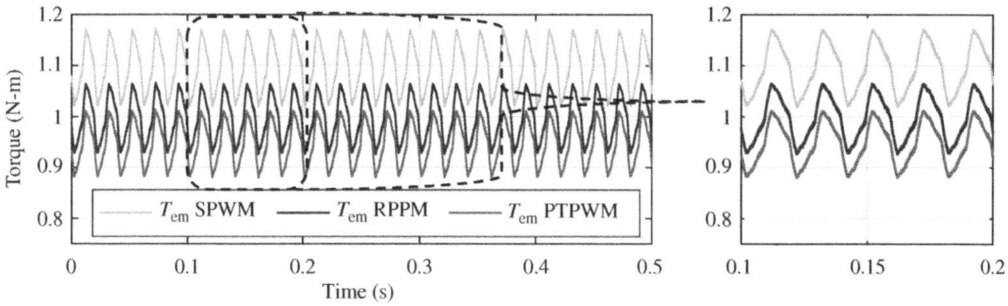

Figure 5.13 Steady-state torque ripple at motor shaft S_1 corresponding to various control strategies.

where $i = 1,2,3$ corresponds to SPWM, PTPWM, and RPPM control strategies, respectively. $(T_{avg})_i$ is the average or mean value of torque, and $(T_a)_i$ is the amplitude of torque waveform. All the findings for (5.10) from Figure 5.13 are shown in Table 5.4.

A control strategy with high average torque and low ripple is considered to be the most effective. Torque ripple magnitude directly influences torsional vibrations. Therefore, ripples are selection criteria for control strategy. This torsional vibration leads to cyclic loading over the shaft and torsional shear stress is induced and maybe stress becomes greater than permissible stress and causes shaft failure [39, 40]. Therefore, it is very important to do a torsional vibration analysis of the system. For mathematical modeling of torsional vibration considering computational and time complexity, lumped modeling with multi-degrees of freedom is the most appropriate [41–44]. For the selection of optimum degrees of freedom especially in a continuous mass distribution system as in Figure 5.6, MPF is integrated with a lumped model to ensure accuracy with computational complexity. The vibration response at resonance is actually a free vibration response of the system, which can be validated easily by frequency response at resonance. Free vibration response helps us identify whether geometrical and material properties are accurately modeled and also helps identify the natural frequency, to avoid situation of resonance. Also, in the modal analysis approach, that is going to be implemented in this chapter, the output of free vibration response like orthonormalized mode shape is useful for dynamic vibration response calculation corresponding to each stated control strategy. The various steps involved in mathematical modeling of analysis of torsional vibration are discussed in the following sections.

Table 5.4 Parameters of steady-state torque for various control strategies.

	Various control strategies		
Parameters	**SPWM**	**PTPWM**	**RPPM**
T_{avg} (N-m)	1.06	0.9984	1.0075
T_{max} (N-m)	1.177	1.064	1.072
T_{min} (N-m)	0.943	0.9328	0.943
$T_a = \frac{(T_{max} - T_{min})}{2}$ (N-m)	0.117	0.0656	0.0645

5.7 Implementation of MPF Accuracy Enhancement Technique in Lumped Model for Number of Modes or DoF Selection

5.7.1 Mathematical Modeling of Torsional Vibration Equation for All Lumped Elements

This technique states that the contribution of any mode to the vibration response is indicated by corresponding effective mass, which usually decreases for higher modes. So, only those modes or only that number of DoFs are to be considered up to which effective mass is 95% of actual mass [45, 46]. For effective mass calculation, orthonormalized mode shapes are required corresponding to the initially assumed DoF. So, initially, five DoFs are assumed and analyzed in the following steps, whether the guess number of DoF is optimum or not. For this, PMSM as the first element, S_1 shaft along with disc C_1 of claw coupler A as the 2nd element, C_2-S_2-C_3 as the 3rd, C_4-S_3 as 4th, and generator as 5th element, respectively [47]. The 5 DoF lumped model is shown in Figure 5.14.

In the lumped model, any structure or component is replaced by a disc representing lumped mass of equivalent polar moment of inertia (PMoI), equivalent stiffness and damping coefficient (Table 5.5) [44].

The modeling presented in this chapter is in such a way that it can be used for any number of DoF varying as per power train present in EVs and HEVs. So, a free body diagram of the generalized jth element along with the various possible torques acting over it is shown in Figure 5.15.

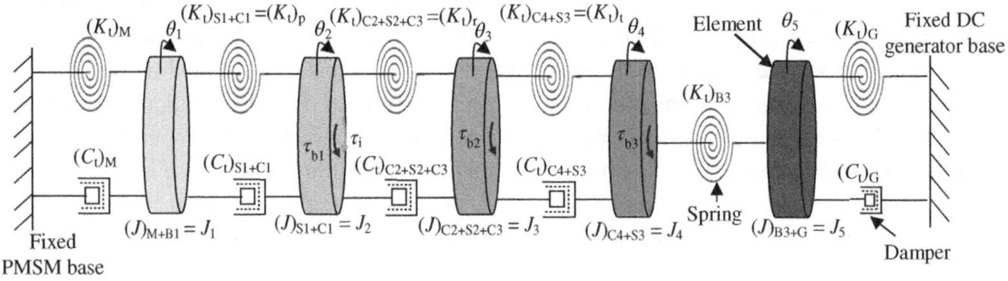

Figure 5.14 Five DoF lumped model for torsional vibration analysis.

Table 5.5 Nomenclature used in lumped modeling representation of experimental setup.

Symbols	Meaning
θ_1 to θ_5	Angular displacement or DoF for each lumped mass
J_1 to J_5	Polar moment of inertia (PMoI) for each lumped mass
$(K_t)_M$ and $(C_t)_M$	Equivalent stiffness and torsional damping coefficient for PMSM
$(K_t)_{S1+C1}$ ($(K_t)_p$) and $(C_t)_{S1+C1}$ ($(C_t)_q$)	Equivalent stiffness and torsional damping coefficient of second element
$(K_t)_{C2+S2+C3}$ ($(K_t)_r$) and $(C_t)_{C2+S2+C3}$ ($(C_t)_s$)	Equivalent stiffness and torsional damping coefficient of third element
$(K_t)_{C4+S3}$ ($(K_t)_t$) and $(C_t)_{C4+S3}$ ($(C_t)_u$)	Equivalent stiffness and torsional damping coefficient of fourth element
$(K_t)_{B3}$	Equivalent stiffness bearing of generator
$(K_t)_G$ and $(C_t)_G$	Equivalent stiffness and torsional damping coefficient of generator
τ_i	Torque corresponding to SPWM, PTPWM, and RPPM techniques
τ_{b1}, τ_{b2}, and τ_{b3}	Resisting torque offered by bearings

Using the D'Alembert principle, the generalized equation of motion for the *j*th element is given as (5.12).

$$J_j\ddot{\theta}_j + (K_t)_j(\theta_j - \theta_{j-1}) + (C_t)_j(\dot{\theta}_j - \dot{\theta}_{j-1}) - (K_t)_{j+1}(\theta_{j+1} - \theta_j) - (C_t)_{j+1}(\dot{\theta}_{j+1} - \dot{\theta}_j) = \tau_j - \tau_{bj} \tag{5.11}$$

This generalized equation can be used for any number of lumped elements using Table 5.6, which is generated using the lumped model shown in Figure 5.14.

Using Table 5.7 in (5.11) for the present setup of 5 DoF, the torsional vibration equation in matrix form corresponding to each lumped element considering the forced damped system is shown in (5.12).

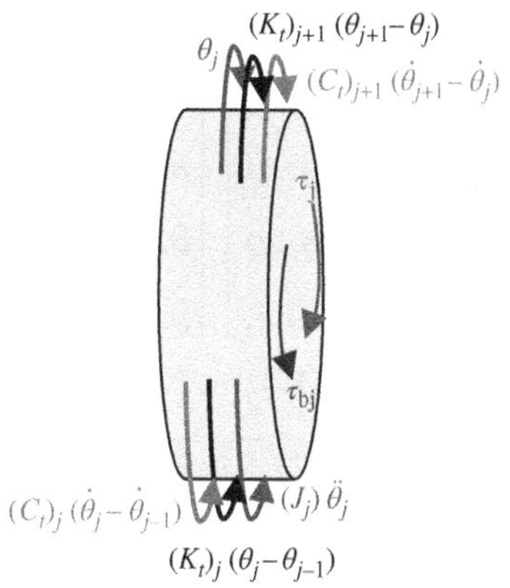

Figure 5.15 Generalized lumped model.

Table 5.6 Nomenclature of generalized element.

Symbols	
θ_j, θ_{j-1}	Angular displacement of the *j*th and (*j*–1)th elements
J_j	PMoI of the *j*th element
K_{tj}, $(K_t)_{j+1}$	Torsional stiffness of the *j*th and (*j*+1)th elements
C_{tj}, $(C_t)_{j+1}$	Damping coefficient of the *j*th and (*j*+1)th elements
τ_j	Torque at the *j*th element
τ_{bj}	Resisting torque by bearing at the *j*th element
$(C_t)_j\ (\dot{\theta}_j - \dot{\theta}_{j-1})$	Damping torque over the *j*th element by the (*j*–1)th element
$(K_t)_j\ (\theta_j - \theta_{j-1})$	Stiffness torque over the *j*th element by the (*j*–1)th element
$(C_t)_{j+1}\ (\dot{\theta}_{j+1} - \dot{\theta}_j)$	Damping torque over the *j*th element by the (*j*+1)th element
and $(K_t)_{j+1}\ (\theta_{j+1} - \theta_j)$	Stiffness torque over the *j*th element by the (*j*+1)th element
$J_j\ \ddot{\theta}_j$	Inertia force over the *j*th element

Table 5.7 Nomenclature used in lumped modeling representation of experimental setup.

Element number	1	2	3	4	5
J_j	J_1	J_2	J_3	J_4	J_5
θ_j	θ_1	θ_2	θ_3	θ_4	θ_5
$(K_t)_j$	$(K_t)_M$	$(K_t)_p$	$(K_t)_r$	$(K_t)_u$	$(K_t)_{b3}$
θ_{j-1}	0	θ_1	θ_2	θ_3	θ_4
$(C_t)_j$	$(C_t)_M$	$(C_t)_q$	$(C_t)_s$	$(C_t)_v$	0
$(K_t)_{j+1}$	$(K_t)_p$	$(K_t)_r$	$(K_t)_u$	$(K_t)_{b3}$	$(K_t)_G$
θ_{j+1}	θ_2	θ_3	θ_4	θ_5	0
$(C_t)_{j+1}$	$(C_t)_q$	$(C_t)_s$	$(C_t)_v$	0	$(C_t)_G$
τ_j	0	τ	0	0	0
τ_{bj}	0	τ_{b1}	τ_{b2}	τ_{b3}	0

$$\begin{bmatrix} J_1 & 0 & 0 & 0 & 0 \\ 0 & J_2 & 0 & 0 & 0 \\ 0 & 0 & J_3 & 0 & 0 \\ 0 & 0 & 0 & J_4 & 0 \\ 0 & 0 & 0 & 0 & J_5 \end{bmatrix} \begin{bmatrix} \ddot{\theta}_1 \\ \ddot{\theta}_2 \\ \ddot{\theta}_3 \\ \ddot{\theta}_4 \\ \ddot{\theta}_5 \end{bmatrix} + \begin{bmatrix} (C_t)_M+(C_t)_q & -(C_t)_q & 0 & 0 & 0 \\ -(C_t)_q & (C_t)_q+(C_t)_s & -(C_t)_s & 0 & 0 \\ 0 & -(C_t)_s & (C_t)_s+(C_t)_v & -(C_t)_v & 0 \\ 0 & 0 & -(C_t)_v & (C_t)_v & 0 \\ 0 & 0 & 0 & 0 & (C_t)_G \end{bmatrix} \begin{bmatrix} \dot{\theta}_1 \\ \dot{\theta}_2 \\ \dot{\theta}_3 \\ \dot{\theta}_4 \\ \dot{\theta}_5 \end{bmatrix}$$
$$+ \begin{bmatrix} (K_t)_M+(K_t)_p & -(K_t)_p & 0 & 0 & 0 \\ -(K_t)_p & \left((K_t)_p+(K_t)_r\right) & -(K_t)_r & 0 & 0 \\ 0 & -(K_t)_r & \left((K_t)_r+(K_t)_u\right) & -(K_t)_u & 0 \\ 0 & 0 & -(K_t)_u & \left((K_t)_u+(K_t)_{b3}\right) & -(K_t)_{b3} \\ 0 & 0 & 0 & -(K_t)_{b3} & \left((K_t)_{b3}+(K_t)_G\right) \end{bmatrix} \begin{bmatrix} \theta_1 \\ \theta_2 \\ \theta_3 \\ \theta_4 \\ \theta_5 \end{bmatrix} = \begin{bmatrix} 0 \\ \tau-\tau_{b1} \\ -\tau_{b2} \\ -\tau_{b3} \\ 0 \end{bmatrix} \tag{5.12}$$

5.7.2 Calculation of Parameters Required in Resonance Response of Torsional Vibration

For resonance response, the system is considered to be undamped and under free vibration. So, (5.12) can be rewritten as (5.13):

$$[J]_{5\times 5}\{\ddot{\theta}\}_{5\times 1} + [K_t]_{5\times 5}\{\theta\}_{5\times 1} = \{0\} \tag{5.13}$$

where $[J]$ is PMoI matrix and $[K_t]$ is the torsional stiffness matrix. Various parameters required in (5.13) for resonance response of torsional vibration are PMoI (J_j) and torsional stiffness ($(K_t)_j$) corresponding to all lumped elements. For motor and generator, these parameters are taken from the manufacturer catalog [48] shown in Table 5.8.

For remaining elements 2, 3, and 4, the torsional stiffness is obtained by spring series formulation, i.e. inverse addition of stiffness of each stepped part in each element and similarly equivalent PMoI is obtained using the addition of PMoI of each stepped part in each element. The geometry and corresponding detailed calculation of PMoI and equivalent stiffness for element 2, 3, and 4 is given in Tables 5.9–5.11, respectively (Figures 5.16–5.18).

5.7.3 Natural Frequency, Mode Shape, and Orthonormalization of Modes

Using (5.13), multiplying with $[J]^{-1}$, we get,

$$[I]\{\ddot{\theta}_0\} + [J]^{-1}[K_t]\{\theta_0\} = \{0\} \tag{5.14}$$

Table 5.8 PMSM and generator parameter value.

Element number	Inertia (J)	Inertia (kg-m^2)	Stiffness (K_t)	Stiffness (N-m/rad)
1 (PMSM)	J_1	1.6	$(K_t)_M$	0.0434
5 (DC generator)	J_5	2.4	$(K_t)_G$	0.642

Table 5.9 Calculation of PMOI and equivalent stiffness of lumped element 2.

Parameters	Values (m) × 10^{-3}	Mass (kg) × 10^{-3}	J (kg-m^2) × 10^{-7}	$(K_t)_p$ N-m/rad
D11, L11	22, 10	22.56	17.88	4.36×10^5
D12, L12	49, 13	190.63	572.11	
D13, L13	54, 12	213.7	778.95	
L14, H14, W14	12, 20, 54	100.8	278.59	
$J_2 = \sum J = 1647.53 \times 10^{-7}$				

Table 5.10 Calculation of PMOI and equivalent stiffness of lumped element 3.

Parameters	Values (m) × 10^{-3}	Mass (kg) × 10^{-3}	J (kg-m^2) × 10^{-7}	$(K_t)_p$ N-m/rad
L21, H21, W21 (cavity)	10, 40, 54	−168.048	−632.42	0.146×10^5
D22, L22	54, 23	409.60	1492.99	
D23, L23	49, 12	175.963	528.10	
L24, D24	60, 19	132.28	59.69	
L25, D25	20, 57	396.85	1611.70	
L26, D26	28, 65	722.99	3815.65	
L27, H27, W27	18, 40, 65	−364.104	−1767.4	
$J_3 = \sum J = 5108.29 \times 10^{-7}$				

Table 5.11 Calculation of PMOI and equivalent stiffness of lumped element 4.

Parameters	Values (m) × 10^{-3}	Mass (kg) × 10^{-3}	J (kg-m^2) × 10^{-7}	$(K_t)_p$ N-m/rad
L31, H31, W31	15, 20, 65	151.71	584.71	2.56×10^5
D32, L32	65, 13	335.44	1771.54	
D33, L33	57, 20	396.85	1611.70	
L34, D34	50, 38	440.94	795.89	
$J_4 = \sum J = 4763.84 \times 10^{-7}$				

Replacing $[J]^{-1}[K_t] = [D]$, where $[D]$ is dynamic matrix. Initially assumed harmonic solution for (5.14) with ω_n as frequency, (5.14) can be re-written as:

$$[D] - \lambda[I])\{\theta_0\} = \{0\} \tag{5.15}$$

where $\lambda = \omega_n{}^2$. Using values from Tables 5.9–5.11 in (5.15), all-natural frequencies are obtained and listed in Table 5.12.

Natural frequency gives an indication of the number of significant modes. Only those modes are significant until natural frequency does not have abrupt incremental change. From Table 5.12, it is evident that only two modes are significant, which was further validated using MPF method.

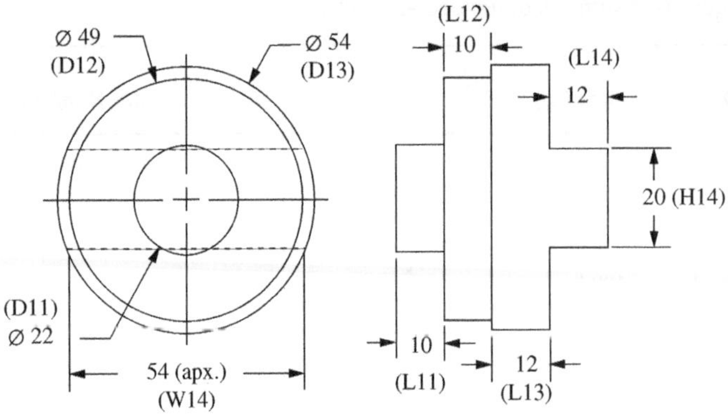

Figure 5.16 Geometry of element 2 (S_1-C_1).

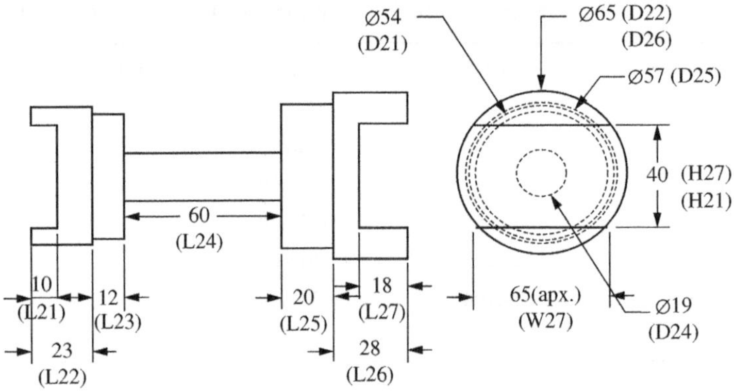

Figure 5.17 Geometry of element 3 (C_2-S_2-C_3).

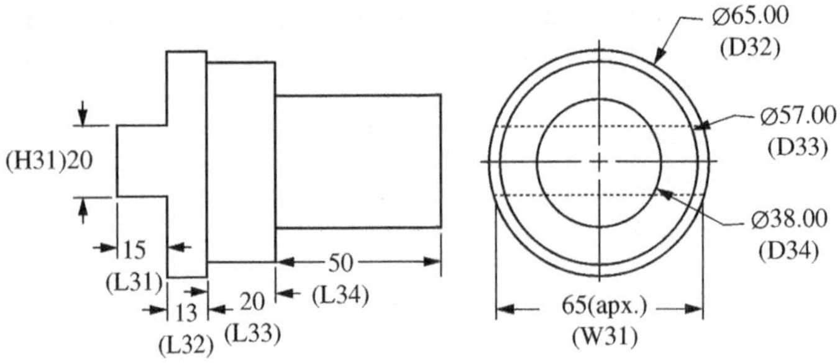

Figure 5.18 Geometry of element 4 (C_4-S_3).

Table 5.12 Natural frequencies for a small-scale experimental setup.

Mode number	1	2	3	4	5
Natural frequency (Hz)	23.78	34.6	3.133×10^3	7.855×10^3	8.33×10^3

Mode shapes are relative amplitude of vibration at the resonance of different lumped elements. Putting these natural frequencies from Table 5.12 in (5.15), mode shape (θ_{0j}) can be obtained. For plotting, mode shapes are displacement normalized (θ_{Ni}), i.e. reduced to scale 1 and shown in Figure 5.19. This is done by dividing with the maximum term in the same mode shape.

Further orthonormalization of mode shape is done. Orthonormalized modes are required in finding out a significant number of modes and also orthonormalization leads to making coupled differential equations uncoupled, which represents the dynamic response of torsional vibration that can be easily solvable. For this, a modal mass is multiplied corresponding to each mode shape given by (5.17).

$$A_j = \sqrt{\frac{1}{\left[\theta_{Nj}\right]^{T}[J]\left[\theta_{Nj}\right]}} \tag{5.16}$$

So, modal mass is multiplied with each mode shape as in (5.17)

$$\{\theta_{ONj}\} = A_i \times \{\theta_{Nj}\} \tag{5.17}$$

Hence, orthonormalized modal matrix

$$[\theta_{ON}] = [\theta_{ON1}, \theta_{ON2}, \theta_{ON3}, \theta_{ON4}, \theta_{ON5}] \tag{5.18}$$

5.7.4 Calculation of Computationally Optimum Number of Lumped Elements

5.7.4.1 Calculation of Coefficient Vector {L}

$$\{L\} = [\theta_{ON}]^{T}[J]\{r\} \tag{5.19}$$

where $\{r\} = [1; 1; 1; 1; 1]$, whose element represents the displacement of each DoF when a unit displacement is given as input to the system and while the system is assumed to be rigid. So, in this case, the whole system gets displaced by 1 unit. Hence, all terms are 1.

5.7.4.2 Calculation of Model Participation Factor (MPF)

The MPF for the jth mode is given as

$$\mathrm{MPF}_j = \frac{L_{j1}}{\left(J_g\right)_{jj}} \tag{5.20}$$

For orthonormalized mode, the generalized PMoI matrix is the identity matrix. So, $(J_g)_{jj} = 1$.

5.7.4.3 Calculation of Effective Mass

The effective modal mass for jth mode ($(J_{\mathrm{eff}})_j$) is given as

$$\left(J_{\mathrm{eff}}\right)_j = \frac{\mathrm{MPF}_j^2}{\left(J_g\right)_{jj}} \tag{5.21}$$

The total PMOI of the system is given as (5.22)

$$J_{\mathrm{actual}} = \sum_{J=1}^{5} J_J = J_{11} + J_{22} + J_{33} + J_{44} + J_{55} \tag{5.22}$$

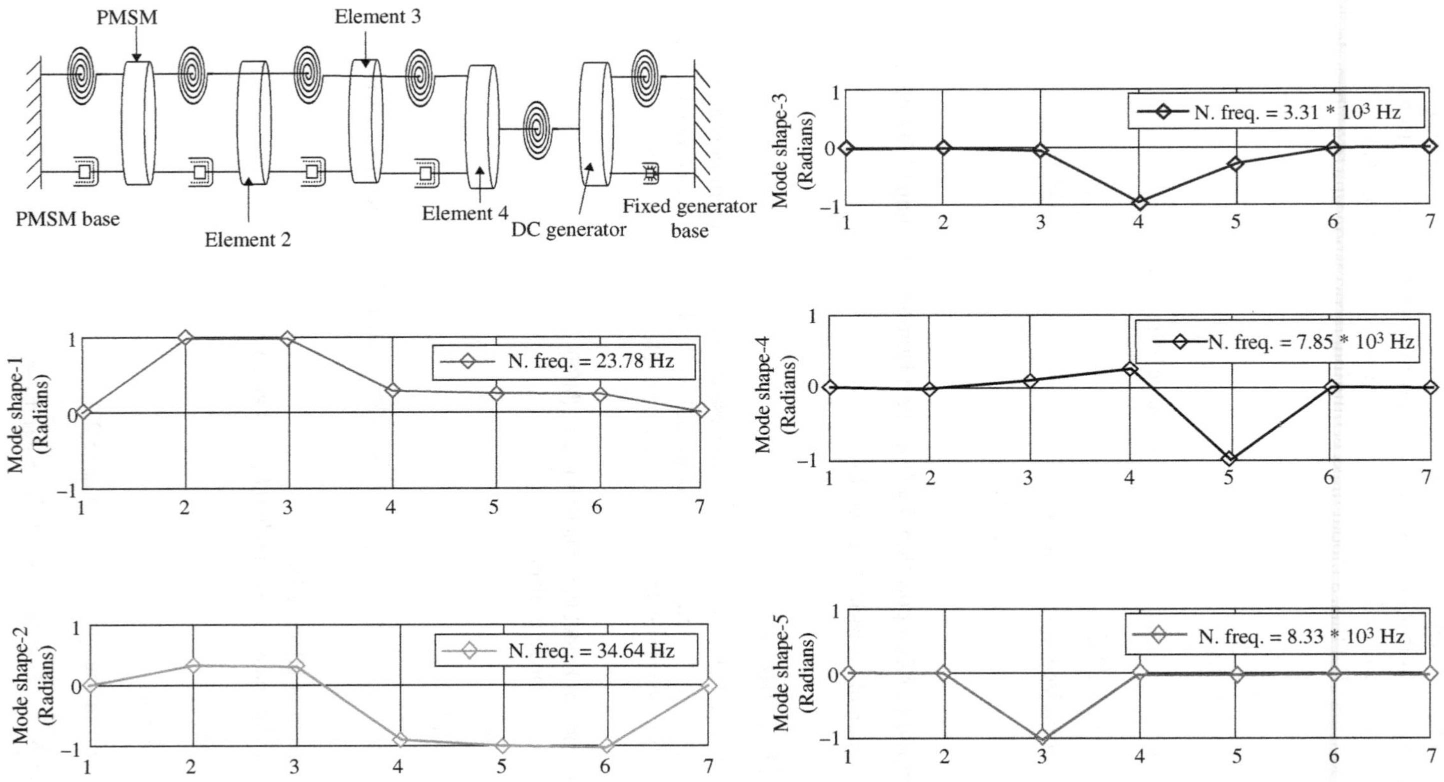

Figure 5.19 Displacement-normalized mode shape corresponding to all-natural frequencies.

Table 5.13 Percentage of effective mass corresponding to the number of modes under consideration.

Mode no.	Effective PMoI	% Effective PMoI
1	2.71	67.4
2	3.9	97
3	4.01	99.8
4	4.015	99.9
5	4.018	100

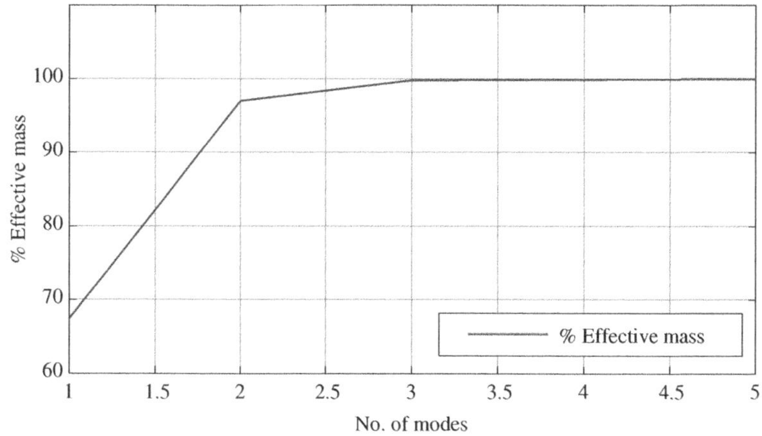

Figure 5.20 Graphical representation of % effective mass vs no. of mode to be considered.

The percentage of effective mass for the *i*th mode or their contribution in total response is given by (5.23)

$$\%(J_{\text{eff}})_i = \frac{\sum_{i=0}^{i} (J_{\text{eff}})_i}{J_{\text{actual}}} \times 100 \tag{5.23}$$

For the present setup, the percentage of effective mass corresponding to the mode number is given in Table 5.13 and Figure 5.20.

The number of modes are considered up to which effective mass is 95%. So, significant modes are only 2 and for dynamic vibration analysis, 2.5 times the significant number of modes are considered. Hence, for the present setup, 5 DoFs are optimum.

5.8 Extended Mathematical Modeling for the Effectiveness of Control Strategies Over Torsional Vibration Reduction

In this section, the simulation study of the vibration of each DoF corresponding to each control strategy is presented. As modes are orthonormalized, so, they follow the condition as in (5.24)

$$\left[\theta_{ONj}\right]^{T}[J]\,[\theta_{ONi}] = 0 \text{ and } \left[\theta_{ONj}\right]^{T}[K]\,[\theta_{ONi}] = 0 \tag{5.24}$$

Now,

$$[J]\{\ddot{\theta}\} + [C_t]\{\dot{\theta}\} + [K_t]\{\theta\} = \{\tau\} \tag{5.25}$$

By using principal coordinate $q_i(t)$, where $\{\theta\} = [\theta_{ON}]\,\{q\}$, these five coupled differential equations become uncoupled. By putting it in (5.25) and multiplying with $[\theta_{ON}]^T$, (5.25) can be rewritten as (5.26)

$$[\theta_{ON}]^T\left\{[J][\theta_{ON}]\{\ddot{q}\} + [\theta_{ON}]^T[C_t][\theta_{ON}]\{\ddot{q}\} + [\theta_{UN}]^T[K_t][\theta_{ON}]\{q\}\right\} - [\theta_{ON}]^T\{\tau\}_i \tag{5.26}$$

Also, modal matrix is inertia normalized. So, $[\theta_{ON}]^T\,[J]\,[\theta_{ON}]$ becomes an identity matrix called generalized inertia matrix ($[J_g]$) and $[\theta_{ON}]^T\,[K]\,[\theta_{ON}]$ becomes a diagonal matrix, whose diagonal element is $\omega_{nj}^{\;2}$ and called a generalized stiffness matrix ($[K_g]$) where ω_{nj} is natural frequency. $[\theta_{ON}]^T\,[C_t]\,[\theta_{ON}]$ is the generalized damping matrix ($[C_g]$). $[\theta_{ON}]^T\{\tau\}_i$ is the generalized torque corresponding to various control strategies.

5.8.1 Calculation of Generalized Damping Matrix ($[C_g]$)

Using Rayleigh damping and the property of orthonormal modes [45, 46], $[C_g]$ is given as (5.27)

$$\left[C_g\right] = \text{diag}(2 \times \mathfrak{z}_i \times \omega_{ni}) \tag{5.27}$$

where $\mathfrak{z}_i$ is the modal damping ratio corresponding to each mode. Its value increases for higher modes as they have less effective mass and cause a low critical damping value. Usually, the constant value of damping ratio leads to error in results, especially for higher modes. Here, the methodology opted for the calculation of modal damping ratio is taken from references [45, 46], considering carbon steel (Grade 304) as a material whose modal damping ratio varies from 0.001 to 0.002 for significant modes. In [45, 46], basically four methods are illustrated, i.e. linear interpolation method, second method considering only significant modes, third method considering full-length modes, and the last one is average of significant and full-length mode method. The overall findings from these methods are shown in Table 5.14.

From the graphical representation in Figure 5.21, it seems that the significant mode method gives a close result to linear interpolation only up to a significant mode and the full-length mode method gives a close result to linear interpolation after significant mode. By [45, 46], that strategy is opted, which gives a close result with linear interpolation up to significant modes. So, finally, the modal damping ratio value corresponding to the significant mode method is used in this chapter.

Table 5.14 Variation of modal damping ratio corresponding to mode number.

Mode no.	Linear interpolation	Significant mode method	Full-length mode	Average method
1	0.001	0.001	0.0009	0.0009
2	0.002	0.002	0.0023	0.0022
3	0.289	0.226	0.289	0.257
4	0.725	0.565	0.725	0.645
5	0.769	0.599	0.769	0.684

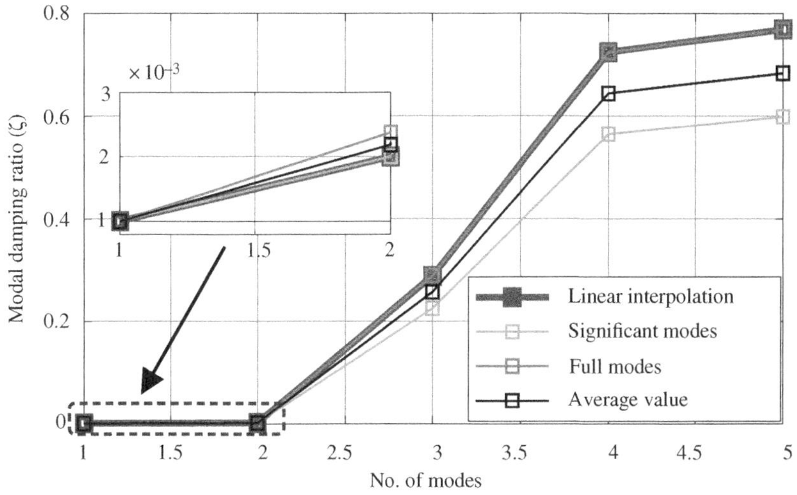

Figure 5.21 Graphical representation of relative error in various methods of approximation.

5.8.2 Calculation of Generalized Torque Corresponding to Each Control Strategy

The generalized torque needs two inputs, one is resisting torque offered by bearing and the other is torque corresponding to various control strategies. The resisting torque by bearing is given by (5.28) taken from the bearing manufacturer [49] and tabulated in Table 5.15.

$$\tau_b = 0.5 \times 0.0015 \times \text{radial force (N)} \times \text{bearing bore diameter (m)} \tag{5.28}$$

And torque corresponding to various control strategies is taken from Table 5.4 and (5.10). All these values are used in (5.26) and uncoupled equations of the form (5.29) are obtained, representing the dynamic torsional vibration response of the system corresponding to various control strategies.

$$\ddot{q}_i + C_i\dot{q}_i + D_i q_i = E_{ij} + B_{ij}\sin\omega_i t \tag{5.29}$$

whose solutions are given as (5.30)

$$q_{ij}(t) = \frac{E_{ij}}{D_i} + \frac{B_{ij}/D_i \times \sin(\omega_i t - \emptyset_i)}{\sqrt{(1 - r_i^2)^2 + (2\xi_i r_i)^2}} \tag{5.30}$$

where j varies from 1 : 3, representing control strategy and i from 1 : 5 representing DoF under consideration.

Table 5.15 Resisting torque offered by bearing.

S. no.	Name of bearing	Radial force (N)	Bore diameter (m)	Resisting torque (τ_{bj}) (N-m)
1	B_1	13.37081	22×10^{-3}	2.265×10^{-4}
2	B_2	1.25	19×10^{-3}	1.7281×10^{-5}
3	B_3	16.06351	38×10^{-3}	4.5781×10^{-5}

5.9 Results and Discussion

5.9.1 Validation of Torsional Vibration Response at Resonance

For resonance response validation, an accelerometer is attached over PMSM and DC generator and the motor is driven up to the rating speed and then stopped and allow to come to a standstill condition. An accelerometer at PMSM and DC generator measures the peak of vibration when motor speed matches with the resonance frequency of PMSM and DC generator, respectively. The result shown in Figure 5.22 reveals the natural frequency of PMSM as 24 Hz and for DC generator 36.5 Hz, which are in close vicinity of simulation result shown in Table 5.11, hence validated and can be used for further simulation of dynamic response.

5.9.2 Analysis of Dynamic Response Corresponding to Various Control Strategies

Using (5.30), the dynamic response of torsional vibration under various control strategies is shown in Figure 5.23 and various findings are shown in Tables 5.16–5.20.

5.9.3 Simulation Results of SPWM, RPPM, and PTPWM Techniques for PMSM Drive

Three modulation techniques are implemented to observe and analyze speed response, torque ripple, and stator current of PMSM drive coupled with DC generator. The SPWM, RPPM, and PTPWM techniques are implemented for investigation and analysis of reduction of torque ripple and ANV of PMSM drive coupled with DC generator. The magnitude of ANV directly depends on the

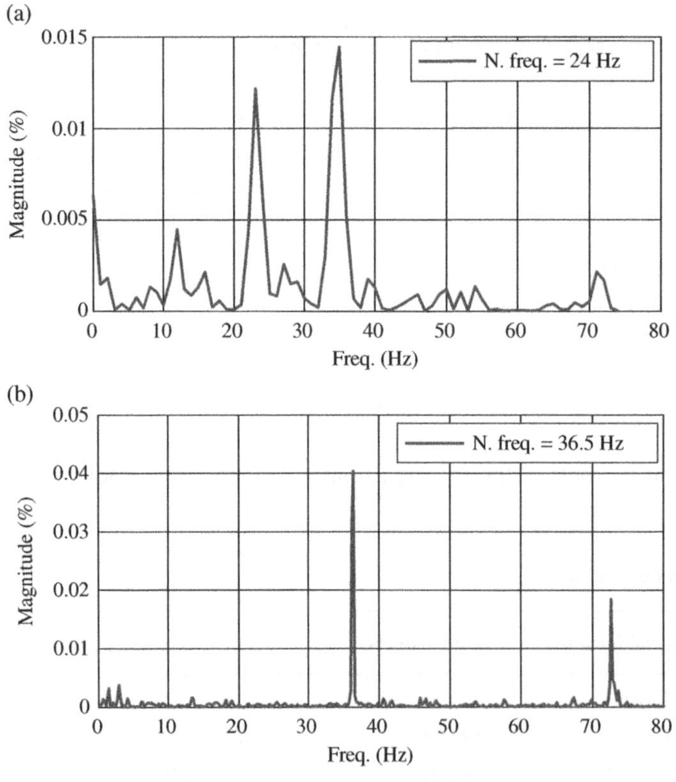

Figure 5.22 Experimental result of the natural frequency response: (a) PMSM drive and (b) DC generator.

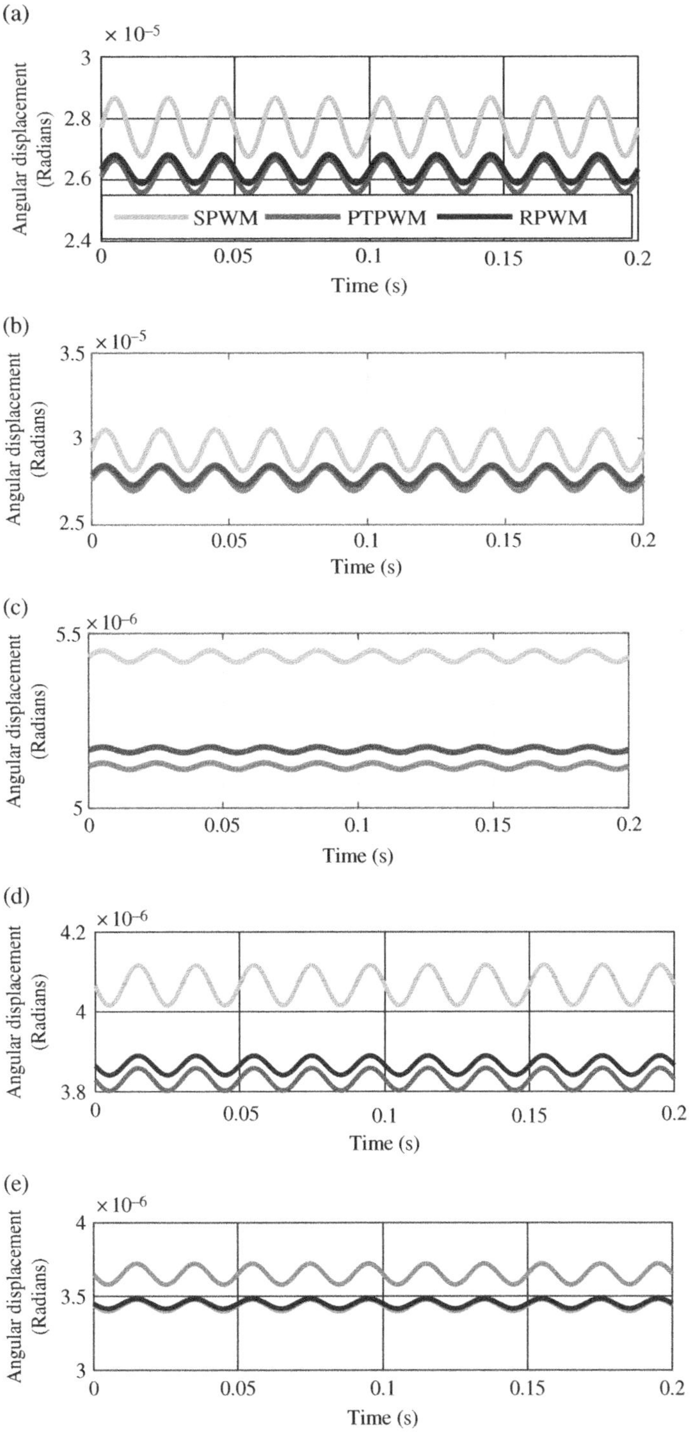

Figure 5.23 Dynamic response of torsional vibration at (a) PMSM, (b) element 2, (c) element 3, (d) element 4, (e) DC generator.

Table 5.16 Dynamic torsional vibration response of PMSM.

Control strategy	θ_{avg}	θ_a
SPWM	2.7×10^{-5}	9.6×10^{-7}
PTPWM	2.61×10^{-5}	5.39×10^{-7}
RPPM	2.63×10^{-5}	4.64×10^{-7}

Table 5.17 Dynamic torsional vibration response of element 2.

Control technique	θ_{avg}	θ_a
SPWM	2.93×10^{-5}	1.19×10^{-6}
PTPWM	2.76×10^{-5}	6.67×10^{-7}
RPPM	2.78×10^{-5}	5.75×10^{-7}

Table 5.18 Dynamic torsional vibration response of element 3.

Control technique	θ_{avg}	θ_a
SPWM	5.43×10^{-6}	1.66×10^{-8}
PTPWM	5.12×10^{-6}	9.31×10^{-9}
RPPM	5.17×10^{-6}	8.02×10^{-9}

Table 5.19 Dynamic torsional vibration response of element 4.

Control technique	θ_{avg}	θ_a
SPWM	4.07×10^{-6}	5.07×10^{-8}
PTPWM	3.83×10^{-6}	2.83×10^{-8}
RPPM	3.86×10^{-6}	2.45×10^{-8}

Table 5.20 Dynamic torsional vibration response of generator.

Control technique	θ_{avg}	θ_a
SPWM	3.64×10^{-6}	7.14×10^{-8}
PTPWM	3.43×10^{-6}	3.98×10^{-8}
RPPM	3.46×10^{-6}	3.43×10^{-8}

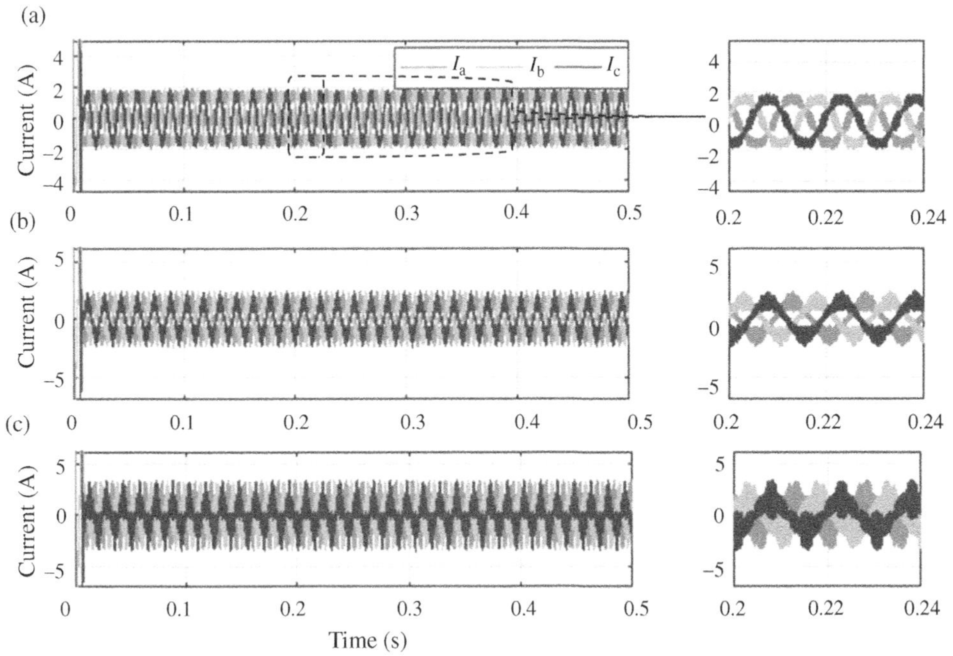

Figure 5.24 Simulation results of PMSM drive: (a) three-phase stator current response by PTPWM technique, (b) three-phase stator current response by RPPM technique, and (c) three-phase stator current response by SPWM technique.

magnitude of the torque ripples. Therefore, the analysis of torque ripples for the different modulation techniques has been done in MATLAB/Simulink software. Simulations of PMSM drive coupled with DC generator are carried out for drive speed of 1000 rpm and mechanical loading T_m of 1 N-m and results are shown in Figure 5.24.

Three-phase stator current of converter-fed PMSM drive coupled with DC generator operated by PTPWM, RPPM, and SPWM techniques is shown in Figure 5.24a–c, respectively. The effects of the different modulation techniques on the current ripples are shown in Figure 5.24a–c. The current ripples produced by PTPWM, RPPM, and SPWM are 2.5%, 4.8%, and 9.3%, respectively. It means current ripples produce higher in SPWM compared to PTPWM and RPPM techniques.

The steady-state speed response of the PMSM drive coupled with the DC generator for a speed reference of 1000 rpm is shown in Figure 5.25a. The steady-state speed response of SPWM, RPPM, and PTPWM techniques for PMSM drive coupled with DC generator is shown in Figure 5.25a. Torque response for PTPWM and RPPM techniques exhibits excellent performance with almost 38.5% and 41.7% reduction in torque ripples as compared to the SPWM technique, as shown in Figure 5.25b. Reduction in torque ripple response is evidently related to the reduction in stator current harmonics. Torque ripple reduction will, in turn, lead to a reduction in vibration and acoustic noise. Table 5.21 presents numeric comparative results for torque ripples.

5.9.4 Experimental Results of SPWM, RPPM, and PTPWM Techniques for PMSM Drive

Experimental results for the operation of PMSM drive coupled with DC generator for SPWM, RPPM, and PTPWM techniques are shown in Figures 5.26–5.28.

Three-phase stator current of converter-fed PMSM drive coupled with DC generator operated by PTPWM, RPPM, and SPWM techniques is shown in Figure 5.26a–c, respectively. The effects of the

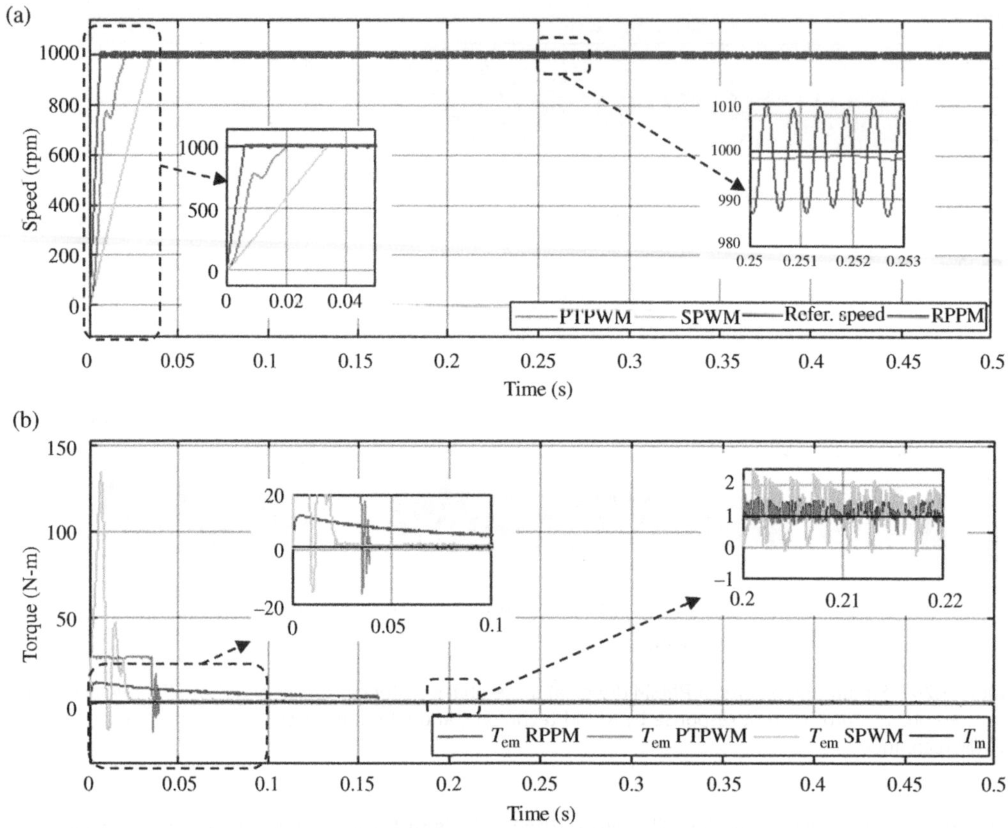

Figure 5.25 Simulation results of PMSM drive: (a) steady-state speed response and (b) steady-state torque response.

Table 5.21 Comparison of torque ripples of PMSM drive for SPWM and PTPWM.

Control strategy	SPWM technique	PTPWM technique	RPPM technique
T_{max} (N-m)	2.01	1.8	1.571
T_{min} (N-m)	-0.1	0.9	1
T_{avg} (N-m)	0.95	1.35	1.285
T_{ripple} (%)	1.97	0.66	0.44

different modulation techniques on the current ripples are shown in Figure 5.26a–c. The current ripples produced by PTPWM, RPPM, and SPWM are 3%, 5%, and 8%, respectively. It means current ripples are produced higher in SPWM compared to PTPWM and RPPM techniques.

Experiments are carried out for reference drive speed of 1000 rpm and mechanical loading T_m of 1 N-m. The steady-state speed response of SPWM, RPPM, and PTPWM techniques for PMSM drive coupled with a DC generator is shown in Figure 5.27. The speed controller forces the motor to follow the reference speed. Speedy responses for the three techniques are well behaved with comparable performance under steady-state conditions. Torque response is measured using an in-line

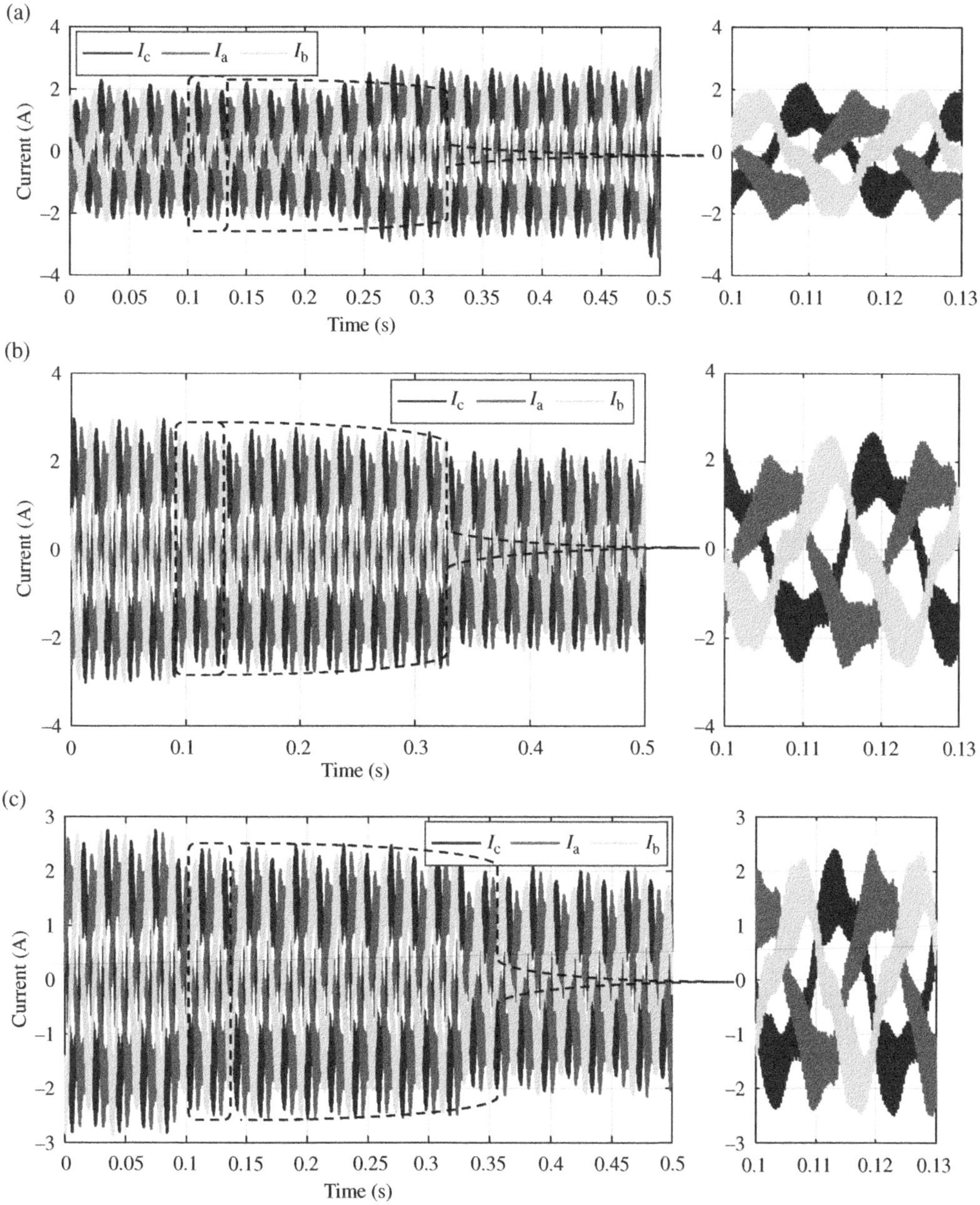

Figure 5.26 Experimental results of PMSM drive coupled with DC generator: (a) three-phase stator current by PTPWM technique, (b) three-phase stator current by RPPM technique, and (c) three-phase stator current by SPWM technique.

torque sensor (torque constant = 0.49 N-m, range 0–5 N-m for 12,000 rpm, make FUTEK), which is attached to the shaft of PMSM and DC generator (used as load).

Experimentally obtained ANV of SPWM, RPPM, and PTPWM techniques for PMSM drive coupled with DC generator is shown in Figure 5.28. The vibration of the PMSM drive was measured using an accelerometer sensor with an output sensitivity of 10 mV/g. An acoustic noise of the PMSM drive coupled with a DC generator was measured using a microphone with a sensitivity

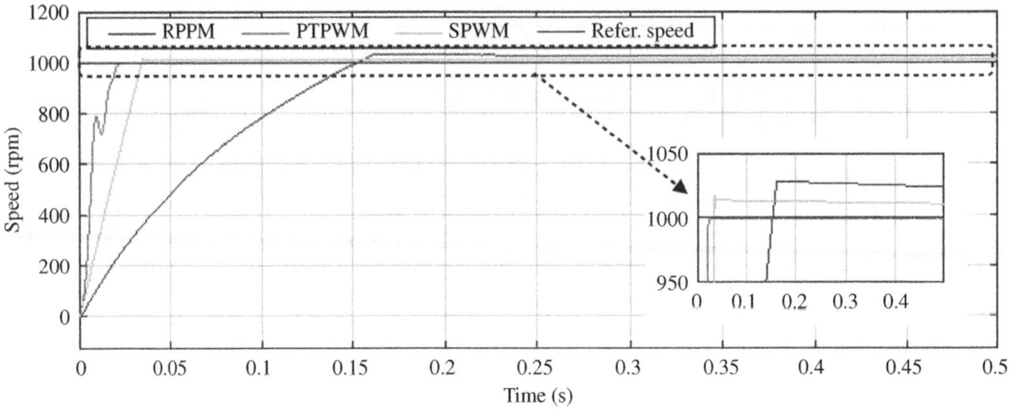

Figure 5.27 Steady-state speed response of PMSM drive coupled with DC generator.

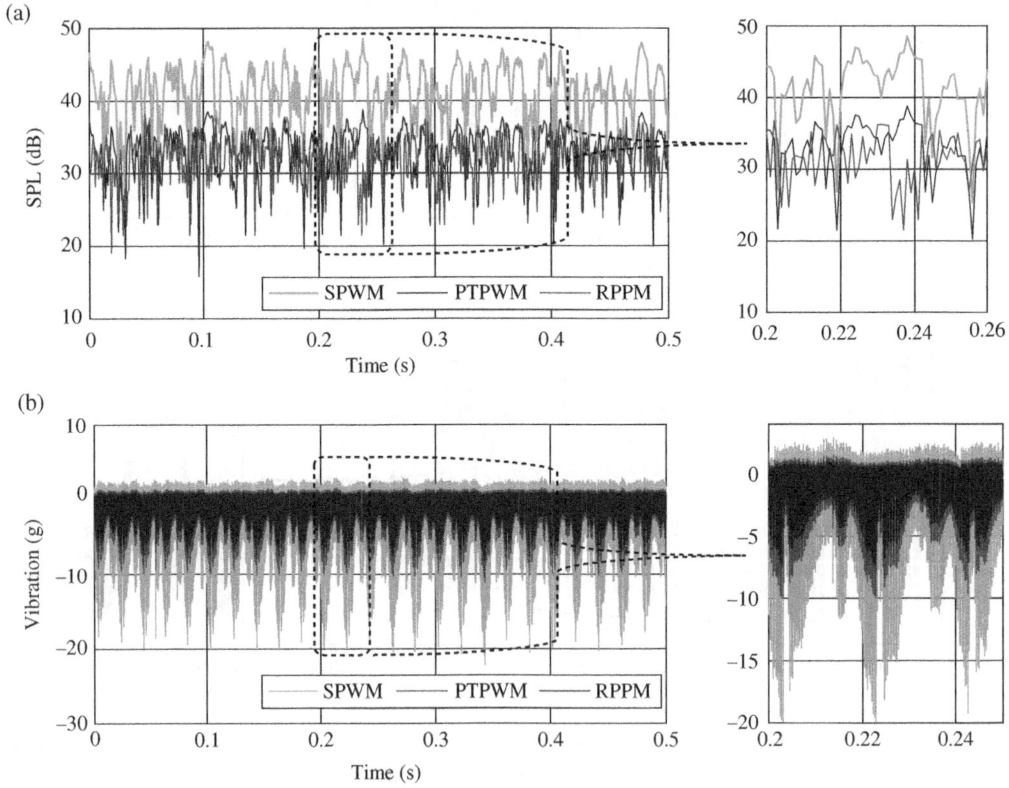

Figure 5.28 Experimental results of acoustic noise and vibration during steady-state speed response of PMSM drive: (a) Time domain spectrum of sound and (b) Time domain spectrum of vibration.

of 10 mV/Pa. To measure the acoustic noise of the PMSM drive coupled with the DC generator, a half-inch free-field microphone provided by National Instruments is used. The sampling rate of the sound and vibration data recorded through DAQ is 50,000. Once data of acoustic noise is recorded in Pascal by microphone, the data is converted into decibel (dB) using $10\log_{10}\frac{P}{P_{\mathrm{ref}}}$ where P is measured data of acoustic noise in Pascal and P_{ref} is reference sound pressure in Pascal ($P_{\mathrm{ref}} = 20\,\mu\mathrm{Pa}$).

The sound spectrum of the PMSM drive coupled with the DC generator is measured in free-field environments, i.e. inside the acoustic chamber. Time-domain sound spectrum of SPWM, RPPM, and PTPWM techniques for PMSM drive coupled with the DC generator (during steady-state speed response) is shown in Figure 5.28a. The maximum amplitude of the sound spectrum is observed at around 48, 37, and 36 dB for SPWM, PTPWM, and RPPM techniques, respectively as shown in Figure 5.28a. That means acoustic noise survived is around 10% and 12% by using PTPWM and RPPM as compared to SPWM techniques, respectively.

The maximum amplitude of vibration is observed around 1.8, 0.9, and 0.81 gravitation for SPWM, PTPWM, and RPPM techniques, respectively, as shown in Figure 5.28b. PMSM drive coupled with the DC generator exhibits twice the magnitude of vibration for SPWM as compared to PTPWM and RPPM techniques for PMSM drive.

Comparison of acoustic noise of PMSM drive coupled with the DC generator (in percentage) with SPWM, RPPM, and PTPWM techniques are given in Table 5.22. Substantial reduction of acoustic noise is observed with PTPWM and RPPM techniques. For applied load torque of 1 N-m, the reduction in acoustic noise is found to be 22.92% and 35.26% less in PTPWM and RPPM techniques as compared to SPWM, respectively.

A comparison of vibration of PMSM drive coupled with DC generator with SPWM, RPPM, and PTPWM techniques is given in Table 5.23. A substantial reduction of vibration of PMSM drive is observed with PTPWM and RPPM techniques. For applied load torque of 1 N-m, the reduction in vibration of PMSM drive is found to be 40.12% and 42.85% less in PTPWM and RPPM techniques as compared to SPWM, respectively.

The PTPWM and RPPM techniques prove to be effective for the substantial reduction of torque ripples and in-turn ANV in the PMSM drive coupled with a DC generator. However, both techniques, i.e. PTPWM and RPPM may pose a few limitations such as challenges in device loss calculation, and thermal and loop control design, and, hence, thermal subsystems cannot be optimally designed.

Table 5.22 Comparison of acoustic noise of the PMSM drive.

		SPL (dB)				
S. no.	T_m (N-m)	SPWM technique	PTPWM technique	RPPM technique	% reduction of acoustic noise by PTPWM	% reduction of acoustic noise by RPPM
1	1	48	37	0.9	22.92	35.26
2	1.5	52	42.23	1.6	18.79	30.43
3	2	59.5	48.45	1.9	18.57	26.08

Table 5.23 Comparison of vibration of the PMSM drive.

		Vibration (g)				
S. no.	T_m (N-m)	SPWM technique	PTPWM technique	RPPM technique	% reduction of vibration by PTPWM	% reduction of vibration by RPPM
1	1	1.8	0.9	0.8	40.12	42.85
2	1.5	2.8	1.2	1.4	53.4	36.36
3	2	3.2	1.7	1.8	46.2	33.33

5.10 Conclusions and Future Scope

This chapter presented power electronics modulation techniques such as PTPWM and RPPM for PMSM drive coupled with a DC generator to reduce the overall ANV. The well-known conventional sinusoidal pulse width modulation (SPWM) technique was compared with PTPWM and RPPM to check the effect on the ANV of the PMSM drive coupled with the DC generator. A small-scale experimental setup is developed to validate modeling and control strategies experimentally. Further, lumped parametric modeling along with MPF-based torsional vibration analysis is presented, which is experimentally validated over resonance and further extended mathematical modeling for dynamic response of torsional vibration under various control strategies is also presented. It is obtained that the RPPM technique is the most effective in ANV reduction followed by PTPWM and SPWM. In the present chapter, the electrical way of controlling vibration is presented. The present vibration analysis will act as a base for mechanical ways of controlling vibration like active vibration control using smart material. The present analysis will help predict the condition of catastrophic failure of components like resonance and will provide guidelines or suggestions to mitigate those circumstances, hence enhancing the system reliability to a significant extent.

References

1 Gieras, J.F., Wang, C., and Lai, J.C.S. (2006). *Noise of Polyphase Electrical Motors*. Taylor & Francis Group.

2 Prasad, S.R. and Sekhar, A.S. (2018). Life estimation of shafts using vibration-based fatigue analysis. *J. Mech. Sci. Technol.* 32: 4071–4078.

3 Han, H., Lee, K., and Park, S. (2015). Estimate of the fatigue life of the propulsion shaft from torsional vibration measurement and the linear damage summation law in ships. *Ocean Eng.* 107: 212–221.

4 Chen, Y., Liang, S., Li, W. et al. (2019). Faults and diagnosis methods of permanent magnet synchronous motors: a review. *Appl. Sci.* 9 (10): 2116.

5 Vijayraghavan, P. and Krishnan, R. (1999). Noise in electric machines: a review. *IEEE Trans. Ind. Appl.* 35 (5): 1007–1013.

6 Chan, C.C. (2007). The state of the art of electric, hybrid, and fuel cell vehicles. *Proc. IEEE* 95 (4): 704–718.

7 Zhu, Z.Q., Howe, D., and Ackermann, B. (1992). *Electrical Machines & Power System*, 661–678. Springer.

8 Le-Huy, H., Perret, R., and Feuillet, R. (1986). Minimization of torque ripple in brushless DC motor drives. *IEEE Trans. Ind. Appl.* IA-22 (4): 748–755.

9 Bolton, H.R. and Ashen, R.A. (1984). Influence of motor design and feed-current waveform on torque ripple in brushless DC drives. *IEE Proc. B Electr. Power Appl.* 131 (3): 82–90.

10 Pindoriya, R.M., Gautam, G., and Rajpurohit, B.S. (2020). A novel application of pseudorandom-based technique for acoustic noise and vibration reduction of PMSM drive. *IEEE Trans. Ind. Appl.* 56 (5): 5511–5522.

11 Li, H., Chen, W.L., Xu, F. et al. (2010). A numerical and experimental hybrid approach for the investigation of aerodynamic forces on stay cables suffering from rain-wind induced vibration. *J. Fluids Struct.* 26 (7–8): 1195–1215.

12 Zhang, Y., Zhang, J., Li, T. et al. (2016). Research on aerodynamic noise reduction for high-speed trains. *Shock Vibr.* 2016, Article ID 6031893: 1–21.

13 Lupi, F., Niemann, H.J., and Hoffer, R. (2018). Aerodynamic damping model in vortex-induced vibrations for wind engineering applications. *J. Wind Eng. Ind. Aerodyn.* 174: 281–295.

14 Vasques, C.M.A. and Rodrigues, J.D. (2006). Active vibration control of smart piezoelectric beams: comparison of classical and optimal feedback control strategies. *Comput. Struct.* 84 (22–23): 1402–1414.

15 Kandasamy, R., Cui, F., Townsend, N. et al. (2016). A review of vibration control methods for marine offshore structures. *Ocean Eng.* 127: 279–297.

16 Liu, Z., Wanyou, L., and Ouyang, H. (2016). Structural modifications for torsional vibration control of shafting systems based on torsional receptance. *Shock Vibr.* 2016: 1–8.

17 Denhartog, J.P. (2013). *Mechanical Vibrations*. MA, USA: *Courier Corporation*.

18 Shivashankar, P. and Gopalakrishnan, S. (2020). Review on the use of piezoelectric materials for active vibration, noise, and flow control. *Smart Mater. Struct.* 29 (5): 053001.

19 Kumar, T., Kumar, R., and Jain, S.C. (2020). Finite element analysis of an actively controlled heavy rotor using a PVDF piezoelectric layer as a sensor and actuator. *Eng. Rep.* 2 (10): 1–17.

20 Huang, Y., Xu, Y., Zhang, W., and Zou, J. (2019). Hybrid RPWM technique based on modified SVPWM to reduce the PWM acoustic noise. *IEEE Trans. Power Electron.* 34 (6): 5667–5674.

21 Lee, K., Shen, G., Yao, W., and Lu, Z. (2017). Performance characterization of random pulse width modulation algorithms in industrial and commercial adjustable-speed drives. *IEEE Trans. Ind. Appl.* 53 (2): 1078–1087.

22 Ruiz-Gonzalez, A., Vargas-Merino, F., Heredia-Larrubia, J.R. et al. (2013). Application of slope PWM strategies to reduce acoustic noise radiated by inverter-fed induction motors. *IEEE Trans. Ind. Electron.* 60 (7): 2555–2563.

23 Zhang, W., Xu, Y., Huang, H., and Zou, J. (2020). Vibration reduction for dual-branch three-phase permanent magnet synchronous motor with carrier phase-shift technique. *IEEE Trans. Power Electron.* 35 (1): 607–618.

24 Qu, J., Zhang, C., Jatskevich, J., and Zhanga, S. (2021). Deadbeat harmonic current control of permanent magnet synchronous machine drives for torque ripple reduction. *IEEE J. Emerg. Sel. Top. Power Electr.* 10 (3): 3357–3370.

25 Qu, J., Jatskevich, J., Zhang, C., and Zhang, S. (2021). Torque ripple reduction method for permanent magnet synchronous machine drives with novel harmonic current control. *IEEE Trans. Energy Convers.* 36 (3): 2502–2513.

26 Fang, Y. and Zhang, T. (2018). Sound quality of the acoustic noise radiated by PVM-fed electric powertrain. *IEEE Trans. Ind. Electron.* 65 (6): 4534–4541.

27 Zou, J., Lan, H., Xu, Y., and Zhao, B. (2017). Analysis of global and local force harmonics and their effects on vibration in permanent magnet synchronous machines. *IEEE Trans. Energy Convers.* 32 (4): 1523–1532.

28 Kang, B.J. and Liaw, C.M. (2001). Random hysteresis PWM inverter with robust spectrum shaping. *IEEE Trans. Aerosp. Electron. Syst.* 37 (2): 619–629.

29 Nami, A. and Zare, F. (2015). A comparison between random hysteresis current control technique with bipolar and unipolar modulations. *Aust. J. Electr. Electron. Eng.* 4 (1): 1–8.

30 Pindoriya, R.M., Yadav, A.K., Rajpurohit, B.S., and Kumar, R. (2020). A novel application of random hysteresis current control technique for acoustic noise and vibration reduction of PMSM drive. *2020 IEEE Industry Applications Society Annual Meeting*, Detroit, MI, USA, pp. 1–8.

31 Zhang, W., Xu, Y., Huang, H., and Zou, J. (2020). Vibration reduction for dual-branch three-phase permanent magnet synchronous motor with carrier phase-shift technique. *IEEE Trans. Power Electron.* 35 (1): 607–618.

32 Xu, J. and Zhang, H. (2020). Random asymmetric carrier PWM method for PMSM vibration reduction. *IEEE Access* 8: 109411–109420.

33 Lee, K., Shen, G., Yao, W., and Lu, Z. (2017). Performance characterization of random pulse width modulation algorithms in industrial and commer-cial adjustable-speed drives. *IEEE Trans. Ind. Appl.* 53 (2): 1078–1087.

34 Lim, Y.C., Jung, Y.G., Oh, S.Y., and Kim, J.G. (2012). A two-phase separately randomized pulse position PWM (SRP-PWM) scheme with low switching noise characteristics over the entiremodulation index. *IEEE Trans. Power Electron.* 27 (1): 362–369.

35 Gamoudi, R., Elhak Chariag, D., and Sbita, L. (2018). A review of spread-spectrum-based pwm techniques—a novel fast digital implementation. *IEEE Trans. Power Electron.* 33 (12): 10292–10307.

36 Lai, Y., Chang, Y., and Chen, B. (2013). Novel random-switching PVM technique with constant sampling frequency and constant inductor average current for digitally controlled converter. *IEEE Trans. Ind. Electron.* 60 (8): 3126–3135.

37 Trzynadlowski, A.M., Blaabjerg, F., Pedersen, J.K. et al. (1994). Random pulse width modulation techniques for converter-fed drive systems – a review. *IEEE Trans. Ind. Appl.* 30 (5): 1166–1175.

38 Lin, Y.C., Jung, Y.G., and Kim, J. (2010). A pseudorandom carrier modulation scheme. *IEEE Trans. Ind. Appl.* 25 (4).

39 Huang, Q., Liu, H., and Cao, J. (2019). Investigation of lumped-mass method on coupled torsional-longitudinal vibrations for a marine propulsion shaft with impact factors. *J. Mar. Sci. Eng.* 7 (4): 95.

40 Huang, Q., Yan, X., Wang, Y. et al. (2017). Numerical modeling and experimental analysis on coupled torsional-longitudinal vibrations of a ship's propeller shaft. *Ocean Eng.* 136: 272–282.

41 Chengsheu, H. and Wenchen, L. (2000). A lumped mass model for parametric instability analysis of cantilever shaft–disk systems. *J. Sound Vib.* 234 (2): 331–348.

42 Yin, J., Xu, L., Wang, H. et al. (2019). Accurate and fast three-dimensional free vibration analysis of large complex structures using the finite element method. *Comput. Struct.* 221: 142–156.

43 Cho, D.S., Kim, B.H., Kim, J.H. et al. (2016). Free vibration analysis of stiffened panels with lumped mass and stiffness attachments. *Ocean Eng.* 124: 84–93.

44 Rao, S.S. (2016). *Mechanical Vibration*, 6e. Pearson India Education Services, Pvt. Ltd.

45 Rahman, M.H. and Gupta, C. (2020). Computation of Rayleigh damping coefficient of a rectangular submerged floating tunnel. *SN Appl. Sci.* 2: 936.

46 Chowdhury, I. and Dasgupta, S.P. (2003). Computation of Rayleigh damping co -efficient for large systems. *Electron. J. Geotech. Eng.* 8: 1–11.

47 Pindoriya, R.M., Thakur, R.K., Rajpurohit, B.S., and Kumar, R. (2022). Numerical and experimental analysis of torsional vibration and acoustic noise of PMSM coupled with DC generator. *IEEE Trans. Ind. Electron.* 69 (4): 3345–3356.

48 ABB Manual. (2015). AC Servo Motors BSM Series, pp. 1–99. https://library.e.abb.com/public/.

49 SMB Bearings. (2022). Empirical relation to calculate frictional torque offered by radial ball bearing. https://www.smbbearings.com/.

6

Challenges and Applications of Blockchain Technology in Electric Road Vehicles

Nabeel Mehdi

North Carolina State University, Department of Industrial and Systems Engineering, Raleigh, NC 27695, USA

6.1 Mobility and Electric Vehicles

Mobility can be regarded as one of the basic human rights of the modern society. It is estimated that by the year 2050, the total worldwide road travel through freight, passenger vehicles, motorbikes, and buses would be 50 trillion miles annually [1]. This will have a massive impact on the industrial, societal, economic, and environmental aspects of the civilized world. To realize and sustain such an enormous infrastructure judiciously would require a concerted effort from government bodies, private players, and the general population. Fossil fuels continue to deplete and the side effects of global warming due to man-made pollution are becoming more prominent each day. The last decade was a pivotal period for sustainability both in terms of the introduction of disruptive technologies (e.g., electric vehicles and renewable energy) [2, 3] (Figure 6.1) and novel sharing-based economic models, such as Uber [4, 5]. Electrification of mobility can accelerate the transition toward sustainable energy and reduce the toxic carbon emissions from gasoline vehicles in the environment. The overall well-to-wheel (W2W) efficiency of a conventional internal combustion engine (ICE) vehicle is around 15% and greenhouse gas emissions can reach up to 11,500 lbs of CO_2 equivalent annually (US national average) [6]. In contrast, for a fully electric vehicle charged by the traditional electricity grid, the total annual CO_2-equivalent emissions are 3780 lbs with a W2W efficiency of 70%. Factors like these have stimulated government interest in research and development in the vehicle electrification domain.

Rapid proliferation and important developments in technologies of energy storage devices (Figure 6.2), renewable energy systems (RES), power electronics, and artificial intelligence (AI) have made it possible to manufacture EVs that are highly efficient, environmentally friendly, and competitively priced. Blockchain is a digital technology that can support EV infrastructure by enabling decentralized, resilient, and secure applications. Some of the EV industry blockchain applications will be summarized in the following sections. These applications can have a considerable impact on a range of aspects like economic, logistic, interactive, energy, security, and privacy for the EV owner.

Transportation Electrification: Breakthroughs in Electrified Vehicles, Aircraft, Rolling Stock, and Watercraft,
First Edition. Edited by Ahmed A. Mohamed, Ahmad Arshan Khan, Ahmed T. Elsayed, and Mohamed A. Elshaer.

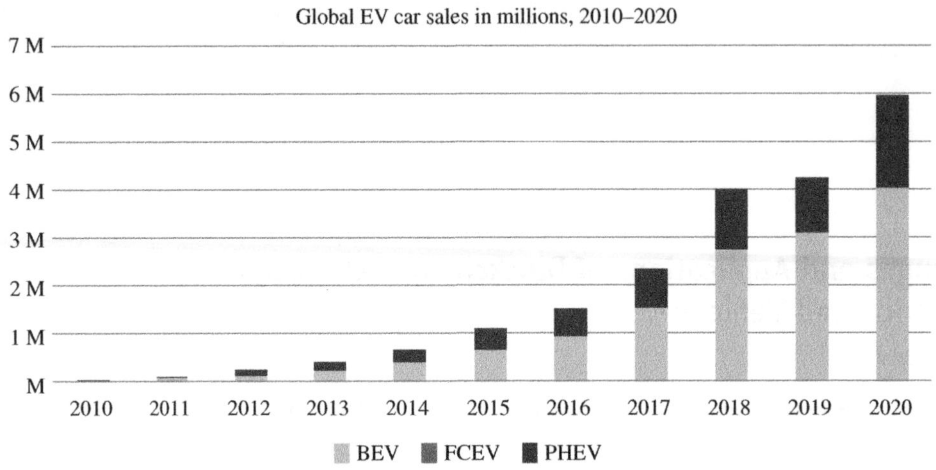

Figure 6.1 Global electric car sales. *Source:* Data from IEA 2022.

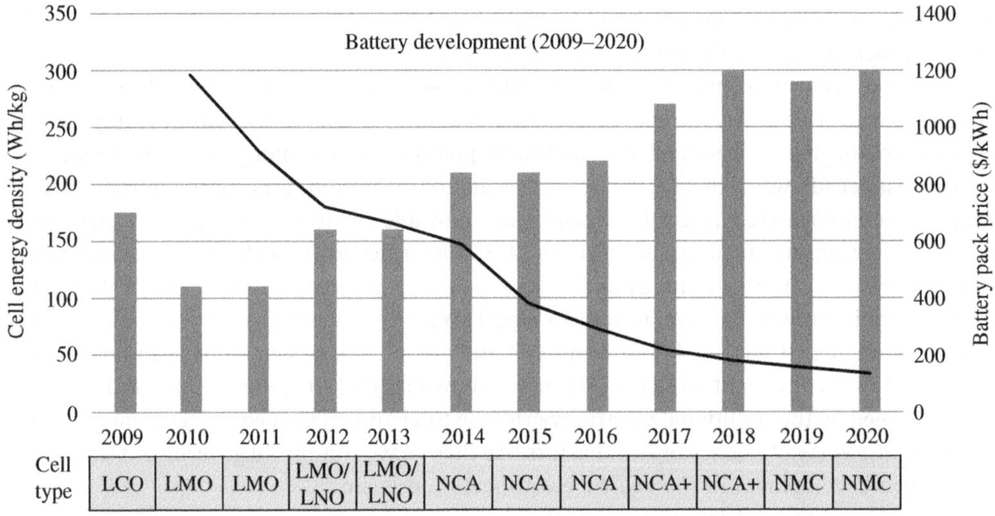

Figure 6.2 EV battery energy density, cost over the years. *Source:* Bloomberg New Energy Finance.

6.2 Electric Vehicle Overview

In the simplest terms, an EV can be described as a vehicle that is partially or fully propelled by an electric motor and powered by an electric battery. This may include four-wheelers, two-wheelers, cargo trucks, rail vehicles, underwater vehicles, aircrafts, or boats. Compared to conventional ICE vehicles, EVs have a much simpler design and far less components (Figure 6.3). This chapter's discussion revolves mainly around four-wheeled passenger electric vehicles.

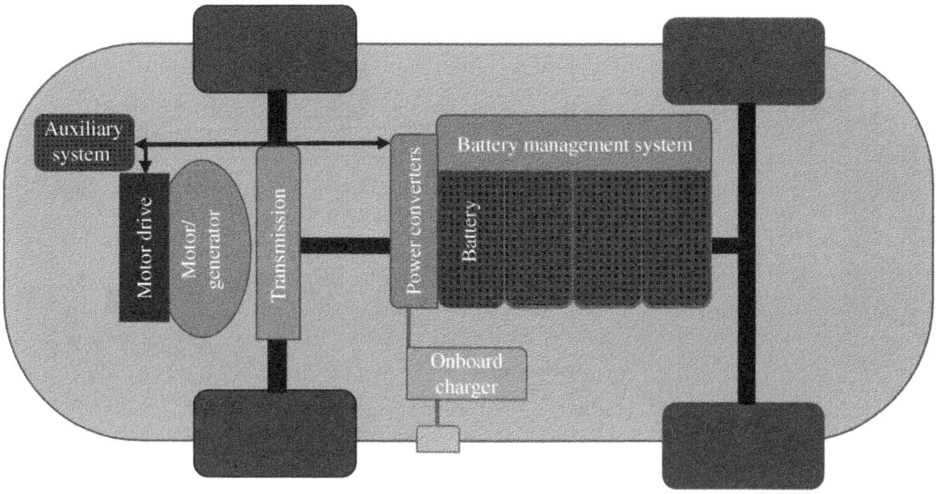

Figure 6.3 High-level internal schema of EV.

According to their energy usage and battery configuration, EVs can be further categorized as shown in Table 6.1.

Table 6.1 Categorization of electric vehicles.

	Type	Description
1	Battery electric vehicle (BEV)	Propelled by battery energy with external charging
2	Hybrid electric vehicle (HEV)	Propelled by gasoline and battery, no external charging
3	Plug-in hybrid electric vehicle (PHEV)	Propelled by battery and gasoline, with external charging
4	Fuel cell electric vehicle (FCEV)	Propelled by energy from burning fuel like hydrogen

6.3 Challenges of the Electric Vehicle Industry

Despite the numerous advantages offered by the EVs, there are some bottlenecks that must be overcome to enable their mass adoption and integration with the current infrastructure. These challenges and concerns are summarized in the following sections.

6.3.1 Range Anxiety

Compared to ICE vehicles, most BEVs have a shorter driving range and long recharge times resulting in range anxiety for the drivers. Although battery cell technology continues to improve in terms of features like energy density (kWh/kg) and power density (kW/kg), fossil fuels are still 50 times more energy dense per unit weight. Thus, to achieve an equivalent driving range as an ICE vehicle, the corresponding weight of the battery pack in an EV would be significantly high. However, the higher tank-to-wheel (T2W) efficiency of the EV and reduced engine weight negates the adverse effects of the extravehicular overall mass. Another factor that contributes to range anxiety is the lack of enough fast-charging EV-compatible infrastructure worldwide. For instance, currently

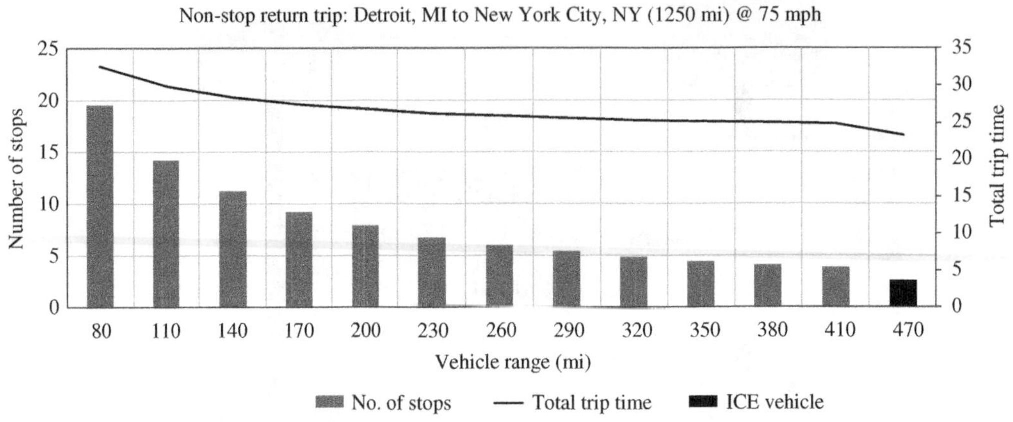

Figure 6.4 Number of stops vs. EV driving range.

the number of EV charging stations in the US is roughly one-third of the gas stations [7] and would be insufficient to meet any sudden surge in the demand. Figure 6.4 shows the number of recharging stops (0–80% battery) and total trip time on a simulated return trip from Detroit, MI to New York city and 30 minutes charge-time and 30 minutes recreational stop every 100 mi. There is a significant reduction in the number of stops (80%) and total trip time (24%) for EVs with a longer driving range.

6.3.2 Lengthy Charging Times

An EV battery charging (0–80%) time can range anywhere from 30 minutes on a high-voltage direct current (HV-DC) offboard fast-charger to 40 hours through a residential alternating current (AC) onboard charger. This is several magnitudes higher than an ICE vehicle's 5-minute full-tank refilling at a gas station. Furthermore, slow chargers and long wait times at the charging station during peak times can aggravate the situation. The charging profile till about 80% capacity is linear while it slows down afterward. This implies that during a long trip, partial-capacity charging at multiple stops would take overall less time than full-capacity charging at fewer stops [7]. Figure 6.5 shows the time it takes for an EV battery to charge till 80% at different charging levels (3, 11, 50 kW power). Assuming external factors to be negligible (battery degradation, weather, charging loss, etc.), it is evident from the simulation that the charging time reduces significantly on a fast-charging station (50 kW). However, research has shown that fast charging can be detrimental to battery life as it has higher charging currents and heat dissipation.

6.3.3 Battery Safety Concerns

EV battery pack comprises hundreds or thousands of individual cell units and a complex battery management system (BMS) that balances the load and monitors thermal runaway. Even one single malfunctioning cell unit can start a fire that propagates through the entire battery pack and can pose a severe threat to vehicle safety. It is difficult to suppress EV fires because the internal battery pack is inaccessible to external fire-extinguishing agents and may reignite if not cooled sufficiently [8]. Additionally, BMS software is also vulnerable to malicious attacks from adversaries due to the digital communication nature of the vehicle charging system. These factors necessitate additional measures and comprehensive and transparent monitoring of battery health. Ideally, the current

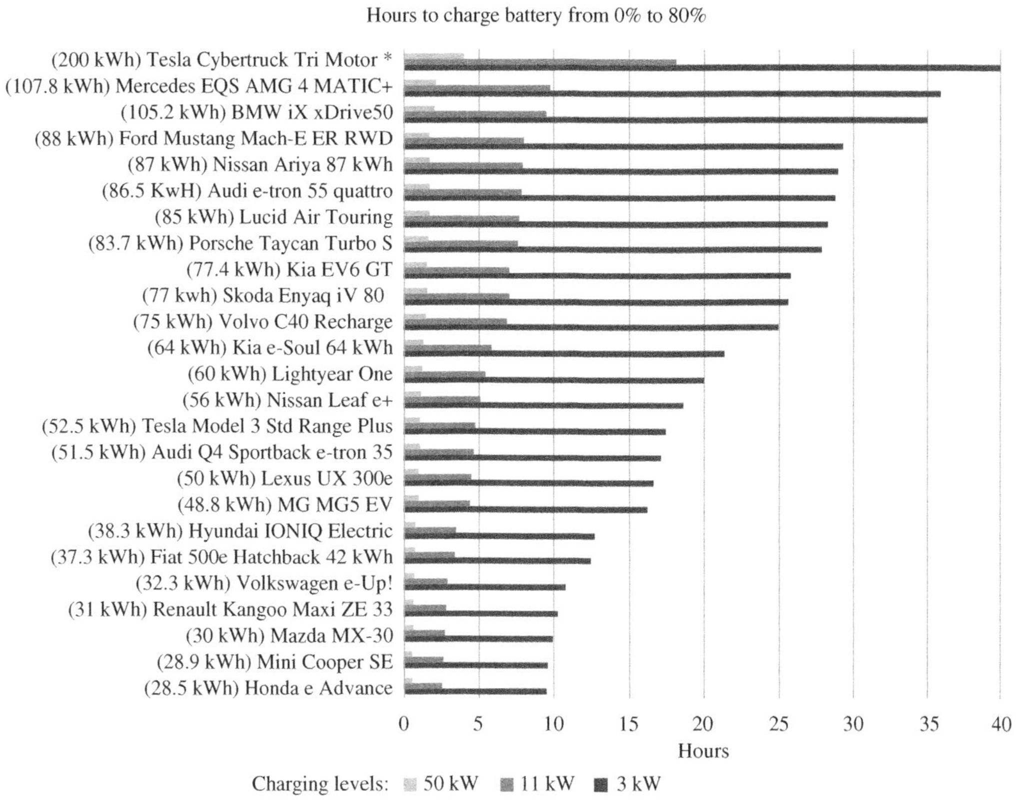

Figure 6.5 EV vs. charge time at different modes.

safety and health of EVs must be publicly accessible with the ability to trace its key metrics (such as temperature and load) and past events (battery swaps, repairs, etc.).

6.3.4 Lack of Standardization

A significant impediment toward building truly scalable EV infrastructure is the lack of standardization in the industry. An ICE vehicle can be refueled at any gas station across the world, whereas the same cannot be said for an electric vehicle. There are multiple layers of complexities that should be addressed to achieve interoperability and compatibility between an electric vehicle and the charging infrastructure.

Firstly, the charging infrastructure must be capable to supply and satisfy the power level requirements of the EV. The charging levels are mainly classified as follows:

- Level 1: (0–10 kW AC) (slow charging)
- Level 2: (10–50 kW AC/DC)
- Level 3: (>50 kW DC) (fast charging)

Secondly, the connector types of the charger and EV should pair with each other mechanically and electrically. Due to different standards of distribution voltage and current across various countries, power electronic devices are required for necessary conversions. Some of them are inbuilt into the EV while others (like high-voltage devices) require a separate off-board charging device. So far,

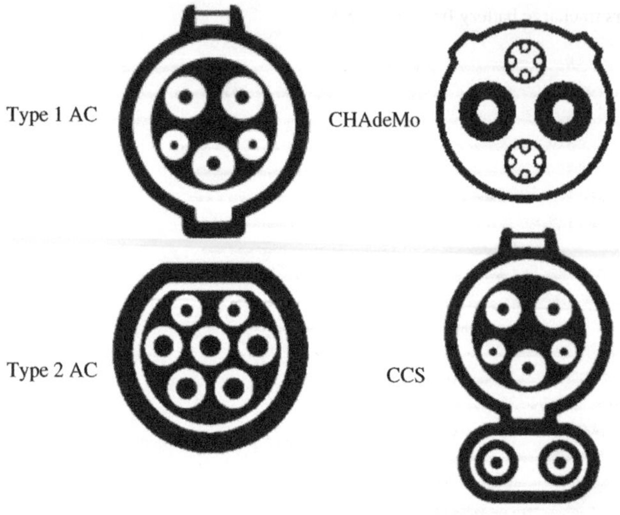

Figure 6.6 Examples of common EV charging connector types.

there has been limited consensus amongst the auto-industry leaders and other stakeholders on any universal EV charging component design and principles. The main EV charging connector types (Figure 6.6) currently being used in the industry are:

- Type 1: AC, common in the United States, Japan, single-phase plug.
- Type 2: AC, common in European and Asian vehicles, triple-phase plug.
- Type 3: AC, used in Europe, but getting replaced by Type 2 connectors.
- CHAdeMo: DC, common in Asian cars, bidirectional and fast-charging capability.
- CCS: DC, like Type 2 with fast charging capability.
- Tesla: AC and DC both from the same plug.
- Combo Plugs: A combination of AC and DC plugs.

Thirdly, an EV's BMS should be able to communicate with the offboard charging system and regulate, monitor, or trigger charging with the necessary cable. EVs can have different communication modes that are summarized as:

- Mode 1: Slow AC charging on regular electrical sockets without communication.
- Mode 2: Slow AC charging on regular electrical sockets with communication.
- Mode 3: Slow and semi-fast charging with dedicated sockets and communication.
- Mode 4: Fast charging on a dedicated DC charger with bi-directional control and communication (Figure 6.7).

6.3.5 Electricity Grid Disruption

A single EV with a 24 kWh battery takes as much electricity as an average US household. The number of EVs on the road is expected to grow to 230 million by the year 2030 [3] along with the battery energy capacity. Unregulated EV charging will have a negative impact on the supply–demand dynamics of the electricity grid. It may cause issues like an increase in the peak demand, voltage deviation, phase unbalance, harmonics distortion, power system equipment overloading, increased power losses, decreased life of transformers, and higher electricity bills for all users [9].

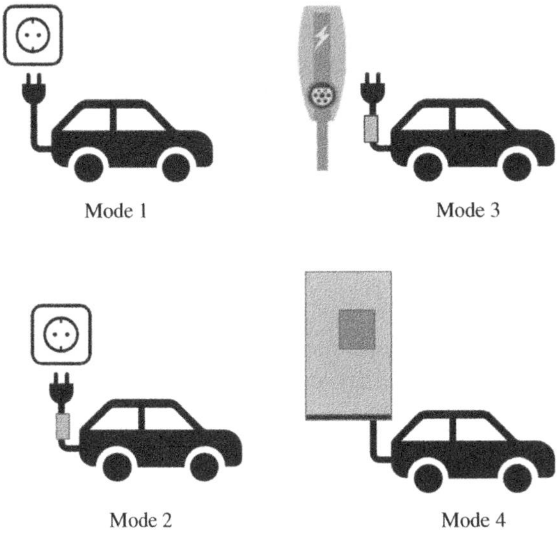

Figure 6.7 Different EV charging modes.

Electrical utility companies would need to spend a lot of capital to revamp/upgrade their energy infrastructure.

A strategic approach is required to minimize the negative impact of unregulated charging on the electrical grid by harnessing the power storage and bi-directional power-flow capabilities of an EV. For instance, a simulation of uncontrolled EV charging in Figure 6.8 shows the impact on the regular demand curve. Assuming there is a limited supply capacity during different times of the day, unregulated EV charging will saturate the demand and deprive other consumers of power. On the other hand, the capacity problem could be eliminated through a controlled EV charging process. The green dotted line shows the total demand if the EVs are charged in a regulated manner throughout the day.

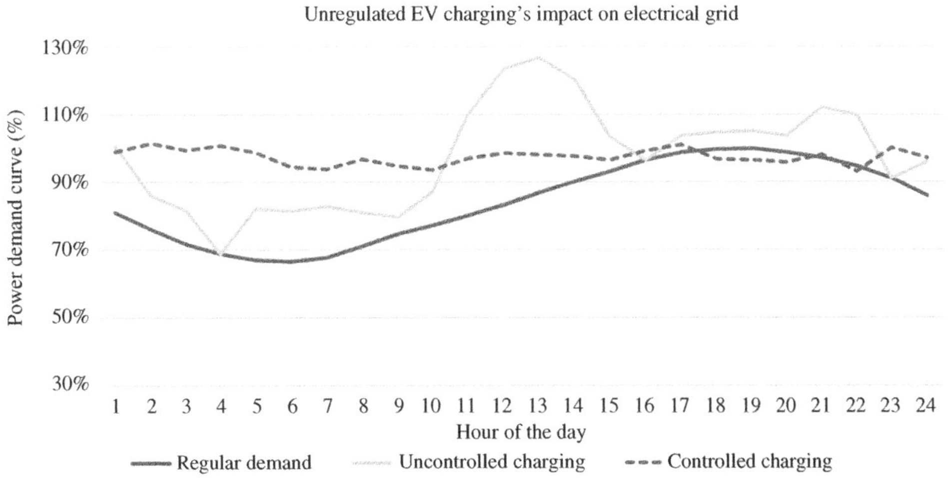

Figure 6.8 Simulation of a city's daily power demand curve.

6.3.6 Battery Waste

EV batteries have a finite life that depends on a multitude of external and internal factors. Once the end of life (EOL) is reached, the battery can be recycled or discarded. Due to the hazardous and explosive nature of batteries, these processes require technical capabilities and tactical planning. This issue will be amplified in the next decade as EV adoption increases, and new battery technologies are developed [10].

6.3.7 Cyber-Security Hazard

EVs comprise multiple cyber-physical systems (CPSs) that rely on internet-connected software controls to optimize motor drive, battery management, and auxiliary systems. There is bidirectional communication of power and information through multiple channels (charging, auxiliary systems, software updates, and vehicle-to-vehicle interaction). This makes EVs vulnerable to potential cyber-security risks from various types of malicious actors [11].

6.4 Applications of Blockchain Technology

As described in Section 6.3, there are significant hurdles before mass adoption of EVs can ensue. Blockchain is a promising technology that can alleviate some of the challenging aspects of the EV industry. It can act as an information and communications technology (ICT) platform to create decentralized or semi-centralized digital ecosystems that are self-sustaining and resilient. EV-related public applications can be built on top of a blockchain framework and can be accessed by state utilities, RES, electric vehicle software, and the charging infrastructure. Various types of users with an encrypted public address can co-exist and operate on a set of predefined logics in a blockchain-enabled democratized application. Such a system can play a transformative role in democratizing the green energy sector and channeling the actual potential of electric vehicles as power storage devices.

Blockchain is a distributed, decentralized, and immutable public set of records that are cryptographically linked with each other [12]. Network users with public–private key pairs propose transactions/data that are bundled together in records. Each record known as a block contains data in the form of a Merkle tree and a cryptographic hash of the contents from the prior block forming a chain-like pattern. Figure 6.9 shows the Merkle tree representing a single block in the blockchain network. "L1" and "L2" represent transactional data and hash is a one-way cryptographic function that outputs a string of fixed length from any variable length input data.

Figure 6.10 depicts the linkage between each block in the blockchain and the hashing mechanism that imparts immutability to this ledger. Hence, if any content ("L1" and "L2") of the previous blocks is tampered, all the subsequent hashes in the ledger blocks mutate.

The blockchain concept was designed to be inherently decentralized where the participants, also known as Nodes, adhere to a consensus mechanism and append new blocks to the ledger. For every additional block, a reward is generated as an incentive for the node that validated it. These node(s), also known as miners, solve a cryptographic challenge using brute-force computing and present proof in the form of a hash. Whichever miner presents this hash first gets the reward from the blockchain and, subsequently, all the nodes update their records with the new block. Due to this, powerful feature blockchain can be self-sufficient to execute transactions using a peer-to-peer network of nodes without the involvement of any intermediaries. This feature has applications in scenarios

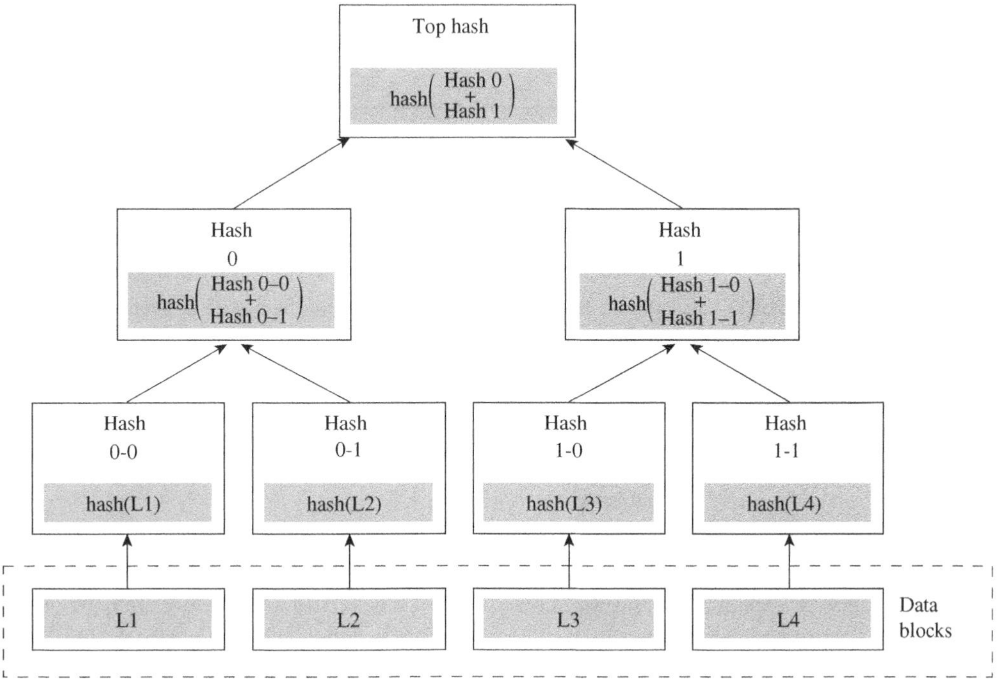

Figure 6.9 Merkle tree showing data layer and top hash.

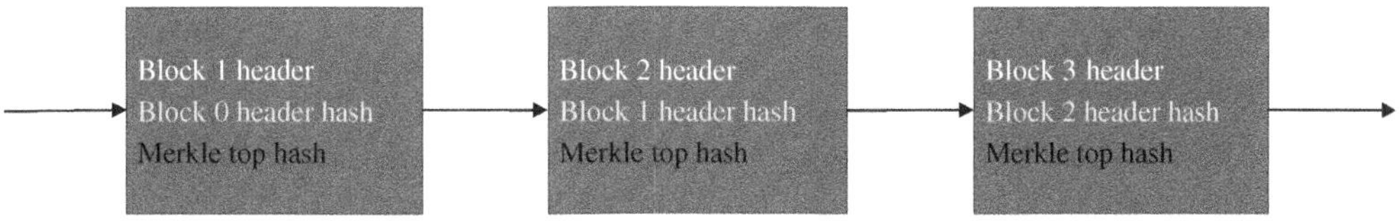

Figure 6.10 Blockchain ledger.

where traceability, trust is indispensable, and auditing is expensive (such as in the finance sector and food supply chains).

Blockchain successfully solves the "double-spending" problem in the payment sector without the need for a third party to verify it. The heart of a decentralized network is the consensus protocol that all the nodes agree upon for validating and adding a block to the ledger. Once a block addition is validated, all the nodes update their blockchain and the network is said to be in consensus. Another popular protocol is "proof of authority," which gives special nodes (regulatory bodies) the authority to validate transactions. A few additional consensus mechanisms are described in Table 6.2.

The transaction speed, energy consumption, and security of a blockchain network depend on the choice of consensus mechanism, underlying algorithms and overall network topology. For instance, Bitcoin's PoW applies "SHA-256 hash function" as its mining algorithm while Litecoin PoW employs a simplified version of "scrypt key-derivation function." The computational complexity of mining a block in Bitcoin keeps getting harder as the chain gets larger while there is a finite number of blocks in it. Due to the difference in complexity of these two algorithms, mining a new block in the Bitcoin chain is about 10 times slower and much more energy intensive [13].

Table 6.2 Various types of blockchain consensus mechanisms.

Consensus mechanism	Approach
Proof of Work (PoW)	Miners solve brute-force mathematical puzzles to produce a new block
Proof of Stake (PoS)	Nodes put up their tokens as collateral to produce the next block
Delegated Proof of Stake (DPoS)	Nodes vote and elect delegates to validate the next block
Proof of Capacity (PoC)	Nodes use their available hard drive space to decide mining rights
Proof of Elapsed Time (PoET)	Randomly allot waiting time to users and the first responder is the producer
Proof of Identity (PoI)	Based upon private key allowing only a certain user to perform a transaction
Proof of Authority (PoA)	Consensus mechanism based on identity as a stake such as regulatory body
Proof of Activity (PoAc)	Combines aspects of Proof of Work and Proof of Stake consensus mechanisms

Depending upon the application requirements, different blockchain architectures may be used. Ethereum blockchain network allows the participating peers to define and deploy custom business logic in the form of programmable functions called "Smart Contracts" that can interact with the other nodes. These Smart-Contracts reside on the blockchain and can be fetched through a hash address. They can be triggered through a network transaction and can also execute other smart contracts [14]. A blockchain network can be categorized as permissionless if it is open to the public or permissioned if it is private and only authorized nodes can read, write, or validate transactions.

6.4.1 Energy Blockchain Ledger

Distributed RES will be ubiquitous in the smart communities of the future. A myriad of IoT devices and smart vehicles will need to interact with the electricity grid in a secure and efficient manner. This should happen seamlessly where data integrity can automatically be verified. Blockchain enables this by reducing latency in transactions and securing energy devices. Its data immutability characteristics will furnish the necessary trust and visibility to vet out malicious players and reduce the chances of financial exploitation. Smart contracts executing on nodes (energy generators) would manage the supply of excessive power to the grid without the need for any external auditing from regulators, thus saving both cost and time. If required, a permissioned or semi-decentralized energy blockchain type of architecture can be utilized to enforce regulations by delegating special privileges to institutional nodes in the network. Blockchain architecture can serve as a public ledger shared across different energy stakeholders and EV owners to record their transactions of energy generation, storage, and trading (Figure 6.11) in a transparent and foolproof manner [15, 16].

6.4.2 Blockchain-Powered Billing in E-mobility Systems

Robust, swift, and foolproof payment systems are required to process micro-payments at low-transaction fees. The current payment systems are highly centralized and opaque, which makes them less suitable for distributed renewable energy trading processes. Due to the involvement of several intermediary financial institutions, a transaction through a traditional banking process would be slower than a similar blockchain-enabled payment channel. International payment accompanies a high fee, and the transaction can take several days to complete (Figure 6.12).

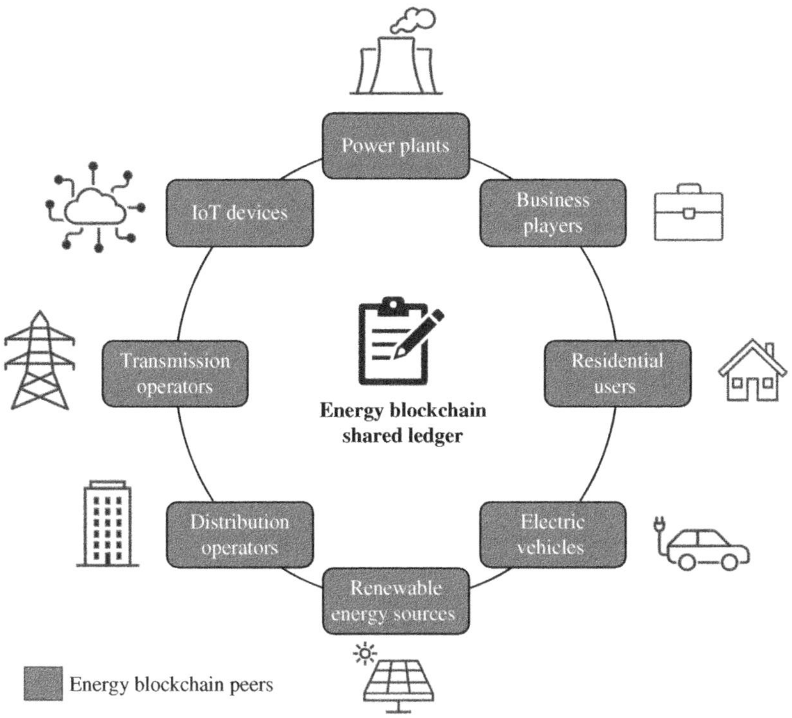

Figure 6.11 Blockchain as an electrical energy ledger.

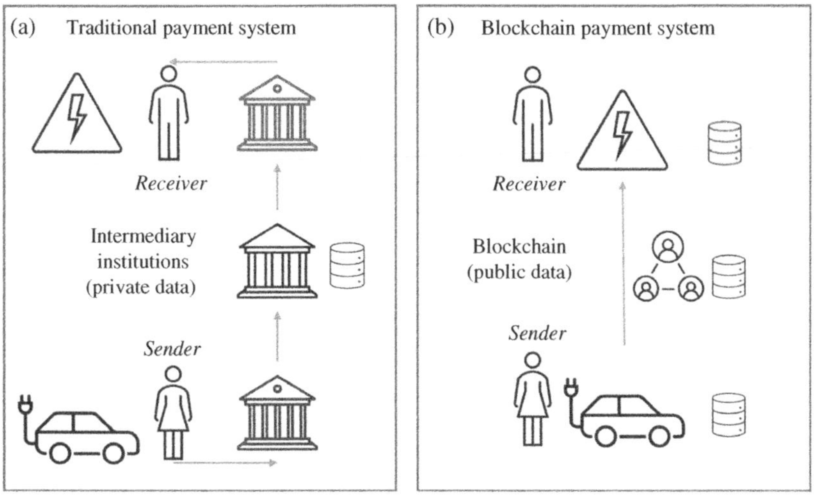

Figure 6.12 Traditional payment system (a) vs. blockchain payment system (b).

Another major disadvantage of a traditional payment system is the lack of user privacy that can be concerning in scenarios like V2V (vehicle to vehicle) charging. Blockchain-based payment systems can be anonymized through pseudonym addresses that are untraceable and can be dynamically changing. Lastly, blockchain integration with CPSs can enable automatic verification and triggering of transactions through smart contracts after a physical event like battery recharging or tire replacement. Researchers have established the benefits of blockchain in eliminating the shortcomings in current financial infrastructure and making the payment process more efficient [17].

6.4.3 Charging-as-a-Service (CaaS) Ecosystem

The automotive and energy sector workforce will be disrupted because of electrification and the transition to clean energy. This will engender an economic void that can be filled by creating new opportunities for the displaced workforce. Renewable energy sources like solar energy are freely available and can be easily harnessed through affordable and accessible solar panels. Producers can independently store and sell this energy to consumers like EV drivers in their region. This type of service can be called a Charging-as-a-Service (CaaS) and will be highly distributed, decentralized, and variable in nature [18, 19]. Blockchain-based applications can mitigate trust and privacy issues that are key concerns in such unregulated ecosystems (Figure 6.13).

A service provider can configure RES-powered EV charging units on mobile (off-board chargers) or stationary structures (V2V) to automate the energy trading and billing process through a blockchain interface. The available capacity will be advertised live on a high throughput blockchain network so that EV users can find and schedule the charging event in advance. The smart contract will act as an intermediary agent between the EV driver and the EV charging station. Based upon the current location, battery capacity, state of charge, charging time, and other parameters, the smart contract will recommend available charging options to the driver. Through dynamic pricing and bidding mechanism, the smart contracts will reserve a "charging – transaction" for the EV on the ledger. This will assert that the EV charger is unable to "double-spend" the capacity or "over-promise" to the driver. The EV and the charging station both should contain a smart-metering system that provides physical confirmation of energy transfer and triggers a payment. If the charge-service provider fails to deliver the service, then that event information would appear permanently on the network. This will be insightful for other users and assist them in choosing decent service providers. If there are regulatory authorities on the network, they can audit the historical data and penalize misbehaving peers. These processes can be performed by smart contracts and digital agents on behalf of the participating nodes. Government institutions can directly issue "Renewable Energy Credits" (RECs) to producers that can be traded amongst users. On the one hand, a truly decentralized CaaS ecosystem can allow energy producers to charge desirable rates

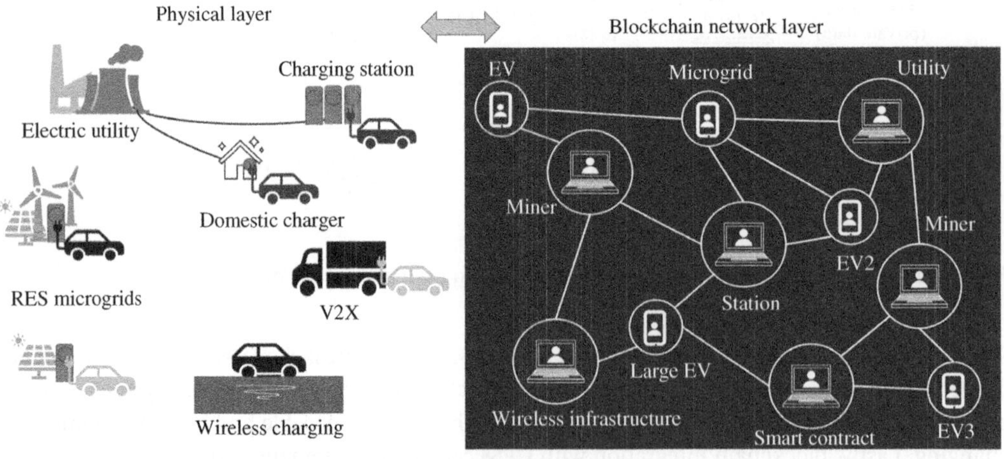

Figure 6.13 Charging-as-a-Service enabled by the blockchain network.

while, on the other hand, it provides EV users with the flexibility to analyze public data and choose the most suitable service provider. Smart-Contracts-based dynamic pricing can also be computed from the live demand, geographical location, competitors, and network fee [16, 20–22].

6.4.4 Electric Vehicle Battery Management with Blockchain

Battery swapping is a method to replace a discharged electric car battery with a fully charged one like any battery-operated consumer device. This can eliminate long charging times, which are a major limitation of electric vehicles [23]. Ideally, the battery should be interchangeable even by the driver; however, most EVs do not presently support this feature and it requires a dedicated service provider. In the future, this service can potentially be very common as it is highly reliable, efficient, and time-saving. Provided that there are sufficient battery-swapping stations, this technique can eliminate range anxiety and waiting times for EV drivers [23].

Some key considerations for battery swapping are:

1. EV modular design to enable quick full-battery or partial-battery swapping.
2. Battery and EV compatibility.
3. Both batteries should have a comparable state of health (SoH) and wear and tear.
4. The difference in the degraded quality of the battery must be offset by a cost.
5. Data authenticity and fair pricing.

Battery life, performance, and available capacity depend on several factors like age, usage, number of past charging cycles, temperature, and the charging mechanism. With the assistance of a blockchain network, an EV's BMS can make this data publicly available on an immutable ledger after each "swapping event" (Figure 6.14).

Once EV broadcasts a battery swap request on the network, the smart contract can trigger the following functions:

I Extract the battery metadata from the EV blockchain address's last transaction block.
II Get the EV's physical location (ZIP) from the broadcasted message.
III The battery swapping service providers bid to that request.
IV Smart-Contract filters the bidders and verifies the compatibility, region, etc.
V Each battery has a blockchain hash address with past records and other key metadata.
VI Smart-Contract computes the tariff by comparing the "State of Health" (SoH) of EV's current battery and the bidder's battery. The tariff also considers service charges and blockchain transaction fees.
VII If the bidder's battery has worse SoH than the EV battery, the bidder pays tariff to the EV owner. The swapping station also charges a fee for the services.
VIII After the battery is swapped, the EV's BMS validates the transaction on the blockchain network.

6.4.5 Vehicle to Grid (V2G)

Electric vehicles are an energy reservoir that can supplement the electric grid with bidirectional electricity flow. A single EV battery's capacity (24 kWh) is worth the energy consumption of a US household. This can be useful to electric utilities in managing peak shaving (eliminating short-term electricity demand spikes) and managing outages to name a few.

The price per unit of electricity fluctuates daily and seasonally with the variability in green energy sources and demand. EV owners can charge batteries and supply energy to the electric grid during

BLOCK 12	6be87e139df7e22fcb64ea5ee2fce70d9417dd1bc20062fccb30
REGION	48.7345, 47.9234
BATTERY_ADDRESS	60303ae22b998861bce3b28f33eec1be758a213c86c93c076db
CURRENT_OWNER	fd61a03af4f77d870fc21e05e7e80678095c92d808cfb3b5c279
LAST_TXN	a4e624d686e03ed2767c0abd85c14426b0b1157d2ce81d27bb
TIMESTAMP	2021-07-01 05:16:13
CHARGE_CYCLES_COUNT	46
AVG_TEMP_C	94
AGE_DAYS	600
CAPACITY_KWH	80
HIGH_AMP_CYCLES	37
LOW_AMP_CYCLES	9

BLOCK 40	7f92e3b0fd253b5fc95ab7c30db940beb6fe1c84732dd0d06101
REGION	48.6411, 46.9888
BATTERY_ADDRESS	60303ae22b998861bce3b28f33eec1be758a213c86c93c076db
CURRENT_OWNER	ed0cb90bdfa4f93981a7d03cff99213a86aa96a6cbcf89ec5e88
LAST_TXN	6be87e139df7e22fcb64ea5ee2fce70d9417dd1bc20062fccb30
TIMESTAMP	2021-09-04 03:28:20
CHARGE_CYCLES_COUNT	51
AVG_TEMP_C	95
AGE_DAYS	663
CAPACITY_KWH	80
HIGH_AMP_CYCLES	40
LOW_AMP_CYCLES	11

Figure 6.14 Example of battery-swapping blockchain network blocks.

peak demand. Blockchain can facilitate this ecosystem by confidentially and securely letting prosumers and state utilities trade energy in a decentralized fashion [16, 22].

Moreover, utilities issue renewable energy credits (RECs) to renewable energy producers that certify that 1 MWh of electricity was generated. Once this energy is injected into the grid, the RECs can be freely traded over an online platform [24]. Entities that produce greenhouse emissions or other pollutants can purchase their RECs and offset their carbon emissions. A solar-powered EV (e.g., lightyear [25]) owner can be awarded RECs or similar tokens by the utilities once the grid receives this energy from the EV battery. A blockchain-based decentralized trading platform would be an ideal network where this token functions like a cryptocurrency that is issued by the state utilities and traded by peers [22]. RECs can be easily traced from generation to ownership trades to final reclamation. According to researchers [24], this leaves a simple audit trail, significantly reducing the associated time and cost, and enables producers to monetize their credits immediately after generation.

6.4.6 Blockchain-Enabled Security in Electric Vehicles Computing

In the V2X (Vehicle to Anything) paradigm, electric vehicles can communicate with any other "thing" like other vehicles, IoT devices, passengers, and the electricity grid. This is achieved through an intelligent system of sensing devices, an inbuilt computing system, cloud technologies, and a wireless telecommunication network. The security requirements of such a system should address the data confidentiality, integrity, and availability (CIA); authentication, authorization, and accounting (AAA); and user privacy concerns (e.g., location, charge status, and identity). The repercussions of a centralized EV computing system being compromised by an adversary far outweigh the shortcomings accompanied by blockchain technology. EVs can function as a pool of intelligence that share data and perform complex interactions through edge-computing devices. Some major limitations of EV communication channels like data tampering, privacy theft, and identity spoofing can be alleviated through blockchain technology by using public–private keys for authentication and data encryption methods [26]. Cryptographic challenges like "zero-knowledge proof" (ZKP) [27] and user responses can be applied to establish communication sessions between different vehicles or devices securely [20] (Figure 6.15).

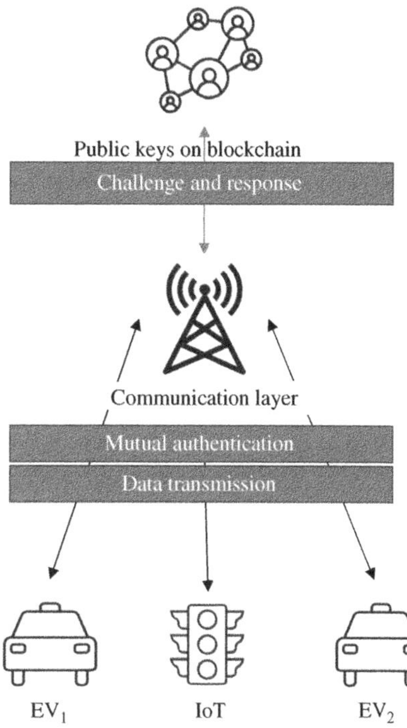

Figure 6.15 Blockchain-based security and authentication system.

6.4.7 Privacy-Preserving Blockchain-Based EV Charging

Blockchain-enabled privacy applications will be crucial in building a truly democratic and decentralized EV-charging infrastructure. ZKP is a cryptographic protocol where different parties can prove that a given statement is true, without communicating any additional information apart from the fact that the statement is indeed true [27]. ZKP is combined with blockchain to develop decentralized applications where the user's personal information (like GPS location) can be safeguarded while still proving their geographical region. Researchers have proposed a system where EV users raise a charging request and various charging stations in a region send a bid to that request [20, 28]. After the user finalizes a bid, only the accepted charging station and the EV user knows the EV GPS location on the network. This mechanism is depicted in Figure 6.16.

6.4.8 Battery Analytics

SoH is a KPI that denotes the condition of a battery (or a cell or a battery pack), compared to its ideal conditions (brand new state) as a percentage. This can reflect the remaining life of the battery and its suitability in hardware applications. It can be computed by the BMS from a single factor or a combination of factors like the age, charge cycles, temperature, discharge time, capacity, resistance, etc. This data resides inside the BMS memory and can be digitally tampered by a malicious actor. A blockchain-based cloud-sourcing battery data collection platform can enable advanced analytics to analyze SoH and identify anomalous behavior [29]. These battery analytics on a public blockchain can serve multiple benefits in areas ranging from dynamic pricing of EV charging to vehicle safety and insurance premiums. This will add transparency, trust, and fairness to any decentralized applications that involve trading of EV or its battery. As seen in Figure 6.17, the manufacturer creates a new record for a battery and gives it a unique address on block B1 of the blockchain network. B2 depicts the "transfer of ownership" blocks that will tract the battery's journey till EV installation.

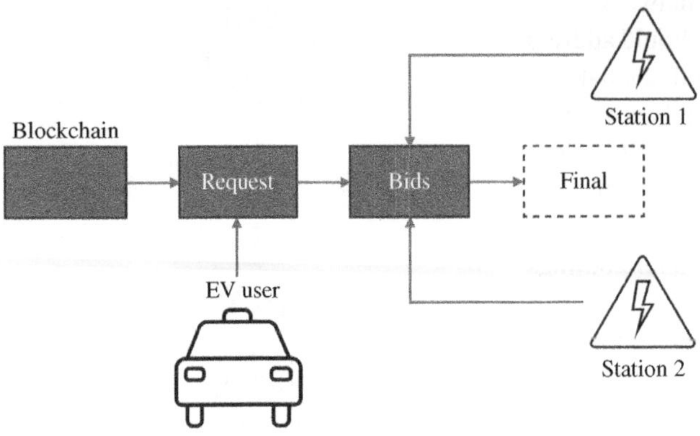

Figure 6.16 Privacy-preserving blockchain network.

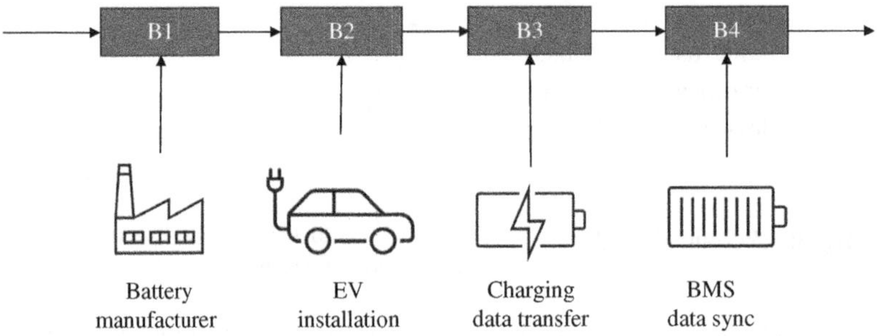

Figure 6.17 EV battery operations' life cycle.

The blocks B3 and B4 represent update of key-battery related metrics (example: charging cycle) on the blockchain network by the BMS.

6.4.9 Supply-Chain Traceability and Provenance

Counterfeiting is a major security and integrity challenge in the supply chain industry. Due to globalization, the complexity of tracking and tracing electronic parts globally has grown by leaps and bounds due to the widespread infiltration of counterfeit, cloned, and sub-quality hardware. Batteries are a dynamic part of an EV that can be swapped by the owner for economic incentives or EOL replacement. This leaves the EV potentially vulnerable to a counterfeit or compromised battery part that can significantly affect the performance and safety of the vehicle. Thus, hardware provenance of the battery and full traceability of its operation are essential to safeguard the EV safety and minimize the risk of adversarial attacks. A blockchain-based framework can track and trace every battery that is circulating in the supply chain. This can be achieved by a permissioned-blockchain network where original equipment manufacturers (OEMs) would have the authority to create and validate unique identities using a non-resource intensive consensus algorithm such as "PoA" [30]. A quasi set of blockchain transactions for part traceability can be summarized as follows:

- Part is manufactured, the OEM generates a unique ID (UID) for the part on blockchain.
- Part gets sold to distributor D1; transfer-ownership verified on the network.

```
{
"User_Address": '9F86D081884C7D659A2FEAA0C55AD015A3BF4F1B2B0B822CD15D6C15B0F00A08',
"MetaData": {
   "Driver_Id" : '1B4F0E9851971998E732078544C96B36C3D01CEDF7CAA332359D6F1D83567014',
   "EV":{
      "EV_ID":"60303AE22B998861BCE3B28F33EEC1BE758A213C86C93C076DBE9F558C11C752",
      "Parts":{
         "Part1_ID":"FD61A03AF4F77D870FC21E05E7E80678095C92D808CFB3B5C279EE04C74ACA13",
         "Part1_Inst":"2020-03-22 12:03:33",
         "Part1_Prev":"8A1CF282F08265CF1A38EF470D9D56D5BF4D01E6F3B973A2E824453B67400DB2"
            }}}}
```


Figure 6.18 Block data representation for part traceability.

- EV OEM purchases the part and installs in EV, UID gets linked to EV-UID address metadata.
- EV purchased by driver X, OEM dealership transfer EV-UID to driver X-UID on blockchain.
- At every step, part is cryptographically linked to its "previous" ownership state and the timestamp of physical transfer (Figure 6.18).

6.5 Vehicle Insurance Management

Data transparency and traceability through blockchain can improve the insurance experience and premium for users and insurance companies. User-driving activity, vehicle health, and performance insights will enable the companies to tailor the insurance plan accordingly. The benefits of blockchain-based systems are huge in terms of speed, transparency, and trust for this sector [31]. Individual drivers, vehicle dealers, lawyers, law enforcement agencies, insurance companies, and law and motor vehicle agencies will all be stakeholders of this blockchain-based solution. Smart-Contracts can trigger alerts to stakeholders and create automated records of events like accidents and assist the authorities to investigate better through this evidence. The insurers will be able to track claims in an unprecedented way by analyzing the shared trusted ledger. Fraudulent claims would be eliminated through data provenance and anomaly algorithms executing on the smart contract as described in the research [32]. The underwriting process can be real-time or automated in blockchain-based insurance mechanism while it can take several weeks to months to present an "insurance quote" through the state-of-the-art process. The EV industry concepts described in the sections above like V2G charging will introduce additional constraints and liabilities to the buyer and seller that will require reforms in the present insurance coverages. Blockchain would be a key element to operate this large, dynamic, and decentralized type of insurance network.

6.5.1 Electric Vehicle Crypto Mining

Blockchain applications consume computing power and depend on the activity of participating nodes for transactions. In a futuristic scenario, highway tolls, parking tickets, drive-through orders, EV charging fees, etc., will be paid in cryptocurrency by the EV itself. Modern EVs are equipped with a powerful computer processing unit that lays idle or low usage while the vehicle is parked or charging. This computing power can be utilized to facilitate transactions on the blockchain network through the process also known as mining. This will convert the EV into a self-sufficient revenue-generating instrument that will be capable of paying off its investment costs [33]. This would be a

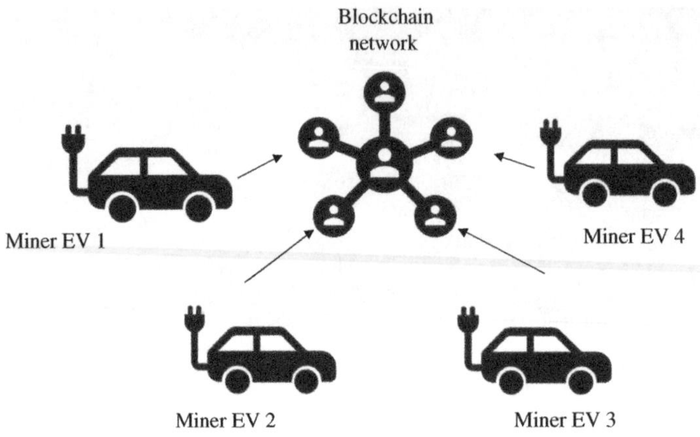

Figure 6.19 Cluster of EV miners.

win-win scheme both for the EV owner who earns the mining reward and the blockchain network that gets an active node. EV crypto mining will create novel opportunities like pooled-mining clusters for consumers to profit from their EVs (Figure 6.19).

6.6 Summary

The rapid growth of electric vehicles in the last few years is very promising for the transition to sustainable energy goals. This chapter described various challenges related to the electric vehicle industry and some applications of blockchain to tackle them. Since it is a fledgling technology, most solutions are currently at a conceptual stage and would require an integrated effort from regulatory bodies, state institutions, EV industry leaders, and consumers for fruition.

The blockchain framework can enable the development of applications related to EVs that are unattainable or infeasible with state-of-the-art systems. EVs are sophisticated CPSs with game-changing features and interactivity. First and foremost, EVs will disrupt the electrical energy market by supporting clean energy sources that are inherently decentralized, independent, and uncontrolled (e.g., solar energy). The additional energy requirements of EVs would need to be met in a strategic and controlled manner with minimal impact on the electricity grid. With its decentralized and distributed characteristics, blockchain will be key to building and operating such ecosystems. Secondly, blockchain will also enable numerous other applications in areas like battery swapping, V2V charging, privacy, dynamic pricing, analytics, and the insurance industry by offering immutable, transparent, and pseudo-anonymous data-storage systems. Finally, blockchain enables the true potential of EV as a high-value asset that can sell energy, trade batteries, and mine cryptocurrencies on a blockchain network.

Major infrastructural concerns like electricity grid disruption, data integrity, and user privacy are key to the future of electric vehicles and clean energy. Ongoing research and development in these areas using blockchain have demonstrated several applications to realize viable, resilient, and secure systems that will form the crux of energy ecosystems of the future. As the mass adoption of electric vehicles and renewable energy continues to expand, the need for the blockchain-based applications summarized in this chapter becomes imperative in solving these challenges. However, additional research is required to assess the reliability, network threats, and innovative fail-safe systems before the production implementation of blockchain in critical infrastructure [34, 35].

References

1 Dulac, J. (2014). International Energy Agency. https://sustainabledevelopment.un.org/content/documents/23490411Globaltransport.pdf (accessed 10 February 2021).
2 Sun, X., Li, Z., Wang, X., and Li, C. (2019). Technology development of electric vehicles: a review. *Energies* 13 (1): 90.
3 IEA. (2020). Global electric car sales by key markets, 2010–2020. https://www.iea.org/data-and-statistics/charts/global-electric-car-sales-by-key-markets-2015-2020 (accessed 10 February 2021).
4 Cramer, J. and Krueger, A.-B. (2016). Disruptive change in the taxi business: the case of uber. *American Economic Review* 106 (6): 177–182.
5 Aloui, A., Hamani, N., Derrouiche, R., and Delahoche, L. (2021). Systematic literature review on collaborative sustainable transportation: overview, analysis and perspectives. *Transportation Research Interdisciplinary Perspectives* 9: Article 100291.
6 United States Department of Energy. Electric emissions. https://afdc.energy.gov/vehicles/electric_emissions.html (accessed January 2021).
7 Sullivan, B. and Taylor, H. (2021). CNBC road test: the U.S. EV charging network isn't ready CNBC. https://www.cnbc.com/2021/08/24/cnbc-road-test-the-us-ev-charging-network-isnt-ready-for-your-family-road-trip-let-alone-the-expected-wave-of-new-cars.html (accessed 23 August 2021).
8 Sun, P., Bisschop, R., and Niu, H. (2020). A review of battery fires in electric vehicles. *Fire Technology* 56: 1361–1410.
9 Nour, M., Chaves-Ávila, J., Magdy, G., and Sánchez-Miralles, Á. (2020). Review of positive and negative impacts of electric vehicles charging on electric power systems. *Energies* 13 (4675).
10 Morse, I. (2021). Millions of electric cars are coming. What happens to all the dead batteries? https://www.science.org/news/2021/05/millions-electric-cars-are-coming-what-happens-all-dead-batteries (accessed 10 February 2021).
11 Acharya, S.a.D., Pandžić, H., and Karri, R. (2020). Cybersecurity of smart electric vehicle charging: a power grid perspective. *IEEE Access* 8: 214434–214453.
12 Srivastava, G., Dhar, S., Dwivedi, A.D., and Crichigno, J. (2019). Blockchain education. *EEE Canadian Conference of Electrical and Computer Engineering (CCECE)*, Los Alamitos, CA, 10 October 2019. IEEE.
13 Gallersdörfer, U., Klaaßen, L., and Stoll, C. (2020). Energy consumption of cryptocurrencies beyond bitcoin. *Joule* 4 (9): 1836–1846.
14 Christidis, K. and Devetsikiotis, M. (2016). Blockchains and smart contracts for the Internet of Things. *IEEE Access* 4: 2292–2303.
15 Imbault, F., Swiatek, M., de Beaufort, R., and Plana, R. (2017). The green blockchain: managing decentralized energy production and consumption. *IEEE International Conference on Environment and Electrical Engineering,* Milan, Italy, 6–9 June 2017, pp. 1–5. IEEE.
16 Luo, L., Feng, J., Yu, H., and Sun, G. (2021). Blockchain-enabled two-way auction mechanism for electricity trading in internet of electric vehicles. *IEEE Internet of Things Journal* 9 (11): 8105–8118.
17 Cocco, L., Pinna, A., and Marchesi, M. (2017). Banking on blockchain: costs savings thanks to the blockchain technology. *Future Internet* 9: Article 3.
18 Ngo, H., Kumar, A., and Mishra, S. (2020). Optimal positioning of dynamic wireless charging infrastructure in a road network for battery electric vehicles. *Transportation Research Part D: Transport and Environment* 85: Article 102385.
19 Arif, S., Lie, T., Seet, B. et al. (2021). Review of electric vehicle technologies, charging methods, standards and optimization techniques. *MDPI Electronics* 10: Article 10161910.
20 Knirsch, F., Unterweger, A., and Engel, D. (2018). Privacy-preserving blockchain-based electric vehicle charging with dynamic tariff decisions. *Computer Science - Research and Development* 33: 71–79.

21 Zhang, T., Pota, H., Chu, C.-C., and Gadh, R. (2018). Real-time renewable energy incentive system for electric vehicles using prioritization and cryptocurrency. *Applied Energy* 226: 582–594.

22 Asfia, U., Kamuni, V., Sheikh, A. et al. (2019). Energy trading of electric vehicles using blockchain and smart contracts. *18th European Control Conference (ECC)*, Naples, Italy, 25–28 June 2019, pp. 3958–3963. IEEE.

23 Shao, S., Guo, S., and Qiu, X. (2017). A mobile battery swapping service for electric vehicles based on a battery swapping van. *Energies* 10: Article 10101667.

24 Ashley, M.J. and Johnson, M.S. (2018). Establishing a secure, transparent, and autonomous blockchain of custody for renewable energy credits and carbon credits. *IEEE Engineering Management Review* 4: 100–102.

25 Lightyear. (2020). Lightyear research vehicle — Tesla Model 3 with solar roof. https://www.youtube.com/watch?v=bJm2m4lbVWo (accessed July 2021).

26 Liu, H., Zhang, Y., and Yang, T. (2018). Blockchain-enabled security in electric vehicles cloud and edge computing. *IEEE Network* 32: 78–83.

27 Yang, X. and Li, W. (2020). A zero-knowledge-proof-based digital identity management scheme in blockchain. *Computers & Security* 99: Article 102050.

28 Firoozjaei, M.D., Ghorbani, A., Kim, H., and Song, J. (2019). EVChain: a blockchain-based credit sharing in electric vehicles charging. *17th International Conference on Privacy, Security and Trust (PST)*, Fredericton, NB, Canada, 26–28 August 2019. IEEE.

29 Jin, R., Wei, B., Luo, Y. et al. (2021). Blockchain-based data collection with efficient anomaly detection for estimating battery state-of-health. *IEEE Sensors Journal* 21 (12): 13455–13465.

30 Cui, P., Dixon, J., Guin, U., and Dimase, D. (2019). A blockchain-based framework for supply chain provenance. *IEEE Access* 7: 157113–157125.

31 Demir, M., Turetken, O., and Ferworn, A. (2019). Blockchain based transparent vehicle insurance management. *Sixth International Conference on Software Defined Systems (SDS)*, Rome, Italy, 10–13 June 2019. IEEE.

32 Oham, C., Jurdak, R., Kanhere, S.S. et al. (2018). B-FICA: BlockChain based Framework for Auto-Insurance Claim and Adjudication. *2018 IEEE International Conference on Internet of Things (iThings) and IEEE Green Computing and Communications (GreenCom) and IEEE Cyber, Physical and Social Computing (CPSCom) and IEEE Smart Data (SmartData)*, Halifax, NS, Canada, 30 July - 03 August 2018. IEEE.

33 Cormack, R. (2021). The Daymak Spiritus is the world's first EV that mines cryptocurrency. *Robb Report*. https://robbreport.com/motors/cars/daymak-futuristic-new-ev-mines-cryptocurrency-while-charging-1234616964/ (accessed 2021).

34 Kuang-Lo, S., Xu, X., Staples, M., and Yao, L. (2020). Reliability analysis for blockchain oracles. *Computers & Electrical Engineering* 83: Article 106582.

35 Zheng, K., Liu, Y., Craig, P. et al. (2018). Evaluating the reliability of blockchain based Internet of Things applications. *IEEE International Conference on Hot Information-Centric Networking*, Shenzhen, China, 15–17 August 2018. IEEE.

7

Starter/Generator Systems and Solid-State Power Controllers

Tao Yang, Xiaoyu Lang, and Zhen Huang

The University of Nottingham, Faculty of Engineering, University Park, Nottingham NG7 2RD, UK

7.1 Background

Air travel has become an inevitable part of today's world. Going through turbulent times with rising fuel prices, the industry has been re-engineering itself to survive and counter the impact [1]. Given the fact that fuel prices are unlikely to remain stable in the years to come, governments worldwide are urging the industry to develop more efficient and cleaner solutions. The total contribution of aircraft emissions to total anthropogenic carbon dioxide (CO_2) emission was considered to be about 2% in 1990 and this figure is estimated to increase to 3% by 2050 [2, 3]. The recently released "Flight2050" updates the ACARE targets and sets out some very tough long-term environmental goals for the industry. Compared to a new aircraft in 2000, by 2050, the aim is to reduce CO_2 emissions per passenger kilometer by 75%, emissions of oxides of nitrogen (NO_x) by 90% and the perceived noise emission of flying aircraft by 65% [4]. With more and more concern about aircraft CO_2 emission, airlines are under pressure to reduce their carbon emissions by governments concerned about global warming.

Research on alternative fuels has been ongoing for decades, but none has the energy density of aviation fuel, which is essential since minimizing the gross weight of aircraft is critical for efficient operations. Synthetic fuels have been demonstrated on aircraft and are mentioned as an alternative. However, producing these fuels is far more energy-intensive than common aviation fuels. Alternatives to fossil fuels are more practical with other forms of transportation; hence, the airline industry must focus on increasing efficiency [1].

Increasingly moving toward more-electric aircraft (MEA) is one of the few existing solutions available for the development of more efficient and environment-friendly aircraft. Driven by the development of power electronics, electric machines and advanced control technologies, many functions which are conventionally driven by hydraulic, pneumatic, and mechanical power are being replaced by electrical subsystems in the MEA [5–7]. Compared with conventional aircraft, the MEA offers significant cost benefits with lower recurring costs due to fewer parts, integration of key sub-systems, and multi-use of components. It also reduces the overall cost of operation and ownership because its more-electric architecture helps reduce fuel consumption per passenger per mile, and increase overall aircraft performance and energy usage.

Electrical systems have made significant advances over the years with the development of power electronics and electrical drive systems. The use of electrical power structure in a conventional aircraft has been illustrated by an electrical power system structure shown in Figure 7.1. Each

Transportation Electrification: Breakthroughs in Electrified Vehicles, Aircraft, Rolling Stock, and Watercraft,
First Edition. Edited by Ahmed A. Mohamed, Ahmad Arshan Khan, Ahmed T. Elsayed, and Mohamed A. Elshaer.

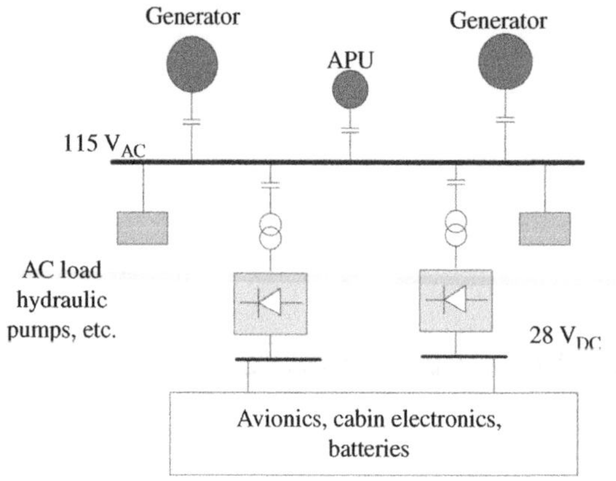

Figure 7.1 Conventional aircraft electrical system architecture.

generator delivers 115 VAC/400 Hz electrical power to the main AC bus and is controlled by its own generator control unit (GCU). The 115 VAC power is transformed to 28 VDC power using Transformer Rectifier Units (TRUs). TRUs consist of a multiphase transformer and an n-pulse diode rectifier, where $n = 12$ or 18, to reduce the ripple on the DC-link and to achieve the power quality requirements. The electrical loads supplied by the 28 VDC bus are the avionics, cabin electronics and the back-up batteries. Other AC electrical loads, such as lighting, galley loads, entertainment system and auxiliary hydraulic pumps are directly fed by the AC bus.

7.2 Future Design Options

The application of the MEA concept will lead to the use of a large number of motor drives for functions such as fuel pumping, cabin pressurization, air conditioning, engine start, and flight control actuation, making power electronics an essential part of MEA technologies [5].

Due to the increase of electrical power as well as the application of power electronics in future aircraft, different electrical power system architectures are emerging recently including 540 Vdc (±270 V), 230 Vac at 400 Hz and 230 Vac at variable frequencies (320–800 Hz) [7]. As a result, the main bus will not necessarily be the 115 Vac HVAC bus. Figure 7.2 shows a possible MEA electrical power system developed within the EU more-open electrical technologies (MOET) project [8]. The electrical system shown is split into two main AC buses, AC bus 1 and AC bus 2, which are fed by two synchronous generators, SG1 and SG2, respectively. These generators can be driven by one engine or by different engines. The voltages at the outputs of these generators are controlled by the GCUs. Large loads on these AC buses include the wing ice protection system and the autotransformer rectifier units, which feed the high-voltage direct current (HVDC) buses, DC bus 1 and DC bus 2. In addition to the two main AC buses, there is an essential bus, which can be fed from either generator for redundancy consideration. This essential bus is used for the flight critical actuation systems, represented in this diagram by two electromechanical actuators (EMA1 and EMA2), which are driven by permanent magnet motor (PMM) drives. The most significant loads found on the DC buses are the environmental control systems (ECSs), which maintain the temperature and pressure of the passenger cabin of civilian aircraft.

Figure 7.3 shows a potential MEA electrical power system layout with HVDC buses. The AC power from the two main starters/generators is transformed into DC power through bidirectional AC/DC converters. The flight control actuation system, the environment control system (ECS), and the actuators are driven electrically through DC/AC converters. The main advantage of the DC

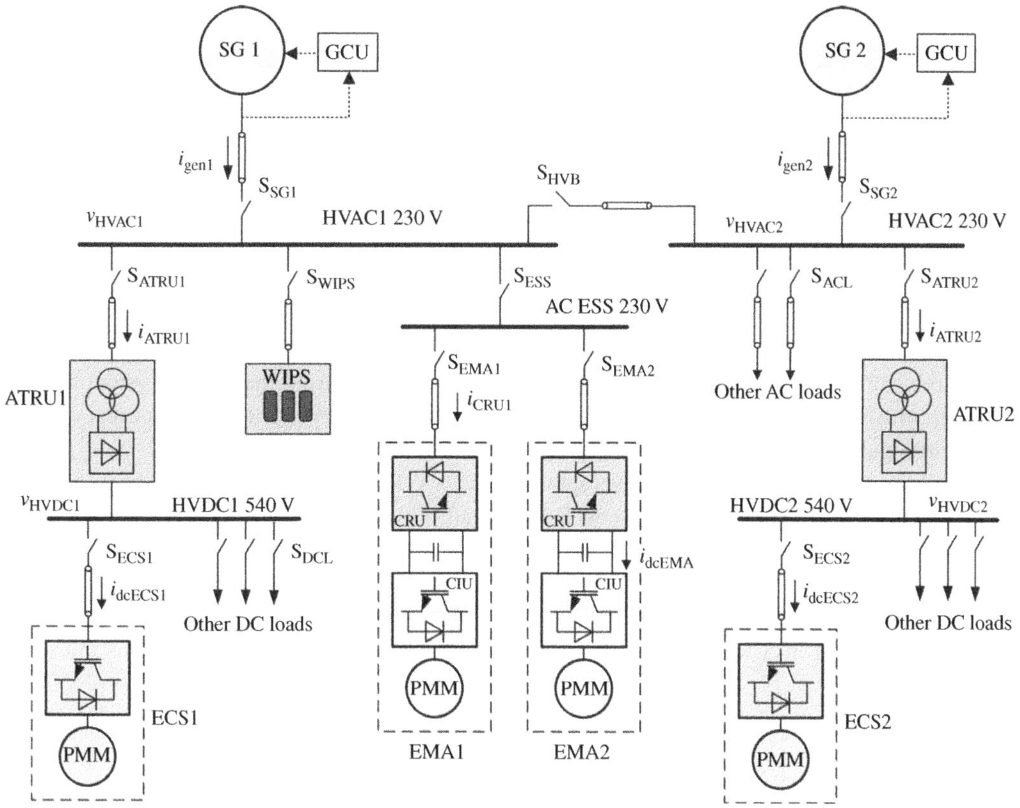

Figure 7.2 MOET aircraft electrical power system architecture. *Source:* 2014 IEEE.

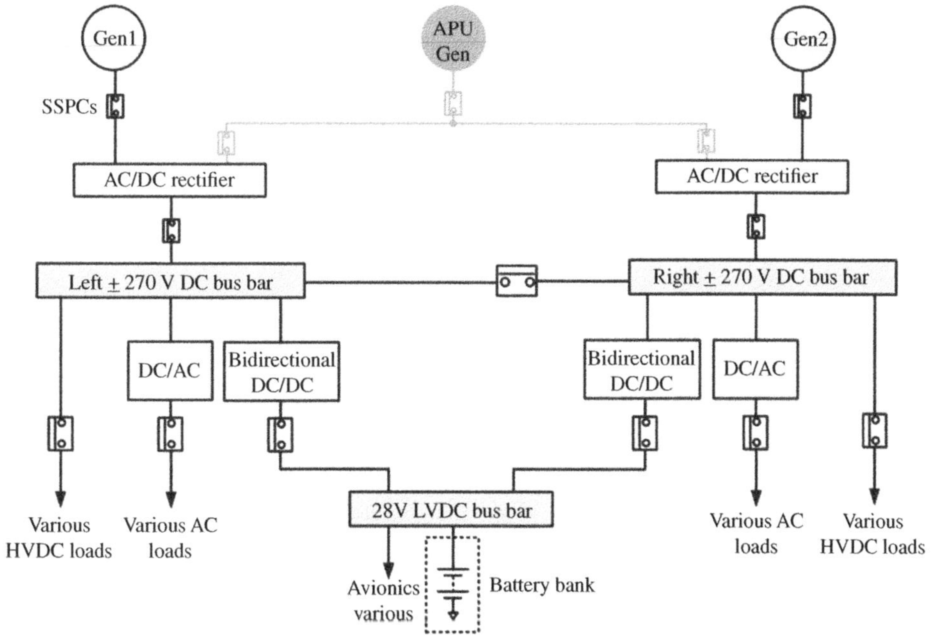

Figure 7.3 Architecture of a DC electrical power system for MEA. *Source:* 2018 IEEE [9].

system includes less cable weight as it only requires two cables instead of three cables in the AC systems. It also decouples the generator frequency from that of the main distribution system [10]. Furthermore, DC distribution readily permits the paralleling of multiple generators onto a single bus [11] and enables the application of variable-frequency power to be more convenient. The choice of DC power distribution, however, is also associated with some issues, for example, the safety aspect requires safe isolation of power buses carrying fault currents [12]; the demanding protection requirements under fault conditions [13]; the transient voltage disturbance issues during regeneration [14]. DC power system architectures for aircraft are important areas of study.

To design an optimal electrical system for the future MEA, four candidates, including constant-speed constant-frequency (used by Boeing 777 and Airbus A340 [15]), variable-speed 115 VAC with 270 VDC (used by Airbus A380), variable-speed 230 VAC with ±270 VDC (used by Airbus A380), and purely ±270 VDC, have been investigated in [9]. After comparing the overall system weight together with stability analysis, it can be concluded that the purely ±270 VDC electrical system shown in Figure 7.3 is the most promising option [9].

US military has emphasized the development of 270 VDC systems, which have been already adopted by the Lockheed Martin F-22 and F-35 aircraft [16]. This approach is more efficient with no reactive power and a lightweight solution by reducing the amount of required power conversion, allowing parallel asynchronous operation of electric motors, compared to the previous AC distribution system. To further reduce aircraft weight, a higher voltage, i.e. 540 VDC (±270 VDC), is recommended in an electrical system to decrease the current and, hence, weights of cables and cooling system equipment [7, 17].

7.3 The Starters/Generators and Their Power Electronics Control

As designs move toward the MEA concept, more power needs to be extracted from the primary source, the engine, and converted into electrical power to feed onboard loads. Electrical generators can be coupled to both the low-pressure spool (LPS) and the high-pressure spool (HPS) of the engine to exploit the power supply capability of the engine. Generally, the LPS and the HPS have different speeds. This means different AC frequencies from the LP generator (LPG) and the HP generator (HPG). This prevents these generators to be connected in parallel due to conflict in frequencies. Various power-generation centers (PGCs) have been proposed for this application [18–20]. Among them, a single-bus dual-channel PGC, shown in Figure 7.4, can handle the parallel operation of multiple generators and has potential benefits in terms of weight, efficiency, availability, and cost [21]. Moreover, the twin-generator architecture meets the redundancy and requirements for extended-range twin-engine operations [22].

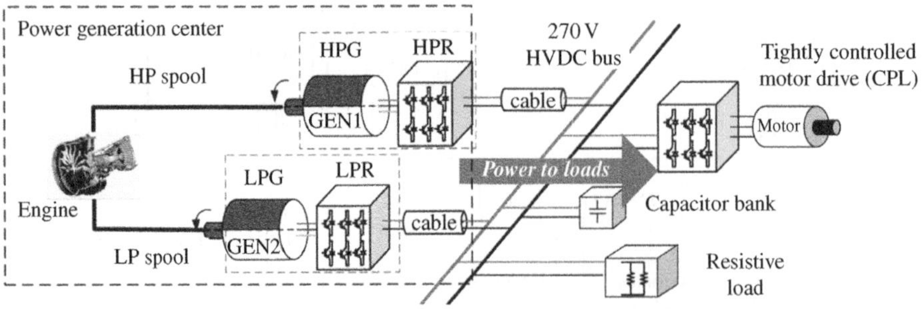

Figure 7.4 A single-bus dual-channel power-generation center. *Source:* 2021 IEEE [18].

In this architecture, the rectifiers convert the output of the generators into a single HVDC bus through the supervisory controller to share the load between the power sources. Then, the DC power can be routed by the distribution system around the airframe.

7.4 System Analysis and Control Design

To ensure normal operation of the PGC in Figure 7.4, different objectives should be met:

1) Regulating a stable DC bus voltage.
2) Appropriate load power sharing between the HP shaft and LP shaft generators.
3) Regulating the stator currents and torque of the generators.

To fulfil this control objective, Figure 7.5 shows the control block diagram for the generator-rectifier system. By controlling the magnetic flux on the *d*-axis and the active power on the *q*-axis, the PMSG can operate in starter or generator mode. The conventional proportional-integral (PI) controller is used to control the excitation component (*d*-axis) and the torque component (*q*-axis). When the rotary speed is lower than the base speed (constant torque region), the *d*-axis current component is equal to zero, ensuring the maximum torque to current ratio. In the high-speed operation area, the back electromotive force (EMF) of the PMSG, which is proportional to the speed, may be higher than the AC voltage of rectifier. The maximum value V_{cmax} of the reference voltage of the rectifier defines the constant power area where the flux weakening (FW) is implemented. The

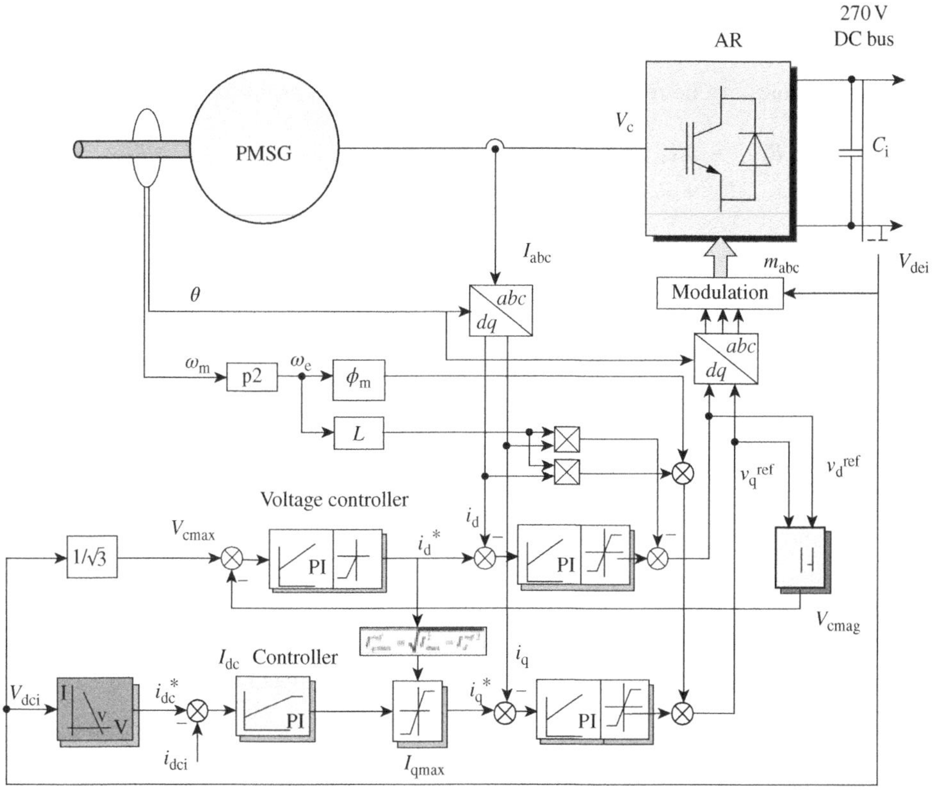

Figure 7.5 Control block diagram for the generator-rectifier system. *Source:* 2017 IEEE.

stator current references on the d and q axes are obtained from the output of the FW controller and the I_{dc} controller, respectively. The reference of the AC voltage limit V_{cmax} depends on the DC-link voltage. The DC current reference i_{dc}^* is determined by the droop characteristic. This control scheme is applied to both LP and HP generator-rectifier systems.

In the control block diagram, there are two control objectives: the main HVDC bus voltage and the stator currents of the generator. Instead of directly adjusting the DC bus voltage, the DC current I_{dc} is a control target for adjusting the appropriate active power injected into the DC bus. The output of I_{dc} controller determines i_q^*. The FW loop regulates the stator voltage magnitude V_{cmag}, whose output is d-axis current reference i_d^*. The control design for each controller, i.e. current control, FW control, and DC voltage control, is discussed in the following sections.

7.4.1 Current Control Design

The PMM-based generators are considered for both HP and LP power-generation channels. The electrical dynamics of PMM in the dq rotary frame are given as follows:

$$\begin{cases} \dfrac{di_d}{dt} = \dfrac{1}{L_d}\left(v_d - R_s i_d + \omega_e L_q i_q\right) \\ \dfrac{di_q}{dt} = \dfrac{1}{L_q}\left(v_q - R_s i_q - \omega_e L_d i_d - \omega_e \psi_m\right) \end{cases} \tag{7.1}$$

where v_d and v_q are stator voltages in d and q axes, respectively; i_d and i_q are stator currents in d and q axes, respectively; L_d and L_q are stator inductance; R_s is the stator resistance; ψ_m is the flux linkage of permanent magnet; ω_e is the electrical angular speed in rad/s. For a surface-mounted PMM, $L_d = L_q = L_s$.

The electrical dynamics can be rewritten using Laplace transform as follows:

$$\begin{bmatrix} v_d \\ v_q \end{bmatrix} = \begin{bmatrix} L_s s + R_s & -\omega_e L_s \\ \omega_e L_s & L_s s + R_s \end{bmatrix} \begin{bmatrix} i_d \\ i_q \end{bmatrix} + \begin{bmatrix} 0 \\ \omega_e \psi_m \end{bmatrix} \tag{7.2}$$

Based on (7.2), the control plant and respective current loops can be derived as shown in Figure 7.6. It can be seen that for the current control plant, the dq-axes variables are cross coupled. The control performance of i_d will be affected by i_q and vice versa. To eliminate the cross-coupling effect, a feedforward decoupling strategy is applied [18]. With a classical PI structure [23], the AC voltages of a generator are given as follows:

$$\begin{cases} v_d = \left(K_{iP} + \dfrac{K_{iI}}{s}\right)\left(i_d^* - i_d\right) - \omega_e L_s i_q \\ v_q = \left(K_{iP} + \dfrac{K_{iI}}{s}\right)\left(i_q^* - i_q\right) + \omega_e L_s i_d + \omega_e \psi_m \end{cases} \tag{7.3}$$

where K_{iP} and K_{iI} are the P and I gains of the current controller, respectively. i_d^* and i_q^* are the current references.

From (7.3) and Figure 7.6, the feedforward decoupling terms in current loops are chosen as below:

$$v_{d\_decouple} = -\omega_e L_s i_q \qquad v_{q\_decouple} = \omega_e L_s i_d + \omega_e \psi_m \tag{7.4}$$

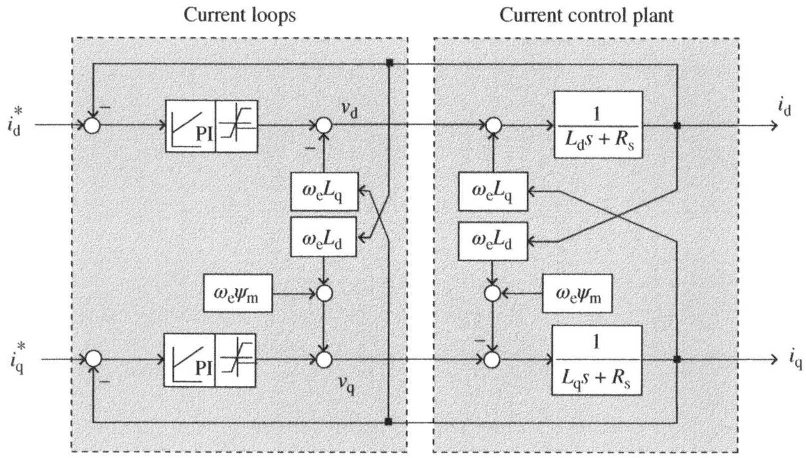

Figure 7.6 Control plant and control loops. *Source:* 2020 IEEE.

Applying (7.3) into (7.2), relationships between current references and feedback can be derived as follows:

$$\begin{bmatrix} \dot{i}_d \\ \dot{i}_q \end{bmatrix} = \begin{bmatrix} g & 0 \\ 0 & g \end{bmatrix} \begin{bmatrix} i_d \\ i_q \end{bmatrix} + \begin{bmatrix} n & 0 \\ 0 & n \end{bmatrix} \begin{bmatrix} i_d^* \\ i_q^* \end{bmatrix} \tag{7.5}$$

where $g = -\left[\left(K_{iP} + \frac{K_{iI}}{s}\right) + R_s\right]/L_s, n = \left(K_{iP} + \frac{K_{iI}}{s}\right)/L_s$. Equation (7.5) indicates that the variation of dq currents is only affected by themselves and their references. Therefore, a decoupled control is achieved. The *dq* axes closed-loop transfer function can be derived as follows:

$$G_{c.1}(s) = \frac{i_d(s)}{i_d^*(s)} = \frac{i_q(s)}{i_q^*(s)} = \frac{K_{iP}s + K_{iI}}{L_s s^2 + (K_{iP} + R_s)s + K_{iI}} \tag{7.6}$$

The denominator of the closed-loop transfer function mainly determines the characteristic of the current loop. Hence, the controller parameters can be tuned by comparing (7.6) to a desired second-order system with the characteristic equation given by:

$$c(s) = s^2 + 2\zeta\omega_n s + {\omega_n}^2 \tag{7.7}$$

Then the controller parameters can be derived:

$$K_{iP} = 2\zeta\omega_n L_s - R_s, K_{iI} = {\omega_n}^2 L_s \tag{7.8}$$

where ω_n is the natural frequency, ζ is the damping ratio. Generally, the damping ratio is set as 0.707, and the natural frequency can be determined by the targeted control bandwidth (ω_b) [24]:

$$\omega_n = \frac{\omega_b}{\sqrt{1 - 2\zeta^2 + \sqrt{4\zeta^4 - 4\zeta^2 + 2}}} \tag{7.9}$$

Although the tuning method is easy to implement, it has disadvantages because of the zero in the closed-loop transfer function (7.6). To analyze the impact of this additional zero, (7.6) can be rearranged as follows:

$$G_{c.1}(s) = \underbrace{\frac{K_{iP}s}{L_s s^2 + (K_{iP} + R_s)s + K_{iI}}}_{G_1(s)} + \underbrace{\frac{K_{iI}}{L_s s^2 + (K_{iP} + R_s)s + K_{iI}}}_{G_2(s)} \tag{7.10}$$

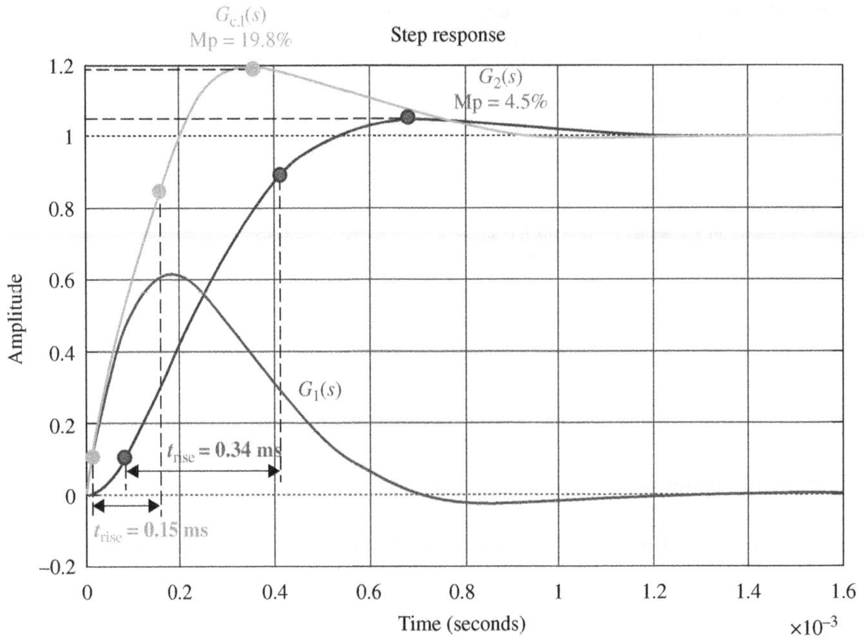

Figure 7.7 Step response of $G_1(s)$, $G_2(s)$, and $G_{c.l}(s)$. *Source:* 2020 IEEE.

It can be seen in (7.10) that the system response has two components, $G_2(s)$ represents the desired response and $G_1(s)$ refers to the responses due to zero. To study the effect of the zero on control performance, step response tests are carried out for $G_1(s)$, $G_2(s)$, and $G_{c.l}(s)$, as shown in Figure 7.7. The PI gains are calculated from (7.8) and (7.9) using $\zeta = 0.707$, $\omega_b = 2\pi \times 1000$ Hz. $L_s = 100\ \mu$H, $R_s = 53$ mΩ as shown in Table 7.1. It can be seen that the overshoot (Mp) of $G_2(s)$ is 4.5%, while with the effect of zero, the Mp significantly rises to 19.8%. The rise time decreases from 0.34 to 0.15 ms due to the zero. Therefore, it can be concluded that the zero can improve dynamic performance but indicate a lower stability margin.

In general, the selection of ω_b for the current loop privileges a fast dynamic response. However, this will introduce a large overshoot and may lead to steady-state oscillation [25]. Thus, the

Table 7.1 Parameters of the AEGART project PMM.

Parameter	Rated value
Motor power	45 kW @270VDC
Rated speed	8000 rpm
Rated current	170 A
Poles	6
Phases	3
Stator resistance	53 mΩ
Stator inductance (in *dq* axes)	100 μH
Flux linkage of magnet	0.0365 Wb

Source: 2019 IEEE.

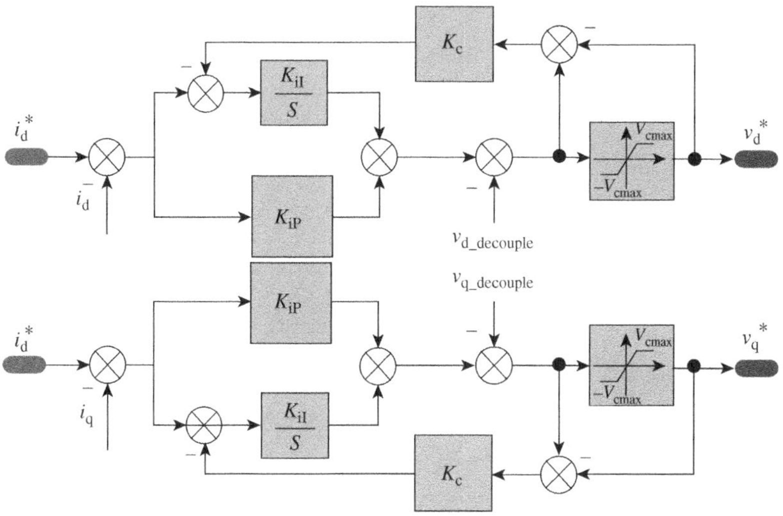

Figure 7.8 Current controller with anti-windup and decoupling terms in (7.4). *Source:* 2020 IEEE.

selection control bandwidth should be a trade-off. According to the machine parameters in Table 7.1, where $L_s = 100\ \mu\text{H}$, $R_s = 53\ \text{m}\Omega$, and the time constant is $L_s \div R_s = 1.88$ ms, equivalent to 530 Hz. The switching frequency is $f_{sw} = 16$ kHz, according to [26], ω_b should be within $0.18 f_{sw}$, namely 2.8 kHz. Hence, the control bandwidth f_b should be within the following range:

$$530\ \text{Hz} < \omega_b < 2.8\ \text{kHz} \tag{7.11}$$

For meeting the condition in (7.11), the current controller is designed with 1 kHz bandwidth as a trade-off between dynamic and steady-state performance. Then the *P* and *I* parameters can be obtained from (7.8).

The overall structure of current controller for both HP and LP power-generation channels is demonstrated as shown in Figure 7.8. The *dq*-axes voltage commands, i.e. the outputs of current controllers are limited. When the output of the PI controller goes into a limitation, the current controller will disengage, but its integrator will continue to accumulate without affecting any control actions. This may lead to a high current overshoot, slow down the settling time, and eventually lead to instability.

To minimize the impact of oversaturation on control performance, an anti-windup technique is adopted [23]. Figure 7.8 shows that in the used anti-windup scheme, the deviation between the limited voltage commands and the unbounded voltage commands is passed through a proportional gain (K_c), feeding to limit the error to the integrator, thus avoiding windup. The anti-windup parameters can be tuned as follows [23]:

$$K_c = \frac{K_{iI}}{K_{iP}} \tag{7.12}$$

7.4.2 Field-Weakening Control Design

In the starter mode of the SG, it performs as a motor to crank the HPS until reaching the self-sustaining speed. While in the majority of flight time, the SG operates as a generator, supplying electric power to the downstream loads. The requirements of engine start (cranking the engine to a self-sustaining speed by the motoring mode of the SG) and power generation (generator mode of the SG)

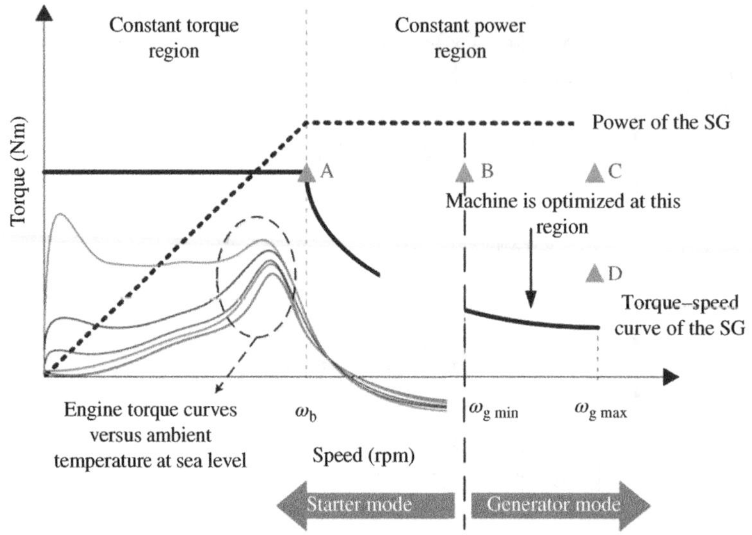

Figure 7.9 Torque and speed characteristics of the SG-converter system. Point A: base speed, maximum torque. Point B: minimum generator speed, maximum torque. Point C: maximum speed, maximum torque. Point D: maximum speed, minimum requirements imposed by the engine torque. *Source:* 2020 IEEE.

make the design of the SG system include coordinated consideration of machine-converter interactions. Since it operates as a generator in most flight time, the converter must therefore be able to deal with high speeds and associated large-induced back EMF. This requires a high-voltage rating for the converter. However, the same converter must also be able to handle the starting mode, which requires high torque, hence a high current rating. Overall, it means that the kVA rating of the converter should be selected at the maximum value based on point C in Figure 7.9.

However, unfortunately, at point C, it means that the converter will be significantly underutilized under normal operating scenarios. Furthermore, the converter kVA rating is related to the weight and volume of the converter, which is critical for aircraft applications. Although these can be reduced by moving the machine's electromagnetic design point down the saturation curve, this will increase the size and weight of the SG [27]. A solution to reduce size and weight of the SG is to implement a set of reconfigurable winding, such as switching from series to parallel connection, in different operation modes (for example, selecting point B or D in Figure 7.9 for design). However, this will lead to an increased number of switches and less efficiency due to additional losses in switching devices [28].

To reduce the kVA rating of the converter, another option is to operate the SG in the FW mode [29]. In this case, when above the base speed shown by point A in Figure 7.9, a negative d-axis current is injected into the machine and the flux linkage is gradually reduced, limiting the back EMF. This means that the voltage rating of a converter can be substantially reduced at point A compared with B, C, and D. Therefore, FW control becomes a must which brings benefits for the aircraft SG system since the 270V DC link voltage is typically low compared to the induced back EMF in practical high-speed generation applications.

According to (7.1), the steady-state voltage limit equation of SG can be derived as follows:

$$v_d^2 + v_q^2 \le V_{cmax}^2 \Rightarrow \left(R_s i_d - \omega_e L_s i_q\right)^2 + \left(R_s i_q + \omega_e L_s i_d + \omega_e \psi_m\right)^2 \le V_{cmax}^2 \tag{7.13}$$

The aircraft electrical generation system with active rectification and health monitoring (AEGART) project has developed a beyond state-of-the-art type of starter-generator system for MEA [30]. Based on the AEGART machine parameters in Table 7.1, the operational limits and

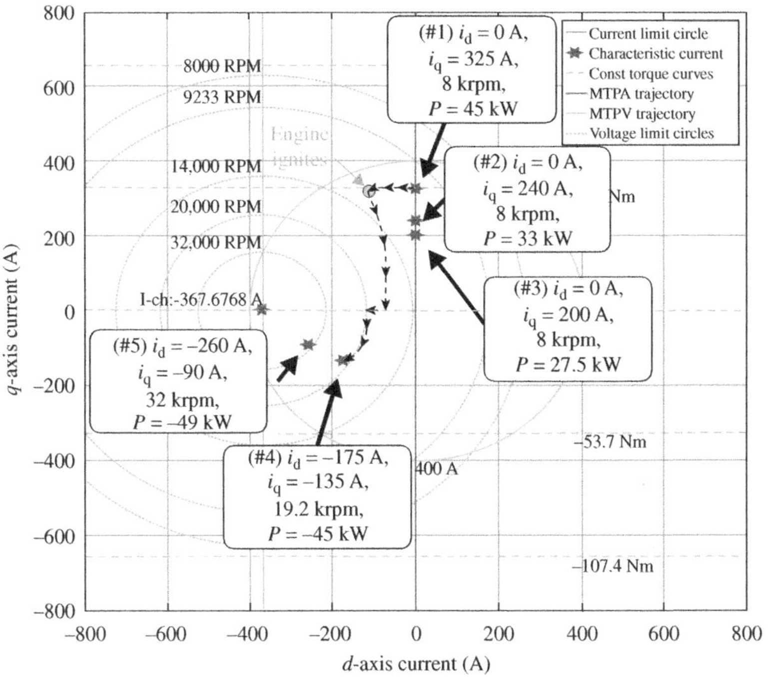

Figure 7.10 Operating trajectory of the SG drive system. *Source:* 2020 IEEE.

trajectory of the SG drive system are presented in Figure 7.10, where the voltage limit circles are plotted based on (7.13). In the starter mode, the machine goes into point #1, cranking the engine shaft at its maximum torque. When the speed increases beyond the base speed (8 krpm), the FW operation is automatically activated since the voltage reference magnitude is larger than the actually available voltage as shown in Figure 7.5. A negative *d*-axis current is injected into the machine according to the error between the reference voltage and the voltage limit V_{cmax} set by the inverter (Figure 7.5). The magnetic field generated by the *d*-axis current is opposite to the magnetic field of the permanent magnet on rotor, diminishing the induced back EMF. At 10 krpm, the engine ignites, and the SG moves to standby mode, where the *q*-axis current falls to zero. The generation mode starts at 20 krpm and outputs up to 45 kW power indicating by point #4 with significantly negative *d*-axis current flowing into the machine.

From this analysis, the control logic of the FW controller can be derived as follows: when the stator voltage is less than the maximum value V_{cmax}, the output of the FW controller, i.e. *d*-axis current reference should be zero. When the stator voltage exceeds the maximum voltage limit, the controller starts to operate, generating a negative d-axis current reference. The magnitude of current reference needs to increase as the increase of speed to maintain constant stator voltage. A classical PI control structure is applied for FW purpose [31]:

$$\begin{cases} i_d{}^* = \dfrac{K_{FWp}s + K_{FWi}}{s}(V_{cmax} - v^*) \\ v^* = \sqrt{v_d{*}^2 + v_q{*}^2}, \quad i_d{}^* \in [-I_{max},\ 0] \end{cases} \tag{7.14}$$

where K_{FWp} and K_{FWi} is the P and I gain of the FW controller, respectively. $V_{cmax} = v_{dc}/v_{dc}\sqrt{3}$when using the space-vector pulse width modulation (SVPWM). I_{max} is the machine's current limitation.

The d-axis current set-point given in (7.14) is automatically adjusted by voltage feedback through tracking the voltage constraints as the speed varies [31]. Unlike model-based FW controls, such as model-predictive control [32], the tuning of control gains in (7.14) does not require accurate machine parameters. It can maintain consistent control performance when the motor parameters change due to temperature rise or magnetic saturation. Hence, it is considered robust against machine parameter variation [31].

7.4.3 Analysis and Control Design of the DC Voltage Loop

As shown in the system schematic diagram in Figure 7.4, the DC bus can be regarded as the "energy interface" between the PGC and the various onboard loads. The onboard loads acquire power from the DC bus, whilst the PGC supplies power to the DC bus. To ensure the normal operation of the generators and loads, the DC bus voltage should be actively regulated. In this subsection, a DC voltage control loop is established to fulfil the regulation of the DC bus voltage, where the control plant and control design are specifically analyzed.

7.4.4 DC Bus Voltage Control: The Control Plant

Since multiple power sources are involved in the PGC, appropriate power sharing between power sources is required. A current-mode droop control method presented in [33] is adopted due to its advantages including the absence of a communication link, high modularity, and immunity from the impact of cable impedance. The power-sharing is achieved by splitting the total load current i_{dc} into currents i_{dcLP} and i_{dcHP}. Using droop control characteristic, the DC current reference can be derived as follows:

$$i_{dc}{}^{*} = \frac{v_{dc}{}^{*} - v_{dc}}{g_D} \tag{7.15}$$

where g_D is the droop gain. With the current-mode droop control, the output of control plant should be the DC current i_{dc} to regulate DC voltage changes caused by the load currents. The DC link equation can be formulated as:

$$C\frac{dv_{dc}}{dt} = i_{dc} - i_L \tag{7.16}$$

The control plant for i_{dc} control can be derived from the electrical dynamics of PMM shown in (7.1) and the DC link dynamics shown in (7.16) using small signal analysis, given as follows:

$$\begin{aligned}&\left[\overline{v}_{dc} - \frac{3\left(\overline{v}_d\overline{i}_d + \overline{v}_q\overline{i}_q\right)}{2\overline{v}_{dc}Cs}\right]\Delta i_{dc} = -\frac{3}{2}\left(\overline{v}_d\Delta i_d + \overline{v}_q\Delta i_q\right) - \\ &\frac{3}{2}\left\{\overline{i}_d\left[(R_s + L_s s)\Delta i_d - \omega_e L_s\Delta i_q\right] + \overline{i}_q\left[(R_s + L_s s)\Delta i_q + \omega_e L_s\Delta i_d\right]\right\}\end{aligned} \tag{7.17}$$

where the superscript "-" indicates the selected operating points. i_d and i_q are the stator currents of the PMM generator in dq frame. v_d and v_q are the terminal voltages of PMM generator in *dq* frame. The relation shown in (7.17) can be rearranged as follows:

$$\Delta i_{dc} = -\frac{3\overline{v}_{dc}Cs\left[\overline{v}_d + \omega_e L_s\overline{i}_q + (R_s + L_s s)\overline{i}_d\right]\Delta i_d}{2\overline{v}_{dc}{}^2Cs - 3\left(\overline{v}_d\overline{i}_d + \overline{v}_q\overline{i}_q\right)} - \frac{3\overline{v}_{dc}Cs\left[\overline{v}_q - \omega_e L_s\overline{i}_d + (R_s + L_s s)\overline{i}_q\right]\Delta i_q}{2\overline{v}_{dc}{}^2Cs - 3\left(\overline{v}_d\overline{i}_d + \overline{v}_q\overline{i}_q\right)} \tag{7.18}$$

Based on (7.18) and the current loop transfer function shown in (7.6), the control plant can be derived as shown in Figure 7.11, where the expression of f_d and f_q are given as follows:

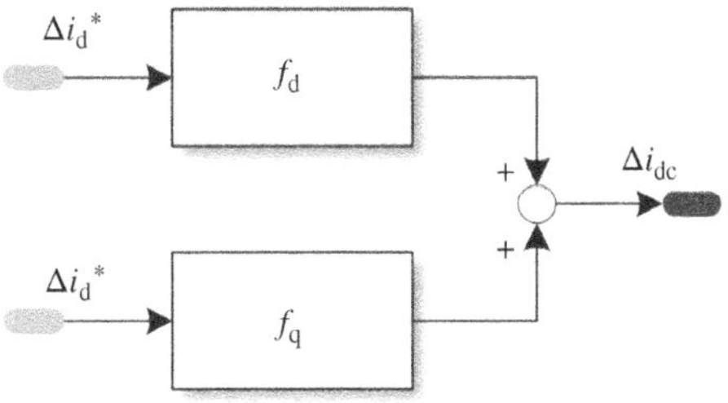

Figure 7.11 Control plant for the i_{dc} control. *Source:* 2020 IEEE.

$$\begin{cases} f_{\mathrm{d}} = -\dfrac{K_{\mathrm{iP}}s + K_{\mathrm{iI}}}{L_{\mathrm{s}}s^2 + (K_{\mathrm{iP}} + R_{\mathrm{s}})s + K_{\mathrm{iI}}} \dfrac{3\overline{v}_{\mathrm{dc}}Cs\left[\overline{v}_{\mathrm{d}} + \omega_{\mathrm{e}}L_{\mathrm{s}}\overline{i}_{\mathrm{q}} + (R_{\mathrm{s}} + L_{\mathrm{s}}s)\overline{i}_{\mathrm{d}}\right]}{2\overline{v}_{\mathrm{dc}}{}^2Cs - 3\left(\overline{v}_{\mathrm{d}}\overline{i}_{\mathrm{d}} + \overline{v}_{\mathrm{q}}\overline{i}_{\mathrm{q}}\right)} \\ f_{\mathrm{q}} = -\dfrac{K_{\mathrm{iP}}s + K_{\mathrm{iI}}}{L_{\mathrm{s}}s^2 + (K_{\mathrm{iP}} + R_{\mathrm{s}})s + K_{\mathrm{iI}}} \dfrac{3\overline{v}_{\mathrm{dc}}Cs\left[\overline{v}_{\mathrm{q}} - \omega_{\mathrm{e}}L_{\mathrm{s}}\overline{i}_{\mathrm{d}} + (R_{\mathrm{s}} + L_{\mathrm{s}}s)\overline{i}_{\mathrm{q}}\right]}{2\overline{v}_{\mathrm{dc}}{}^2Cs - 3\left(\overline{v}_{\mathrm{d}}\overline{i}_{\mathrm{d}} + \overline{v}_{\mathrm{q}}\overline{i}_{\mathrm{q}}\right)} \end{cases} \tag{7.19}$$

Before moving to the control design, the control plant is verified with a nonlinear equivalent model in the Matlab/Simulink environment as presented in Appendix. The derived control plant is verified by comparison with the non-linear model in Simulink. The characteristics of the derived control plant depend on the operating point. The operating point used for control plant verification is obtained in the steady state, as shown in Table 7.2.

With the operating points shown in Table 7.2, the expressions of f_{d} and f_{q} can be derived as follows:

$$\begin{cases} f_{\mathrm{d}} = \dfrac{0.87s + 3908}{10^{-4}s^2 + 0.928s + 3908} \dfrac{0.0128s^2 + 14.88s}{174.96s + 37621} \\ f_{\mathrm{q}} = \dfrac{0.87s + 3908}{10^{-4}s^2 + 0.928s + 3908} \dfrac{0.0058s^2 - 216.14s}{174.96s + 37621} \end{cases} \tag{7.20}$$

The step responses of the derived control plant and the nonlinear model are presented in Figure 7.12. It can be seen that the step responses of the control plant and nonlinear model are very close in terms of transient and steady-state performances. Hence, the derived control plant is verified. Moreover, it can be seen from (7.19) that there is a positive zero in f_{q}. This positive zero will lead to a non-minimum phase characteristic, delaying the response and causing an opposing response to the input [34]. This is the reason why an initial undershoot occurs in Figure 7.12b.

To study how the poles and zeros of the control plant will move with different operating points, the closed-loop response of the plant was investigated, where a proportional (P) controller is used since the P controller will not introduce additional poles or zeros. The proportional gain is denoted

Table 7.2 Operating point used for control plant verification.

Parameter	Value	Parameter	Value
$\overline{v}_d$	30 V	$\overline{i}_d$	−128 A
$\overline{v}_q$	143 V	$\overline{i}_q$	−60 A
$\overline{v}_{dc}$	270 V	ω_{e}	6283 rad/s (20 krpm)
K_{iP}	0.87	K_{iI}	3908

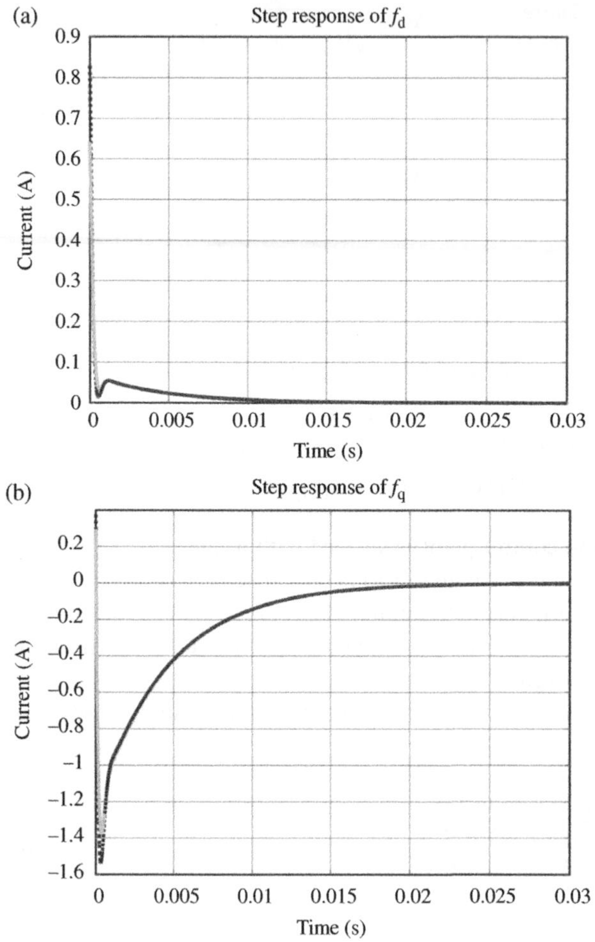

Figure 7.12 Step response of the derived control plant (light grey curves) and the nonlinear equivalent model in Simulink (dark grey curves). (a) step response of f_d with $\Delta i_d = 1$ A. (b) step response of f_q with $\Delta i_q = 1$ A. *Source:* 2020 IEEE.

as K_{vP}. Figures 7.13 and 7.14 show the closed loop root locus at different speeds of a generator and different load powers. It can be seen that the closed loop poles move toward the right half plane (RHP) as the increase of K_{vP}. The value of K_{vP} must be carefully selected to ensure that the poles do not move across the imaginary axis into the RHP, otherwise the system will become unstable.

It can be seen from Figure 7.13 that as the speed increases, the allowable value for K_{vP} decreases from 1.14 to 0.59. Also, from Figure 7.14, it can be seen that as the load power increases, the allowable value for K_{vP} decreases from 0.85 to 0.66. These results show that the worst operating point corresponds to the highest speed and full power. Hence, this operating point is specifically considered when designing the i_{dc} PI controller.

7.4.5 DC Bus Voltage Control Design

In the proposed APGC (Advanced Power Generation Center), the DC-link voltage within the back-to-back converter can be set to a high value, ensuring the HP generator operates without field weakening. Thus, i_d can be controlled to be zero for the surface-mounted PMM in use. Hence, a

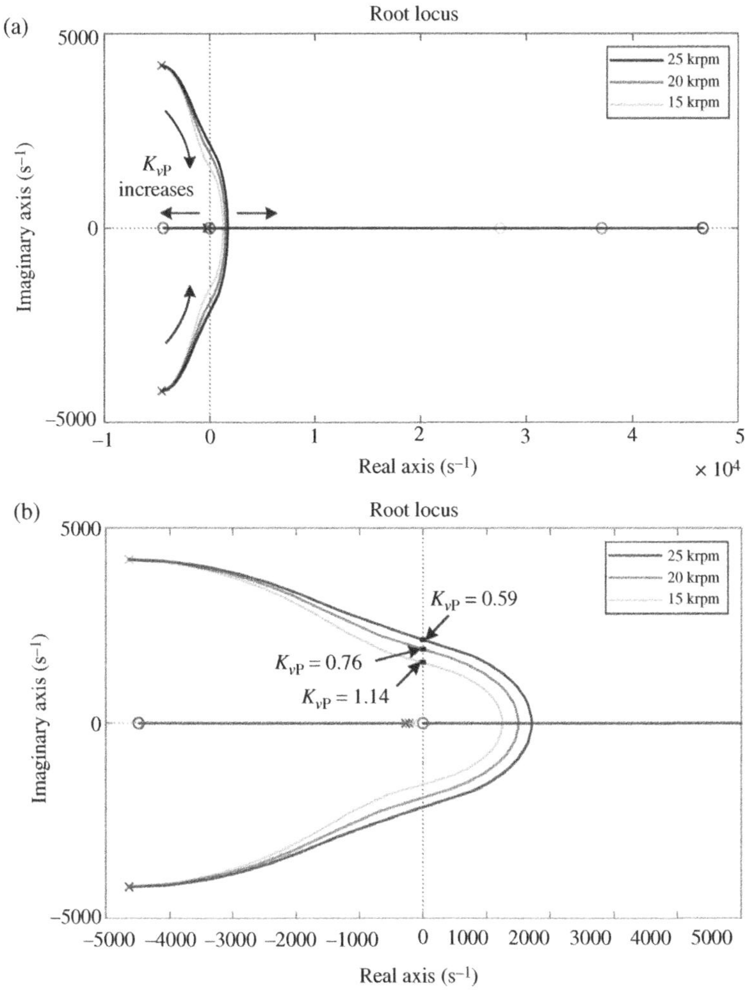

Figure 7.13 Closed loop root locus of control plant at different speeds. (a) overview. (b) zoomed area. Load power is 15 kW. *Source:* 2020 IEEE.

reasonable assumption was made for the following control design, which is that the influence of i_d is neglected. The dynamics of Δi_{dc} are dominantly contributed by Δi_q. With this assumption, the block diagram of the DC bus voltage control can be derived as shown in Figure 7.15.

As shown in Figure 7.15, a PI controller is applied to regulate the inner i_{dc} control loop. The controller can be expressed as:

$$G_{pi}(s) = K_{vP} + \frac{K_{vI}}{s} \tag{7.21}$$

With the f_q in (7.19) and (7.21), the open loop transfer function of the i_{dc} control loop can be written as follows:

$$G_{idc}(s) = -\frac{(K_{iP}s + K_{iI})(K_{vP}s + K_{vI})}{L_s s^2 + (K_{iP} + R_s)s + K_{iI}} \frac{3\overline{v}_{dc}C\left[\overline{v}_q - \omega_e L_s \overline{i}_d + (R_s + L_s s)\overline{i}_q\right]}{2\overline{v}_{dc}^2 Cs - 3\left(\overline{v}_d\overline{i}_d + \overline{v}_q\overline{i}_q\right)} \tag{7.22}$$

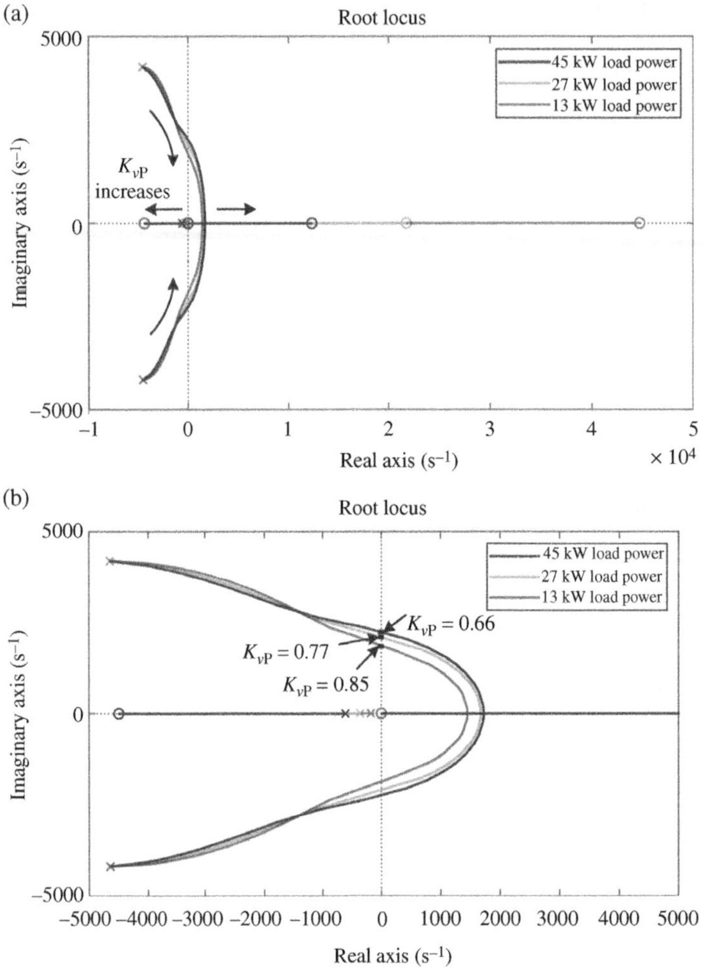

Figure 7.14 Closed loop root locus of control plant at different load powers. (a) overview. (b) zoomed area. Operating speed is 20 krpm. *Source:* 2020 IEEE.

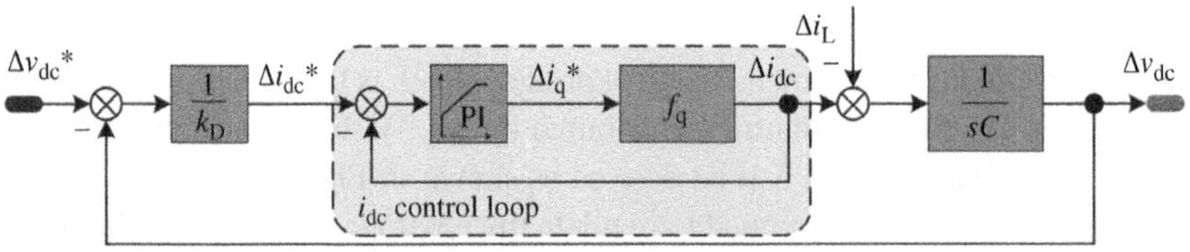

Figure 7.15 Block diagram of the DC bus voltage control. *Source:* 2020 IEEE.

As can be seen in (7.22), there are three poles in $G_{idc}(s)$, and two out of the three are conjugate poles brought by the current loop transfer function. To study the effect of the three poles, their locations on the complex plane are plotted with respect to different speeds and load power, as shown in Figure 7.16. It can be seen from Figure 7.16 that the location of the two conjugate poles is fixed since it only depends on the current loop control parameters and generator parameters. The dominant pole depends on the operating point. With the increase of speed and load power, it moves toward the RHP, indicating degraded stability.

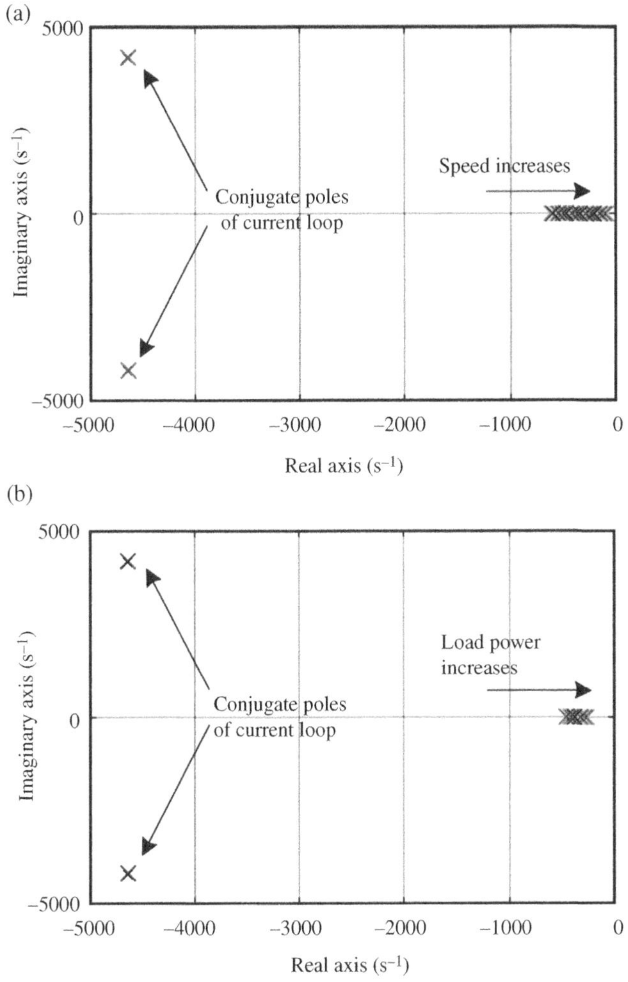

Figure 7.16 Locations of poles of $G_{idc}(s)$ at different speeds and load powers. (a) Poles are obtained at 20 kW load power, speed increases from 15 to 25 krpm. (b) Poles are obtained at 20 krpm, load power increases from 5 to 35 kW. *Source:* 2020 IEEE.

The i_{dc} PI controller shown in (7.21) provides a zero. To cancel the dominant pole with this zero, the control parameters can be set as follows:

$$\begin{cases} K_{vP} = 2\overline{v}_{dc}{}^{2}C\cdot\gamma \\ K_{vI} = -3\left(\overline{v}_{d}\overline{i}_{d} + \overline{v}_{q}\overline{i}_{q}\right)\cdot\gamma \end{cases} \tag{7.23}$$

where γ is a gain used to tune the control parameters.

From (7.23) it can be seen that the PI parameters depend on operating points and a gain γ. By applying (7.23) into (7.22), the open loop transfer function of the i_{dc} control loop can be derived as:

$$G_{idc}(s) = -\frac{3\gamma\overline{v}_{dc}C\left[\overline{v}_{q} - \omega_{e}L_{s}\overline{i}_{d} + (R_{s} + L_{s}s)\overline{i}_{q}\right](K_{iP}s + K_{iI})}{L_{s}s^{2} + (K_{iP} + R_{s})s + K_{iI}} \tag{7.24}$$

As concluded in the above section, the highest speed and heaviest load power is identified as the worst operating condition in terms of control stability. The root locus of $G_{idc}(s)$ in (7.24) is analyzed

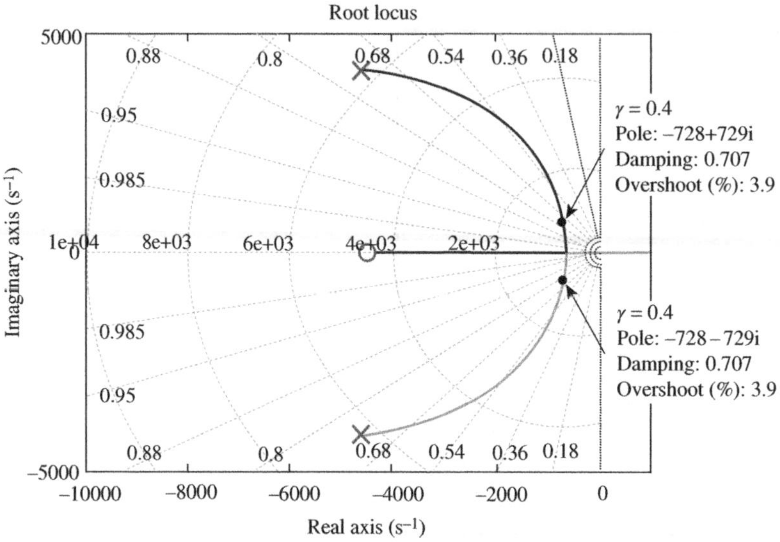

Figure 7.17 Root locus of $G_{idc}(s)$ at 32 krpm, 45 kW. *Source:* 2020 IEEE.

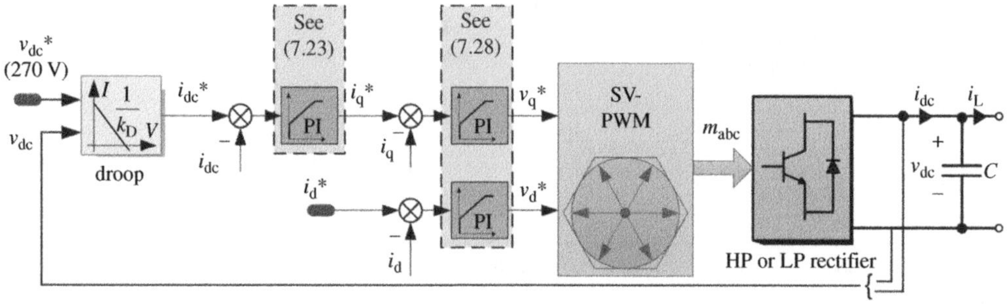

Figure 7.18 Control block diagram of the electrical power-generation channels. *Source:* 2021 IEEE.

at the operating point (32 krpm, 45 kW), as shown in Figure 7.17. The trajectory means the locations of closed loop poles with respect to different values of the gain γ. It can be seen when γ is set as 0.4, the damping ratio is 0.707 and the overshoot is 3.9%. This is an acceptable performance considering the worst operating condition. Hence, the gain γ is set as 0.4 in the following analysis. With a known γ, the control parameters can be adaptively tuned based on the relation in (7.23).

To conclude, the control block diagram can be built as shown in Figure 7.18, where the design criteria of control parameters are highlighted. DC current reference comes from the predefined V-I droop characteristic. Then with a closed-loop calculation, optimal q-axis current reference is derived. Subsequently, the inner current loop generates voltage references in the dq frame. With the aid of the SVPWM technique, the voltage references are transformed into certain PWM-switching sequences for driving the power electronics modules of the HP or LP rectifier.

7.4.6 Simulation Results of the Single-Bus Power-Generation Center

A PGC model is built in the Simulink in the continuous domain, where averaged model for power converters is adopted. The transmission cable impedance is omitted. Subsystems of the model can

Table 7.3 Parameters of the studied power-generation center.

Category	Parameter	Value
Electrical machines	Speed of LP machine	7000 rpm
	Speed of HP machine	20,000 rpm
Rectifiers	Topology	Two-level, bidirectional
	Maximum current	400 A
DC link	Rated voltage	270 V
	Local shunt capacitor	1 mF
	Main bus capacitor	1.2 mF
CPL	Load power	From 10 to 30 kW
Control parameters	Current loop PI gains	0.87, 3908
	Voltage loop PI gains	Adaptive tuned by setting the gain $\gamma = 0.4$ in (7.23)
	FW control PI gains	1.5, 2000
	Droop gains	$g_{LP} = 1/8$, $g_{HP} = 1/4$

be found in Appendix. Control performance for the PGC is investigated. The HP and LP machines both perform as generators, supplying power to the loads. Power-sharing ratio depends on droop gains. The main parameters of two PMMs have been presented in Table 7.1. More details about design of the machine can be found in [35, 36].

As discussed earlier, in MEA many conventional onboard systems are replaced by electrically powered ones, which are regulated by power electronic converters. For example, for the ECS, conventionally, bleed air from one or two of the compressor stages of the main engine is used to regulate the cabin temperature and pressure. However, in MEA such as Boeing 787, a set of electrical compressors driven by power converters is used to regulate the temperature and pressure [37]. Tightly controlled power electronic converters and motor drives often behave as constant power load (CPL), which show the constant power characteristic. Therefore, an inverter-controlled PMM that performs as a CPL is used as DC bus load in the simulation. The system parameters are given in the Table 7.3.

The simulation results are demonstrated in Figure 7.16. The cruise mode of engine is considered where the speed of HP generator is set as 20,000 rpm and that of LP generator is 7000 rpm. The findings in [22] reveal that extracting more power from the LP spool than HP spool of the engine is beneficial for the compressor surge margin. Hence, in this study, the droop gains for the LP and HP rectifier are set as 1/8 and 1/4, respectively. Then the power-sharing ratio between the LP and HP rectifier is 2 : 1. During simulations, the power demand of CPL is changing at 0.05, 0.1, 0.15, 0.2 seconds, and takes values 10, 20, 30, and 20 kW, respectively. The main findings are listed as follows:

1) Figure 7.19a shows that as the CPL power demand increases, the DC bus voltage will slightly drop due to the droop effect.
2) Figure 7.19b exhibits the output power of the LP and HP rectifiers. It can be seen that the power ratio is kept to 2 : 1 in the whole process.
3) Figure 7.19c shows the *d*-axis currents of LP and HP generators. Since the speed of LP generator is relatively low, there is no need for FW and its *d*-axis current remains zero. While a significant negative *d*-axis current is injected into the HP generator for FW purpose, the *dq* currents can also

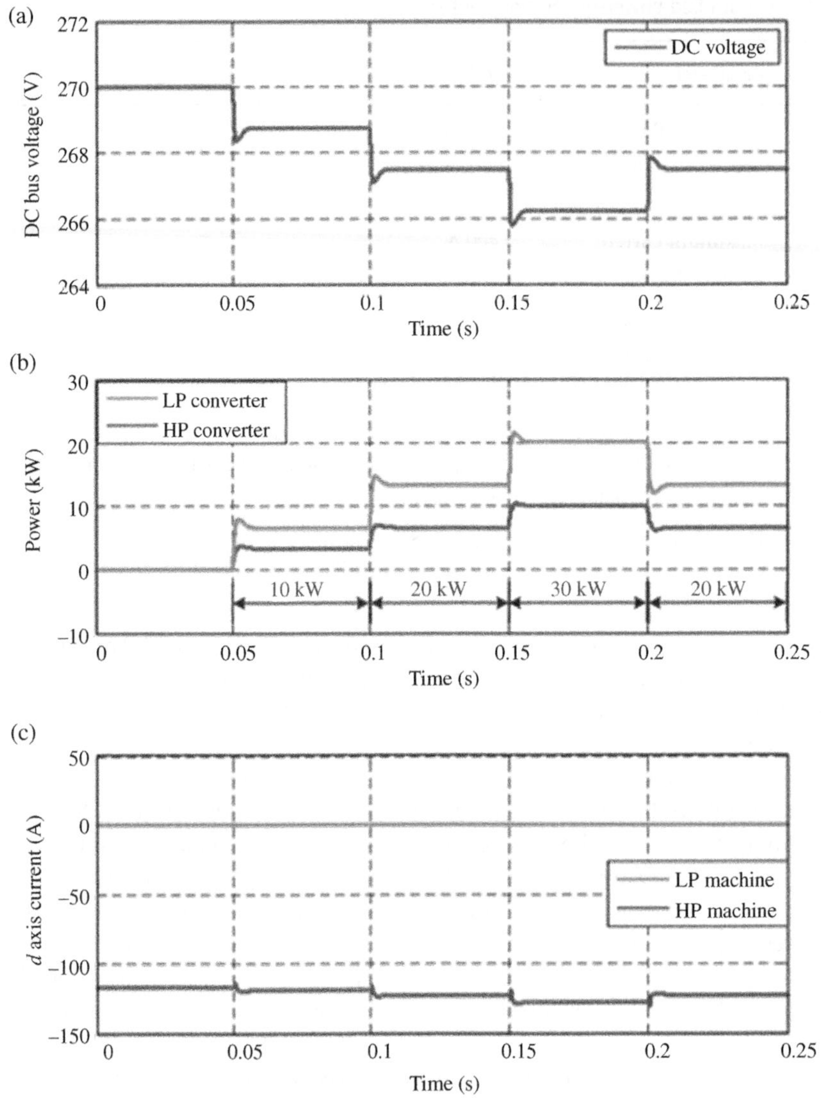

Figure 7.19 Simulation results in the CPO mode. (a) DC bus voltage; (b) output power of LP and HP rectifiers; (c) *d*-axis currents of LP and HP machines. *Source:* 2020 IEEE.

be seen from the operating trajectory in Figure 7.10. Moreover, as the power increases, the magnitude of i_d will slightly increase. This can be explained by Figure 7.20 from which it can be seen that the magnitude of i_d will increase as the operating points (the dark grey dots) moving on the boundary of voltage limit circle in the 3rd quadrant. This significant i_d current is expected to lead to considerable power losses in both the HP machine and the HP rectifier.

7.4.7 Appendix

Simulation analysis in point 2 was primarily conducted in the Matlab/Simulink environment. The following are the nonlinear equivalent models of the PGC created in the Simulink to support the study in point 3. The overall model of the PGC is demonstrated in Figure 7.21.

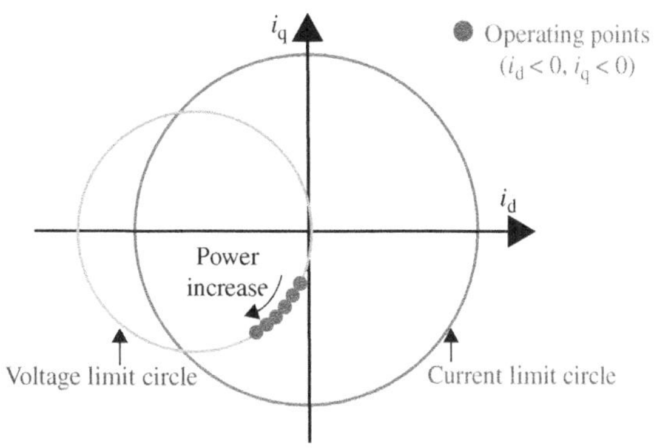

Figure 7.20 Operating points of the HP generator with increased power. *Source:* 2020 IEEE.

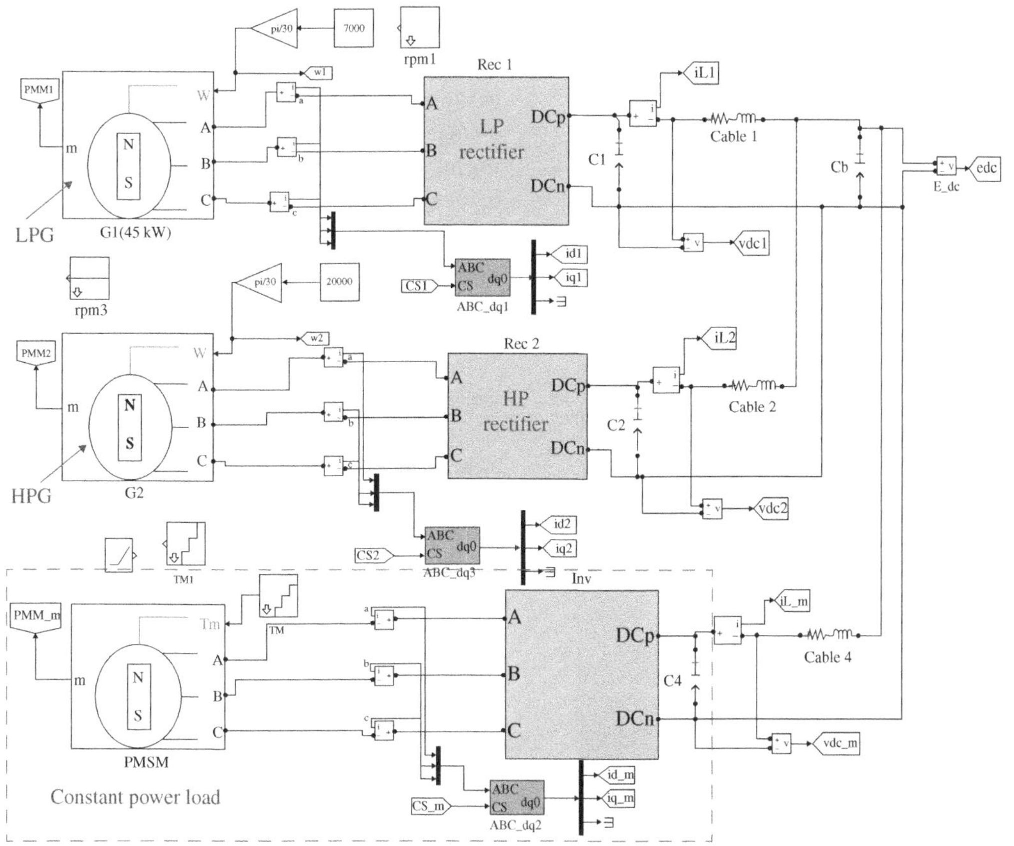

Figure 7.21 The PGC simulation model.

The control structure for the LP and HP rectifiers is shown in Figure 7.22. The functionalities of core blocks are highlighted. The schematic diagram of the control structure has been shown in Figure 7.18. To be specific, the error between the voltage reference and actual voltage is amplified through a droop gain, giving the DC current reference. Then through a closed-loop calculation of the DC currents, the q-axis current reference is generated. It is correlated to the d-axis current, to make sure that the current references are within the current limits. Then through the inner current loop calculation, voltage references are generated and also limited within the voltage limits. The voltage references are transformed into gate drive signals to drive the HPR and LPR.

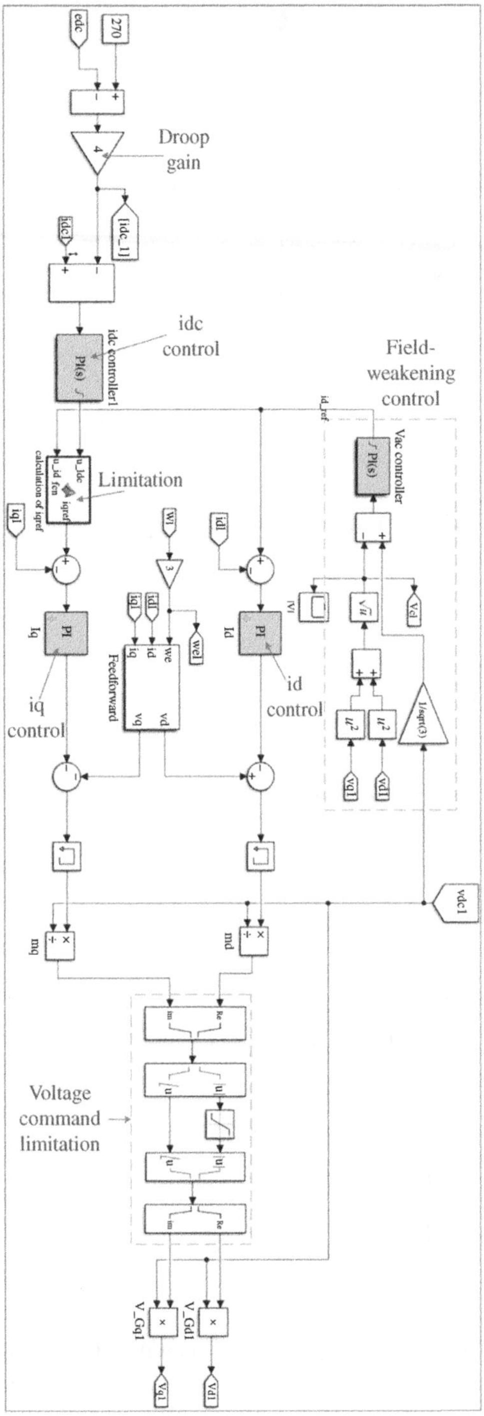

Figure 7.22 Control structure for the rectifiers.

7.5 The Solid-State Power Controllers and the Protection Features

7.5.1 Background of Solid-State Power Controllers

Whilst many potential benefits could be brought by the DC electrical system, there are serious concerns related to DC fault protection [17]. Mechanical circuit breakers (MCBs) are commonly applied in an AC electrical system to protect the cables and associated loads from excess current. However, in a DC power system rather than an AC system, the electric arcs would appear more easily and with greater effect as there is no zero-voltage crossing [38–41]. Consequently, the arcing may damage the MCB, posing a serious threat to the system safety. Besides, the moving parts, i.e. contacts in an MCB and relatively long tripping time are challenging for safety-critical aircraft applications associated with harsh environmental condition, such as high vibration [42].

To enhance the reliability of the protection design, it can be achieved by two aspects: (i) accurate monitoring of the electrical power system; and (ii) instant tripping once a fault happens. The SSCB gains increasing attention recently as it could be an attractive alternative to replace the MCB. Instead of using electromagnetic components in an MCB, an SSCB adopts fully solid-state devices, where the switching could be arcless and considerably fast. The detailed comparison between an solid-state power controller (SSPC) and an MCB in terms of reliability, availability, and power performance is shown in Table 7.4.

Similar to the MCB, the contactors on aircraft controlling their associated loads are also mechanical. Therefore, an SSPC integrating the functionalities of both a circuit breaker and a contactor emerges recently [9]. In particular, the switching losses of an SSPC can be significantly low with the assistance of a soft switching mechanism [40]. In summary, the features of an SSPC are:

1) Suitable for DC power system due to arc-less switching.
2) Configuring and protecting the system at the same time.
3) Protection in a more accurate and instant approach.
4) High power/weight and power/volume density.

Table 7.4 Comparison of an SSPC and an MCB.

	SSBC	MCB
Reliability	High due to arc-less switching, no moving parts	Limited service lifetime
Availability	Instant trip (<0.1 ms)	Slow trip (>10 ms)
	Programmable tripping	Adjustable tripping only for high currents
	System-level diagnostics and prognostics	NA
Power performance	Low switching losses	High switching losses from coils, contactors, strips, solenoids, etc.
	High weight/volume density	Bulky and heavy at high current

7.5.2 Design of Solid-State Power Controllers

Figure 7.23 shows the basic schematic of an SSPC comprised of the solid-state devices and their protection circuits. For an SSPC, one single-power module is normally not adequate as its voltage/current class is constrained. Hence, several power modules can be series-connected, enhancing the withstand voltage or be parallel-connected, strengthening the current capability and thus increasing the overall power rating. If connecting *n* IGBTs or MOSFETs in series, the total conduction losses can be expressed in (7.25).

$$\begin{aligned} P_{\text{con,IGBT}} &= nI_{\text{CE}}V_{\text{CE}} \\ P_{\text{con,MOSFET}} &= nI_{\text{DS}}{}^2R_{\text{DS(on)}} \end{aligned} \tag{7.25}$$

where I_{CE} is the IGBT collector-emitter current, V_{CE} is the IGBT collector-emitter voltage, I_{DS} is the MOSFET drain-source current, and $R_{\text{DS(on)}}$ is the MOSFET on-state resistance.

Equation (7.25) shows the overall conduction losses will be n times larger than series connection, regardless of IGBTs or MOSFETs. Otherwise, if connecting n IGBTs or MOSFETs in parallel, the total conduction losses can be expressed (7.26).

$$\begin{aligned} P_{\text{con,IGBT}} &= I_{\text{CE}}V_{\text{CE}} \\ P_{\text{con,MOSFET}} &= I_{\text{DS}}{}^2\frac{R_{\text{DS(on)}}}{n} \end{aligned} \tag{7.26}$$

From (7.26), it is seen that parallel-connected IGBTs are unchanged in the conduction losses in comparison to a single IGBT module. On the other side, the conduction losses decrease as greater n

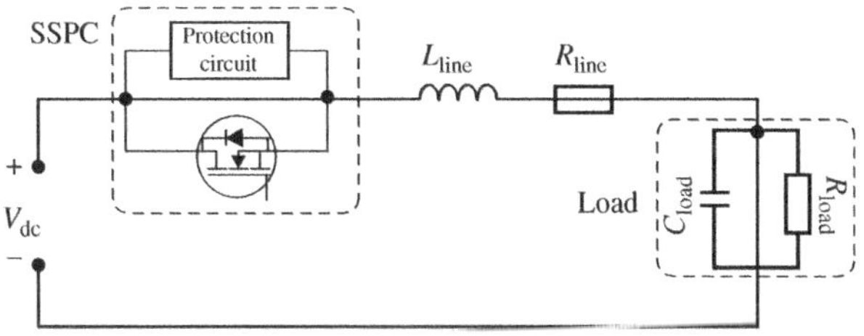

Figure 7.23 Schematic of an SSPC. *Source:* 2021 IEEE.

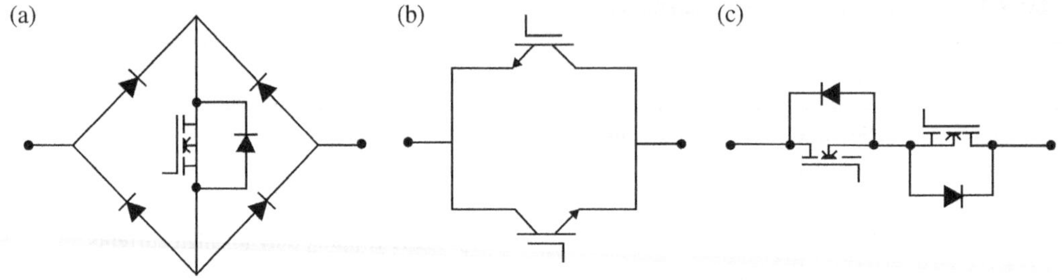

Figure 7.24 Several bidirectional breakable SSPC topologies. *Source:* 2021 IEEE.

of parallel-connected MOSFET. The power losses P_{loss} associated with the SSPCs, as defined in (7.27), pose a significant challenge for thermal management where those losses directly impact the size, weight, and lifetime of SSPCs [40].

$$P_{loss} = P_{sw} + P_{con} \tag{7.27}$$

where P_{sw} and P_{con} are the switching losses and condition losses, respectively.

At the steady state, the junction temperature T_j rise of the power devices can be expressed in (7.28).

$$T_j = P_{loss}R_{th,ja} + T_a \tag{7.28}$$

where $R_{th,\,ja}$ is the device's junction-to-case thermal resistance, and T_a is the ambient temperature.

For the SSPC comprised of parallel-connected MOSFETs, the required n can be derived if it only considers the conduction losses, as shown in (7.29). This equation can be a practical guideline determining the suitable number n to meet the thermal requirement of the maximum allowable junction temperature $T_{j,max}$.

$$\begin{aligned} P_{loss} &< \frac{T_{j,\,max} - T_a}{R_{th,ja}} \\ n &> \frac{R_{th,ja}I_{DS}{}^2R_{DS(on)}}{T_{j,\,max} - T_a} \end{aligned} \tag{7.29}$$

Figure 7.24 shows three applicable bidirectional breakable SSPC topologies, in which both forward current (from the left to the right of topologies) and negative currents (from the right to the left of topologies) can be tripped [11]. For the topology in Figure 7.24a, the conduction losses are contributed by two diodes and one IGBT for the forward current conduction, while three diodes are for the reverse current conduction. Figure 7.24b results in the identical conduction losses of one IGBT when current flows are forward and reverse. In comparison, the conduction losses of Figure 7.24c are caused by two MOSFETs for the forward current, while two MOSFETs and their anti-parallel diodes at higher currents for the reverse current flow.

SSPCs are the essential elements protecting the whole electric power distribution system, particularly protecting the cables from over-current burnout. However, whilst SSPCs secure the safety of electric systems, some necessary circuits should be designed to protect SSPCs themselves. As the SSPCs are comprised of solid-state devices, those devices have certain safe operation areas with limitations of current, voltage as well as thermal temperature.

7.5.3 Protection of Solid-State Power Controllers

In light of the protection of SSPCs, over-voltage protection is one of the most critical considerations. There are four potential sources attributing to the over-voltage threats on SSPCs:

1) Electrostatic discharge
2) Inductive load switching
3) Lightning-induced transient
4) Automotive load dump

The electrostatic discharge (ESD) is normally induced when the static electrical charge is transferred from a body to an electronic circuit. Its damages include faulty circuit operation, latent defects, and even catastrophic failure of sensitive components. For those SSPCs protecting transformers, generators, motors, and relays with potentially high inductance, switching can cause voltage transients up to hundreds of volts and amps, lasting as long as 400 ms [43]. The lightning threat is commonly applied to aircraft as they are struck by lightning twice a year on average [44]. Lightning strikes will cause an electromagnetic disturbance on electrical and communication lines, posing threats of voltage spikes on SSPCs. Load dump refers to when a load is removed, such as when the battery is disconnected while the engine is running. In that case, the voltage may surge before stabilizing and damaging electronic components. For example, for a typical 12 V circuit in automotive applications, load dump can ramp up as high as 120 V and take 400 ms to decay [43].

To suppress the over-voltage thus protecting the SSPCs, there are several applicable options:

1) Diodes based on PN junction, such as avalanche diode, transient voltage suppression (TVS) diode, Zener diode;
2) Varistors, such as metal oxide varistor (MOV);
3) Snubbers, such as resistor, resistor-capacitor, resistor-capacitor-diode;
4) Free-wheeling diodes.

Zener diode is a diode with a highly doped p–n junction that allows the current passing from its anode to its cathode, while the reverse current (from its cathode to its anode) can also flow if the reverse voltage reaches the Zener voltage. This effect is known as the Zener effect. A similar breakdown exists in a general-purpose diode, but the voltage is not clearly defined as in Zener diodes. Similar to Zener diodes, avalanche diodes and TVS diodes have a clear breakdown voltage. However, the working principle of those diodes is avalanche breakdown, which can define a much higher breakdown voltage than the Zener voltage. On the other side, TVS diodes can absorb excessive voltage/power to protect other devices. In contrast, Zener diodes provide a considerably stable voltage that is more suitable for voltage regulation than over-voltage protection.

Figure 7.25 shows the equivalent circuit and I–V characteristics of unidirectional (clamping voltage in one direction), bidirectional (clamping voltage in both directions) TVS diodes [45]. Referring to Figure 7.25, a TVS diode is characterized as follows:

Peak pulse current I_{pp}: the maximum current for a defined pulse waveform that the diode can sustain before failure. This parameter is crucial as TVS diodes will only fail due to excess current rather than excess voltage [46];

Stand-off voltage V_r: the maximum voltage that can be applied to the TVS diodes without conduction;
Breakdown voltage V_{br}: the voltage measured between the TVS diode at a specified test current I_t, normally 1 or 10 mA;
Clamping voltage V_c: the peak voltage measured across the TVS diodes at the I_{ppm};
Reverse leakage current I_r: current measured at V_r;
Forward voltage drop (only applicable for unidirectional diodes) V_f.

As mentioned earlier, the I_{pp} of a TVS is given for a pre-defined pulse waveform, such as the waveform shown in Figure 7.26. To characterize it, two time widths, the rise time t_r and the decay time t_d,

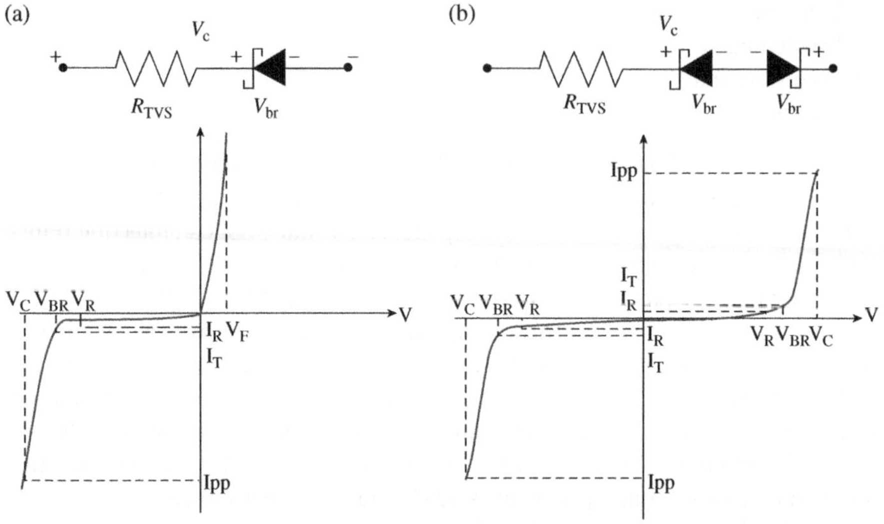

Figure 7.25 Equivalent circuit and *I*–*V* characteristics of (a) unidirectional; (b) bidirectional TVS diodes. *Source:* 2021 IEEE.

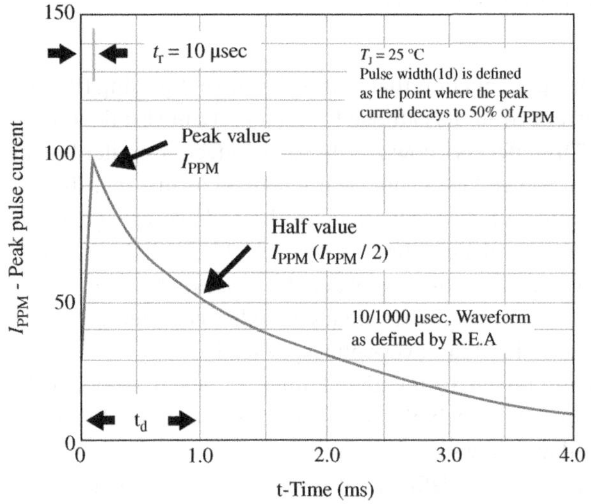

Figure 7.26 A typical 10/1000 μs pulse waveform. *Source:* [47]/Littlefuse Inc.

need to be specified. t_r refers to the time taken to reach the peak pulse current I_{ppm}, as shown in Figure 7.26, while the time taken to half of the I_{ppm} is denoted as t_d. For instance, a pulse of 10/1000 μs implies its current waveform arrives at the maximum point at $t = 10$ μs, decreasing to half of the maximum value at $t = 1000$ μs.

MOVs are voltage-dependent, non-linear devices that are composed primarily of zinc oxide (Z_NO) with small additions of other metal oxides. Those conductive Z_NO grains are separated by grain boundaries, providing similar electrical characteristics to PN junction semiconductors. As each Z_NO grain works as a P–N junction with non-linear electrical behavior occurring at its boundary. The MOV can be considered as a multi-junction device comprised of many series and parallel

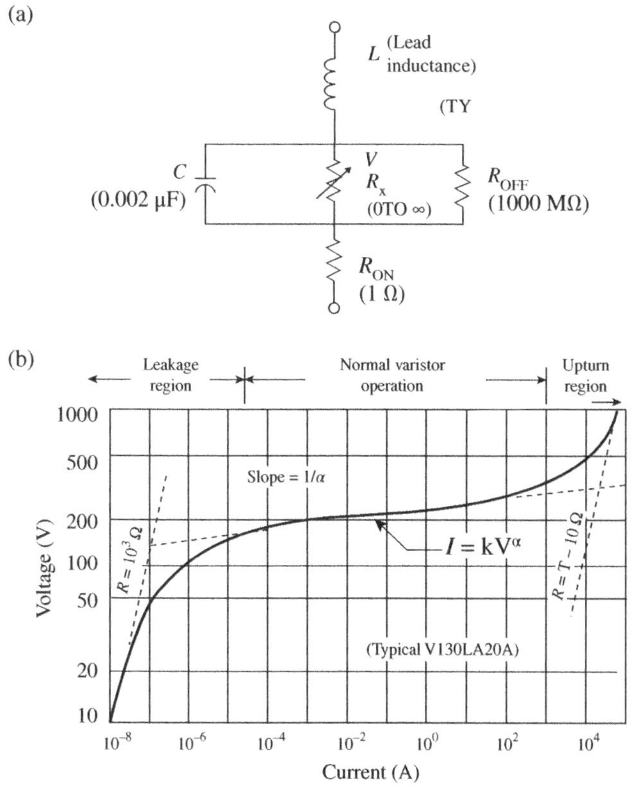

Figure 7.27 Equivalent circuit and typical *V–I* curve of an MOV. *Source:* [48]/Littlefuse Inc.

connections of grain boundaries. During the manufacturing operation, the MOV is sintered into a ceramic semiconductor resulting in a crystalline microstructure that allows MOVs to dissipate very high levels of transient energy across the device.

Figure 7.27 shows the equivalent circuit and typical *V–I* curve of an MOV. Figure 7.27a shows the MOV can be modelled by a leakage inductor L, a leakage capacitor C, a variable resistor R_x, an off resistor R_{off}, and an on resistor R_{on}. At low current levels, the MOV nearly appears as an open circuit because R_x is in a very high resistance mode and the R_{off} of 10 MΩ will be pre-dominant. This operation region is called the leakage region as shown in Figure 7.27b, which has a distinct temperature dependence. As the current increases, the MOV will enter the normal varistor region, and the varistor characteristic follows the equation: $I = kV^a$, where k is a constant and a determines the degree of non-linearity. In this region, the R_x pre-dominates other resistors. When the current approaches the maximum rating of the MOV, the resistance of R_x will reduce to around 1–10 Ω [48]. The MOV can be regarded as a short circuit, and this operation region is named as upturn region.

Table 7.5 compares the performance of MOVs and TVS diodes in terms of power dissipation capability, lifetime, operating temperature, leakage current, response time, and clamped voltage. It can be seen from Table 7.5 that TVS diodes feature with minimal leakage current, extremely fast response time, and fixed clamped voltage. Apart from that, TVS diodes are more suitable for protecting SSPC on aircraft in comparison to the MOVs because they are durable and reliable in very high working temperatures, which are vital for safety-critical aerospace applications. Table 7.6 summarizes the over-voltage events and recommended protection devices for each case.

Table 7.5 Protection devices' comparison.

Device	MOV	TVS diode
Protection principle	Mass of metal grains that create diodes like breakdown	Avalanche breakdown diode behavior to shunt current
Power dissipation capability	High	Moderate to high
Lifetime	Shorten after taking power pulses	Long
Operating temperature	Up to 125 °C	Up to 175 °C
Leakage current	Low (5–25 μA)	Very low (1 μA)
Response time	Fast (1–5 ns)	Very fast (<1 ps)

Table 7.6 Common transient voltage events and protection solutions.

Over-voltage events	Occasions	Protection criteria	Protection devices
Lightning	Any electronic or electrical equipment with connections to the outside environment	Fast response, proper switching threshold, surge current rating	TVS diodes, varistors
Inductive load switching	Any system that has inductive loads	High energy rating	Varistors, TVS diodes
ESD	Any electronic equipment with a human interface	Fast response, high peak voltage rating	ESD suppressors TVS diode array

Source: Adapted from [43, 48].

References

1 Srimoolanatha, A. (2008). Aircraft electrical power systems - charged with opportunities. *Aerospace and Defense Executive Briefing of Frost & Sullivan*. Mountain View, CA: Frost & Sullivan.

2 Rypdal, K. (2000). Aircraft emission. *Good Practice Guidance and Uncertainty Management in National Greenhouse Gas Inventories*. Japan: The Institute for Global Environmental Strategies.

3 IATA (2012). Fact Sheet: Environment. http://www.iata.org/pressroom/facts_figures/fact_sheets/pages/environment.aspx.

4 E. Commision (2011). Flightpath 2050 Europe's Vision for Aviation.

5 Rosero, J.A., Ortega, J.A., Aldabas, E., and Romeral, L. (2007). Moving towards a more electric aircraft. *Aerospace and Electronic Systems Magazine*. IEEE, vol. 22, pp. 3–9.

6 Degoutte, C., Sanchez, O., Renaudin, J. et al. (2016). Aircraft 270VDC power distribution improvements using wide band gap semi-conductors. *2016 International Conference on Electrical Systems for Aircraft, Railway, Ship Propulsion and Road Vehicles & International Transportation Electrification Conference (ESARS-ITEC)*, pp. 1–6.

7 Wheeler, P. and Bofigko, S. (2014). The more electric aircraft: technology and challenges. *IEEE Electrification Mag.* 2: 6–12.

8 More Open Electrical Technologies. (2006–2009). TRIMIS. http://trimis.ec.europa.eu/project/more-open-electrical-technologies (accessed 3 November 2022).

9 Chen, J., Wang, C., and Chen, J. (2018). Investigation on the selection of electric power system architecture for future more electric aircraft. *IEEE Trans. Transport. Electrification* 4 (2): 563–576.

10 Rakhra, P., Norman, P.J., Galloway, S.J., and Burt, G.M. (2011). Modelling and Simulation of a MEA Twin Generator UAV Electrical Power System. *Universities' Power Engineering Conference (UPEC), Proceedings of 2011 46th International*, pp. 1–5.

11 Norman, P.J., Galloway, S.J., Burt, G.M. et al. (2008). Transient analysis of the more-electric engine electrical power distribution network. *Power Electronics, Machines and Drives, 2008. PEMD 2008. 4th IET Conference on*, pp. 681-685.

12 Raimondi, G.M., Sawata, T., Holme, M. et al. (2002). Aircraft embedded generation systems. *Power Electronics, Machines and Drives, 2002. International Conference on (Conf. Publ. No. 487)*, pp. 217–222.

13 Fletcher, S.D.A., Norman, P.J., Galloway, S.J., and Burt, G.M. (2011). Determination of protection system requirements for dc unmanned aerial vehicle electrical power networks for enhanced capability and survivability. *Electr. Syst. Transport.*, IET 1: 137–147.

14 Louganski, K.P. (1999). *Modeling and Analysis of a DC Power Distribution System in 21st Century Airfilters*. Virginia: Virginia Polytechnic Institute and State University.

15 F. Gao (2016). Decentralised control and stability analysis of a multi-generator based electrical power system for more electric aircraft. PhD thesis. University of Nottingham.

16 Moir, I. and Seabridge, A. (2008). *Aircraft Systems: Mechanical, Electrical, and Avionics Subsystems Integration*. Wiley.

17 Sztykiel, M., Fletcher, S., Norman, P. et al. (2015). Ac/dc converter with dc fault suppression for aircraft+/-270 vdc distribution systems. *SAE 2015 AeroTech Congress & Exhibition*.

18 Lang, X., Yang, T., Huang, Z. et al. (2021). Stability improvement of onboard HVDC grid and engine using an advanced power generation center for the more-electric aircraft. *IEEE Trans. Transport. Electrification* 8 (1): 660–674. https://doi.org/10.1109/TTE.2021.3095256.

19 Jia, Y. and Rajashekara, K. (2017). An induction generator-based AC/DC hybrid electric power generation system for more electric aircraft. *IEEE Trans. Ind. Appl.* 53 (3): 2485–2494.

20 Taneja, D.N. (2011). Method and apparatus for extracting electrical power from a gas turbine engine. US patent US20130062885A1, 14 March 2013. https://patents.google.com/patent/US20130062885A1/en (accessed 3 November 2022).

21 Buticchi, G., Bofigko, S., Liserre, M. et al. (2019). On-board microgrids for the more electric aircraft-technology review. *IEEE Trans. Ind. Electron.* 66 (7): 5588–5599.

22 Enalou, H.B., Lang, X., Rashed, M., and Bofigko, S. (2020). Time-Scaled Emulation of Electric Power Transfer in the More Electric Engine. *IEEE Trans. Transport. Electrification* 6 (4): 1679–1694.

23 Burgos, R., Kshirsagar, P., Lidozzi, A. et al. (2006). Mathematical model and control design for sensorless vector control of permanent magnet synchronous machines. Proceedings of IEEE Workshops COMPEL, pp. 76–82.

24 Kuo, B.C. (1962). *Automatic Control System*. London, UK: Prentice-Hall.

25 Diab, A.M., Bofigko, S., Galea, M., and Gerada, C. (2020). Stable and robust design of active disturbance-rejection current controller for permanent magnet machines in transportation systems. *IEEE Trans. Transport. Electrification* 6 (4): 1421–1433.

26 Diab, A.M., Bozhko, S., Guo, F. et al. (2021). Fast and simple tuning rules of synchronous reference frame proportional-integral current controller. *IEEE Access* 9: 22156–22170.

27 Bofigko, S., Rashed, M., Hill, C.I. et al. (2017). Flux-weakening control of electric starter? generator based on permanent-magnet machine. *IEEE Trans. Transport. Electrification* 3 (4): 864–877.

28 Lang, X., Yang, T., Li, C. et al. (2021). An enhanced feedforward flux weakening control for high-speed permanent magnet machine drive applications. *IET Power Electron.* https://doi.org/10.1049/pel2.12170.

29 Figang, Z., Huang, J., Jiang, Y. et al. (2017). Overview and analysis of PM starter/generator for aircraft electrical power systems. *CES Trans. Electr. Mach. Syst.* 1 (2): 117–131.

30 Final Report Summary - AEGART (Aircraft Electrical Generation System with Active Rectification and Health Monitoring) | FP7 | CORDIS | European Commission. https://cordis.europa.eu/project/id/296090/reporting.

31 Bolognani, S., Calligaro, S., Petrella, R., and Pogni, F. (2011). Flux-weakening in ipm motor drives: Comparison of state-of-art algorithms and a novel proposal for controller design. *Proceedings of 2011-14th Eur. Conf. Power Electron. Appl. (EPE 2011)*, pp. 1–11.

32 Preindl, M. and Bolognani, S. (2013). Model predictive direct torque control with finite control set for PMSM drive systems part 2: field weakening operation. *IEEE Trans. Ind. Inf.* 9 (2): 648–657.

33 Gao, F., Bofigko, S., Asher, G. et al. (2016). An improved voltage compensation approach in a droop-controlled DC power system for the more electric aircraft. *IEEE Trans. Power Electron.* 31 (10): 7369–7383.

34 Begum, K., Ghousiya, A., Rao, S., and Radhakrishnan, T.K. (2017). Enhanced IMC based PID controller design for non-minimum phase (NMP) integrating processes with time delays. *ISA Trans.* 68: 223–234.

35 Fernando, W., Arumugam, P., and Gerada, C. (2018). Design of a stator for a high-speed turbo-generator with fixed permanent magnet rotor radius and volt–ampere constraints. *IEEE Trans. Energy Convers.* 33 (3): 1311–1320.

36 Fernando, W., Arumugam, P. and Gerada, C. (2016). Volt-ampere constrains and its influence on inductance limits in high speed PM machine design. *8th IET International Conference on Power Electronics, Machines and Drives (PEMD 2016)*, Glasgow, pp. 1–6.

37 Sayed, E., Abdalmagid, M., Pietrini, G. et al. (2021). Review of electric machines in more/hybrid/turbo electric aircraft. *IEEE Trans. Transport. Electrification* 7 (4): 2976–3005. https://doi.org/10.1109/TTE.2021.3089605.

38 Gu, C., Wheeler, P., Castellazzi, A. et al. (2017). Semiconductor devices in solid-state/hybrid circuit breakers: current status and future trends. *Energies* 10 (4): 495.

39 Adhikari, J., Yang, T., Figang, J. et al. (2018). Thermal analysis of high-power high voltage dc solid state power controller (sspc) for next generation civil tilt rotor-craft. *2018 IEEE International Conference on Electrical Systems for Aircraft, Railway, Ship Propulsion and Road Vehicles & International Transportation Electrification Conference (ESARS-ITEC)*, pp. 1–6. IEEE.

40 Molligoda, D.A., Chatterjee, P., Gajanayake, C.J. et al. (2016). Review of design and challenges of dc sspc in more electric aircraft. *2016 IEEE 2nd Annual Southern Power Electronics Conference (SPEC)*, pp. 1–5. IEEE.

41 Feng, X. (2007). *Sic Based Solid State Power Controller.* University of Kentucky.

42 Glass, M. (2010). Performance comparison: Solid state power controllers vs. electromechanical switching. Technical Report. Data Device Corporation.

43 Littelfuse. (2015). Electronics Circuit Protection Product Selection Guide.

44 Clark, M. and Walters, K. (2020). Aircraft lightning protection. *Micronote 132.*

45 Phillips, C. (2018). Tvs surge rating: Power vs. current. No. SLVAE37, pp. 1–3, Texas Instruments, 36.

46 Forbes, A. (2019). *How to select a Surge Diode, Application Report SLVAE37.* Texas Instruments.

47 Littelfuse. (2013). *TVS Diode Arrays Transient Voltage Suppression Diode Devices*, 1–174. Littelfuse, Inc.

48 Littelfuse. (2017). *Metal-Oxide Varistor (MOV)*, Product Catalog & Design Guide, 1–254. Littelfuse, Inc.

8

DC–DC Converter and On-board DC Microgrid Stability

Giampaolo Buticchi[1] *and Jiajun Yang*[2]

[1] *University of Nottingham Ningbo China, Faculty of Science and Engineering, Ningbo, Zhejiang 315000, China*
[2] *University of Nottingham Ningbo China, China Beacons Institute, Ningbo, Zhejiang 315000, China*

8.1 Introduction

The physical and functional requirements for onboard aircraft electronics have been constantly increasing, in order to realize a more power-dense, more reliable, and safer power distribution system. Although there are different kinds of electrical power distribution system (EPDS) (with AC or DC lines), it is a general result that the DC bus should be made available (either high- or low-voltage) to supply the DC loads.

The traditional solution, which has been successfully employed in aircraft with low-to-medium DC power requirements [1], is to employ a transformer rectifier unit (TRU) or the auto-transformer-based version in case the galvanic isolation is not needed. A TRU is built up from an (auto)transformer and a diode bridge rectifier. The TRU which offers voltage control capability also has a DC–DC converter connected in series to the rectifier.

This solution has proven to be reliable (due to the reduced component count and the absence of controlled switches), but it offers neither the bidirectional power-transfer capability nor it has a good power density (due to the low-frequency transformer).

For the aforementioned reasons, researchers have been actively investigating power electronics-based solutions, which offer an increased degree of control and, at the same time, can achieve improved power-density figures due to the employment of high-frequency transformers. Among the several options researched in literature, the DC–DC converter based on soft-switching appears to have attracted the most interest [2].

8.2 The Dual Active Bridge Converter

The dual active bridge (DAB) converter is formed by two full bridges, two capacitors, and a high-frequency (HF) transformer for galvanic isolation. Figure 8.1 shows the structure of the DAB converter, where V_1 is the input voltage, V_2 is the output voltage, C_1 is the input capacitor, C_2 is the output capacitor, L_lk is the leakage inductance of transformer, and i_L is the current going through the leakage inductance. With such a structure, the converter takes advantage of high efficiency, high power density, and low stresses for devices and components [3].

Transportation Electrification: Breakthroughs in Electrified Vehicles, Aircraft, Rolling Stock, and Watercraft, First Edition. Edited by Ahmed A. Mohamed, Ahmad Arshan Khan, Ahmed T. Elsayed, and Mohamed A. Elshaer.

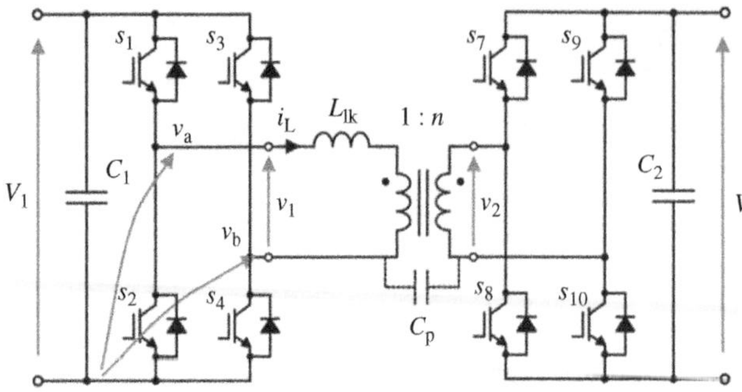

Figure 8.1 The dual active bridge converter. *Source:* Buticchi et al. [4]/IEEE.

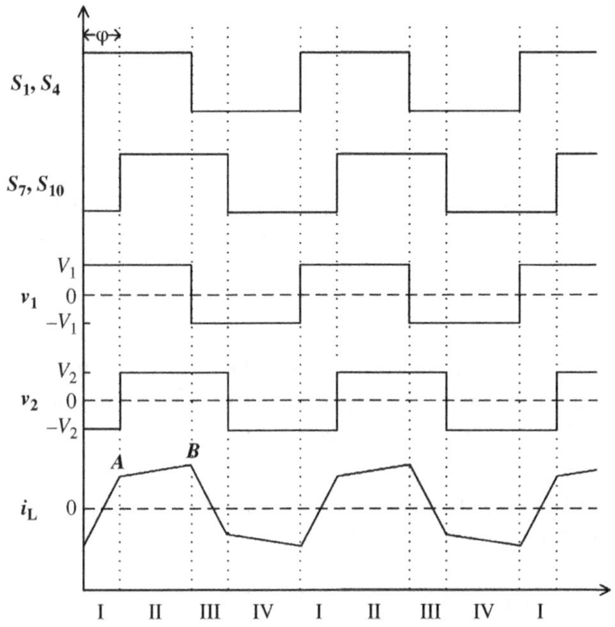

Figure 8.2 Single-phase shift modulation.

By employing phase shift modulation (PSM) to full bridges, the bidirectional power flow can be enabled. Figure 8.2 shows the waveforms by applying single PSM and Figure 8.3 shows the waveforms by applying dual PSM. When the switching signals of the full bridge at the primary side lead to the switching signals of the full bridge on secondary side, the power will flow from primary side to secondary side. Conversely, if the switching signals of the full bridge at primary side lag, the power will flow from secondary side to primary side. The power equation of DAB converter is given as

$$P = \frac{V_1 V_2}{2n\pi f_s L_{lk}} \varphi\left(1 - \left|\frac{\varphi}{\pi}\right|\right) \tag{8.1}$$

One of the main advantages of DAB converter is that it can achieve zero voltage switching (ZVS) at both full bridges under certain conditions. The peak value of i_L is denoted as A and B, it exists

$$A = \frac{1}{4L_{lk}f_s}\left(\frac{2\varphi V_1}{\pi} + \frac{V_2}{n} - V_1\right) \tag{8.2}$$

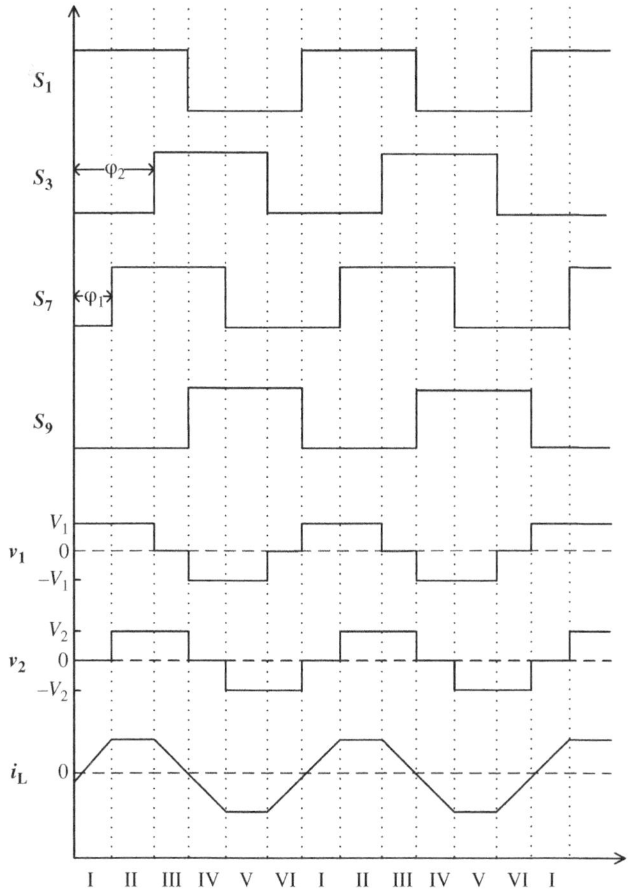

Figure 8.3 Dual-phase shift modulation.

$$B = \frac{1}{4L_{\text{lk}}f_s}\left(\frac{2\varphi V_2}{n\pi} - \frac{V_2}{n} + V_1\right) \tag{8.3}$$

A ratio of input voltage and output voltage can be defined as:

$$X = \frac{V_2}{nV_1} \tag{8.4}$$

The equations of A and B can be rewritten as:

$$A = \frac{V_1}{4L_{\text{lk}}f_s}\left(\frac{2\varphi}{\pi} + X - 1\right) \tag{8.5}$$

$$B = \frac{V_1}{4L_{\text{lk}}f_s}\left(\frac{2\varphi}{\pi}X - X + 1\right) \tag{8.6}$$

Therefore, to achieve ZVS for both full bridges, peak values A and B must be larger than zero, which gives the constraints as:

$$\frac{\varphi}{\pi} > \frac{X-1}{2X}, \text{ for leading full bridge} \tag{8.7}$$

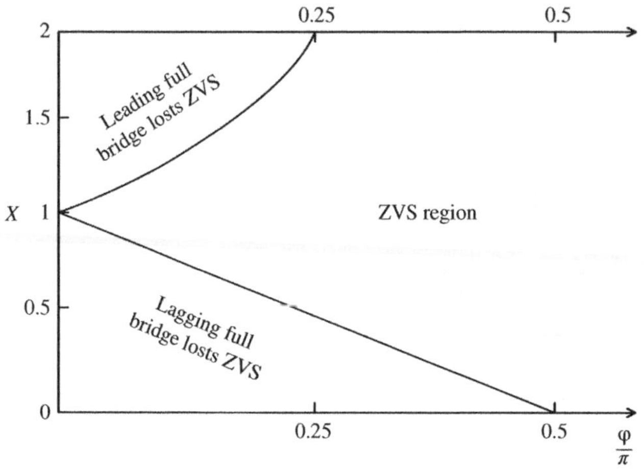

Figure 8.4 Constraints for enabling ZVS of DAB converter.

$$\frac{\varphi}{\pi} > \frac{1-X}{2}, \text{ for lagging full bridge} \tag{8.8}$$

The constraints are presented in Figure 8.4.

The DAB has been actively researched for aerospace application, such as reducing the common-mode current and increasing the transformer lifetime [4]. Also, considering the improved performance in terms of power density and the reduced requirement in terms of DC link capacitance, the three-phase DAB seems a very attractive option for aerospace application [5]. Besides, buck converters have been employed if isolation is not required [6–8].

8.3 The LLC Series-Resonant Converter

In order to achieve better soft-switching performance than the DAB converter, a resonant tank can be added to the primary stage circuit [2, 9]. If the resonant tank is composed of a capacitor connected in series to the transformer (the inductance can either be an external inductor or the leakage inductor), the converter is named LLC or series-resonant converter, shown in Figure 8.5. Another advantage compared to the DAB converter is the simplification of the secondary stage, which is composed of a diode bridge rectifier.

The principle of operation resides in the transfer function of the resonant circuit, which depends on the design parameter and on the load. Figure 8.6 shows the normalized frequency (i.e. the ratio of converter switching frequency to resonant frequency) of excitation of the resonant circuit determines the amplitude of the output voltage, which is then rectified. And it can be depicted by an expression as follows [10]:

$$K(Q, m, F_x) = \left|\frac{V_{o_ac}(s)}{V_{in_ac}(s)}\right| = \frac{F_x^2(m-1)}{\sqrt{(mF_x^2-1)^2 + F_x^2(F_x^2-1)^2(m-1)^2Q^2}} \tag{8.9}$$

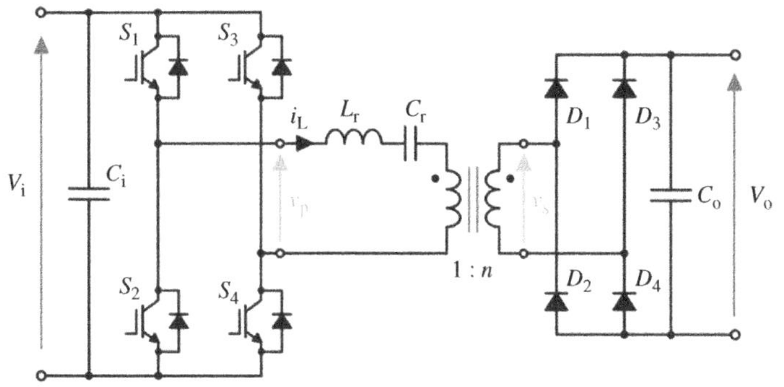

Figure 8.5 The series resonant converter. *Source:* Costa et al. [11]/IEEE.

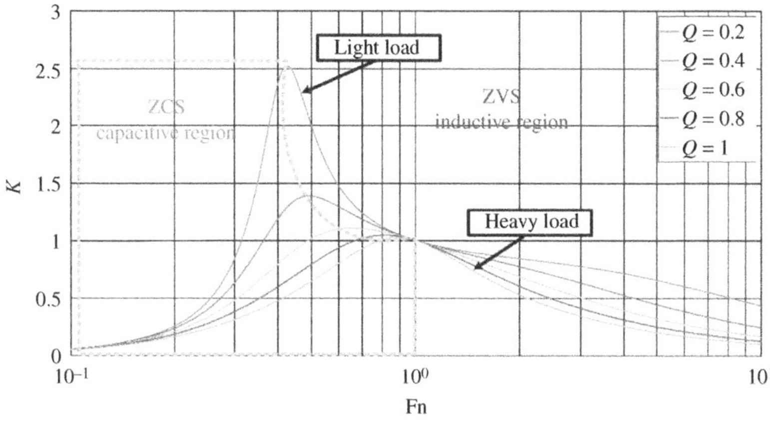

Figure 8.6 The operating region of the series resonant converter.

where $Q = \frac{\sqrt{L_r/C_r}}{R_{ac}}$, $R_{ac} = \frac{8{N_P}^2 R_o}{\pi^2 {N_S}^2}$, $F_x = \frac{f_s}{f_r}$, $f_r = \frac{1}{2\pi\sqrt{L_r C_r}}$ and $m = \frac{L_r + L_m}{L_r}$. Q is the quality factor, R_{ac} is the resistance reflecting the secondary side circuit, F_x is the normalized switching frequency, f_r is the resonant frequency, m is the ratio of total primary inductance to resonant inductance. Thereinto, R_o is the DC output resistance, L_m is the magnetizing inductance of the transformer, N_P is the number of turns of primary winding of transformer, N_S is the number of turns of secondary winding of transformer, f_s is the switching frequency, and L_r and C_r are the resonant inductance and capacitance. In addition, it can be seen from Figure 8.6 that different values of Q can determine the soft switching region for the converter loads. The selection of Q depends on which converter mode (buck or boost) the user wants to adopt, referring to the load and which soft switching mode (i.e. ZVS or zero current switching) can properly bring more efficiency in the case. Typical values are usually between 0.5 and 0.7.

Depending on the power level and the device choice, it may be more advantageous to operate in either of the regions. For high-power solutions which do not need a wide voltage ratio range, operating in ZCS in the proximity of the resonant frequency has the advantage of the ease of control (since the converter maintains a stable output voltage regardless of the load) [11]. For lower-power solutions (for example, the 270–28V for the avionic), operating in ZVS would allow for a very compact solution with some degrees of controllability due to the high switching frequency. The latter solution is usually adopted for more electric aircraft applications [9].

8.4 Constant Power Load

As the onboard power electronics increase, the scale of EPDS becomes much larger and the system structure becomes more complex. The combination of source and load subsystems with different behaviors could make the system unstable, especially for those load subsystems, which perform as constant power loads (CPL). The existence of CPL will destabilize the system due to its negative incremental impedance characteristic [12].

Figure 8.7 shows the I–V curve of CPL. Assume P is the power, I and V are the current and voltage of CPL, and i and v are the current and voltage of CPL at the operating point. It exists

$$I = \frac{P}{V} \tag{8.10}$$

By doing derivative of (8.10), it can have

$$\partial I = -\frac{P}{v^2}\partial V \tag{8.11}$$

By rearranging (8.19), the slope of CPL curve at the operating point can be obtained as

$$\frac{\partial I}{\partial V} = -\frac{P}{v^2} = -\frac{i}{v} \tag{8.12}$$

8.5 Stability Criteria

To avoid the instability introduced by the negative incremental impedance characteristic of CPL, the design of system needs to be well considered while the stability analysis needs to be carried out for predicting system stability. As for the approaches to analyzing the system stability, one of the

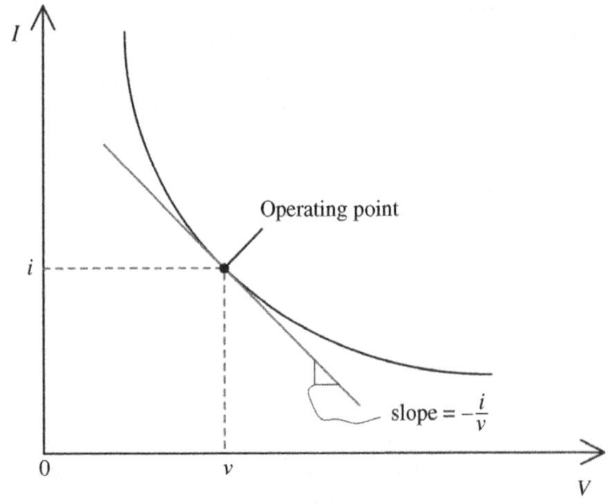

Figure 8.7 *I*–*V* curve of CPL.

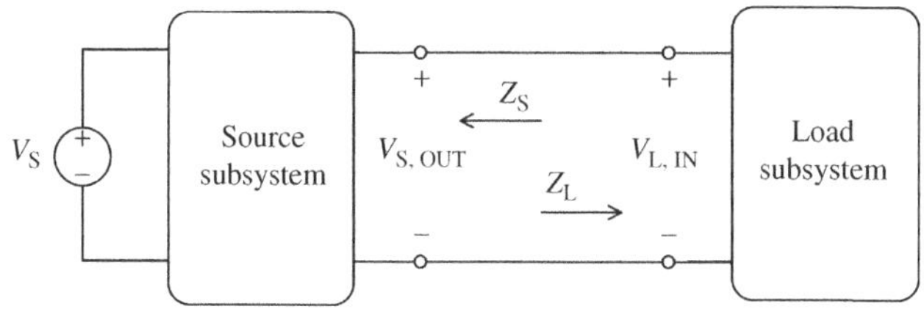

Figure 8.8 The impedance model of a DC microgrid. *Source:* 2021 IEEE.

mainstreams is based on the impedance of source subsystems and load subsystems. Figure 8.8 shows the impedance model of a DC microgrid in which V_s is the source voltage, Z_S is the output impedance of source subsystem and Z_L is the input impedance of load subsystem. Hence, if the source is an ideal voltage source, the load voltage can be given as

$$V_L = V_S \frac{Z_L}{Z_S + Z_L} = V_S \frac{1}{\frac{Z_S}{Z_L} + 1} \tag{8.13}$$

Here, a concept called "minor loop gain," which is defined as the ratio of the source impedance Z_S to the load impedance Z_L, is introduced into the equation, denoted as G_{MLG}. This equation can be represented as:

$$V_L = V_S \frac{1}{G_{MLG} + 1} \tag{8.14}$$

However, if the source is an ideal current source, the minor loop gain becomes reciprocal [13], which is explained by Eq. (8.15). Assuming that the source current is I_S, the output admittance of source subsystem is Y_S and the input admittance of load subsystem is Y_L, the load current can be given as

$$I_L = I_S \frac{Y_L}{Y_S + Y_L} = I_S \frac{1}{\frac{Y_S}{Y_L} + 1} = I_S \frac{1}{\frac{Z_L}{Z_S} + 1} = I_S \frac{1}{G_{MLG} + 1} \tag{8.15}$$

It can be seen that the minor loop gain G_{MLG} in (8.13) and (8.15) are reciprocal to each other. The system stability can be determined by the transfer function which has unity gain on the feedforward path and has G_{MLG} on the feedback path. With this feature, several stability criteria were proposed based on the precondition that the subsystems are individually stable. The most basic one is the Nyquist Criterion, which states that the system will be stable if the Nyquist contour of G_{MLG} does not encircle (−1, 0). The Nyquist Criterion is a sufficient and necessary condition for system stability. Based on that, other criteria were proposed. The most conservative is the Middlebrook Criterion [14], and the constraint is given as

$$|G_{MLG}| = \left|\frac{Z_S}{Z_L}\right| < \frac{1}{\text{GM}}, \text{ where GM} > 1 \tag{8.16}$$

where GM is the desired gain margin. The Middlebrook Criterion not only guarantees the system stability but also considers the dynamics of load converter with influence from the input filter. However, it could result in overdesign of input filter, making the size of filter circuit large. To lose the design requirement, the gain margin phase margin (GMPM) criterion was proposed [15]. The constraint of GMPM criterion is given as

$$|G_{\mathrm{MLG}}| = \left|\frac{Z_{\mathrm{S}}}{Z_{\mathrm{L}}}\right| < \frac{1}{\mathrm{GM}}, \text{ if not, } |\angle G_{\mathrm{MLG}}| \leq 180^{\circ} - \mathrm{PM} \tag{8.17}$$

where GM and PM are the desired gain and phase margin. By introducing the phase margin, the system design can be liberated.

8.6 Impedance Modeling and Stability Analysis

Before analyzing the system stability, the impedance model of source subsystem and load subsystem needs to be worked out. In this case, the source subsystem is a permanent magnet synchronous generator (PMSG) and the load subsystem consists of a DAB converter and a tightly controlled motor drive, which can be regarded as a CPL.

8.6.1 Impedance Model of PMSG

In this subsection, the output impedance of PMSG is derived [16, 17]. Figure 8.9 shows the basic control structure of PMSG. The open loop model in *d-q* frame is presented in Figure 8.10. R_s and L_s are the stator resistance and inductance, ω_e is the electrical speed which is equal to $p\omega m$ (p is the pole-pair and ω_m is the mechanical speed), D_d and D_q are the duty cycle of V_d and V_q to V_{dc}, I_{dc} is the DC bus current, and ψ_{PM} is the flux linkage of a permanent magnet. Hence, the machine equation in *d-q* frame can be given as

$$\begin{cases} V_d = -(R_s + sL_s)I_d + \omega_e L_s I_q \\ V_q = -(R_s + sL_s)I_q - \omega_e L_s I_d + \omega_e \psi_{PM} \end{cases} \tag{8.18}$$

The small signal equations can be obtained as

$$\begin{cases} \hat{V}_d = -(R_s + sL_s)\hat{I}_d + \omega_e L_s \hat{I}_q \\ \hat{V}_q = -(R_s + sL_s)\hat{I}_q - \omega_e L_s \hat{I}_d \end{cases} \tag{8.19}$$

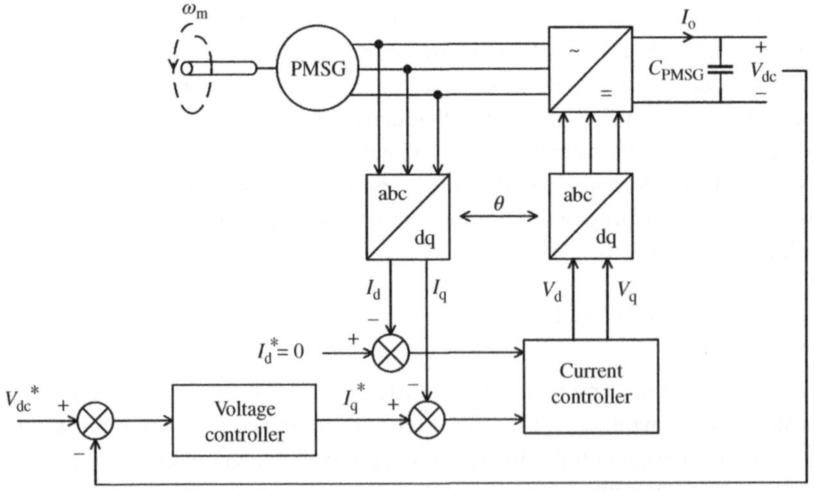

Figure 8.9 The basic control structure of PMSG. *Source:* From Ref. [16]/2021 IEEE.

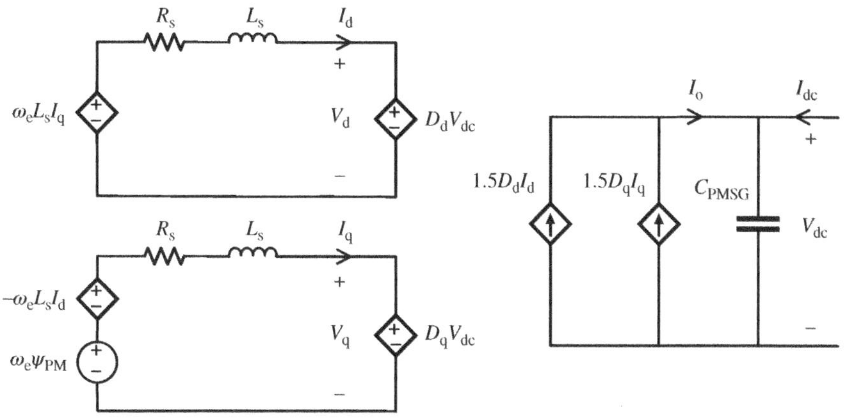

Figure 8.10 The open loop model of PMSG. *Source:* From Ref. [17]/2021 IEEE.

Assuming that d and q axes are well decoupled, so only the components of q axis need to be considered. The control of q axis can be given as

$$\begin{cases} \hat{I}_q^* = \left(\hat{V}_{dc}^* - \hat{V}_{dc}\right) G_v(s) \\ \left(\hat{I}_q^* - \hat{I}_q\right) G_c(s) = (R_s + sL_s)\hat{I}_q \end{cases} \tag{8.20}$$

where $G_v(s)$ and $G_c(s)$ are the transfer functions of the voltage controller and current controller. Besides, the dynamics during power transfer also have influence on system impedance. Considering I_d is zero in steady state and with the constant amplitude transformation, the power equation of PMSG at steady state can be given as

$$V_{dc} I_o = 1.5 V_q I_q \tag{8.21}$$

The small signal equation of I_o can be derived as

$$\hat{I}_o = \frac{1.5 I_q}{V_{dc}} \hat{V}_q + \frac{1.5 V_q}{V_{dc}} \hat{I}_q - \frac{I_o}{V_{dc}} \hat{V}_{dc} \tag{8.22}$$

Hence, the transfer function block scheme of PMSG can be worked out as Figure 8.11. The output impedance of PMSG can be derived as

$$Z_{o,PMSG} = \frac{V_{dc}[G_c(s) + R_s + sL_s]}{sC_{PMSG} V_{dc}[G_c(s) + R_s + sL_s] + G_v(s) G_c(s) [1.5 V_q - (R_s + sL_s) 1.5 I_q] G_{d,PMSG}(s) + I_o[G_c(s) + R_s + sL_s] G_{d,PMSG}(s)} \tag{8.23}$$

8.6.2 Controller Design

In this section, the design processes of current controller and voltage controller for three-phase voltage source rectifier are presented. Considering that the stator of PMSG has inductance L_s and resistance R_s, the admittance of stator is given as

$$G_s(s) = \frac{1}{L_s s + R_s} \tag{8.24}$$

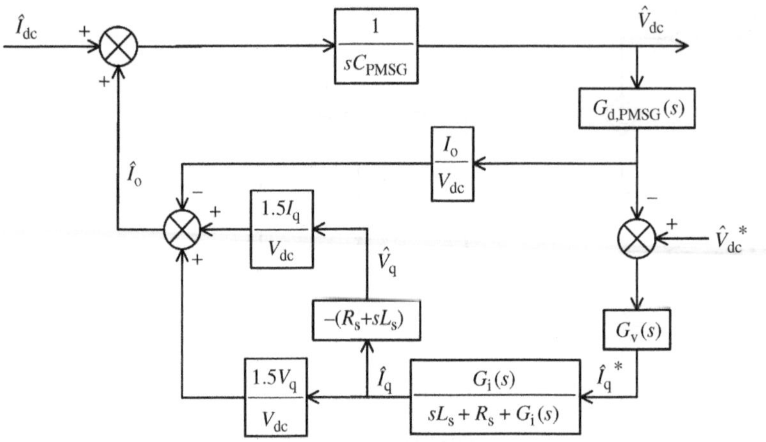

Figure 8.11 The transfer function block scheme of PMSG. *Source:* From Ref. [16]/2021 IEEE.

The current controller can be expressed as:

$$G_c(s) = k_{pc} + \frac{k_{ic}}{s} = \frac{k_{pc}s + k_{ic}}{s} \tag{8.25}$$

where k_{pc} and k_{ic} are the proportional gain and integral gain. Then, the transfer function of current control loop can be given as

$$G_{c,cl}(s) = \frac{G_c(s)G_s(s)}{G_c(s)G_s(s) + 1} = \frac{\dfrac{k_{pc}s + k_{ic}}{L_s s + R_s}}{\dfrac{k_{pc}s + k_{ic}}{L_s s + R_s} + s} \tag{8.26}$$

It is easy to observe that this transfer function is first-order. Hence, by canceling the poles and zeros in numerator and denominator, the equation can be simplified as

$$G_{c,cl}(s) = \frac{\omega_c}{\omega_c + s} = \frac{1}{1 + s/\omega_c} \tag{8.27}$$

where ω_c is the desired bandwidth of current controller. The parameters of current controller can be determined as

$$G_c(s) = k_{pc} + \frac{k_{ic}}{s} = \omega_c L_s + \frac{\omega_c R_s}{s} \tag{8.28}$$

The voltage controller can be expressed as

$$G_v(s) = k_{pv} + \frac{k_{iv}}{s} \tag{8.29}$$

The transfer function block scheme of voltage control loop is shown in Figure 8.12. The open loop gain of voltage control can be obtained as

$$G_{v,ol}(s) = \frac{1.5V_q k_{iv}}{V_{dc}C_{PMSG}} \frac{\dfrac{k_{pv}}{k_{iv}}s + 1}{s^2(1 + s/\omega_c)} = A\frac{\dfrac{k_{pv}}{k_{iv}}s + 1}{s^2(1 + s/\omega_c)} \tag{8.30}$$

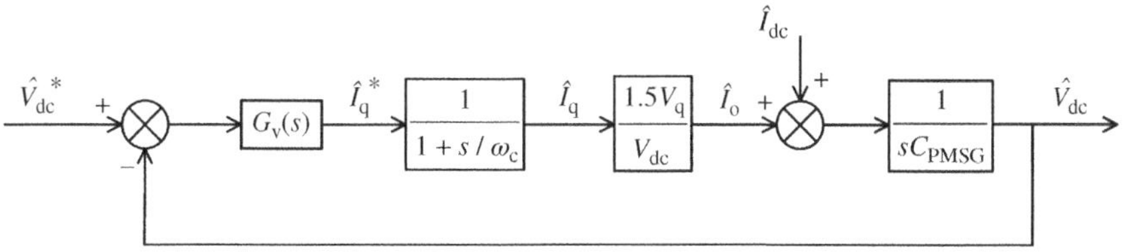

Figure 8.12 The transfer function block scheme of voltage control. *Source:* From Ref. [17]/2021 IEEE.

where A represents the constant. In this case, the technique of "symmetrical optimum" [18] can be used to maximize the phase margin of open loop gain, the frequency at maximum phase margin can be given as

$$\omega_o = \sqrt{\omega_c \omega_v} \tag{8.31}$$

where ω_o is the frequency at maximum phase margin and ω_v is the desired bandwidth of voltage controller. Considering that the open loop gain will be unity at ω_o, it exists

$$|G_{v,ol}(\omega)| = 1 = \frac{A}{\omega_c \omega_v}\sqrt{\frac{1+\dfrac{\omega_c}{\omega_v}}{1+\dfrac{\omega_v}{\omega_c}}} \tag{8.32}$$

Hence, the integral gain of voltage controller can be calculated as

$$k_{iv} = \frac{V_{dc}C_{PMSG}}{1.5V_q}\omega_v\sqrt{\omega_c\omega_v} \tag{8.33}$$

And the proportional gain of voltage controller can be calculated as

$$k_{pv} = k_{iv}/\omega_v = \frac{V_{dc}C_{PMSG}}{1.5V_q}\sqrt{\omega_c\omega_v} \tag{8.34}$$

Besides, it is worth mentioning that for a given phase margin θ, it exists

$$\frac{\omega_c}{\omega_v} = \left(\frac{1+\cos\theta}{\sin\theta}\right)^2 \tag{8.35}$$

8.6.3 Impedance Model of DAB Converter

In this subsection, the input impedance of DAB converter is derived [19]. Figure 8.13 shows the basic control structure of DAB converter, where the single PSM and load voltage control are applied. V_i and V_o are the input voltage and output voltage, C_1 and C_2 are the input capacitor and output capacitor, L_{lk} is the leakage inductance, and n_1 and n_2 are the number of turns of a transformer. The power equation of DAB converter is given as

$$P = \frac{V_i V_o}{2Nf_s L_{lk}} d(1-|d|) \tag{8.36}$$

where f_s is the switching frequency, N is the turn ratio of transformer $\frac{n_2}{n_1}$, and d is the phase shift $d_2 - d_1$. The input and output currents can be obtained as

$$I_1 = \frac{P}{V_i} = \frac{V_o}{2Nf_s L_{lk}} d(1-|d|) \tag{8.37}$$

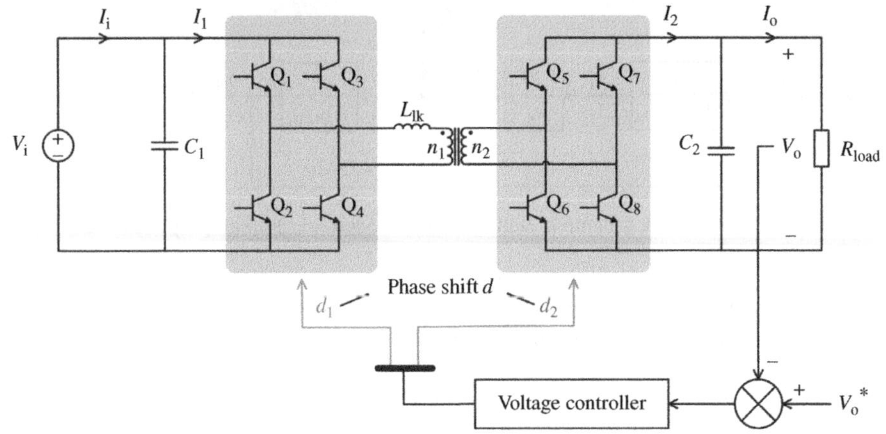

Figure 8.13 The basic control structure of DAB converter.

$$I_2 = \frac{P}{V_o} = \frac{V_i}{2Nf_sL_{lk}}d(1-|d|) \tag{8.38}$$

Hence, the small signal equations can be derived as

$$\hat{I}_1 = \frac{1}{2Nf_sL_{lk}}d(1-|d|)\hat{V}_o + \frac{V_o}{2Nf_sL_{lk}}(1-|2d|)\hat{d} \tag{8.39}$$

$$\hat{I}_2 = \frac{1}{2Nf_sL_{lk}}d(1-|d|)\hat{V}_i + \frac{V_i}{2Nf_sL_{lk}}(1-|2d|)\hat{d} \tag{8.40}$$

According to these small signal equations, the small signal open loop model of DAB converter can be worked out in Figure 8.14. The small signal gains G_1, G_2, G_3, and G_4 are given as

$$G_1 = \frac{V_o}{2Nf_sL_{lk}}(1-|2d|),\ G_2 = \frac{1}{2Nf_sL_{lk}}d(1-|d|) \tag{8.41}$$

$$G_3 = \frac{V_i}{2Nf_sL_{lk}}(1-|2d|),\ G_4 = \frac{1}{2Nf_sL_{lk}}d(1-|d|) \tag{8.42}$$

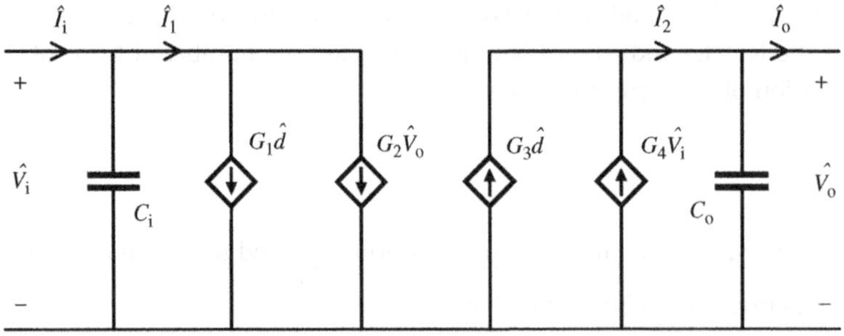

Figure 8.14 The small signal open-loop model of DAB converter.

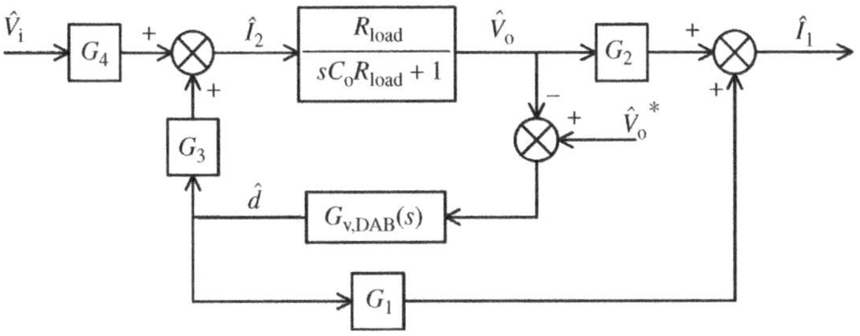

Figure 8.15 The transfer function block scheme of DAB converter.

Considering the controller is input by load voltage and outputs the phase shift, the transfer function block scheme can be obtained as Figure 8.15, where $G_{v,DAB}(s)$ is the voltage controller. By considering the input capacitor, the input impedance of DAB converter can be obtained as

$$Z_{i,DAB} = \frac{G_3G_{v,DAB}R_{load} + sC_oR_{load} + 1}{G_4R_{load}(G_2 - G_1G_{v,DAB}) + sC_i(G_3G_{v,DAB}R_{load} + sC_oR_{load} + 1)} \tag{8.43}$$

8.6.4 Impedance-Based Stability Analysis

In this section, a switching model of a DC microgrid formed with a PMSG, a DAB converter and a tightly controlled motor drive is created in PLECS, where the line impedance is also considered. The impedance models are validated with the switching model, and the system stability is predicted by using the impedance models and stability criteria. Referring to Airbus A380, the power capacity of DC microgrid in following analysis is set as 500 kW. Considering the real on-board situation that the power of tightly controlled motor drives is much larger than DC-DC converters, it is assumed that the tightly controlled motor drives occupy 90% of system power and the DAB converter occupies 10% in the following analysis.

Although the Nyquist Criterion is the sufficient and necessary condition for system stability, it is discussed in [20] that different grouping of subsystems can influence how they will be designed and also the accuracy of stability analysis based on the bode diagram. In this case, the PMSG, cable, input capacitor of DAB converter are grouped to be the source impedance, and the rest components are grouped to be the load impedance. Figure 8.16 shows the grouping details of source and load. Meanwhile the proportion of load power is presented.

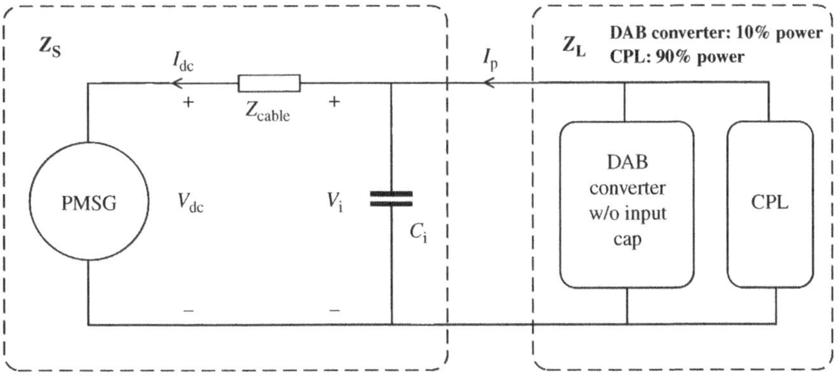

Figure 8.16 The grouping details of source and load. *Source:* From Ref. [16]/2021 IEEE.

8.6.5 Specifications

Before analysis, the feasible specifications of PMSG, DAB converter, and the line impedance are given. Table 8.1 gives the parameters of PMSG and Table 8.2 gives the parameters of DAB converter.

Considering that the wingspan of Airbus A380 is 80 m, it is assumed that the DC bus cable has the same length as the wings. To handle the large current, the cable is chosen to be a combination of seven wires with type of ASNE0438-YV AWG 4/0 in [21]. Assuming that the relative permeability of the conductor is 1, with the known diameter and length of cable, the inductance of each wire can be calculated using the formula in [22]. The parameters of DC bus cable are given in Table 8.3.

Table 8.1 Simulation parameters of PMSG.

Parameters of PMSG		
Symbol	**Definition**	**Value**
R_s	Stator resistance	4 mΩ
L_s	Stator inductance	40 μH
p	Number of pole pairs	1
f_e	Carrier (switching) frequency	20 kHz
ω_e	Current controller bandwidth	1256.6 rad/s
V_{de}	DC bus voltage	540 V
ω_m	Mechanical rotating speed	15,000 rpm
ψ_{PM}	Flux linkage of the permanent magnet	0.127 Wb
C_{PMSG}	Output capacitor	10 mF
θ	Desired phase margin	60°

Source: From Ref. [16]/2021 IEEE.

Table 8.2 Simulation parameters of DAB converter.

Parameters of DAB converter		
Symbol	**Definition**	**Value**
V_i	Input voltage	540 V
V_o	Output voltage	270 V
N	Turn ratio of transformer	270/540
L_{lk}	Leakage inductance	10 μH
f_s	Switching frequency	50 kHz
$k_{p,DAB}$	Proportional coefficient	0.02
$k_{i,DAB}$	Integral coefficient	1
C_i	Input capacitor	6 mF
C_o	Output capacitor	6 mF

Source: From Ref. [16]/2021 IEEE.

Table 8.3 Parameters of DC bus cable.

Parameters of DC bus cable		
Symbol	**Definition**	**Value**
l	Length of cable	80 m
d	Diameter of cross section of each wire	14.1 mm
μ	Relative permeability of conductor	1
R_{cable}	Resistance of cable	3.3 mΩ
L_{cable}	Inductance of cable	21.2 μH

Source: From Ref. [20]/2021 IEEE.

$$L_{\text{wire}} = 2l\left\{\ln\left[\frac{2l}{d}\left(1+\sqrt{1+\left(\frac{d}{2l}\right)^2}\right)\right]-\sqrt{1+\left(\frac{d}{2l}\right)^2}+\frac{\mu}{4}+\frac{d}{2l}\right\} \tag{8.44}$$

8.6.6 Impedance Model Validation

To validate the output impedance model of the PMSG, an ideal current source containing DC current and AC sinusoidal current is connected to the output of PMSG. By performing Fast Fourier Transformation (FFT) of the output voltage and current of PMSG at steady state, the output impedance of PMSG at specific frequency can be calculated. Figure 8.17 shows the output impedance of PMSG at 500 kW. It can be seen that the proposed impedance model is consistent with simulation

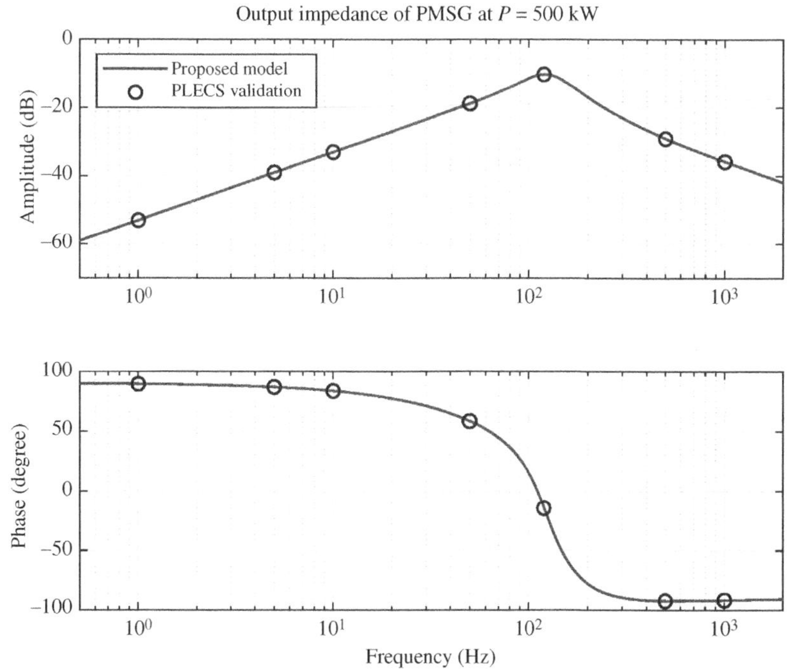

Figure 8.17 The output impedance of PMSG at power of 500 kW. *Source:* From Ref. [20]/2021 IEEE.

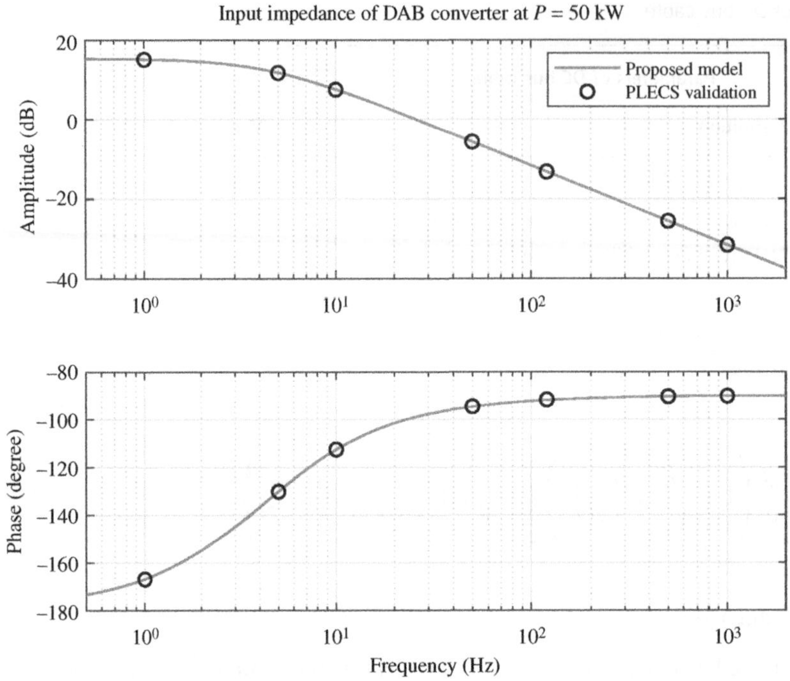

Figure 8.18 The input impedance of DAB converter at power of 50 kW. *Source:* From Ref. [16]/2021 IEEE.

measurements. The validation of the input impedance model of DAB converter is same, except that an ideal voltage source is connected to the input of DAB converter. Figure 8.18 shows the input impedance of DAB converter at 50 kW. It can be seen that the proposed impedance model is consistent with simulation measurements.

8.6.7 System Instability

For a CPL-dominated system, the increase in power will make the system tend to instability. Figure 8.19 shows the bode diagram of impedance of source and load at different power levels. It can be seen that the source impedance and load impedance have intersection at a frequency of approximately 570 Hz when system powers are 400 and 500 kW, respectively. By recalling the GMPM Criterion, if the $|Z_S|$ is larger than $|Z_L|$, then the phase information is needed, i.e. the difference of $\angle Z_S$ and $\angle Z_L$ must be smaller than 180 to guarantee the system stability. It can be seen that the difference of $\angle Z_S$ and $\angle Z_L$ is over 180° at intersection point indicating that the system will become unstable when system power is 400 and 500 kW, respectively. Also, the system instability can be seen in the Nyquist contour of minor loop gain shown in Figure 8.20. The contour of minor loop gain encircles (−1, 0) at system power of 400–500 kW, respectively. Besides, the bode diagram can indicate the frequency of instability harmonics. Figure 8.21 shows the output voltage of PMSG. It can be observed that the output voltage is unstable and contains a harmonic component with frequency of 570 Hz.

8.6.8 Proposed Control Techniques for Stabilization

As mentioned before, the intersection by the impedance models in bode diagram indicates system instability. To solve this problem and extend the stable region of system, the impedance models

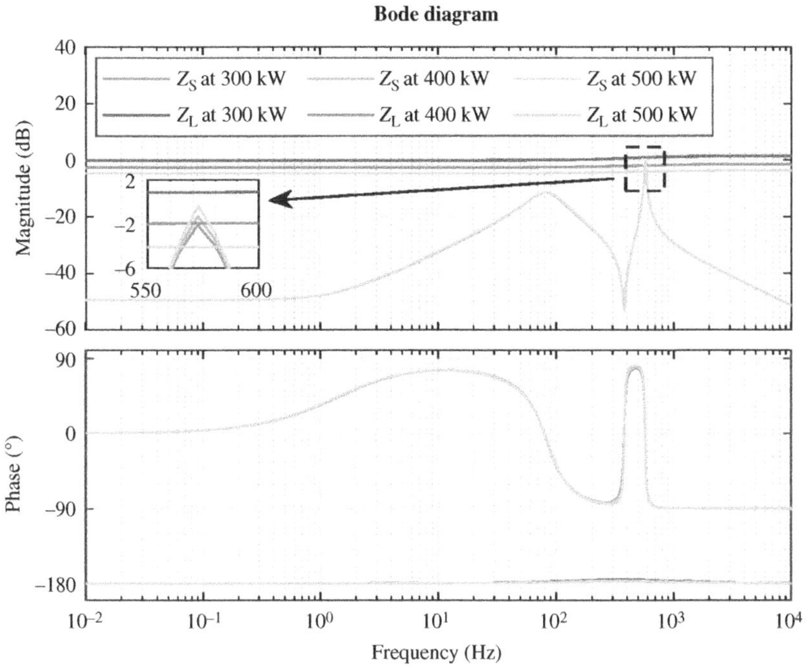

Figure 8.19 The bode diagram of impedance of source and load at different power levels. *Source:* From Ref. [16]/2021 IEEE.

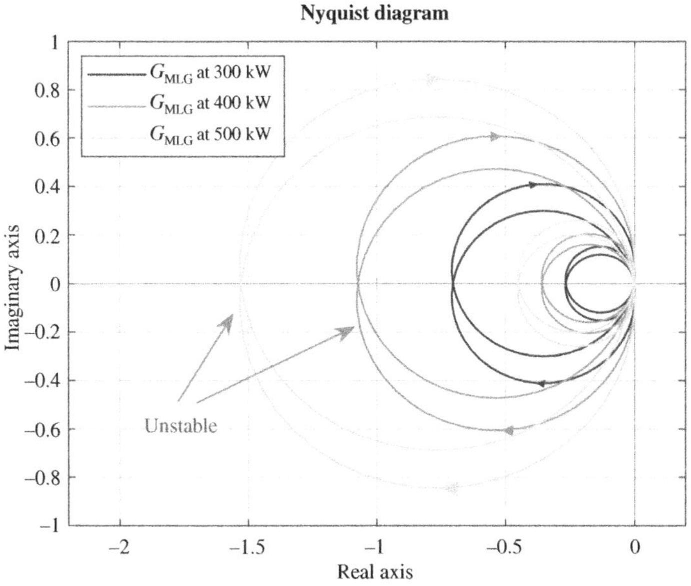

Figure 8.20 The Nyquist contour of minor loop gain at different power level. *Source:* From Ref. [16]/2021 IEEE.

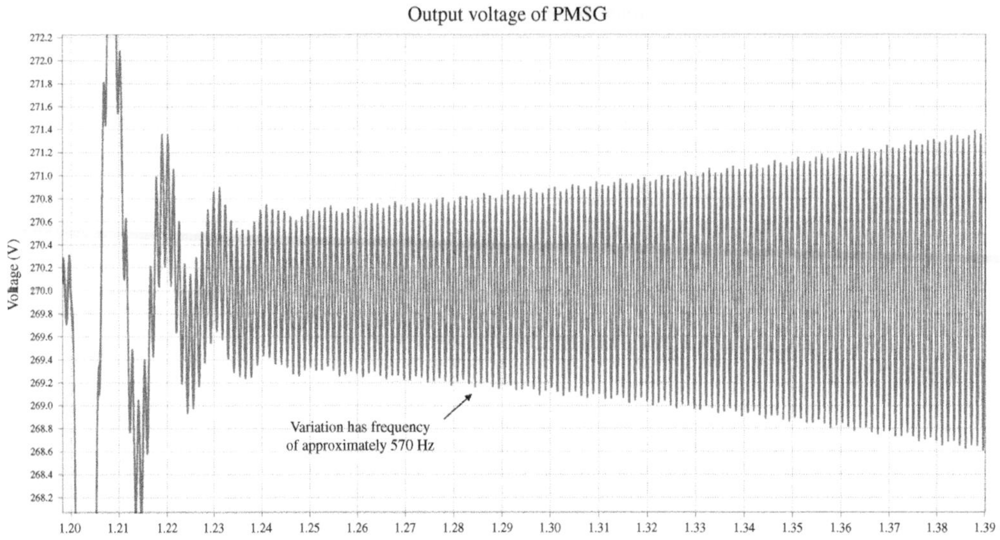

Figure 8.21 The output DC voltage of PMSG with instability harmonic of 570 Hz.

need to be reshaped, and one main approach is reshaping the impedance by modifying control strategies. The voltage sag pass-through control (VSPTC) for reshaping the Z_L and the point of load control (PLC) for reshaping the Z_S were proposed in [16]. It is seen that by applying VSPTC, the phase difference of $\angle Z_S$ and $\angle Z_L$ is largely diminished lower than 180°. And by applying PLC, the $|Z_S|$ at intersection point is reshaped to avoid intersecting with $|Z_L|$. The simulation and experiment were also carried out. Both simulation results and experimental results were successfully predicted with the proposed impedance models, and the effect of VSPTC and PLC on stabilizing system was verified, too.

8.7 Conclusion

With the impedance models of PMSG and DAB converter, the system instability introduced by the bus cable is successfully predicted by Nyquist contour. According to the bode diagram, a harmonic component with frequency of 570 Hz in DC bus voltage can be foreseen when system is unstable. To stabilize the system, two control strategies, VSPTC and PLC are proposed by reshaping the load impedance and source impedance. Moreover, what is more significant is the system design can be effectively optimized in terms of volume, kVA rating etc., while the system stability is guaranteed with the impedance models. Besides, the impedance models also enable the stability prediction for a system with more complex structure (e.g. multi-source multi-load system) during the design.

References

1 Madonna, V., Giangrande, P., and Galea, M. (2018). Electrical power generation in aircraft: review, challenges, and opportunities. *IEEE Trans. Transport. Electrification* 4 (3): 646–659. https://doi.org/10.1109/TTE.2018.2834142.

2 Buticchi, G., Costa, L., and Liserre, M. (2017). Improving system efficiency for the more electric aircraft: a look at dc–dc converters for the avionic onboard dc microgrid. *IEEE Ind. Electron. Mag.* 11 (3): 26–36. https://doi.org/10.1109/MIE.2017.2723911.

3 Kheraluwala, M.N., Gascoigne, R.W., Divan, D.M., and Baumann, E.D. (1992). Performance characterization of a high-power dual active bridge DC-to-DC converter. *IEEE Trans. Ind. Appl.* 28 (6): 1294–1301.

4 Buticchi, G., Barater, D., Costa, L.F., and Liserre, M. (2018). A PV-inspired low-common-mode dual-active-bridge converter for aerospace applications. *IEEE Trans. Power Electron.* 33 (12): 10467–10477. https://doi.org/10.1109/TPEL.2018.2801845.

5 De Doncker, R.W., Divan, D.M., and Kheraluwala, M.H. (1998). A three-phase soft-switched high power density DC/DC converter for high power applications. *Conference Record of the 1988 IEEE Industry Applications Society Annual Meeting*, pp. 796–805, Vol 1. https://doi.org/10.1109/IAS.1988.25153.

6 Gao, F. and Bozhko, S. (2016). Modeling and impedance analysis of a single DC bus-based multiple-source multiple-load electrical power system. *IEEE Trans. Transport. Electrification* 2 (3): 335–346. https://doi.org/10.1109/TTE.2016.2592680.

7 Gao, F., Bozhko, S., Costabeber, A. et al. (2017). Comparative stability analysis of droop control approaches in voltage-source-converter-based DC microgrids. *IEEE Trans. Power Electron.* 32 (3): 2395–2415. https://doi.org/10.1109/TPEL.2016.2567780.

8 Gao, F., Bozhko, S., Costabeber, A. et al. (2017). Control design and voltage stability analysis of a droop-controlled electrical power system for more electric aircraft. *IEEE Trans. Ind. Electron.* 64 (12): 9271–9281. https://doi.org/10.1109/TIE.2017.2711552.

9 Chen, G., Chen, L., Deng, Y. et al. (2019). Topology-reconfigurable fault-tolerant LLC converter with high reliability and low cost for more electric aircraft. *IEEE Trans. Power Electron.* 34 (3): 2479–2493. https://doi.org/10.1109/TPEL.2018.2848297.

10 Abdel-Rahman, S. (2012). *Resonant LLC Converter: Operation and Design*, 19. North America: Infineon Technologies.

11 Costa, L., Buticchi, G., and Liserre, M. (2017). A fault-tolerant series-resonant DC–DC converter. *IEEE Trans. Power Electron.* 32 (2): 900–905. https://doi.org/10.1109/TPEL.2016.2585668.

12 Emadi, A., Khaligh, A., Rivetta, C.H., and Williamson, G.A. (2006). Constant power loads and negative impedance instability in automotive systems: definition, modeling, stability, and control of power electronic converters and motor drives. *IEEE Trans. Veh. Technol.* 55 (4): 1112–1125. https://doi.org/10.1109/TVT.2006.877483.

13 Sun, J. (2011). Impedance-based stability criterion for grid-connected inverters. *IEEE Trans. Power Electron.* 26 (11): 3075–3078. https://doi.org/10.1109/TPEL.2011.2136439.

14 Middlebrook, R.D. (1976). Input filter considerations in design and application of switching regulators. *Conf. Rec. IEEE IAS Annu. Meeting*, pp. 366–382.

15 Wildrick, C.M., Lee, F.C., Cho, B.H., and Choi, B. (1995). A method of defining the load impedance specification for a stable distributed power system. *IEEE Trans. Power Electron.* 10 (3): 280–285.

16 Yang, J., Yan, H., Gu, C. et al. (2021). Modeling and stability enhancement of a permanent magnet synchronous generator based DC system for more electric aircraft. *IEEE Trans. Ind. Electron.* 1–1. https://doi.org/10.1109/TIE.2021.3066934.

17 Yang, J., Buticchi, G., Gu, C., and Wheeler, P. (2021). Impedance-based stability analysis of permanent magnet synchronous generator for the more electric aircraft. *2021 IEEE Workshop on Electrical Machines Design, Control and Diagnosis (WEMDCD)*, pp. 181–185. https://doi.org/10.1109/WEMDCD51469.2021.9425679.

18 Teodorescu, R., Liserre, M., and Rodrguez, P. (2011). *Grid Converters for Photovoltaic and Wind Power Systems*. Chichester, UK: Wiley. https://doi.org/10.1002/9780470667057.

19 Yang, J., Buticchi, G., Yan, H. et al. (2019). Impedance-based sensitivity analysis of dual active bridge DC–DC converter. *2019 IEEE 13th International Conference on Compatibility, Power Electronics and Power Engineering (CPE-POWERENG)*, pp. 1–5. https://doi.org/10.1109/CPE.2019.8862418.

20 Sudhoff, S.D., Glover, S.F., Lamm, P.T. et al. (2000). Admittance space stability analysis of power electronic systems. *IEEE Trans. Aerosp. Electron. Syst.* 36 (3): 965–973. https://doi.org/10.1109/7.869516.

21 Nexans. (2007). Aircraft Wires and Cables.

22 Grover, F.W. (2004). *Inductance Calculations: Working Formulas and Tables*. Courier Corporation.

9

Packed U-Cell Inverter and Its Variants with Fault Tolerant Capabilities for More Electric Aircraft

Haroon Rehman[1], Mohd Tariq[1,2], Hasan Iqbal[1], Arif I. Sarwat[2], and Adil Sarwar[1]

[1] *Aligarh Muslim University, Department of Electrical Engineering, ZHCET, Aligarh, Uttar Pradesh 202002, India*
[2] *Florida International University, Department of Electrical and Computer Engineering, Energy Power Sustainability & Intelligence (EPSi) Lab, Miami, FL 33174, USA*

9.1 Introduction

The rise in power efficiency, the drop in operating costs, and most importantly, the reduction in smog, are driving significant changes in mobility. The electrification of the transportation sector is receiving considerable attention. The power electronics (PE) and electric machines required for this goal must be more dependable, effective, compact, and quiet, and have much higher power densities than they are now. A vehicle that is electrically propelled depends heavily on its PE to mitigate problems like specific power generation, energy storage capacity, and thermal management [1].

Civil aircraft have conventionally been powered by three general types of secondary power: hydraulic, pneumatic, and electrical. As manufacturers move forward, they are seeking electrical alternatives to traditional secondary hydraulic and pneumatic power systems [2]. In order to accomplish anti-icing, air conditioning, cabin pressurization, as well as fuel supply, traditional aircraft normally use hydraulic, pneumatic, or mechanical actuators and aviation fuel is used for the propulsion system. The significance of PE is increasing in the research and development of a cleaner sky, due to the MEAs, which includes several electrically powered equipment, along with propellers [3]. In addition to switching all hydraulic and pneumatic power systems (surface actuation, air conditioning, and wing deicing) over to electric power, MEA reduces the power required by load sharing and optimizing the electrical generation source [4, 5].

Since general aviation is becoming more competitive, MEAs are becoming more and more popular owing to their environmental friendliness and low cost [6–10]. An MEA replaces its mechanical, hydraulic, and pneumatic systems with electrical systems. This results in lighter aircraft and lower maintenance costs due to the simplified structure. Furthermore, it reduces its fuel consumption, noise, and emission, which has an advantageous effect on the environment [11]. Table 9.1 illustrates the comparison of an electrical system with hydraulic, mechanical, and pneumatic systems.

The number of electrically driven aircraft has grown by roughly 30% across the world in the last year as more than 200 electric aircraft demonstrators have been developing this system using a variety of methodologies [13]. Among the electric-powered eVTOLs developed by CityAirbus is the 4-seater full-electric, powered by eight motors of 100 kW each, allowing it to operate at 800 V [13]. The company is also pursuing large aircraft like the series hybrid e-fan X featuring a power

Transportation Electrification: Breakthroughs in Electrified Vehicles, Aircraft, Rolling Stock, and Watercraft, First Edition. Edited by Ahmed A. Mohamed, Ahmad Arshan Khan, Ahmed T. Elsayed, and Mohamed A. Elshaer.

Table 9.1 Comparative study of power distribution systems in aircraft.

System	Complexity	Maintenance	Technological maturity
Electrical	Complicated	Easy	System – developed New technologies – developing
Hydraulic	Easy	Complicated and risky	Developed
Mechanical	Complicated	Regular	Developed
Pneumatic	Easy	Complicated	Developed

Source: Adapted from Ounis et al. [12].

converter with a 2 MW capacity and 3 kV bus voltage [14]. All through this race, Bell also developed a hybrid electric propulsion system that uses six 100-kW motors to generate 600-kW of power. There are four 250 kVA generators on a Boeing 787, along with two 225 kVA secondary power units on board for onboard power [3, 15]. All these electrical systems require converters in order to transform power from one form to another. Typical two-level inverters have issues such as high dv/dt, high switching voltage, and common mode voltage, which can cause motors to overheat and fail. The output voltage profile produced by MLIs can be sinusoidal, with diminished dv/dt, resulting in less stress on the electronic devices, and windings, and with reduced harmonic content. Cascaded H-bridge (CHB) and other topologies like packed U-cell (PUC) provide a better alternative for an electrified powertrain due to the various benefits, like enhanced efficiency, improved power quality, and the use of lower-rated power devices [16–18].

An aircraft hybrid propulsion system with superior efficiency and enhanced density is introduced in [19] that uses a power converter based on active neutral point clamped (ANPC) for large-scale power. With the use of a five-level CHB converter, the paper [20] provides a detailed analysis of different modulation techniques, along with the LCL filters required for each modulation technique. Many industries including NASA are trying to use MLI for MEA application. Electric machines are the backbone of powertrain applications. One of the four jet engines of Airbus was swapped for an electric engine of 2 MW in 2017 [14]. Power is generated for the engine by a variety of machines, including switch reluctance machines [21], permanent magnet (PM) [22], and induction machines with special designs [23]. Multi-phase machines with more than three phases are frequently constructed with more than three phases. The literature shows that PM machines have significant advantages compared to other machines for fitting the MEA requirement since they have high power densities, reliability, and toughness [21]. The paper [24] shows the advantage of the use of PM machines.

In this chapter, firstly power system architecture in MEA is discussed. Secondly, different power converters employed in MEA are explained. Thirdly, different PUC-based inverter topologies are discussed with their controlling action. Fourthly, fault tolerant operation of PUC-based inverter is discussed for MEA application. Lastly, simulation results of PUC inverter in normal and faulty cases are discussed.

9.2 Power System Architecture in MEA

The power system architecture in the MEA includes generation and distribution architecture. An MEA power system is categorized based on the way power is generated and distributed.

The power generation architecture (PGA) is classified into two types, i.e. constant speed (or constant frequency) power generation architecture (CSFPGA) and variable speed (or variable frequency) power generation architecture (VSFPGA), depending upon the speed at which the generator shaft is rotating [25]. In CSFPGA, the generator shaft rotates at constant speed with the help of constant speed drive (a mechanical drive) connected to the engine. This results in a constant electrical frequency on the bus connected to the generator. In VSFPGA, the generator shaft rotates at speed proportional to the engine-rotating speed. There is no constant speed drive present between the engine and the generator. As a result, the bus connected to the generator can have variable electrical frequency. For instance, the frequency can range from 350 up to 800 Hz.

The power distribution architecture (PDA) is classified into three types, i.e. constant frequency power distribution architecture (CFPDA), variable frequency power distribution architecture (VFPDA), and DC power distribution architecture (DCPDA), depending upon the frequency of generator's output voltage. In CFPDA, there is distribution of electrical power at constant frequency as shown in Figure 9.1a [25]. This constant frequency can be provided using one or more methods. These methods are as follows:

- Directly from generator's output under CSFPGA
- VSFPGA output is connected to the AC/AC converter

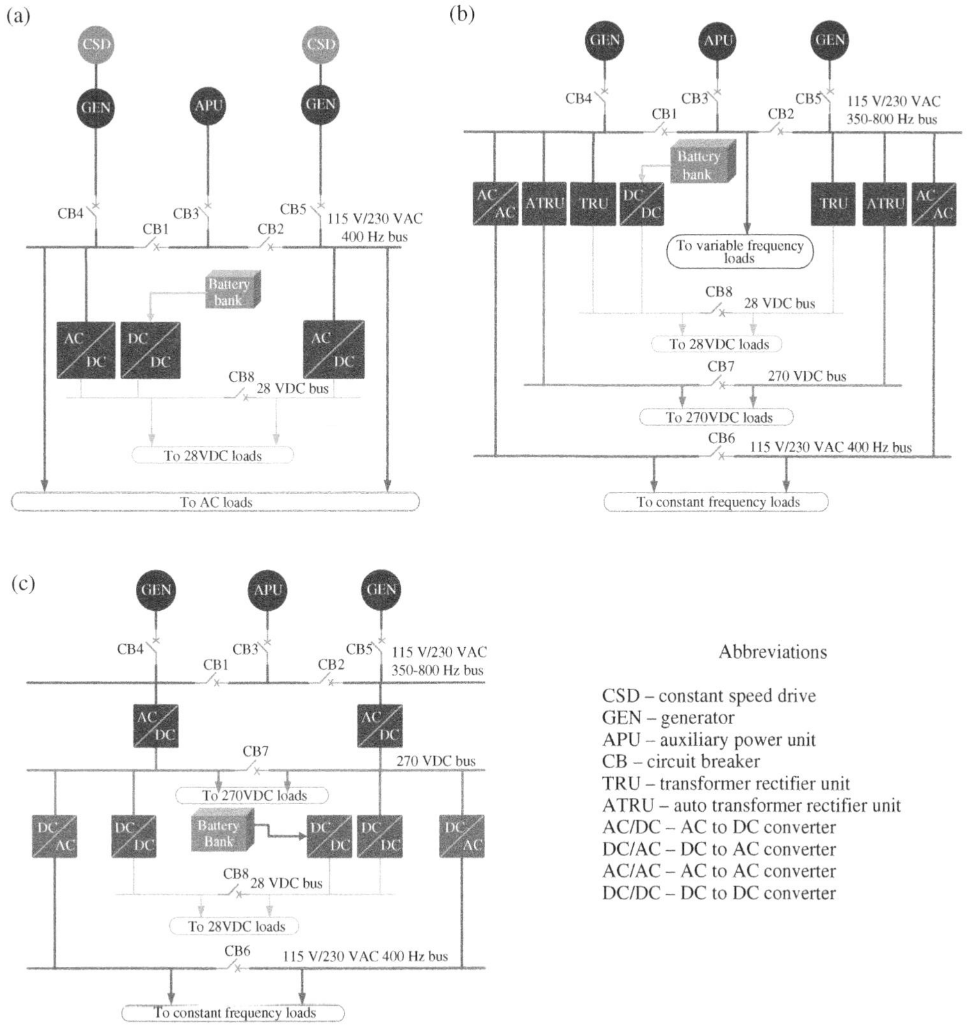

Figure 9.1 (a) Constant frequency power distribution architecture. (b) Variable frequency power distribution architecture. (c) DC power distribution architecture.

In VFPDA, there is distribution of electrical power at variable frequency directly from generator's output under VSFPGA [25]. Depending upon the type of load, the power is converted to AC or DC via different PE converters. Figure 9.1b shows the VFPDA [25]. In DCPDA, there is distribution of electrical power at 270 V high voltage DC (HVDC) voltage. The HVDC voltage is provided with AC/DC converters connected with generator's output under VSFPGA. Figure 9.1c shows the DCPDA [25].

9.3 Power Converters in MEA

The electrical power system in every airplane consists of both AC and DC components. Due to this, flight operations are significantly impacted by the use of PE converters. Also, this provides required power management between various networks. The voltage and power ratings of these converters depend on the type of aircraft being under consideration. Therefore, the system architecture should be tailored to the requirements, which is done by selecting the right configuration, parts, or cooling system. Additionally, PE have also opened up new applications in protection systems because of electrification. In this following subsection, the review of different converter topologies for MEA application is presented and discussed. Within the MEA power system, power conversion stages vary according to the distribution architecture, which is summarized here [25].
Fixed Frequency Conversion:

- AC-to-DC: (115 or 230 V) 400 Hz AC/28 V DC

Variable Frequency Conversion:

- AC-to-AC: (115 or 230 V) 350–800 Hz AC/(115 or 230 V) 400 Hz AC
- AC-to-DC: (115 or 230 V) 350–800 Hz AC/28 V DC
- AC-to-DC: (115 or 230 V) 350–800 Hz AC/270 V DC

DC Conversion:

- DC-to-AC Conversion: Load determines the magnitude and frequency of the AC voltage
- DC-to-DC Conversion: 270 V DC/28 V DC

AC–AC Converters

The AC–AC conversion systems fall into two categories. The indirect type uses a DC link to convert the AC power to DC, and the direct type does not use an energy storage component. For an AC–AC converter to be suitable for MEA applications, its input and output should be sinusoidal, have low harmonic content, bidirectional, and have a high-power factor [26]. Conversion from AC–AC is indirect using a back-to-back converter. Its DC link capacitor is bulky, reducing the lifespan of the component, making the design more complicated, and decreasing overall performance [26]. So, direct AC–AC conversion, being advantageous for MEA application, can be of the following two types – cycloconverter and matrix converter.

a. *Cycloconverter:* The switching circuit of this converter regulates the frequency and voltage of the AC bus. Cycloconverters are characterized by one-stage conversion, varying load conditions, and the ability to produce power both in and out. They have a disadvantage that they have bulky output filters based on their input voltage frequency [27].
b. *Matrix Converter:* In this converter, the conversion is done without any energy storage components. To reduce switching ripple, this converter only requires small filters. Its advantage

includes low total harmonic distortion (THD) output voltage, in and out power flow, longer life, and an adjustable power factor [27].

Apart from these two, in [28], indirect matrix converter is presented having a dual-output single-input system with integrated dc.

AC–DC Conversion

In MEA applications, six-pulse rectifiers are the simplest three-phase rectifiers. But this configuration has disadvantages of having poor THD. The transformer rectifier unit (TRU) and the autotransformer rectifier unit (ATRU) are other configurations that can be applied to the variable frequency distribution system. A 28VDC bus must be supplied by the TRU, and a 270VDC bus must be supplied by the ATRU in a variable frequency distribution. MEA applications typically use 12- or 18-pulse configurations. However, the ATRU's weight, volume, and cost are lower than the TRU's due to its lower equivalent KVA as well as its low copper and iron necessities [29]. MEA AC–DC power converters are designed in three major ways as discussed here:

a. *Two-Level Rectifier:* Power can be bidirectional with this converter, depending on the control. Since this rectifier is connected by DC link capacitor rather than neutral point, its control is simpler than that of rectifiers that use neutral point connections. Active damping elimination can minimize the possibility of resonance between the power distribution system and the filters [30].
b. *Neutral Point Clamped (NPC) Three-Level Rectifier:* For the aerospace industry, three-level topologies are preferred to eliminate corona issues and safety concerns. Due to the ability to achieve three-step voltages in this topology, the ripple of the ac input current is almost minimized in the NPC three-level rectifier. In addition to being able to be bidirectional, the NPC three-level rectifier can be used to reduce EMI effect [30].
c. *Vienna Rectifier:* Only three switches are required in the Vienna rectifier, thus having lower gate drives control, simpler protection design, as well as fewer cooling necessities and heat sinks [30–32]. Furthermore, active semiconductor switches have a lower blocking voltage, and the power factor can also be controlled to operate at unity [32]. In [33–35] a boosted rectifier for three-phase with a single switch along with the other popular topologies is presented. There are, however, some concerns regarding the input current harmonic content, which remains higher and should be further improved to achieve prerequisites of the MEA standards [29]. Furthermore, in [36, 37], there are two approaches to power factor correction (PFC) converters for three-phase – buck-boost for three-phase and semi-controlled for three-phase in MEA.

DC–DC Conversion

Different isolated bidirectional DC–DC power converters can provide DC–DC conversion. For DC–DC converters in aerospace applications, higher power and efficiency, more consistency and more versatility along with rapid recovery after faults are essential [38]. The simplest DC–DC converters in the MEA are flyback, forward, and forward-flyback converters but having poor efficiency [38]. Here is given a summary of five other configurations used in MEA:

a. *Resonant Converters:* In MEA application, the LLC and CLLC along with series resonant (SR) converters [39–42] are used as these have better efficiency while used in various conditions. This requires additional passive devices and complex control [42].
b. *Dual-Active-Bridge (DAB) Converter:* This converter employs zero voltage switching to reduce switching losses without significantly increasing controller complexity when the switches are

turned on [43]. Three-phase DAB allows for the reduction of current THD and DC link capacitors [44]. Due to the complex transformer design and the large number of components, high-current applications may pose problems for this converter [45].

c. *Interleaved Boost with Coupled Inductors (IBCI) Converter:* This converter replaces the H-Bridge on the input side with a clamping circuit, and inductors are used to replace the high frequency transformer in this converter [46].
d. *Active Bridge Active Clamp (ABAC) Converter [47, 48]:* In this converter, a high frequency transformer supplants the two coupled inductors of the IBCI and two output inductors are used instead of the coupled inductors in the IBCI [38].
e. *Multi-port Converters:* The overall power density can be increased by triple-active bridge (TAB) [8] and the quadruple-active bridge (QAB) converter [49]. Having multiple converters, integrated into one unit, makes communication, synchronization, control, and exchange of power much simpler while increasing the control complexity of HF transformer [45]. As well, multiple separate DC–DC converters show better performance when compared to multi-port converters when a single component fails [49]. In addition to converter, in [50], a fault tolerant configuration can be implemented to improve voltage regulation. Moreover, in [51], a topology is proposed to supply avionic loads.

DC–AC Conversion

In DC distribution system, there's a need for DC–AC conversion to provide AC power. These inverters can be two-level voltage-source inverters or multilevel inverters like NPC inverters, CHB, T-type converters, or Z-source inverters or even higher-level inverters like PUC, etc. VSIs have a simple structure; however, they are less efficient and less power dense than the NPC three-level inverters, since the three-level NPCs use a lower switching voltage. Lighter EMI filters are required for the NPC [52], while there is no need for EMI filters in the case of the T-type inverter, thus being more efficient [53]. A shoot-through mode has been developed in Z-source inverters [54–56] that provides increased output voltage and enhanced reliability. When there is no output load, the Z-link capacitor voltage will increase [57]. In [58–60] a quasi-Z-source inverter (QZI) along with the previously mentioned topologies is described. Through the addition of a step up between the input filter and the VSI, this inverter attempts to reduce the size of the Z-source inverter's VSI and heat sink [58].

Besides all these topologies, there is another kind of converters for DC–AC conversion, i.e. the multilevel family converters. These inverters equipped with switches and floating capacitors can generate symmetrical voltage levels when controlled properly. Harmonic contents are reduced significantly when the voltage levels are increased. Traditional MLI becomes bulkier and costlier when voltage levels are increased due to the increase in the number of switches and capacitors [61].

To overcome these shortcomings of these inverters, new topologies are proposed with many effective results. A very capable topology is the PUC inverter, which includes the merits of CHBs and a flying capacitor. Further in this work, PUC5 is used as it provides the self-balancing of the capacitor voltage. PUC employs a single stand-alone DC source. However, the other DC buses are maintained to obtain appropriate voltage levels. It also makes use of a smaller number of components in comparison with the traditional type. The multilevel inverter topology used in this work is PUC inverter, which has a very low switch and capacitor count in comparison with conventional inverters. PUC topology as depicted in Figure 9.2 was invented in early 2008 by Al-Haddad and Ounejjar. It is an improved version of CHB and flying capacitor topology and can be used in both the rectifier and inverter mode [61].

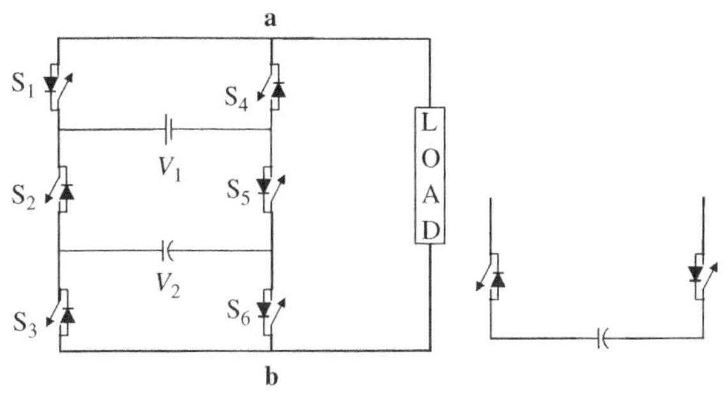

Figure 9.2 Packed U-cell topology and single U-cell.

9.4 PUC Topologies and Control

The basic structure of PUC5 is presented in Figure 9.2. There are six switches and one DC source along with one capacitor. For five-level inverters, the capacitor voltage (V_2) is retained at 50% of the DC voltage (V_1) [62].

In Table 9.2, the output voltage levels of PUC inverters are listed. It should be mentioned that switches S_4, S_5, and S_6 are operating in the opposite of S_1, S_2, and S_3. So, each pair of (S_1, S_4), (S_2, S_5), and (S_3, S_6) cannot conduct simultaneously.

In the situation of using a capacitor along with a DC voltage source, the output voltage level is, therefore, attained from the following five levels (V_1, $V_1/2$, 0, $-V_1/2$, $-V_1$). In sequence to achieve these levels, the capacitor voltage (V_2) must be regulated to $V_1/2$.

The sinusoidal level-shifted PWM technique, as shown in Figure 9.3, is used for the closed loop control of the inverter. Here four-triangular carrier wave is compared to a sinusoidal reference wave as described in Figures 9.4 and 9.6 depending on the topologies being used.

Table 9.2 All possible switching states of PUC inverter.

Switching states	S_1	S_2	S_3	V_{ab}	Output voltage	Capacitor voltage
1	1	0	0	V_1	V_1	No effect
2	1	0	1	V_1-V_2	$V_1/2$	Charging
3	1	1	0	V_2	$V_1/2$	Discharging
4	1	1	1	0	0	No effect
5	0	0	0	0	0	No effect
6	0	0	1	$-V_2$	$-V_1/2$	Discharging
7	0	1	0	V_2-V_1	$-V_1/2$	Charging
8	0	1	1	$-V_1$	$-V_1$	No effect

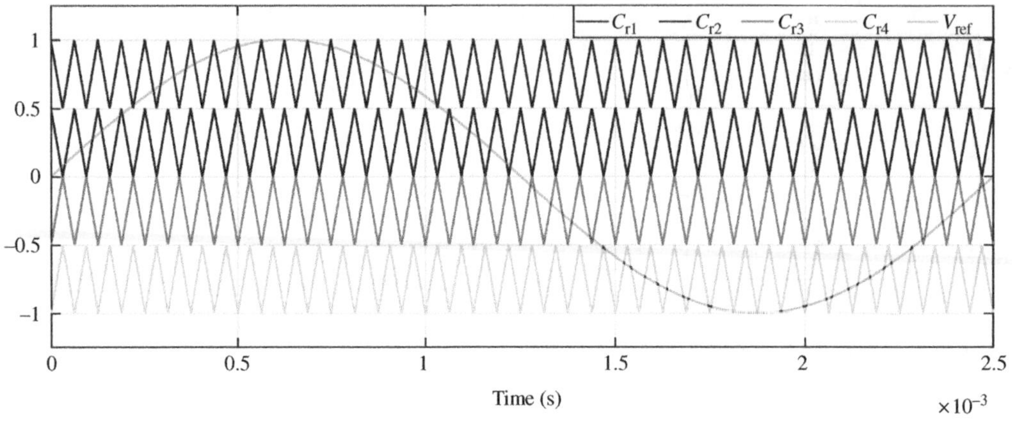

Figure 9.3 Sinusoidal modulation.

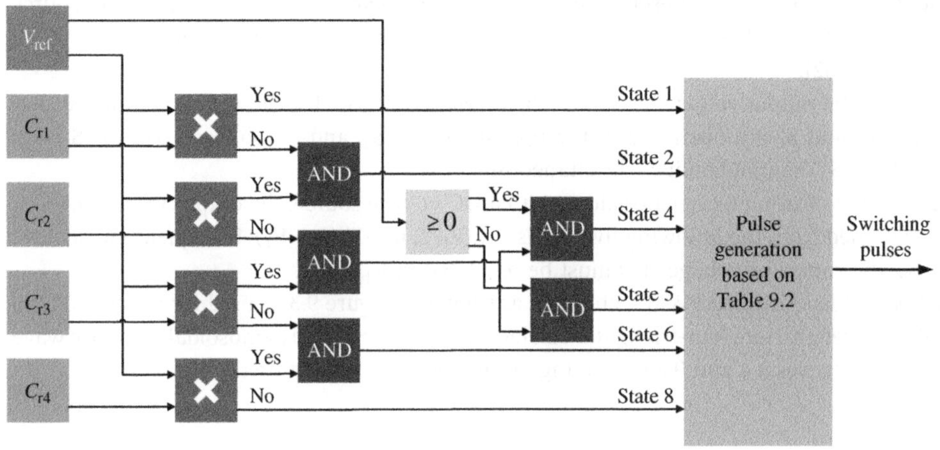

Figure 9.4 Open-loop switching algorithm for self-voltage balancing PUC5 inverter.

Five-Level Packed U-Cell (PUC5) Inverters – Open Loop

In this topology, the switching algorithm is such that the capacitor voltage is self-balancing at 50% of the DC supply voltage (V_1). A five-level voltage waveform can be generated by the algorithm described in Figure 9.4 with a minimum switching frequency while simultaneously maintaining the capacitor voltage at a target level using no feedback sensor. Here, the 3 and 7 switching states are omitted, as in Table 9.2; it can be seen that these two states are redundant.

Through the use of this algorithm, the PUC5 inverter can generate a waveform that has five levels of voltage without relying on voltage sensors or complex calculations. The capacitor voltage, at start-up as well as under load change conditions, will always be at 50% of the DC supply (V_1). It is much faster to control the algorithm than other controllers used on PUC inverters. System models (e.g. average modeling), feedback sensors, modulation indices, switching frequencies, or grid frequencies do not have effect on this technique. It allows operation of the system starting at zero voltage and reaching arbitrary output amplitudes and can operate in variable DC source voltage situations as well.

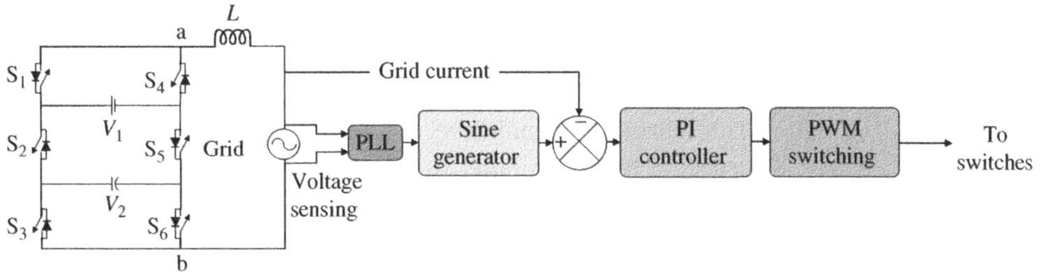

Figure 9.5 Five-level PUC-based grid-tied inverter.

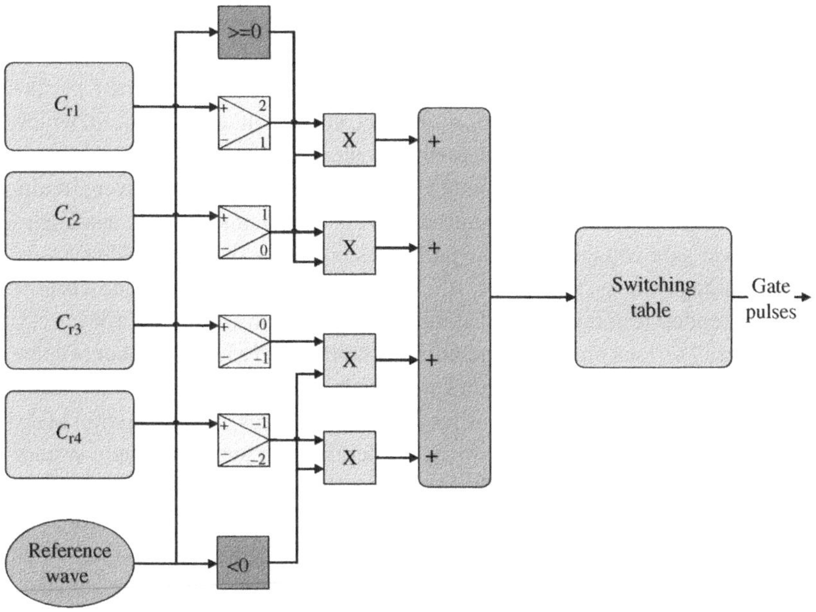

Figure 9.6 Switching strategy for five-level inverter.

Five-Level Packed U-Cell (PUC5) Inverters – PI Control

In Figure 9.5, the closed loop PI control is presented with the help of sinusoidal level-shifted PWM technique. The PI controller is operating on the current injected while maintaining the capacitor voltage (V_2) at 50% of the DC voltage (V_1). The detailed switching technique is discussed in Figure 9.6. Here, the switching states are decided using the states mentioned in Table 9.2.

Five-Level Modified Packed U-Cell (MPUC5) Inverters – Open Loop

Figure 9.7 shows the MPUC5 topology. A total of six switches are involved, along with two DC sources in this topology. Now these DC sources can be supplied by flyback converter with one output at 1 : 1 ratio for V_1 and the other output at 2 : 1 ratio for V_2. So, in this topology, there is no need for the balancing of the auxiliary source voltage. Thus, the control of this topology becomes much simpler than the previous ones.

9.5 Fault Tolerant Capability of PUC Inverter

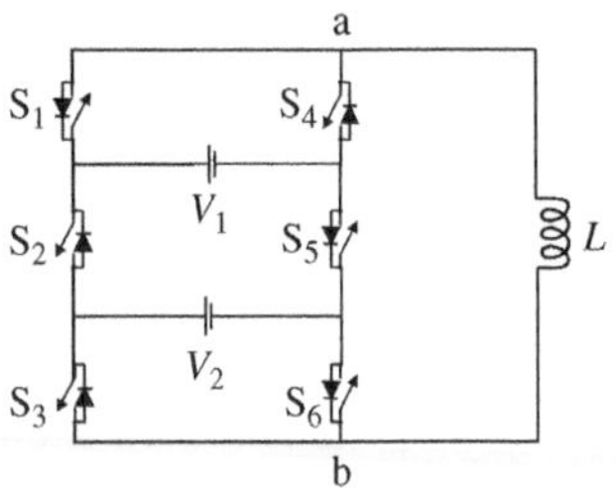

Figure 9.7 Modified packed U-cell inverter topology.

The reliability of multilevel inverter (MLI) is a very important aspect of MEAs. In order to maintain the stability and safety of the electrical distribution system, MEA requires highly reliable MLIs with the ability to maintain their performance when faced with unexpected faults. One of the major concerns associated with MLI is their lower reliability, which can be ascribed to the presence of more switches and capacitors than conventional two-level inverters [63]. Literature has paid considerable attention to achieving fault tolerant operation when switches fail. For such environments, MLIs must be highly capable in order to supply continuous power to the loads. So, there is a need for fault tolerant MLIs (FT MLIs) to have optimal number of components or switches with fault tolerance capability [64]. An analysis of fault tolerance in MLI with regard to their performance is highly significant. In cases of OC or SC faults, multilevel output voltage is distorted, which causes malfunctioning of the MLI that can propagate further to other systems and cause the load/system to fail [65]. Loss of gating signals, thermal cycling, and malfunctioning of gate driver circuits are possible causes of OC faults. SC faults can be caused by very high temperatures, avalanche stress, overvoltage, or incorrect gate voltages. The OC fault does not cause damage to the system. In contrast, SC fault can damage the system due to the flow of the higher-magnitude current [66–68]. Hence, it is recommended to remove the SC fault as soon as fault is detected [66].

MLIs used in MEAs face many serious problems, including reliability and a large number of components. The failure of power switches on these inverters may result in severe damage to other MEA equipment since failure of power switches will affect other components. Thus, it is crucial that MLIs used in MEAs should be fault tolerant. MLIs that are fault tolerant help avoid complete system shutdown, thus promoting higher reliability. All critical loads should be able to continue receiving power even when the system is malfunctioning. Several fault tolerant solutions are presented in literature and reviewed in [69]. The MLI FT solutions are classified as switch-level, leg-level, module-level, and system-level. In switch-level MLI FT solution, extra-redundant switches are added to the original MLI structure to achieve fault tolerance. In leg-level MLI FT solution, extra-redundant leg(s) are added to the original MLI topology in series or parallel to achieve fault tolerance. In module-level MLI FT solution, either faulty module is replaced by redundant healthy module or faulty module is bypassed to achieve fault tolerance. A fault tolerant MLI system is constructed by connecting a healthy MLI parallel or series to the faulty MLI [69, 70].

A fault tolerant structure of five-level MPUC5 is presented in [70]. Figure 9.8 shows the MPUC5 MLI and its fault tolerant variant. Three redundant switches (S_7, S_8, and S_9) are added to the original MPUC5 topology to make it fault tolerant. These redundant switches provide redundant paths that provide different conduction paths to replace lost paths under faulty conditions. There are total nine switches and two DC supplies. The magnitudes of DC supplies are V and $V/2$. There is complete shutdown of MPUC5 MLI under switch S_1 or S_4 failure. The MPUC5 MLI continues to supply with three levels in output voltage under switch S_2 or S_3 failure. The FT MPUC5 provides five-level output voltage under any single switch failure in group (S_1, S_4, S_5, and S_6) and three-level output voltage under any single switch failure in group (S_2 and S_3). Table 9.3 shows the switching table of FT MPUC5.

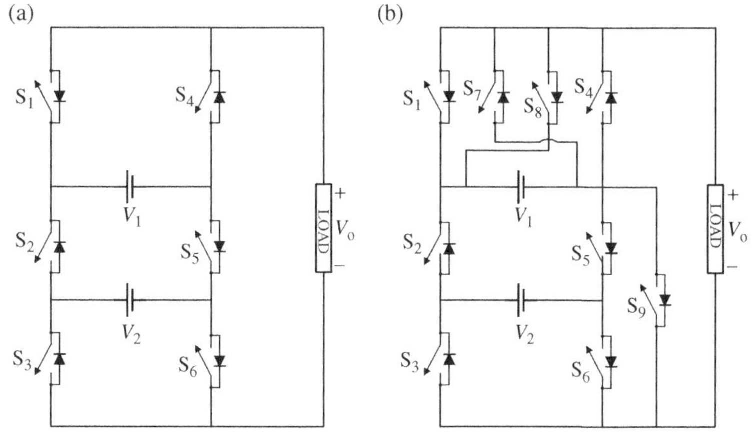

Figure 9.8 (a) MPUC5 MLI and (b) FT MPUC5 MLI.

Table 9.3 Switching table of FT five-level MPUC5 MLI ($V_1 = V$, $V_2 = V/2$).

States	S_1	S_2	S_3	S_4	S_5	S_6	S_7	S_8	S_9	V_{ab}	V_{out}
β_1	1	0	0	0	0	0	0	0	1	V_1	V
β_2	1	0	0	0	1	1	0	0	0		
β_3	0	0	0	0	1	1	0	1	0		
β_4	0	0	0	0	0	0	0	1	1		
β_5	1	0	1	0	1	0	0	0	0	V_1-V_2	$V/2$
β_6	0	0	1	0	1	0	0	1	0		
β_7	1	1	0	0	0	1	0	0	0	V_2	$V/2$
β_8	0	1	0	0	0	1	0	1	0		
β_9	1	1	1	0	0	0	0	0	0	0	0
β_{10}	0	0	0	1	1	1	0	0	0		
β_{11}	0	1	1	0	0	0	0	1	0		
β_{12}	0	0	0	0	0	0	0	1	1		
β_{13}	0	0	0	0	1	1	1	0	0		
β_{14}	0	1	0	0	0	1	1	0	0	$-V_2$	$-V/2$
β_{15}	0	0	1	0	1	0	1	0	0		
β_{16}	0	0	1	1	1	0	0	0	0		
β_{17}	0	1	0	0	0	1	1	0	0		
β_{18}	0	1	0	1	0	1	0	0	0	$-(V_1-V_2)$	$-V/2$
β_{19}	0	1	1	1	0	0	0	0	0	$-V_1$	$-V$
β_{20}	0	1	1	0	0	0	1	0	0		

9.6 Results and Discussion

The simulation parameters for these topologies in the case of MEA application are presented in Table 9.4 using MATLAB/Simulink® 2021b.

PUC5 with Self-Balancing Capacitor Voltage Sensor-Less Inverter

In this topology, Figure 9.9 shows the five-level voltage waveform obtained by a sensor-less open loop modulation technique aimed at self-voltage regulation as described in Figure 9.4. Thus, here the capacitor voltage is maintained at 50% of the DC supply with the help of this switching algorithm. Figure 9.9 also depicts the current waveform for PUC5. In Figure 9.10, the capacitor voltage in reference to the required capacitor voltage is shown.

PI Control of PUC5 with Balanced Capacitor

In Figure 9.11, the five-level inverter output is plotted with the help of PI controller. PI controller is used for current control as illustrated in Figure 9.5. Figure 9.11 presents the current waveform using PI controller. Using PI controller, the capacitor voltage is balanced at a voltage level of half of the DC supply as shown in Figure 9.12.

Table 9.4 Simulation parameters.

Parameters	Values
DC voltage (V_1)	270 V
DC capacitor (C_1)	2500 μF
Line inductor (L)	50 mH
Stand-alone RL load	40 Ω, 20 mH
Grid voltage (RMS)	220 V
Line frequency	400 Hz
Modulating frequency	16 kHz

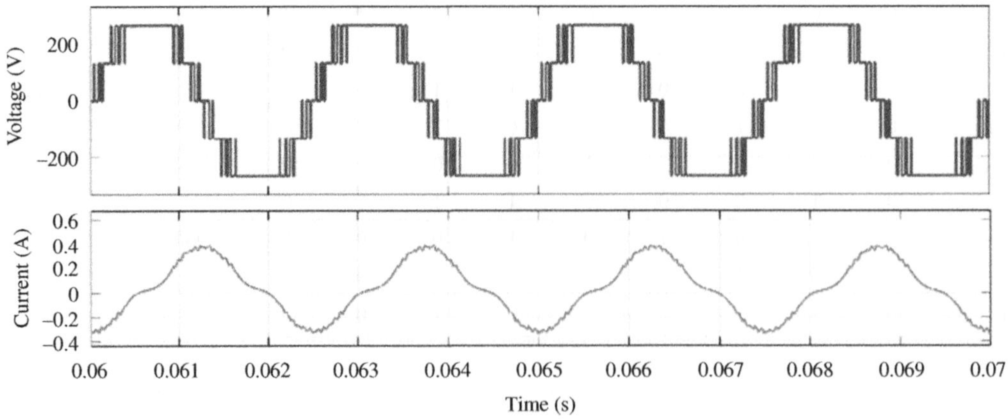

Figure 9.9 Output voltage and current of PUC5 – open loop.

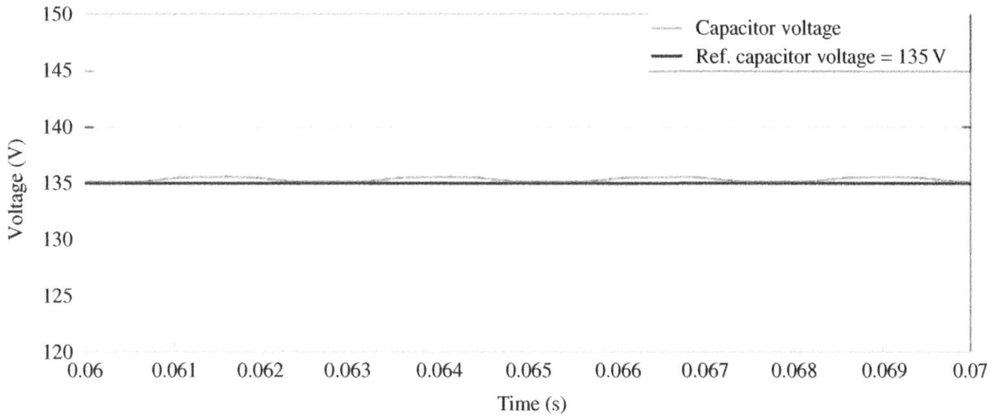

Figure 9.10 Capacitor voltage and reference capacitor voltage – open loop.

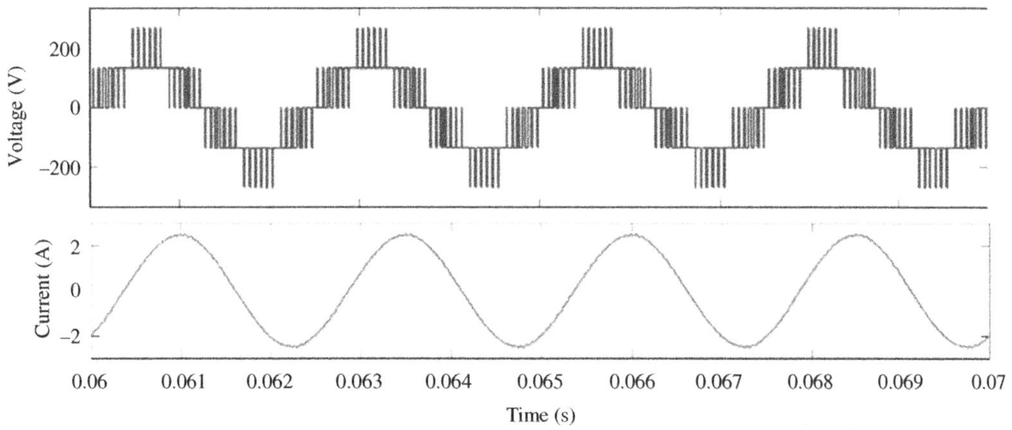

Figure 9.11 Output voltage and current of PUC5 – PI control.

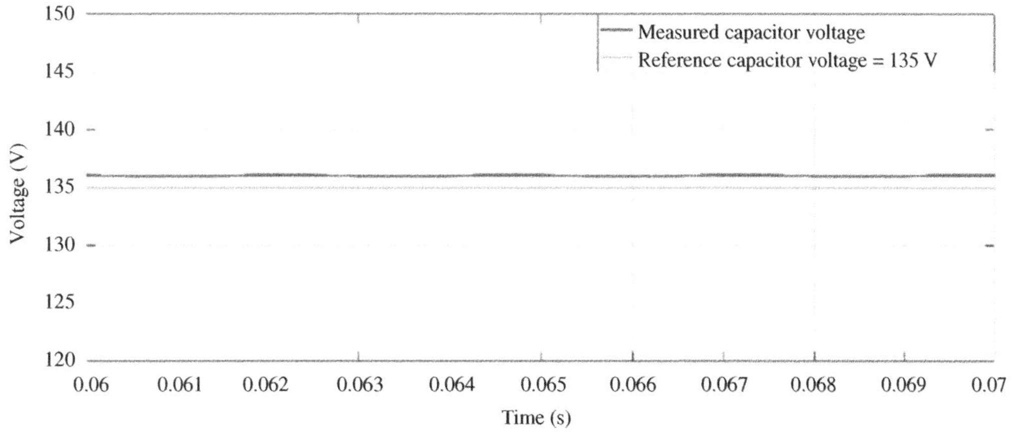

Figure 9.12 Measured and reference capacitor voltage – PI control.

Modified PUC5 by Using Flyback Converter

In this simulation, there is no capacitor used. Here, with the help of flyback converter, V_1 and V_2 voltages are supplied while maintaining 2 : 1 ratio in these two. Figure 9.13 shows the inverter output voltage of a modified packed U-cell (MPUC5). Figure 9.13 also shows the current waveform.

Fault Tolerant MPUC5

The five-level FT MPUC5 is simulated in MATLAB/Simulink 2021b. The magnitudes of DC supplies used for simulation purposes are 270 V and 135 V. The load (RL) with $R = 40\ \Omega$ and $L = 50$ mH is used for simulation purposes. The modulation index chosen is 0.5465. Figure 9.14 shows output voltage and load current under pre-fault, during-fault, and post-fault periods for all single-switch OC faults in FT MPUC5. Pre-fault is the period before the occurrence of fault, during-fault is the period between the instance of fault (FI) and instance of reconfiguration (RI), and the post-fault period is the period after the RI. The OC fault is simulated on the switch. The OC fault occurs at 0.01 second. Fault detection and fault reconfiguration are assumed to occur after a time interval of 0.01 second.

Figure 9.14a shows the simulation results under switch S_1 OC fault. There is a loss of $+V$, $+V/2$ voltage levels under S_1 fault. Hence, there is shutdown of the MPUC5. At RI, redundant switching states are utilized to provide output voltage levels that are lost under faulty conditions in FT modified PUC5. The available switching states under S_1 fault are β_3, β_4, β_6, β_8, β_{10}, β_{11}, β_{12}, β_{13}, β_{14}, β_{15}, β_{16}, β_{17}, β_{18}, β_{19}, and β_{20}. Hence, after fault reconfiguration, five-level operation is achieved.

Figure 9.14b shows the simulation results under switch S_2 OC fault. There is a loss of −V voltage level under S_2 fault. Hence, three-level operation is possible in MPUC5. The available switching states under S_2 fault are β_1, β_2, β_3, β_4, β_5, β_6, β_{10}, β_{12}, β_{13}, β_{15}, and β_{16}. At RI, $+V$ voltage level is not considered to avoid DC-offset in FT modified PUC5. Hence, after fault reconfiguration, three-level operation is achieved.

Figure 9.14c shows the simulation results under switch S_3 OC fault. There is loss of $-V$ voltage level under S_3 fault. Hence, three-level operation is possible in MPUC5. The available switching states under S_3 fault are β_1, β_2, β_3, β_4, β_7, β_8, β_{10}, β_{12}, β_{13}, β_{14}, β_{17}, and β_{18}. At RI, $+V$ voltage level is not considered to avoid DC-offset in FT modified PUC5. Hence, after fault reconfiguration, three-level operation is achieved.

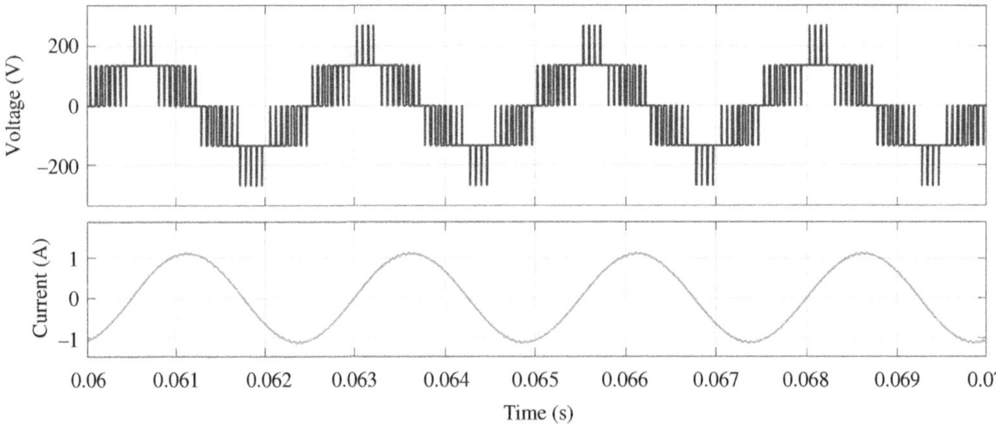

Figure 9.13 Output voltage and current of MPUC5.

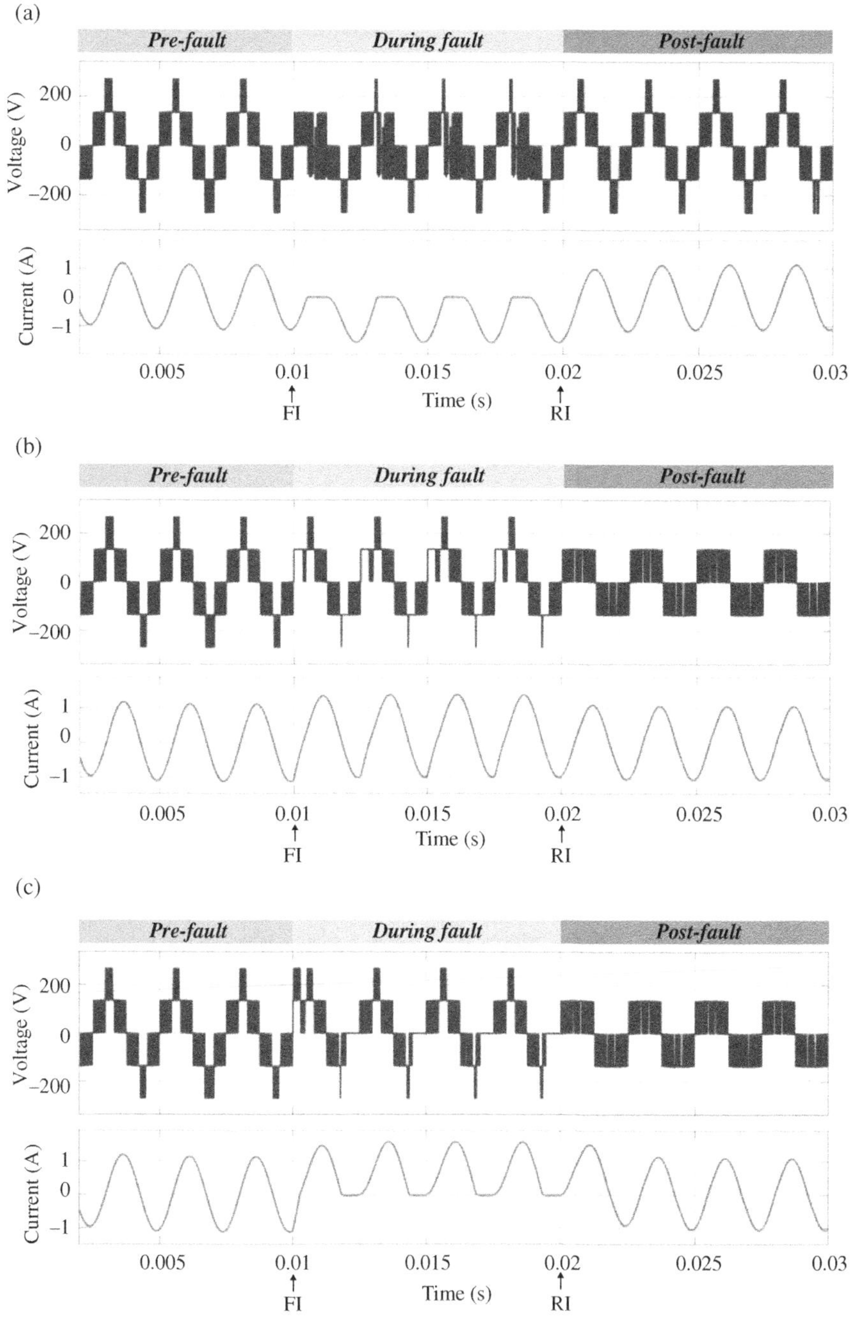

Figure 9.14 Output voltage and load current under pre-fault, during-fault, and post-fault condition for single switch OC fault (a) S_1, (b) S_2, (c) S_3, (d) S_4, (e) S_5, (f) S_6.

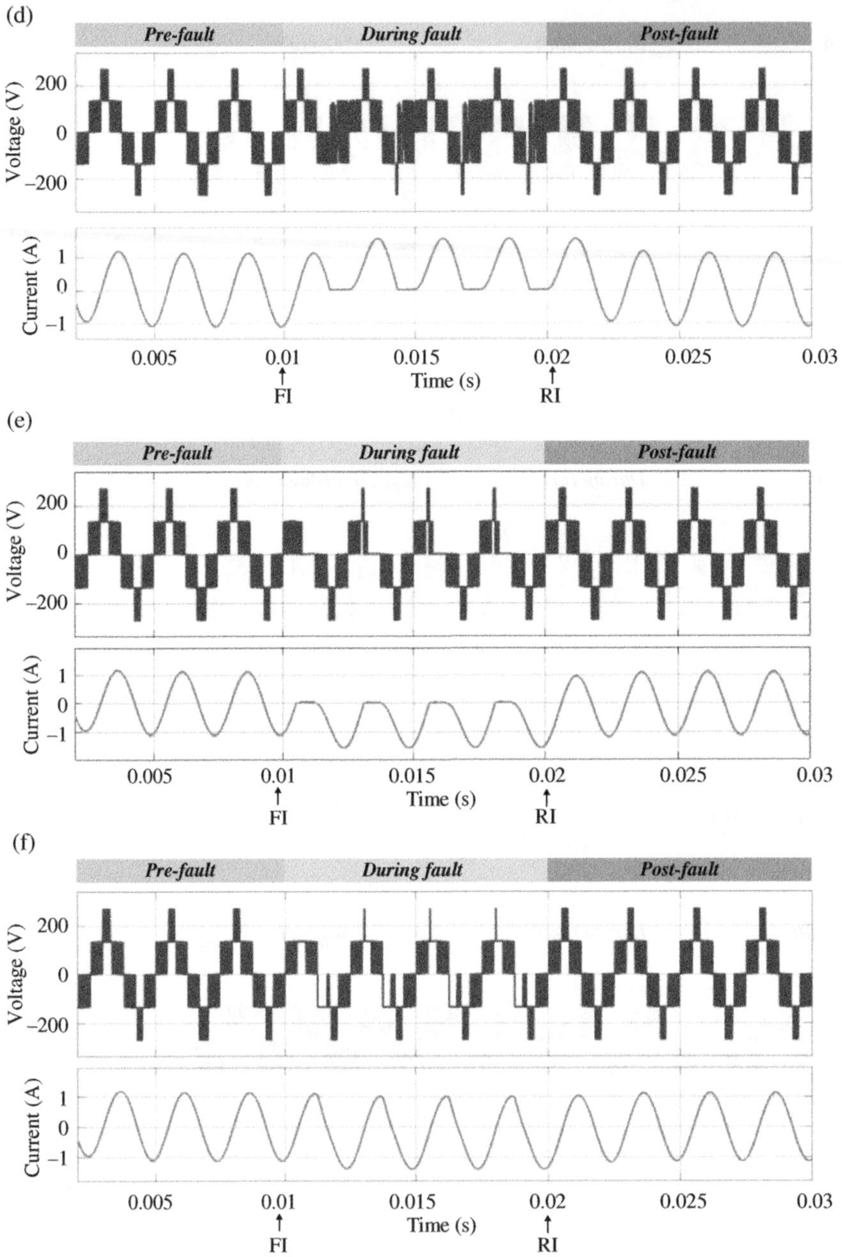

Figure 9.14 (Continued)

Figure 9.14d shows the simulation results under switch S_4 OC fault. There is loss of $-V$, $-V/2$ voltage levels under S_4 fault. Hence, there is shutdown of the MPUC5. At RI, redundant switching states are utilized to provide output voltage levels that are lost under faulty conditions in FT modified PUC5. The available switching states under S_4 fault are β_1, β_2, β_3, β_4, β_5, β_6, β_7, β_8, β_9, β_{11}, β_{12}, β_{13}, β_{14}, β_{15}, β_{17}, and β_{20}. Hence, after fault reconfiguration, five-level operation is achieved.

Table 9.5 Comparison of operation under faulty conditions for MPUC5 and its FT variant.

Faulty switch	MPUC5 (operation)	FT MPUC5 (operation)
S_1	Shutdown	Five-level
S_2	Three-level	Three-level
S_3	Three-level	Three-level
S_4	Shutdown	Three-level
S_5	Three-level	Three-level
S_6	Three-level	Three-level

Figure 9.14e shows the simulation results under switch S_5 OC fault. There is loss of $+V$ voltage levels under S_5 fault. Hence, three-level operation is possible in MPUC5. At RI, redundant switching states are utilized to provide output voltage levels that are lost under faulty conditions in FT modified PUC5. The available switching states under S_5 fault are β_1, β_4, β_7, β_8, β_9, β_{11}, β_{12}, β_{14}, β_{17}, β_{18}, β_{19}, and β_{20}. Hence, after fault reconfiguration, five-level operation is achieved.

Figure 9.14f shows the simulation results under switch S_6 OC fault. There is loss of $+V$ voltage levels under S_6 fault. Hence, three-level operation is possible in MPUC5. At RI, redundant switching states are utilized to provide output voltage levels that are lost under faulty conditions in FT modified PUC5. The available switching states under S_6 fault are β_1, β_4, β_5, β_6, β_9, β_{11}, β_{12}, β_{15}, β_{16}, β_{19}, and β_{20}. Hence, after fault reconfiguration, five-level operation is achieved.

Table 9.5 shows the comparison of operation under OC faulty condition for MPUC5 and its FT variant. The FT variant can tolerate all single switch OC faults, which is absent in MPUC5. The FT variant also restores all output voltage levels except for switches S_2 and S_3. The FT variant provides continuous supply at reduced voltage levels under switch S_2 or S_3 failure. Hence, the FT MPUC5 provide fault tolerance in MEAs.

9.7 Conclusions

In this chapter, different MEA power system architectures have been discussed. Different types of converters that are used in MEA have been categorized. The control and fault tolerance operation of PUC inverter for MEA application has been analyzed and verified with MATLAB/Simulink results. The fault tolerance operation of MPUC5 shows that if the fault occurs in switch S_1, then the normal MPUC5 shuts down its operation, whereas the FT MPUC5 continues to operate at the same level. Similarly, different fault cases were successfully analyzed and reported in the chapter.

Acknowledgments

The authors would like to acknowledge the support from the collaborative research grant scheme (CRGS) project, Hardware-In-the-Loop (HIL) Lab, Department of Electrical Engineering, Aligarh Muslim University, India, having project numbers CRGS/MOHD TARIQ/01 and CRGS/MOHD TARIQ/02.

References

1 Dorn-Gomba, L., Ramoul, J., Reimers, J., and Emadi, A. (2020). Power electronic converters in electric aircraft: current status, challenges, and emerging technologies. *IEEE Transactions on Transportation Electrification* 6 (4): 1648–1664. https://doi.org/10.1109/TTE.2020.3006045.

2 Wileman, A.J., Aslam, S., and Perinpanayagam, S. (2021). A road map for reliable power electronics for more electric aircraft. *Progress in Aerospace Sciences* 127: 100739. https://doi.org/10.1016/j.paerosci.2021.100739.

3 Wheeler, P. (2016). Technology for the more and all electric aircraft of the future. *2016 IEEE International Conference on Automatica (ICA-ACCA)*, pp. 1–5. https://doi.org/10.1109/ICA-ACCA.2016.7778519.

4 Provost, M.J. (2002). The more electric aero-engine: a general overview from an engine manufacturer. *International Conference on Power Electronics Machines and Drives*, pp. 246–251. https://doi.org/10.1049/cp:20020122.

5 Bennett, J.W., Mecrow, B.C., Atkinson, D.J. et al. (2011). Fault-tolerant electric drive for an aircraft nose wheel steering actuator. *IET Electrical Systems in Transportation* 1 (3): 117. https://doi.org/10.1049/iet-est.2010.0054.

6 Wheeler, P. and Bozhko, S. (2014). The more electric aircraft: technology and challenges. *IEEE Electrification Magazine* 2 (4): 6–12. https://doi.org/10.1109/MELE.2014.2360720.

7 Sarlioglu, B. and Morris, C.T. (2015). More electric aircraft: review, challenges, and opportunities for commercial transport aircraft. *IEEE Transactions on Transportation Electrification* 1 (1): 54–64. https://doi.org/10.1109/TTE.2015.2426499.

8 Karanayil, B., Ciobotaru, M., and Agelidis, V.G. (2017). Power flow management of isolated multiport converter for more electric aircraft. *IEEE Transactions on Power Electronics* 32 (7): 5850–5861. https://doi.org/10.1109/TPEL.2016.2614019.

9 Gao, F., Bozhko, S., Asher, G. et al. (2015). An improved voltage compensation approach in a droop-controlled DC power system for the more electric aircraft. *IEEE Transactions on Power Electronics* 31 (10): 1–1. https://doi.org/10.1109/TPEL.2015.2510285.

10 Bifaretti, S., Zanchetta, P., and Lavopa, E. (2014). Comparison of two three-phase PLL systems for more electric aircraft converters. *IEEE Transactions on Power Electronics* 29 (12): 6810–6820. https://doi.org/10.1109/TPEL.2014.2307003.

11 Chen, G., Chen, L., Deng, Y. et al. (2019). Topology-reconfigurable fault-tolerant *LLC* converter with high reliability and low cost for more electric aircraft. *IEEE Transactions on Power Electronics* 34 (3): 2479–2493. https://doi.org/10.1109/TPEL.2018.2848297.

12 Ounis, H., Sareni, B., Roboam, X., and de Andrade, A. (2016). Multi-level integrated optimal design for power systems of more electric aircraft. *Mathematics and Computers in Simulation* 130: 223–235. https://doi.org/10.1016/J.MATCOM.2015.08.016.

13 Electric propulsion is finally on the map | Roland Berger. https://www.rolandberger.com/en/Insights/Publications/Electric-propulsion-is-finally-on-the-map.html (accessed 17 March 2022).

14 E-Fan X - Electric Flight - Airbus. https://www.airbus.com/en/innovation/zero-emission/electric-flight/e-fan-x (accessed 17 March 2022).

15 lo Calzo, G., Zanchetta, P., Gerada, C., et al. (2015). Converter topologies comparison for more electric aircrafts high speed starter/generator application. *2015 IEEE Energy Conversion Congress and Exposition (ECCE)*, pp. 3659–3666. https://doi.org/10.1109/ECCE.2015.7310177.

16 Patel, V., Buccella, C., Saif, A.M. et al. (2018). Performance comparison of multilevel inverters for E-transportation. *2018 International Conference of Electrical and Electronic Technologies for Automotive*, pp. 1–6. https://doi.org/10.23919/EETA.2018.8493209.

17 Tolbert, L.M., Peng, F.Z., and Habetler, T.G. (1999). Multilevel converters for large electric drives. *IEEE Transactions on Industry Applications* 35 (1): 36–44. https://doi.org/10.1109/28.740843.
18 Rodriguez, J., Lai, J.-S., and Peng, F.Z. (2002). Multilevel inverters: a survey of topologies, controls, and applications. *IEEE Transactions on Industrial Electronics* 49 (4): 724–738. https://doi.org/10.1109/TIE.2002.801052.
19 Zhang, D., He, J., Pan, D. et al. (2018). Development of megawatt-scale medium-voltage high efficiency high power density power converters for aircraft hybrid-electric propulsion systems. *2018 AIAA/IEEE Electric Aircraft Technologies Symposium (EATS)*, pp. 1–5. https://ieeexplore.ieee.org/document/8552782 (accessed 17 March 2022).
20 Viola, F. (2018). Experimental evaluation of the performance of a three-phase five-level cascaded H-bridge inverter by means FPGA-based control board for grid connected applications. *Energies (Basel)* 11 (12): 3298. https://doi.org/10.3390/en11123298.
21 Jack, A.G., Mecrow, B.C., and Haylock, J.A. (1996). A comparative study of permanent magnet and switched reluctance motors for high-performance fault-tolerant applications. *IEEE Transactions on Industry Applications* 32 (4): 889–895. https://doi.org/10.1109/28.511646.
22 Mitcham, A.J. (2002). Permanent magnet generator options for the more electric aircraft. *International Conference on Power Electronics Machines and Drives*, 2002, pp. 241–245. https://doi.org/10.1049/cp:20020121.
23 Jia, Y. and Rajashekara, K. (2017). An induction generator-based AC/DC hybrid electric power generation system for more electric aircraft. *IEEE Transactions on Industry Applications* 53 (3): 2485–2494. https://doi.org/10.1109/TIA.2017.2650862.
24 Wenping, C., Mecrow, B.C., Atkinson, G.J. et al. (2012). Overview of electric motor technologies used for more electric aircraft (MEA). *IEEE Transactions on Industrial Electronics* 59 (9): 3523–3531. https://doi.org/10.1109/TIE.2011.2165453.
25 Rahrovi, B. and Ehsani, M. (2019). A review of the more electric aircraft power electronics. *2019 IEEE Texas Power and Energy Conference, TPEC 2019*. https://doi.org/10.1109/TPEC.2019.8662158.
26 Zhang, J., Li, L., and Dorrell, D.G. (2018). Control and applications of direct matrix converters: a review. *Chinese Journal of Electrical Engineering* 4 (2): 18–27. https://doi.org/10.23919/CJEE.2018.8409346.
27 Khlebnikov, A.S. and Kharitonov, S.A. (2008). Application of the Z-source converter for aircraft power generation systems. *2008 9th International Workshop and Tutorials on Electron Devices and Materials*, pp. 211–215. https://doi.org/10.1109/SIBEDM.2008.4585900.
28 Shi, B., Zhou, B., and Hu, D. (2017). A novel dual-output indirect matrix converter for more electric aircraft. *IECON 2017 - 43rd Annual Conference of the IEEE Industrial Electronics Society*, pp. 4083–4087. https://doi.org/10.1109/IECON.2017.8216700.
29 Sarlioglu, B. (2012). Advances in AC-DC power conversion topologies for More Electric Aircraft. *2012 IEEE Transportation Electrification Conference and Expo (ITEC)*, pp. 1–6. https://doi.org/10.1109/ITEC.2012.6243487.
30 Gong, G., Heldwein, M.L., Drofenik, U. et al. (2005). Comparative evaluation of three-phase high-power-factor AC–DC converter concepts for application in future more electric aircraft. *IEEE Transactions on Industrial Electronics* 52 (3): 727–737. https://doi.org/10.1109/TIE.2005.843957.
31 Kolar, J.W. and Drofenik, U. (1999). A new switching loss reduced discontinuous PWM scheme for a unidirectional three-phase/switch/level boost-type PWM (VIENNA) rectifier. *21st International Telecommunications Energy Conference. INTELEC '99 (Cat. No.99CH37007)*, p. 572. https://doi.org/10.1109/INTLEC.1999.794128.
32 Wang, B., Venkataramanan, G., and Bendre, A. (2007). Unity power factor control for three-phase three-level rectifiers without current sensors. *IEEE Transactions on Industry Applications* 43 (5): 1341–1348. https://doi.org/10.1109/TIA.2007.904433.

33 Zhang, R. and Lee, F.C. (1998). Optimum PWM pattern for a three-phase boost DCM PFC rectifier. *Proceedings of APEC 97 - Applied Power Electronics Conference*, pp. 895–901. https://doi.org/10.1109/APEC.1997.575751.

34 Jang, Y. and Jovanovic, M.M. (1998). A comparative study of single-switch, three-phase, high-power-factor rectifiers. *APEC '98 Thirteenth Annual Applied Power Electronics Conference and Exposition*, pp. 1093–1099. https://doi.org/10.1109/APEC.1998.654033.

35 Barbosa, P., Canales, F., and Lee, F. (2001). Passive input current ripple cancellation in three-phase discontinuous conduction mode rectifiers. *2001 IEEE 32nd Annual Power Electronics Specialists Conference (IEEE Cat. No.01CH37230)*, pp. 1019–1024. https://doi.org/10.1109/PESC.2001.954253.

36 Sivanagaraju, G., Rathore, A.K., and Fulwani, D.M. (2018). Discontinuous conduction mode three phase buck-boost derived PFC converter for more electric aircraft with reduced switching, sensing and control requirements. *2018 IEEE Applied Power Electronics Conference and Exposition (APEC)*, pp. 1467–1472. https://doi.org/10.1109/APEC.2018.8341210.

37 Gangavarapu, S. and Rathore, A.K. (2017). Three-phase interleaved semi-controlled PFC converter for aircraft application. *IECON 2017 - 43rd Annual Conference of the IEEE Industrial Electronics Society*, pp. 6652–6657. https://doi.org/10.1109/IECON.2017.8217161.

38 Tarisciotti, L., Costabeber, A., Linglin, C. et al. (2017). Evaluation of isolated DC/DC converter topologies for future HVDC aerospace microgrids. *2017 IEEE Energy Conversion Congress and Exposition (ECCE)*, pp. 2238–2245. https://doi.org/10.1109/ECCE.2017.8096437.

39 Shakib, S.M.S.I. and Mekhilef, S. (2017). A frequency adaptive phase shift modulation control based LLC series resonant converter for wide input voltage applications. *IEEE Transactions on Power Electronics* 32 (11): 8360–8370. https://doi.org/10.1109/TPEL.2016.2643006.

40 Li, Z. and Wang, H. (2016). Comparative analysis of high step-down ratio isolated DC/DC topologies in PEV applications. *2016 IEEE Applied Power Electronics Conference and Exposition (APEC)*, pp. 1329–1335. https://doi.org/10.1109/APEC.2016.7468040.

41 Walter, J. and de Doncker, R.W. (2003). High-power galvanically isolated DC/DC converter topology for future automobiles. *IEEE 34th Annual Conference on Power Electronics Specialist, 2003. PESC '03*, pp. 27–32. https://doi.org/10.1109/PESC.2003.1218269.

42 Liu, B., Qiu, M., Jing, L. et al. (2018). Design of high-performance bidirectional DC/DC converter applied for more electric aircraft. *The Journal of Engineering* 2018 (13): 520–523. https://doi.org/10.1049/joe.2018.0044.

43 Rodriguez Alonso, A., Sebastian, J., Lamar, D.G. et al. (2010). An overall study of a dual active bridge for bidirectional DC/DC conversion. *2010 IEEE Energy Conversion Congress and Exposition*, pp. 1129–1135. https://doi.org/10.1109/ECCE.2010.5617847.

44 de Doncker, R.W.A.A., Divan, D.M., and Kheraluwala, M.H. (1991). A three-phase soft-switched high-power-density DC/DC converter for high-power applications. *IEEE Transactions on Industry Applications* 27 (1): 63–73. https://doi.org/10.1109/28.67533.

45 Buticchi, G., Costa, L., and Liserre, M. (2017). Improving system efficiency for the more electric aircraft: a look at dc\/dc converters for the avionic onboard dc microgrid. *IEEE Industrial Electronics Magazine* 11 (3): 26–36. https://doi.org/10.1109/MIE.2017.2723911.

46 Spiazzi, G. and Buso, S. (2015). Analysis of the interleaved isolated boost converter with coupled inductors. *IEEE Transactions on Industrial Electronics* 62 (7): 4481–4491. https://doi.org/10.1109/TIE.2014.2362496.

47 Chen, L., Tarisciotti, L., Costabeber, A. et al. (2017). Parameters mismatch analysis for the active-bridge-active-clamp (ABAC) converter. *2017 IEEE Southern Power Electronics Conference (SPEC)*, pp. 1–6. https://doi.org/10.1109/SPEC.2017.8333619.

48 Chen, L., Tarisciotti, L., Costabeber, A. et al. (2017). Advanced modulation for the active-bridge-active-clamp (ABAC) converter. *2017 IEEE Southern Power Electronics Conference (SPEC)*, pp. 1–6. https://doi.org/10.1109/SPEC.2017.8333618.

49 Buticchi, G., Costa, L., and Liserre, M. (2017). DC/DC conversion solutions to enable smart-grid behavior in the aircraft electrical power distribution system. *IECON* 2017 - *43rd Annual Conference of the IEEE Industrial Electronics Society*, pp. 4369–4374. https://doi.org/10.1109/IECON.2017.8216752.

50 Costa, L.F., Buticchi, G., and Liserre, M. (2017). Highly efficient and reliable SiC-based DC–DC converter for smart transformer. *IEEE Transactions on Industrial Electronics* 64 (10): 8383–8392. https://doi.org/10.1109/TIE.2017.2696481.

51 Forest, F., Martiré, T., Flumian, D. et al. (2018). A nonreversible 10-kW high step-up converter using a multicell boost topology. *IEEE Transactions on Power Electronics* 33 (1): 151–160. https://doi.org/10.1109/TPEL.2017.2662224.

52 Wang, Q., Turriate, V., Burgos, R. et al. (2017). Towards a high performance motor drive for aerospace applications: topology evaluation, converter optimization and hardware verification. *IECON 2017 - 43rd Annual Conference of the IEEE Industrial Electronics Society*, pp. 1622–1628. https://doi.org/10.1109/IECON.2017.8216275.

53 Schweizer, M. and Kolar, J.W. (2013). Design and implementation of a highly efficient three-level T-type converter for low-voltage applications. *IEEE Transactions on Power Electronics* 28 (2): 899–907. https://doi.org/10.1109/TPEL.2012.2203151.

54 Peng, F.Z. (2002). Z-source inverter. *Conference Record of the 2002 IEEE Industry Applications Conference. 37th IAS Annual Meeting (Cat. No.02CH37344)*, pp. 775–781. https://doi.org/10.1109/IAS.2002.1042647.

55 Trinh, Q.-N., Lee, H.-H., and Chun, T.-W. (2011). A new Z-source inverter topology to improve voltage boost ability. *8th International Conference on Power Electronics - ECCE Asia*, pp. 1981–1986. https://doi.org/10.1109/ICPE.2011.5944462.

56 Shen, M., Wang, J., Joseph, A. et al. (2006). Constant boost control of the Z-source inverter to minimize current ripple and voltage stress. *IEEE Transactions on Industry Applications* 42 (3): 770–778. https://doi.org/10.1109/TIA.2006.872927.

57 Khlebnikov, A.S., Kharitonov, S.A., Bachurin, P.A. et al. (2011). Modeling of dual Z-source inverter for aircraft power generation. *2011 International Conference and Seminar on Micro/Nanotechnologies and Electron Devices Proceedings*, pp. 373–376. https://doi.org/10.1109/EDM.2011.6006976.

58 Cuenot, J., Zaim, S., Nahid-Mobarakeh, B. et al. (2017). Overall size optimization of a high-speed starter using a quasi-Z-source inverter. *IEEE Transactions on Transportation Electrification* 3 (4): 891–900. https://doi.org/10.1109/TTE.2017.2738022.

59 Asghari-Gorji, S. and Ektesabi, M. (2015). Input current ripples cancellation in bidirectional switched-inductor quasi-Z-source inverter using coupled inductors. *2015 Australasian Universities Power Engineering Conference (AUPEC)*, pp. 1–6. https://doi.org/10.1109/AUPEC.2015.7324876.

60 Battiston, A., Miliani, E.-H., Pierfederici, S., and Meibody-Tabar, F. (2016). A novel quasi-Z-source inverter topology with special coupled inductors for input current ripples cancellation. *IEEE Transactions on Power Electronics* 31 (3): 2409–2416. https://doi.org/10.1109/TPEL.2015.2429593.

61 Ounejjar, Y., Al-Haddad, K., and Grégoire, L.-A. (2011). Packed U cells multilevel converter topology: theoretical study and experimental validation. *IEEE Transactions on Industrial Electronics* 58 (4): 1294–1306. https://doi.org/10.1109/TIE.2010.2050412.

62 Vahedi, H., Labbé, P.A., and Al-Haddad, K. (2016). Sensor-less five-level packed U-cell (PUC5) inverter operating in stand-alone and grid-connected modes. *IEEE Transactions on Industrial Informatics* 12 (1): 361–370. https://doi.org/10.1109/TII.2015.2491260.

63 Chappa, A., Gupta, S., Sahu, L.K., and Gupta, K.K. (2021). A fault-tolerant multilevel inverter topology with preserved output power and voltage levels under pre- and postfault operation. *IEEE Transactions on Industrial Electronics* 68 (7): 5756–5764. https://doi.org/10.1109/TIE.2020.2994880.

64 Dewangan, N.K., Gupta, K.K., and Bhatnagar, P. (2020). Modified reduced device multilevel inverter structures with open circuit fault-tolerance capabilities. *International Transactions on Electrical Energy Systems* 30 (1): e12142. https://doi.org/10.1002/2050-7038.12142.

65 Mhiesan, H., Wei, Y., Siwakoti, Y.P., and Mantooth, H.A. (2020). A fault-tolerant hybrid cascaded H-bridge multilevel inverter. *IEEE Transactions on Power Electronics* 35 (12): 12702–12715. https://doi.org/10.1109/TPEL.2020.2996097.

66 Choi, U.M., Blaabjerg, F., and Lee, K.B. (2015). Study and handling methods of power IGBT module failures in power electronic converter systems. *IEEE Transactions on Power Electronics* 30 (5): 2517–2533. https://doi.org/10.1109/TPEL.2014.2373390.

67 Kumar, G.K. and Elangovan, D. (2020). Review on fault-diagnosis and fault-tolerance for DC–DC converters. *IET Power Electronics* 13 (1): 1–13. https://doi.org/10.1049/IET-PEL.2019.0672.

68 Dewangan, N.K., Prakash, T., Tandekar, J.K., and Gupta, K.K. (2020). Open-circuit fault-tolerance in multilevel inverters with reduced component count. *Electrical Engineering* 102 (1): 409–419. https://doi.org/10.1007/S00202-019-00884-9.

69 Zhang, W., Xu, D., Enjeti, P.N. et al. (2014). Survey on fault-tolerant techniques for power electronic converters. *IEEE Transactions on Power Electronics* 29 (12): 6319–6331. https://doi.org/10.1109/TPEL.2014.2304561.

70 Asif, M., Tariq, M., Sarwar, A. et al. (2021). A robust multilevel inverter topology for operation under fault conditions. *Electronics* 10: 3099. https://doi.org/10.3390/ELECTRONICS10243099.

10

Standards and Regulations Pertaining to Aircraft

Lujia Chen, Prem Ranjan, Qinghua Han, Abir Alabani, and Ian Cotton

The University of Manchester, Department of Electrical and Electronic Engineering, Manchester, M13 9PL, UK

10.1 Introduction

The increasingly strict environmental regulations and the drive for improved energy efficiency have led to the development of aerospace systems operating at higher voltages and the continued move toward hybrid and all-electric propulsion. Design and testing in accordance with standards and regulations are crucial to the aviation industry from the initial design stage to the daily operation of the aircraft. Relevant standards have been formulated and published by international standard bodies that provide the necessary guidance that facilitates quality control and assurance. There is a myriad of standards that require a simple overview as per their end use for industrialists and academics working or having an interest in entering this sector. Given the rapid development of high-power electrical systems in aircraft, it is certain that standards will also need to develop and evolve.

The Federal Aviation Administration (FAA) divided the Federal Aviation Regulations (FARs) into the different sub-chapters A to N [1]. Sub-chapter D is the most relevant to aircraft and its associated part 21 details the regulations related to "certification procedures for products and parts." This document is further divided into other sub-parts from A to O covering different certification requirements, changes in certificates, and approval of different parts including export and authorization procedures for the Technical Standard Order (TSO). The Environmental Protection Agency (EPA) defines the regulations on the CO_2 emission from aircraft and their engines in accordance with greenhouse gas emission standards with Docket ID no. EPA-HQ-OAR-2018-0276 [2]. The aforementioned standards are equivalent to the airplane CO_2 standards adopted by the International Civil Aviation Organization (ICAO) in 2017 and are applicable to airplanes that are in-production and in-design phases. The majority of emissions generated by aircraft are CO_2 and nitrous oxide (N_2O) with fuel efficiency-based metric controls in place for both gases since January 2021 [2].

European Commission published a report on "Electrification of the Transport System" detailing (i) EU policy targets and challenges facing aeronautics industry, (ii) baseline of the state of the art on the electrification technological development, barriers, and competitiveness, and (iii) strategic implementation plan including the aeronautics mode of transport [3]. The European Flightpath 2050 vision has been set up with goals for reduced emission. It is envisaged that by 2050, technologies and procedures will be able to facilitate a 75% reduction in CO_2 emissions per passenger kilometer and a 90% reduction in nitrogen oxide (NOx) emissions.

Transportation Electrification: Breakthroughs in Electrified Vehicles, Aircraft, Rolling Stock, and Watercraft,
First Edition. Edited by Ahmed A. Mohamed, Ahmad Arshan Khan, Ahmed T. Elsayed, and Mohamed A. Elshaer.

The Carbon Offsetting and Reduction Scheme for International Aviation (CORSIA) is an EU-led initiative that aims to stabilize CO_2 emissions at 2020 levels by requiring airlines to offset the growth of their emissions post-2020. Airlines will be required to monitor emissions on all international routes and offset emissions from routes included in the scheme. This will be achieved by purchasing eligible emission units generated by projects that reduce emissions in other industrial sectors. The participation in the scheme is mandatory for EU countries and the anticipated participation will offset estimated emissions of around 80% above 2020 levels in the period of 2021–2035 [4].

Standards are categorized in sub-sections based on aircraft components, viz. power generation, cables, connectors and contacts, switching devices, and materials. Generators, motors, fuel cells, batteries, and power electronic converters are grouped into the power-generation section. Standards for cables can be divided into electric cable and fiber optics, both for electric power and data transmission. This will include different test schemes and standards related to checking on flammability, flexibility, and toxicity. Connectors and contacts in Section 10.4 will cover the standards related to design and testing of different parts including plugs, lugs, nut bolts, and contacts. The sub-section for switching devices will comprise all standards related to different switches used in aircraft. In the case of materials, standards on metallic and non-metallic parts used in aircraft will be described.

10.2 Power Generation

This section summarizes the existing standards related to power generation on aircraft, which include the specifications on power source, electrical machine, and transformer.

10.2.1 Characteristics of Aircraft Electrical Systems

Three distribution voltage systems are specified in ISO 1540 on both small and large transport aircrafts: (i) primary AC power generation provides a nominal voltage of either 115/200 V_{rms} or 230/400 V_{rms} with a nominal frequency of 400 Hz for constant frequency sources, (ii) power sources with a frequency higher than 360 Hz variable frequency, and (iii) DC power sources that provide a nominal voltage of 14, 28 or 42 V [5]. Table 10.1 specifies the operating limits under normal, abnormal, and emergency operating conditions [5]. The respective definitions of normal, abnormal, and emergency electrical system operations are detailed in ISO 1540. Furthermore, transient conditions in which the system experiences momentary variation before returning to a steady-state limit are also defined. It is clear that these voltage ranges will need to be extended to accommodate higher-power aircraft.

BS 4G 220 and ISO 7137 standards provide environmental test conditions and procedures for all the onboard equipment, which can be categorized in four different types of test environments: atmospheric, mechanical, electromagnetic, and explosion and fire environments [6]. ISO 6858 specifies the ground support electrical supply based on electrical characteristics, electrical protection, control circuit and supply, test requirements, safety requirements, general design features, and installation, operation, and maintenance. The electrical characteristics can be measured at the aircraft attaching connector with an output that is either AC or DC [7]. In the case of ISO 12384, the requirements for digital measuring methods and equipment are specified for AC power system characteristics with constant or variable frequency, DC power system characteristics, current parameter, power transfer characteristics, voltage spike characteristics, and power factor [8].

Table 10.1 Operation conditions of DC and AC (115/200 V_{rms}) at 400 Hz, narrow range variable frequency, and wide range variable frequency.

			Normal		Abnormal		Emergency	
	Category	Named value	Lower limit	Upper limit	Lower limit	Upper limit	Lower limit	Upper limit
400 Hz 115/200 V AC	Three phase average V_{rms}	115	104	120.5	95.5	132.5	104	120.5
	Individual phase V_{rms}	115	100	122	94	134	100	122
	Frequency Hz	400	390	410	360	440	360	440
Narrow range variable frequency	Three-phase average V_{rms}	115	104	120.5	98.5	132.5	104	120.5
	Individual phase V_{rms}	115	100	122	97	134	100	122
Wide range variable frequency	Three-phase average V_{rms}	115	101.5	120.5	98.5	132.5	101.5	120.5
	Individual phase V_{rms}	115	100	122	97	134	100	122
DC	28 V Categories A and B	28	22	30	20.5	32.2	18	32.2
	14 V Category B	14	11	15	10.25	16	–	–
	28 V Category R	28	26.5	28.5	22	30.5	–	–
	42 V Category R	42	40	45.5	33	48	–	–

Source: Reproduced based on Ref. [5].

10.2.2 Electrical Machines

BS G 102 summarizes the general requirements of the rotating machine including requirements for the bearing, lubrication, rotor balances (<2.5 mm/kg), brush gear (>200 mA), commutators, slip rings, assessment of commutation, air, and liquid cooling, mounting flanges, and drive shafts [9]. Dimensions of brushes and brush holders for electrical machines are detailed in EN 4999 [10]. Specifications on the mounting can be found in BS M 32, which provides specifications for aircraft accessory drives and mounting pad dimensions [11].

Three-phase constant frequency and variable frequency squirrel-cage induction motors are two types of motors used in aircraft [11, 12]. The testing requirement for the motor are specified in BS 3G 100 [13] and BS G 102 standards [9], whereas BS 2G 146 is the specification for DC motor [14]. It has been defined that stall, altitude, compass (magnetic) interference, and radio interference are the type tests of DC motor.

Constant frequency and variable frequency are the two types of AC generators with their corresponding test requirements specified in BS 3G 100 [13] and BS G 102 [9], which include construction, cooling, mounting and drive, and terminal arrangement. There are specific requirements on voltage frequency phases and waveform, rating and overload requirements, efficiency, extrication, overspeed, vibration acceleration, climatic proofing ratio interference and compass interference, and declaration of performance [15, 16].

As specified in BS 2G 134, the DC generator shall be able to maintain 125% of the rated load current when operated at the rated voltage, remain operational at a speed 125% lower than the rated minimum speed for at least 5 minutes, and produce at least 200% of rated current for 5 seconds' duration [17]. The machine shall comply with a range of environmental and stress conditions that include vibration, acceleration, temperature/pressure, ratio interference suppression, and compass safe distance as detailed in BS 2G 100-2, BS 3G 100-2.3.1, and BS 3G 100-2.3.2 standards [13]. The type tests of DC generator shall be conducted at an ambient temperature of $20 \pm 5\,°C$ and are listed as follows [17]:

- Weight, center of gravity, and rotor inertia;
- Initial check on the brush;
- Cold resistance of windings measured under room temperature and corrected to 20 °C;
- Commutation as per BS G 102 [9];
- Open-circuit magnetization and load excitation characteristics;
- Ripple voltage of <4.5 $V_{pk\text{-}pk}$ (this can be affected by the feeder cables);
- Brush performance at altitude;
- Temperature tests that include a shortened standard test;
- Insulation resistance measurement of the machine shortly after the temperature tests with the windings still hot. The resistance between live parts and the frame shall not be <2 MΩ at 250 V_{DC} and with suppression capacitors disconnected;
- Overspeed test is followed immediately after the insulation resistance test with the machine to run for 3 minutes at 120% of its maximum rated speed or at a higher speed to be agreed between the manufacturer and end user;
- Other tests include containment, endurance, overload and radio interference suppression, vibration, acceleration, temperature-pressure tests, mold growth, compass interference, waterproofness, salt corrosion, dust and sand, and fluid contamination.

Following on from the specified type tests, all the components and sub-assemblies of the machine shall comply with production routine tests that include component check, the resistance of windings, brush bedding and adjustment, temperature, overspeed, loading, and insulation. Further examination as part of product quality tests is undertaken on a selection of samples from the current production that passed the previous production routine tests.

10.2.3 Power Conversion

Auto-transformer rectifier units (ATRUs) can convert AC to DC using multi-pulse phase shifting auto-transformers to provide cancelation of harmonic current [18]. ATRUs have demonstrated numerous advantages over conventional harmonic suppression devices, such as weight reduction, simple equipment design structure, and higher reliability. ISO 24071 provides information on general requirements, environmental conditions and test procedures, and qualification inspection methods. The nominal steady-state DC output voltage of two output terminals of the ATRUs to the neutral of the system shall be ±135 or ±270 V. The output voltage shall remain between ±125 and ±160 V_{DC}. The output voltage differential mode ripple amplitude is less than 160 $V_{pk\text{-}pk}$. Voltage distortion factor shall be less than 3%. Environmental conditions and test procedures are defined in BS 3G 100 and ISO 7137 [6, 13].

BS 2G 127 provides the specifications on power and current transformers, which are intended for use in applications operating at altitudes less than 60,000 ft (18,288 m) and with temperatures from −65 to 70 °C as well as a maximum temperature of 125 °C [19]. The rating should be obtained at sea level and 70 °C. For efficiency and regulation, at least 90% efficiency is required on full load with voltage regulation of less than 3%. Environmental conditions and stresses including vibration, average acceleration, crash acceleration, climatic proofness, waterproofness, resistance to fire, and explosion proofness must be considered. The operation life shall be 1000 hours at maximum temperature and 30,000 hours at 25 °C. The acoustic noise should be below 60 dB at a power less than 1 kVA and 75 dB at a power higher than 1 kVA.

BS G 174 covers the general design and test procedure for an inverter that derives AC at controlled voltage and frequency from the DC supply [20]. There are two inverter classes: (i) Class 1 that is suitable for use up to an altitude of 20,000 ft (6096 m) or equivalent air pressure of 37–53 kPa, and (ii) Class 2 that is suitable for use up to a specific altitude specified by manufacturers. Each class has two groups with group A for use where closely controlled output is necessary and group B is for general purpose. The input voltage is either 28 V_{DC} (+1, −3 V) or 112 V_{DC} (+4, −2 V), whereas the output voltage is 115 and 200 V_{rms} at a frequency of 400 Hz.

10.2.4 Batteries

The EN 60952 standard series defines the minimum environmental and performance requirements for establishing a qualification standard for the airworthiness of lead–acid and nickel–cadmium aircraft batteries, which contain corrosive electrolytes [21–23]. EN 60952-1 defines test procedures for evaluating, comparing, and qualifying the batteries and states the minimum environmental performance levels for airworthiness [21]. EN 60952-2 details the design requirements for aircraft batteries, shape and size, and the range of aircraft interface connectors used [22]. EN 60952-3 defines the product specification with specific requirements for an application and a declaration of design and performance, which details the performance of a battery when tested in accordance with EN 60952-1 [23]. Electrical requirements of the battery include the following aspects [21]:

- Capacity of −30 to 50 °C;
- Constant voltage at −30 to 23 °C;
- Constant voltage discharge at 14 V and −30 to 23 °C;
- Rapid discharge capacity at both 23 and −30 °C;
- Charge retention at either 23 or 50 °C;
- Short-circuit test, charge acceptance at 23 °C,
- Low temperatures at −18 and −40 °C;
- Insulation resistance (measured at 250 V_{DC}) and dielectric strength (1500 V_{rms} 50 Hz);
- Overcharge endurance with 1.45 V $\pm$ 0.005 per cell for nickel-cadmium and 2.417 V $\pm$ 0.01 per cell for lead-acid;
- Other tests include duty cycle performance, water consumption test, cyclic endurance, deep discharge test, induced destructive overcharge, and electrical emission.

The type tests for lithium batteries are specified in BS 2G 239 with a tolerance of $\pm$2 °C, $\pm$1 kPa, and $\pm$10% humidity [24]. The use of nickel-cadmium batteries is also specified in the following categories [25–30]:

- EN 2570 for technical specification;
- EN 2985 for Type A, 40 kg, and 34 Ah (300∗268∗262 mm);
- EN 2986 for Type B type, 29 kg, 22 Ah;
- EN 2987 for Type C type 27.5 kg 22 Ah;
- EN 2988 for Type D type, 16.5 kg, 11 Ah;
- EN 2991 for Type E type 40 kg 34 Ah (240∗216∗270 mm).

10.2.5 Challenges for Higher Voltage Aerospace Systems

Most conventional aircrafts are operated at a voltage of 28 V_{DC} and AC voltage 115 V_{AC}, 400 Hz. At these voltages, there is little risk of flashover and consequent arc fault the higher voltage aerospace systems are seeing in the replacement of mechanical and hydraulic apparatus with their electric equivalent to reducing the overall weight of the system. Such new concepts would contribute to enhancing the reliability of the system, facilitate maintenance, scale costs down, and reduce environmental impact. For this to be achieved, certain measures have to be introduced including frequency increase and raising the voltage levels well beyond the conventional. As such, the higher voltage aerospace systems are using a combination of AC and DC systems. Voltages in existing aerospace systems have increased to 540 V_{DC} and 230 V_{AC} with frequencies of ≥400 Hz. This creates another challenge, which was not of concern in low-voltage operations. Commercial aircrafts fly at a pressure of ~20 kPa, which corresponds to ~12 km above sea level. Therefore, at such a low pressure, and with the increasing voltage, the probability of flashover and electrical discharges is increased. All electric aircraft and next-generation higher-voltage aerospace systems are proposed to operate on voltage level of 1 to 2 kV and frequency range of 1–2 kHz. Relevant standards need to be updated or new standards need to be formulated to address the challenges in design, testing, operation, and safety of aircraft equipment.

10.3 Cable

This section details the standards and practices adopted on cables used in the aircraft wiring system. All related cable terms used are defined as per ISO 8815 [31]. Cables can be constructed as single-core or multi-core assembly in accordance with respective standards. Corresponding standards to different applications of cables were formulated to streamline the production and installation processes. The main standards can be broadly summarized in the following categories.

10.3.1 Cable Component and Type

EN 3475-100 is a general-purpose document that defines terminologies including components such as strand, plated-strand, jacket, insulation, sheath, screen, and concentric, bunched, rope-stranded, and insulated conductors. The different cable types were categorized as electrical, airframe, fire-resistant, and fire-proof. Furthermore, cross-sectional area and size of conductor and cable lay formation were also specified [32].

Relevant standards on different cable types can be classified based on the specified operating temperature. EN 2266 and EN 2713 cover a temperature range of −55 to 200 °C, whereas the temperature range of −55 to 260 °C is covered by EN 2267 and EN 2714 [33–36]. In the case of fire-resistant and fire-proof cables, the respective EN 2346 and EN 4608 standards specify an operating temperature range of −65 to 260 °C [37, 38]. Note that the maximum specified temperature accounts for any potential temperature rise in the conductor in addition to the ambient temperature. Fire-resistant and fire-proof cables must maintain a defined dielectric performance when subjected to a flame of 1100 °C for 5 and 15 minutes respectively. Note that EN 2713, EN 2714, and EN 4608 standards are intended for screened and jacketed cables.

Technical specification on fire-resistant cables and cables with screen and jacket are covered in EN 2234 and EN 2235 respectively [39, 40]. In the case of EN 2084, the document specifies the characteristics, test methods, qualification, and acceptance conditions of single-core and multi-core electrical cables with copper or copper alloy conductors and without outer jackets. The cable insulation is designed to withstand the required operating aircraft voltages of 115/200 V unless specified otherwise and at a system frequency that is below 2 kHz [41].

10.3.2 Digital Data and Signal Transmission

EN 3375-001 details the technical specifications related to cables used to transmit digital data [42]. The lay length of the outer cable layer shall be between 8 and 16 times the nominal diameter of the cable core. For screening, it can be in different combinations of one or several spiral layers, braids, metallic strips, and layers of extruded conductive or non-conductive materials. However, the minimum angle γ of spiral screening or braiding is at least 10°, which is measured against the longitudinal axis of the cable. For ethernet quad single braided and screened cables, EN 3375-011 and EN 3375-012 have defined maximum operating temperatures of 125 and 260 °C for cables with lightweight and fire-resistant properties respectively [43, 44].

EN 3745 defines the usage of optical fibers and cable in terms of core, cladding, fiber coating, optical cable, multiple fiber cable, buffer, strength member, jacket, refractive index profile, core diameter, cladding diameter, concentricity error core/cladding, non-circularity of core, non-circularity of cladding, attenuation, numerical aperture, and bandwidth [45]. Terms are defined as per IEC 50 (731) International Electro-technical Vocabulary – Chapter 731: Optical fiber communication [46]. Optical cables with 125 µm outer diameter cladding are specified in EN 4641, which defined its construction as a tight and semi-loose structure cable. The term "tight" refers to no movement between different component layers including fiber, inner jacket and outer strength member [47].

Coaxial and micro-coaxial cables used in signal transmission are covered in EN 4604. Note that "micro-coaxial" refer to cables with an outer cable diameter of approximately 1 mm or less. Cables are classified in accordance to different conditions such as an operating frequency of 3 to 8 GHz, an impedance of 50 and 75 Ω with a tolerance of ±2 to ±5 Ω, an attenuation at 1 GHz with 15 to 140 dB per 100 m, an external diameter of 2.4 to 7.7 mm, a screen in foil and/or braid, and temperature ranges of either −55 to 180 or −65 to 200 °C [48].

10.3.3 Cable Identification Marking

Identification marking for cables is specified in ISO 2574, with cable code identification defined in TR 6058, marking specification detailed in both EN 2084 and EN 3838, and the color code is specified in SAE-AMS-STD-595 [41, 49–52]. Generally, the relevant standards provide the number of cores, color codes for single-core and multi-core cables, and the number of cores from 1 to 10 are assigned with the corresponding code of A to K except for I.

Marking designation is specified in the corresponding product standard that includes the standard number, markability, type/gauge/manufacture code. As an example, for EN3375-00XXABB WJ C GG FR F YY with 4 spaces between each group, XX is standard part number, A is the nature of braid, BB is UV laser markability, WJ is the type code, C is the specific code, GG is the gauge code, FR is the manufacturer's country code, F is the manufacture code, and YY is the year of manufacturing. The maximum length of the marking is 60 mm, and this marking is repeated every 300 mm of cable length with a tolerance of ±50 mm. There are more detailed standards that define the printing technique, for instance, EN 2714-015 specifies UV laser printing for marking of cables to be used in a low-pressure environment [53]. Quality management is conducted in accordance with EN 9133, where routine tests are required immediately post-production and prior to the delivery. Periodic tests are recommended by sampling at least every three years [54].

10.3.4 Cable Test Specifications

Test conditions for cables are defined in EN 3475-100 [32], with an ambient temperature of 20 ± 5 °C, atmospheric pressure of 86–106 kPa, and relative humidity of 45–75%. The temperature and humidity shall remain constant during a series of consecutive measurements. There are many sub-standards as part of EN 3475 [32] series that contain different test methods, which can be grouped as follows:

- General (EN 3475-20X, X = 1-3), visual examination, and measurements of fiber/cable dimensions;
- Electrical (EN 3475-30X, X = 1-7), measurements of ohmic resistance per unit length, and insulation, surface, and overload resistances, and voltage proof, continuity, and corona extinction voltage tests;
- Environmental (EN 3475-40X, X = 1-18), accelerated aging, fire resistant, thermal endurance, and bending tests;
- Mechanical (EN 3475-5XX, XX = 01-15), solderability, abrasion, porosity, deformation resistance, and tensile tests;
- Sundry (EN 3475-60X, X = 1-5), smoke density, toxicity, resistance to wet arc tracking, resistance to dry arc propagation, and wet short-circuit tests;
- Handling (EN 3475-70X, X = 1-5), flexibility and laser markability;
- High frequency (EN 3475-8XX, XX = 01-13), measurements of capacitance, impedance, attenuation, and power rating.

Other than the aforementioned tests, burning behavior of non-metallic materials under radiating heat and flames with determination of smoke density defined in EN 2825 [55] and gas components as per EN 2826 [56] must also be considered.

10.4 Connectors and Contacts

This section provides an overview of standards for a wide range of electrical and optical connectors as well as their contacts and elements used to establish onboard connections. ISO/DIS 2100-100 and EN 2591 define an element of electrical and optical connectors as a component such as a connector or a module with the purpose of ensuring the connection of circuits [57, 58]. Other terms and definitions relevant to connectors, contacts, or elements of electrical and optical connections are defined in ISO/DIS 2100-100, EN 2591, and EN 3155-001 [57–59].

10.4.1 Classification

In accordance with EN 3197, the selection of appropriate connectors in the Electrical Wiring Interconnection System (EWIS) and the Optical Fiber Interconnection System (OFIS) should adhere to the safety, system requirement, ease of maintenance, and cost approved by the Design Authority [60]. The component selection should consider the service lifetime of connectors, contacts, and elements of electrical and optical connections, which shall not fall below the lifetime of a civil aircraft structure, that being approximately 60,000 flying hours or 20 years. Regular maintenance and inspection of components should not wait for a fault occurrence or to take place near its end of life. Visual inspection, testing, and examination as well as conditions of quality assurance should be applied according to relevant standards and technical specifications for each component.

Electrical and optical connectors used on aircrafts can be classified as follows:

- *Installation environment*: surface finish, material, and plating;
- *Geometry*: circular and rectangular;
- *Functionality*: electrical and optical.

10.4.2 Connectors

Coaxial, triaxial, twinax, and quadrax are all variants of coaxial connectors that are used with compatible electrical cables to transmit high-frequency digital data. These variants fall under the circular geometry classification. Circular connectors are characterized by variant coupling mechanisms including coupling ring with self-lock, single and multi-way coaxial, and bayonet or screw-on mechanism [61–65]. In the case of rectangular connectors, there are fewer variations such as rack and panel or modular [65, 66].

The choice of manufacturing material for a connector housing is found to be common across rectangular and circular connectors. Housings can be made of aluminum alloys, passivated stainless steel, composite, nickel-copper or copper alloys, which are plated with nickel, aluminum, cadmium, or tin to offer corrosion protection. Connector plugs can mate with their receptacles through different fixing/mounting features. The main mounting features for circular connectors are square flanges, bulkhead, nuts, jam-nut with O-ring seal, oval flange, or round solder and brazed flange, which are usually only used with Y and YE class receptacles [63]. For rectangular connectors, mounting configurations depend on the connector but they are either rectangular flange or two-end flange [67].

10.4.3 Contacts

EN 3155-001 further describes the types of contacts that are present in connectors. It clearly identifies the terms and definitions used to understand contacts and the electrical, mechanical, environmental, and dimensional characteristics of different contact types including removable crimp, wrap, and solder contacts. EN 3155-001 can be a general reference to all contacts unless there is a contradiction with the product's specific standard. In this case, the specific product standards will overrule the general standard. Contacts can be categorized based on their working mechanism or temperature. The working mechanism could be sub-divided into removable for general-purpose, non-removable, removable contacts for thermo-couple, and removable contact with screening features. They could also be sub-categorized based on the maximum operating temperature, which ranges from 125 to 350 °C [59].

10.4.4 Testing of Tools, Contacts, and Connectors

Unless specified otherwise, the general test temperature is 23 ± 5 °C, the atmospheric pressure is from 86 to 106 kPa and the relative humidity is in the range of 45–75%. Based on the specific test conditions for each component, optical and electrical connectors have an operating temperature that lies from −65 to 260 °C [57]. The types of crimping tools and accessories used with electrical and optical connectors are specified in the individual product standards of EN 4008-001 [68]. Customized technical specifications of what crimping tools to be used, crimping set, positioners, and other accessories of different products can be found within each component standard. Generally, marking crimping tools and other appliances should by default include the identity block, date of manufacture, and manufacturer's identity. All relevant information should be implemented in legitimate manner with a permanent marker to avoid mixing. Testing of contacts, connectors, and elements of electrical and optical connectors is summarized in ISO 2100-100, EN 2591, and its sub-sections [57, 58]. Each set of test programs is then customized to each individual component and explained in their designated standard. The tests highlighted within this standard can be subdivided into nine main categories: environmental impact, visual inspection and physical properties, electrical performance, mechanical stress/resiliency, aging, measurements, signals and shielding, insulation, and assembly.

10.5 Switching Device

This section discusses the standards/practices related to circuit breakers and relays used in the aircraft wiring system. Standards and practices on circuit breakers and relays used in the aircraft wiring system are presented. Specific terms for circuit breakers are provided in EN 2350 whereas all other terms used in these standards are defined by IEC 50 (441) [69, 70].

10.5.1 Circuit Breaker Classification

Circuit breakers are categorized as single-pole, double-pole, or three-pole devices used for both AC and DC supplies in accordance with EN 2282 [71]. The normal operating current (I_n) is used to select the circuit breaker type and usually there is a designated white marking on the black actuator button for the equipment. The short-circuit or prospective current varies in the range of $65 \cdot I_n$ up to 1 to 5 kA depending on I_n or as specified in the relevant product standard. Most of the circuit breakers produced have been temperature compensated for thermal protection and can operate between −55 and 125 °C up to an altitude of 15 km. The upper temperature limit reduces to 90 °C for I_n exceeding 15 A for three-pole devices. For Arc Fault Circuit Breaker (AFCB), the operational ambient temperature lies between −40 and 85 °C with a switching capacity of $65 \cdot I_n$ or 1 kA [72].

10.5.2 Design of Circuit Breakers

For the design of circuit breakers, electrical connection components that are suitable to fit the cable lugs or contacts, actuator button, calibration safety device, leakage paths, non-flammable insulants, corrosion resistant, and fasteners (screws and nuts need to be locked) are defined in EN 2350 [69]. These are operated by a "push-pull" type of single non-removable actuator button incorporated with a delayed action of "trip-free" tripping. The metallic materials used for the circuit breaker components must be resistant to corrosion or coated to protect against corrosion. For dissimilar materials in the contact, the electromagnetic force of the galvanic couple does not exceed 0.25 V to ensure that the use of bimetals will not affect the operation of circuit breakers [72]. Based on relevant standards, circuit breakers for aircraft applications can be broadly divided into three types: (i) low-current range of 0.5–25 A, (ii) high-current range of 20 to 50 A, and (iii) AFCB.

10.5.2.1 Low-Current Range

A I_n current range of 0.5–25 A is specified in EN 2495 and EN 2592, which define the characteristics of circuit breakers for single-pole and three-pole, respectively [73, 74]. Both standards specify the following parameters for the low-current range of 0.5–25 A:

- *Physical*: dimension, mount, and specified maximum mass of 26 and 65 g for single-pole and three-pole devices, respectively;
- *Mechanical*: force, load, and torque;
- *Environmental*: vibrations, and external mechanical shock;
- *Electrical*: a leakage current of less or equal to 1 mA, and a voltage drop of 0.2–2.2 V.

The fault establishment time is 2 to 5 seconds. The prospective current varies as (i) single-pole 5 kA (28 V_{DC}) and 1 to 2 kA (115 V_{AC}), and (ii) for three-pole 1 kA (0.5–3 A), 2 kA (5–25 A), and with a power factor in the range of $0.8 \leq \cos\varphi \leq 1$.

10.5.2.2 High-Current Range

A comparatively higher current range of 20–50 A is defined in EN 2794 and EN 2665 standards for single-pole and three-pole circuit breakers respectively [75, 76]. With technological development, circuit breakers with polarized and non-polarized signal contacts were developed for 20–50 A defined in EN 3661 and EN 3662 for single-pole and three-pole circuit breakers respectively [77, 78].

10.5.2.3 Arc Fault Circuit Breaker (AFCB)

AFCBs are designed to protect aircraft wiring system from potential circuit overload and arc fault. These are specified in EN 4838 and EN 4839 for single-pole and three-pole with a current rating of 3–25 A used in aircraft on-board circuits [72, 79]. To avoid circuit overload and arc faults, the design of a "trip-free" tripping to ensure protection of aircraft wiring system in any operating system. For any AFCB containing software or complex hardware, the design of such software and hardware systems must be developed in accordance with RTCA DO-178B and RTCA DO-254 respectively [80, 81].

10.5.3 Circuit Breaker Testing Specifications

The test conditions for circuit breakers are defined in EN 3841 with an ambient temperature of 23 ± 5 °C, atmospheric pressure of 84–107 kPa, and relative humidity of less than 85% [82]. For all the tests, cables with at least 0.5 m length should be used and the cross-section varies in the range of 0.6–5 mm^2 for all types of circuit breakers with a current range of 1–50 A, as defined in EN 2083 [83]. For the EN 3841 standard series, there are 29 sub-standards that detail different test methods as follows [82]:

- Physical (EN 3841-20X, X = 1, 2), visual examination, dimensions, and masses;
- Electrical (EN 3841-30X, X = 1-8), voltage drop, insulation resistance, dielectric strength, tripping point, short-circuit performance, service life, and lightning;
- Environmental (EN 3841-40X, X = 1-7), sand and dust, corrosion, humidity, explosion proofness, fuel resistance, and flammability;
- Mechanical (EN 3841-5XX, XX=01-11), actuator button travel, operating forces, strength of different parts – actuating components, mounting elements, and main terminals.

Similar to tests defined in Section 10.3.4, EN 2825 defines the burning behavior of non-metallic materials under radiating heat and flames with determination of smoke density and EN 2826 defines the required check of gas components [55, 56]. Furthermore, arc fault detection test is carried out in accordance with Sections 4.7.7.6 and 4.7.7.7 of AS 5692 for AFCBs [84].

Qualification and inspection tests are carried out as per their corresponding product standards, EN 3841 and EN 9133 (EN 2000) respectively [54, 82, 85]. After testing, circuit breakers can be stored for 3 to 5 years with main contacts in closed position for an ambient temperature of 5–40 °C, a maximum allowable relative humidity of 80% and protected from UV rays and dust under non-explosive and non-corrosive atmosphere. After the specified storage duration, voltage drop at I_n and tripping time at $2{\cdot}I_n$ tests need to be repeated. Periodic tests are recommended to be performed every two years or after the production of 50,000 circuit breakers of all ratings.

10.6 Material

Continued advances in materials are crucial to the design and manufacturing of airframe and propulsion in modern aircraft. The metallic structure requires properties such as mechanical strength, durability, damage tolerance, safety and fatigue, and corrosion resistance. There is an increasing usage of composite materials for airframe applications as a substitute for metallic material due to advantages of lightweight, fatigue, and corrosion resistance, and possessing tailorable mechanical strength and stiffness. However, the two drawbacks of composite materials are cost and long-term maintainability and reparability. This section details the relevant standards on existing specifications for materials in aircraft [86].

10.6.1 Metallic Materials

Typical metallic materials used in aerospace applications include aluminum, magnesium, and titanium alloys as well as heat-resistant alloys that are mixed with nickel, cobalt, or iron base. The EN standards on metallic materials can be broadly categorized as [87–95]:

- EN 4258 [87] is a general standard that connects different EN standards and their applications;
- Standards that define general terms related to metallic materials, for example, EN 2078 [88] defines manufacturing and inspection schedules, and inspection and test reporting, EN 2032-1 [89] for material designations, and EN 2032-2 [90] for coding of metallurgical condition of metallic semi-finished products in delivery condition;
- Standards covering the specific rules for drafting and presentation of material standard in EN 4500 [91], technical specification in EN 4260 [91], test method in EN 4261 [91], dimensional standard in EN 4000 [92] and general requirements for qualification in EN 2043 [93];
- Supplementary standards on quality assurance requirements concerning personnel facilities and processes. For example, qualification and approval of personnel for non-destructive tests are specified in EN 4179 [94] and heat treatment facilities are defined in EN 4268 [95].

The development of aluminum alloy has attained higher fracture toughness and specific yield strength that is relatively low-cost and lightweight. Aluminum and aluminum alloy can be used for conductors, nuts, bolts, rivets, bushes, clamps, and coatings for fasteners [96–102]. There are numerous standards pertaining to aluminum alloy since it is the main choice of material for aircraft accounting for approximately 80% of structure structural material due to the advantages of lightweight and cost. It is envisaged that aluminum alloy usage will be reduced from 80% down to 20% in overall aircraft structure [86]. In the case of magnesium alloy, it possesses a high strength-to-density ratio with excellent machinability, which is used in aircraft engines and gearbox casing [103]. Nickel alloys can resist extreme high temperature, corrosion resistance, and wear, which can be used for applications in bolts, fasteners, inserts, and nuts [103]. Ferrous alloys such as steels offer the highest strengths for commercial metallic materials and with a limited number of applications in landing gear, flap tracks, and actuation components. Corrosion of steel landing gear structure is a major issue, and the landing gear must be refurbished every 7–10 years to remove any rust or pits [86]. Titanium alloy, especially Ti-6Al-4V, possesses low density, good strength, and superior resistance to corrosion, which make it suitable for the construction of airframes and engines [86].

There are three types of conductors for electrical cable: (i) copper and copper alloy in EN 2083, (ii) copper-clad aluminum alloy in EN 4651, and (iii) aluminum or aluminum alloy in EN 3719 [83, 98, 104]. To improve the performance of the conductor, metal plating materials such as tin, silver, and nickel are introduced. Physical property requirements are also specified in these standards including elongation, tensile strength, resistivity, breaking load, and mass. EN 2853 specifies the current rating of electrical cables: (i) continuous rating and (ii) duty cycle rating operating in a short period below the maximum [105].

According to EN 3197, connectors and elements made of metallic materials must be corrosion-resistant or adequately protected against corrosion throughout their service life to minimize the risk posed by electrolytes at joints or mutual connections [60]. Unless otherwise specified, the maximum potential difference between any two joints should not exceed 0.25 V with a maximum potential difference of 0.3 V in a limited range of components such as the high-density miniature connectors referenced in EN 4857-1 [106].

10.6.2 Non-metallic Material

Non-metallic materials include polymer, glass, textiles, structural adhesive, and composition, which are mainly used for electrical insulation, sleeving, seal, and accessories. Polytetrafluoroethylene (PTFE) has a wide range of applications with EN 3572 specifies the use of in flexible hose, EN 4708-107 for heat-shrinkable sleeving, ISO 23933 for hose assemblies, and EN 4166 for bushes [107–110]. Polyolefin is used for sleeving at an operating temperature from −55 to 135 °C whereas adhesive-lined polyolefin is used for sleeving at an operating temperature from 30 to 105 °C as defined in EN 4708. EN 3001 defines the use of tempered glass mainly for cockpit glazing. The elastomer material is used for cable outlet, cap, seal in hydraulic system and plug [111–115]. For all components of non-metallic material, they shall be immune to micro-organic and fungus formation and should maintain adequate performance under severe weather, changing ambient conditions and in applicable fluids.

Composite materials are considered viable replacements for aluminum alloy for aircraft structure. There are several composites identified in the standards that include glass fiber-reinforced resin in EN 2374, polyetherketone with 55% continuous carbon fiber by volume (PEEK-CF55) in EN 4717, polyetheretherketone with 55% continuous glass fiber by volume (PEEK-GF55) in EN 4718, and continuous fiber reinforced-PEEK composite specified in EN 4711, EN 4712, and EN 4713 [116–121]. Other composition types are also being investigated for enhanced performances such as titanium foil in carbon fiber epoxy, and aluminum sheet in either glass fiber or epoxy resin [86].

References

1 Federal Aviation Regulations – List of FAR Parts. https://www.risingup.com/fars/info/ (accessed 3 September 2021).
2 Environmental Protection Agency (2021). Control of air pollution from airplanes and airplane engines: GHG emission standards and test procedures, Federal Register, Vol. 86, No. 6.
3 European Commission (2017). Directorate-General for Research and Innovation. Electrification of the Transport System – Studies and Reports.
4 Reducing emissions from aviation. https://ec.europa.eu/clima/policies/transport/aviation_en (accessed 3 September 2021).
5 ISO 1540 (2006). Aerospace – characteristics of aircraft electrical systems.
6 BS 4G 229:1996, ISO 7137 (1995). Schedule for environmental conditions and test procedures for airborne equipment.
7 ISO 6858 (2017). Aircraft – ground support electrical supplies – general requirements.
8 ISO 12384 (2010). Aerospace – requirements for digital equipment for measurements of aircraft electrical power characteristics.

9 BS G 102 (1971). Specification for general requirements for rotating electrical machinery.
10 BS 4999-147:1988, IEC 60136:1986 (1988). General requirements for rotating electrical machines – specification for dimensions of brushes and brush-holders for electrical machinery.
11 BS 2G 147-1 (1966). Specification for a.c. motors for aircraft – three-phase constant frequency squirrel-cage induction motors.
12 BS 2G 147-2 (1966). Specification for a.c. motors for aircraft – three-phase variable frequency squirrel-cage induction motors.
13 BS 3G 100 (1980). Specification for general requirements for electrical equipment and indicating instruments for aircraft – introduction.
14 BS 2G 146 (1966). Specification for d.c. motors for aircraft
15 BS 2G 124-1 (1966). Specification for a.c. generators for aircraft – constant frequency.
16 BS 2G 124-2 (1966). Specification for a.c. generators for aircraft – variable frequency.
17 BS 2G 134 (1970). Specification for d.c. generators.
18 ISO 24071 (2021). Aircraft – Autotransformer Rectifier Units (ATRUs) – general requirements.
19 BS 2G 127 (1967). Specification for power and current transformers for use in aircraft electrical power supply systems.
20 BS G 174 (1959). Invertors for secondary electrical supplies for aircraft.
21 EN 60952-1 (2013). Aircraft batteries – general test requirements and performance levels.
22 EN 60952-2 (2013). Aircraft batteries – design and construction requirements.
23 EN 60952-3 (2013). Aircraft batteries – product specification and declaration of design and performance (DDP).
24 BS 2G 239 (1992). Specification for primary active lithium batteries for use in aircraft.
25 BS EN 2570 (1996). Nickel-cadmium batteries – technical specification.
26 EN 2985 (1996). Nickel-cadmium batteries of format A type.
27 EN 2986 (1996). Nickel-cadmium batteries of format B type.
28 EN 2987 (1996). Nickel-cadmium batteries of format C type.
29 EN 2988 (1996). Nickel-cadmium batteries of format D type.
30 EN 2991 (1996). Nickel-cadmium batteries of format E type.
31 ISO 8815 (1994). Aircraft – electrical cables and cable harnesses – vocabulary.
32 EN 3475-100 (2010). Aerospace series – cables, electrical, aircraft use – test methods – Part 100: general.
33 EN 2266-002 (2005). Cables, electrical, for general purpose – operating temperatures between -55 °C and 200 °C – Part 002: general.
34 EN 2713-002 (2006). Aerospace series – cables, electrical, single and multicore for general purpose – operating temperatures between -55 °C and 200 °C – Part 002: screened and jacketed – general.
35 EN 2267-002 (2015). Aerospace series – cables, electrical, for general purpose – operating temperatures between -55 °C and 200 °C – Part 002: general.
36 EN 2714-002 (2016). Aerospace series – cables, electrical, single and multicore for general purpose – operating temperatures between -55 °C and 260 °C – Part 002: Screened and jacketed – general.
37 EN 2346-002 (2006). Aerospace series – cable, electrical, fire resistant – operating temperatures between –65 °C and 260 °C – Part 002: general.
38 EN 4608-001 (2019). Aerospace series – cable, electrical, fire resistant – single and twisted multicore assembly, screened (braided) and jacketed – operating temperatures between -65 °C and 260 °C Part 001: technical specification.
39 EN 2234 (2018). Aerospace series – cable, electrical, fire resistant – technical specification.
40 EN 2235 (2015). Aerospace series – single and multicore electrical cables, screened and jacketed – technical specification.
41 EN 2084 (2018). Aerospace series – cables, electrical, general purpose, with conductors in copper or copper alloy – technical specification.

42 EN 3375-001 (2018). Aerospace series – cable, electrical, for digital data transmission – Part 001: technical specification.
43 EN 3375-011 (2017). Aerospace series – cable, electrical, for digital data transmission – Part 011: Single braid – Star Quad 100 ohms – Light weight – Type KL – Product standard.
44 EN 3375-012 (2013). Aerospace series – cable, electrical, for digital data transmission – Part 012: Single braid – Star Quad 100 ohms – 260 °C – Type KH – roduct standard.
45 EN 3745-100 (2008). Aerospace series – fibres and cables, optical, aircraft use – test methods, Part 100: general.
46 IEC 50(731) (1991). International electro-technical vocabulary – Chapter 731: optical fiber communication.
47 EN 4641-001 (2018). Aerospace series – cables, optical, 125 μm diameter cladding – Part 001: technical specification.
48 EN 4604-001 (2019). Aerospace series – cable, electrical, for signal transmission – Part 001: technical specification.
49 ISO 2574 (1994). Aircraft – electrical cables – identification marking.
50 ASD-STAN TR 6058 (2019). Aerospace series – cable code identification list.
51 EN 3838 (2010). Aerospace series – requirements and tests on user-applied markings on aircraft electrical cables.
52 SAE-AMS-STD-595 (2017). Colours used in government procurement.
53 EN 2714-015 (2016). Aerospace series – cables, electrical, single and multicore for general purpose – operating temperatures between -65 °C and 260 °C – Part 015: xx family, screened (spiral) and jacketed, UV laser printable for use in low pressure atmosphere – product standard.
54 EN 9133 (2018). Aerospace series – Quality Management Systems – qualification procedure for Aerospace Standard products.
55 EN 2825 (2011). Aerospace series – burning behaviour of non metallic materials under the influence of radiating heat and flames – determination of smoke density.
56 EN 2826 (2011). Aerospace series – burning behaviour of non metallic materials under the influence of radiating heat and flames – determination of gas components in the smoke.
57 ISO 2100-100 (2017). Aerospace elements of electrical and optical connection – test methods – Part 100: general.
58 EN 2591:1992 (1992). Aerospace series – elements of electrical and optical connection – test methods – general.
59 EN 3155-001 (2016). Aerospace series – electrical contacts used in elements of connection – technical Specification, .
60 EN 3197 (2010). Aerospace series. Design and installation of aircraft electrical and optical interconnection systems.
61 EN 3645-001 (2019). Aerospace series – connectors, electrical, circular, scoop-proof, triple start threaded coupling, operating temperature 175 °C or 200 °C continuous – Part 001: technical specification.
62 EN 3733-001 (2009). Aerospace series – connector, optical, circular, single channel, coupled by self-locking ring, operating temperature up to 150 °C continuous – Part 001: technical specification.
63 EN 4857-001 (2017). Aerospace series – iniature connectors, high density, electrical, circular, scoop-proof, triple start threaded coupling, operating temperatures 175 °C continuous or 200 °C continuous – Shell sizes : 05 to 07 – Part 001: technical specification.
64 EN 4652-001 (2015). Aerospace series – connectors, coaxial, radio frequency – Part 001: technical specification.
65 EN 3716 (2006). Aerospace series – connectors, single-way, with triaxial interface, for transmission of digital data – Part 001: technical specification.

66 EN 4830-001 (2015). Aerospace series – connectors, optical, rectangular, modular, operating temperature 125 °C, for EN 4639-10X contacts – Part 001: technical specification.

67 EN 4644-001 (2017). Aerospace series – connector, electrical and optical, rectangular, modular, rectangular inserts, operating temperature 175 °C (or 125 °C) continuous – Part 001: technical specification.

68 EN 4008-001 (2006). Aerospace series – elements of electrical and optical connection – crimping tools and associated accessories – Part 001: technical specification.

69 EN 2350 (1991). Circuit breakers – technical specification.

70 IEC 50 (441) (1984). International Electrotechnical Vocabulary (IEV) – Part 441: switchgear, controlgear and fuses.

71 EN 2282 (1992). Characteristics of aircraft electrical supplies.

72 EN 4838-001 (2018). Aerospace series – Arc Fault Circuit breakers, single pole, temperature compensated, rated current 3 A to 25 A – 115 V a.c. 400 Hz Constant Frequency Part 001: technical specification.

73 EN 2495 (1991). Single-pole circuit breakers, temperature compensated, rated currents up to 25 A – product standard.

74 EN 2592 (1991). Three-pole circuit breakers, temperature compensated, rated currents up to 25 A – product standard.

75 EN 2794-001 (2014). Aerospace series – circuit breakers, single-pole, compensated, rated current 20 A to 50 A – Part 001: technical specification.

76 EN 2665-001 (2013). Aerospace series – circuit breakers, three-pole, temperature compensated, rated current 20 A to 50 A – Part 001: technical specification.

77 EN 3661-001 (2006). Aerospace series – circuit breakers, single-pole, temperature compensated, rated current 20 A to 50 A – Part 001: technical specification.

78 EN 3662-001 (2006). Aerospace series – circuit breakers, three-pole, temperature compensated, rated current 20 A to 50 A – Part 001: technical specification.

79 EN 4839-001 (2018). Aerospace series – Arc fault circuit breakers, three poles, temperature compensated, rated current 3 A to 25 A – 115 V a.c. 400 Hz constant frequency Part 001: technical specification.

80 RTCA DO-178B (1992). Software considerations in airborne systems and equipment certification.

81 RTCA DO-254 (2000). Design assurance guidance for airborne electronic hardware.

82 EN 3841-100 (2004). Circuit breakers – test methods – Part 100: general.

83 EN 2083 (2001). Copper and copper alloy conductors for electrical cables – product standard.

84 AS 5692 (2004). ARC Fault Circuit Breaker (AFCB), aircraft, trip-free single phase 115 VAC, 400 Hz – constant frequency.

85 EN 2000 (1992). Quality assurance – EN aerospace products – approval of the quality system of manufacturers.

86 Boyer, R.R., Cotton, J.D., Mohaghegh, M. et al. (2015). Materials considerations for aerospace applications. *MRS Bull.* 40: 1055–1066.

87 EN 4258 (2021). Metallic materials – general organization of standardization – links between types of EN standards and their use.

88 EN 2078 (2002). Aerospace series – metallic materials – manufacturing schedule – inspection schedule, inspection and test report – definition, general principles, preparation and approval.

89 EN 2032-001 (2014). Aerospace series – metallic materials – conventional designation.

90 EN 2032-2 (1994). Metallic materials – coding of metallurgical condition in delivery condition.

91 EN 4500-001 (2012). Aerospace series – metallic materials – rules for drafting and presentation of material standards.

92 EN 4000 (2002). Aerospace series – metallic materials – rules for the drafting and presentation of dimensional standards for metallic semi-finished products.
93 EN 2043 (2013). Aerospace series – metallic materials – general requirements for semi-finished product qualification (excluding forgings and castings).
94 EN 4179 (2017). Aerospace series – qualification and approval of personnel for non-destructive testing.
95 EN 4268 (2012). Aerospace series – metallic materials – heat treatment facilities – general requirements.
96 EN 3719 (2018). Aerospace series – aluminium or aluminium alloy conductors for electrical cables – product standard.
97 EN 2876 (2019). Aerospace series – nuts, hexagon, plain, reduced height, normal across flats, in aluminium alloy, anodized – classification: 450 MPa (at ambient temperature)/120 °C.
98 BS 4A 169:2002+A2 (2019). Aerospace series – specification for bolts, hexagonal heads (unified threads) in aluminium alloy, anodized.
99 EN 6104 (2018). Aerospace series – rivets, solid, in aluminium or aluminium alloy – inch series – technical Specification.
100 EN 2285 (2017). Aerospace series – bushes, plain, aluminium alloy, with self-lubricating liner – dimensions and loads.
101 EN 3730 (2009). Aerospace series – clamps, saddle fixed and sliding version in aluminium alloy with rubber cushioning – dimension, masses.
102 EN 4473 (2010). Aerospace series – aluminium pigmented coatings for fasteners.
103 Prasad, N.E. and Wanhill, R.J.H. (2016). *Aerospace Materials and Material Technologies. Volume 2, Aerospace Material Technologies.* Singapore: Springer.
104 EN 4651 (2014). Aerospace series – copper-clad aluminium alloy conductors for electrical cables – product standard.
105 EN 2083, EN 2853 (2005). Current ratings for electrical cables with conductor.
106 EN 4857-1 (2017). Aerospace series – miniature connectors, high density, electrical, circular, scoop-proof, triple start threaded coupling, operating temperatures 175 °C continuous or 200 °C continuous – shell sizes:05 to 07 – Part 001.
107 EN 3572 (2021). Aerospace series – PTFE flexible hose assembly with convoluted inner tube of a nominal pressure up to 6800 kPa and 8°30' fitting in titanium – product standard.
108 EN 4708-107 (2019). Aerospace series – sleeving, heat-shrinkable, for binding, insulation and identification – polytetrafluoroethylene (PTFE) – operating temperatures – 65 °C to 260 °C – product standard.
109 ISO 23933 (2006). Aerospace series – aramid reinforced lightweight polytetrafluoroethylene (PTFE) hose assemblies, classification 135 °C/20 684 kPa (275 °F/3 000 psi) and 135 °C/21 000 kPa (275 °F/3 046 psi) – procurement specification.
110 EN 4166 (2004). Aerospace series – clips, spring tension, three parts – PTFE bushes.
111 EN 3001 (2019). Aerospace series – tempered float glass plies for aircraft applications – technical specification.
112 EN 3660-037 (2009). Aerospace series – cable outlet accessories for circular and rectangular electrical and optical connectors – bushing strip, elastomer, for cable outlet – style Z.
113 EN 6139 (2020). Aerospace series – cap, protective, non-metallic, for EN 6123 fitting ends.
114 EN 6111 (2020). Aerospace series – ethylene-propylene elastomer (EPM/EPDM) – Hardness 80 IRHD for static seal elements in hydraulic systems for long-term application – material standard.
115 EN 6140 (2020). Aerospace series – plug, protective, non-metallic, for NAS1760 fitting ends and AS33649 boss ports.

116 EN 2374 (1991). Specification for glass fibre reinforcing mouldings and sandwich composites – production of test panels.
117 EN 4717 (2021). Aerospace series – polyetheretherketone with 55% continuous carbon fibre by volume (PEEK-CF55) stock shape material – material specification.
118 EN 4718 (2021). Aerospace series – polyetheretherketone with 55% continuous glass fibre by volume (PEEK-GF55) – stock shape material – material specification.
119 EN 4711 (2012). Aerospace series – screw, 12-point flange head (bi hexagonal head bolt), long thread, for tensile applications, continuous fibre reinforced-PEEK composite, temperature range -65°C to 135°C – inch series.
120 EN 4712 (2012). Aerospace series – nuts, 12-point flange, self-locking, continuous fibre reinforced-PEEK composite, temperature range -65% °C to 135 °C – inch series.
121 EN 4713 (2012). Aerospace series – nuts, anchor, self-locking, floating. Continuous fibre reinforced PEEK composite, temperature range -65°C to 135°C – inch series.

11

Overview of Rolling Stock

Deepak Ronanki

Indian Institute of Technology Delhi, Department of Energy Science and Engineering, New Delhi, Delhi 110016, India

11.1 Introduction

The demand for sustainable transportation infrastructure is continuously increasing due to the rise in environmental pollution and traffic congestion caused by conventional vehicles. The term rolling stock is typically used in the rail transport industry, which represents rail wheels including both powered and unpowered vehicles. Electrification of the rail system is considered to be a more energy-efficient way of mass transportation. Furthermore, innovations and tremendous developments in power semiconductor devices, cooling systems, power converter topologies, materials, improved drive systems, and drive control techniques could rise the rolling stock. Also, researchers have carried out extensive research in terms of gravimetric density, effective utilization of regenerative braking energy, and the design and control of electrical/electronic equipment. However, significant efforts are needed to achieve reliable, highly efficient and compact rolling stock systems, which effectively utilize the energy and promote a faster mode of operation.

Until the 1990s, railway systems are equipped with tap control or Ward-Leonard control-based DC propulsion systems. Squirrel-cage induction motors (IMs)-based AC propulsion systems employing pulse width modulated power converters have become popular because of their robustness, maintenance-free and low cost [1]. These traction converters (TRCs) utilize gate turn-off (GTO) thyristors and snubber circuits for efficient operation and control. However, they have bulky value sets, and complex gate driver circuits and their operation is limited to 500 Hz switching frequencies [2]. The use of insulated gate bipolar transistors (IGBTs) allows the converter to operate at up to 1.5 kHz, reducing traction transformer losses and current harmonics while retaining 98% efficiency [3, 4].

Currently, IGBT-based AC propulsion systems are standardized and the current trend toward high-speed trains has better comfort and efficiency. Therefore, it is essential to know the fundamentals and recent technological improvements of the rolling stock. The current trend is to build rapid electric traction systems that can drive longer distances and give improved passenger comfort while being efficient. To accomplish the aforementioned requirements, energy consumption throughout the route must be reduced, which can be done by reducing the weight and running of rolling stock, implementing effective regenerative braking, and designing energy-efficient electrical equipment. This chapter mainly covers the fundamentals, overview and latest technologies in rolling stock including traditional and modern railway propulsion systems.

Transportation Electrification: Breakthroughs in Electrified Vehicles, Aircraft, Rolling Stock, and Watercraft,
First Edition. Edited by Ahmed A. Mohamed, Ahmad Arshan Khan, Ahmed T. Elsayed, and Mohamed A. Elshaer.

This chapter is organized as follows:

- Section 11.2 provides an overview of railway power systems, focusing on power supply types and rolling stock categorization based on fuel and power supply.
- Various components and rolling stock standards are covered in detail in Section 11.3.
- Section 11.4 describes modern rail systems that use solid-state transformer (SST) technology.
- Section 11.5 depicts advancements and challenges in modern rolling stock, such as drive train systems, traction motors, advanced control techniques, and semiconductor technology.
- This chapter's concluding observations are found in Section 11.6.

11.2 Rolling Stock Architectures

11.2.1 Railway Traction Power Systems

Railways use varied amounts of catenary voltage depending on geographical area, power network availability, and the kind of rolling stock for historical reasons [5]. Two different traction power systems (TPS) that exist today are DC and medium voltage AC (MVAC) systems. The most common TPSs of different voltage amplitude and frequencies along with estimated coverage in the world are summarized in Table 11.1.

The global standard catenary voltages are DC levels (600 VDC, 750 VDC, 1500 VDC, 3000 VDC), and AC levels (25 kV/50/60 Hz and 15 kV/16.7 Hz) [5]. The voltage used varies depending on the geographical area and the rolling stock class. It is to be noted that all AC voltage levels and DC voltage levels (1500 V, 3000 VDC) are usually employed for main-line (long-distance) trains. However, urban short railway lines such as trams, metros or city trains, use either 600, or 750 V as the main power supply. The primary motivation for lowering voltage levels is to ensure people's safety. In some countries, AC power supply (15 kV/16.7 Hz and 25 kV, 50/60 Hz) is also employed for sub-urban and urban lines.

11.2.2 Classification of Rolling Stock

The most frequent way to classify rolling stock is by its power output as indicated in Figure 11.1. As illustrated in Figure 11.2, rolling stock based on power distribution can be characterized as concentrated or distributed traction systems. A train set with one, two, or more engines pulling the coaches

Table 11.1 Summary of the traction power systems (TPSs).

TPS	Voltage level	Estimated coverage (km)	Selected countries
DC	600 and 750 V	7650	United Kingdom
	1.5 kV	20,440	Japan, France, The Netherlands
	3 kV	68,890	Russia, Spain, South Africa
MVAC (Single phase)	15 kV, 16 2/3 Hz	32,940	Germany, Switzerland, Norway
	25 kV, 50/60 Hz	72,110	Russia, France, China, South Africa, India
	11–12 kV, 25 Hz, and others	3000	USA

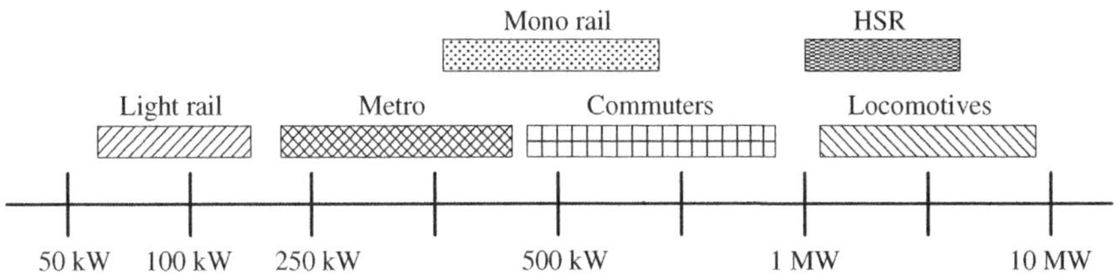

Figure 11.1 Categorization of rolling stock based on power levels.

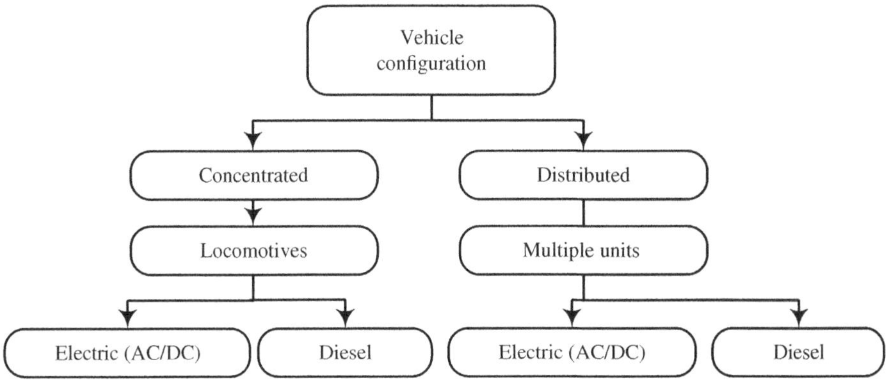

Figure 11.2 Classification of rolling stock based on the architecture, fuel, and source of supply.

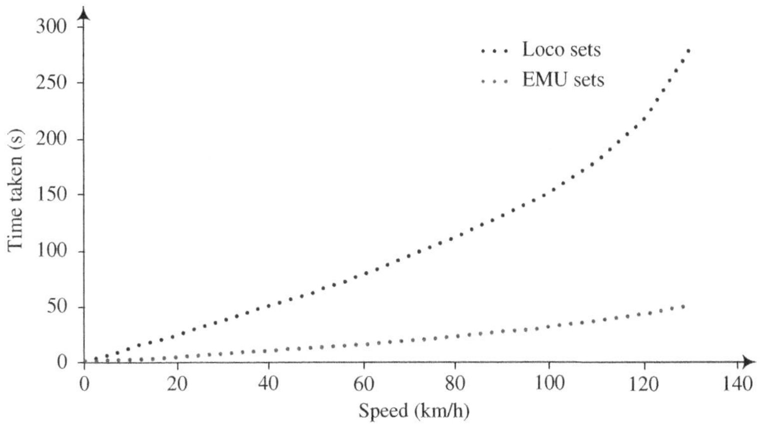

Figure 11.3 Field trial comparison between distributed and concentrated traction systems. *Source:* [7]/IEEE.

can be used to demonstrate concentrated traction systems. On the contrary, distributed traction systems feature many electric units made up of motor cars, referred to as electric multiple units (EMUs) [6]. Figure 11.3 shows a field trial comparison of the same power-level traction systems to attain 130 km/h [7]. Due to large axle loads, transport capacity, and superior adhesion effort, the offered research suggests that distributed-based rolling stock performs better in achieving higher acceleration and deceleration. However, it has certain reservations about maintenance, passenger comfort, installation, and high-speed pantograph functioning.

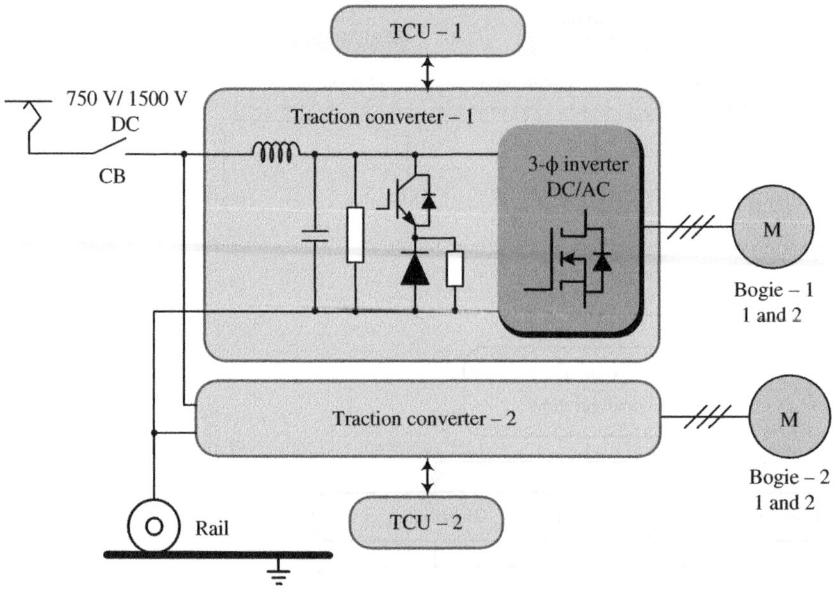

Figure 11.4 Typical configuration of LRV rolling stock.

11.2.2.1 Light Rail Vehicle (LRV)

DC electrification is best for this type of rolling stock as it necessitates rapid acceleration and frequent stops. Because there is no huge transformer and accompanying rectifier portion onboard, these systems are less expensive. As a result, the rail vehicle's weight is reduced, and passenger capacity is increased. Figure 11.4 shows a powertrain for light rail vehicles (LRVs) that uses a low-voltage DC of 600–750 V from the catenary or a third rail [8]. In several parts of the globe, a third rail is used since it has fewer electromagnetic interference (EMI) issues, requires less maintenance, and is more efficient than an overhead line [7]. Third rails are more compact than overhead wires and may be utilized in tunnels with smaller diameters, making them ideal for subway rolling stock. However, operational voltages are restricted to 1 kV for safety concerns [7]. The DC/AC TRC receives electricity from the DC supply grid via line filters and delivers the needed voltage and frequency wave shape for the propulsion motors' performance [9]. To decouple the differential power and absorb the harmonic currents, the line filter serves as a resonance filter. The number of traction motors and drive converters is determined by the power-rating requirement. Each TRC is controlled by a traction control unit (TCU). When the DC-link voltage exceeds the rated value, the braking chopper activates and discharges through braking resistors. When just a DC overhead line is available, this configuration is utilized for DC EMUs, metros, trams, and small-scale rail vehicles [9].

11.2.2.2 Heavy Rail-Diesel Locomotive

The diesel locomotive has self-propelled electrical systems that include two synchronous generators, several electric motors for propulsion, and a diesel engine as the principal power source as shown in Figure 11.5. A power conversion system converts electrical power generated by the alternator and sends it to an electric traction motor. The conventional power design of 4500-hp diesel

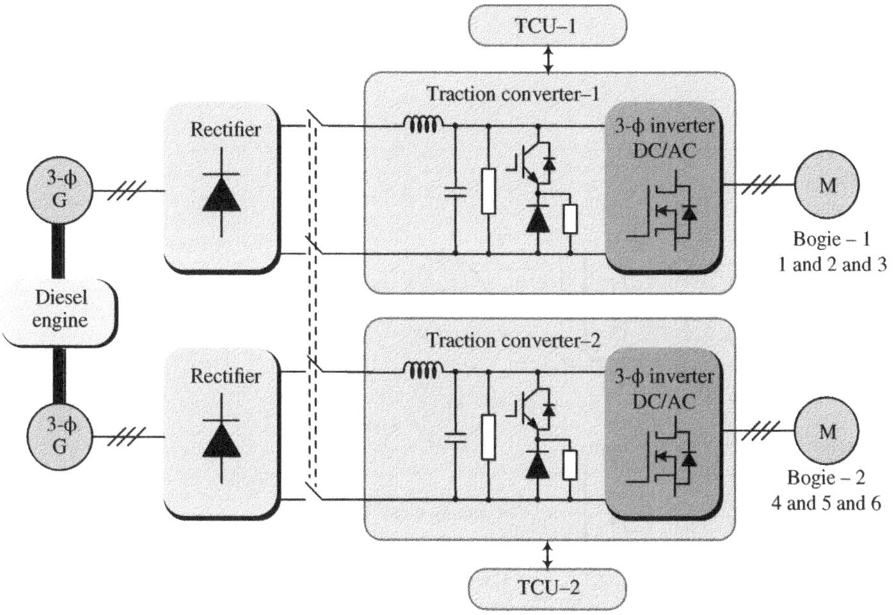

Figure 11.5 Typical rolling stock configuration of 4500 hp diesel locomotive.

locomotives is seen in Figure 11.5, in which six motors are pushed by two separate TRCs, each with a TCU for control operations. Field control of synchronous generators can provide DC-link voltage flexibility. In non-electrified sectors with long-distance uses, diesel locomotives are preferable [10]. This rolling stock architecture has the problem of being inefficient and relying on non-renewable fossil fuels as its primary source of energy. During downhill running and braking, this train does not present any opportunities for regeneration. The US Environmental Protection Agency (EPA) enacted requirements to reduce line-haul NOx from 5.5 to 1.5 (g/hp-h) to achieve Tier IV (2015-present) pollution limits [11]. General Electric (GE) ET44C4 (4400 hp), Electromotive Diesel (EMD) F125 (4700 hp), and Motive Power and Equipment Solutions MP-1500 "Greenville" are all developing low-emission diesel locomotives. Different firms, such as GE, EMD, and Bombardier are developing new diesel locomotives with power ratings of above 6000 hp. Further research is being conducted to have the technology for regenerative braking, to reduce carbon footprint according to regulations required by the US EPA [7].

11.2.2.3 Heavy Rail-Electric Locomotive

IGBT-based IM drives are the most common in heavy rail applications, and their configuration is shown in Figure 11.6. A pantograph collects current from the AC overhead catenary, and the traction transformer's lowered secondary voltage is pushed into a 4-quadrant converter (FQC) [5]. Two FQCs are connected in parallel by each group of TRCs. The FQC supplies variable-voltage and variable-frequency AC power to the traction motors through an intermediate circuit of a series resonant circuit that works as a second harmonic filter [12]. Many manufacturers are interested in using individual dedicated inverters to control parallel machines rather than installing a larger inverter for cost and reliability reasons [7]. Instead of bogie control, this enables axle control via tailored slip

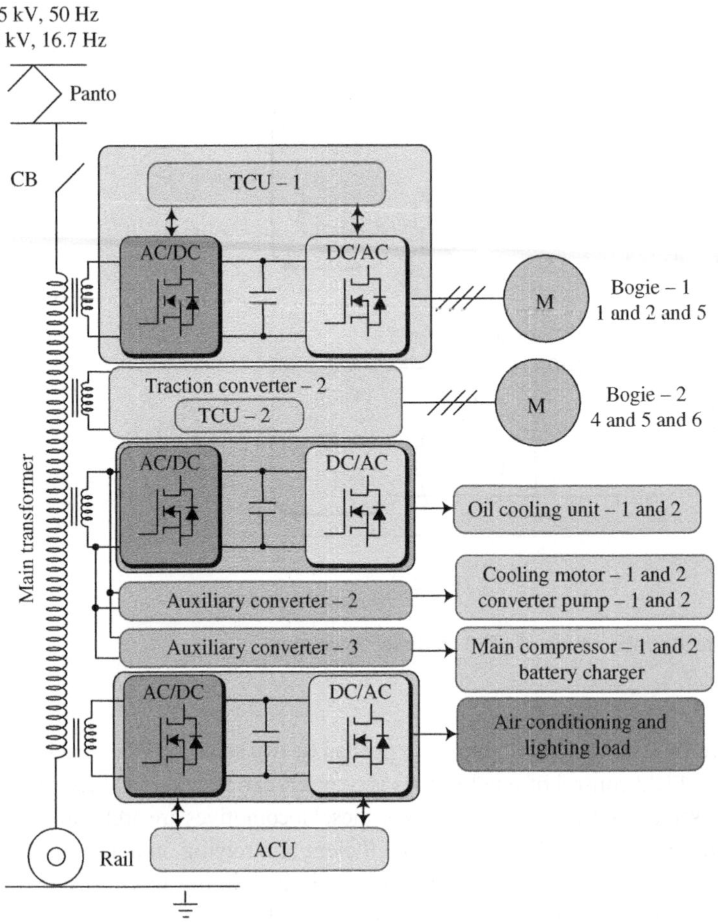

Figure 11.6 Typical rolling stock configuration of 6000 hp electric locomotive.

control of the motor. The braking chopper's function is similar to LRV. The choice of the cooling system varies significantly across different rolling stocks, depending on the needs of the rail car. In heavy rail systems today, water-cooling systems for IGBT-based power converters are prevalent [12].

11.2.2.4 Electric Multiple Units [EMUs] (AC or DC)

A distributed traction system is exemplified by EMU. It is made up of several motorized rail cars that make up the entire train. In contrast to locomotives, which take at least 30 minutes to reverse at the terminal, EMUs can do it in under two minutes. One motor coach (MC) and two trailing coaches (TCs) make up a typical unit. Figure 11.7 shows the arrangement of EMUs with 6, 9, and 12 car rake patterns. All of the electrical equipment in the MC is housed in a high-tension chamber, and motors are installed on the respective axles (powered axles). In TC, the axles are not motorized (powered). This indicates that these axles do not have motors. As shown in Figure 11.8, the DC chopper stage is placed in the traction drive train to give power to auxiliary loads unlike locomotives [7]. Two traction motors are powered by each TRC (two axles per bogie). This arrangement is common in high-speed, suburban, and metro trains.

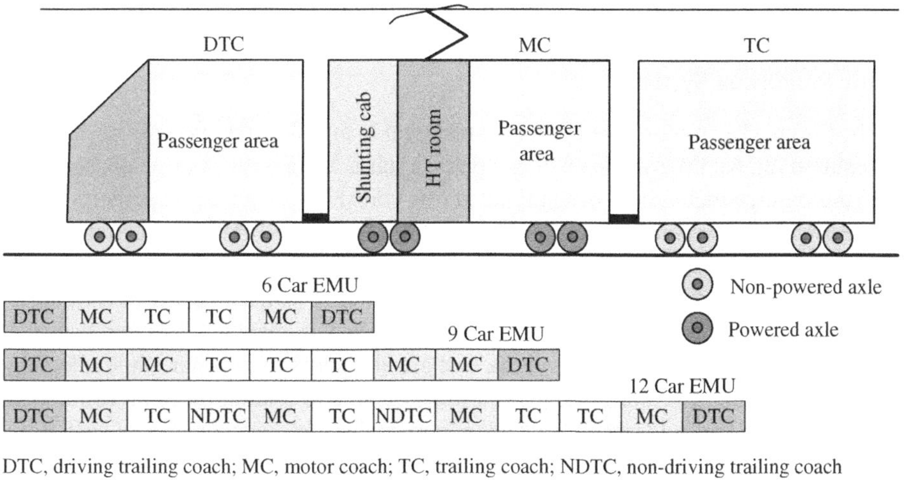

Figure 11.7 Typical rolling stock configuration of 1600 hp AC EMU.

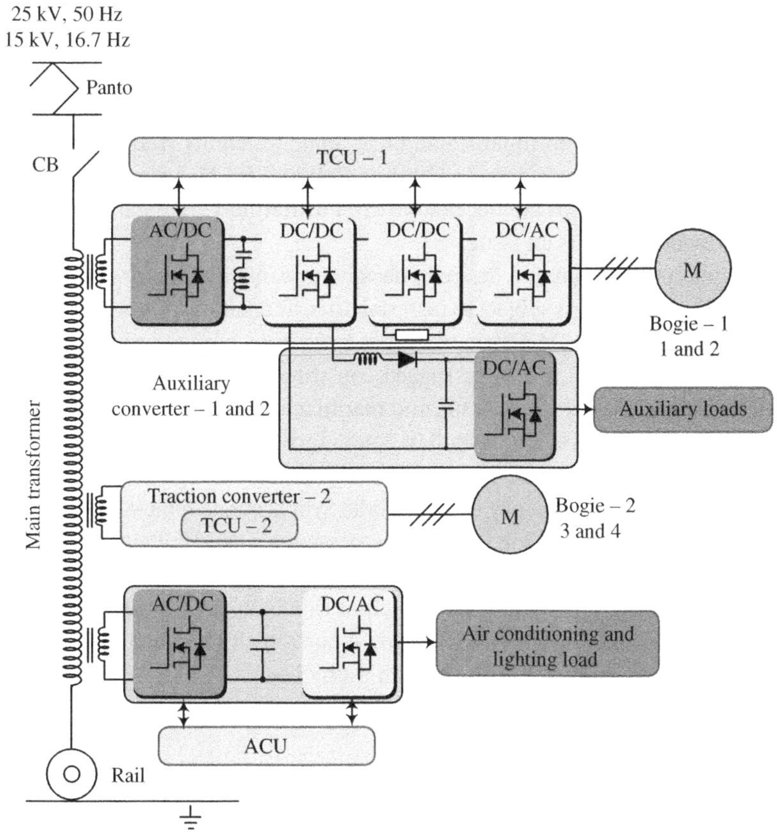

Figure 11.8 Typical rolling stock configuration of 1600 hp AC EMU.

11.3 Sub-Systems and Components of Rolling Stock Architectures

11.3.1 Electric Propulsion Systems

Railways switched from a DC motor-based propulsion system to an IM system because of the little maintenance required [2]. An IM is made up of two parts: a stator and a rotor. A three-phase winding is housed in the stator, which is built of a laminated iron core. Form-wound coils constructed of rectangular copper wires are separated by an insulating material in the three-phase winding. The stator also has an open-slot layout to accommodate wound winding. As a result, the rotor is likely to have a lot of slot ripple and surface losses. A laminated core material with squirrel-cage winding is also used in the rotor. Electrical loading, power density, efficiency, magnetic loading, acoustic noise, fault tolerance, mean time to failure and the cost are all factors to consider while selecting and building traction motors. Furthermore, motors must be able to tolerate extreme conditions such as dampness, vibrations, and shocks. Traction motors have a power rating of less than 1000 kW and a speed range of 2000–3000r/min (without gear). Synchronous motors (SMs) are utilized on occasion, such as in the TGV in France [5]. IMs have been widely used in traction propulsion systems for the past two decades [13]. Furthermore, because of the inherent slip of IMs, which equalizes discrepancies in the motor's rotational speed induced by wheel diameter differences, it permits a group drive (more than two motors) powered by a single converter.

Permanent magnet SMs (PMSMs) have taken the place of IMs in newer rail vehicles such as the Tokyo subway for the last 15–20 years [14]. PMSMs eliminate the intrinsic rotor losses, resulting in increased power density, efficiency and power factor. On the contrary, PMSMs have some drawbacks, such as magnet demagnetization due to high currents as well as high temperatures, the need for one TRC per motor, and special attention to fault handling. Some feasibility studies on commuter trains are conducted for forming group drives (with two PMSMs) fed by a single inverter, and it was discovered that extra control loops are necessary [15]. Furthermore, rare-earth magnet prices are variable, and worldwide supply is an issue.

Due to the volatility of magnet prices, there is a desire to design more efficient motor configurations with a higher power density that are less subject to material price fluctuations. Switched reluctance machines (SRMs) are one of the PMSM alternatives that have been tried for rolling stock. SRMs have the merits of being economical, having a rugged construction, being able to operate at high temperatures, being straightforward to assemble and manufacture, and having a constant power speed ratio of up to seven [16]. However, SRMs have some intrinsic limitations, including high torque ripple, high vibrations, the requirement for a six-lead connection, high acoustic noise levels, complex management, and poorer efficiency than PMSMs. Synchronous reluctance motors (SynRMs) are thought to be a good substitute for IMs as they possess a high torque density and efficiency. However, the power factor is poor and the speed of operations is limited [17]. Multiphase motors have also received some interest due to their fault-tolerant capability and lower space harmonics. Hub motors such as PMSMs and SRMs are ideal for low-floor vehicles because they have a low moment of inertia and negligible gear losses, resulting in a tiny bogie.

11.3.2 Power Converter Systems and its Components

A TRC is an essential component of the rolling stock, which is used to drive the traction motor based on speed, torque, and flux requirements. As discussed earlier, it comprises FQC, a DC chopper and an inverter. The configuration of the TRC depends on the TPS and the type of the rolling stock. The decision to propel multiple traction motors in parallel with each power converter or to assign a separate power converter to each machine is one of the key discrepancies in this fundamental configuration of the rolling stock. While there are cost savings to be had by using a larger power converter to excite paralleled machines, the use of individual dedicated converters allows for improved traction characteristics by allowing each traction motor to have its slip control. The number of paralleled traction motors generated by each three-phase inverter, and the number of single-phase DC/AC inverter units paralleled to supply the DC link, are examples of these variances. The FQC utilizes the full-bridge (h-bridge) converter, which is connected in parallel to meet the current requirements [7].

The power semiconductor device is a building block of TRCs. During the early 1990s, GTO power switches with ratings of 4.5 kV and 4 kA turnoff current were the favored power semiconductor devices for railway traction systems. However, the GTO-based TRCs are often limited to converter switching frequencies of 500 Hz or less as they demand very complex gating circuits, which can lead to reliability issues [3]. Over the past years, due to technological advancement of high-voltage (HV) IGBTs, traction drive manufacturers have chosen IGBTs for their newest semiconductor device to benefit from greater switching frequencies and simpler gate drives. Furthermore, these devices exhibit optimal characteristics, low power losses, ease of gate control, and snubber-less operation [5]. The IGBTs with ratings of 3.3 kV and 1.5 kA are readily accessible, while new devices with greater voltage ratings up to 6.5 kV at 1.2 kA are now being commercially produced by semiconductor manufacturers. These IGBTs enable pulse width modulation (PWM) switching frequencies to be increased from 500 to 1500 Hz for railway applications, reducing harmonic currents at both the input and output while attaining more than 98% inverter efficiencies [2]. As a result of these technological advances, various manufacturers have developed tiny IGBT-based traction inverter units with power ratings ranging from 1000 to 1500 kVA. Depending on the power converter topology chosen, these units' DC bus voltage is normally between 1.5 and 3.5 kV. Because of their known durability, traction inverters have historically preferred rugged press-pack power switch packages. In recent times, wire-bonded power devices were used in the popular plastic power module packaging. This significant shift was made possible by significant improvements in wire-bond ruggedness, which were proven through thorough extensive reliability testing of IGBT modules [2, 7].

Over the past few years, several traction power converters are developed including voltage source converters (VSCs), current source converters (CSCs), and power converters without DC link (e.g. matrix converters and cyclo converters). Among them, the VSCs are quite popular due to their flexibility in design and control and ease to meet the drive requirements compared to other topologies. Traction drive manufacturers have developed the inverter power circuit topology to take one of two forms: the standard two-level (2L) or the multilevel power converters [18]. The 2L-VSC may require devices in series to attain desired medium-voltage range depending on the semiconductor blocking voltage availability. In such cases, equal voltage sharing among devices necessitates the use of voltage equalization circuits during the blocking mode and switching transients. The power losses in power converters are increased by these additional circuits [18].

On the contrary, the traction power converters, which use low blocking voltage active switches, known as multilevel converters (MCs), constructively meet the requirements including low dv/dt, low common-mode voltage (CMV), low voltage harmonic distortion, and high efficiency owing to low switching losses and elimination of output filters. The most popular and commercially available MCs are neutral-point clamped (NPC), flying capacitor converter (FCC), and cascaded H-bridge for an operating voltage of 2.3–4.16 kV. The capacity to almost double the DC-link voltage for a given power semiconductor device voltage rating is the main attraction of the three-level NPC arrangement. This allows 3.3-kV IGBTs to be used in NPC converters with a 3-kV DC-link voltage, allowing two-level GTO inverters constructed with 4.5-kV devices to be directly replaced [2].

The chosen cooling system has a significant impact on the weight and dimensions of the TRCs. In recent years, improved cooling technology on a variety of heat pipe configurations, water cooling

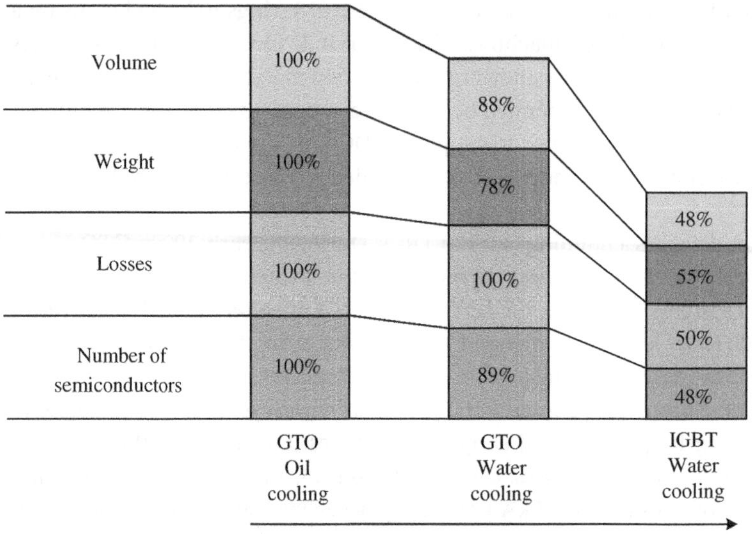

Figure 11.9 Advancement of propulsion converters in rail systems.

with an anti-freezing agent (glycerol), circulating coolant, evaporative cooling, natural-air and forced-air convection cooling has gained a lot of research attention. The choice of a cooling method is usually made based on system difficulties that are specific to the rail car requirements and the type of rolling stock. The advancement of electric propulsion converters in rail systems with the adoption of IGBTs and water cooling systems from traditional GTO-based converters is illustrated in Figure 11.9. It is noticed that the losses are reduced by half while reducing the volume and weight requirements.

11.3.3 Auxiliary Power Systems

Both heavy rail locomotives and EMUs have auxiliary loads including oil-cooling units, compressors, traction motor scavenger blowers, control power supply units, converter water pumps, and cooling motors for main traction motors [19]. To energize all control cards, TCUs, driver display units (DDUs), gate drivers, and a power supply unit with an output voltage range from 36–110 VDC are required. In addition, for passenger comfort, consistent power is needed for hotel loads such as air conditioning, indication lamps, and interior lighting equipment [19]. The quantity of auxiliary converters needed is determined by the kind of rolling stock used and the auxiliary functions it performs. Each locomotive has three auxiliary converters that are directly connected main transformer's auxiliary winding (Figure 11.6), or to the traction unit through a down chopper (Figure 11.8) in EMUs and are controlled through an auxiliary control unit (ACU).

Contractors switch loads from one converter to another, resulting in equal load distribution and redundancy among auxiliary converters [19]. Compressors, battery chargers, motor loads, and dedicated auxiliary converters for hotel loads typically demand two or three auxiliary converters. Due to pantograph bounce or gap, auxiliary systems require quick recovery, neutral section traversal, DC-link voltage control and adequate v/f control when initiating compressor loads [20]. These loads are extremely important, and their reliability is paramount. For instance, control power supplies are critical to the performance and dependability of all control cards [20]. These auxiliary power supplies are constructed with galvanic separation between the catenary line and low-voltage AC and DC grids, either using a power converter with low-frequency transformer architecture.

11.3.4 Traction Drive Control

Effective wheel slip-slide control, electronic stability with rapid torque control across the entire speed range, and good field weakening control are typical railway traction drive requirements. Particularly, IMs are widely used in these traction drives due to their rugged construction and maintenance-free operation. Therefore, this section focuses primarily on the control of speed and torque/flux of IMs. However, the fundamental functions of these control techniques also apply to PMSMs. For GTO-based IM drives, direct self-control techniques were used [21]. The most popular traction drive control techniques are broadly categorized into scalar, vector, and predictive control, as shown in Figure 11.10. Without knowing the position of the voltage/current/flux space vectors, scalar control methods can adjust their magnitude and frequency. The volt per hertz (v/f) control is a simple and most commonly used method, where the ratio of stator voltage vector to frequency is maintained constant. However, these methods exhibit poor dynamic performance [7].

Vector control or field orientation control (FOC) approaches including stator-flux-oriented control, rotor field-oriented control, or indirect stator-quantities control are now used in IGBT-based IM traction drives owing to a good steady-state and transient response. These methods function by separating the flux and torque-producing current components from the motor stator currents, allowing the motor flux and torque to be adjusted independently. Since FOC is highly sensitive to motor characteristics, flux measurement and estimation must be precise [7]. There is a control function known as repowering control, which requires detecting rotor speed before starting the inverter. In such circumstances, the rotor position must be calculated, which can be done using a mechanical simulator method for beginning IMs at zero speed. In [22], a rotor-induced voltage is used to estimate the initial rotor position and rotor frequency in normal conditions. Furthermore, several sensorless techniques for railway traction are presented in the literature to improve reliability, eliminate sensor interface noise, and extend the motor temperature range [23]. FOC requires

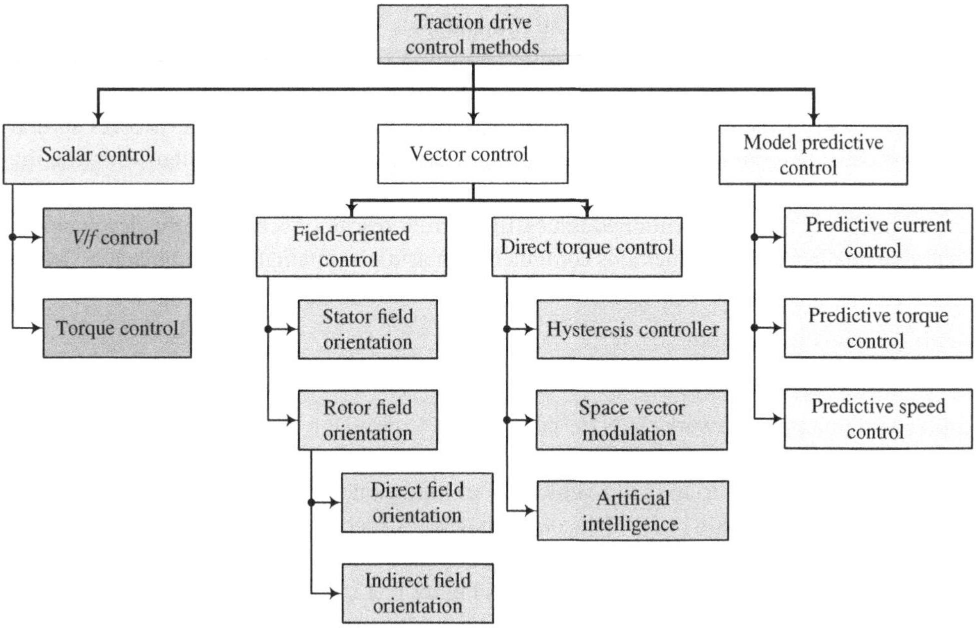

Figure 11.10 Classification of traction drive control strategies.

a fast microprocessor because the computation algorithm takes longer to perform and must be sensitive to the system as well as load variations.

On the other hand, direct torque control (DTC), as the name implies, controls the torque and stator flux directly without the use of intermediate current controllers [24]. As a result, the DTC outperforms the FOC in terms of dynamic performance. DTC characteristics include ease of implementation in digital control platforms and robust system operation. However, DTC performance suffers from high torque and flux ripple, the violence of polarity consistency rules, and variable switching frequency and is heavily influenced by motor speed, particularly at low speeds [24]. Furthermore, the digital implementation of hysteresis comparators necessitates a high sampling frequency [24]. To overcome the aforementioned concerns, new variants of DTC techniques including space vector modulation (SVM), deadbeat control, sliding mode controllers (SMCs), self-tuned linear controllers, and artificial intelligence-based controllers are introduced.

Model predictive control (MPC) is an advanced control approach employed in traction drives that can directly generate switching signals without the use of a modulator for the power converters with simple and intuitive nature [25, 26]. MPC is distinguished by the use of a system model to forecast the behavior of the control variables to be managed, which includes reference variable generation and extrapolation, control variable prediction using a system model, and system optimization utilizing cost function. Generally, MPC methods are divided into direct and indirect MPCs based on the methodology used to achieve the control objectives. A single cost function is sufficient to achieve all control objectives in direct MPC. To achieve all control objectives, the indirect MPC employs the cost function in conjunction with the classical control methods. The switching states are either directly or indirectly manipulated in the direct and indirect MPC methods. The former method results in variable switching frequency operation, whereas the latter method results in fixed switching frequency operation [27]. The MPC methods are further classified based on the primary control variables used in the optimization process. The MPC methods employed for traction systems, in particular, are derived from FOC and DTC methods and are listed in Figure 11.10.

The stator current controller loops in FOC are replaced with a predictive algorithm, and the resulting method is known as predictive current control (PCC) [27]. Instead of torque and flux controllers in DTC, the predictive algorithm is used to manage stator flux and torque. Hence, the resultant method is referred to as predictive torque control (PTC) [28, 29]. Similarly, a predictive algorithm can be used to control motor speed, and the resulting method is known as predictive speed control (PSC). Despite its popularity, the digital implementation of MPC involves several challenges such as appropriate tuning of weighting factors, the accuracy of systems models, the effect of system parameters variation, variable switching frequency and computational complexity [25–29]. Considering the aforementioned issues, the current research focused on the development of model-free predictive control methods combined with artificial intelligence and machine learning techniques for rolling stock.

11.3.5 Control Hierarchy of Rolling Stock

The train communication network (TCN) as per IEC 61375 specifies a communication architecture as well as the protocols required for non-vital communication on the train and at the vehicle level [30]. To meet the needs of inter-and intra-vehicle communication, it has a two-layered, hierarchical design. It comprises a wire train bus (WTB) connecting the rail vehicles and a multifunction vehicle bus (MVB) connecting the equipment on a train or group of rails as shown in Figure 11.11 [30]. The main control hierarchy of rolling stock is depicted in Figure 11.12. Vehicle control units (VCUs) are master controllers that interface with all other rolling stock components through a common bus, including the TRC, auxiliary converter, DDU, and remote input and output modules (RIOM).

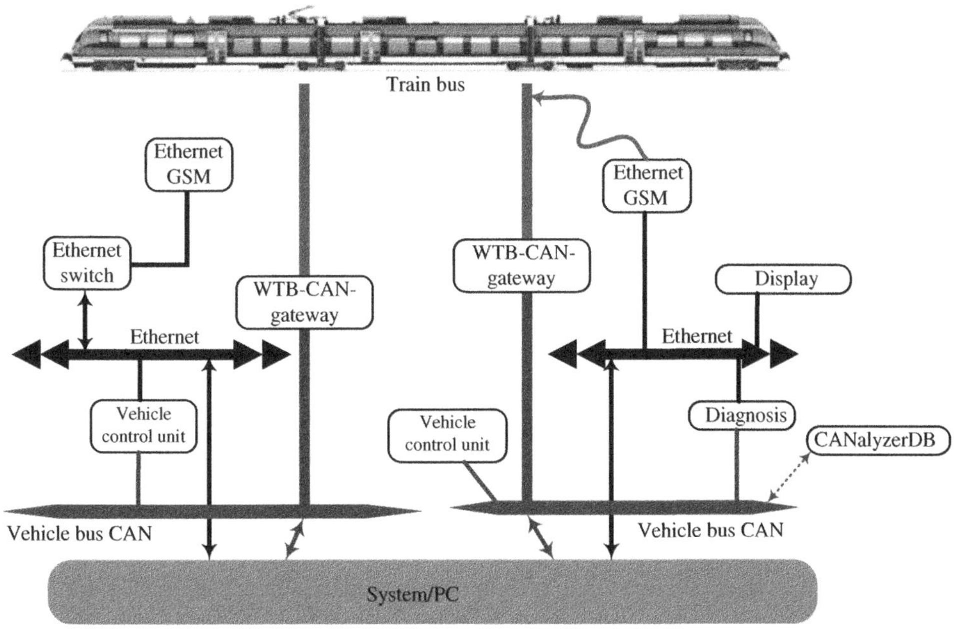

Figure 11.11 Communication links between various components of rolling stock.

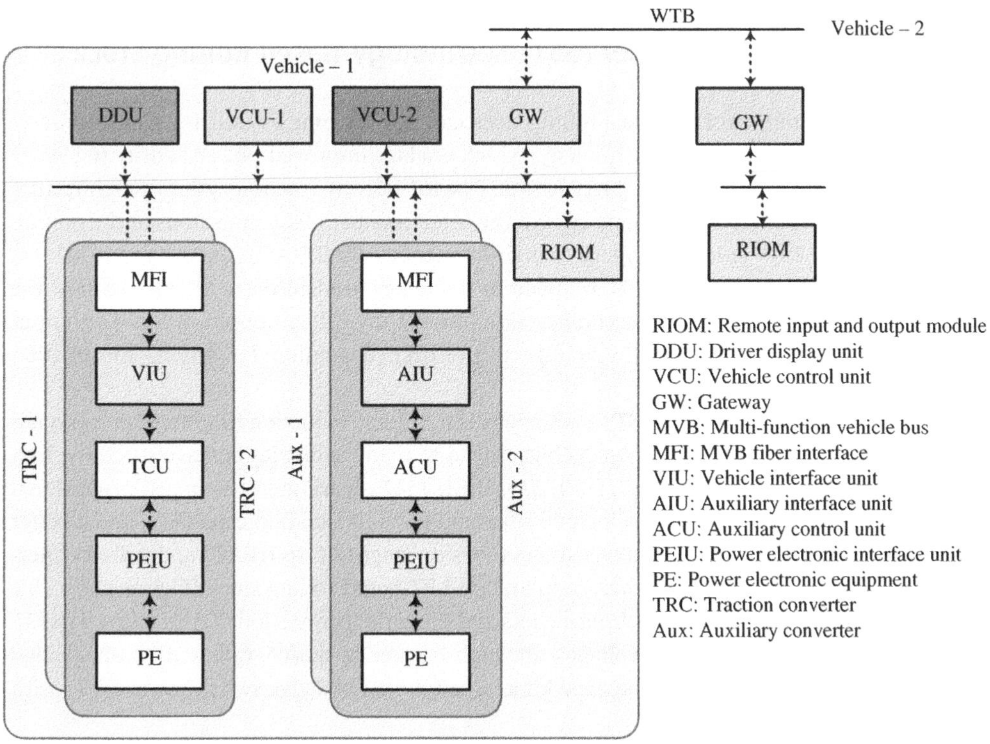

Figure 11.12 General control hierarchy of rolling stock. *Source:* [7]/IEEE.

Through fiber-optic connection, this bus may operate on the MVB protocol, controller area network (CAN) protocol, or Ethernet protocol within the car. For multi-vehicular operations, a WTB interacts between two rail vehicles via a gateway circuit [31]. Through the vehicle interface unit (VIU), VCU sends command signals to the TCU. The TCU gives proper PWM signals to power electrical (PE) equipment to manage the driving and braking operations of a motor. Similarly, the ACU manages the auxiliary operations of the vehicle. Figure 11.13 depicts the TRC's line, drive, and braking control actions in a systematic block diagram. The VCU's fault produces a diagnostic data record in the DDU, which contains information on the value and instance of the occurrence.

11.3.6 Standards and Regulations

Even though these systems are designed with high efficiency and are compactable, however, some specific regulations and standards are indispensable to achieving safe, punctual, and efficient operation of rolling stock. Therefore, it is essential to impose safety standards on the federal government and railway operators so that they can align specific requirements by meeting the standards. The main purpose of establishing the standards and regulations for ensuring safety, convenience, environmental countermeasures, reduction of production cost, and efficient operation of rolling stock. The standards and regulations are measured by the authorities and are classified for the systems as shown in Figure 11.14.

Table 11.2 summarizes the appropriate IEEE, European Norm (EN), and International Electrotechnical Commission (IEC) standards and technical codes for rolling stock design, safety, maintenance, communications, electric power supplies, train control, and testing [7, 32].

11.4 Solid State Transformer (SST) Technology-Based Rolling Stock

The constraints of traditional LFT-based rolling stock can be overcome by utilizing a new technology known as SST traction (SSTT) technology, which enables improved power density (0.5–0.75 kVA/kg) [9]. As illustrated in Figure 11.15, the SST-based railway traction system's architecture comprises cascaded connections of many cells at the input and parallel connections at the output. The AFE converts the HV catenary voltage to high-frequency AC (HF-AC) (>400 Hz to 10 kHz) at line frequency (16.7 or 50 or 60 Hz). A medium-frequency transformer (MFT) transfers the medium-frequency AC voltage to the secondary side without any voltage amplification. The output of this converter is used to adjust the voltage to the desired amount for driving the traction motors [33].

The most important advantage of SSTT technology is harmonic reduction (improved power quality) due to higher operating frequency, which is critical as many countries do have strict line harmonics requirements as per standards (IEEE 519, IEEE 1547 (distributed resources)). Additional benefits of SST-based traction systems include adaptability, fault isolation capability, and a better multi-voltage interface. It necessitates an extra conversion stage (AC to HFAC) and a direct interconnection of AFE to HV catenary when compared to LFT-based rolling stock (Figure 11.6). This necessitates the use of HV power-switching devices and the series connection of AFE converters [3]. Furthermore, MFT's terminal side voltages are high-frequency pulsed rather than sinusoidal. SST based systems are costly due to the need for a large number of active switches and MFTs with sophisticated cores.

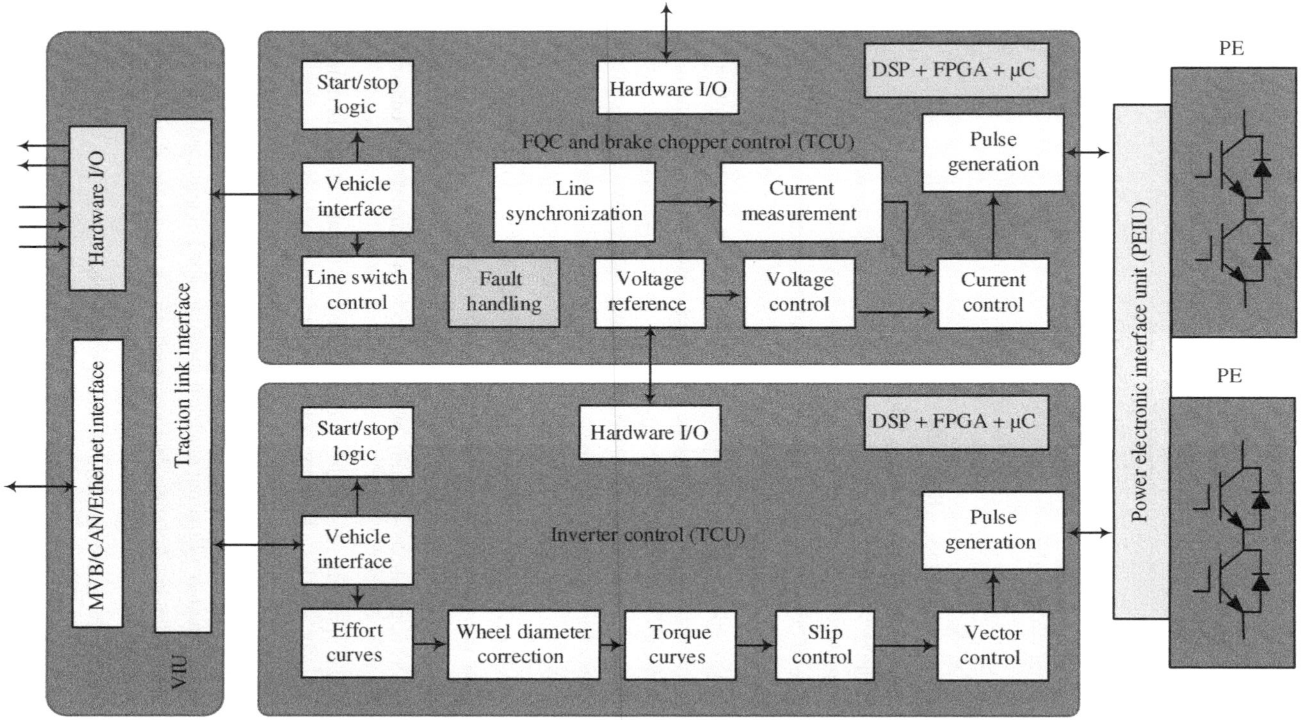

Figure 11.13 Traction control unit (TCU) with vehicle interface unit (VIU) of rolling stock. *Source:* [7]/IEEE.

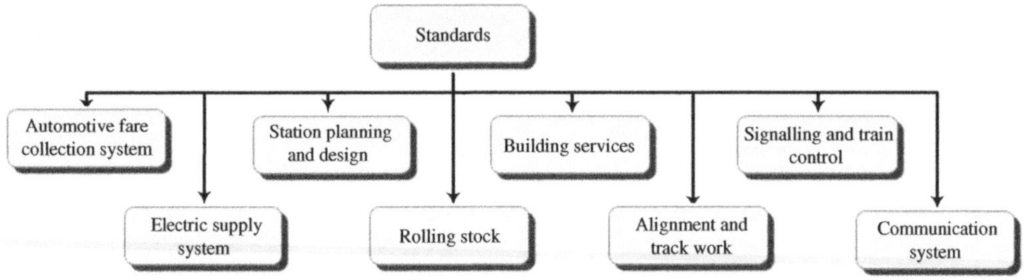

Figure 11.14 Classification of standards for rolling stock.

Table 11.2 Technical standards applicable for rolling stock.

Standard	Description
IEC 62278	Railway application's specification and demonstration of reliability, availability, maintainability, and safety
IEC 61508	Functional safety of electrical/electronic/programmable electronic safety-related systems
EN 14752	Railway applications – Body-side entrance systems for rolling stock
EN 13272	Railway applications – Electrical lighting for rolling stock in public transport systems
EN 45545	Fire protection on railway vehicles
EN 14750	Air conditioning for urban and suburban rolling stock
EN 16186	Railway applications – Driver's cab
BS EN 13262	Railway applications – Wheel sets and bogies, wheels, product requirements
EN 13749	Railway applications – Wheelsets and bogies – Method of specifying the structural requirements of bogie frames
EN 13103	Railway applications – Wheelsets and bogies – Non-powered axles – Design methods
EN 13452	Railway applications – Braking – Mass transit brake systems
EN 50155	Railway applications – Electronic equipment used on rolling stock
EN 50207	Railway applications – Electronic power converters for rolling stock
EN 60529	Degree of protection provided by enclosures (IP code)
EN 12663	Railway applications – Structural requirements of railway vehicle bodies
IEC 61133	Railway applications – Testing of rolling stock on completion of construction and before entry into service
EN 14363	Railway applications – Testing and simulation for the acceptance of running characteristics of railway vehicles – Running behavior and stationary tests
IEC 61377	Railway applications – Rolling stock – Combined test method for traction systems
IEC 61287	Electronic power converters onboard for rolling stock
IEC 60349-1	Electric traction – Rotating electrical machines for rail and road vehicles
IEC 60349-2	Electric traction – Rotating electrical machines for rail and road vehicles – Part 2: Electronic converter-fed alternating current motors
IEC 60571	Rules for the electronic control part of converters
IEC 61375	The train communication network (TCN) for rolling stock

Table 11.2 (Continued)

Standard	Description
IEC 61373	Stock and vibration test for onboard equipment
EN 50125-1	Environmental conditions for the equipment onboard
IEC 60077	Electronic equipment for rolling stock
IEC 61881	Capacitors for power electronics for rolling stock
IEC 60310	Traction transformers and inductors onboard
IEC 61991	Protective provisions against electrical hazards
IEC 62236	Electromagnetic compatibility for traction equipment
IEC 61709	Reliability of electronic components for rolling stock
IEC 62928	Onboard lithium-ion traction batteries for rolling stock
IEEE 1476	Passenger train auxiliary power system interface
IEEE 11	Rotating electric machinery for rail and road vehicles
IEEE 1482.1	Rail transit vehicle event recorders
IEEE 16	Electrical and electronic control apparatus on rail vehicles
IEEE 1473	Communications protocol abroad passenger trains
IEEE 1478	Environmental conditions for electronic equipment
IEEE 1483	Verification of virtual functions in processor-based systems
IEEE 1475	Functioning of interfaces among propulsion, friction brake, and rain-borne master control
IEEE 1474.1	Communications-based train control performance and functional requirements
IEEE 1662.3	Design and application of power electronics in electrical power systems
EN 62864-1	Power supply with hybrid energy storage systems onboard

Source: Adapted from [7, 32].

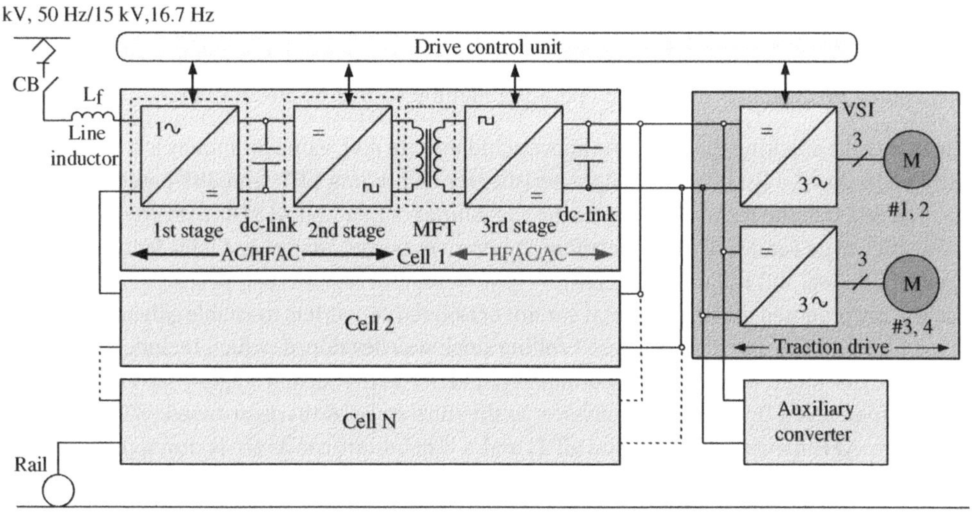

Figure 11.15 SST-based rolling stock architecture of modern rail vehicle (two-stage).

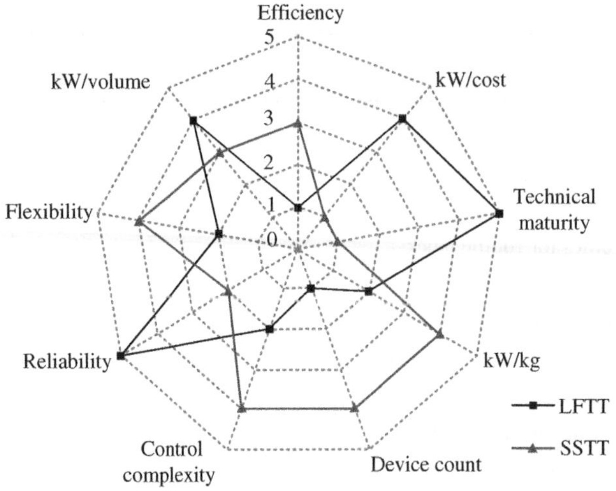

Figure 11.16 Comparison between LFTT and SSTT technology.

Along with cutting-edge developments in magnetic materials, advanced converter topologies and power-switching devices, further essential improvement requirements are summarized as follows [34]:

- More than 1 kVA/kg power density and 2–3% efficiency improvement of SST when compared to LFT
- Unity power factor at grid side with advanced power factor control (PFC) strategies
- Able to feed hotel and auxiliary loads apart from traction loads
- DC voltage stabilization at the inverter input during pantograph bounce or in neutral sections
- Reliability improvement and demands for no less than 150 kilo-hour mean time between failures (MTBF)
- Integrated cooling technologies for SSTT power conversion systems
- Effective improvements in regenerative braking are mandatory to save energy
- Power converters with a reconfigurable modular construction and control approach in the event of a failure in one of the sub-assemblies
- Depending on customer needs, the operating lifetime should be 20–30 years.

Finally, while ensuring that all of the above standards are met, value for money is a critical factor. Figure 11.16 shows a list of critical characteristics for comparing LFT- and SST-based traction systems. The SST traction technology covers the capabilities of LFT and AC/DC converter systems that provide higher power density. Furthermore, it improves power quality and adds features like fault current limitation and isolation [33–35].

McMurray proposed the fundamental variant of isolated converters to enable galvanic separation in 1968 [36]. In 1985, the first SST-based rolling stock was developed, which included a diode rectifier as the first stage, a thyristor-based matrix converter as the second stage, a 400 Hz MFT, and a diode rectifier, and a DC/DC boost converter as the third stage. A thyristor-based SSTT system that includes an AFE matrix converter, an MFT, and a 4-quadrant converter is developed [37]. High switching loss, high line harmonics, and low switching frequency operation plague this thyristor-based converter. The AFE power converters must be connected to the HV side (15-kV, 25-kV), and a series connection of these structures is required due to the maximum blocking capability of power switches (6.5-kV until now). An input series and output parallel (ISOP) arrangement are formed by connecting AFEs in series at the input and parallel at the output. Assuming the same drive design, a maximum of three conversion stages are required from single-phase AC to DC power

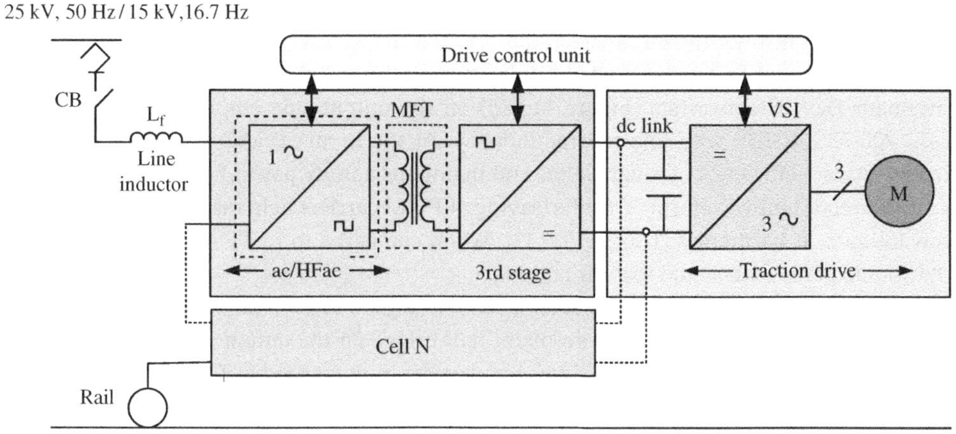

Figure 11.17 Single-stage AC/HFAC power conversion architecture for SSTT technology.

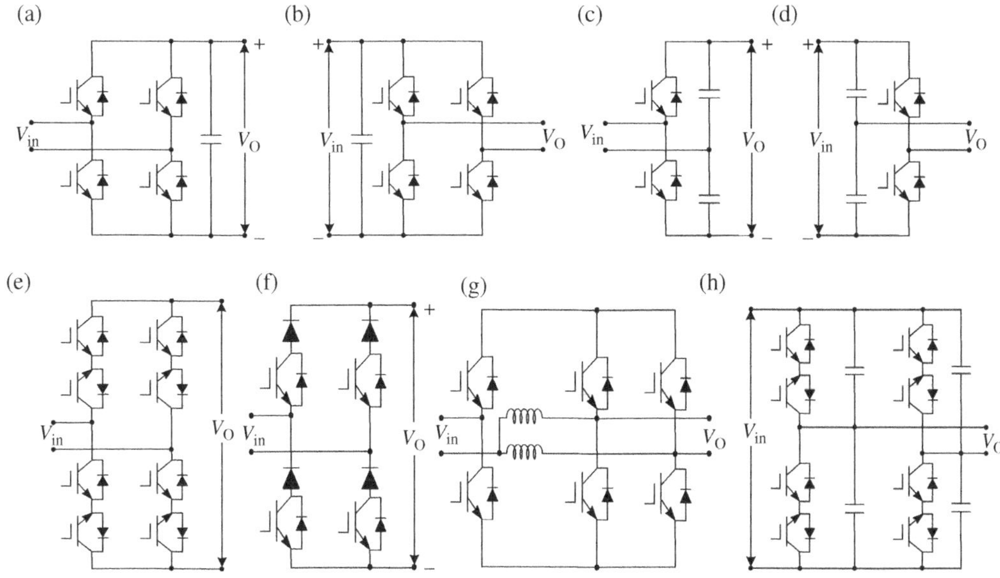

Figure 11.18 Power conversion topologies: (a) Full-bridge AC/DC or HFAC/DC, (b) full-bridge DC/HFAC, (c) half-bridge HFAC/DC, (d) half-bridge DC/HFAC, (e) AC/HFAC matrix converter, (f) AC/HFAC or HFAC/DC CSC, (g) AC/DC three-state switching cell (3SSC), (h) hybrid dual-active bridge AC/HFAC.

conversion (Figure 11.15). They can be classed as two-stage (AC/DC/HFAC) (Figure 11.15) or single-stage (AC/HFAC) (Figure 11.17) based on conversion stages from AC to HFAC (assumed 3rd step is typical). Figure 11.18 depicts the classification-based grouping of power converter topologies.

11.4.1 Two-Stage (AC/HFAC) Power Conversion Topologies

The two-stage conversion system comprises (i) single-phase AC/DC conversion (AFEs) as stage 1, and (ii) a second harmonic resonant filter and a DC/HFAC conversion as stage 2. The ISOP configuration-based SSTT system was first designed for 15 kV/16.7 Hz railway applications using a cascaded H-bridge arrangement with HV-IGBTs [38].

ABB designed and tested a 1.2 MVA SST-based traction drive on a shunting locomotive operated by Swiss Federal Railways (SBB), Switzerland. A 150 kVA SSTT configuration shown in Figure 11.19 is composed of 9-cell cascaded full-bridges on the AC side (Figure 11.18a) and half-bridge resonant DC/DC converters (Figure 11.18d) on the output side employing 6.5 kV/400 A and 3.3 kV/800 A Si-IGBTs with one cell redundancy. At the input HV-AC side, this prototype has a traction system efficiency of roughly 96% and maintains a unity power factor. The AFE converter's multilevel output waveform and interleaving of PWM carriers help meet rolling stock specifications for current harmonics (IEEE 519) [33]. When compared to LFTT technology, this has higher harmonic performance and extra functions.

ALSTOM designed a 2 MVA SSTT system configuration (Figure 11.19) similar to ABB, with 8-cell cascaded H-bridges on the AC side and a resonant half-bridge on the output side (Figure 11.18a) using 6.5 and 3.3 kV Si-IGBTs respectively [34]. Bombardier built a 5 MW SSTT system with 8-cell full bridges (Figure 11.18a) on a grid side and a full-bridge series resonant DC/DC converter (Figure 11.18b) operating at an 8 kHz switching frequency [38]. The sole difference between these prototype designs is the output stage, which uses a full-bridge topology rather than a half-bridge as in the ABB prototype. Two-stage SSTT-based power conversion topologies are very popular in terms of manufacturability and controllability. This structure enables each converter step to be developed and optimized independently.

Auxiliary power supplies, other sources, and loads can also be assigned to HV and LV ports using this approach. These constructions necessitate bulky aluminum electrolytic capacitors in the DC links, which is one of the most important components that affect the converter's reliability [39]. These topologies have a second harmonic ripple that necessitates the use of an LC resonant filter. As a result, DC capacitors are oversized, and the TRC consumes more energy.

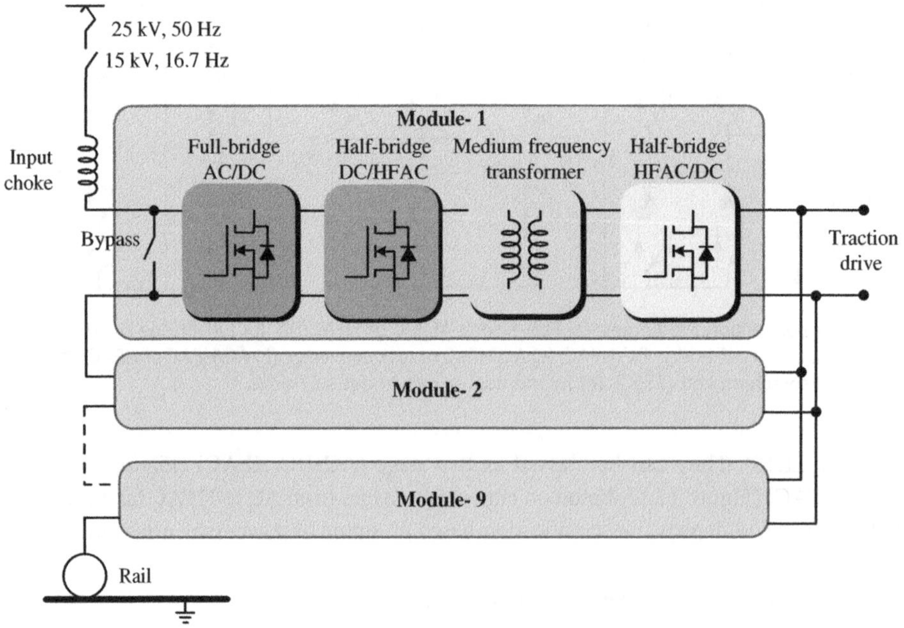

Figure 11.19 The two-stage conversion PETT architecture of 9-cell PETT developed by ABB.

11.4.2 Single-Stage (AC/HFAC) Power Conversion Topologies

The single-stage configuration in SSTT systems will be a viable option for achieving higher efficiency and lower cost owing to fewer power conversion stages. ABB's first 1.2 MVA SSTT prototype, used cyclo-converters for direct power conversion and had a nominal 1.8 kV DC-link voltage for a 15 kV/16.7 Hz EMU application [40]. ABB claims a 3% increase in efficiency, as well as weight and volume reductions of 50% and 20%, respectively. The University of West Bohemia (Czech Republic) implemented a 4 kVA SSTT system comprising a cascaded connection of single-phase matrix converters (Figure 11.18e) to the primary winding of an MFT as illustrated in Figure 11.20 [41]. North Carolina State University in the United States developed SSTT comprised of current source-based full-bridge converters (Figure 11.18f) with a medium-frequency fly-back inductor (MFT) [42]. This inductor functions as a galvanic isolation and voltage adaptation MFT. ALSTOM developed a 2 MVA 12-stage SSTT system with this topology [43]. In this prototype, the phase-controlled converter employs 6.5 and 3.3 kV IGBTs on the primary and secondary sides, respectively.

A new converter variant known as modular MC is made up of several submodules (SMs) called arms (Figure 11.21). The primary features of MMCs are modularity, good adaptability, scalability and fault-tolerant capability. By charging and discharging SM capacitors, these converters enable bidirectional energy flow. SMs having higher capacitor voltages will be used when motoring to achieve the desired arm voltage, and vice versa in braking mode. All SM capacitor voltages or stored energy must be kept constant. or within a tight tolerance zone for MMC to operate reliably and efficiently. This structure is made up of 4N similar arms that contribute to the structure's robustness and ease of maintenance. The frequency can be adjusted to the point where a modulation index is an even number, which aids in the eradication of even harmonics and thus eliminates the need for the second-order harmonic LC filter. Siemens implemented a 5 MW SSTT system that uses the

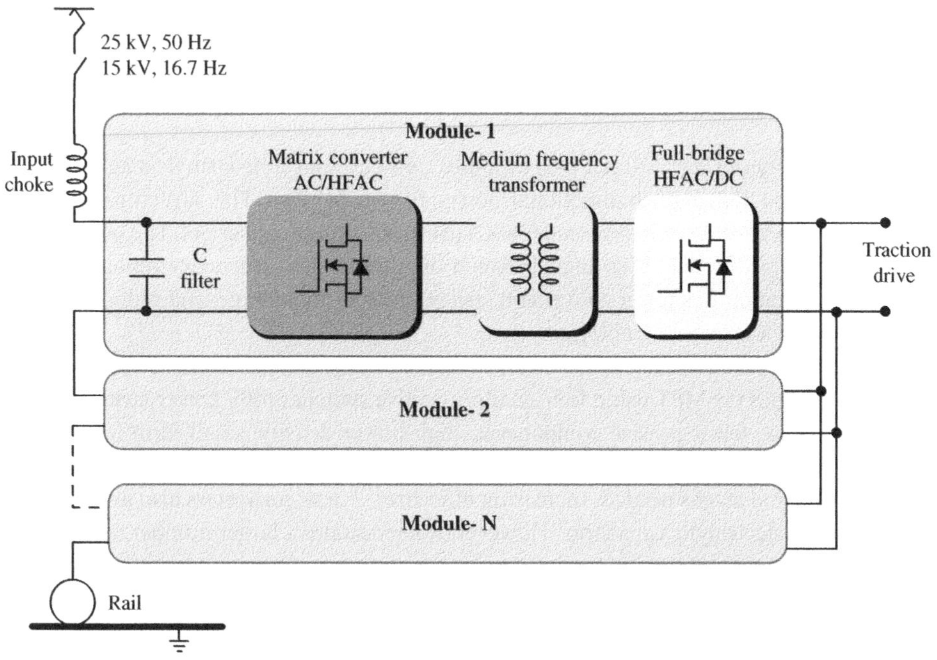

Figure 11.20 Single-stage SSTT using matrix converters developed by West Bohemia.

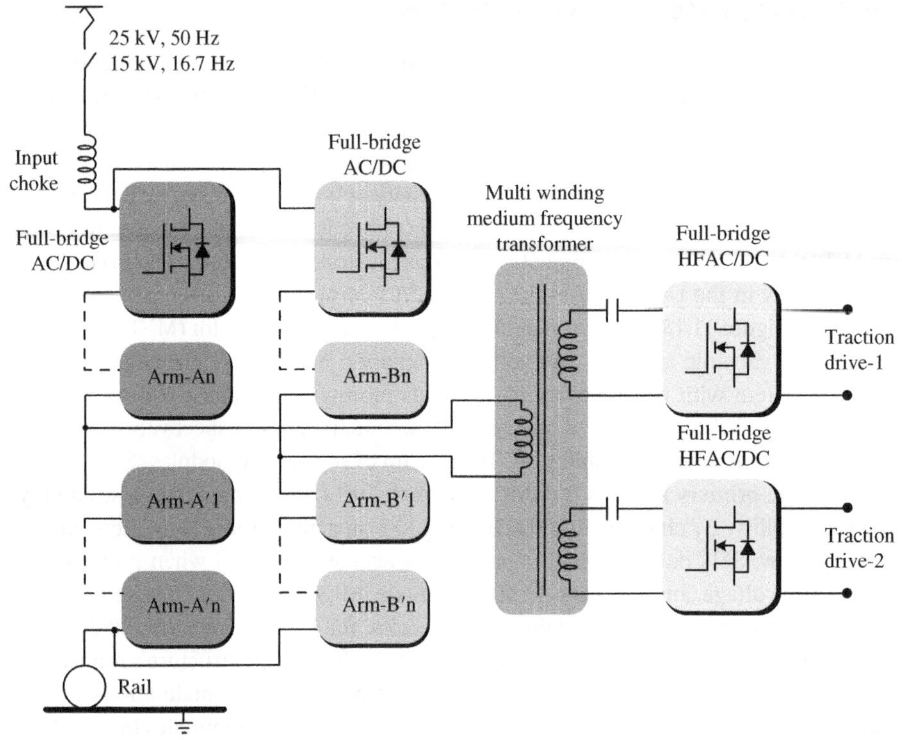

Figure 11.21 Single-stage PETT architecture using MMC developed by Siemens.

MMC architecture in combination with a concentrated multi-winding MFT. It employs 17-level MMCs made up of 1.2 kV/400 A IGBTs, which offers an efficiency of nearly 98% at a 10 kHz switching frequency [44]. However, MMC architecture possesses some serious disadvantages such as high active switch count, limited switching frequency functioning, and non-intuitive modulation and control.

The Federal University of Ceara, Brazil [45] proposes and develops a new single-stage AC/DC cascaded MC (Figure 11.18g) with bidirectional power flow capability. This structure is based on the formation of a three-state switching cell (3SSC) using full-bridge converters. The power flow can be controlled by adjusting the phase angle between the primary and secondary voltages across the MFT. This configuration has fewer conversion stages, is easier to manage, and reduces power loss when compared to alternative topologies.

An AC/AC hybrid dual active bridge (Figure 11.18h) based SST system that consists of full-bridge and half-bridge topologies via MFT using four-quadrant active switches [46]. Lower switch count, bidirectional power flow, fewer passive components, high-power density, small form factor, and high reliability are all advantages of these systems. Finally, single-stage converters cut down on the number of conversion stages needed, increasing efficiency. These converters also avoid bulky life-limited aluminum electrolytic capacitors. However, it necessitates a larger number of switches and a more complicated control system. It also imposes constraints on their optimization and may necessitate the use of specialized switches such as reverse blocking IGBTs (RB-IGBTs). Table 11.3 summarizes a prototype comparison of various SSTT systems implemented by rolling stock manufacturers and suppliers [34, 43].

Table 11.3 Prototype comparison of SSTT systems by equipment manufacturers.

Manufacturer	1st stage	2nd stage	MFT type	3rd stage	Frequency
ABB (1.2 MVA)	16-cell cyclo converters	–	1 : 1 multiple cells	Full-bridge	400 Hz
ABB (1.2 MVA)	8+1-cell cascaded full-bridges	Half-bridge	1 : 1 multiple cells	Half-bridge	1.75 kHz
Siemens-1 (5 MW)	MMC	–	Single multi-winding	Full-bridge	–
Siemens-2 (2 MW)	17-level MMC	–	Single multi-winding	Full-bridge	10 kHz
Bombardier (5 MW)	8-cell cascaded full-bridges	Full-bridge	1:1 multiple cells	Full-bridge	8 kHz
Alstom-1 (2 MVA)	12-cell full-bridge	Half-bridge	Single multi-winding	Half-bridge	2 kHz
Alstom-2 (2 MVA)	CSC	Half-bridge	Single multi-winding	Half-bridge	5 kHz

11.4.3 Auxiliary Systems for SSTT Systems

For reliable functioning, all auxiliary loads must be fed and the vehicle service batteries must be recharged. For charging the vehicle service battery, this needs an onboard generation of 415 V AC with 5–10% fluctuation at the output and a low-voltage DC power supply with 1–2% ripple [1]. The configuration of an SSTT-based auxiliary converter is shown in Figure 11.22. Due to the presence of nonlinear loads, the quality of the auxiliary AC supply voltage waveform is limited, which could lead to low-frequency current harmonics.

A sine wave filter's cost, maintainability, and reliability are always a compromise, according to regulation criteria [19]. A DC/DC buck converter is commonly suggested for a battery charger due to the low voltage of the batteries [20]. Only unidirectional flow and precise coordination management between the traction and auxiliary systems are required by these converters, particularly

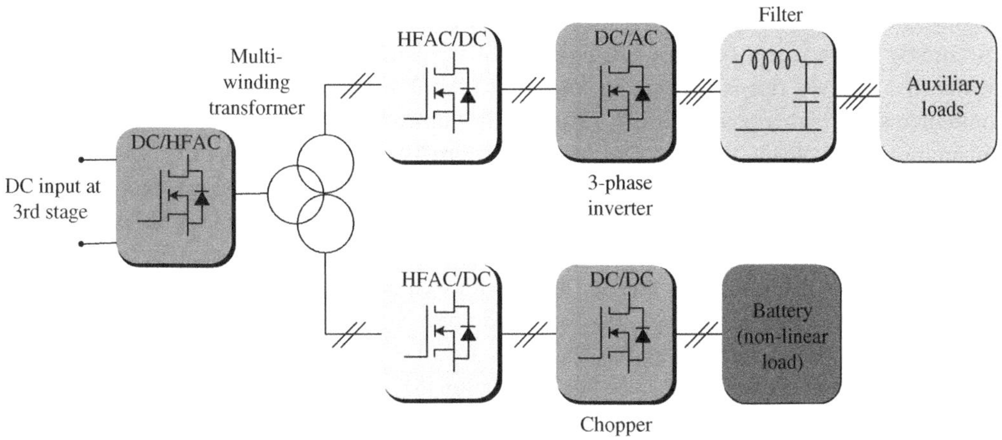

Figure 11.22 Configuration of SSTT-based auxiliary converter.

during neutral section traversals. Even though SST-based auxiliary converter configuration achieves a large boost in power density, there are still issues with increasing stages and switching frequency.

11.5 Advancements and Challenges in Modern Rolling Stock

11.5.1 Semiconductor Technology and Cooling Systems

The silicon (Si)-based power semiconductor devices such as IGBTs at higher power levels suffer from poor thermal properties, resulting in lower efficiency and the need for a better cooling system. Also, these devices have higher on-state resistance leading to higher conduction losses and are capable of operating at a lower switching speed only. Mitsubishi Electric recently introduced a Si-based 6.5-kV/1000-A X-series HV-IGBT module with the lowest thermal resistance, and highest operating temperature (150 °C) of counterparts currently available [47]. However, operating temperatures and power densities of Si-based power devices are insufficient for the current power-switching device requirements, prompting research into the use of wide bandgap (WBG) materials. Recently, Hitachi developed a traction inverter in commuter train Japanese railways using SiC hybrid modules (Si-IGBT+SiC-Schottky barrier diode). This results in a reduction of the mass and the volume of the inverters by 60% [48]. WBG devices such as silicon carbide (SiC) and gallium nitride (GaN) are a superior solution for greater switching frequency operation. However, GaN devices are not seasoned for medium voltage and high-power applications. SiC devices can withstand high voltages, produce less thermal energy during operation, support high-speed switching, have a low voltage drop and improved switching properties, and have fewer switching losses, which assist to improve efficiency and reduce cooling requirements.

The world's first full-SiC-based 3-level diode clamped TRCs using 3.3 kV/1200 A was launched by Mitsubishi Electric Corp for N700 Shinkansen bullet trains. This system minimizes total energy consumption by about 30% when considered along with regenerated energy from braking in a wide speed range [49]. The limitations in design utilizing Si-IGBTs with higher blocking voltages can be conquered by the application of SiC-MOSFET devices for SST-based rolling stock. As the conduction losses of SiC-MOSFET-based converter depend on turn-on drain-source resistance (R_{DSon}) and drain current (i_D), which helps in cascaded connection of devices with lower blocking voltages. These properties cause SiC devices to operate at a higher switching frequency, influencing the heatsink's weight, volume, size, and component size reductions such as the filter and MFT.

According to the recent findings, SiC-based auxiliary converters have high efficiency and the junction temperature rise is minimal at high switching frequencies, allowing for lower cooling requirements and higher power density. Although SiC devices have acquired acceptability for high temperature, high voltage, and high switching operation, there are still several difficulties that prevent widespread production. Due to the formation of positive spurious gate voltage when common-mode (CM) current flows through a miller capacitance, the quicker functioning of SiC devices (10 kV/s) necessitates faster short circuit protection, correct dv/dt, and di/dt management [50]. For HV applications, active gate control in the gate driver with well-designed laminated busbars is usually advised to prevent parasitic ringings and conducted noise. Snubber circuits, key design concerns, HV arcing, and electromagnetic interference issues are all eliminated with a laminated busbar. It also allows for simple, flexible, and error-free installation [51]. Figure 11.23 depicts the potential for all-SiC power modules in various rolling stocks by replacing Si-IGBTs with SiC-MOSFETs. For traction applications, higher current modules are required. Hence, new SiC-MOSFET modules must be

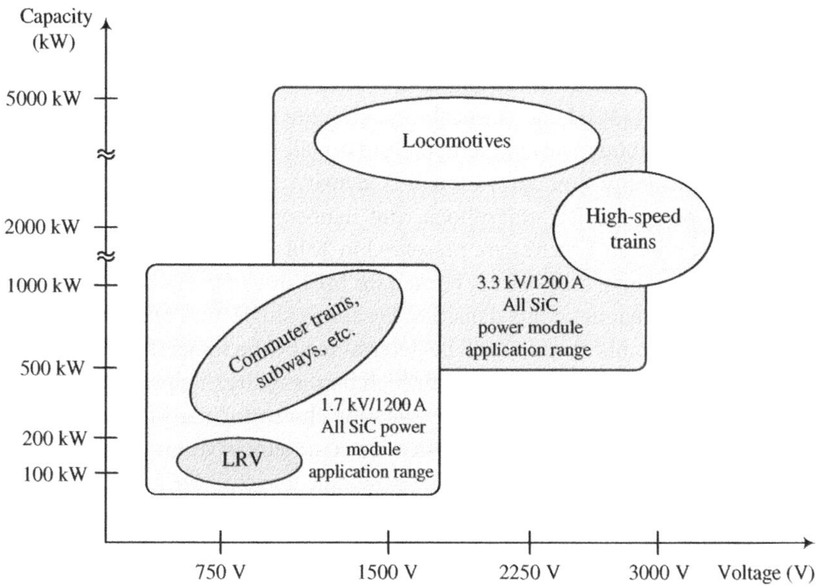

Figure 11.23 Potential replacement of Si-IGBTs with SiC MOSFETs in various rolling stocks.

developed that allow for the parallelization of a larger number of chips while minimizing stray inductance.

Another significant factor to consider is the design of a good cooling system to attain a better power density. Active cooling methods such as forced air-cooled and liquid-cooled-based heat sinks are typically used for rolling stock. These approaches, on the other hand, necessitate complex piping networks with external fans, liquid materials, pumps, and heavy heat exchangers, which generate acoustic disturbance and raise system costs. Overcoming the aforementioned concerns and trends toward higher dissipation with smaller size prompted the researchers to explore innovative passive cooling methods. The current state-of-the-art passive cooling methods such as two-phase cooling methods (heat pipes and phase-change materials) are not mature enough to replace conventional cooling methods in traction applications. However, with the considerable advancements in thermal analysis procedures and manufacturing processes definitely dominate in the near future. Overall, there are some issues with SiC devices in terms of manufacturing, gate driver design, EMI filtering, busbar arrangement, thermal management, and reliability that must be addressed.

11.5.2 Advanced Materials for Passive Components

The MFTs have a greater operating frequency than LFTs, allowing the traction power converter to attain higher power density and efficiency. Furthermore, it eliminates voltage and current distortion at the input side owing to LFT's core saturation. In contrast to LFT, when designing and optimizing the MFT, several technical elements such as proximity effects and eddy currents must be considered. Lower flux density material and higher insulation at higher power and operating frequency are critical criteria for the design of MFT [34]. To perform soft switching, magnetizing, leakage inductances, and the parasitic capacitance of MFT must all be within prescribed limits. High dielectric strength, good heat conductivity, and low dielectric loss are all requirements for MFT insulation [43].

The magnetic material's operation in LFT is very close to saturation, whereas, in MFT, the flux density of the magnetic circuit is frequency-dependent (0.3 T for 20 kHz) and has a fixed value. Due to the high power and insulation requirements, the choice of core material, in addition to winding arrangement and structure, is critical [52]. High permeability, low core loss, high saturation magnetic flux density, and a high continuous working temperature are all requirements for the core material. Despite its high permeability and saturation flux density, Si steel is unsuitable for MFT design due to excessive losses. Ferrite, amorphous, and nanocrystalline materials are the materials to consider for MFT, and their features are presented in Table 11.4. Even though ferrite cores have a low cost and a low core loss, they are not chosen for MFT due to the need for a bigger core size to achieve high saturation magnetic flux densities. Because a single-core material cannot meet all of the requirements, the suitable material will be determined by the tradeoff between efficiency, volume, and cost. Amorphous materials, with their high permeability, high saturation flux density, strong resistance, and low cost, appear to be a superior choice for the MFT core [53]. Because of its higher saturation flux density, lower core loss, a lower amount of vibration, and lower acoustic noise, the nanocrystalline core is also an appealing candidate for small MFT. Nanocrystalline materials, on the other hand, are more expensive and require unique designs, which is a disadvantage when compared to amorphous materials. Because of its higher copper density and better use of a winding area, Litz wires are favored for windings.

Core and shell designs are the most prevalent transformer core configurations for SSTT systems. A comparative analysis of alternative core structures was performed, and core-type transformers are recommended for most MFT designs, making it easier to mass-produce MFTs. The MFT has increased core losses, higher harmonic content, and aberrant eddy losses, and becomes significantly overloaded as a result of nonsinusoidal excitation. Owing to the aforementioned characteristics, the optimal design of MFT is critical when considering transient thermal behavior for an effective cooling system. All magnetic components, such as a pre-charging device, MFTs, and line inductors, are stacked in a tank in ABB prototypes. C-cut nanocrystalline cores, Litz wire windings, and an oil-directed forced air cooling technique are employed in the MFTs [34]. Bombardier developed a 3-MW MFT (18-kg, 8-kHz) prototype composed of nanocrystalline core and windings that are cooled by deionized water through rectangular aluminum profiled tubes [54]. Further research into

Table 11.4 A summary of magnetic core material properties.

Parameter	Ferrite	Amorphous	Nanocrystalline
Saturation flux density	0.3–0.5 T	0.8–10.5 T	1.1–1.3 T
Operating temperature	100–150 °C	120–15 °C	120–180 °C
Curie temperature	220 °C	>350 °C	>550 °C
Magnetostriction	$(20–30) \times 10^{-6}$	$<0.2 \times 10^{-6}$	$(1–6) \times 10^{-6}$
Permeability	1.5–15 k	1–100 k	20–200 k
Loss (20 kHz, 0.2 T)	15–20 W/kg	5–7 W/kg	4–8 W/kg
Composition	MnZn	$(Co)_x(SiB)_y$	FeCuNbSiB
Thermal conductivity	High	Medium	Low
Processing	Easy	Difficult	Difficult
Operating frequency	High	Medium	Medium
Cost	Low	Medium	High

Source: Adapted from [34, 43].

medium-frequency voltage-stress-resistant insulating materials, low acoustic noise emissions, low loss MFT interconnections, MFT construction optimization, and advanced thermal management systems (air and H_2O cooling methods) to replace oil, which can all contribute to the high-power compact MFT design. Much more research is needed to investigate the manufacturing process, life cycle evaluation, recyclability, and environmental consequences of MFT materials. One such initiative was by Hitachi to develop high-efficiency super amorphous transformers (oil-immersed and molded). This is achieved by utilizing recyclable material to reduce no-load losses, which resulted in a long-term impact on the global environment and improves product lifetime [55].

11.5.3 Reversible Substations and Off-Board Energy Storage Systems

Another promising solution is the effective utilization of energy during the braking process. During the braking process, electric traction motors serve as electric generators and deliver a significant amount of energy. To avoid an overvoltage in the catenary, most of the braking energy is dissipated as heat in the onboard brake resistor. If the braking energy could be captured and used in the next accelerating phase, it would be extremely useful. It is determined that a railway vehicle's energy consumption from the overhead catenary can be reduced, allowing the braking energy to be stored and used to accelerate the railway vehicle into the next accelerating phase or transferred to another train nearby.

Reversible substations can be used to do this, with extra DC energy being supplied back into the AC distribution system. As previously stated, rail vehicles are fed by two types of electrification systems: AC voltage (up to 25 kV) and DC voltage (up to 3 kV). Because AC electric substations are connected to the main grid, they have a higher receptivity than DC electric substations. The regenerative energy generated in AC electric substations can be transmitted to the main grid, where it can be used to power other vehicles or loads. Only if the AC power supply is constructed with controlled rectifiers would this be possible. The controlled rectifier is activated when the DC voltage at the substation level rises as a result of regenerative braking, and the braking energy is sent into the AC grid, where it can be transferred to another train in the vicinity.

Reversible substations have become a reality thanks to cutting-edge power semiconductor technologies and cost savings from energy storage devices (ESDs). INGETEAM has illustrated several field implementations at 2–6 of Metro Brussels (750 VDC) and Suburban Malaga-Fuengirola (3300 VDC). ALSTOM planned and executed an alternate system using a thyristor-based rectifier in parallel with an IGBT converter in the London Metro and the Milan Metro [7]. On the other hand, by installing IGBT-based converters in parallel with the uncontrolled rectifier, existing substation equipment, such as transformers and rectifiers, may be repurposed. Reversible substations can recover up to 99% of the maximal braking energy [56]. However, mechanical braking consumes the majority of the remaining energy, making it impossible to combine reversible systems with the catenary-free operation.

As a roadside energy storage system (ESS) infrastructure shown in Figure 11.24, off-board energy storage at substations may be a realistic option. These technologies are put at ground level, where there are no restrictions on size, weight, or area. Both the off-board ESS and the electric substation can send energy to the existing railway vehicles on the line in this situation. However, only the off-board ESS can recover the braking energy that the rest of the rail vehicles do not use. At the West Falls Church substation (Washington DC Metro rail station), a 2-MW/400-kWh battery power system composed of a nickel-metal hydride (Ni-MH) was installed in 2013. ESSs have higher cost-effectiveness and maintenance than reversible substations. Reversible substations and off-board aggregation systems are more efficient, but they still cannot provide catenary-free segments or boost vehicle performance [57].

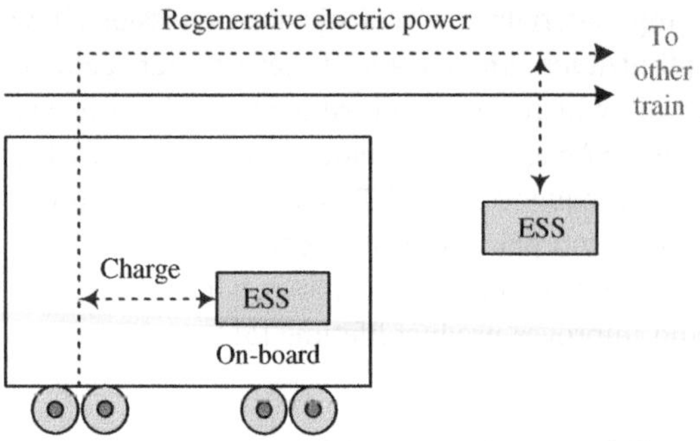

Figure 11.24 Arrangement of both onboard and off-board ESS for energy recovery.

Table 11.5 Comparison of regenerative braking technologies.

Technology	Infrastructure cost	Energy recovery	Catenary free	Modification cost
Reversible substation	High	Medium	No	Low
Off-board ESS	Medium	High	No	Medium
Onboard ESS	Low	High	Yes	High

An onboard ESS in railway vehicles, as depicted in Figure 11.24, is another interesting approach. When a rail vehicle has an onboard ESD, its energy can be utilized to lower the catenary line's energy consumption during acceleration. Table 11.5 shows a comparison of three methods for effective braking and improved traction system performance. This demonstrates that onboard ESS is the optimum option in terms of energy recovery, vehicle performance, safety, and catenary-free operating while remaining independent of a power source. On the other hand, onboard ESS adds weight and increases the cost of rolling stock modification. Section 11.5.4 goes into great length about onboard ESS and the latest technology.

11.5.4 On-Board Energy Storage Systems in Rolling Stock

High-power density, high peak power for a short time, high energy density, high load cycling capacity, and the capture of regenerative energy are all requirements for ESSs. The following are the key benefits of onboard ESS in a railway context [7]:

- Minimizes the power peak during the acceleration of the vehicle
- Enhancement of traction/braking characteristics, particularly in high-speed areas
- A self-contained or catenary-free operation
- Can stabilize the network voltage (voltage regulation) and run the rail vehicle without overhead catenary
- Less strain on the power distribution system
- An ESS can reduce the power peak consumed from the catenary and demand reduction on electric substations

While flywheel storage technologies have advanced significantly, their energy density remains low when compared to other devices. Batteries and supercapacitors (SCs) are employed in commercially available ESDs for railway locomotives. However, electric double-layer capacitors (EDLCs) are more popular than batteries owing to their high-power capabilities (same charge and discharge).

EDLCs feature superior power density, longer life cycles, lower internal resistance, and faster charge and discharge times, making them an excellent choice for storing regenerative braking energy during deceleration and supplying peak power during acceleration. In Mannheim, Bombardier Transportation presented a field trial on one bogie of an LRV equipped with onboard ultracapacitors and a lightweight MITRAC energy saver for operator RNV. These improve acceleration and result in a 30% reduction in energy consumption. A similar approach was also applied to EMUs and diesel EMUs (DEMUs), resulting in fuel savings and lower carbon emissions [58]. Figures 11.25 and 11.26 depict the operation modes of EMUs and DEMUs, respectively. The rolling stock with

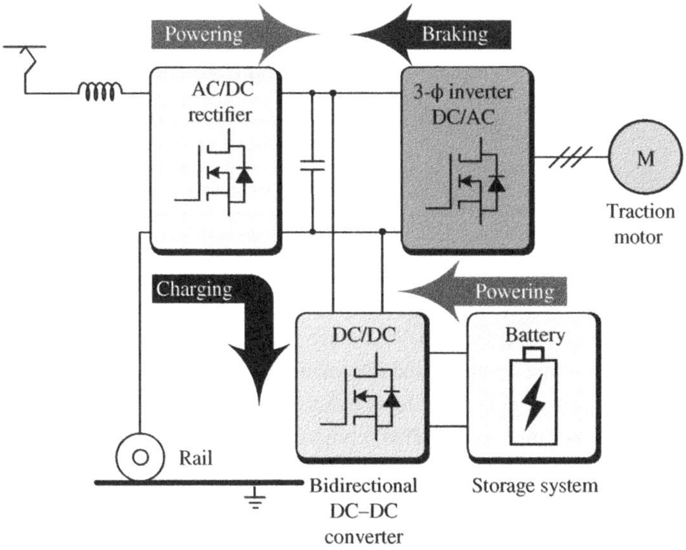

Figure 11.25 Typical configuration of propulsion system with onboard ESS for EMUs.

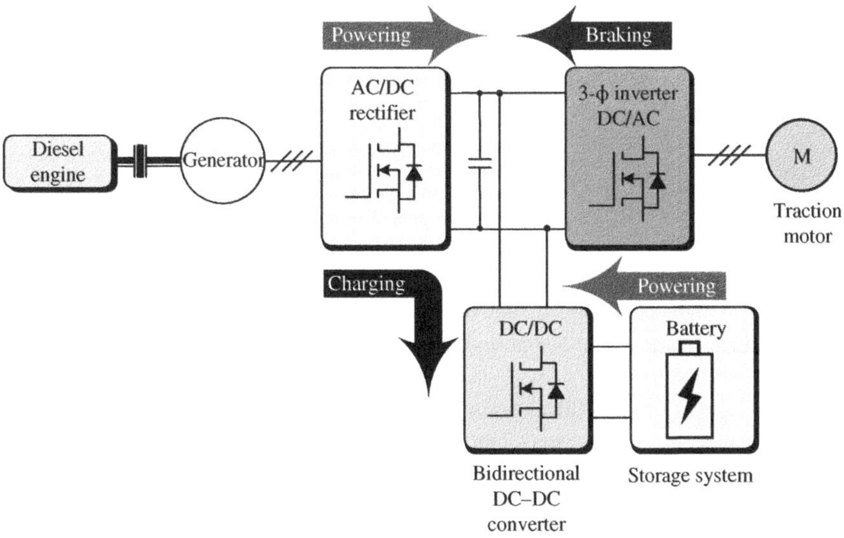

Figure 11.26 Typical configuration of propulsion system with onboard ESS for DEMUs.

onboard ESS in the DEMUs and EMUs ameliorates overhead line discontinuities, neutral sections, and non-electrified lines. The Li-ion batteries were placed in the SEIBU 20000 series by the SEIBU railway company's commuter trains [59]. The IPEMU project was established by Bombardier to build battery-powered EMUs for longer travels.

A correct combination of SCs with batteries with increased safety and efficiency will be the most viable solution for catenary-free operation in the DEMUs and EMUs; the same was developed in series EV-E30 [60]. Hybrid locomotives were developed using the same approach, combining the environmentally favorable qualities of AC locomotives with the lower infrastructure costs of diesel locomotives. As onboard ESS necessitates large spaces on the vehicle, it is typically located on the roof in the case of a low-floor tramway. Depending on the energy storage size, the system's weight increases by 2–4 tonnes. Due to their short-term energy storage capability, flow batteries can be regarded as a replacement for SCs [61]. However, advances in battery materials, characterization, and packaging, combined with cutting-edge chemical technology, may make SCs outdated shortly.

11.6 Concluding Remarks

This chapter examines several rolling stock architectures and their components, both in use and under investigation. The state-of-the-art developments in rolling stock architectures, traction motors, TRCs, auxiliary systems, control techniques, and VCUs are all discussed in detail. Over the two decades, significant developments have been carried out, which resulted in new semiconductor technologies, magnetic materials, power converter topologies, control schemes, communication systems, and other features for rolling stock. This chapter mainly focuses on SSTT technology, which has received a lot of attention from both academia and industry. These technologies have successfully finished field trials and attained a medium level of maturity with technological advancements in power conversion systems, the MFT design, and advanced control techniques. Further investigation is required especially for single-stage active front-end converters, WBG devices, an optimal number of cascaded modules, reliability aspects, and optimal transformer design with advanced magnetic materials. With ongoing advancements in semiconductor technologies, magnetic materials, power converter topologies, control schemes, and cooling systems, as well as reliability-oriented design, advanced manufacturing processes, policies, and regulations, SSTT technology can achieve mass-market penetration in modern rolling stock. Finally, this chapter describes and introduces fresh research opportunities.

References

1 Uzuka, T. (2011). Trends in high-speed railways and the implications on power electronics and power devices. *Proceedings of IEEE International Symposium on Power Semiconductor Devices and ICs, San Diego*, CA, pp. 6–9.

2 Jahns, T.M. and Blasko, V. (2001). Recent advances in power electronics technology for industrial and traction machine drives. *Proc. of the IEEE* 89 (6): 963–975.

3 Stemmler, H. (1993). Power electronics in electric traction applications. *Proceedings of International IEEE Industrial Electronics, Control, and Instrumentation,* Maui, HI, pp. 707–713, Vol. 2.

4 Steimel, A. (1996). Electric railway traction in Europe. A survey of the state-of-the-art. *Proceedings of Internatinal Symposium on Industrial Electronics, Warsaw*, pp. 40–48, Vol. 1.

5 Mermet-Guyennet, M. (2010). New power technologies for traction drives. *Proceedings of International Symposium on Power Electronics and Electrical Drives, Automation and Motion*, Pisa, pp. 719–723.
6 Bhargava, B. (1999). Railway electrification systems and configurations. *Proceedings of IEEE Power Engineering Society Summer Meeting, Edmonton*, Alta, pp. 445–450.
7 Ronanki, D., Singh, S.A., and Williamson, S. (2017). Comprehensive topological overview of rolling stock architectures and recent trends in electric railway traction systems. *IEEE Trans. Transport. Electrification* 3 (3): 724–738.
8 Eckel, H.G., Bakran, M.M., Krafft, E.U., and Nagel, A. (2005). A new family of modular IGBT converters for traction applications. *Proceedings of European Conference on Power Electronics and Applications*, Dresden, Germany, pp. 10–15.
9 Drofenik, U. and Canales, F. (2014). European trends and technologies in traction. *Proceedings of International Power Electronics Conference (ECCE-ASIA)*, Hiroshima, pp. 1043–1049.
10 Bernet, S. (2000). Recent developments of high power converters for industry and traction applications. *IEEE Trans. Power Electron.* 15 (6): 1102–1117.
11 Pinney, C. and Smith, B. (2013). Cost-benefit analysis of alternative fuels and motive designs. *U.S. Dept. Transp., Tech. Rep.* DOT/FRA/ORD-13/21 [Online]. Available: https://www.fra.dot.gov/Elib/Document/3128.
12 Bakran, M.M. and Eckel, H.G. (2007). Power electronics technologies for locomotives. *Proceedings of IEEE International Conference on Power Conversion*, Nagoya, Japan, pp. 1362–1368.
13 El-Refaie, A.M. (2013). Motors/generators for traction/propulsion applications: a review. *IEEE Veh. Technol. Mag.* 8 (1): 90–99.
14 Shikata, K., Kawai, H., Nomura, H. et al. (2012). PMSM propulsion system for Tokyo Metro. *Proceedings of Electrical Systems for Aircraft, Railway and Ship Propulsion*, Bologna, Italy, pp. 1–6.
15 Koerner, O. and Binder, A. (2004). Feasibility of a group drive with two permanent magnet synchronous traction motors for commuter trains. *Eur. Power Electron. Drives J.* 14 (3): 32–37.
16 Ray, W.F., Davis, R.M., Lawrenson, P.J. et al. (1984). Switched reluctance motor drives for rail traction: a second view. *IEEE Proc. B Electr. Power Appl.* 131 (5): 220–225.
17 Germishuizen, J.J., Van der Merwe, F.S., Van der Westhuizen, K., and Kamper, M.J. (2000). Performance comparison of reluctance synchronous and induction traction drives for electrical multiple units. *Proceedings of IEEE International Conference on Industrial Applications*, Vol. 1, pp. 316–323.
18 Kouro, S., Rodriguez, J., Wu, B. et al. (2012). Powering the future of industry: high-power adjustable speed drive topologies. *IEEE Ind. Appl. Mag.* 18 (4): 26–39.
19 Marogy, B.Y.M. (1992). Specification of static auxiliary converters for rolling stock. *Proceedings of IEE Colloquium Auxiliary Power Supplies Rolling Stock*, London, UK, pp. 1-1–1-4.
20 Kotaro, U. Auxiliary Inverters for Traction, Fuji Electric [Online]. Available: http://americas.fujielectric.com/files/FER-58-4-175-2012.pdf.
21 Janssen, M. and Steimel, A. (2002). Direct self-control with minimum torque ripple and high dynamics for a double three-level GTO inverter drive. *IEEE Trans. Ind. Electron.* 49 (5): 1065–1071.
22 Toliyat, H.A., Levi, E., and Raina, M. (2003). A review of RFO induction motor parameter estimation techniques. *IEEE Trans. Energy Convers.* 18 (2): 271–283.
23 Alsofyani, I.M. and Idris, N.R.N. (2013). A review on sensorless techniques for sustainable reliablity and efficient variable frequency drives of induction motors. *Renew. Sustain. Energy Rev.* 24: 111–121.
24 Idris, N.R.N. and Yatim, A.H.M. (2004). Direct torque control of induction machines with constant switching frequency and reduced torque ripple. *IEEE Trans. Ind. Electron.* 51 (4): 758–767.
25 Rodriguez, J. et al. (2022). Latest advances of model predictive control in electrical drives–part I: basic concepts and advanced strategies. *IEEE Trans. Power Electron.* 37 (4): 3927–3942.

26 Rodriguez, J. et al. (2022). Latest advances of model predictive control in electrical drives–part II: applications and benchmarking with classical control methods. *IEEE Trans. Power Electron.* 37 (5): 5047–5061.

27 Yaramasu, V. and Wu, B. (2016). *Model Predictive Control of Wind Energy Conversion Systems.* Hoboken, NJ, USA: Wiley.

28 Zhang, Y., Yang, H., and Xia, B. (2016). Model-predictive control of induction motor drives: Torque control versus flux control. *IEEE Trans. Ind. Appl.* 52 (5): 4050–4060.

29 Vazquez, S., Rodriguez, J., Rivera, M. et al. (2017). Model predictive control for power converters and drives: advances and trends. *IEEE Trans. Ind. Electron.* 64 (2): 935–947.

30 Schifers, C. and Hans, G. (2000). IEC 61375-1 and UIC 556-international standards for train communication. *Proceedings of IEEE 51st Vehicular Technology Conference (Cat. No.CH)*, Vol. 2. Tokyo, Japan, pp. 1581–1585.

31 Neil, G. (2012). On board train control and monitoring systems. *Proceedings of IET 13th Professional Development Course on Electric Traction Systems.* London, UK, 2012, pp. 223–246.

32 List of rail standards used in India. https://www.sesei.eu/wp-content/uploads/2018/02/Railway-standards.pdf.

33 Besselmann, T., Mester, A., and Dujic, D. (2014). Power electronic traction transformer: efficiency improvements under light-load conditions. *IEEE Trans. Power Electron.* 29 (8): 3971–3981.

34 Ronanki, D. and Williamson, S. (2018). Evolution of power converter topologies and technical considerations of power electronic transformer based rolling stock architectures. *IEEE Trans. Transport. Electrification* 4 (1): 211–219.

35 Dujic, D., Chuanhong, Z., Mester, A. et al. (2013). Power electronic traction transformer-low voltage prototype. *IEEE Trans. Power Electron.* 28 (12): 5522–5534.

36 McMurray, W. (1970). Power converter circuits having a high frequency link. *U.S. Patent 3517300.*

37 Huber, J.E. and Kolar, J.W. (2016). Solid-state transformers: on the origins and evolution of key concepts. *IEEE Ind. Electron. Mag.* 10 (3): 19–28.

38 Steiner, M. and Reinold, H. (2007). Medium frequency topology in railway applications. *Proceedings of European Conference on Power Electronics and Applications*, pp. 1–10

39 Ronanki, D. and Williamson, S.S. (2020). Failure prediction of submodule capacitors in modular multilevel converter by monitoring the intrinsic capacitor voltage fluctuations. *IEEE Trans. Ind. Electron.* 67 (4): 2585–2594.

40 Claessens, M., Dujic, D., Canales, F. et al. (2012). Traction transformation: a powerelectronic traction transformer (PETT) [Online]. Available: https://library.e.abb.com.

41 Pittermann, M., Drabek, P. and Bednar, B. (2015). Single phase high-voltage matrix converter for traction drive with medium frequency transformer. *Proceedings of International 41st IEEE Annual Conference of the IEEE Industrial Electronics*, Yokohama, pp. 005101–005106.

42 Roy, S., De, A. and Bhattacharya, S. (2014). Current source inverter based cascaded solid state transformer for AC to DC power conversion. *Proceedings of IEEE International Conference on Power Electronics (ECCE ASIA)*, Hiroshima, pp. 651–655.

43 Feng, J., Chu, W.Q., Zhang, Z., and Zhu, Z.Q. (2017). Power electronic transformer-based railway traction systems: challenges and opportunities. *IEEE J. Emerg. Sel. Top. in Power Electron.* 5 (3): 1237–1253.

44 Glinka, M. (2004). Prototype of multiphase modular-multilevel-converter with 2 MW power rating and 17-level-output-voltage. *Proceedings of IEEE International Conference on Power Electronics Spec. (IEEE Cat. No. 04CH37551)*, pp. 2572–2576.

45 Honrio, D., Oliveira, D. and Barreto, L.H. (2015). An AC-DC multilevel converter feasible to traction application. *Proceedings of IEEE International Conference on Power Electronics and Applications (EPE'15)*, Geneva, 2015, pp. 1–9.

46 Facchinello, G.G., Mamede, H., Brighenti, L.L. et al. (2016). AC-AC hybrid dual active bridge converter for solid state transformer. *Proceedings of IEEE International Symposium on Power Electronics for Distributed Generation Sys*tem, Vancouver, BC, pp. 1–8.

47 Mitsubishi Electric to Launch X-Series HVIGBT Module. (2015). http://www.mitsubishielectric.com.

48 Ishikawa, K., Yukutake, S., Kono, Y. et al. (2014). Traction inverter that applies compact 3.3 kV/1200 A SiC hybrid module. *Proceedings of International Conference on Power Electronics*, Hiroshima, Japan, pp. 2140–2144.

49 Mitsubishi Electric Launching 3.3kV, 1500A Inverter With All-SiC Power Module for High Power Trains. (2013). http://www.mitsubishielectric.com/news/2013/1225.html.

50 She, X., Huang, A.Q., Lucía, Ó., and Ozpineci, B. (2017). Review of silicon carbide power devices and their applications. *IEEE Trans. Ind. Electron.* 64 (10): 8193–8205.

51 Callegaro, A.D., Guo, J., Eull, M. et al. (2018). Bus bar design for high-power inverters. *IEEE Trans. Power Electron.* 33 (3): 2354–2367.

52 Balci, S., Sefa, I., and Bayram, M.B. (2014). Core material investigation of medium-frequency power transformers. *Proceedings of International Power Electronics and Motion Control Conference Expo.*, Antalya, Turkey, pp. 861–866.

53 She, X., Huang, A.Q., and Burgos, R. (2013). Review of solid-state transformer technologies and their application in power distribution systems. *IEEE J. Emerg. Sel. Topics Power Electron.* 1 (3): 186–198.

54 Leibl, M., Ortiz, G., and Kolar, J.W. (2017). Design and experimental analysis of a medium-frequency transformer for solid-state transformer applications. *IEEE J. Emerg. Sel. Topics Power Electron.* 5 (1): 110–123.

55 Inagaki, K., Kuwabara, M., Sato, K. et al. (2011). Amorphous transformer contributing to global environmental protection. *Hitachi Rev.* 60 (5): 250–256.

56 Arboleya, P., Bidaguren, P., and Armendariz, U. (2016). Energy is on board: energy storage and other alternatives in modern light railways. *IEEE Electrification Mag.* 4 (3): 30–41.

57 Zelinsky, M. (2016). Market Advancement of NiMH Batteries for Stationary Applications [Online]. Available: http://www.battcon.com/Papers2016/Zelinsky%20Paper%202016.pdf.

58 Steiner, M., Klohr, M., and Pagiela, S. (2007). Energy storage system with ultracaps on board of railway vehicles. *Proceedings of European Conference on Power Electronics and Applications*, Aalborg, Denmark, pp. 1–10.

59 Ayata, M., Kusano, N., Shinomiya, T. et al. (2015). Traction inverter system with Lithium-ion batteries for EMUs. *Proceedings of 17th European Conference on Power Electronics and Applications*, Geneva, Switzerland, pp. 1–9.

60 Kono, Y., Shiraki, N., Yokoyama, H., and Furuta, R. (2014). Catenary and storage battery hybrid system for electric railcar series EV-E301. *Proceedings of International Conference on Power Electronics*, pp. 2120–2125.

61 Mayet, C., Pouget, J., Bouscayrol, A., and Lhomme, W. (2014). Influence of an energy storage system on the energy consumption of a diesel-electric locomotive. *IEEE Trans. Veh. Technol.* 63 (3): 1032–1040.

12

Electromagnetic Compatibility in Railways

Sahil Bhagat

Hitachi Rail STS, Traction Power and EMC, Abu Dhabi, United Arab Emirates

12.1 Introduction

Electromagnetic compatibility (EMC) is the ability of an equipment or system to function satisfactorily in its electromagnetic (EM) environment without introducing intolerable EM disturbances to anything in that environment [1]. It involves the generation, transmission, and reception of EM energy. There are mainly three elements, which form the basic framework for any EMC design, as shown in Figure 12.1. These elements are the source ("culprit" also indicated as emitter), which emits EM energy; coupling method ("path"), which provides the path for EM energy to flow; and the victim, which receives and processes this EM energy. Interference is considered to occur when the victim behaves in an undesirable manner [2]. Therefore, in order to ensure EMC, the following three steps shall be applied:

- At the source of disturbance: reduction of emissions;
- At the coupling: reduction of coupling;
- At the victim: increase of immunity.

A typical railway system is a complex installation with static and moving sources of EM disturbance and also a wide range of EM susceptible equipment. For the purpose of safe and punctual transport of passengers, the EMC of all subsystems and components is mandatory. Furthermore, there are EM sources in the neighborhood of the railway, such as radio transmitters, power distribution networks, other railways, military installations, airports and sinks like telecommunication networks, and medical institutions. They may interfere with the railway system and, therefore, shall be part of the EMC analysis.

Disturbances have to be limited to reasonable values and the equipment adjacent to sources of disturbances needs adequate level of immunity to get the system function satisfactorily in its environment. As far as safety-relevant subsystems are the potential sinks, like signaling, telecom, where immunity including a safety margin is essential to ensure high availability of this type of subsystems. Furthermore, the system shall not influence its environment in an inadmissible manner.

Transportation Electrification: Breakthroughs in Electrified Vehicles, Aircraft, Rolling Stock, and Watercraft,
First Edition. Edited by Ahmed A. Mohamed, Ahmad Arshan Khan, Ahmed T. Elsayed, and Mohamed A. Elshaer.

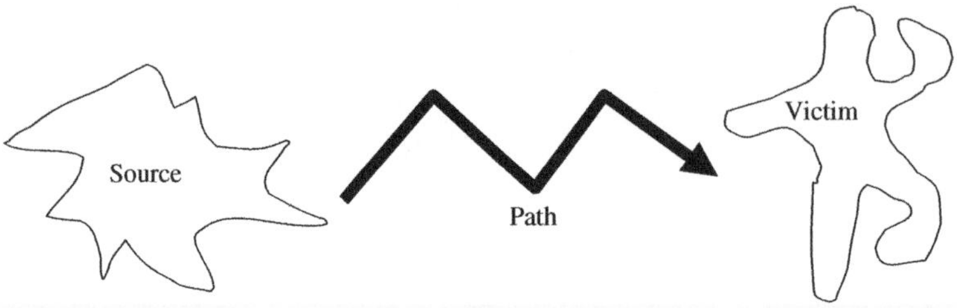

Figure 12.1 Elements of an EMI situation.

In this chapter, we will understand how EMC is managed in railways based on international standards along with guidelines required to be followed in order to avoid any interference.

12.2 The Phenomenon of Electromagnetic Interference

12.2.1 The Interference Model

EM disturbance will be found in general, where EM energy finds a way of undesirable coupling from a source to any other sink than those it is intended for. Figure 12.2 shows the general interference model with its three partitions, source of disturbance, coupling path and sink of disturbance, with some examples.

According to the definition of EMC, the emission of equipment, which is the source of disturbance, has to be at the location of the second device, which is the sink of disturbance, to be lower than its immunity level with respect to the mechanism of the coupling path. The coupling path of the disturbance is never the same as that of the useful signal. Any particular device may be the source of disturbance and also sink of another disturbance. For example, a handheld device (walkie-talkie), if used, can act as a source of disturbance if placed near any numerical relay or digital meter and cause some disturbance in the display or readings. The same device (walkie-talkie) can have interference issues if similar frequency source is used near its operating frequency, typically 430 MHz. Thus, devices have to meet defined requirements, to function satisfactorily in a specific EM environment, which means a space with a defined level of disturbances.

Typical sources of EM disturbance in a railway system are transformers, power cables, current collectors (arcing), converters, radio-based communication, external radio transmission, and electrical charge of human bodies. These sources generate a wide spectrum of disturbances from static and low-frequency magnetic fields at radio frequencies in the range of some GHz.

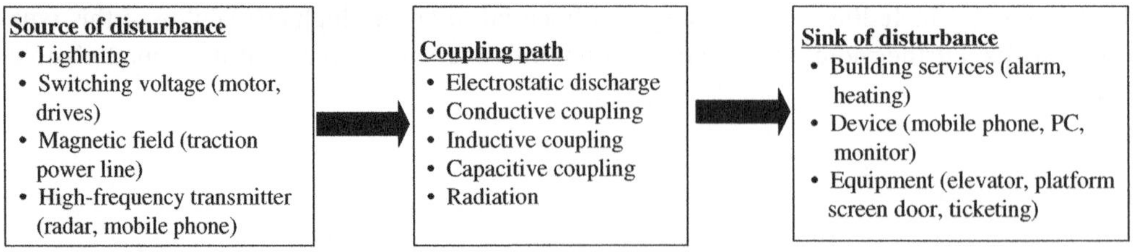

Figure 12.2 Interference model.

Coupling describes the physical process of transmission of energy from a source to a sink. Transmission will occur if a specific form of energy emitted by the source finds a suitable path of coupling, and the sink is susceptible to this form of energy. If the value of energy transmitted exceeds the immunity of the sink it will cause electromagnetic interference (EMI). There are five basic ways of coupling:

- Conductive coupling;
- Inductive coupling;
- Capacitive coupling;
- Direct radiation; and
- Electrostatic coupling

Knowing these ways of coupling is necessary to take suitable measures where EMC problems are identified.

Conductive coupling is the coupling by means of common impedances. Conductive coupling dominates at low frequencies, where circuit impedances are low. For example, a transformer with a large fault current should not be bonded using the same earthing conductor as sensitive electronic equipment. This could lead to sudden voltage rise and cause failure. Thus, it is recommended to connect the apparatuses at the nodes of the earthing network.

Inductive coupling is the transmission of energy between conductors via the magnetic field according to the current linkage.

Capacitive coupling is the transmission of energy between conductors via the electric field.

Inductive and capacitive couplings are limited to conductors with a length of less than 10% of the wavelength to be considered. Example: power frequency 50 Hz, wavelength 6000 km – inductive and capacitive coupling will predominate; communication frequency 1 GHz, wavelength 30 cm – radiation predominates. Inductive and capacitive coupling is usually found between power cables, especially where they run parallel over long distances as shown in Figure 12.3.

Direct radiation has to be considered for conductive structures with geometrical dimensions in the range of the wavelength or bigger because such structures generate EM fields. Direct radiation dominates at high frequencies, where conductor lengths, such as external cables or tracks on circuit boards, are a significant fraction of wavelength at the frequency of interest. Figure 12.4 explains the concept of this type of coupling.

Electrostatic coupling or discharge (ESD) generally may occur between conductive bodies insulated to each other in the case of a high static charge.

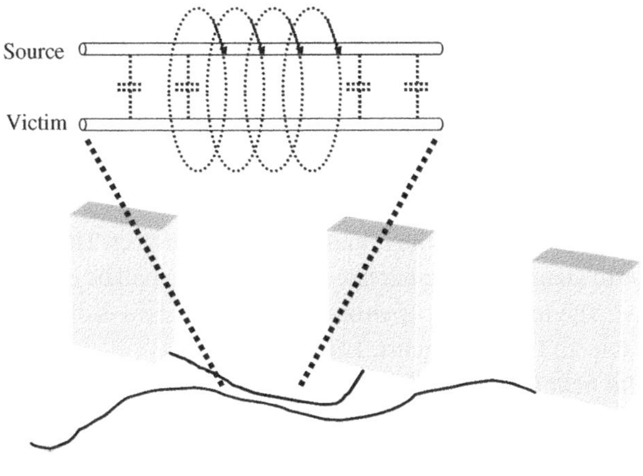

Figure 12.3 Inductive and capacitive couplings between cables.

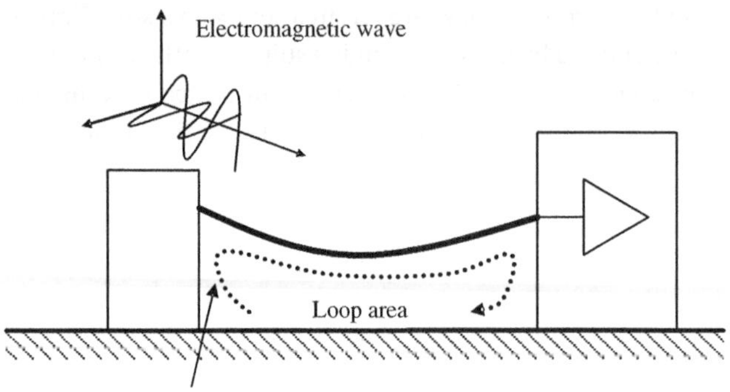

Figure 12.4 Radiated coupling.

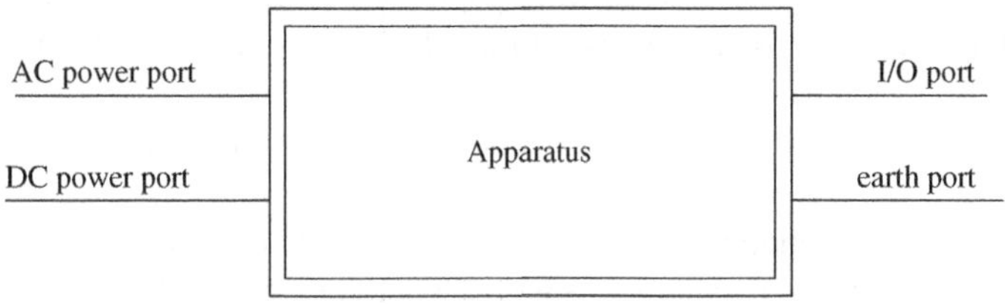

Figure 12.5 Main categories of ports.

From the understanding of coupling mechanisms, it can be derived that the EM waves will enter the equipment either through the ports such as power, signaling, communication, and earth, which have the cables exposed in EM environment or by passing directly through enclosure (after attenuation if metallic). Therefore, all the EMC standards revolve around defining the immunity and emission levels for the ports through EM waves can enter or exit the apparatus. Figure 12.5 [3] provides a typical block diagram that indicates the type of ports as discussed earlier.

The sink of an EM disturbance is the potentially influenced device. Influence may result in degraded performance, unsafe operation or loss of operation of the system, depending on the function of the equipment in the system. Typical sinks of a railway system are signaling system (track circuits, cables, and monitors), Supervisory Control and Data Acquisition (SCADA) system, telecommunication (antennas and cables), passenger information system, building services (control systems), and external telecommunication systems.

12.3 EMC Strategy

The best way to achieve EMC is to start working on it from the initial stage of the project. The primary objective of an EMC engineer must be to identify and demarcate EM zones that shall be implemented throughout the project lifecycle [4]. The installation environment of the system is divided into four different zones as depicted in Table 12.1 and in Figure 12.6.

Based on the zone definition earlier, the next approach shall be to prepare EMC compliance matrix for all subsystems such as signaling, power supply, telecom, etc., and collect the declaration of conformities or EMC type test reports for verification.

Table 12.1 EM zoning concept.

EM zone	Description
Zone 1: Track side [3, 5]	• Vital equipment such as protection devices, interlocking or command and control • Equipment having connection to traction power conductors, • All apparatuses installed within 3 m from the centerline of the outer track • Ports of apparatus installed within 10 m, but interfaced with cables inside the 3 m zone or longer than 30 m
Zone 2: Industrial environment	All apparatus whose installation environment is equivalent to Industrial Environment as per definition of EN 61000-6-2/4 and does not fall within cases covered by zone 1, zone 3, and zone 4. The equipment have to comply with the following standards: 1) Emissions: EN 61000-6-4 2) Immunity: EN 61000-6-2
Zone 3: Commercial and light industrial environment	The installation environment presents the same features as a residential location. The equipment comply with CISPR 32 (or EN 55032) emission limits (class A) and with CISPR 35 (or EN 55035) immunity limits or COTS equipment product standards. As an alternative, emission and immunity provisions can be selected in compliance with EN 61000-6-3 and EN 61000-6-1, respectively
Zone 4: Onboard environment [6]	All the apparatus are installed onboard the rolling stock. Emission and Immunity criteria shall be set as per EN 50121-3-2/EN 50155

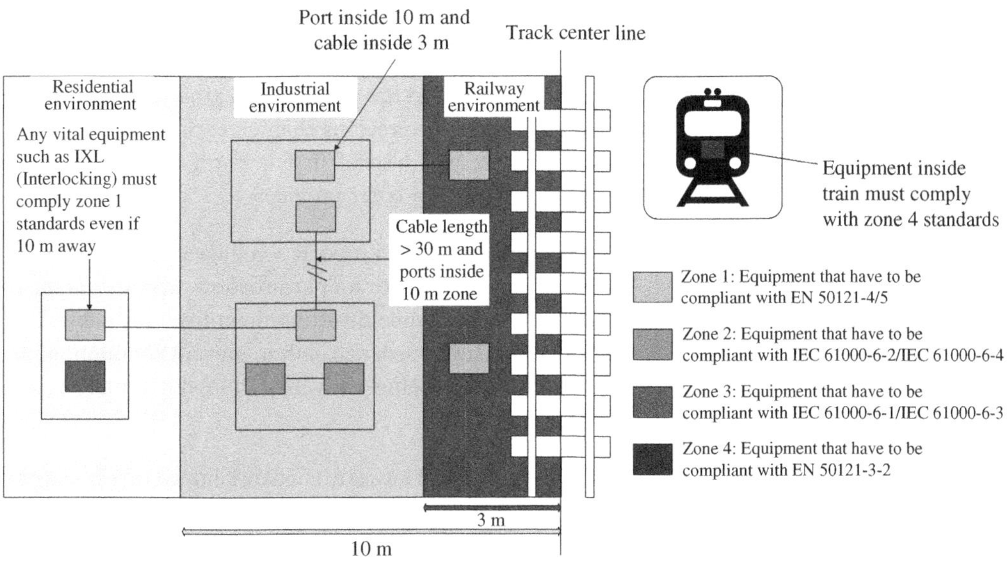

Figure 12.6 EM zones for railway system.

However, conformity to EMC standards based on zones might not be sufficient, or appropriate, when EMC for sensitive apparatus is concerned. Although the installation guidelines might seem not necessary when the equipment themselves have sufficiently high immunity levels, it is preferred to adopt them in most of the situations, especially in complex EM environment such as railways.

12.4 Design and Installation

The main EMC design guidelines are grouped in the following categories discussed later. These requirements have been derived from IEC 61000-5-2 [7], EN 50174-2 [8], IEC 60364-4-44 [9], and other best practices adopted in railway projects.

12.4.1 Equipment Layout

To mitigate coupling paths between noisy and sensitive equipment, noisy equipment should be kept as physically separated as possible from sensitive equipment. Indeed, coupling efficiency of EMI, in particular for radiating coupling, can dramatically reduce with distance, so that the physical separation is an easy means to reach an acceptable level of EMC for a given installation. It concerns both physical separations of equipment frames and cabling (Figure 12.7).

To this extent, specific areas of the buildings (and rooms) should be allocated to different categories of equipment, as much as possible, e.g., a control room's zone for sensitive cubicles separated from the power distribution zone gathering noisy equipment. Both those zones will be as physically separated as reasonably achievable. In the same way, where similar classes of equipment are installed in adjacent rooms, it is advantageous to group each side of the common separating wall. This approach provides a better EMC efficiency and also helps decrease the cabling routing complexity.

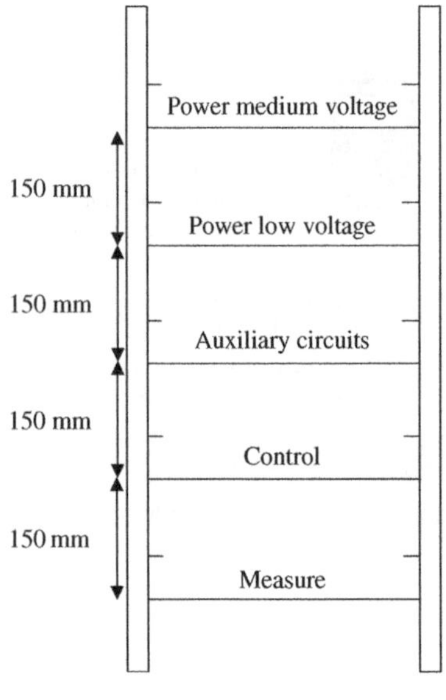

Figure 12.7 Cable segregation between different cable types.

12.4.2 Minimizing the Earth Network Impedance

The primary purpose of the earthing network is to ensure electrical safety, and the earthing and grounding principles are described in the previous chapter, to this extent. But the earthing system is also of first importance for the EMC aspects. Indeed, the latter provides:

- an equipotential area used as a voltage system reference;
- a low impedance path for perturbation currents to return to their source while diverting them from the cables;
- a low transfer impedance path to prevent common mode currents from being converted into differential mode voltages.

In order to obtain low earth network impedance, its structure has to be a meshed-type as shown in Figure 12.8.

For instance, in stations, it is required to create a three-dimensional meshed earthing system. This is to bond every structural and non-structural metalwork together, including concrete reinforcement bars, cable trays, ducts,

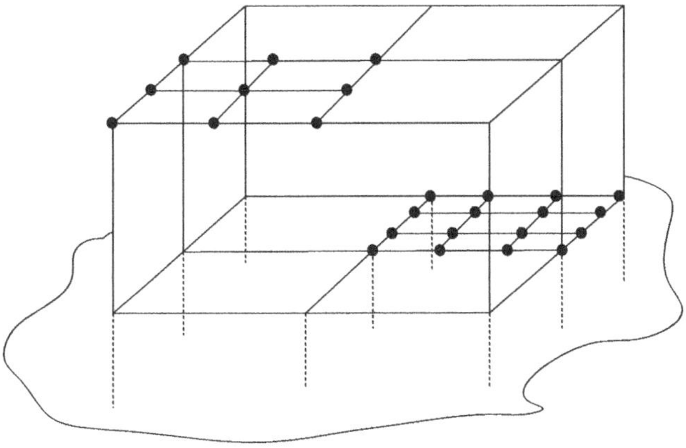

Figure 12.8 Three-dimensional schematic for the earthing network.

frameworks, conduits, elevator structures, and the metallic carriers of services such as gas, water, smoke, air, etc., to make a highly interconnected mesh. The latter is then connected to the earthing system. This highly meshed three-dimensional system is also connected to the cable shields and armors, as well as the equipment frames or chassis. When a false floor or technical void is implemented, the jacks must be connected to the general earth network.

12.4.3 Minimizing the Earth Bond Impedance

The various apparatus, electronic cubicles, shields, filters, etc., shall have adequate low impedance bonding to the earthing network to limit the CM voltages between circuits, in such a way that the CM perturbations will run in a preferential way to the earthing network without passing through or coupling to other sensitive circuits (Figure 12.9). The various apparatus metal enclosures should be bonded to the earthing network through a very low impedance connection. In case of grounding wires, bond straps or braids, their length shall be as short as possible (no "pigtails"). Moreover, the bond connections shall ensure good electric continuity: no painting or isolating surfacing, or protection to prevent corrosion shall impair the electrical contact provided by screws or welding. Particularly, connector receptacles ensuring the grounding of shields shall be mounted on a paint-free surface.

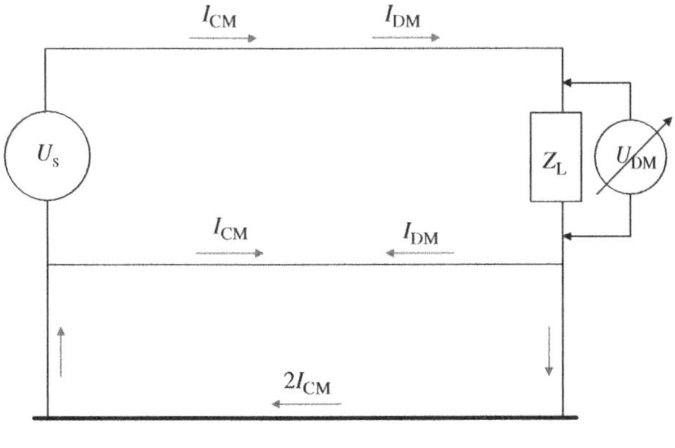

Figure 12.9 Flow of DM and CM currents.

12.4.4 Grounding of Cable Shields

Cables used for noisy and sensitive signals should be shielded, respectively, to limit their emission and their susceptibility to perturbations, in particular, for all but balanced differential twisted pairs. Cable shields act as parallel earth conductors (PECs), i.e., reduce the inductive coupling on inner wires, only when such cable shields are earthed at least at both extremities by short connections (no "pigtails"). The cable shield will be efficient, particularly at high frequencies (i.e., above 20 kHz) only if the following conditions are fulfilled:

- Cable shields should be directly earthed at both ends as a general rule, but at one end only if it is specified by the equipment manufacturer (e.g., for floating DM circuits). In such a case, an over-shield, earthed at both ends, is recommended in order to take the benefit of a PEC as close as possible to the DM circuit.

  Furthermore, there is no other PEC than the shield, such as metallic cable trays or an earth bus. This configuration is not recommended: If no cable tray is routing a cable, at least an earth cable bundled to the routed cable is highly recommended. Moreover, in DC traction systems, earthing of shielding or screening of trackside cables at each end must take into account the presence of stray currents. Connecting to the termination earth (mainly in the cables frames of technical rooms or cubicles) is recommended. Trackside end might be left unconnected, but protected against dangerous touching voltage due to EM coupling, as necessary.
- Cable shields shall be earthed with the lowest possible impedance. So, direct and peripheral connections to the metal chassis should be preferably used (see also the following points). Bonding a cable screen requires a connector back shells which achieve 360° (full circular) electrical contact to the connector, i.e., over the whole circumference of the screen: such connection ensures that the CM currents are diverted from the cables without creating a locally high transfer impedance Z_t.
- Pigtail-type earthing of cable screens are very poor for EMC and should be avoided (Figure 12.10). In cases where such a connection method cannot be avoided, the exposed inner conductors and the pigtail bond itself should be as short as possible, i.e., less than about 30 mm.

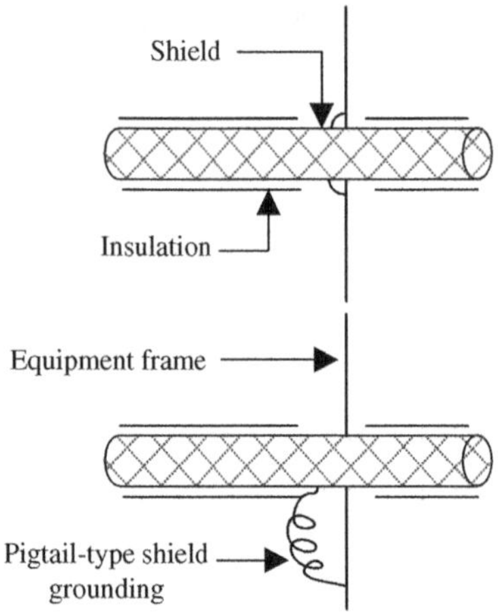

Figure 12.10 Good/bad practices for passing a cable shield through an equipment enclosure.

In relation with the cable tray design, it is advisable to properly bond cable shields to the metallic cable trays at their point of entry on the tray, in order to allow the EMI, which are flowing into the cable shield, being diverted into the cable tray.

12.4.5 Appropriate Design of Cables Routes

The objective of this design is to reduce the magnetic coupling of cables with external or local fields and the radiation of loops, such as common mode (CM) loops which are formed between active cables (i.e., differential mode circuits) and the circuit ground/chassis, or the earthing network and differential mode (DM) loops which are formed between one signal or power wire and its return wire.

All cable lengths shall be minimized. However, when an extra length of cables is deemed necessary (e.g., for maintenance purpose), it is advisable to lay the extra length of cable in snake shape, when possible. Loop-shape areas in circuits of high current variations (i.e., high di/dt) shall be strictly minimized.

12.4.5.1 Minimizing CM Loops

To reduce the earth loop currents flowing into cables due to magnetic fields, cables should always run as close as possible to directly earth metallic structures, such as earthed metallic cable trays, structural metalwork, steel beams or any other PEC, throughout the length of their run. However, if some cables are to be routed between two apparatus without any metallic conduit or trunk, an earth conductor could be routed with them, and properly connected to both equipment chassis, i.e., as close as practically possible to the cables entry. In the same way, each multi-conductor cable or bundle of single wires, if not armored nor screened, should include an earth conductor bonded to the local earth at both the equipment to which the cables are terminated, in order to create a PEC.

12.4.5.2 Minimizing DM Loops

"Signal" and "return" wires shall be kept together in the same bundle (e.g., for a power supply line fitted with a switch, traction feeder and return cables) to achieve a compact DM circuit. All DM circuits pairs should be kept compact, and preferably twisted. In multi-lead cables, using one single return wire for several 2-wires DM circuits should be avoided to limit couplings via a mutual resistance.

12.5 Cable Tray Assembling and Earthing

Conductive cable trays and ducts act as a PEC, provided that metallic cable trays and ducts must be electrically continuous over the whole length, including when routing through a wall. Metallic cable trays, supports and ducts must be made electrically continuous and earthed at both ends as a minimum. They shall be interconnected together with electrically short (i.e., of low impedance) leads and earthed over their length at regular intervals (as practically achievable) 50–100 m [9].

The bonding/earthing technique should cover the control of all frequencies of signals in cables that run inside these trays, i.e., the bonds impedance should be considered at the carried signal frequencies. Hence, the bonding/earthing with one single wire instead of one of the U-bracket evoked above will only cover the disturbances at the power frequency. Furthermore, bonding/earthing with several round cables can be suitable for low frequencies carrying trays. For high frequencies carrying trays, low inductance bonds, such as wide braids, should be envisaged.

12.5.1 Cable Segregation

The main cabling segregation, in every railway installation, aims to limit capacitive or inductive couplings of EMI between parallel cables, regardless of safety concerns. Cables routing constraints can however be minimized, thanks to the decoupling effects associated to constructive choices for the cables ducting system. Cables segregation constraints imply separation distances to be maintained between the cables, based on the cables classes, the type of ducting system, and its proper installation.

The constructive choices for a cables ducting system can provide significant attenuation effects on the EMI coupling. It can help minimizing the separation distances required between cables by taking advantage of the cable route acting as a PEC. This latter capability greatly depends on material employed, i.e., insulating materials (e.g., plastic, air) does not provide any attenuation factor for

EMI. Resistive materials (e.g., concrete, soil) provide weak attenuation factors for EMI. For this reason, the same rules apply to both those types of ducting. Metallic ducts, well installed, bring significant attenuation factor for EMI. Still, the plain metallic ducts will provide a much better attenuation factor than a ladder type.

Several metallic systems are available to duct cables. In addition to the dedicated systems presented below, the structural metalwork of the works can also be used as a metallic ducting system. Because of their open construction, ladder type cable support systems act as a PEC for low frequencies only, because their impedance significantly increases with the frequency, thus decreasing their decoupling effect.

Metallic cable trays are usually perforated with slots to make cable fixing easier, but this can degrade its high frequency performance. The shielding effectiveness is reduced by slots and gaps which interrupt the current flow and therefore increase the transfer impedance of the structure. Slots or holes in the corners of the trays should be avoided. Also, slots perpendicular to the direction of the tray should be avoided.

Further improvements allowing a greater concentration of cables while keeping the necessary decoupling are internal longitudinal separators and covers. Those improvements are effective provided that these additional parts are metallic, and properly connected to the main tray structure. The most sensitive cables should be routed as close as possible to the trays edges, which provide a better decoupling effect than the trays centers.

12.5.2 Cables Classification

Although there are many different standard practices that provide cable segregation rules, this section here below is inspired from the standard IEC 61000-5-2 [7], EN 50174-2 [8], IEC 60364-4-44 [9] and EN 50343 [10]. The cables are split broadly into three main categories according to the kind of signal they carry:

- Sensitive
- Indifferent
- Noisy.

Those categories are then subdivided into several classes, according to Table 12.2.

Each cables class should run along in a different trunking, or at least on one edge of a shared trunking, provided that the separation distances with other classes proposed hereafter are respected. Cables should only be bundled with (or in close proximity to) cables from its own class. Outdoor cables should be segregated from indoor cables, even if they are from the same class. Cables classes would ideally not cross over each other, but where they must cross, they should do so at right angles.

Cables distribution and route direction changes must maintain the separation distances between cable classes, including by improving the cable support type for the most sensitive cables classes 1 and 2, e.g., improving from an open cable tray to a covered cable tray.

12.5.3 Separation Distances

The segregation rules presented in this section are inspired by IEC 61000-5-2 [7] and IEC 60364-4-44 [9]. The golden rule of no segregation requirement for the length of parallelism if less than 35 m has been removed from the latest version of IEC 60364-4-44 [9]. Therefore, it is preferred to perform a detailed analysis if segregation prescribed in Table 12.3 is not achieved.

Table 12.2 Cable classification.

Category	Class	Signal type	Example
Sensitive	1	Analog low-level circuits of ~1 mV sensitivity scale	• Low level analog signals (millivolt output) • Sensor measurement signal • Audio and video analog signal cables
	2	Analog/digital low-level circuits of ~1 V sensitivity scale	• Standard analog signal (4–20 mA, 1–10 V) • HF and UHF transmitter and receiver signal cables (except leaky cables) • Digital communications cables (RS232, Ethernet). • AC/DC control signals ≤ 110 V (no inductive loads, such as contact relays)
Indifferent	3	Discrete control circuits	• AC/DC control signals for relays, contactors, etc. (e.g., relay command, pilot wire) of voltage level > 110 V
		Small power circuits	• AC/DC power supply <1 kV and <20 A • Battery related cables (from charger to battery and from battery to UPS)
Noisy	4	Medium power circuits	• Lighting of discharge type • PWM
	5	High power circuits	• HV power distribution cables, traction feeder cables (25 kV) • AC/DC power supply <1 kV and >20 A

Table 12.3 Separation distance (mm) in the air.

Class	1	2	3	4
2	300			
3	300	300		
4	600	600	600	
5	1200	1200	1200	1200

Furthermore, the distance defined can be reduced by adding a level of segregation. Typical reduction is considered as 50% [9]. A level of segregation is typically considered as follows:

- Metallic cable tray (perforated)
- Reinforced concrete duct
- Metallic conduit
- Cable tray cover.

It must be noted that the values in this table are applicable to cable routings and the related harnesses design of the mainline and depot. However, for cableways of internal buildings all the distances in this table will be reduced by a reducing factor of 33% (e.g., 600–200 mm) [9].

12.5.4 Filtering

Filtering is a method to limit conducted disturbances. The filter enables the frequency of the wanted signal or power and blocks the unwanted signal. A typical situation is the usage of low-pass filters for power frequency to block harmonic disturbances of power converters. Such filters shall be placed near the converter because cables conducting harmonics become a significant source of disturbance. A Common mode choke provides an increase in impedance of the CM circuit, thus reducing the CM current in the circuit. Another typical usage of filters is reducing the disturbance level of conducted disturbances at the border between different EMC-oriented lightning protection zones.

One of the most commonly used methods to increase the impedance of the CM circuit is electrical separation. This is effective at lower frequencies because at higher frequencies the parasitic capacitances come into play. Typically optical fibers, optocouplers, or isolation transformers are installed.

12.6 Overvoltage Arrestors

Overvoltage is a voltage exceeding the nominal voltage of an installation and, therefore, this might cause damage. The scope of an overvoltage arrestor is to limit the overvoltage below the insulation level of the installation. The insulation level of an installation has to be chosen according to the environmental and operational conditions of the installation. Overvoltages are caused by lightning strikes, switching of electrical loads or induced voltages. They vary significantly regarding its maximum level, its duration, and the maximum current driven. Overvoltage arrestors may not limit permanent overvoltages or frequently occurring transients as this would, on one hand, influence the signal or power to be transmitted and on the other hand, excessively stress the arrestor and damage it.

12.7 EMC Analysis

The EMC analysis is a compilation and evaluation of EMC data and EMC requirements to determine the degree of interference with electrical devices. During the EMC analysis, qualitative and quantitative studies are used to analyze the degree of interference and the paths of electrical transmission between the system and its environment and between various components of the system itself. Available methods include theoretical studies, numerical computation, laboratory measurements, estimates, experimental values, and experience. Furthermore, EMI risks are identified and a risk matrix is prepared to define the ranking of each risk and accordingly mitigation measures are prescribed to reduce it to as low as reasonably possible level (ALARP).

Moreover, to understand the existing EM environment, a cartography survey is performed. This is to identify typically special locations which fall along the alignment such as radio transmission equipment, medical installations with magnetic imaging equipment, overhead power lines adjacent to the alignment, military installations, etc. The outcome of this survey is further used to identify EMI risks and it sets the baseline for EM emissions from the whole railway system.

Furthermore, an induced voltage study along with a magnetic field pattern is conducted to understand the impact on installation running along the track. The method applied to estimate the magnitude of the EM interference requires evaluating the magnetic coupling **M** between the rail power system conductors and an adjacent conductor parallel to the tracks (Figure 12.11). Along with **M**, and the noise current (I_f) of the railway power, the resultant is coupled to provide a common mode noise voltage V_{cm} for a particular supply frequency (f), which is developed on the telecom cables, signaling cables, and other installations.

$$V_{cm} = 2\pi f \cdot M \cdot I_f \tag{12.1}$$

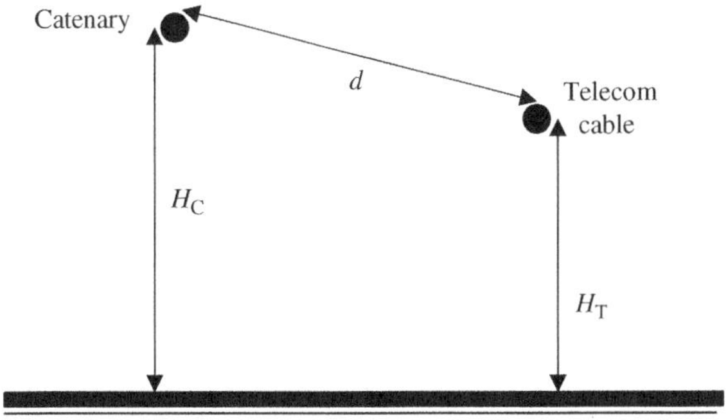

Figure 12.11 Conductor cross-section.

A number of sources provide simple models for rail power system coupling to nearby conductors. The complication in modeling this system is the earth's conductivity which is non-homogeneous and varies with seasonal temperature and humidity, and if affected by underground conductors such as pipes and structures. The two most commonly used formulae to calculate mutual inductance have been presented here.

The CCITT simple model [11]

$$M = 1425 + 45.96x - 1.413x^2 - 198.4 \ln x \ \mu\text{H/m}$$

$$x = d \cdot \sqrt{\frac{\mu_o \cdot \omega}{\rho}}$$

where

d, distance between two conductors
ρ, soil resistivity
μ_o, permeability of free space
ω, angular frequency ($2\pi f$)
f, frequency of power supply

The Tesche Model [11]:

$$M_t = \frac{\mu_o}{2\pi} \cdot \ln \sqrt{\frac{d^2 + (H_C + H_T)^2 + \delta^2}{d^2 + (H_C - H_T)^2}}$$

$$\delta = 1.85 \sqrt{\frac{\rho}{2\pi\mu_o f}}$$

where

H_C, height of conductor (here catenary)
H_T, height of another conductor (here telecom cable)

12.8 EMC Tests

EMC measurements and tests are required to demonstrate the compatibility of the overall system with the outside world. All the tests and examinations are conducted according to the environmental conditions set in the corresponding standard. Sometimes it triggers that once the declaration of conformity or EMC type test reports of equipment is obtained, then there is no requirement of any further testing. However, it must be highlighted again that even if all equipment installed are EMC-compliant, there is still a possibility of EM interference to occur due to arrangement (or layouts), cable management system, etc. Thus, the following are the minimum requirements that are required to be covered during the commissioning of railway system:

- To conduct the measurement of radiated EM field emissions according to EN 50121-2 [12]. The measurements of radiated emissions shall be carried out at a distance of 10 m from the track or traction power substation. The covered frequency range is 9 kHz–1000 MHz. Figures 12.12–12.14 provide a typical arrangement for EMC test as per EN 50121-2.

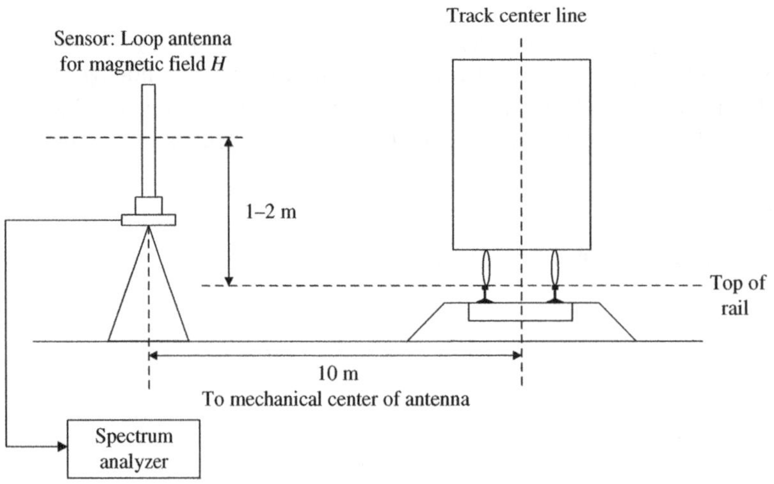

Figure 12.12 Position of antenna for measurement of a magnetic field in 9 kHz to 30 MHz frequency band.

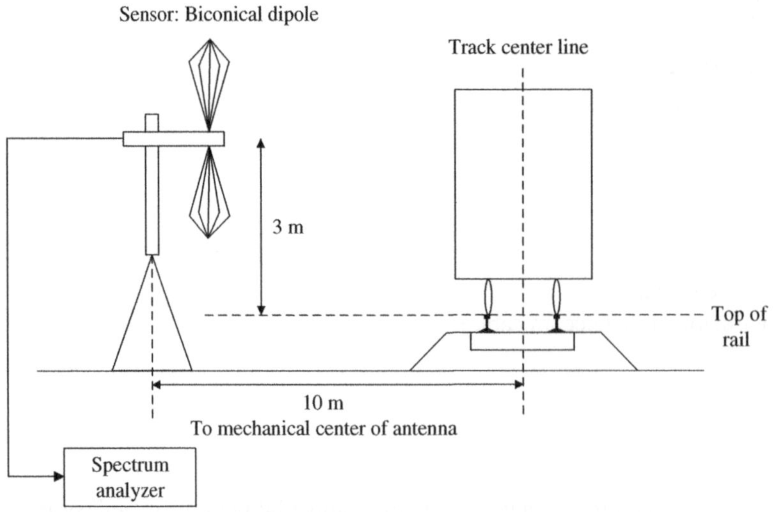

Figure 12.13 Position (vertical polarization) of antenna for measurement of electric field in 30–300 MHz frequency band.

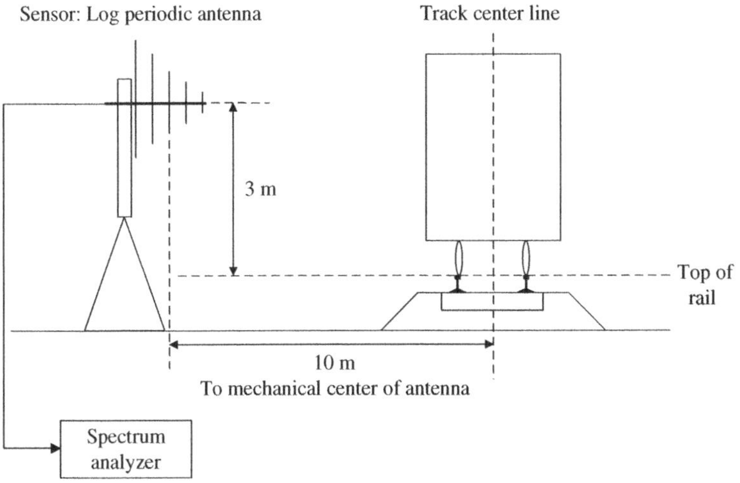

Figure 12.14 Position (vertical polarization) of antenna for measurement of electric field in the 300 MHz-to-1 GHz frequency band.

- Similar to this, radiated EM field emissions according to EN 50121-3-1 [13] are recommended for trainsets (traction stock, hauled stock, and trainsets).
- The measurement of magnetic field levels according to EN 50500. The covered frequency range is 0 Hz–20 kHz. This measurement is expected to be carried out where passengers and maintenance staff are expected such as platforms, rolling stock, substations (traction and auxiliary).
- Compatibility between the train detection system and rolling stock must be measured and complied with EN 50238-2 [14] in case of track circuits or EN 50238-3 [15] (test as per EN 50592 [16]) in case of axle counters.

References

1 IEC 60050-161:1990/AMD9. (1990/2019). *International Electrotechnical Vocabulary (IEV) – Part 161: Electromagnetic compatibility.*
2 Paul, C.R. (2006). *Introduction to Electromagnetic Compatibility*, 2e. Hoboken, NJ: Wiley.
3 EN 50121-4. (2015). *Railway applications — Electromagnetic compatibility — Part 4: Emission and immunity of the signalling and telecommunications apparatus.*
4 Ogunsola, A. and Mariscotti, A. (2013). *Electromagnetic Compatibility in Railways*. Berlin: Springer.
5 EN 50121-5. (2017). *Railway applications — Electromagnetic compatibility — Part 5: Emission and immunity of fixed power supply installations and apparatus.*
6 EN 50121-3-2. (2015). *Railway applications — Electromagnetic compatibility — Part 3-2: Rolling stock - Apparatus.*
7 IEC 61000-5-2. (1997). *Electromagnetic compatibility — Part 5: Installation and mitigation guidelines – Section 2: Earthing and cabling.*
8 EN 50174-2. (2018). *Information technology – Cabling installation – Part 2: Installation planning and practices inside buildings.*
9 IEC 60364-4-44. (2016). *Low-voltage electrical installations – Part 4-44: Protection for safety — Protection against voltage disturbances and electromagnetic disturbances.*
10 EN 50343. (2015). *Railway applications – Rolling stock – Rules for installation of cabling.*

11 CCITT (International Telegraph and Telephone Consultative Committee). (1989). *Directives concerning the protection of telecommunication lines against harmful effects from electric power and electrified railway lines*, vol. 2. Geneva, Switzerland: ITU.

12 EN 50121-2. (2006). *Railway applications — Electromagnetic compatibility — Part 2: Emission of the whole railway system to the outside world.*

13 EN 50121-3-1. (2017). *Railway applications — Electromagnetic compatibility — Part 3-1: Rolling stock – Train and complete vehicle.*

14 EN 50238-2. (2020). *Railway applications — Compatibility between rolling stock and train detection systems – Part 2: Compatibility with track circuits.*

15 EN 50238-3. (2013). *Railway applications — Compatibility between rolling stock and train detection systems – Part 3: Compatibility with axle counters.*

16 EN 50592. (2016). *Railway applications – Testing of rolling stock for electromagnetic compatibility with axle counters.*

13

Stray Current and Rail Potential Control Strategies in Electric Railway Systems

Aydin Zaboli and Behrooz Vahidi

Amirkabir University of Technology (Tehran Polytechnic), Department of Electrical Engineering, Tehran, 1591634311, Iran

13.1 Introduction

In our modern society, time and speed are two necessary factors. Electric trains have a significant role in a wide range of factors such as air pollution and traffic congestion. Hence, governments usually allocate a considerable budget to establish these green transportation systems. Besides, it is imperative to pay attention to the adverse effects of these systems [1]. DC traction power supply system is generally utilized in railway systems with running rails as a return path. One of the most critical issues of electric railway systems is the leakage of traction currents to the earth due to the weak insulation of the rail fastening systems called stray currents. There is no suitable insulation between the running rails and soil. It can lead to non-uniform potential distribution along the running rail [2, 3]. Meanwhile, the new structure of DC railway systems can be adopted to reduce the negative effects of the stray current with presented methods in the next sections [4].

For more clarification, first the current transfers to the train through a third rail or overhead line. Then, the traction currents enter the running rails. These currents should go back to the traction power substation (TPS), but a small percentage leaks into the earth. These are very destructive for underground structures and can cause high expenditures for governments.

Nevertheless, these expenditures can dramatically be decreased with the proper management of railway systems. When companies tend to establish a rail line, it is essential to consider the aforementioned factors and apply the control methods before the establishment. Hazardous stray currents can generally lead to issues including noise pollution and communication's noise production, making faults in electric trains' navigation systems, destruction of equipment insulation, underground infrastructure and pipeline corrosion, weakening of metal structures' cathodic protection and damage to computer sites [5–10].

Transportation Electrification: Breakthroughs in Electrified Vehicles, Aircraft, Rolling Stock, and Watercraft,
First Edition. Edited by Ahmed A. Mohamed, Ahmad Arshan Khan, Ahmed T. Elsayed, and Mohamed A. Elshaer.

13.2 Principle of Stray Current and Corrosion Calculation

13.2.1 Mathematical Calculation of Stray Current

A single-train representation of a DC-electrified railway system is considered in Figure 13.1 in which $i(x)$ is the current in the running rails and $v(x)$ is the rail voltage through Sections x_1–x_2 of the running rails that can be calculated using Kirchhoff's current law as Eqs. (13.1)–(13.3) [11]:

$$i(x) = c_1 e^{\gamma x} + c_2 e^{-\gamma x} \tag{13.1}$$

$$v(x) = -R_0(c_1 e^{\gamma x} - c_2 e^{-\gamma x}) \tag{13.2}$$

$$I_S(x) = i(x + \Delta x) - i(x) \tag{13.3}$$

where $\gamma = \sqrt{RG}\,(\mathrm{m}^{-1})$ is the propagation constant, $R_0 = \sqrt{R/G}\,(\Omega)$ is the characteristic resistance of the rail conductor's earth system, c_1 and c_2 are the constants determined by boundary conditions, R is running rails resistance per rail, G is the rail-to-earth conductance, $I_s(x)$ is the stray current along the track and x is the distance between TPS and train [12, 13].

Effects of running rails and the design of power systems were assessed on the stray current by Schaffer et al. in [14]. They stated that there is no necessity for stray current collection mats if running rails are entirely insulated and power system design can maintain the stray current values at the desired level. Yu and Goodman [15] assumed that the running rails and the rail-to-earth resistance are uniformly distributed, while it was not valid in real life because these parameters are affected by the local values of soil resistivity, construction techniques, and ballast or track-bed conditions.

13.2.2 Corrosion Formulation

Stray currents can only flow through damaged coating areas to initiate electrochemical reactions. The corrosion reactions of the steel under the DC anode can be summarized as follows [16, 17]:

$$\left.\begin{array}{l} \mathrm{Fe} \rightarrow \mathrm{Fe}^{2+} + 2\mathrm{e}^- \\ \mathrm{Fe}^{2+}\, 2\mathrm{H_2O} \rightarrow \mathrm{Fe(OH)_2} + 2\mathrm{H}^+ \\ 4\mathrm{Fe(OH)_2} + \mathrm{O_2} \rightarrow 2\mathrm{Fe_2O_3} + 4\mathrm{H_2O} \end{array}\right\} \tag{13.4}$$

When pipelines or other underground metal structures are corroded by DC stray current, the metal corrosion satisfies Faraday's electrochemical first law; the corrosion w is directly proportional to the stray current as follows [18]:

$$w = KIt, \quad v = {}^{wI}\!/_{A}, \quad K = {}^{M}\!/_{nF} \tag{13.5}$$

where w is the metal corrosion in grams; K is the electrochemical equivalent of the metal in kg/(A s); I is the value of the current flowing out of the metal anode in A; t is the corrosion duration

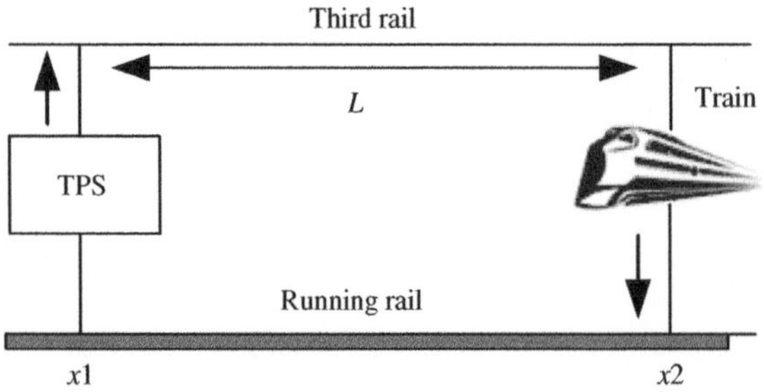

Figure 13.1 A simple view of the single-train system with traction power substation.

in hours; M is the molecular mass of the metal; F is a constant. K value is supposed to be 0.00104 which obtains approximately as 1/96,500 that demonstrates an electric charge of 96,500 coulombs passing through an electrolyte cavity [18]. These equations are useful to estimate the corrosion damages caused by DC stray current $I = 1$ A, steel corrosion amounts lead to 9.13 kg/yr [19].

To clarify, Figure 13.2 demonstrates the corrosion process and the equivalent electrical circuit. As you can observe, traction currents leak into the anodic area (point A) of the soil and then enter the cathodic area of the concrete. In this part, currents penetrate the underground infrastructures and

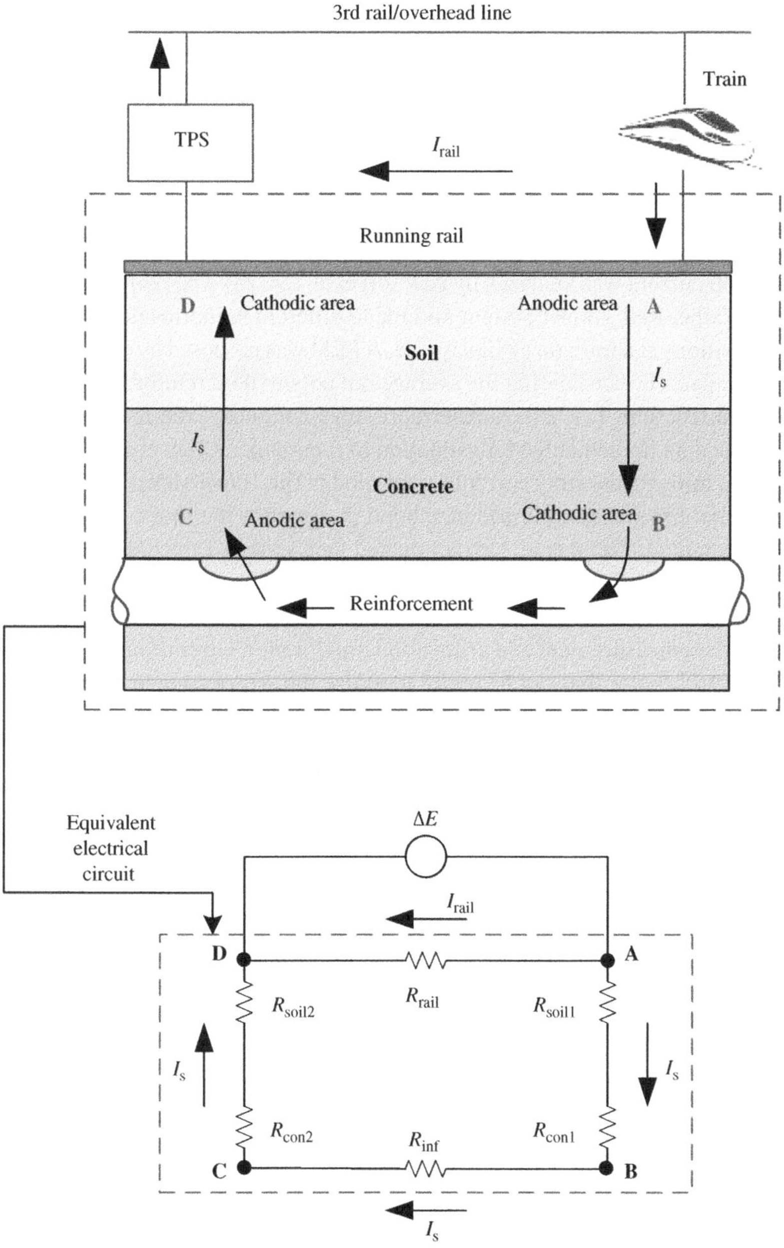

Figure 13.2 Process of the corrosion through stray current penetration to reinforcement and an equivalent electrical circuit.

lead to corrosion. The closed equivalent electrical circuit relevant to this process is illustrated in Figure 13.2 where R_{rail}, R_{soil1}, R_{soil2}, R_{con1}, R_{con2}, and R_{inf} are the electrical resistances of the running rails, soil, concrete, and infrastructures at different positions, correspondingly; I_s is the stray current flowing into the infrastructures. ΔE has been raised because of the potential difference between points A and B as well [19–21]. In [9], Zhichao and Cheng presented a transient finite element method (FEM) to make a comparison between different factors affecting the stray currents that lead to corrosion in underground infrastructures. These significant factors are cathodic protection, the operation condition of the train, soil types, depth of buried metals, and amount of traction currents [17].

13.3 Literature Review of Control Strategies

Cotton et al. [22] presented a comparison between the stray current collection mats and floating systems and then analyzed the efficiency of the collection systems as a method for stray current mitigation. Lee and Lu [23] compared the different methods of grounding systems (GSs) and made computational formulations for their efficiency in a case study. The effect of GSs on the practical subway systems and corrosion current was analyzed by Alamuti et al. [24, 25]. Lee [26] studied the maximum rail potential in Taipei Rail Transit System and made different scenarios due to calculating the rail potential in positions of a train on a railway line. A FEM was proposed by Dolara et al. to reduce the stray current impacts by considering the average rail potential on reinforcement bars and other underground infrastructures [27, 28]. Furthermore, they conducted the real measurements in comparison with the data implemented for validation in their paper. Chen et al. [29] used the diode-grounded system to mitigate the stray current for a period in the Taipei Metro. Using field tests, they figured out that the disconnecting impedance bond at a part of the line can lead to a severe reduction of the rail potential. Tzeng et al. [30] conducted field tests in Taipei Metro by collection mats as a stray current and rail potential limitation method and provided a comparison between the accumulated stray currents in different GSs. Charalambous et al. [31] presented a computer module based on local data measurement and acquisition unit for momentum control of stray currents by the operation control center that can be useful to utilize the sensors due to monitoring the hazardous currents and rail potential as well. Stray current control using lumped parameters in the presence of different structures of soil has been conducted by Fichera et al. [32]. Assessment of the stray current for tunnel sections by using a boundary element platform was done by Charalambous et al. [33, 34] that the evaluation of these currents relied on the tunnel geometry and topology. Research related to the tunnel structure associated with the GSs is performed comprehensively by Lin et al. [35, 36] using the FEM. Research conducted by Park [37] has demonstrated useful results related to stray current decline. He utilized the fault location and detection methods to find the faulty sections instantly, which these faults can be problematic for the railway systems and engenders stray currents at higher levels because of the high amplitude of fault currents. Chen et al. [21] presented a comprehensive review of stray currents' impacts on corrosion of underground infrastructures. Charalambous illustrated an efficient method for the variability of different parameters known as the Monte Carlo method to describe certain parameters related to stray currents between the running rails and the earth [38, 39]. He represented a robust simulation method with uniform and non-uniform soil models to evaluate the stray currents. A comparison between the analytical method and the finite element model of a DC railway system by taking advantage of the stray current collection mats was carried out by Zaboli et al. [12], which covers the simple and precise models of the electric train in different scenarios with modeling of horizontal and vertical soil layers. Recent research conducted by Ibrahem et al. [40, 41] has presented an impedance emulator as variable resistors for the stray current and touch voltage prediction via power electronic devices to reduce the destructive effects. In recent years, many researchers have suggested a wide range of power filters with different configurations to reduce the electromagnetic compatibility impacts on power quality, which is directly effective on stray currents [42–48]. Alnuman et al. [49, 50] developed a fourth rail system by MATLAB/SIMULINK, which can be advantageous for the stray current decrement. This system predominantly served in the UK in recent years. The effect of trains' movement timetable on stray currents control is done by Luo et al. [51] that these currents originate from regenerative braking currents, in which the stray current is a negligible percentage of the total current [36, 52].

A comparison of stray current control methods relies on several factors such as location, soil type, distance between the TPSs, moisture, etc. Some methods are almost impossible to be applied to the existing railway lines, but some can be costly or there will be some limited applications. For instance, decreasing of the distance between the TPSs is only feasible for new railway lines, while it will be complicated and expensive to renovate the existing system. Another example can be the stray current collection system that should be regarded at the initial design. This control method also has some issues such as the corrosion of bars after a long time that will make it difficult to be maintained. Changes in the nominal voltage are also unjustifiable because these can affect the electric equipment of the railway system regarding their voltage threshold. The explanations of different control methods along with their advantages or drawbacks are presented in the next sections.

13.4 Stray Current Control and Limitation Methods

13.4.1 Increase of Rail-to-Earth Resistance

This case is equivalent to the reduction of the rail-to-earth conductance, which is one the most outstanding points in the running rails' design. A simple representation of a single-train DC-electrified railway system with two TPSs is illustrated in Figure 13.3. The distance between TPS1 and TPS2 is L, which can be considered a variable for different simulations. Also, the train contains 28 mutually

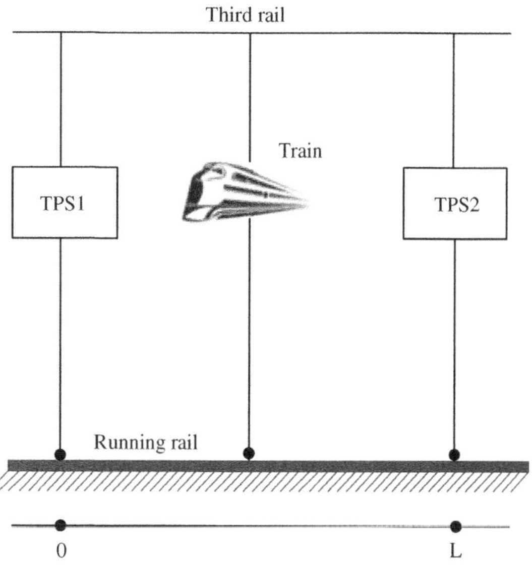

Figure 13.3 Single-train model of DC railway system with two TPSs.

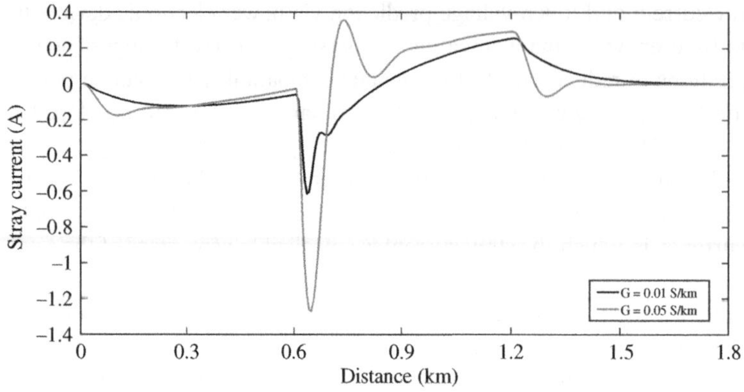

Figure 13.4 Stray current variations with different conductance.

series DC motors and line voltage is 750 V. The train is regarded as a dynamic load to calculate the stray current and rail voltage changes. Some slight fluctuations are due to the dynamic nature of train performance, depicted in Figure 13.4. As shown in this figure, a wet rail with higher conductance leads to the maximum stray current. Therefore, the running rail with less conductance results in less stray current in the traction system [53].

Take railway sleepers or railway ties as an example. It is favorable to have high resistance between rails and sleepers; therefore, concrete sleepers have much better resistance than wooden ones. Because wooden sleepers have a low current transmittable capability on their surface, their lifetime is less than concrete sleepers due to the inner destruction of the wood. Moreover, other factors are impactful as follows: dryness or wetness of the running rails, electric insulation's exhaustion underneath the rail, cleanliness or dirtiness of running rails, and environmental pollution. Investigations demonstrate that composite sleepers have appropriate characteristics including vibration's dampening, low weight, and tolerable against abrasion and severe weather conditions. Furthermore, they are compatible with the environment and lead to less pollution [53–56].

13.4.2 Locating TPSs Adjacent to the Points of Maximum Train Acceleration or Adding TPSs

This control method represents that the lesser the distance between the TPSs, the lesser the traction currents produce. That means the traction current does not exceed from the specific limitation. The most optimal condition is that the stations are located near each other to reduce the acceleration speed. Equation (13.6) depicts the total stray current leaking from an ungrounded system as follows [34]:

$$I_{\text{stray}} = I \times r_t \times l^2 / 8 \times r_c \tag{13.6}$$

where I is the traction current in A; r_t is the track resistance (the two running rails in parallel) in Ω/km; l is the distance between each of the two substations and the train when the train is at the midpoint measured in km; r_c is the resistance to earth of the track [22].

Stray current (I_{stray}) is proportional to the square of length. A series of factors are effective in the distance between TPSs such as the simple possibility of installation, coordination with traffic and the number of trains, less maintenance expenditure, and optimal performance at the beginning of railway design [12, 26]. These factors are impactful on stray current levels. That means the time scheduling for train movement or using the artificial intelligence algorithms in time management

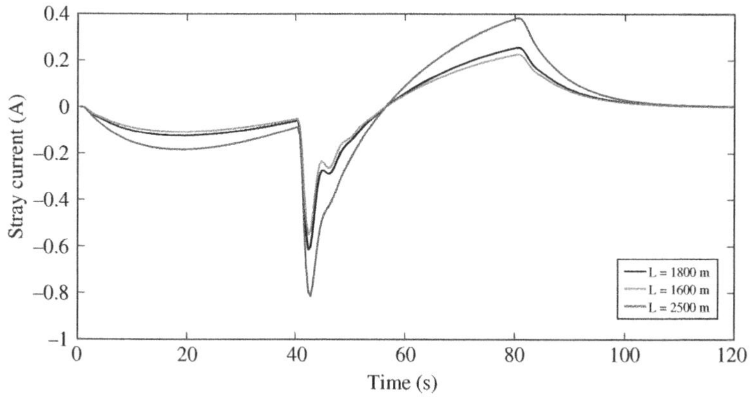

Figure 13.5 Stray current changes between two consecutive TPSs.

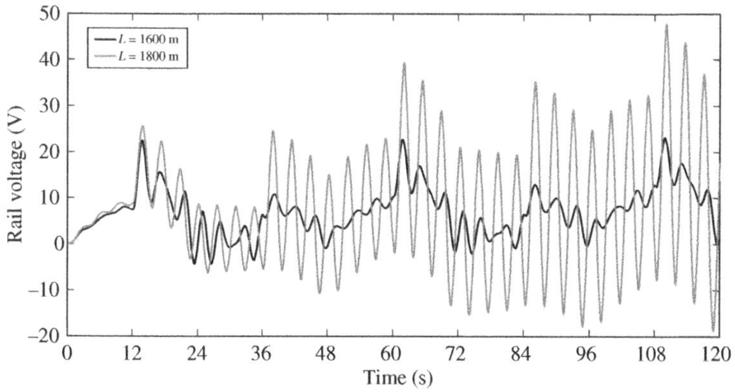

Figure 13.6 Rail voltage changes between two TPSs.

are such factors in stray current levels mitigation. The impact of different distances between the TPSs on stray current and rail potential is demonstrated in Figures 13.5 and 13.6, respectively. Figure 13.5 shows that the more the distance between the TPSs increases, the more the stray currents are. Also, the distance is an efficient factor in the rail potential which is shown in Figure 13.6.

13.4.3 Traction Supply Voltage Increase

One method for the traction current reduction is an increment in supply voltage [57]. The fundamental voltage at the pantograph of each train is analyzed at each simulation time step. The mean voltage is utilized for DC systems, while the rms value of fundamental voltage is used for AC systems. The mean voltage at the pantograph can be expressed mathematically as follows [58]:

$$U_{\text{mean}} = \frac{\sum_{i=1}^{n} \frac{1}{T_i} \int_{0}^{T_i} U_{\text{pi}} \times |I_{\text{pi}}| \times dt}{\sum_{i=1}^{n} \frac{1}{T_i} \int_{0}^{T_i} |I_{\text{pi}}| \times dt} \tag{13.7}$$

Table 13.1 Parameters in mean voltage at the pantograph in DC systems.

Parameter	Definition
T_i	The integration period on train i
N	The number of trains
Up_i	The momentary average DC voltage at the pantograph of train i
$\lvert Ip_i \rvert$	The modulus of momentary average DC current passing through the pantograph of train i

Equation (13.7) represents the relationship between the mean power calculated for the trains during the traction sequences and the corresponding mean current. More information related to this equation for DC systems is given in Table 13.1.

The larger the supply voltage, the smaller the absorbed and the traction current [55], [59–61]. However, not all voltage levels are allowed (standardized levels may be found in EN 50163). It is worth observing that for a given nominal line voltage, there is a wide range of operating voltage levels that can be selected by the turn ratio of the TPS transformers and by means of tap changers: slightly higher voltage levels are compatible with the voltage rating and insulation level of the infrastructure, are accepted by the rolling stock (that is designed for a wide range of variations of the input voltage), and are also helpful for the energy efficiency.

13.4.4 Stray Current Collection Mats

In this control method, designers generally use metallic low-resistance mats to drain and collect the stray current leaving the running rails. These mats locate at certain intervals along the running rails and connect to the negative busbar of the TPS through a cable and a diode [62]. It serves in the paths which have concrete frames. An example of the single-train model with two TPSs including collection mats is presented in Figure 13.7.

As mentioned, these mats are connected to the traction earth busbar, where the bus bar is grounded through resistance for measuring the secondary stray current (R_{ss}). Moreover, R_{ps} is the resistance for measuring the primary stray currents that leak from the running rails. A drainage diode serves as a device to record and isolate the source from an increased stray current. A simple electric circuit of the train is presented in Figure 13.8, where R_c, R_{rr}, R_{rm}, R_{mm}, and R_{mg} are the resistance of the third rail or overhead line, running rails, rail-to-mats, collection mats, and mats to earth. The train is modeled as a constant current source and Norton's equivalent circuit is considered for the TPS [12].

To analyze the electric current and voltage at different segments of the network, we can achieve matrices (13.9)–(13.11) using the assumptions given in (13.8). They calculate the node's voltages, rail potential, mat current or voltage, and earth resistance's voltage [30, 63].

$$G_c = \frac{1}{R_c},\ G_c' = \frac{1}{R_c'},\ G_{rm} = \frac{1}{R_{rm}},\ G_{mm} = \frac{1}{R_{mm}},\ G_{rr} = \frac{1}{R_{rr}},\ G_{mg} = \frac{1}{R_{mg}} \tag{13.8}$$

$$I = Y_{bus} V \tag{13.9}$$

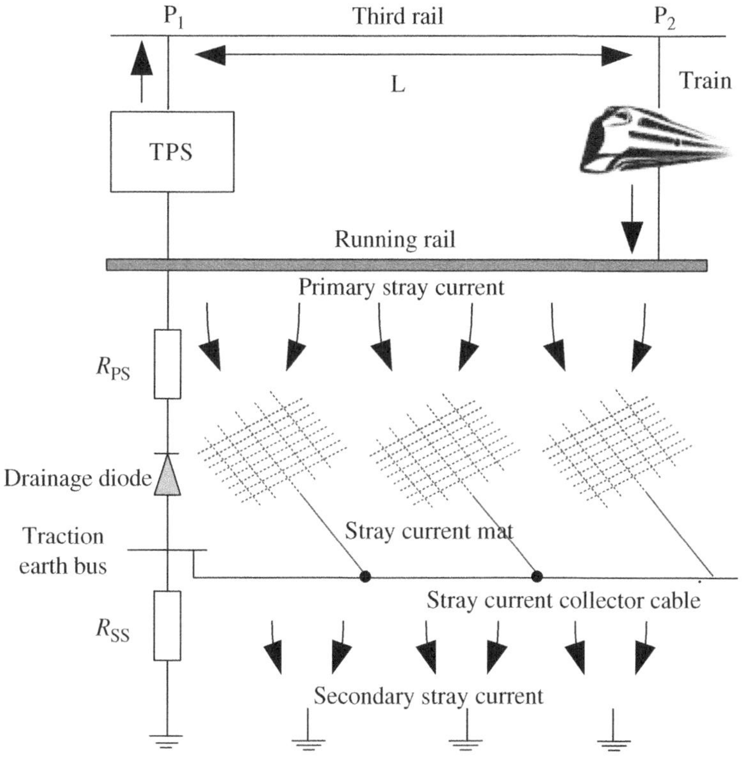

Figure 13.7 Stray current collection mats located beneath the running rails.

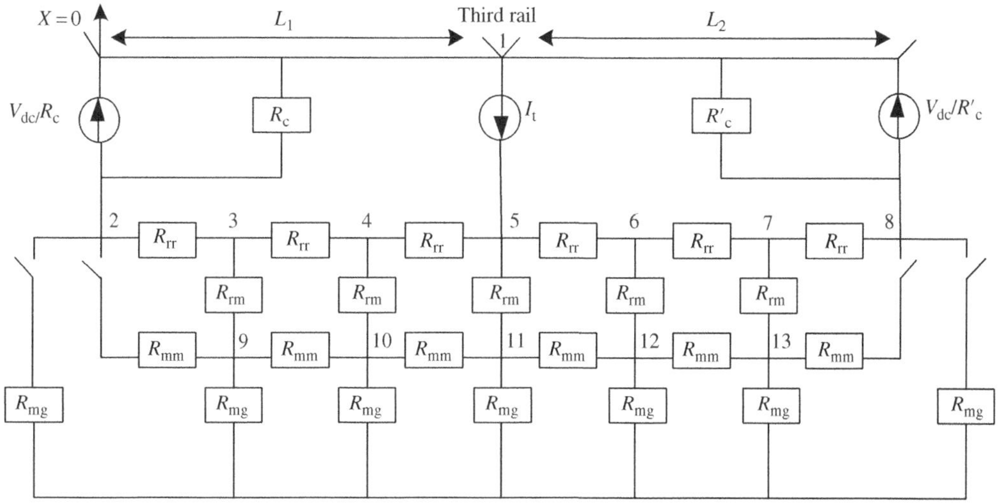

Figure 13.8 The equivalent electrical circuit of the collection mats associated with other resistances.

$$
Y_{\text{bus}} = \begin{pmatrix}
G_c + G_c^{'} & -G_c & 0 & 0 & 0 & 0 & 0 & -G_c^{'} & 0 & 0 & 0 & 0 & 0 \\
-G_c & \begin{matrix} G_{rr} + G_c + \\ G_{mg} + G_{mm} \end{matrix} & -G_{rr} & 0 & 0 & 0 & 0 & 0 & -G_{mm} & 0 & 0 & 0 & 0 \\
0 & -G_{rr} & 2G_{rr} + G_{rm} & -G_{rr} & 0 & 0 & 0 & 0 & -G_{rm} & 0 & 0 & 0 & 0 \\
0 & 0 & -G_{rr} & 2G_{rr} + G_{rm} & -G_{rr} & 0 & 0 & 0 & 0 & -G_{rm} & 0 & 0 & 0 \\
0 & 0 & 0 & -G_{rr} & 2G_{rr} + G_{rm} & -G_{rr} & 0 & 0 & 0 & 0 & -G_{rm} & 0 & 0 \\
0 & 0 & 0 & 0 & -G_{rr} & 2G_{rr} + G_{rm} & -G_{rr} & 0 & 0 & 0 & 0 & -G_{rm} & 0 \\
0 & 0 & 0 & 0 & 0 & -G_{rr} & 2G_{rr} + G_{rm} & -G_{rr} & 0 & 0 & 0 & 0 & -G_{rm} \\
-G_c^{'} & 0 & 0 & 0 & 0 & 0 & -G_{rr} & G_c^{'} + G_{rr} + G_{mm} + G_{mg} & 0 & 0 & 0 & 0 & -G_{mm} \\
0 & -G_{mm} & -G_{rm} & 0 & 0 & 0 & 0 & 0 & 2G_{mm} + G_{mg} + G_{rm} & -G_{mm} & 0 & 0 & 0 \\
0 & 0 & 0 & -G_{rm} & 0 & 0 & 0 & 0 & -G_{mm} & 2G_{mm} + G_{mg} + G_{rm} & -G_{mm} & 0 & 0 \\
0 & 0 & 0 & 0 & -G_{rm} & 0 & 0 & 0 & 0 & -G_{mm} & 2G_{mm} + G_{mg} + G_{rm} & -G_{mm} & 0 \\
0 & 0 & 0 & 0 & 0 & -G_{rm} & 0 & 0 & 0 & 0 & -G_{mm} & 2G_{mm} + G_{mg} + G_{rm} & -G_{mm} \\
0 & 0 & 0 & 0 & 0 & 0 & -G_{rm} & 0 & 0 & 0 & 0 & -G_{mm} & 2G_{mm} + G_{mg} + G_{rm}
\end{pmatrix}_{13 \times 13} \tag{13.10}
$$

$$
I = \begin{pmatrix}
\dfrac{V_{\text{dc}}}{R_{\text{c}}} + \dfrac{V_{\text{dc}}}{R_{\text{c}}^{'}} - I_{\text{t}} \\
-\dfrac{V_{\text{dc}}}{R_{\text{c}}} \\
0 \\
0 \\
I_t \\
0 \\
0 \\
-\dfrac{V_{\text{dc}}}{R_{\text{c}}^{'}} \\
0 \\
0 \\
0 \\
0 \\
0
\end{pmatrix}_{13 \times 1} \tag{13.11}
$$

As you can observe, there are two types of electric current leaking from running rails; first, it is known as the primary stray current that leaks from rails to the collection mats, and another one is the secondary stray current that transfers from the collection cable to the earth. Explicitly, the latter is less than the primary stray current. Hence, there is a relationship between stray current collection mats' efficiency as (13.12) [13]:

$$\eta = \frac{I_{\text{collected}}}{I_{\text{st}}} \times 100 \tag{13.12}$$

This equation demonstrates how a percentage of total currents leaking from the running rails can accumulate in collection mats where $I_{\text{collected}}$ is the accumulated stray current by the collection mats, and I_{st} is the leakage current from the running rails. The more the collection mats, the better the efficiency. In [12, 29, 38, 64], analytical and numerical simulations have demonstrated a wide range of mats' efficiency for different types of soil models.

There is a case study for better clarification of collection mats efficiency in DC railway system. Suppose there is a horizontally two-layer earth with sample collection mats. A characteristic of data related to soil type is in Table 13.2. As you can see in Figure 13.9, there is an intense density in stray current below 0.2 mA/m for different soil types analyzed by FEM.

There is a mixed variety of changes, including soil types with different resistivities and layer placements. Efficiency values vary from 82.9% (for the soil models with 1000 Ωm for the upper layer and 100 Ωm for the lower layer) to 97.9% (for the soil models with 300 Ωm for the upper layer and 100 Ωm for the lower layer). This research also presents a comparison between the FEM and analytical model of the soil [12]. These mats serve merely as a control method for stray current mitigation, not structural goals.

Table 13.2 Different models of horizontally two-layer soil type along with the relevant resistivity.

	Model A	Model B	Model C
Soil type (upper layer/lower layer)	100/300 Ω m	1000/300 Ω m	1000/100 Ω m

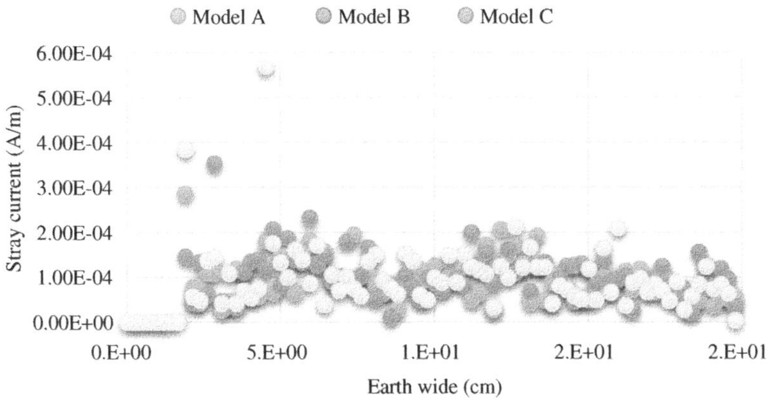

Figure 13.9 Stray current distribution in earth-wide horizontally two-layer soil type.

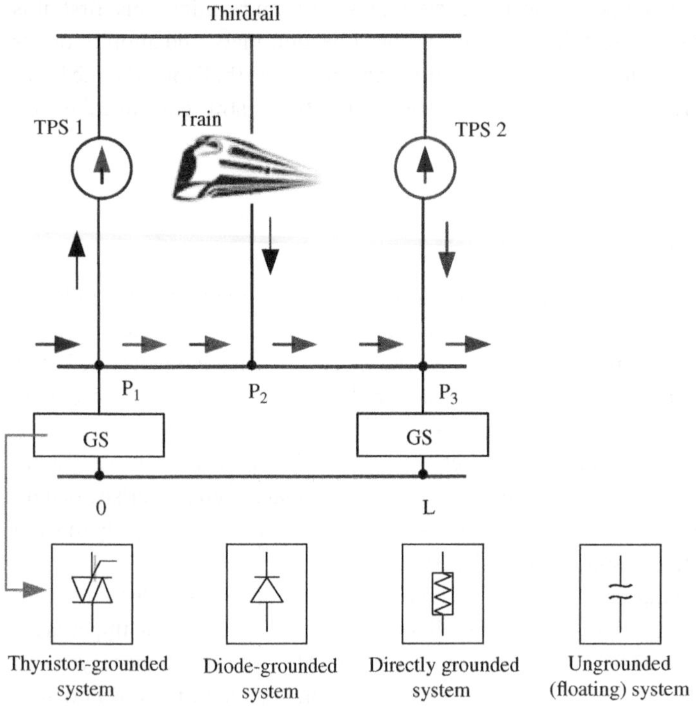

Figure 13.10 Single-train with TPSs accompanied by different types of the grounding system.

13.4.5 Grounding Schemes

GSs are essential for two main goals: stray current and rail potential minimization (in view of electrical safety). Grounding of track is designed and sized for normal and faulty conditions with a wide range of solutions, as a compromise between the two mentioned goals: ungrounded (floating) system, directly grounded system, diode-grounded system, thyristor-grounded system. Each has advantages and drawbacks described later. Figure 13.10 illustrates a single train with TPSs including the GSs that are placed at the substations and common types of grounding schemes [24, 25, 65].

13.4.5.1 Ungrounded System

There is no direct electrical connection between the running rails and the negative busbar or earth. Because the voltage can vary between positive and negative values, it is known as a floating system. Resistance values depend on some factors such as the insulation material between the running rail and earth, weather conditions, line fatigue, and so on. This structure presents a high touch potential that is unfavorable for personnel, but the stray current level is low. This presents the minimum stray current among GSs. Therefore, protection solutions against touch voltage, as well as implementing the insulation of passengers' platforms is vital for the reduction of electric shock risk [66, 67]. For achieving this purpose, a voltage-limiting device (VLD) can be used that can provide an appropriate system associated with the floating system. The total mechanism of this process with VLD is represented in Figure 13.11.

Secheron SA Company presented two configurations as follows: basic VLD, hybrid VLD [66]. In the basic configuration (as shown in Figure 13.12 with light gray lines), an electric current enters the circuit through the negative point, then earthing contractor (1) will switch to on-state by the protection and control unit (2) and current flow will connect to the earth. This structure provides a new path for stray currents to enter the earth instead of the underground infrastructures. Hybrid VLD (as demonstrated in Figure 13.12 with light and dark gray lines) presents an advanced circuit with power electronic devices. In this case, stray currents limit with the limiting resistor that anti-parallel thyristors (6) active by trigger board (7). Basic VLD has benefits such as cost-efficient, easy installation, and no launcher required as the system is supplied, it is ready to operate. Moreover, hybrid VLDs have more benefits than the basic VLDs, that is including low maintenance due to the application of thyristor, and maximum protection for passengers and personnel because of the short reaction time. This configuration is adapted to demand more than 1 kV. VLD limits the touch voltages regardless of whether they are merely DC or combined voltages (both DC and AC voltages) to the allowable limit values defined in EN 50122-1 [30, 68].

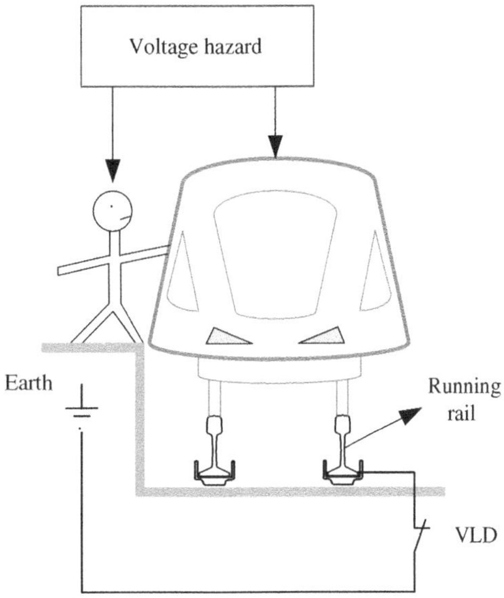

Figure 13.11 A schematic structure of touch voltage regarding VLD position.

13.4.5.2 Directly Grounded System

In this case, we suppose a resistance in each substation (R_g) that the currents I_{g1} and I_{g2} enter the points P1 and P3, correspondingly. Regarding boundary conditions in a grounded system in P1 and P3, the touch voltage is low, but the stray current level is high enough that it intensifies the corrosion of the metallic structures [69]. Hence, designers do not utilize this configuration in modern railway systems, but it is secure for people in stations, reducing the electric shock jeopardy [65].

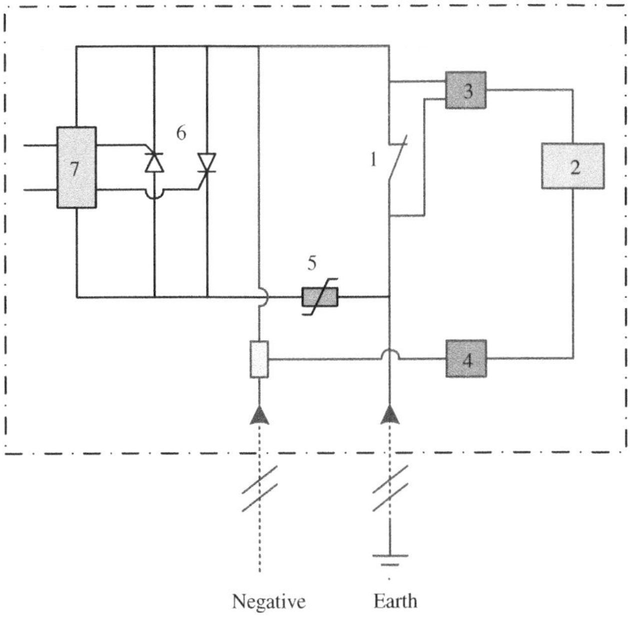

Figure 13.12 An illustration of Basic VLD and Hybrid VLD.

13.4.5.3 Diode-Grounded System

This case holds the stray current and rail potential at an acceptable level, which Figure 13.7 presents as a type of collection mat with the grounded diode. The diode (Figure 13.13a) permits the electric current to flow from the grounding mats to the negative busbar when a certain threshold voltage is reached. This threshold can be variable from 10 to 50 V depending on the TPS conditions [24, 25]. It also provides safe situations for the personnel by disconnecting. However, the stray current corrosion can still happen on diode grounded systems, particularly on running rails and rail fasteners where low rail-to-earth resistance is observed. Moreover, the return rails alternatively discharge the electric current when a threshold voltage is exceeded because of the diode ground circuit path. In this case, the system will be switched to the floating mode when the ground faults occur at the negative point. However, an issue of the system is to maintain periodically against the possibility of infrastructures' corrosion [63, 70].

13.4.5.4 Thyristor-Grounded System

This system is grounded by thyristor and its performance is that when the rail potential exceeds a certain limit, an appropriate thyristor activates depending on the voltage polarity and then it connects the negative busbar to the earth. This thyristor conducts whether the electric current is zero or the voltage polarity is inversed. In thyristor, the voltage level switch relies on the system's features [63].

To achieve a logical compromise between the voltage limit and stray current, it is necessary to determine an appropriate value for a threshold that is predominantly considered 60 V. This configuration was developed to have control over the grounding structures. Figure 13.13 depicts sample configurations of diode-grounded and thyristor-grounded systems. According to Figure 13.13b, an overvoltage relay (R_v) continuously controls the difference between negative bus and ground voltage amplitudes, and it triggers the thyristor gate when it exceeds a predefined value. Moreover, there is a current sensor (R_{cs}) to check out the status of the flowing current. The system can be switched back to the floating system if the level of current was lowered. Another reason for changing the status of DC breakers could be probably a ground fault occurrence [71]. Sustainability of the system as ungrounded is an outstanding feature of the thyristor-grounded system in comparison with the diode-grounded system unless a dangerous voltage is sensed. Therefore, the thyristor-grounded configuration has a remarkable impact on stray current mitigation and its negative consequences [72]. To clarify, Table 13.3 demonstrates a comparison between different GSs for practical purposes.

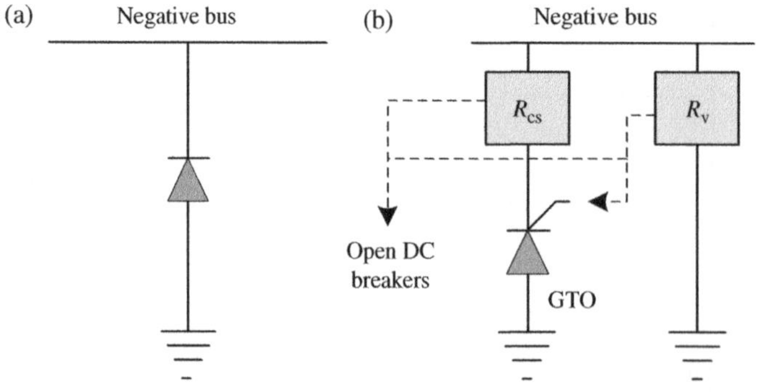

Figure 13.13 Structures of sample: (a) diode-grounded system, (b) thyristor-grounded system.

Table 13.3 A general comparison between diverse grounding systems in terms of stray current and touch voltage level.

Grounding system type	Stray current level	Touch voltage level
Ungrounded (floating)	Low	High
Directly grounded	High	Low
Diode grounded	Average/high	Average/low
Thyristor grounded	Average/low	Average/high

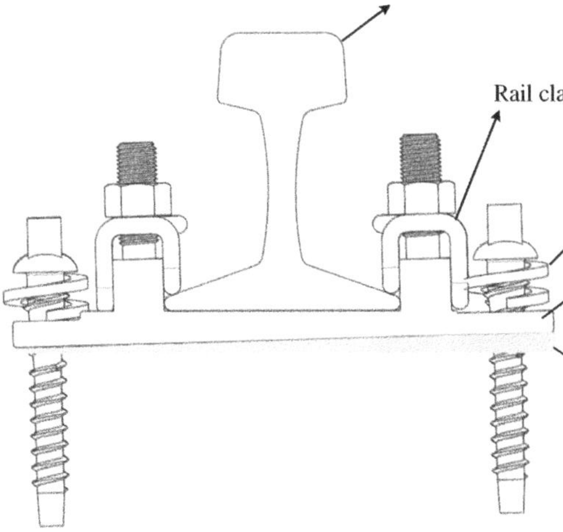

Figure 13.14 A double insulated fastening system.

13.4.6 Insulating Pad

Rail pads are elastomeric mats which are interposed between the running rails and sleepers to preserve the sleeper top. The reason why rail pads make up rubbers or plastics is to dampen the shocks of the vibration of a train. In addition to the earlier benefit, there are some advantages including the reduction of noise and vibration, load distribution over a wider surface, etc. Obviously, the material type of these pads is important in stray current mitigation [73], which means the better the insulation is, the less the stray current will be. It can dramatically decline the current leakage to the earth. The pad design and the material used to make it are chosen to ensure that the insulation pad has favorable characteristics in stiffness and coating. These are necessities for the railway track design to be considered. Figure 13.14 demonstrates different parts of the rail-fastening system in detail [74].

Table 13.4 illustrates different types of sleepers for more information [74]. Other characteristics of sleepers and comparisons between them are out of the main scope of this chapter.

13.4.7 Welding Running Rails

Screw and nut connections usually are not reliable in terms of electrical characteristics. Welding reduces the electric resistance of the connectors between running rails. So, the electric current can

Table 13.4 Types of sleepers.

Type of sleeper	Figures
Rubber/plastic composite sleeper	
Reclaimed European sleeper	
Green eco-treated softwood sleeper	
Brown pressure-treated softwood log lap landscaping sleeper	
Untreated reclaimed tropical hardwood sleeper	

flow out conveniently. Running rails are generally made of carbon steel that is durable, but the extreme weather environment can quickly affect joint failures from time to time. Hence, a suitable welding process is vital for the lines. In [75], a termite-welding process is proposed to join aligned and adjacent sections of a line that can be very effective in the reduction of the stray current.

13.4.8 4th Rail for Returning Current Path

Railway systems can categorize into two types, including AC and DC. In terms of economical, AC systems are preferable because they can step up and down the voltage levels that can reduce the conductor size. However, line losses in low voltage railway systems are considerable compared to the DC systems due to the skin effects. For instance, DC railway systems use an overhead line, and a 3rd rail or 4th rail. Due to some restrictions in urban areas, it is not common to use the

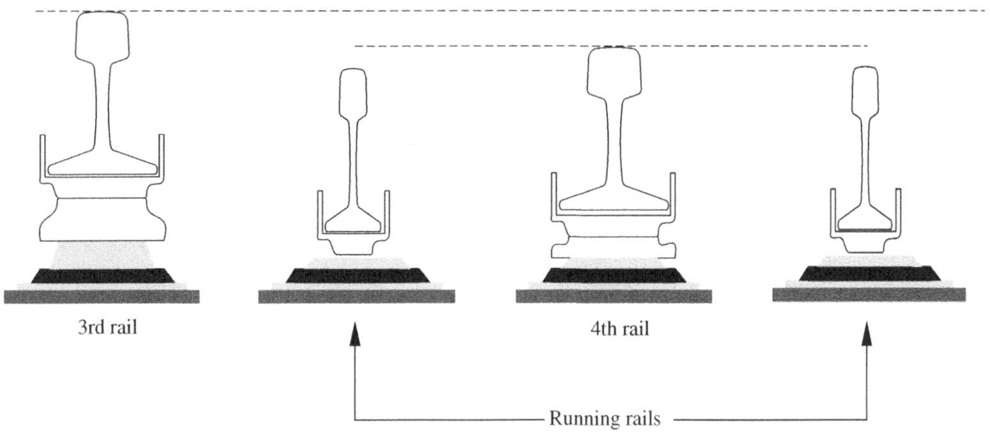

Figure 13.15 A side view of a 4th rail system.

overhead line. Also, the third and fourth rails usually place near the ground. The 4th rail creates a return conductor for the traction currents to go back to the negative busbar of TPSs. It leads to avoiding the return current through the running rails and mitigates the stray current. The 4th rail generally places almost in the middle of running rails at a certain height lower than the 3rd rail [49].

To recap, the 4th rail system is useful for three reasons: (i) it is suitable for the government to limit the voltage drops along the line to 7 V. (ii) Not require too heavy cables and boosters along the route for the returning path. (iii) It is lucrative for signaling purposes, in other words, the railway systems with 4th rail can maintain the service if one pole of supply is earthed, while if this fault occurs in the three-rail supply system, circuit breakers usually detect the short-circuit and the power will be switched off [50]. In Figure 13.15, a side view of this structure in the London Metro is illustrated [59].

13.4.9 Traction Power Substations Based on DC Auto-Transformer

Auto-transformers (ATs) lead to a reduction in loss and voltage drops. The performance is that it locates as a parallel connection between a contact wire and an auto-transformer feeder. A general configuration of the TPS based on a DC auto-transformer is illustrated in Figure 13.16 [76, 77]. In the figure, with conducting a current to negative feeder through an auto-transformer, the rail potential and electric current decrease, resulting in a decline in the stray current from rail-to-earth that is useful for traction systems. It decreases the electromagnetic interferences on adjacent cables' waves. Another main reason for this reduction, that is, electric currents flow through either a positive or negative wire in the opposite directions.

A structure for a DC auto-transformer is presented in [77, 78] that contains two capacitors and an energy transfer module. This module can balance the voltage across two capacitors. DC auto-transformers serve as step-up and step-down converters in the substations and train sides, respectively. Thus, the voltage level rises that the voltage drops, and power losses reduce. A sample of a DC auto-transformer is represented in Figure 13.16. The results of the presented model illustrate that the resonant currents and the transferred energy of the DC auto-transformers are much less than the load current, which shows this system is practical in high current and high-power DC railways. However, this control method has a high cost that includes equipment, repair, and maintenance expenditures that take into account its drawbacks [4].

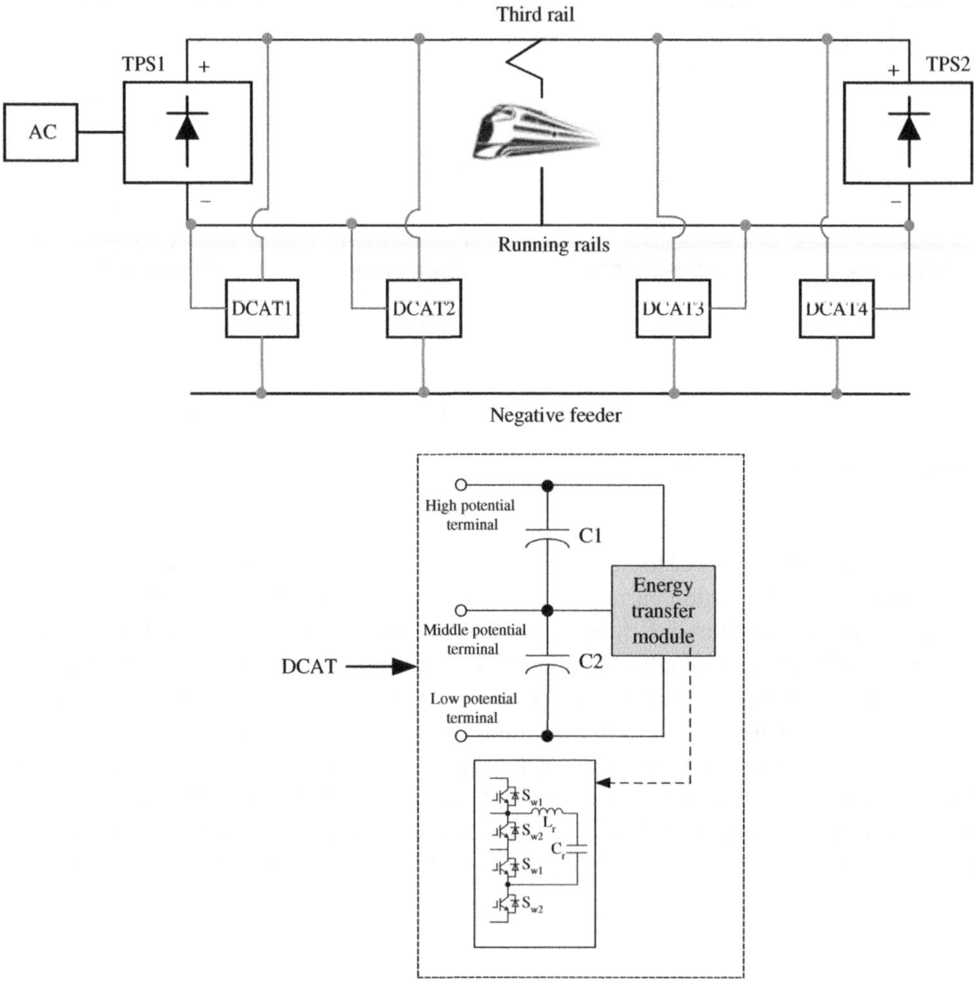

Figure 13.16 General configuration of DCAT-TPS and configuration of DCAT.

13.4.10 Resistance of the Earth Wire to Reinforcing Bar

The stray current interference can be collected by the current return circuit of the AC high-speed line including the earth wire in parallel with rails and flow through the metallic fixed installations. Stray current can be significant for civil structures such as bridges, tunnels, viaducts, etc., in terms of potential corrosion of the reinforcement. Therefore, it is indispensable to provide adequate electrical insulation between the metallic installations and the reinforcing bar (rebar) to mitigate the stray current resulting in the electrolyte. A useful method to increase the insulation resistance is the use of concrete with alkaline properties and free of chlorides [79].

The leakage of current into the rebars depends on some factors such as location, shape, and effectiveness of the diverse insulation materials adopted. Furthermore, it is practical to keep the earth wires at 100 to 200 m at the junctions between the AC and DC lines. When two lines are parallel and close, this is the possible maximum condition. Estimation of earth wire to rebar resistance relies on the resistivity of concrete that can change over a wide range (from 101 to 105 Ωm) influenced by

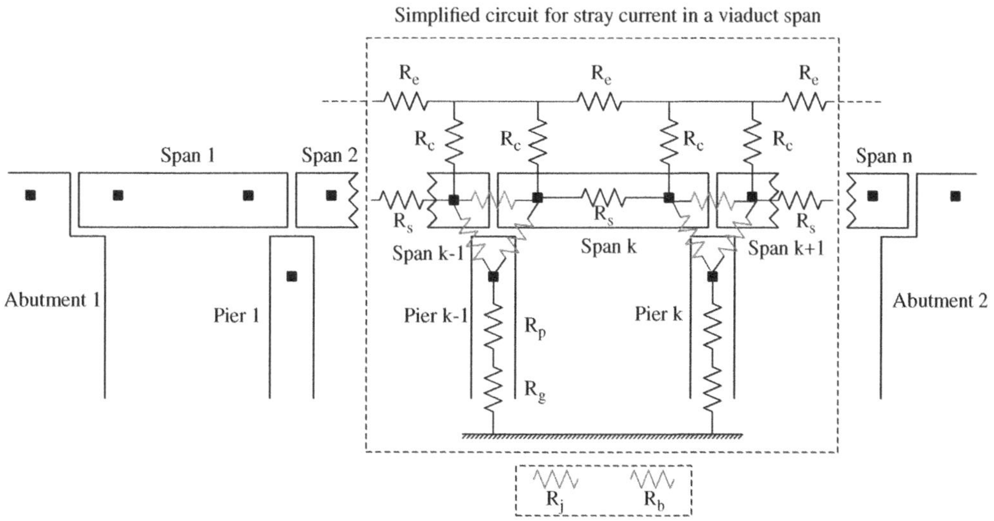

Figure 13.17 A representation of an AC high-speed railway with a concrete viaduct.

various factors including composition, compaction, moisture, type of extension of the metallic installations anchored to the concrete infrastructure, type of fastening system, and aging of the material [80].

A front view of an equivalent circuit of an AC high-speed railway viaduct is demonstrated in Figure 13.17. There is no minimum insulation resistance for each metallic infrastructure, but an estimation of the resistance is conducted by [80]. According to Figure 13.17, the defined parameters are explained in Table 13.5.

There is no concern about the stray current leaking from the earth wire into the concrete reinforcement with a good estimation of various resistances. It is feasible to even the resistance between the earth wire and the reinforcement bars is low [57, 79].

There are some stray current and rail potential control methods, which can be summarized in Table 13.6. Furthermore, a comparison between grounding schemes is shown in Figure 13.18, which the green and orange boxes demonstrate the advantages and deficiencies of GSs, respectively.

Table 13.5 Parameters of the equivalent circuit of a viaduct span.

Parameter	Definition
R_c	Resistances of the earth wire to rebar
R_p	Resistance of pier rebars (Ω)
R_s	Resistance of span and abutment rebars (Ω)
R_j	Resistance of insulated joint between two adjoining spans
R_g	Ground resistance of the piers
R_b	Bearing resistances at the end of each span
R_e	Longitudinal resistance of the earth wire

Table 13.6 A comparison on different stray current control methods.

Control methods	Advantages	Deficiencies
An increase of rail-to-earth resistance [5, 11, 22, 53–56, 58, 60, 61, 81]	Stray current reduction/ isolating the yard track	Destruction of wood sleeper
Locating TPSs near the points of maximum train acceleration/adding TPSs [7, 11–13, 22, 26, 36]	Reduction of traction current	High cost/application limitation/renovating existing railway lines
An increase of nominal voltage [7, 55, 57, 60, 61]	Reduction of traction current	High cost/dangerous for staff
Stray current collection mats [9, 12, 25, 26, 29, 30, 34, 39, 40, 62–64, 82, 83]	Low cost	Worse reduction performance over time
Insulating pad [57, 58, 64, 73, 74, 84]	Insulation improvement/An increase in rail-to-earth resistance	High cost/application limitation/corrosion over time
Welding running rails [60, 64, 75]	More reliability than the screw and nut	Sometimes gaps between steels (tunnels)
Using 4th rail for returning current path [49, 50, 85]	Less corrosion of running rails	Larger cost/insulation deterioration/application limitation
TPSs based on DC auto-transformer [4, 76–78, 82, 86–88]	Transforming the traction currents to other lines	More power quality issues/ application limitations in some TPSs
Resistance of the earth wire to reinforcing bar [57, 79, 80]	Significant contribution in stray current mitigation	Calculation complexity

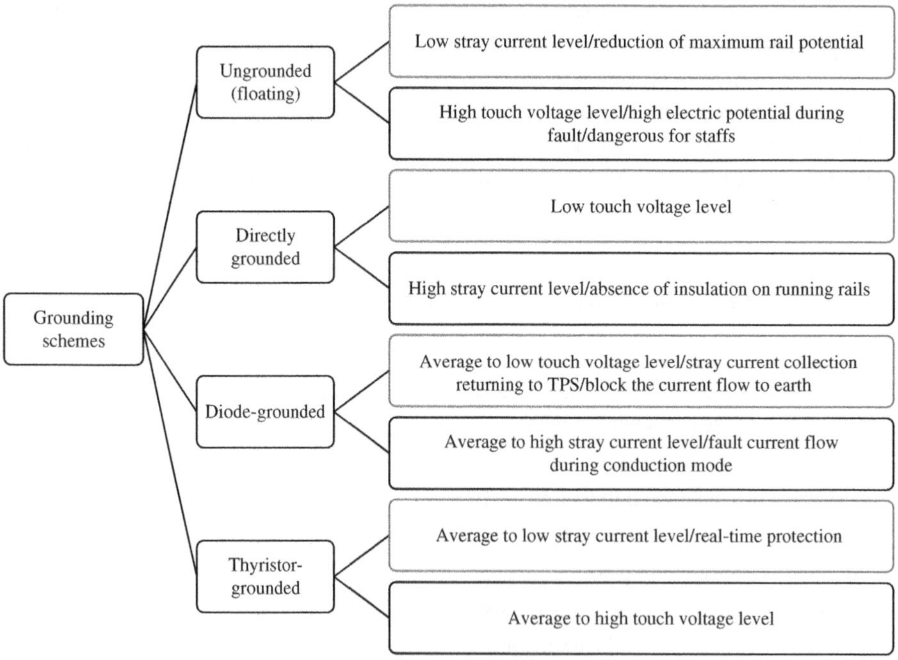

Figure 13.18 A comparison of different grounding systems. *Source:* Adapted from Refs. [5, 6, 8–10, 12, 20, 23, 25, 29, 33, 35, 37, 38, 40, 55, 63, 65, 67, 69–71, 82, 83, 89, 90].

13.5 Conclusion

Stray currents have severe consequences on the electric railway systems and are very problematic for underground infrastructures, communication equipment, pipelines, etc., as well. This paper presents an overview of control methods due to stray currents and rail potential mitigation, which can be beneficial for protecting railway systems from these hazardous currents. It is pivotal to use these control methods for establishing the new railway lines to reduce the extra expenditures due to the stray current increment. Furthermore, it presents a comparison between different control methods and makes a critical review regarding several factors such as location, distance between the TPSs, soil type, insulation resistances, etc.

References

1 Arboleya, P., Mayet, C. et al. (2020). A review of railway feeding infrastructures: mathematical models for planning and operation. *eTransportation* 5: 100063.

2 Du, L., Chang, S., and Wang, S. (2020). The research on DC loop of regional power grid caused by the operation of the subway. *Electronics* 9 (4): 613.

3 Szymenderski, J. et al. (2019). Modeling effects of stochastic stray currents from DC traction on corrosion hazard of buried pipelines. *Energies* 12 (23): 4570.

4 Wang, M., Yang, X., Zheng, T.Q., and Ni, M. (2020). DC auto-transformer based traction power supply for urban transit rail potential and stray current mitigation. *IEEE Trans. Transp. Electrif.* 6 (2): 762–773.

5 Sutherland, P.E. (2011). Stray current analysis. *IEEE IAS Electrical Safety Workshop*, pp. 1–2.

6 Siranec, M., Regula, M., *et al.* (2019). Measurement and analysis of stray currents, in *2019 20th International Scientific Conference on Electric Power Engineering (EPE)*, pp. 1–6.

7 Pires, C.L. (2016). What the IEC tells us about stray currents: guidance for a practical approach. *IEEE Electrif. Mag.* 4 (3): 23–29.

8 Sim, W.M. and Chan, C.F. (2004). Stray current monitoring and control on Singapore MRT system. In: *2004 International Conference on Power System Technology*, vol. 2, 1898–1903. IEEE.

9 Zhichao, C. and Cheng, H. (2019). Evaluation of metro stray current corrosion based on finite element model. *J. Eng.* 2019 (16): 2261–2265.

10 Du, G., Zhang, D., Li, G. et al. (2016). Evaluation of rail potential based on power distribution in DC traction power systems. *Energies* 9 (9): 729.

11 Memon, S.A. and Fromme, P. (2014). Stray current corrosion and mitigation: a synopsis of the technical methods used in dc transit systems. *IEEE Electrif. Mag.* 2 (3): 22–31.

12 Zaboli, A., Vahidi, B., Yousefi, S., and Hosseini-Biyouki, M.M. (2017). Evaluation and Control of Stray Current in DC-Electrified Railway Systems. *IEEE Trans. Veh. Technol.* 66 (2): 974–980.

13 Zaboli, A., Vahidi, B., Yousefi, S., and Hosseini-Biyouki, M.M. (2015). Effect of Control Methods on Calculation of Stray Current and Rail Potential in DC-electrified Railway Systems, *4th International Conference on Recent Advanced in Railway Engineering (ICRARE2015).*

14 Shaffer, R.E. and Fitzgerald, J.H. III (1981). Stray earth current control--washington, DC metro system. *Mater. Performance* 20 (4): 9–15.

15 Yu, J.G. and Goodman, C.J. (1992). Stray curremt design parameters for DC railways. In: *Proceedings of the ASME/IEEE Spring Joint Railroad Conference* (eds. T.S. Larson and A.C. James), 19–28. IEEE.

16 Wang, C., Li, W., Wang, Y. et al. (2018). Stray current distributing model in the subway system: A review and outlook. *Int. J. Electrochem. Sci.* 13: 1700–1727.

17 Wang, X., Wang, Z., Chen, Y. et al. (2019). Effect of a DC stray current on the corrosion of X80 pipeline steel and the cathodic disbondment behavior of the protective 3PE coating in 3.5% NaCl solution. *Coatings* 9 (1): 29.

18 McIntosh, D.H. (1982). Grounding where corrosion protection is required. *IEEE Trans. Ind. Appl.* 6: 600–607.

19 Brenna, A., Lazzari, L. et al. (2017). Stray current control by a new approach based on current monitoring on a potential probe. *Corros. Eng. Sci. Technol.* 52 (5): 359–364.

20 Cai, Z., Zhang, X., and Cheng, H. (2019). Evaluation of DC-subway stray current corrosion with integrated multi-physical modeling and electrochemical analysis. *IEEE Access* 7: 168404–168411.

21 Chen, Z., Koleva, D., and van Breugel, K. (2017). A review on stray current-induced steel corrosion in infrastructure. *Corros. Rev.* 35 (6): 397–423.

22 Cotton, I., Charalambous, C., Aylott, P., and Ernst, P. (2005). Stray current control in DC mass transit systems. *IEEE Trans. Veh. Technol.* 54 (2): 722–730.

23 Lee, C.-H. and Lu, C.-J. (2006). Assessment of grounding schemes on rail potential and stray currents in a DC transit system. *IEEE Trans. Power Delivery* 21 (4): 1941–1947.

24 Alamuti, M.M., Nouri, H. et al. (2011). Effects of earthing systems on stray current for corrosion and safety behaviour in practical metro systems. *IET Electr. Syst. Transp.* 1 (2): 69–79.

25 Jamali, S., Alamuti, M.M., and Savaghebi, M. (2008). Effects of different earthing schemes on the stray current in rail transit systems. In: *2008 43rd International Universities Power Engineering Conference*, 1–5. IEEE.

26 Lee, C.-H. (2005). Evaluation of the maximum potential rise in Taipei rail transit systems. *IEEE Trans. Power Delivery* 20 (2): 1379–1384.

27 Dolara, A., Foiadelli, F., and Leva, S. (2012). Stray current effects mitigation in subway tunnels. *IEEE Trans. Power Delivery* 27 (4): 2304–2311.

28 Brenna, M., Dolara, A., Leva, S., and Zaninelli, D. (2010). Effects of the DC stray currents on subway tunnel structures evaluated by FEM analysis. In: *IEEE PES General Meeting*, 1–7. IEEE.

29 Chen, S.-L., Hsu, S.-C. et al. (2006). Analysis of rail potential and stray current for Taipei Metro. *IEEE Trans. Veh. Technol.* 55 (1): 67–75.

30 Tzeng, Y.-S. and Lee, C.-H. (2010). Analysis of rail potential and stray currents in a direct-current transit system. *IEEE Trans. Power Delivery* 25 (3): 1516–1525.

31 Charalambous, C.A., Aylott, P., and Buxton, D. (2016). Stray current calculation and monitoring in DC mass-transit systems: interpreting calculations for real-life conditions and determining appropriate safety margins. *IEEE Veh. Technol. Mag.* 11 (2): 24–31.

32 Fichera, F., Mariscotti, A., and Ogunsola, A. (2013). Evaluating stray current from DC electrified transit systems with lumped parameter and multi-layer soil models. *Eurocon*, pp. 1187–1192.

33 Charalambous, C.A. (2017). Comprehensive modeling to allow informed calculation of DC traction systems' stray current levels. *IEEE Trans. Veh. Technol.* 66 (11): 9667–9677.

34 Charalambous, C.A., Cotton, I., and Aylott, P. (2008). A simulation tool to predict the impact of soil topologies on coupling between a light rail system and buried third-party infrastructure. *IEEE Trans. Veh. Technol.* 57 (3): 1404–1416.

35 S. Lin, L. Chen, H. Zhang, and Q. Zhou, "FEM with curved hexahedron element and application on tunnel integrated grounding system in high-speed railway," *IEEE Trans. Veh. Technol.*, vol. 68, no. 7, pp. 6441–6452, July 2019.

36 Lin, S. et al. (2019). Research on the regeneration braking energy feedback system of urban rail transit. *IEEE Trans. Veh. Technol.* 68 (8): 7329–7339.

37 Park, J.-D. (2015). Ground fault detection and location for ungrounded DC traction power systems. *IEEE Trans. Veh. Technol.* 64 (12): 5667–5676.

38 Charalambous, C. and Cotton, I. (2007). Influence of soil structures on corrosion performance of floating-DC transit systems. *IET Electr. Power Appl.* 1 (1): 9–16.

39 Charalambous, C.A., Cotton, I. et al. (2012). A holistic stray current assessment of bored tunnel sections of DC transit systems. *IEEE Trans. Power Delivery* 28 (2): 1048–1056.

40 Ibrahem, A., Elrayyah, A. et al. (2016). DC railway system emulator for stray current and touch voltage prediction. *IEEE Trans. Ind. Appl.* 53 (1): 439–446.

41 Ibrahem, A. (2017). Leakage Current Detection and Protection for Electrical Railway Systems. Ph.D. Dissertation, University of Akron.

42 Yousefi, S., Hosseini-Biyouki, M. M., Zaboli, A., and Askarian-Abyaneh, H. (2015). Different hybrid filters configurations impact on an AC 25 kV electric train's harmonic mitigation, in *20th Electrical Power Distribution Conference, EPDC 2015*.

43 Yousefi, S., Zaboli, A., Hosseini-Biyouki, M.M., et al. (2015). Design of a new structure for hybrid filters to eliminate harmonics of 25 kV AC electric railways, in *9th Power Systems Protection and Control Conference, PSPC 2015*.

44 Yousefi, S., Hosseini Biyouki, M.M., Zaboli, A. et al. (2017). Harmonic elimination of 25 kV AC electric railways utilizing a new hybrid filter structure. *AUT J. Electr. Eng.* 49 (1): 3–10.

45 Mariscotti, A., Reggiani, U., Ogunsola, A., and Sandrolini, L. (2012). Mitigation of electromagnetic interference generated by stray current from a DC rail traction system, in *International Symposium on Electromagnetic Compatibility-EMC EUROPE*, pp. 1–6.

46 Jiang, X., Hu, H., Yang, X. et al. (2019). Analysis and adaptive mitigation scheme of low-frequency oscillations in AC railway traction power systems. *IEEE Trans. Transp. Electrif.* 5 (3): 715–726.

47 He, Z., Zheng, Z., and Hu, H. (2016). Power quality in high-speed railway systems. *Int. J. Rail Transp.* 4 (2): 71–97.

48 Brenna, M., Foiadelli, F., Kaleybar, H.J., and Fazel, S.S. (2019). Power quality indicators in electric railway systems: a comprehensive classification. In: *2019 IEEE Milan PowerTech*, 1–6. IEEE.

49 Alnuman, H., Gladwin, D.T., and Foster, M.P. (2018). Development of an electrical model for multiple trains running on a DC 4th rail track," in *2018 IEEE International Conference on Environment and Electrical Engineering and 2018 IEEE Industrial and Commercial Power Systems Europe (EEEIC/I&CPS Europe)*, pp. 1–6.

50 Alnuman, H., Gladwin, D., and Foster, M. (2018). Electrical modelling of a DC railway system with multiple trains. *Energies* 11 (11): 3211.

51 Luo, Z., Yu, X., and Li, X. (2019). An *l0*-norm minimization for energy-efficient timetabling in subway systems. *IEEE Access* 7: 59422–59436.

52 Nunez, F., Reyes, F. et al. (2010). Simulating railway and metropolitan rail networks: From planning to on-line control. *IEEE Intell. Transp. Syst. Mag.* 2 (4): 18–30.

53 Kunpeng, L. et al. (2018). The equivalent circuit and influencing factors of subway rail fastener track-to-earth resistance," in *2018 IEEE Transportation Electrification Conference and Expo, Asia-Pacific (ITEC Asia-Pacific)*, pp. 1–5.

54 Cerman, A., Janicek, F., and Kubala, M. (2015). Resistive-type network model of stray current distribution in railway DC traction system," in *2015 16th International Scientific Conference on Electric Power Engineering (EPE)*,pp. 364–368.

55 Colella, P., Pons, E., and Tortora, A. (2018). Rail potential calculation: impact of the chosen model on the safety analysis, in *2018 AEIT International Annual Conference*, pp. 1–6.

56 Ku, B.-Y. (2016). Rail-to-earth resistance assessment: a medium capacity transit system with continuous negative rails by potential measurement. *IEEE Veh. Technol. Mag.* 11 (4): 29–35.

57 BS EN 50162. (2005). Protection against corrosion by stray current from direct current systems. ISBN: 0580452654.

58 IEC Std. (2009). Railway applications—power supply and rolling stock—technical crite-ria for the coordination between power supply (substation) and rolling stock, vol. 62313.

59 TrainWeb. http://www.trainweb.org/tubeprune/tractioncurr.htm (accessed 25 December 2019).
60 Barlo, T.J. and Zdunek, A.D. (1995). Stray current corrosion in electrified rail systems-final report.
61 (2013). IEEE standard for performance of DC overhead current collectors for rail transit vehicles. *IEEE Std.* 1629-2013: 1–55.
62 Wang, A. et al. (2021). Evaluation model of DC current distribution in AC power systems caused by stray current of DC metro systems. *IEEE Trans. Power Delivery* 36 (1): 114–123.
63 Niasati, M. and Gholami, A. (2008). Overview of stray current control in DC railway systems. In *International Conference on Railway Engineering 2008*, pp. 1–6.
64 Coves, J., Aguilar, J.S., and Rull-Duran, J. (2018). Modelled, simulation and design of collecting grid of stray currents in slab track in DC electrified railway systems," in *2018 IEEE International Conference on Electrical Systems for Aircraft, Railway, Ship Propulsion and Road Vehicles & International Transportation Electrification Conference (ESARS-ITEC)*, pp. 1–7.
65 Colella, P., Pons, E., and Tommasini, R. (2017). A comparative review of the methodologies to identify a global earthing system. *IEEE Trans. Ind. Appl.* 53 (4): 3260–3267.
66 Secheron S.A. (2016). Voltage Limiting Device. https://www.secheron.com/wp-content/uploads/docs/SG825867BEN_D01_Brochure_ESTRA-VGUARD_09-2018.pdf (accessed 20 December 2019).
67 Pons, E., Tommasini, R., and Colella, P. (2017). Fault current detection and dangerous voltages in dc urban rail traction systems. *IEEE Trans. Ind. Appl.* 53 (4): 4109–4115.
68 ABB Voltage Limiting Device HVL 120-0.3 (accessed 28 December 2019).
69 Ogunsola, A. and Mariscotti, A. (2012). *Electromagnetic Compatibility in Railways: Analysis and Management*, vol. 168. Springer Science & Business Media.
70 Zhang, L., Tai, N., Huang, W. et al. (2018). A review on protection of DC microgrids. *J. Mod. Power Syst. Clean Energy* 6 (6): 1113–1127.
71 Mirsaeidi, S., Dong, X. et al. (2017). Challenges, advances and future directions in protection of hybrid AC/DC microgrids. *IET Renew. Power Gener.* 11 (12): 1495–1502.
72 Popescu, M. and Bitoleanu, A. (2019). A Review of the energy efficiency improvement in DC railway systems. *Energies* 12 (6): 1092.
73 Cox, R.J.H.S.J. (2010). Railway Rail Pad, US20100206958A1 (Filed: October 8, 2008; Issued: August 19, 2010).
74 UK Sleepers. https://www.uksleepers.co.uk/product-catalogue/Railway_Sleepers (accessed 28 December 2019).
75 BJF (2010). Welding Railway Lines, US20100243715A1 (Filed: November 21, 2008; Issued: September 30, 2010).
76 Wang, M., Yang, X., Wang, L., and Zheng, T.Q. (2017). DC auto-transformer traction power supply system for DC railways application, in *International Conference on Electrical and Information Technologies for Rail Transportation*, pp. 175–184.
77 Wang, M., Yang, X., Wang, L., and Zheng, T.Q. (2018). Resonant switched capacitor converter based DC auto-transformer for urban rail transit. In: *2018 IEEE Applied Power Electronics Conference and Exposition (APEC)*, 1441–1446. IEEE.
78 Wang, M., Yang, X., Zheng, T.Q., and Gu, J. (2018). Power analysis on DCAT traction power supply system for DC railways. In: *2018 IEEE International Power Electronics and Application Conference and Exposition (PEAC)*, 1–5. IEEE.
79 Sandrolini, L. (2013). Analysis of the insulation resistances of a high-speed rail transit system viaduct for the assessment of stray current interference. Part 2: Modelling. *Electr. Power Syst. Res.* 103: 248–254.
80 Sandrolini, L. (2013). Analysis of the insulation resistances of a high-speed rail transit system viaduct for the assessment of stray current interference. Part 1: Measurement. *Electr. Power Syst. Res.* 103: 241–247.

81 Pham, K.D., Thomas, R.S., and Stinger, W.E. (2001). Analysis of stray current, track-to-earth potentials and substation negative grounding in DC traction electrification system. In: *Proceedings of the 2001 IEEE/ASME Joint Railroad Conference (Cat. No. 01CH37235)*, 141–160. IEEE.

82 G. Du, J. Wang, X. Jiang, D. et al. (2020). Evaluation of rail potential and stray current with dynamic traction networks in multitrain subway systems. *IEEE Trans. Transp. Electrif.* 6 (2): 784–796.

83 Lin, S., Zhou, Q., Lin, X. et al. (2020). Infinitesimal method based calculation of metro stray current in multiple power supply sections. *IEEE Access* 8: 96581–96591.

84 A. Group. Rail Pad. http://www.railway-fasteners.com/rail-rubber-plate-and-rail-pad.html (accessed 28 December 2019).

85 Elliott, R. 3rd and 4th rail dimensions and settingshttp://www.clag.org.uk/3rd-4th.html (accessed 20 December 2019).

86 Wang, L., Yang, X., Xu, J., and Zheng, T.Q. (2018). DC traction system hardware emulator for rail potential distribution in DCAT traction power supply system," in *2018 IEEE International Power Electronics and Application Conference and Exposition (PEAC)*, pp. 1–6.

87 Gu, J., Yang, X., Zheng, T.Q. et al. (2020). Negative resistance converter traction power system for reducing rail potential and stray current in the urban rail transit. *IEEE Trans. Transp. Electrif.* 7 (1): 225–239.

88 Guo, W., Yang, X., Gu, J., and Zheng, T.Q. (2019). Negative impedance converter for reducing rail potential in urban rail transit, in *International Conference on Electrical and Information Technologies for Rail Transportation*, pp. 569–577.

89 Chuchit, T. and Kulworawanichpong, T. (2019). Stray current assessment for DC transit systems based on modelling of earthing and bonding. *Electr. Eng.* 101 (1): 81–90.

90 Sutherland, P.E. (2015). *Grounding of Distribution Systems in Principles of Electrical Safety*. Wiley.

14

Earthing, Bonding, and Stray Current

Sahil Bhagat

Hitachi Rail STS, Traction Power and EMC, Abu Dhabi, United Arab Emirates

In this chapter, the concept of earthing and bonding (E&B) for the railway system (traction and auxiliary), including design provisions for stray current in direct current (DC) traction is briefly discussed. The understanding of these concepts is important particularly for modern systems with high traction currents and extensive use of electrical energy for many purposes.

The traction substation (TSS) supplies the power to the railway traction vehicle through the contact line system. For the current to flow, there must be a return path. The running rails serve as return conductors for the return current. Along with this the resistance between the rails and earth is finite, which causes some portion of the return current to flow to earth and back to the substation via earth. Due to the flow of this return current through running rails, accessible voltage is developed, which can be dangerous and bridged by the passengers.

Thus, apart from the hazardous voltages at accessible parts during fault conditions (like conventional three-phase power distribution system), electrified railways require special provisions to ensure safety of people and protection of installations.

Although some common earthing provisions are applied to both DC and alternating current (AC) traction systems, there are some fundamental differences between the two. While in AC traction system, the return circuit (running rail and return conductors) is earthed, in the case of DC railway system, the return circuit is kept floating. This is due to the fact that if some portion of DC return current is flowing through the earth, it can lead to stray current corrosion.

Thus, in both normal and fault conditions, the potential difference between rails and earth must not exceed acceptable values as specified in EN 50122-1 [1].

The key objectives of E&B are:

- The Safety of the Person: This is characterized by the value of touch and step voltages.
- The Protection of Installations: Damage to installations may arise from overheating of conductors by arcing and electrical corrosion.
- The Intended Operation of the System: For the intended operation of the system, the aspect of electromagnetic compatibility (EMC) has to be considered.

The safety of the person is considered to have the highest priority. We will understand first the earthing and bonding design measures required for DC and AC traction, then some measures required for station building or workshops and, finally, we will briefly discuss stray current in DC traction along with design measures to reduce its impact.

Transportation Electrification: Breakthroughs in Electrified Vehicles, Aircraft, Rolling Stock, and Watercraft, First Edition. Edited by Ahmed A. Mohamed, Ahmad Arshan Khan, Ahmed T. Elsayed, and Mohamed A. Elshaer.

14.1 E&B provisions for Traction Power Supply

14.1.1 DC Traction Return System

The traction return system of the DC railway shall be completely insulated from all earthing installation, conductive structures, and soil. The sole connection to the earthing system shall be via a voltage-limiting device for safety reasons. Neither earth nor structure earth shall become part of the return circuit of DC railways in order to avoid stray current corrosion.

Figure 14.1 shows the return circuit of the DC traction system as well as the relevant area of possible stray current corrosion in the case of an accelerating vehicle. Stray current corrosion may occur in the area where DC leaves the metal parts, such as running rails and structure earth. In case of regenerative braking, this area changes as the current will enter the running rails instead of leaving them.

14.1.2 Wayside Earthing and Bonding in DC Traction System

As the traction return system of the DC railway is completely insulated from the earth, it is not suitable for earthing purposes. The structure earth is the connection point for all earthing purposes. As shown in Figure 14.2, the metal reinforcement of wayside structures and buildings forms the structure earth of the system. This includes the segments, piers, columns, and bored piles of

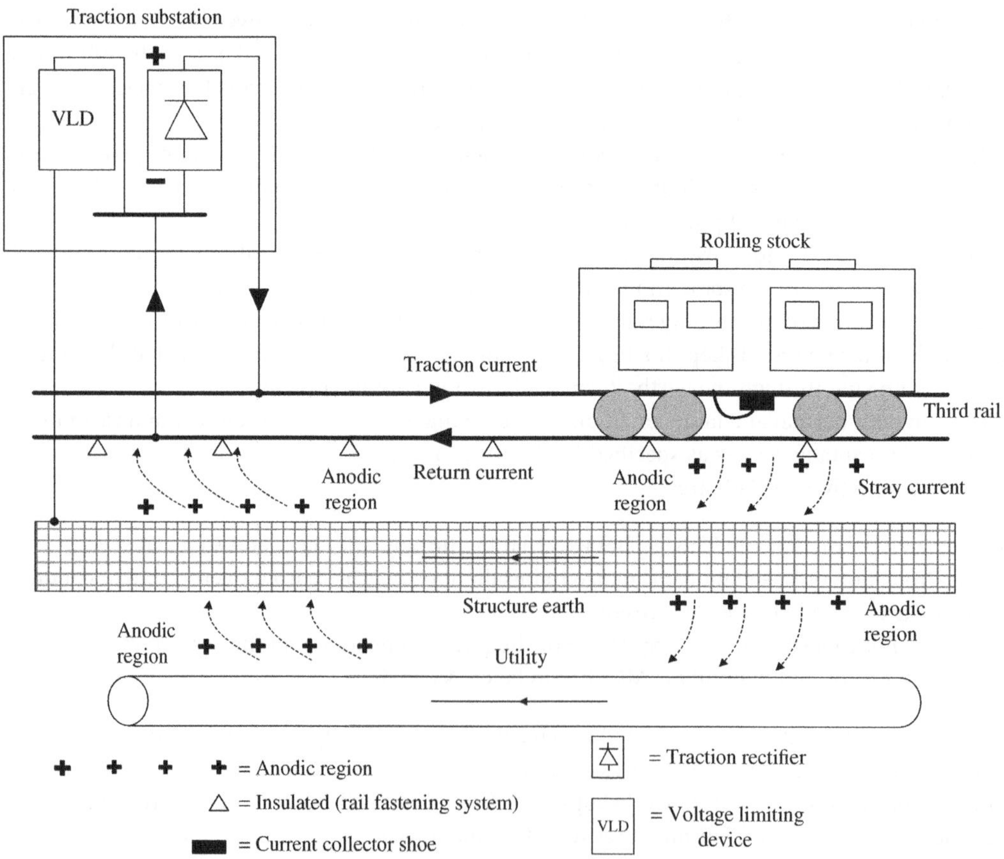

Figure 14.1 Return circuit of third rail DC traction power supply.

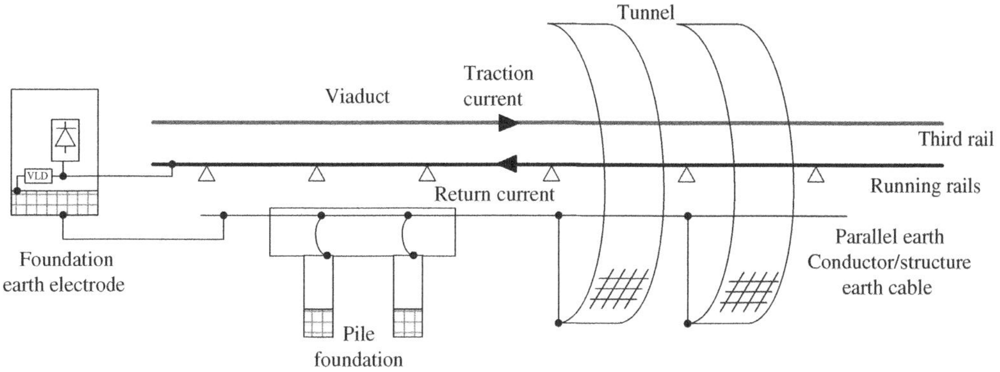

Figure 14.2 Schematic earthing diagram of a DC traction power system.

viaducts and bridges, tunnel segments, slab tracks, retaining walls, technical buildings, stations, and structure earth cable/parallel earth cable.

Any equipment that is 2.5 m away from the outermost running rail of either track must be bonded to the earth cable installed along the alignment or to the structure earth. This distance is derived from annexure B.3 of IEC 60364-4-41 [2], which recommends the value of the zone of arm's reach. Equipment that are within this distance, such as point machines, platform screen doors, signals, etc., shall be connected to running rails as shown in Figure 14.3. Furthermore, these devices shall be carefully insulated and shall not have electrical connections to the structure earth. This insulation shall not decrease the rail insulation and shall not degrade due to different environmental conditions over the lifetime.

14.1.2.1 Rail Potential and Return Circuit

As the flow of return current is through running rails, it gives rise to rail potential. The rail potential is the voltage between running rails and earth either during normal operating conditions or under fault. One of the measures to reduce its value is by reducing the resistance of return circuit. Thus, we provide rail-to-rail bonds and track-to-track bonds, which provide parallel paths, thereby reducing the impedance. It must be noted that running rails are insulated with respect to earth in DC traction system, which causes the rail potential gradient to be very steep. Therefore, the whole rail potential can act as a touch voltage. In some cases, if the voltage does not limit below the touch voltage specified in Section 9 of [1], i.e., 120 V DC, then we need to provide insulated return conductors along the route bonded to running rails at regular intervals to carry some portion of return current. Some

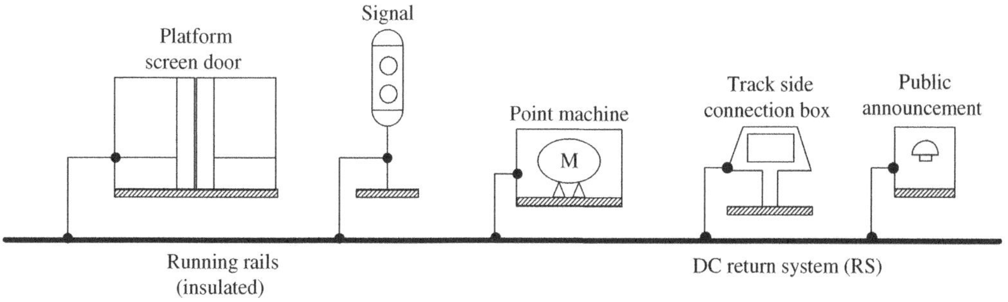

Figure 14.3 Outdoor equipment earthing.

of the less preferred measures to reduce the rail potential are to decrease the spacing between TSS, to change the operational headway and/or increase the operating voltage. Apart from this, the running rails are connected to the negative pole of the traction rectifier by means of return cables, which have to be insulated against the earth and placed close to the feeder cables. Thus, the running rails, rail-to-rail bonds, track-to-track bonds, return current conductor (if used) and return cables from the part of return circuit.

14.1.3 Earthing and Bonding in DC Traction Power Substations

Generally, the TSS shall accommodate MV (medium-voltage) switchgear, LV AC installations, traction transformers, rectifiers, DC switchgear, and other installations to connect to the traction return circuit.

14.1.3.1 Equipment Frames

Equipment frames of AC power supply installations, e.g., the traction transformer and the AC switchgear, shall be connected to the equipotential busbar, which is connected to the structure earth. However, DC switchgear and rectifier frames are insulated from the structure earth and, thus, shall have only one central low-resistance connection to structure earth in order to detect frame fault as indicated in Figure 14.4.

14.1.3.2 Voltage-Limiting Device (VLD)

Depending on the system design, non-permissible touch voltages can occur between the return circuit and structure earth or between the return circuit and earth as a result of abnormal operating conditions and short circuit or fault currents in the DC railway systems. A VLD shall be installed between the running rails and the structure earth (via the main equipotential busbar).

A VLD-O (voltage-limiting device – operation) [3] protects against impermissible voltage caused by rail potential in case of operation and short circuits. In case of short circuit, the current path is identical to the operational one. The VLD-O acts as an equipotential bonding device in order to limit the touch voltage. Only a part of the return current flows through it. The permissible touch voltage according to EN 50122-1 shall not exceed. Tripping of the line circuit breakers caused by the VLD-O is not intended. The VLD-O is normally connected between the return circuit and structure earth, e.g., in passenger stations or substations.

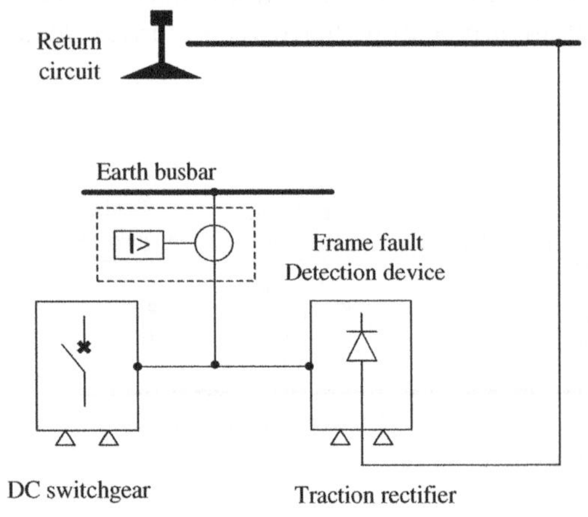

Figure 14.4 Frame leakage detection circuit.

A VLD-F (voltage-limiting device – fault) is deployed if exposed conductive parts are positioned within the current collector zone (CCZ). In this case, the exposed conductive parts, e.g., steel girders of a bridge, are connected to the return circuit by means of VLD-F. VLD-F protects against faults with a connection between a live part of the traction power supply system and a conductive part not intentionally bonded to the return circuit. Due to such a fault, the VLD-F becomes conductive and causes tripping of the DC traction power supply. Generally, the VLD-F is connected between the part to be protected and the return circuit.

14.2 AC Traction Return System

For AC railway systems, the return circuit is connected to the earthing system, e.g., via pole foundations and the structure earth of viaducts, tunnels and other structures. For this reason, a part of the traction return current flows via the earthing systems and the earth back to the TSS. The design of the earthing installation and the return circuit requires protective provisions for the safety of persons and measures for protection of installations.

The operating current as well as the fault current of AC railway systems flows according to Figure 14.5 from the substation through the feeder cables and the overhead contact line system to the vehicles. The traction return current flows from the vehicles through the return circuit and partly through the earth back to the substation.

14.2.1 Wayside Earthing and Bonding in AC Traction

The earthed traction return system forms the connection point for all earthing purposes, as shown in Figure 14.6 for:

- HV, MV, and LV system
- Lightning protection
- Signaling and telecommunication system.

The reinforcement of wayside structures and buildings forms the structure earth of the system in metros.

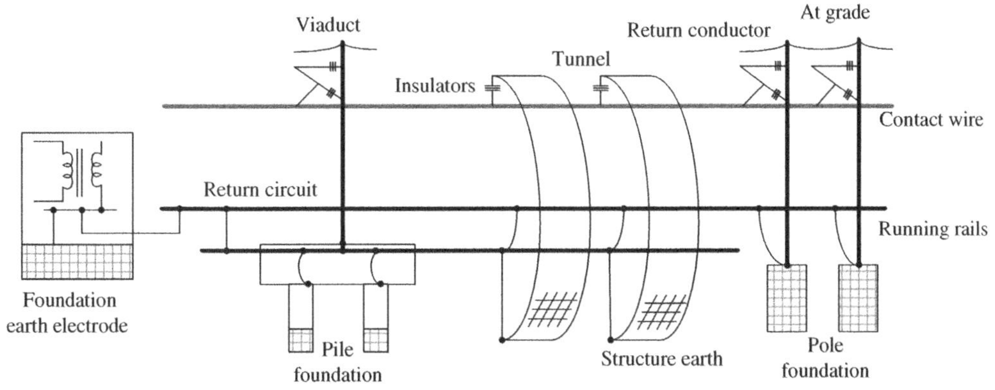

Figure 14.5 Schematic earthing diagram of an AC traction power system.

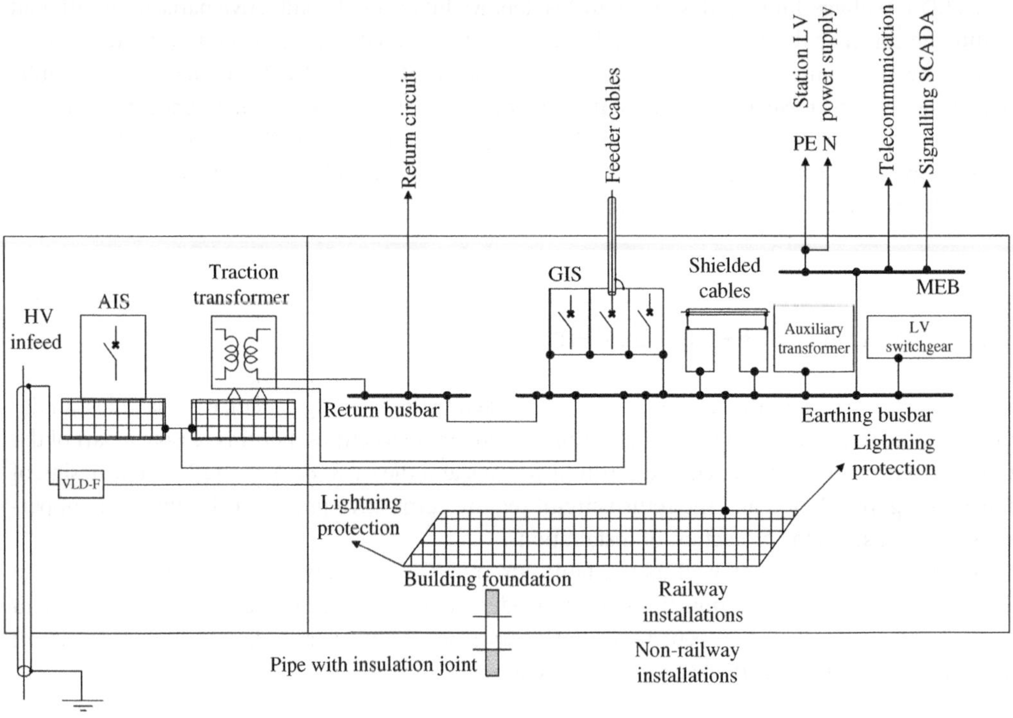

Figure 14.6 Overall earthing scheme – AC traction.

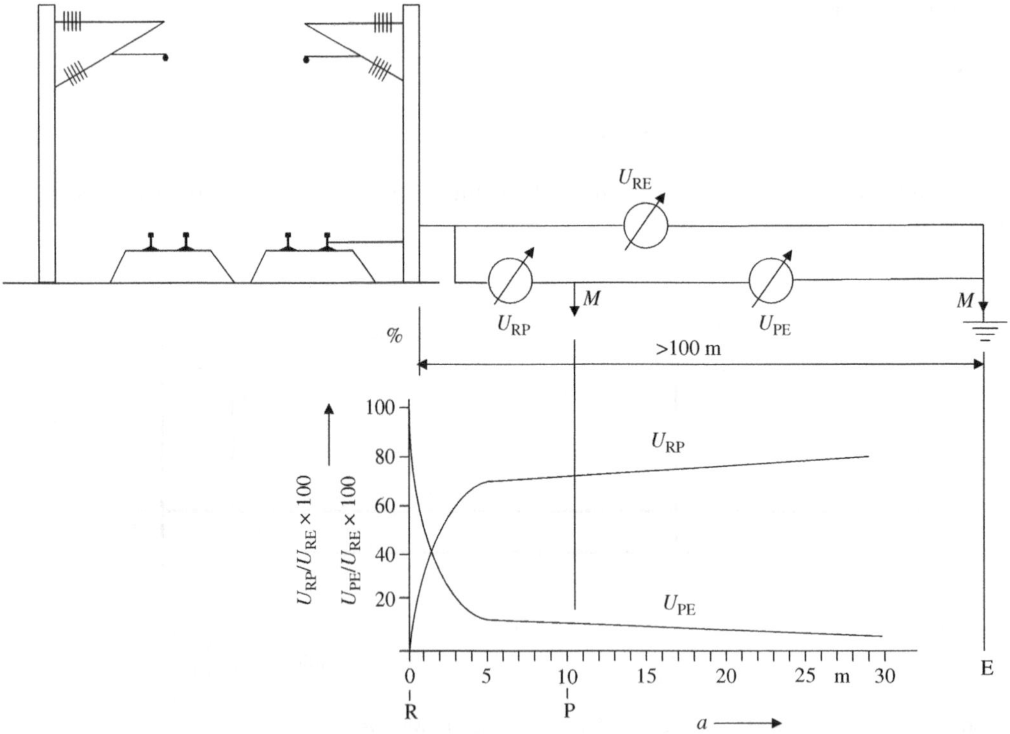

Figure 14.7 Guidance values for the rail potential gradient measured at the mast at a right angle to the track in an AC traction system.

14.2.1.1 Rail Potential and Return Circuit

As the viaduct and tunnels structures of AC railway systems are bonded to the running rails, low voltage difference occurs between running rails and viaduct or tunnel structures. The rail potential (§1.1.1) mainly depends on the conductance per unit length between running rails and earth. A typical potential decrease cross to the line caused by the rail potential is given in Figure 14.7 [1]. Only a part of the maximum potential can be bridged at a distance of 1 minute AC traction. For this reason, the permissible touch voltage is considered satisfied if the rail potential rise, determined by measurement or calculation, does not exceed double the value of the permissible touch voltage (considering bridged distance of 2 m to be more conservative). According to Section 9 of EN 50122-1 [1], the permissible touch voltage is 60 V AC for a period longer than 300 seconds, i.e., for permanent conditions and depending on time duration of current flow for temporary and short-term conditions, e.g., running rails are 645 V AC for 200 ms.

14.3 E&B Provisions for Station and Technical Buildings

14.3.1 Electrical Safety of Persons

14.3.1.1 Direct Contact

Direct contact means an electric contact of a person to live parts. Protection against direct contact of live parts shall be carried out according to IEC 60364-4-41 [2] for LV installations and EN 50522 [4] for HV installations. For installations connected to the traction power supply, the rules of EN 50122-1 [1] shall be considered. Concerning provisions against direct contact, one of the following measures presented in Table 14.1 shall apply. In extraordinary circumstances or for special purposes, further protective provisions against direct contact may be applied like safety extra low voltage (SELV), protective extra low voltage (PELV), functional extra low voltage (FELV), and electrical separation.

The electrical clearance shall be maintained under all circumstances. For traction voltages, the clearances are given in Table 14.2 based on EN 50119 [5] and EN 50124-1 [6]. Minimum clearance to live parts for standing surfaces are given in Figures 14.3 and 14.4 of EN 50122-1 [1], which apply to public areas as well as restricted areas. Obstacles shall be designed according to the relevant standard depending on their use in restricted or public areas. Further measures, which may be applied additionally or alternatively taking into account restrictions, arising from the applicable standards, are:

- Non-conductive locations
- Earth-free local bonding
- Electrical separation for the supply of more than one item (IT system).

Table 14.1 Standards specifying protective provisions against direct contact.

IEC 60364-4-41	EN 50522	EN 50122-1
Basic insulation	Basic insulation	Basic insulation
Placing out of reach	Placing out of reach	Placing out of reach
Obstacles	Obstacles	Obstacles
Barriers	Barriers	Barriers
Enclosures	Enclosures	

Table 14.2 Electrical clearances.

	Recommended clearances	
Voltage	**Static (mm)**	**Dynamic (mm)**
750 V DC	100	50
1500 V DC	100	50
15 kV AC	150	100
25 kV AC	270	150

Table 14.3 Standards specifying protective provisions against indirect contact.

IEC 60364-4-41	EN 50522	EN 50122-1
Equipotential bonding and automatic disconnection including use of RCD	Equipotential bonding and automatic disconnection	Equipotential bonding and automatic disconnection
Placing out of reach	Placing out of reach	Double or reinforced insulation ($U \leq 1$ kV AC or $U \leq 1.5$ kV DC)
Separation		

14.3.1.2 Indirect Contact

Indirect contact means an electric contact of a person to conductive parts, which are not normally live but which may become live under fault conditions. Table 14.3 presents the provisions against indirect contact.

Safety for persons has to be ensured against:

- Voltages in electrical installations and conductive structures under normal and fault conditions;
- Lightning strikes.

Exposed conductive parts within the overhead contact line zone (OCLZ) and CCZ, marked in Figure 14.8 have the risk of becoming alive in case of fault situations. In the case of AC traction system, typically all such parts are connected to the return circuit in accordance with EN 50122-1, which leads to same potential as that of running rails. Whereas in DC system, rails are kept floating in order to avoid flow of stray current, the connection of the equipment in OCLZ/CCZ is made to running rails via a voltage-limiting device (VLD-F) [1]. In tunnels and third rail system, OCL zone can be avoided as per EN 50122-1 due to lower span length. Typical values according to [1] for Figure 14.8 are $X = 4$ m; $Y = 2$ m; $Z = 2$ m.

14.3.1.3 Touch Voltages

Danger to human beings and livestock can arise from currents passing through their bodies. The value of such currents depends amongst others on the voltage bridged by the body and by the body's impedance. IEC 60479-1 [7] describes the body impedance and the effects of AC and DC depending on the time duration. Table 14.4 describes the standards to be referred to for a particular type of installation. Table 14.5 indicates the touch voltage limits as per different standards.

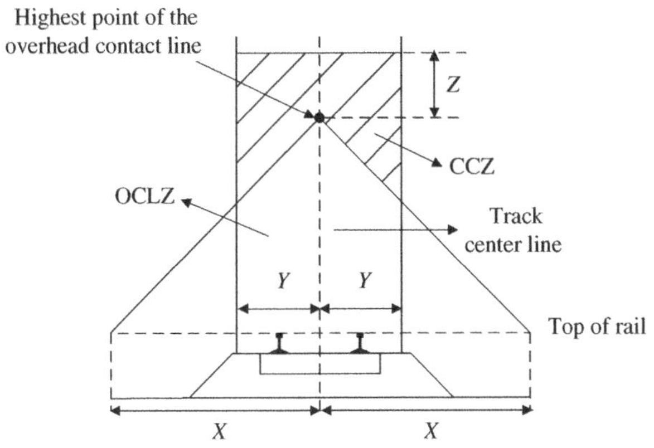

Figure 14.8 OCLZ and CCZ as per EN 50122-1. *Source:* Ref. [1]/British Standards Institution.

Table 14.4 List of standards for touch voltage.

Standard	Installation
EN 50122-1 [1]	Traction power installations
EN 50522 [4]/IEEE 80 [8]	Power installations exceeding 1 kV AC
IEC 60364	Low voltage (LV) systems of service power supply in LV rooms

Table 14.5 Permissible touch voltage limits.

t (seconds)	U_{te}, maximum permissible effective touch voltage for AC traction as per EN 50122-1	U_{te}, maximum permissible effective touch voltage for DC traction as per EN 50122-1	U_{TP}, permissible touch voltage as per EN 50522
>300	60	120	–
300	65	150	–
10	–	–	85
5	–	–	86
2	–	–	96
1	75	160	117
0.9	80	165	–
0.8	85	170	–
0.7	90	175	–
<0.7	155	350	–
0.6	180	360	–
0.5	220	385	220
0.4	295	420	–
0.3	480	460	–
0.2	645	520	537
0.1	785	625	654
0.05	835	735	–
0.02	865	870	–

14.4 Protection

14.4.1 Protection Against Thermal Stress

Live wires/conductors, neutral wires/conductors, and earthing wires/conductors shall be designed to carry operational currents permanently without overheating. Protective earthing and bonding conductors shall be designed to withstand the maximum fault current under short-term conditions according to the relevant tripping time of the backup protection. It needs to be taken into account that even low faults may lead to high thermal load due to longer tripping times depending on the protective device used.

Table 54.2 of IEC 603064-5-54 [9] defined the minimum cross-sectional area of protective conductors, which is the worst case. However, to calculate the exact size refer to section 543.1.2 of [9].

14.4.2 Protection Against Overvoltage

Overvoltages are often caused by switching inductive circuits and by lightning strikes. The magnitude and intensity of lightning-borne overvoltages generally exceed those generated by switching actions. Lightning strikes may not only cause overvoltages in installation by direct hit but may induce high voltage levels in separate conductor loops.

Concerning overvoltages, standards like IEC 62305 on lightning protection and EN 50124 on insulation coordination specify requirements, which depend, amongst others, on the environmental conditions and on voltage rating of the installation. Further mitigations against overvoltages in low voltage installations are given in IEC 60364-4-44 [10].

14.5 Structure Earthing and Bonding System

The earthing and bonding system within railway boundary shall be in accordance with IEC 61000-5-2 [11], IEC 60364-4-44 [10], and IEC 62305-3/4 [12, 13]. These standards strictly recommend use of interconnection of all earth electrodes to a common earthing system for all purposes (See Figure 14.9). This includes the high voltage protective earthing, the low voltage protective

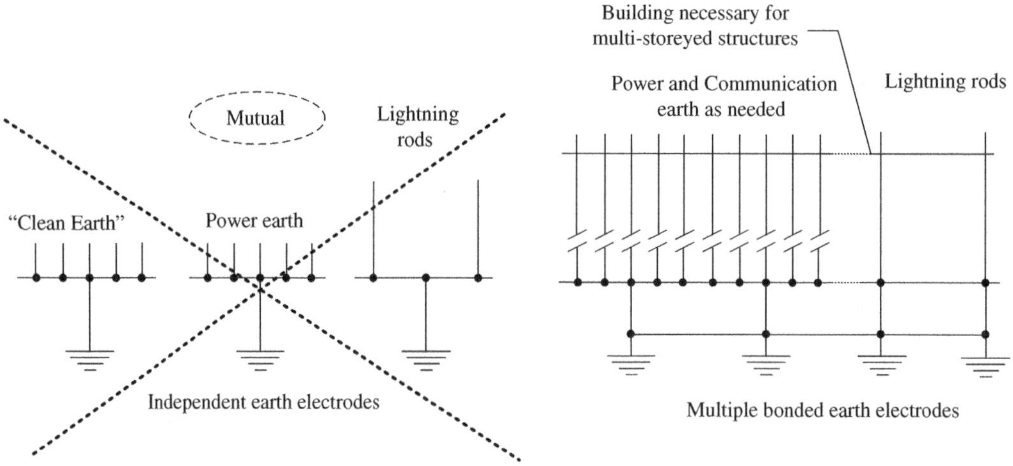

Figure 14.9 Types of earthing configurations.

earthing, signaling and communications earthing, and the lightning protection earthing system. As there are always links by the soil or by parasitic elements (capacitances and mutual inductances) in the installation, in the case of independent earth electrodes at the time of lightning or power system fault, dangerous transient voltages (for personnel safety and for EMC) can occur.

Structures and buildings which are electrically conductive are generally equipped with a structure earthing system. Conductive structures and buildings are made from, e.g., steel. Examples of such structures and buildings are viaducts, tunnels, technical buildings, passenger stations, workshops in depots, etc.

The advantages of structure earthing are:

- The multiple and parallel paths have different resonance frequencies. So, if there is for a given frequency a path with a high impedance, this path is certainly shunted by another which does not have the same resonance frequency. Globally, over a large frequency spectrum, a multitude of paths gives a low impedance system.
- Moreover, equipotential planes are provided in the entire building, thus dangerous touch voltage is not accessible. Since the entire structure has the same electrical potential, there is no voltage difference between different points in the structure, which might cause electromagnetic disturbance. Furthermore, it resembles a cage that screens against magnetic fields and the effects of voltages induced by lightning current pulses. Figure 5 of IEC 62305-4 [13] indicates the utilization of existing concrete reinforcement by bonding to dedicated earthing re-bars.

14.6 Earthing and Equipotential Bonding

With respect to the earthing and equipotential bonding measures, it may be distinguished between indoor and outdoor equipment.

14.6.1 Indoor Equipment

Equipotential bonding ties interconnected metallic frames of electrical devices/apparatus and other conductive parts on the same or nearly the same electrical potential. The aim of this equipotential bonding is generally to avoid dangerous touch voltages and to avoid differences in potentials within an installation. According to IEC 60364-4-54 [9], the minimal cross sections of equipotential bonding conductors or installation with nominal voltage larger than AC 1 kV and traction power installation have to be at least 16 mm^2 of copper or 35 mm^2 of aluminum or 50 mm^2 of steel.

14.6.2 Outdoor Equipment

For DC traction, the outdoor equipment (Figures 14.2 and 14.3) must not be connected to the structure earth. If earthing is required, it may be connected to the structure earth. With respect to structure earthing and bonding, exceptions exist for rail-mounted equipment, parts of which do not meet the clearance requirements (Section 14.1.2). For AC traction (Figure 14.6), the outdoor equipment are bonded to the return system.

14.7 Stray Current

The traction return current uses the running rails as the path back to the traction rectifier. Due to the electrical resistance of the rails, a longitudinal voltage appears between the train location and the TSS. This longitudinal voltage is transferred as a transverse voltage between the rails and the earth and is the driving factor of stray currents.

As no current is "lost" in an electrical circuit and the negative pole of the rectifier is normally insulated from earth, stray currents have to return into the rails in order to return to the negative pole of the rectifier.

Contrary to the electric current with a flow of electrons, stray current may also appear as a flow of metallic ions. The metallic ions lying at the border between a metal and an electrolyte (earth or concrete) tend to leave the metallic mass (rail, pipeline, etc.) and join the electrolyte. Stray current results therefore in the progressive elimination of the metallic mass. This dissolution (mass loss) follows Faraday's electrolytic law, which means for a bare metal surface that an anodic current (current leaving the surface) of 1 A DC dissolves 9.1 kg per year of iron (3 kg of aluminum, or 10 kg of copper, or 12 kg of zinc, or 33 kg of lead). Expressed as corrosion (average penetration of the structure surface), the anodic current density of 1 A/m^2 results in the corrosion rate of 1.1 mm thick iron per year [14]. This corrosion caused by a current originating from an external DC source is known as stray current corrosion.

The location and ampacity of stray current depend on various parameters, among which are as follows:

- The rail-to-earth insulation
- The longitudinal resistance of the track
- The traction return current
- The position of the trainset.

In practice, stray current preferably flows into:

- The metallic utility pipes
- Concrete reinforcement of buildings, bridges, retaining walls, as well as earthing systems and reservoirs.

14.7.1 Stray Current Corrosion

A freely corroding, i.e., not electrically influenced, metallic structure which is surrounded by an electrolyte (e.g., soil, water, or concrete) establishes an electrochemical potential versus the electrolyte. The structure potential can be measured versus a reference electrode, e.g., a saturated copper/copper-sulfate electrode placed in contact with the electrolyte. When electrically influenced by a foreign DC (stray current), the potential of the structure shifts in the negative (cathodic) direction where the current enters or in the positive (anodic) direction where the current leaves the metal surface. At the point where the stray DC leaves the metal surface, an anodic corrosion reaction takes place at the metal/electrolyte interface, resulting in oxidation (dissolution) of the metal.

The anodic reaction results in a positive polarization (positive potential shift) of the metal surface and, thus, stray current can be identified by potential measurement.

At the point where the stray DC enters the metal structure, the cathodic part of the reaction takes place at the metal/electrolyte interface. The cathodic reaction at the surface, where the current enters, generally lowers the corrosion rate.

Regarding the anode consumption, the produced ions, which tend to migrate away from the anode may readily react with ions existing in the electrolyte to form an adherent film on the surface of the metal. Such a film can prevent further corrosion of the metal by restricting the access of the electrolyte to metallic ions. This is called passivation, which is observed on stainless steel or steel in concrete (under specific conditions). For instance, Figure 14.1 shows a simplified case where a buried pipe is running adjacent to the tracks.

Besides corrosion generation, stray current leads to other adverse effects such as:

- Metallic structure heating and consequences
- Influence on signaling and telecom systems
- Influence of existing cathodic protection systems.

14.7.2 Parameters to Control Stray Current

In order to understand the parameters that influence stray current, we shall refer to Figure 14.10, which provides the electric model and equations controlling them.

$$U_{\mathrm{RE}}\ (\text{rail potential}) = I_r{}^{*}R_r \tag{14.1}$$

$$I_{\mathrm{stray}} = \frac{U_{\mathrm{RE}}}{R_u + R_s} \tag{14.2}$$

From Eq. (14.2), we can get that I_{stray} is directly proportional to the rail potential (U_{RE}) and track-to-earth conductance (G'_{RE}). Thus, in order to reduce the stray current, we need to reduce the U_{RE} and/or G'_{RE}. Furthermore, U_{RE} is dependent on the return current and return circuit resistance. Thus, the parameters that control the stray current are:

- The conductance per length between track and the earth;
- The distance between the substations as it governs both return circuit resistance and magnitude of return current;
- The longitudinal resistance of running rails which affects the rail potential;
- Spacing between the cross bonds which affects the rail potential.

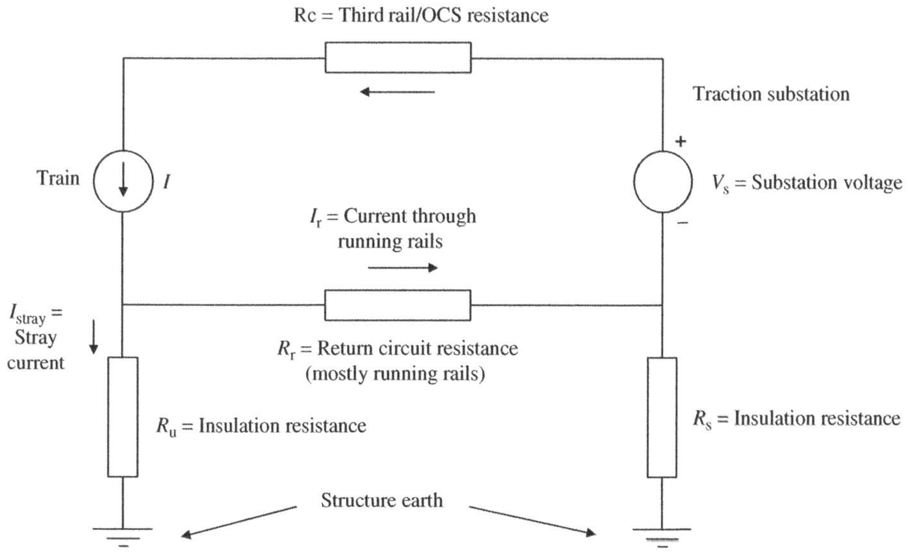

Figure 14.10 Stray current electric model.

14.7.3 Criteria for Stray Current Assessment

There are three main criteria defined by EN 50122-2 [15] for stray current assessment:

a) Insulation Criteria: The rail-to-earth insulation shall be such that average stray current (only positive values) per unit length for a single line track is less than 2.5 mA/m or 2.5 A/km. This ensures the track life for 25 years. The below ensures that average stray current is within the prescribed limits:
   - $G_{RE}' \leq 0.5$ S/km per track and $U_{RE} \leq +5$ V for open formation;
   - $G_{RE}' \leq 2.5$ S/km per track and $U_{RE} \leq +1$ V for closed formation.

NOTE: For the average rail potential shift U_{RE}, only positive values of the rail potential are considered. The average period shall be 24 hours or multiples.

b) Positive shift between the steel in concrete structure and the earth in the hour of highest traffic (headway) $U_{SE} \leq +200$ mV.
c) Longitudinal voltage drop shall be smaller than 200 mV for any two points of the through-connected metal-reinforced structure.

14.7.4 Design Provisions to Reduce Stray Current

The main design principles adopted for the reduction and control of stray current will affect the following disciplines:

- Trackwork and traction return circuit;
- Stray current collection system;
- Power supply;
- Earthing and bonding design principles.

14.7.5 Trackwork

14.7.5.1 Maximum Longitudinal Resistance of the Rail

The electromotive force of the stray current emission is the rail-to-earth potential. In consequence, it is therefore important to limit the traction return circuit resistance by:

- Using Low Resistive Rails: The rail shall have a longitudinal resistance of not more than 50 mΩ/km at 20 °C with 10% wear limit.
- Performing exothermic welding (such welding shall not exceed 5% of the return circuit resistance).
- Ensuring continuity by insulated cables where there are any discontinuities such as expansion joints, switches, and crossings.
- Crossbonding the return circuit: In order to further enhance the return circuit conductivity, negative crossbonds.
- Limiting the rail-to-earth voltage.

14.7.5.2 Insulation Measures

The insulation level of the track shall be designed and maintained during normal operation at a maximal rail-to-earth conductibility of values given in Section 14.7.3, during the whole lifetime of the track. The insulation level will be reached by using appropriate materials and installation methods under the rail. Testing shall be carried out to verify that the designed rail-to-earth resistance of the return circuit is achieved.

14.7.6 Stray Current Collection System (SCCS)

The SCCS is aimed to catch as much stray current as possible to limit its propagation to the rest of the environment. Generally speaking, all reinforced track sections shall meet the corrosion criteria of 200 mV maximum and the calculations must be performed based on annexure C.2 of EN 50122-2. This criterion is understood for the voltage taken between any two points of the track or structure reinforcement as well as on an operating average of the railway traffic.

There are two types of SCCS defined as follows:

- Floating SCCS: In this type, the track slab reinforcement is separated from the deck slab by providing insulation layers to shear connectors bonding both structures as shown in Figure 14.11. This will reduce the amount of stray current leaving from the track reinforcement to deck and further to earth. However, the drawback to this type of arrangement is that there are thousands or millions of shear connectors. The probability of failure of some of the shear connectors (one, two, or even more out of millions) is high. In this case, that failure point will become the hot spot and lead to corrosion in that location.
- Earthed SCCS: In this type, as depicted in Figure 14.12, the track slab reinforcement is bonded to the global earthing system which includes PEC (parallel earth conductor), deck reinforcement, piers, etc. By this method, we are trying to reduce the net impedance of our system, thereby limiting the stray current in this low impedance path.

14.7.7 Power Supply Design

There must be full segregation between the earthing system (including any structure which is not insulated from the earth) and the traction return circuit. No direct and permanent conduction shall be allowed between them. In particular, the negative pole of the traction rectifiers shall not be earthed (except for traction rectifier in the depots, which may be earthed or floating). Return circuit pole shall be made available in the substation to allow electrical drainage if needed. A sufficient

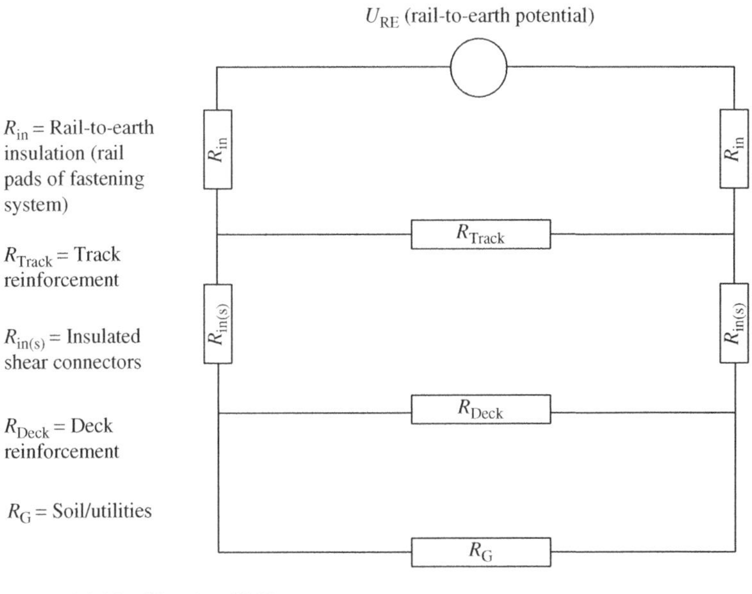

Figure 14.11 Floating SCCS.

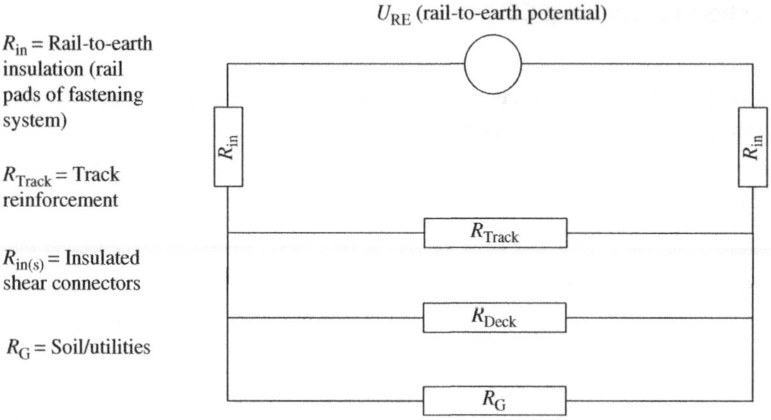

Figure 14.12 Earthed SCCS.

place shall be provided in each substation for installing a stray current drainage cubicle in the case where an active protection system is necessary, as well as a room dedicated to stray current protection. Return circuit cables shall be of insulated type. The number of return circuit cables shall be such that they can withstand the return current despite the loss of one return cable.

14.7.8 Earthing and Bonding

Any permanent and direct connection between the rails and any conductive structure, which is not insulated from ground, shall not be allowed. In particular, all intentional connections to the rails (e.g., return current cable connections, cross-bonding cable connections, point machines, etc.) shall not impair the rail-to-earth insulation.

Particular attention shall be paid to LV electrical system directly in contact with the rails and provided with a PE conductor. Such systems shall comply with Section 7 of EN 50122-1. VLDs may be used to avoid direct connection between the return circuit and earth.

References

1 EN 50122-1. (2017). *Railway applications — Fixed installations — Electrical safety, earthing and the return circuit — Part 1: Protective provisions against electric shock*, 2011 + A4.
2 IEC 60364-4-41. (2005). *Low-voltage electrical installations — Part 4-41: Protection for safety – Protection against electric shock.*
3 EN 50526-1. (2012). *Railway applications — Fixed installations — D.C. surge arresters and voltage limiting devices – Part 1: Surge Arrestors.*
4 EN 50522. (2010). *Earthing of power installations exceeding 1 kV a.c.*
5 EN 50119. (2020). *Railway applications — Fixed installations — Electric traction overhead contact lines.*
6 EN 50124. (2017). *Railway applications — Insulation Coordination — Part 1: Basic requirements – Clearances and creepage distances for all electrical and electronic equipment.*
7 IEC 60479-1. (2018). *Effects of current on human beings and livestock – Part 1: General aspects.*
8 IEEE Std. 80. (2013). *IEEE Guide for Safety in AC Substation Grounding.*
9 IEC 60364-5-54. (2011). *Low-voltage electrical installations — Part 5-54: Selection and erection of electrical equipment – Earthing arrangements and protective conductors.*

10 IEC 60364-4-44. (2005). *Low-voltage electrical installations — Part 4-41: Protection for safety – Protection against electric shock.*

11 IEC 61000-5-2. (1997). *Electromagnetic compatibility (EMC) — Part 5: Installation and mitigation guidelines — Section 2: Earthing and cabling.*

12 IEC 62305-3. (2011). *Protection against lightning - Part 3: Physical damage to structures and life hazard.*

13 IEC 62305-4. (2006). *Protection against lightning - Part 4: Electrical and electronic systems within structures.*

14 EN 50162. (2004). *Protection against corrosion by stray current from direct current systems.*

15 EN 50122-2. (2010). *Railway applications — Fixed installations — Electrical safety, earthing and the return circuit — Part 2: Provisions against the effects of stray currents caused by d.c. traction systems.*

15

Regenerative Braking Energy in Electric Railway Systems

Mahdiyeh Khodaparastan[1], Ahmed A. Mohamed[2], and Constantine Spanos[3]

[1] *EnBW North America, Boston, MA 02210, USA*
[2] *City University of New York, City College, Department of Electrical Engineering, New York, NY 10031, USA*
[3] *Consolidated Edison Company of New York, Distribution Engineering, New York, NY 10003, USA*

15.1 Introduction

Electric rail transit systems are a relatively large consumer of electricity. For instance, New York City Transit (NYCT) consumes about 1700 GWh annually (about 3% of annual electricity demand). Even though electric transportation systems provide relatively low energy consumption per passenger, there is significant potential for energy efficiency enhancement, as well as for peak power and carbon footprint reduction.

In trains with regenerative braking capability, a fraction of the energy used to power a train is regenerated during braking. This regenerated energy, if not properly captured, is typically dumped in the form of heat to avoid overvoltage. Finding ways to recuperate regenerative braking energy can result in substantial economic as well as system benefits. Regenerative braking energy can be effectively recuperated using wayside energy storage, reversible substations, or hybrid storage/reversible substation systems. In this research study, we compare these recuperation techniques. As an illustrative case study, we investigate their applicability to NYCT systems, where most of the regenerative braking energy is currently being wasted.

15.2 Regenerative Braking Energy

Electric trains generally have four modes of operation including acceleration, cruising, coasting, and braking [1]. During the acceleration mode, trains accelerate and draw energy from the power source (i.e. a third rail located next to the traction rails or an overhead catenary). Upon reaching their maximum speed, trains usually switch to a cruising mode (i.e. constant power operation) or a coasting mode, whereby the speed of a train is nearly constant, and it draws a negligible amount of power. A train trip ends with braking mode. When the distance between passenger stations is short, as in the case of light rail traction systems in urban areas, the cruising mode is often omitted.

There are several types of train braking systems, including regenerative braking, resistive braking and air braking. In regenerative braking, which is the most common, a train decelerates by reversing the operation of its motors. During braking, the motors act as generators and convert mechanical energy to electrical energy. This produced electrical energy is referred to as "regenerative

Transportation Electrification: Breakthroughs in Electrified Vehicles, Aircraft, Rolling Stock, and Watercraft,
First Edition. Edited by Ahmed A. Mohamed, Ahmad Arshan Khan, Ahmed T. Elsayed, and Mohamed A. Elshaer.

braking energy" or "regenerative energy." Regenerative braking energy can be utilized to partially supply auxiliary loads onboard the train while the remaining energy is fed back to the third rail. When this energy reaches the third rail, it can be utilized by other nearby accelerating trains. In dense cities, the distance between passenger stations is typically short, and train acceleration/deceleration cycles repeat frequently; therefore, considerable amounts of regenerative energy can potentially be harnessed and reutilized [1].

Currently, there are some challenges that delimit the exchange of regenerative braking energy between trains. Firstly, there needs to be an accelerating train at the same time and location when/where regenerative energy is injected into the third trail. The amount of energy that can be reused by the neighboring trains depends on several factors, e.g. train headway and the age of the system. If there are no nearby trains to use this regenerated energy, which is typically the case, the voltage of the third rail rises, and over-voltage protection systems electrically disconnect the train, preventing it from injecting more regenerative energy. The excess energy is then dissipated in the form of heat in onboard or wayside dumping resistors. Not only does this operation waste useful energy but it also causes heating in underground tunnels [2].

Several solutions have been proposed in the literature to maximize the reuse of regenerative braking energy: (i) train timetable optimization, (ii) energy storage systems (ESS), (iii) reversible substations, and (iv) hybrid systems consisting of a reversible substation and a wayside energy storage system.

In train timetable optimization, synchronizing operations of multiple trains have been investigated. Synchronicity is achieved when the feeding of regenerative energy back to the third rail from train braking coincides with the scheduled acceleration of another train capable of absorbing that energy. In energy storage-based solutions, the regenerative braking energy is stored in an electric storage device, such as a supercapacitor, battery, or flywheel, and released back to the third rail when needed. The storage device can be placed onboard the train or offboard, beside the third rail, i.e. wayside. In the reversible substation solution, a path is provided for regenerative energy to flow in a reverse direction and feed power back to the main AC grid. In the hybrid solution, the reversible substation and wayside energy storage techniques are combined. The ESS is shared between the transit system and the utility company.

In this chapter, regenerative braking energy is described, and the aforementioned solutions for the utilization of regenerative braking energy are discussed in detail. A case study on recuperation of regenerative braking energy in the New York City Subway system is also presented.

15.3 Regenerative Braking Energy Recuperation Methods

15.3.1 Train Timetable Optimization

One of the approaches to maximizing reuse of regenerative braking energy is through optimizing train timetables. Here, the braking and acceleration of two neighboring trains are scheduled to occur simultaneously; hence, some of the energy produced by the decelerating train is used by an accelerating one. Some studies show that up to 14% of energy saving can be achieved through timetable optimization [3–5].

Most train timetable optimization studies focus on two objectives: minimizing peak power demand, and maximizing the utilization of regenerative braking energy [6]. In the early stages of research on timetable optimization (i.e. the early 1960s), the emphasis was on peak power demand minimization. Most of this research proposed methods to shift the acceleration time of some trains to off-peak time (here, the time synchronization between trains was not targeted) [7, 8]. For instance, in [7], a genetic algorithm is used to optimize train scheduling tables to limit the number of trains accelerating at the same time. In [9], a control algorithm for coordinating the movement of multiple trains is proposed to reduce peak power demand. In [10], peak power was reduced by controlling train running time using dynamic programming.

More recent research has aimed at using train timetable optimization to synchronize the acceleration/deceleration intervals of neighboring trains. Most of this research aims at optimizing the dwell time (i.e. stop time at each station) of the trains to increase the chance of synchronizing between accelerating and decelerating trains [3, 5, 11–13]. Other research has focused on determining the optimal time overlap between multiple trains [10, 14, 15] to maximize utilization of regenerative braking energy.

One of the conventional approaches used to improve the energy efficiency of the electric rail transit system is speed profile optimization. In this approach, the speed profile of a single train is optimized such that it consumes less energy during the trips between stations. Currently, ongoing research focuses on combining train timetable optimization and speed profile optimization, which can provide better energy savings when compared to each method individually. In [16], an integrated optimization method has been proposed based on actual operation data from Beijing Metro. The results show that the proposed method can reduce energy consumption of the overall system by 21.17% more than the timetable optimization method [17], and 6.35% more than the speed profile optimization method, for the same system and headways [6, 14].

Generally, timetable and speed profile optimization problems are tied to each other. Timetable optimization provides the best running time that can be used as an input in speed profile optimization. Simultaneously, speed profile optimization determines the optimal acceleration/deceleration rates, which can be used as an input for timetable optimization. Another example of the integrated optimization method is presented in [18], whereby the optimal dwell time at each station and maximum train speed at each section is determined. Here, the results show that 7.31% of energy can be saved using this approach. In [19], other real-world factors, such as the number of performing trains and their cycle times are considered as part of an integrated optimization. Here, better utilization of regenerative braking energy is obtained when all trains supplied by the same electric section are considered in the timetable. In [20], substation energy consumption is minimized by finding optimal train movement mode sequences, inter-station journey times, and service intervals. An overview of some of the studies carried out in this area is presented in Table 15.1.

15.3.2 Storage-Based Solutions

One of the key solutions for better recuperation of regenerative braking is through an energy storage system. An ESS can capture the energy produced during the train braking phase and discharge it when needed [11, 12, 21]. ESS solutions can reduce the overall system energy consumption and the peak power demand, which benefits both the rail transit system and the power utility. ESS may also be used to provide services to the main grid, such as peak shaving [22]. ESS can be implemented in two different ways, onboard or wayside. In onboard, the ESS is mostly located on the roof of each train. On the other hand, wayside ESS is located trackside and can be shared among multiple trains.

Selection of the most suitable storage technology for a given site and application is key. Several essential factors must be considered, including the energy capacity and specific energy, the rate of charge and discharge, and durability and cycle life [23]. The common energy storage technologies that have been utilized in rail transit systems are batteries, supercapacitors, and flywheels:

1) *Batteries:* The oldest and most popular type of electric energy storage. A battery consists of multiple electrochemical cells connected in parallel and series to form a module. Cells consist of two electrodes of different chemical potential (i.e. an anode and a cathode) in contact with a solid or liquid electrolyte. Batteries work based on reversible redox reactions of the electrode material to output power [24, 25]. The most commonly used technology in rail transit systems are lead–acid

Table 15.1 Examples of train timetable optimization implementation.

Method	Saving (% of consumed energy)	Implemented in real system?	Comment	References
Dwell time optimization with GA algorithm	14%	No, but data are gathered from Tehran Metro	The impact of both headway and dwell time on reusing regenerative energy has been studied	[3]
Running time reserve optimization with GA	4%	No, but data are gathered from Berlin Metro	The headway is considered to be constant. At each stop, the amount of reserve time to be spent on the next section of the ride is decided	[4]
Dwell time optimization by greedy heuristic method	5.1%	No	A mathematical model of metro timetable has been defined	[5]
Departure time through multi-criteria mixed integer programing	—	No, but data are gathered from a Korean subway system	Around 40% of peak energy has been reduced and utilization of regenerative braking energy is improved by 5%	[11]
Departure and dwell time optimization	7% in simulation, 3.52% in reality	Proposed models were used to design a timetable for Madrid underground system	85% of braking and accelerating processes are synchronized	[12]
Fuzzy logic control Dwell time	Not presented	No	Train operation is specified by a set of indices, and the aim of fuzzy control is to find the best performance among these indices	[13]
Running time Optimization by GA	7%	No, but data are gathered from a substation in the UK	Two objective functions are considered: energy consumption and journey time. The best possible compromise between them is searched using GA	[26]
Direct climbing optimization	14%	No, but data are gathered from a substation in Italy	The optimal set of speed profile and timetable variables has been found in order to minimize energy consumption	[27]
Dwell time and run-time optimization/Genetic algorithm and dynamic programming model	6.6%	-	From energy saving point of view, run-time control is superior to dwell time control. More flexible train control can also be achieved with run-time control	[28]
Substation energy consumption optimization	38.6%	No, but the case study was based on Beijing Yizhuang Metro	Substation energy consumption optimized by modifying the speed profile and the dwell time	[29]

Table 15.2 A comparison of different battery technologies.

Types	Advantages	Disadvantages	Comment	References
PbA	• Low cost per Wh • Long history • Wide deployment • High reliability • High power density	• Low number of cycles • Low charging current • Environmental concerns • Poor performance at low temperatures	Recently, extensive research has been carried out on hybridizing lead with other materials, such as carbon, to increase power and energy densities	[24, 30, 31]
NiMH	• Long service life • High energy density • High charge/discharge current • High cycle durability • Low environmental concern	• High cost per Wh • High maintenance cost • High self-discharge rate	The main disadvantage of a high self-discharge rate may be addressed by using novel separators	[24, 31–33]
Li-ion	• High energy density • Low maintenance • High cycle life • High efficiency	• High cost per Wh • Requires cell balance and control to avoid overcharge • Requires special packing and protection circuitry • Poor performance at low temperatures	Large research community investigating electrode improvements and new solid-state designs that promise to improve the performance of Li-ion batteries	[24, 31, 34]
NaS	• High energy density • High power density • High energy efficiency	• High capital cost • High operating cost • Safety concerns • Needs external heating	Researchers are investigating ways to reduce operating temperatures and mitigate heating requirements.	[24, 30, 32, 33, 35]

(PbA), lithium-ion (Li-ion), nickel-metal hydride (NiMH) and sodium-sulfur (NaS). A comparison of the advantages and disadvantages of each type has been summarized in Table 15.2.

2) *Supercapacitors*: Like batteries, supercapacitors consist of two electrodes immersed in a liquid electrolyte. However, there is generally no chemical conversion reaction involved, and the energy is stored electrostatically. Applying a voltage across the two electrodes polarizes the electrolyte, and the ions in the solution accumulate only one molecule of distance from each electrode's surface, as part of a capacitive double layer. The combination of short separation distances and very high surface areas of supercapacitor electrodes result in very high capacitance per unit volume when compared to conventional film or ceramic capacitors [24, 30, 36–40]. The electrical characteristics of supercapacitors highly depend on the selection of the electrolyte and electrode materials [37]. Some of the advantages of supercapacitors are high energy efficiency (~95%), large charge/discharge current capacity, long lifecycle (>50,000), high power density (>4000 W/kg) and low internal resistance [30, 37, 39, 41, 42]. However, the maximum operating voltage of supercapacitors is very low, and they suffer from high leakage current [36]. Recently, Li-ion capacitors have been developed with less leakage current and higher energy and power densities than some common batteries and standard supercapacitors [36, 43, 44].
3) *Flywheels*: Flywheels are an electromechanical type of ESS that works based on the storage and release of kinetic energy. A flywheel ESS is composed of an electrical motor-generator driving the rotation of the flywheel rotor, a heavy mass rotating at high speed. The angular momentum of the flywheel rotor determines the amount of energy that can be stored and delivered.

> The electrical machine is coupled to a variable frequency power converter. During charging the electrical machine functions as a motor, accelerating the flywheel rotor. During discharge, the electrical machine transitions into a generator, releasing power and decreasing the speed of the rotor. To reduce friction losses, flywheels use magnetic bearings, often contain the rotor in a vacuum chamber [24, 25, 30, 45, 46].

Flywheel ESS presents some unique advantages, such as high energy efficiency (~95%), high power density (5000 W/kg), high energy density (>50 Wh/kg), low maintenance, high cycling capacity (more than 20,000 cycles), and low environmental concern [42]. Some of their drawbacks include a high self-discharge current, risk of explosion during a failure event, heavy weight, and high cost. However, through predictive design and smart protection schemes, system safety is believed to be improved. According to some publications, if flywheel system costs are decreased, they can be extensively used across a number of industries and play a significant role in worldwide energy sustainability plans [25, 30, 47]. Based on the simulation results presented in [48], a flywheel ESS is capable of achieving 31% energy saving in light rail transit systems.

15.3.2.1 Onboard Energy Storage

In onboard ESS, the storage system is placed on the train, either on the roof or underneath the floor. Figure 15.1 illustrates a schematic of onboard ESS. The efficiency of onboard ESS is highly dependent on the characteristics of the vehicle, which can directly affect the amount of energy produced and consumed during braking and acceleration, respectively [48]. Onboard energy storage systems provide other advantages, such as peak power reduction, voltage stabilization, catenary-free operation, and energy loss reduction [49]. On the other hand, the cost of implementation, maintenance, and safety concerns are higher than that of wayside energy storage, as an ESS is placed on every train, as opposed to at a few targeted stations.

Onboard ESS is already in use by some rail transit agencies. In addition, several agencies around the world are either considering or testing it. Various technologies have been used for onboard ESS; among them, supercapacitors have been most widely implemented in transit systems. Batteries have not been able to compete with supercapacitors due to their relatively short lifetime and low power density. Additionally, flywheels have not received much attention due to size and cost limitations; however, there are ongoing efforts in this area. For instance, construction of a prototype hybrid electric vehicle has been reported in [50], and an agreement was reportedly structured between Alstom Transport and Williams Group to install onboard flywheels on trams in [51].

Important examples of real-world implementation of the onboard ESS configuration are the Brussels metro and tram lines, and the Madrid metro lines in Europe that each realized 18.6–35.8% and 24% in energy savings, respectively [52–54]. Additionally, the Japan metro realized 8% savings from regenerative braking energy recuperation, and the Mannheim tramway achieved a 19.4–25.6% increase in overall system energy efficiency through implementation of onboard ESS [55, 56]. Table 15.3 provides an overview of various applications of onboard ESS all over the world.

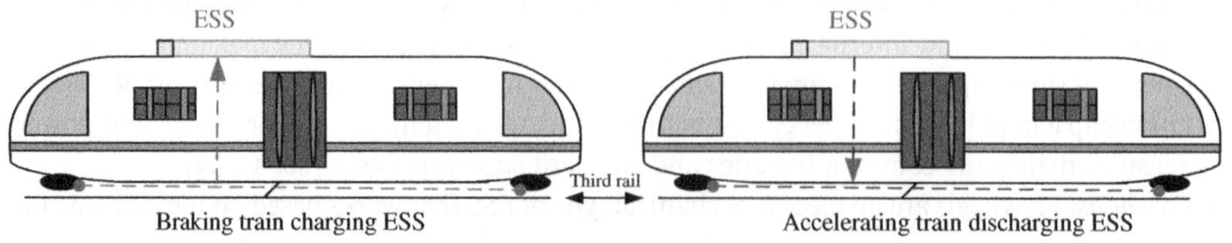

Figure 15.1 Onboard energy storage systems.

Table 15.3 Examples of onboard energy storage implementation.

Type	Location	Purpose	Comment	References
Ni-MH	Sapporo	Energy saving Catenary free operation	Giga-cell™ NiMH batteries provided by Kawasaki have been used. Fully charged in five minutes through a 600 V DC overhead catenary	[57]
Li-ion	Charlotte, NC	Energy saving Catenary free operation	S700 streetcar from Siemens Mobility for the Charlotte Area Transit System	[57–59]
Ni-MH	Lisbon	Operation without overhead contact line	The SITRAS HES (hybrid energy storage) energy storage system has been used	[60, 61]
Ni-MH	Nice	Catenary free operation	—	[53, 62]
Supercapacitor	Mannheim	Reduction of energy consumption and peak power demand Catenary free operation	A 400 V system with 1 kWh energy storage from 640 Ultracapacitors, 1800F each	[63–65]
Supercapacitor	Innsbruck	Energy saving	—	[61]
Supercapacitor	Seville, Saragossa	Energy saving, Catenary free operation	—	[59]
Supercapacitor	Paris	Energy saving, Catenary free operation	Could also be recharged from the overhead contact system in about 20 seconds during station stops	[59, 66]
Flywheel	Rotterdam (France)	Energy saving Catenary free operation	Flywheel located on the roof. Flywheel system was developed and installed by ALSTOM. However, the project stopped due to technical issues	[57, 67]
Li-ion	Brookville, MI	Catenary free operation	Light rail trains built by Brookville Equipment Corporation using 750 V lithium-ion batteries	[59]

15.3.2.2 Wayside Energy Storage

In wayside ESS, the storage system is placed trackside and usually connected to the third rail through a power control unit. A schematic overview of a wayside ESS is shown in Figure 15.2. The main concept of wayside ESS is to absorb the energy regenerated during train braking and deliver it back to the third rail when needed for acceleration. A wayside ESS can be shared among all trains running within the same electrical section [68].

In addition to the general advantages that were previously mentioned for energy storage systems, wayside ESS can provide other benefits, such as peak shaving, load shifting, emergency backup, and frequency and voltage regulation [69]. Wayside ESS can also help minimize problems related to voltage sag (temporary reductions in voltage that occur during times of high throughput over long spans of track between supply points), which can damage electronic equipment in a rail car and affect the performance of trains during acceleration. To overcome this condition, the ESS can be designed to discharge rapidly and help regulate the third rail voltage level [48, 70–72].

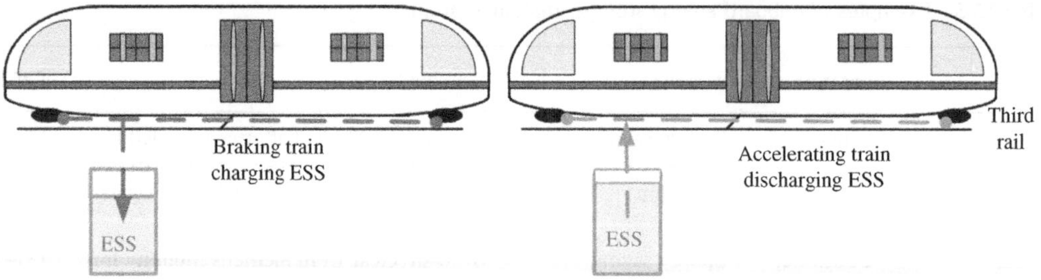

Figure 15.2 Wayside energy storage systems.

Real-world implementation of wayside ESS has a reported energy savings of up to 30%. The amount of energy saved by an ESS highly depends on the system characteristics and storage technology. As an example, the wayside ESS Sitras SES (static energy storage) system formerly commercialized by Siemens was marketed as a solution that could save nearly 30% of energy. The proposed ESS uses a supercapacitor technology that can provide 1MW peak power and is capable of discharging 1400 A DC current into the third rail over 20–30 seconds. Sitras ESS is implemented in different cities in Germany (Dresden, Cologne, Koln, and Bochum), Spain (Madrid), and China (Beijing). Bombardier has developed a system based on supercapacitors, the EnerGstor, which can offer 20–30% reduction in grid power consumption. An Energstor prototype, sized 1 kWh per unit, has been designed, assembled, and tested at Kingston (Ontario) [61].

Another supercapacitor-based system that is commercially available is Capapost, developed by Meiden and marketed by Envitech Energy, a member of the ABB Group, which is scalable from 2.8 to 45 MJ of storable energy. This system has been reported to be installed in the Hong Kong and Warsaw metro systems [73, 74]. Table 15.4 provides an overview of various applications of wayside ESS all over the world. This information is mostly published by manufacturers of wayside ESS like Siemens [61], ABB [75], VYCON [76, 77], and Pillar [57].

15.3.3 Reversible Substation

Reversible substations are another technique for recuperating regenerative braking energy. A schematic of this technique is shown in Figure 15.3. This technique provides a path through an inverter for regenerative braking energy to feed back to the upstream AC grid. This energy can be consumed by other electric AC equipment in the substation, such as escalators and lighting systems, or it can be fed back to the main grid [78]. Reversible substations must follow the rules and legislations of the electricity distribution network, i.e. they must maintain an acceptable power quality level for the power fed back to the grid by minimizing harmonic distortion [78].

There are two common ways to provide a reverse path for the energy: (i) using a DC/AC inverter in combination with a diode rectifier; or (ii) using a reversible thyristor-controlled rectifier (RTCR). In the first approach, the DC/AC converter can either be a pulse width modulation (PWM) converter, or a thyristor line commutated inverter (TCI) [31]. In the first approach, the existing diode rectifier and transformer can be kept. In the second approach, the diode rectifiers need to be replaced with RTCRs, and the rectifier transformers need to be changed, which makes this approach more expensive and complex [31]. However, RTCRs have advantages, such as voltage regulation and fault current limitation [79].

Table 15.4 Examples of wayside energy storage implementation.

Type	Location	Voltage	Purpose	Comment	References
Li-ion	Philadelphia	660 V	– Energy saving – Optimize SEPTA's power and voltage quality – Frequency Regulation Market Revenues	ENVILINE™ ESS provided by ABB has been used	[75, 80]
NaS	Long island	6 kV AC	– Peak shaving	Several challenges have been reported, including the sizing of ESS, safety issues, and unexpected costs	[30, 33, 34, 81]
Li-ion	West Japan	640 V	– Energy saving – Voltage stabilization	—	[79]
Li-ion	Nagoya	640 V	– Voltage stabilization	—	[79]
Li-ion	Kagoshima	640 V	– Voltage enhancement	The ESS was far from the substation and controlled remotely via the internet	[79]
Ni-Mh	Osaka	640 V	– Energy saving	The battery was connected directly to the electric line	[79]
Li-ion	Kobe	640 V	– Voltage enhancement	—	[79]
Super Capacitor	Seibu	640 V	– Energy saving	—	[79]
Ni-MH	New York	670 V	– Voltage enhancement	The battery was directly connected to the third rail	[82]
Flywheel	London	630 V	– Energy saving – Voltage enhancement	Flywheel system provided by URENCO	[30, 83, 84]
Flywheel	Los Angeles	—	– Energy saving	Flywheel system provided by VYCON	[76, 77]
Flywheel	Hanover	—	– Energy saving	Flywheel system provided by Pillar	[57]
Flywheel	New York	670 V	– Energy saving	Flywheel system was provided by KINETIC TRACTION, and successfully tested at Far Rockaway. However, the project was stopped due to budget constraints	[57]
Super capacitor	Madrid	750 V	– Voltage stabilization	Sitras SES has been used	[61]
Super capacitor	Cologne	750 V	– Energy saving, voltage stabilization	Sitras SES has been used	[61]
Super capacitor	Beijing	750 V	– Energy saving	Sitras SES has been used	[61]
Super capacitor	Toronto	600 V	– Energy saving	Sitras SES has been used	[61]

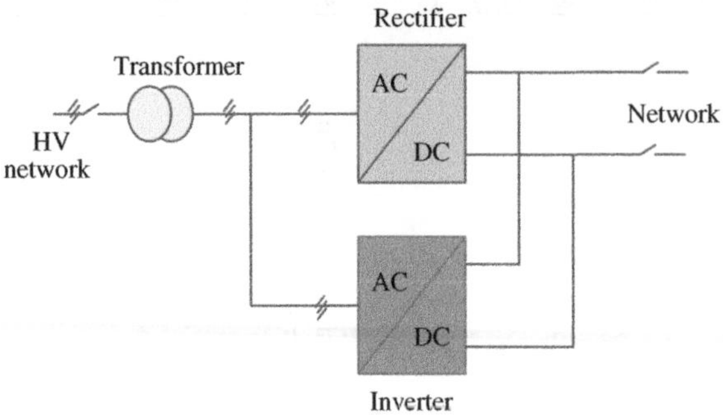

Figure 15.3 Block diagram of a reversible substation.

As mentioned earlier, a diode rectifier can also be combined with a PWM converter to provide a reverse path for the energy. PWM converters have the advantage of working at unity power factor, with the disadvantages of high cost and switching losses. In order to use PWM converters for reversible substation purposes, a step-up DC/DC converter should be added between the PWM converter and the DC bus. To reduce the harmonics and avoid current circulation, a DC filter is also needed at the output of the converter. Some of the currently available reversible substation systems are summarized in Table 15.5.

15.3.4 Hybrid Reversible Substation and Wayside Energy Storage Modeling

A hybrid reversible substation/WESS, as illustrated in Figure 15.4, is a combination of a reversible substation and a wayside energy storage system. Regenerative braking energy can be absorbed by both the ESS and the reversible substation based on a control signal received from the energy management unit.

A key aspect of this technique is the energy management unit that provides setpoints/settings for both the energy storage system and the reversible substation. Setpoints are determined based on predefined modes of operation, and can be classified into grid applications and rail transit applications.

The following is a simplified example of what the different modes of operations may look like. In the rail transit application modes, priority is given to the rail system. The reversible substation and ESS provide the following applications:

- Maximum recuperation of regen braking energy (Mode = 1)
- Peak demand reduction of the rail transit system (Mode = 2)

In the grid application modes, priority is given to the grid. The reversible substation and ESS collaborate to provide the desired output for each of the following applications:

- Feeder peak demand reduction (Mode = 3)
- Voltage regulation and reactive power support (smart inverter) (Mode = 4)

Besides this application, a hybrid reversible substation can provide other services such as demand response, grid black start, and emergency backup for the rail transit system. For these services, special modes and settings need to be designed.

Table 15.5 Examples of reversible substation implementation.

Company	Location	Voltage level	Technology	Energy saving	Comment	References
Alstom-HESOP	Paris Tramway Utrecht–Zwolle London Milan Metro	750 1500	Thyristor rectifier bridge associated with an IGBT converter	7% of traction consumed energy	– Recuperated more than 99% of recoverable braking energy. – Reduced train heat dissipation, which led to reducing energy consumption used for ventilation. – No need for onboard braking resistors; therefore, train mass was reduced leading to less energy consumption during acceleration.	[85–87]
Simense-Sitras TCI	Oslo, Singapore Zugspitze, Germany	750 1500	Inverter, B6 thyristor bridge, autotransformer in parallel with a diode rectifier	—	– Remote control is possible through a communication interface. – Reduced the number of braking resistors on the train. – In the 750 V version, an autotransformer is integrated with the inverter, but in the 1500 V version or high power 750 V, an autotransformer is installed separately.	[57]
Ingeber-Ingteam	Bilbao-Spain Malaga-Spain Bielefeld-Germany Metro Brussels	1500 3000 750 750	Inverter in parallel with existing rectifier and transformer	13% of the substation annual energy consumption	– The current injected into the AC grid is high in quality (THD < 3%). – A DC/DC converter and an inverter are connected in series, and their combination is connected to an existing grid. – Reduced carbon emissions by 40%.	[57, 88–90]

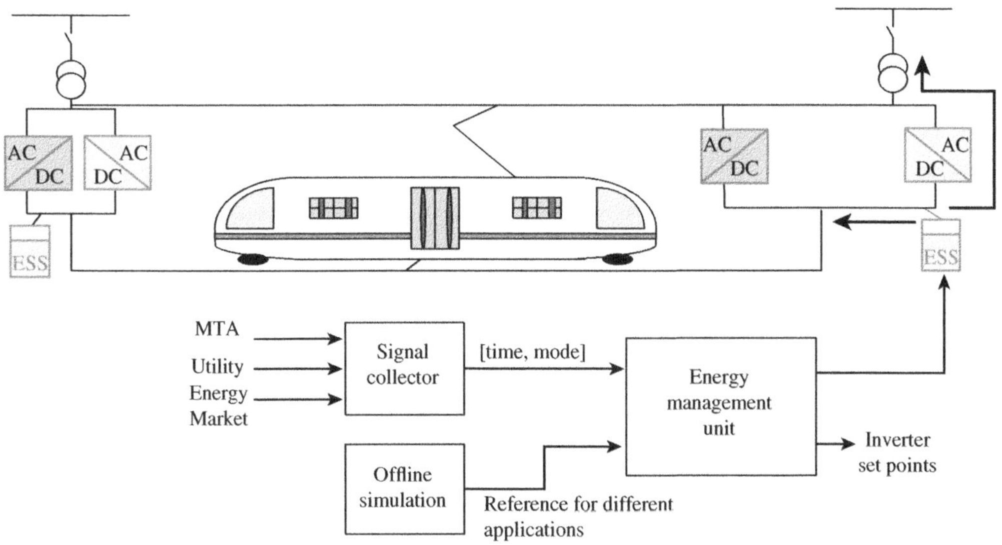

Figure 15.4 Block diagram of a hybrid reversible substation.

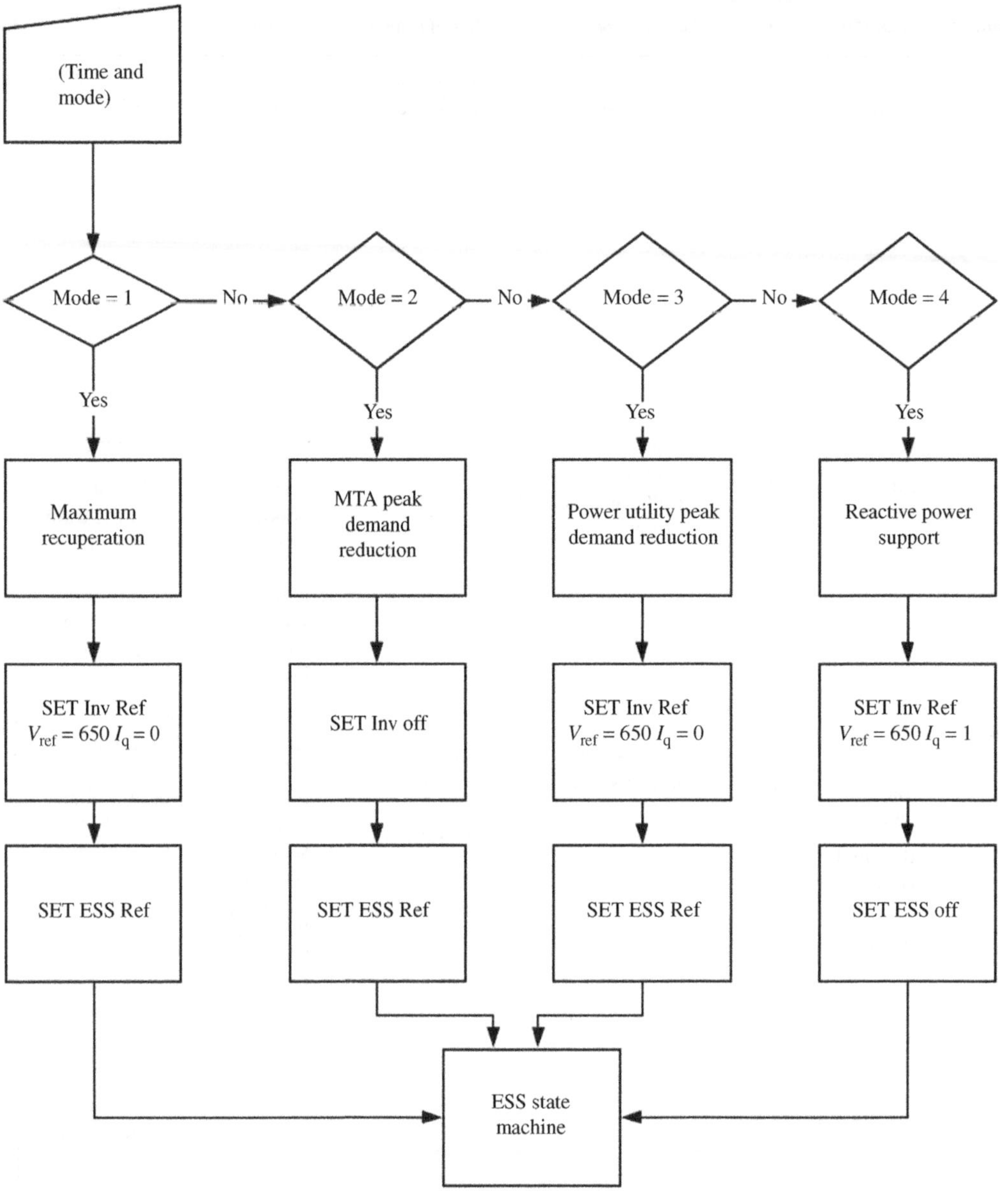

Figure 15.5 Simplified flowchart for an example energy management unit.

ESS and inverter settings and reference points are determined based on offline simulation to make sure the operation of a hybrid reversible substation does not create any conflict with the normal operation of a transit system.

Reversible substation setpoints are sent directly to the inverter's controllers, and ESS setpoints feed into a state machine controller that controls charge and discharge of the energy storage. A flowchart of simplified energy management is presented in Figure 15.5. More detailed algorithms will be presented in the following chapters.

15.3.5 Choosing the Right Application

Different technologies/techniques have been proposed to reuse regenerated braking energy. A comparison between the general pros and cons associated with these alternatives is presented in Table 15.6 [57, 31]. Choosing the right regenerative energy recuperation technique/technology requires careful consideration of various influential parameters, such as [42]:

- Need for catenary-free operation
- Electric network characteristics and ownership
- Vehicle specifications
- Train headway

Catenary-free operation allows vehicles, especially tramways, to run without needing a connection to an overhead supply line. It is a good solution for operating trams in historic areas, for instance. In this case, only onboard energy storage can be used, charged through regenerative braking and/or fast charging infrastructure at the stations [42]. When considering different applications of energy recovery, electric network ownership plays an important role. In many cases, local energy providers own the power system infrastructure, and transportation companies have limitations around exporting power back to the network. In cases where it is possible to sell the recovered energy for a high price, a reversible substation may be the most economic option. If the local energy company charges for power demand, whereby higher prices are charged over several hours of the day due to greater electricity demand, then a hybrid configuration of reversible substations and wayside energy storage may make the most sense. Otherwise, wayside energy storage in isolation should be selected [42].

The characteristics of the network represent another important parameter that should be considered when selecting the energy recovery technique. For example, in some stations with a high level of traffic, the energy exchange between braking and accelerating trains occurs naturally. Therefore, the energy recovery application should be designed in a way that gives priority to natural energy

Table 15.6 A comparison between different recuperation techniques.

Application	Advantages	Disadvantages
Onboard ESS	- Provides possibility for catenary-free operation. - Reduces voltage drop. - Reduces third rail losses and increases efficiency.	- High cost due to placement of ESS on the vehicle. - High safety constraints due to onboard passengers. - Standstill vehicles for maintenance and repair.
Wayside ESS	- Mitigates voltage sag. - Can be used by all vehicles running on the line (within the same section). - Maintenance and repair do not impact train operation.	- Increases overhead line losses due to the absorption and release of energy over the traction line. - Analysis is needed to choose the right sizing and location.
Reversible substation	- Provides possibility for selling electricity to the main grid. - Can be used by all vehicles running on the line. - Maintenance and repair do not impact train operation. - Lower safety constraints.	- No voltage stabilization capability. - Analysis is needed for choosing the right location.

exchange between the trains. In stations that face considerable voltage drop when several trains accelerate simultaneously, wayside energy storage may be used to help sustain the third rail voltage [42].

The type of vehicle used in a transportation system is also an important factor. For example, in some systems, old and new vehicles may be running together, while old vehicles may not have the capability to regenerate energy during braking. Therefore, investing in energy recovery may not yield as much value. In addition, train occupancy needs to be factored in when designing the ESS, since a heavier vehicle produces more energy during braking [42].

15.4 New York City Transit - Case Study

In order to investigate the impact of different solutions on the recuperation of regenerative braking energy, we studied the NYCT system as a case study. A portion of NYCT line 7 consisting of 3 rectifier substations and 5 passenger stations is considered. The NYCT map and the portion under study are presented in Figure 15.6. A schematic of this system is presented in Figure 15.7. This system has been simulated, validated with real measurements, and used for several studies [91, 92]. These case studies are summarized in the following subsections.

15.4.1 NYC Transit Systems

An example of 24-hour power profile of a substation in the NYCT System is illustrated in Figure 15.8. Here, the power demand reduces gradually at night and then starts to increase in the early morning. There is a local maximum at around 8:30 AM when the express and local trains run together and the time interval between trains is reduced to serve a large number of passengers commuting. After this period, the headway between running trains increases and the express train service stops. Therefore, the power demand decreases and remains at an almost constant value until 3 PM, when the express service starts again. The headway between running trains decreases again and the power demand increases until it reaches its maximum at around 6:30 PM. At this time, the power demand of an individual substation exceeds 2 MW.

15.4.2 Wayside Energy Storage

In this section, the impact of installing each of the three wayside energy storage technologies (i.e. battery, supercapacitor, and flywheel) for recuperation of regenerative braking energy and peak demand reduction is investigated. The authors demonstrated in [92] that energy savings and peak demand reduction resulting from wayside energy storage can reach about 35–45%; therefore, the case studies were selected within this range. The scope of the study includes three substations, namely 50 Ave, Jackson, and Queensboro Substations, and five passenger stations, from Hunters Point Ave to 40th St, as presented in Figure 15.9.

The current provided by rectifier substations during the charging process is critical. Generally, when there is no ESS in the system, the power provided by rectifier substations during train deceleration is almost zero. However, when an ESS is added to the system, the substations see that ESS is a new load, and if the ESS requires more current than what is available from the train, the substations will unwantedly contribute to providing it. Therefore, in the first step, we investigated the

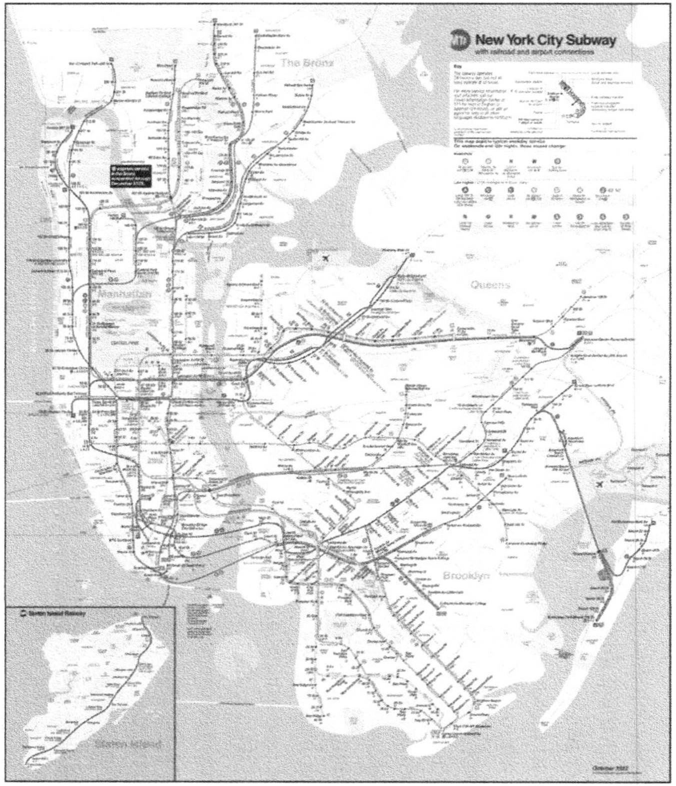

New York Subway Map

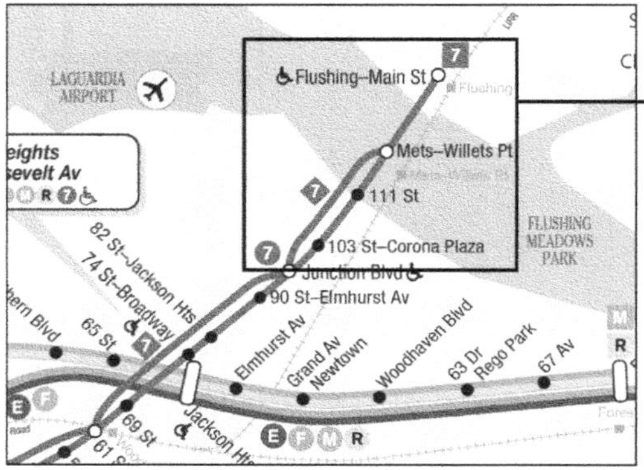

A portion of Line 7

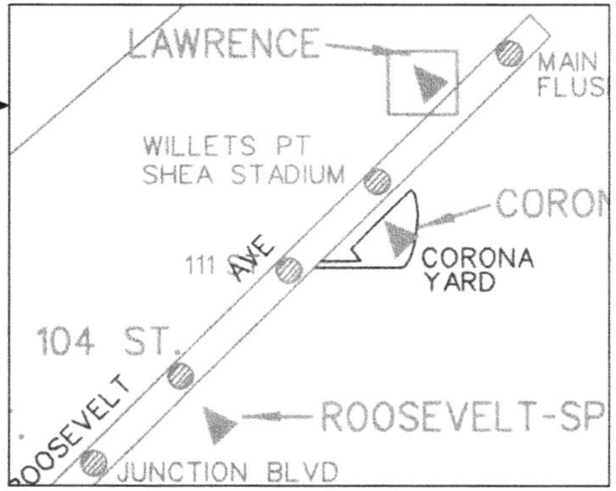

Electric supply network
(triangles represent rectifier substations, and circles represent passenger stations)

Figure 15.6 NYCT map of the system under study.

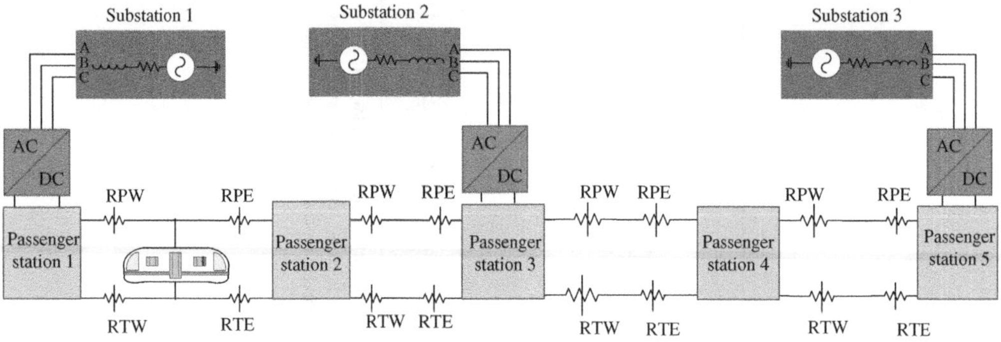

Figure 15.7 Schematic of electric rail transit systems.

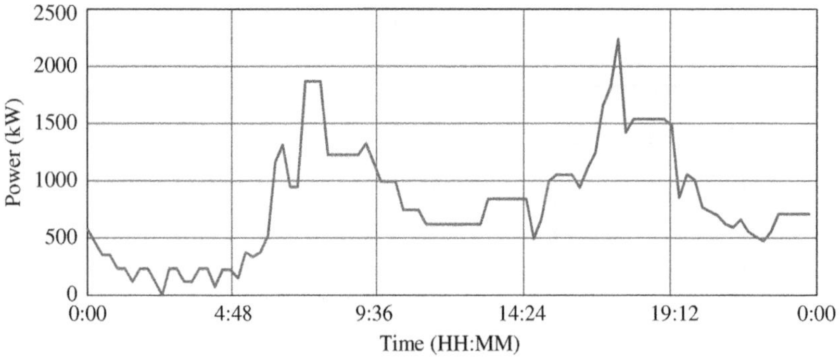

Figure 15.8 24-hour power profile of a substation.

contribution of substations for supplying a train when starting a journey from Hunters Point passenger station to Court Sq. passenger station, and then from Court Sq. to Queensboro Plaza. The power at each substation and the percentage of their contributions are presented in Table 15.7. It can be concluded that the third substation (Queensboro here) has a very small contribution compared to the other two substations. Therefore, a case study between 50 Ave and Jackson Ave substations can neglect the Queensboro substation. The range of the simulation model can be limited to two substations without compromising accuracy of the results.

A typical current profile for a local train running on Line 7 of the NYCT, based on real measurements, is shown in Figure 15.10. Initially, the train is at standstill at the origin station. At about 6 seconds, the train starts accelerating at a constant rate of about 2.5 MPH/s until it reaches its maximum speed at about 13 seconds. During this time, the current rapidly increases. Once the train reaches its maximum speed and stops accelerating, the current drops. Finally, the current becomes negative (flowing back to the third rail) when the train starts to decelerate at a rate of about −3 MPH/s. In the case studies presented here, the regenerative braking energy available for recapture is about 10 kWh.

The wayside energy storage systems will be installed at substation 2 (Figure 15.9). The ESS sizing requirements for achieving 10%, 18%, and 24% energy saving and peak demand reduction at substation 2 are presented in Table 15.8. These percentages represent the amount of regenerative braking energy that can be recuperated out of the total available energy (about 10 kWh in this case). This analysis assumes that wayside energy storage systems will be continuously operational for the entire day. Therefore, the percentage of energy savings observed from a single train trip occurs

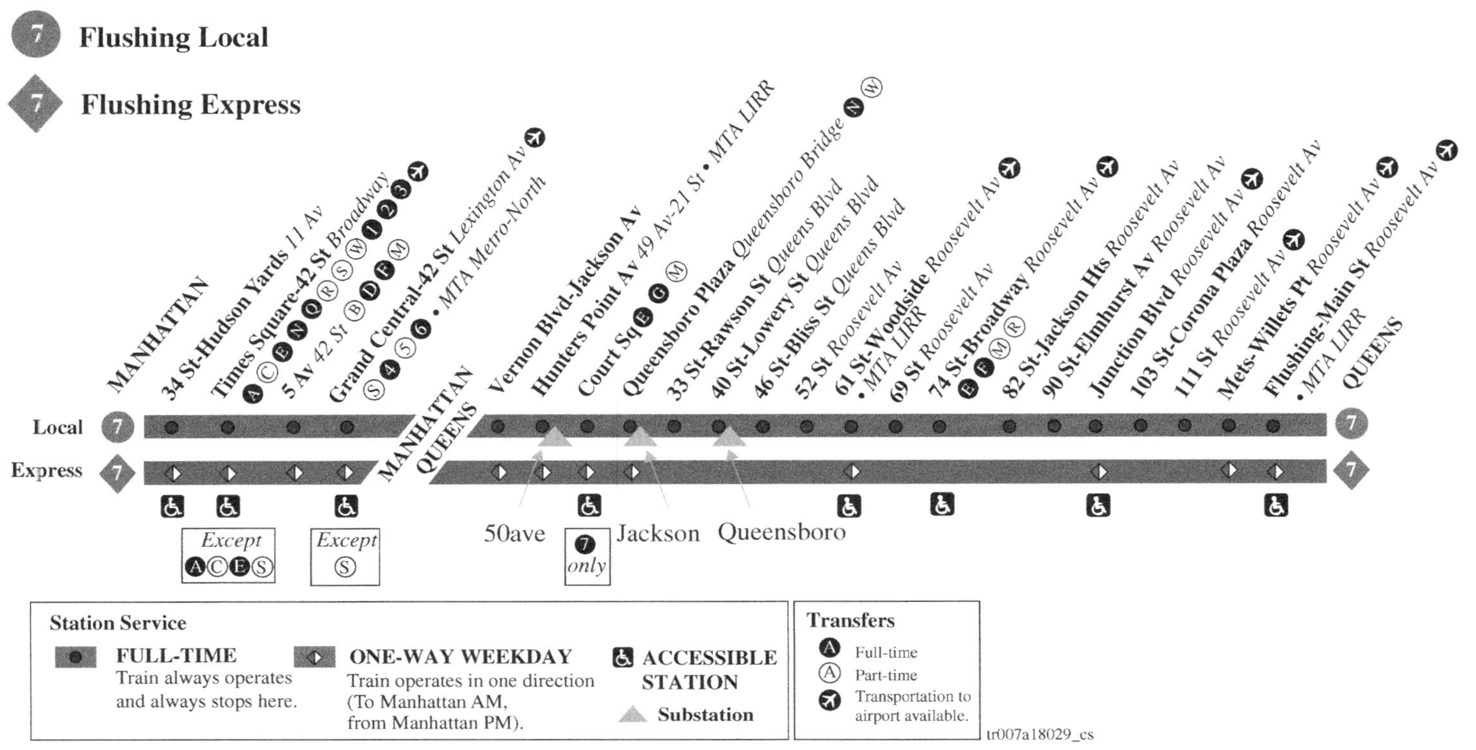

Figure 15.9 Schematic of NYCT line 7.

Table 15.7 Substation Power Contribution

Name	Peak power (w)	% Contribution
Train	3.3×10^6	—
50 Ave 1st stop	2.8×10^6	83
Jackson 1st stop	4.67×10^5	14
Queensboro 1st stop	1.2×10^5	3
50 Ave 2nd stop	1.56×10^6	46
Jackson 2nd stop	1.56×10^6	46
Queensboro 2nd stop	2.87×10^5	8

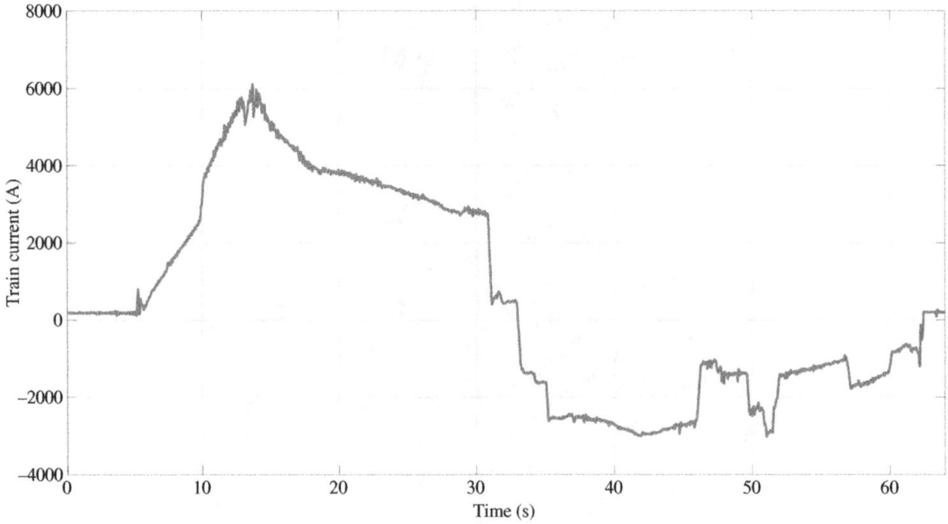

Figure 15.10 Train current profile.

Table 15.8 Sizing requirement for different targets of energy saving and peak demand reduction.

	Charging/ discharging rate	Sizing for 10% demand reduction		Sizing for 18% demand reduction		Sizing for 24% demand reduction	
		Energy (kWh)	Power (kW)	Energy (kWh)	Power (kW)	Energy (kWh)	Power (kW)
Battery	1C/1C	1111.6	570	1990	995	2072	1036
Supercapacitor (capacitance)		1.5 (62 F)	570	2.6 (106 F)	995	3.3 (133 F)	1036
Flywheel		3.8	570	6.6	995	7	1036

Table 15.9 Capital cost for different sizes of wayside ESS.

Capital cost ($) Type of ESS		Peak demand reduction 10%	18%	24%
Battery	1C/1C	1333,914–5002,180	2387,872–8954,521	2486,680–9325,053
Supercapacitor		57,750–250,500	100,800–437,000	105,250–463,900
Flywheel		93,100–247,000	162,450–431,000	169,400–449,400

for all trains associated with the same substation. In addition, since peak demand is computed based on interval metering data on a 24-hours basis, energy savings and peak demand reduction are equivalent. In other words, 10%, 18%, and 24% energy savings will lead to almost 10%, 18%, and 24% peak demand reduction. The estimated capital cost for installing each size of ESS is presented in Table 15.9, based on information available in the literature.

15.4.3 Reversible Substation

For efficient recuperation of regenerative braking energy, whenever the voltage attempts to rise due to the energy produced by an approaching decelerating train, the inverter must rapidly inject as much power as possible to the AC grid. In ideal conditions, it would inject almost all the power that is produced by the decelerating train into the main grid, unless some of the energy is utilized on the third rail (i.e. by a nearby accelerating train). The controller must monitor the voltage and autonomously yield an appropriate setpoint for the current to regulate the third rail voltage. To investigate the impact of a reversible substation on the recuperation of regenerative braking energy, two cases were simulated. In the first case, a train is running from Court Sq. passenger station to the Queensboro passenger station. Jackson substation, close to Queensboro passenger station, is assumed to have a reverse path to the grid (4 units of 0.5 MW inverters connected in parallel). In this case, the reversible substation will attempt to collect the regenerative braking energy by pushing the power back to the grid. The inverter is commanded to regulate the third rail voltage to 650 V. In the second case, the same portion of NYCT's line 7 is also considered. Here, Jackson substation is considered to have a reverse path (2 MW inverter) and the train starts running from Hunters Point Ave passenger station to 40th St passenger station (4 stops).

Based on our simulations, the inverter is capable of injecting a major portion of the energy produced by the train during deceleration to the main AC grid. The distance between a decelerating train and location of the reversible substation dictates the extent of energy saved. When a train is decelerating at a station where a reversible path exists, the energy recuperation is maximum. The savings drop as the train continues and makes stops at the following stations. Our analysis shows that, on average, a reversible substation can provide about 72.26% energy savings.

15.4.4 Hybrid Reversible Substation and Wayside Energy Storage

Results of this section represent the case when an inverter provides a reverse path for the train power to flow back to the main AC grid (as in the previous case), along with the addition of a wayside ESS connected to the third rail. An energy management algorithm (a more advanced version of that presented in Figure 15.5) then determines when and how much energy should be injected into the main AC grid at any given time. The exact type of ESS was not the focus of this task, but rather

Table 15.10 Energy recuperation potential for different configurations.

Application	ESS	Reversible	Hybrid
Energy recuperated	44.1%	63.23%	88.3%
Feeder relief	Yes	No	Yes
Reactive power support	—	—	Yes
Demand response	—	—	yes

the goal was to examine the impact of the ESS controller. Simulation results presented in Table 15.10 indicate that the extent of energy recuperation, presented as a percentage of the total available regenerative braking energy (~10 kWh), reaches almost 90% in the Hybrid configuration.

References

1 Ogasa, M. (2008). Energy saving and environmental measures in railway technologies: example with hybrid electric railway vehicles. *IEEJ Trans. Electr. Electron. Eng.* 3 (1): 15–20.
2 Gunselmann, W. (2005). Technologies for increased energy efficiency in railway systems. *2005 European Conference on Power Electronics and Applications*, pp. 1–10.
3 Nasri, A., Fekri Moghadam, M., and Mokhtari, H. (2010). Timetable optimization for maximum usage of regenerative energy of braking in electrical railway systems. *SPEEDAM 2010 - International Symposium on Power Electronics, Electrical Drives, Automation and Motion*, pp. 1218–1221.
4 Albrecht, T. (2004). Reducing power peaks and energy consumption inrail transit systems by simultaneous train running time control. *WIT Transactions on The Built Environment*, Vol. 74, p. 10, UK.
5 Fournier, D., Mulard, D., Fournier, D., et al. (2015). Agree dyheuristic for optimizing metro regenerative energy usage. *Proceedings of the Second International Conference on Railway Technology: Research, Development and Maintenance.*
6 Yang, X., Li, X., Ning, B., and Tang, T. (2016). A survey on energy-efficient train operation for urban rail transit. *IEEE Trans. Intell. Transp. Syst.* 17 (1): 2–13.
7 Chen, J., Lin, R., Member, S., and Liu, Y. (2005). Optimization of an MRT train schedule: reducing maximum traction power by using genetic algorithms. *IEEE Trans. Power Syst.* 20 (3): 1366–1372.
8 Sansand, B. and Girard, P. (1995). Train scheduling desynchronization and power peak optimizationin a subway system. *Proceedings of the 1995 IEEE/ASME Joint Railroad Conference*, pp. 75–78.
9 Gordon, S.P. (1998). Coordinated train control and energy management control strategies. *Proceedings of the 1998 IEEE/ASME Joint Railroad Conference*, pp. 165–176.
10 Ramos, A., Peña, M.T., Fernández, A., and Cucala, P. (2007). Mathematical programming approach to underground timetabling problem for maximizing time synchronization. *XI Congreso de Ingeniería de Organización.*
11 K. Kim, K. Kim, and M. Han. *A Model and Approaches for Synchronized Energy Saving in Timetabling,"Korea Railroad Research Institute*, http//www.railw.org/IMG/pdf/a4_kim_kyungmin.pdf, 2011.
12 Pena-Alcaraz, M., Fernandez, A., Cucala, A.P. et al. (2011). Optimal underground timetable design based on power flow for maximizing the use of regenerative-braking energy. *Proc. Inst. Mech. Eng. F J. Rail Rapid Transit* 226 (4): 397–408.

13 Chang, C.S., Phoa, Y.H., Wang, W., and Thia, B.S. (1996). Economy/regularity fuzzy-logic control of DC railway systems using event-driven approach. *IEE Proc. Electr. Power Appl.* 143: 9.

14 Yang, X., Li, X., Gao, Z. et al. (2013). A cooperative scheduling model for timetable optimization in subway systems. *IEEE Trans. Intell. Transport. Syst.* 14 (1): 438–447.

15 Zhao, L., Li, K., and Su, S. (2013). A multi-objective timetable optimization model for subway systems. *Proceedings of the 2013 International Conference on Electrical and Information Technologies for Rail Transportation (EITRT 2013)*, Vol. I, pp. 557–565.

16 Li, X. and Lo, H.K. (2014). An energy-efficient scheduling and speed control approach for metro rail operations. *Transp. Res. B* 64: 73–89.

17 Su, S., Li, X., Tang, T., and Gao, Z. (2013). A subway train timetable optimization approach based on energy-efficient operation strategy. *EEE Trans. Intell. Transp. Syst.* 14 (2): 883–893.

18 Yang, T.T., Xin, X.L., and Ning, B. (2015). An optimisation method for train scheduling with minimum energy consumption and travel time in metro rail systems. *Transp. B Transp. Dyn.* 3: 5919–5924.

19 Yang, X., Chen, A., Li, X. et al. (2015). An energy-efficient scheduling approach to improve the utilization of regenerative energy for metro systems. *Transp. Res. C Emerg. Technol.* 57: 13–29.

20 Zhao, N., Roberts, C., Hillmansen, S. et al. (2017). An integrated metro operation optimization to minimize energy consumption. *Transp. Res. C Emerg. Technol.* 75: 168–182.

21 Miyatake, M. and Ko, H. (2007). Numerical analyses of minimum energy operation of multiple trains under DC power feeding circuit. *2007 European Conference on Power Electronics and Applications*. EPE.

22 Holmes, K. (2008). Smart grids and wayside energy storage. *Passenger Transp.* 66 (40).

23 Schroeder, M., Pj, Y., and Teumim, D. (2010). *Guiding the Selection and Application of Wayside Energy Storage Technologies for Rail Transit and Electric Utilities*. Transportation Cooperative Research Program (TCRP).

24 Gonzalez-Gil, A., Palacin, R., and Batty, P. (2013). Sustainable urban rail systems: strategies and technologies for optimal management of regenerative braking energy. *Energy Convers. Manag.* 75: 374–388.

25 Luo, X., Wang, J., Dooner, M., and Clarke, J. (2015). Overview of current development in electrical energy storage technologies and the application potential in power system operation. *Appl. Energy* 137: 511–536.

26 Bocharnikov, Y.V., Tobias, A.M., and Roberts, C. (2010). Reduction of train and net energy consumption using genetic algorithms for trajectory optimisation. *IET Conf. Railw. Tract. Syst.* 32–32.

27 Firpo, P. and Savio, S. (1995). Optimized train running curve for electrical energy saving in autotransformer supplied AC railway systems. *Int. Conf. Electr. Railw. a United Eur.*, pp. 23–27.

28 Van Mierlo, J. and Maggetto, G. (2004). Innovative iteration algorithm for a vehicles imulation program. *IEEETrans. Veh. Technol.* 53 (2): 401–412.

29 Tian, Z., Weston, P., Zhao, N. et al. (2017). System energy optimisation strategies for metros with regeneration. *Transp. Res. C Emerg. Technol.* 75: 120–135.

30 Radcliffe, P., Wallace, J.S., and Shu, L.H. (2010). Stationary applications of energy storage technologies for transit systems. *2010 IEEE Electrical Power and Energy Conference*, pp. 1–7.

31 Vitaly, G. (2013). Energy storage that may be too good to be true. *IEEE Veh. Technol. Mag.* 8 (4): 70–80.

32 Sliker, G. (2005). Long Island bus NaS battery energy storage project. *Electrical Energy Storage Application and Technologies (EESAT), Sanfransisco*, CA.

33 Kishinevsky, Y. (2005). Long Island bus Sodium Sulfur (NaS) battery storage project. *Electrical Energy Storage Application and Technologies (EESAT), Sanfransisco*, CA.

34 Takahashi, H. and Kim, Y.I. (2012). Current and future applications for regenerative energy storage system. *Hitachi Rev.* 61 (7): 336–340.

35 Eckroad, S. (2009). Long Island bus NaS battery energy storage project-EPRI. *DOE Peer Review. Energy Storage & Power Electronics Systems Research Program*, Seattle, WA, no. 704.

36 Ciccarelli, F. and Iannuzzi, D. (2012). A novel energy management control of wayside Li-ion capacitors-based energy storage for urban mass transit systems. *International Symposium on Power Electronics Power Electronics, Electrical Drives, Automation and Motion*, pp. 773–779.

37 Sharma, P. and Bhatti, T.S. (2010). A review on electrochemical double-layer capacitors. *Energy Convers. Manag.* 51 (12): 2901–2912.

38 Ibrahim, H., Ilinca, A., and Perron, J. (2008). Energy storage systems—characteristics and comparisons. *Renew. Sustain. Energy Rev.* 12 (5): 1221–1250.

39 Chen, H., Cong, T.N., Yang, W. et al. (2009). Progress in electrical energy storage system: a critical review. *Prog. Nat. Sci.* 19 (3): 291–312.

40 Maher, B. (2006). Ultracapacitors provide cost and energy savings for public transportation applications. *Batter. Power Prod. Technol. Mag.* 10 (6): 1–4.

41 Khodaparastan, M. and Mohamed, A. (2017). Supercapacitors for electric rail transit system. *6th International Conference on Renewable Energy Research and Application*, Vol. 5, pp. 1–6.

42 X. Tackoen and F.O. Devaux (2014). WB2P energy recovery guidelines for braking energy recovery systems in urban rail networks. *Ticket to Kyoto Project-final Report.* https://www.tickettokyoto.eu/sites/default/files/downloads/T2K_WP2B_Energy%20Recovery_Final%20Report_2.pdf (accessed 5 January 2016).

43 Barcellona, S., Ciccarelli, F., Iannuzzi, D., and Piegari, L. (2014). Modeling and parameter identification of lithium-ion capacitor modules. *IEEE Trans. Sustain. Energy* 5 (3): 785–794.

44 Barcellona, S., Ciccarelli, F., Iannuzzi, D., and Piegari, L. (2016). Overview of lithium-ion capacitor applications based on experimental performances. *Electr. Power Compon. Syst.* 44 (11): 1248–1260.

45 Rossow, M. (2003). *Flywheel Energy Storage.* Washington, DC: Federal Energy Management Program U.S. Department of Energy.

46 Richardson, M.B. (2002). Flywheel energy storage system for traction applications. *International Conference on Power Electronics Machines and Drives*, vol. 2002, pp. 275–279.

47 Liuand, H. and Jiang, J. (2007). Flywheel energy storage – an upswing technology for energy sustainability. *Energy Build.* 39 (5): 599–604.

48 Rupp, A., Baier, H., Mertiny, P., and Secanell, M. (2016). Analysis of a flywheel energy storage system for light rail transit. *Energy* 107: 625–638.

49 Arboleya, B.P. and Armendariz, U. (2016). Energy is on board: energy storage and other alternatives in modern light railways. *IEEE Electrification Mag.* (September): 30–41.

50 Henning, U., Thoolen, F., Berndt, J., and Lohner, A. (2006). Ultra low emission traction drive system for hybrid light rail vehicles. *SPEEDAM 2006. International Symposium on. IEEE*, pp. 12–16.

51 Daoud, M.I. and Ahmed, S. (2013). DC bus control of an advanced flywheel energy storage kinetic traction system for electrified railway industry. *Industrial Electronics Society, IECON 2013-39th Annual Conference of the IEEE*, pp. 6596–6601.

52 Barrero, R., Van Mierlo, J., and Tackoen, X. (2008). Energy saving in public transport. *IEEE Veh. Technol. Mag.* 3 (3): 26–36.

53 Barrero, R., Tackoen, X., and Van Mierlo, J. (2010). Stationary or onboard energy storage systems for energy. *Proc. Inst. Mech. Eng. Part F J. Rail Rapid Transit* 224: 207–225.

54 Domínguez, M., Cucala, A.P., Fernández, A., et al. (2011). Energy efficiency on train control: design of metro ATO driving. *9th World Congress on Railway Research WCRR 2011*, pp. 1–12.

55 Sekijima, Y., Kudo, Y., Inui, M., et al. (2006). Development of energy storage system for DC electric rolling stock applying electric double layer capacitor. *Proceedings of Seventh World Congress on Railway Research*, pp. 1–7, 4–7.

56 Destraz, B., Barrade, P., Rufer, A., et al. (2007). Study and simulation of the energy balance of an urban transportation network- Bombardier Transportation. *12th European Conference on Power Electronics and Applications EPE 2007.*

57 Boizumeau, N.E., Jr., Leguay, P. (2011). Overview of braking energy recovery technologies in the public transport field," In workshop on braking energy recovery systems-Ticket to Kyoto Project.

58 Kinkisharyo (2012). The 100% Low-Floor Streetcar Engineered for North America North.

59 J. Pringle. OCS-free light rail vehicle technology. http://www.miamidade.gov/citt/library/summit/jcpringle.pdf. (accessed 5 January 2016).

60 Meinert, M. (2009). New mobile energy storage system for rolling stock. *2009 13th European Conference on Power Electronics and Applications*, pp. 110.

61 Siemens. (2011). *Increasing Energy Efficiency Optimized Traction Power Supply in Mass Transit Systems.* https://docplayer.net/89948259-Siemens-com-mobility-increasing-energy-efficiency-optimized-traction-power-supply-in-mass-transit-systems.html (accessed 26 October 2022).

62 Van Mierloand, J. and Maggetto, G. (2004). Innovative iteration algorithm for a vehicles imulation program. *IEEE Trans. Veh. Technol.* 53 (2): 401–412.

63 Steiner, M. and Scholten, J. (2004). Energy storage on board of DC fed railway vehicles PESC 2004 conference in Aachen, Germany. 2004 IEEE *35th Annual Power Electronics Specialists Conference (IEEE Cat. No.04CH37551)*, pp. 666–671. doi: 10.1109/PESC.2004.1355828.

64 Steiner, M. and Scholten, J. (2006). *Improving Overall Energy Efficiency of Traction Vehicles.* Mannheim, Germany: Bombardier Transportation.

65 Steiner, M., Klohr, M., and Pagiela, S. (2007). Energy storage system with Ultracaps on board of railway vehicles. *2007 European Conference on Power Electronics and Applications EPE.*

66 Moskowitz, J.P. and Cohuau, J.L. (2010). STEEM: ALSTOM and RATP experience of supercapacitors in tramway operation. *2010 IEEE Conference on Vehicle Power and Propulsion VPPC 2010.*

67 Lacôte, F. (2005). Alstom – Future trends in railway transportation. *Jpn. Railw. Transp. Rev.* (December): 4–9.

68 Yu, J.G., Schroeder, M.P., and Teumim, D. (2010). Utilizing wayside energy storage substations in rail transit systems – some modelling and simulation results association. *PTA Rail Conference.*

69 Rail Transportation Committee (2016). *P 1887 ™/Draft Guide for Wayside Energy Storage for DC Traction Applications. IEEE Vehicular Technology Society.*

70 Holmes, K. (2008). Smart grids and wayside energy storage. *Passenger Transp.* 66 (40).

71 Iannuzzi, D., Pagano, E., and Tricoli, P. (2013). The use of energy storage systems for supporting the voltage needs of urban and suburban railway contact lines. *Energies* 6 (4): 1802–1820.

72 MITRAC Energy Saver (brochure), Bombardier Inc. http://www.bombardier.com/content/dam/Websites/bombardiercom/supporting-documents/BT/Bombardier-Transportation-ECO4-EnerGstor-EN.pdf (accessed 24 February 2016).

73 Fernandez-Rodriguez, A., Fernandez-Cardador, A., De Santiago-Laporte, A. et al. (2017). Charging electric vehicles using regenerated energy from urban railways. *2017 IEEE Vehicle Power and Propulsion Conference (VPPC)*, pp. 1–6. doi: 10.1109/VPPC.2017.8330998.

74 Adetel Group. (2000). NeoGreen power system. France. http://www.adetelsolution.com/app/uploads/2017/06/Flyer-NeoGreen.pdf (accessed 05 January 2016).

75 SEPTA's (Southeastern Pennsylvania Transit Authority) wayside energy storage project. https://library.e.abb.com/public/421a296a68790f53c1257cfa0040c43f/Septa_WhitePaper_V1.pdf (accessed 24 February 2016).

76 O. Solis, F. Castro, L. Bukhin, et al. (2015). Saving money every day: LA metro subway wayside energy storage substation. *JRC2015-5691*, pp. 3–6.

77 Romo, L., Turner, D., Ponzio, J., and B. N. Engineering (2005). Return on investment from rail transit use of wayside energy storage systems. *Rail Transit Conference*, pp. 1–9.

78 González-gil, R., Palacin, P.B., and Powell, J.P. (2014). Energy-efficient urban rail systems: strategies for an optimal management of regenerative braking energy. *Transp. Res. Arena, Paris* 44 (0): 374–388.

79 Konishi, T., Morimoto, H., Aihara, T., and Tsutakawa, M. (2010). Fixed energy storage technology applied for DC electrified railway. *IEEJ Trans. Electr. Electron. Eng.* 5 (3): 270–277.

80 Gillespie, A.J., Johanson, E.S., and Montvydas, D.T. (2014). Energy storage in Pennsylvania: SEPTA's novel and innovative integration of emerging smart grid technologies. *IEEE Veh. Technol. Mag.* 9 (2): 76–86.

81 Campillo, J., Ghaviha, N., Zimmerman, N., and Dahlquist, E. (2015). Flow batteries use potential in heavy vehicles. *2015 International conference on electrical systems for aircraft, railway, ship propulsion and road vehicles*, pp. 1–6.

82 Ogura, K., Nishimura, K., Matsumura, T. et al. (2011). Test results of a high capacity wayside energy storage system using Ni-MH batteries for DC electric railway at New York City Transit. *2011 IEEE Green Technologies Conference (IEEE-Green)*, pp. 1–6.

83 Jackson, D. (2001). High-speed flywheels cut energy bill. *Railw. Gaz. Int.* 1: 7–9.

84 Tarrant, C. Kinetic energy storage wins acceptance. *Railw. Gaz. Int.* 1: 212–213.

85 Cornic, D. (2010). Efficient recovery of braking energy through a reversible dc substation. *International Conference on Electrical Systems for Aircraft, Railway and Ship Propulsion, ESARS 2010.*

86 Benefits, K.E.Y. (2014). *HESOP Energy Saver London Underground-Case Study*. France: Alstom Rail Transport Company.

87 Preda, A. and Suru, V. (2015). Series synchronous reference frame method applied in the indirect current control for active DC traction substation. *Athens: ATINER'S Conference Paper Series, No: TRA2015-1552*, pp. 1–14.

88 Romo, A. (2015). Reversible substation in heavy rail. *Energy Recovery Workshop*, Madrid.

89 Ortega, I.H. (2011). Kinetic energy recovery on railway systems with feedback to the grid. *9th World Congress on Railway Research–WCRR.*

90 Ortega, J.M. (2011). Ingeber system for kinetic energy recovery & metro Bilbao experience. *Rail Technological Forum for Internationlization*, Madrid. https://www.vialibre-ffe.com/pdf/Estaciones_reversibles.pdf (accessed January 2019).

91 Khodaparastan, M. and Mohamed, A. (2018). Modeling and Simulation of Regenerative Braking Energy in DC Electric Rail Systems. *2018 IEEE Transportation Electrification Conference and Expo (ITEC)*, no. 1, pp. 1–6.

92 Khodaparastan, M., Dutta, O., Saleh, M., and Mohamed, A. (2018). Modeling of DC electric rail transit systems with wayside energy storage. *IEEE Trans. Veh. Technol.* 68 (3): 2218–2228.

16

Flywheel Wayside Energy Storage for Electric Rail Systems

Ahmed A. Mohamed[1], Rohama Ahmad[1], William Franks[2], Brian Battle[3], and Robert Abboud[3]

[1] City University of New York, City College, Department of Electrical Engineering, New York, NY 10031, USA
[2] IPRE, Independent Power and Renewable Energy LLC, Wells, ME 04090, USA
[3] Beacon Power, LLC, Tyngsborough, MA 01879, USA

16.1 Introduction

In April of 2020, a Group including Independent Power and Renewable Energy LLC, Scout Economics and Beacon Power LLC, a developer, operator, and manufacturer of kinetic energy storage devices, was awarded a $1 million grant by the New York State Energy Research and Development Authority (NYSERDA) to develop, design, and operate (demonstrate) a 1 MW flywheel-based wayside energy storage facility on the New York City Transit (NYCT) #7 line (NYSERDA Contract 138903). The purpose of this facility would be to capture and reuse regenerative braking energy from subway trains, thereby saving energy and reducing peak demand.

In consultation with NYCT, the Group has identified a site close to the 61st St-Woodside stop and is in the process of vetting that site. The Group also contracted with City University of New York (CUNY) to incorporate a flywheel model into the #7 line model which CUNY had already developed with NYSERDA and ConEd support. This work was carried out by Prof. Ahmed Mohamed's Group at the City University of New York.

This chapter consists of three parts. The first (Section 16.2) is a technical description of Beacon Power's kinetic storage technology and its proposed configuration at 61st Street. The second (Section 16.3) reports on Prof. Mohamed's modeling of the #7 line and the kinetic storage facility's capturing and reinjecting of energy to the trains. The third (Section 16.4) describes the financial results (savings) based on the results of Prof. Mohamed's physical model.

16.2 Beacon Power's Kinetic Energy Storage System

There are four primary technologies that make up the Beacon Kinetic Power System:

1) Ultra-high strength material for the flywheel rim, typically glass and carbon fiber.
2) High-strength magnetics for the bearing systems and motor/generator drive.
3) High-power, fast-switching electronics to create the load and flywheel drive inverters.
4) High-speed computing and data communications.

Transportation Electrification: Breakthroughs in Electrified Vehicles, Aircraft, Rolling Stock, and Watercraft,
First Edition. Edited by Ahmed A. Mohamed, Ahmad Arshan Khan, Ahmed T. Elsayed, and Mohamed A. Elshaer.

These are illustrated in Figure 16.1

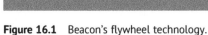

Beacon's high-energy, high-power flywheel has been in commercial operation for over 10 years.

- **Approximately 7 ft tall, 3 ft in diameter**
- **2500 lb. rotor mass**
- **Spinning at up to 16,000 rpm**
- **Lifetime energy throughout is over 5000 MWh**
- **Capable of charging or discharging at full rated power without restriction**
- **Beacon flywheel technology is protected by over 60 patents**
- **U.S. Patents**
  6,614,132; 6,710,489; 6,747,378; 6,817,266; 6,824,861; 6,852,401; 6,884,039; 6,959,756; 7,034,420; 7,174,806; 7,365,461; 7,679,247; 8,314,527 (other U.S. and international patents pending).

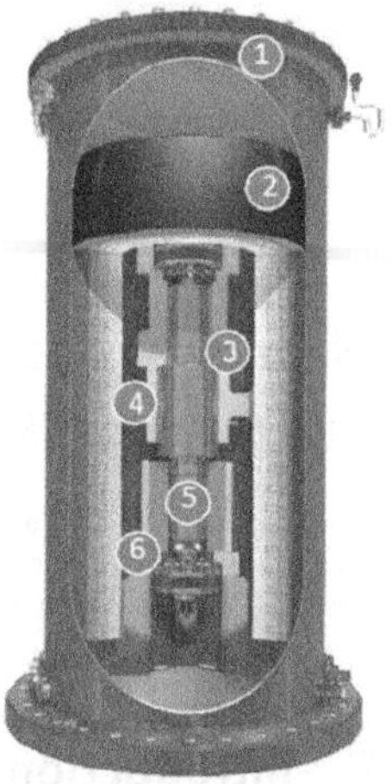

Figure 16.1 Beacon's flywheel technology.

16.2.1 Key Features of Beacon Flywheels

1) *Vacuum Chamber:* A strong vacuum provides a nearly frictionless environment and the containment protects all components from the atmosphere and contamination. This eliminates deterioration of the internal components.
2) *Patented Composite Rim:* A rotating carbon and fiberglass composite cylinder stores the energy. It also optimizes mass and strength to provide energy storage safely and at the best price.
3) *Magnetic Lift System:* A non-contacting magnetic field lifts and supports the rotor, further eliminating wear and extending the life of the parts while minimizing friction. Beacon's patented bearing system ensures the spinning rim maintains its axis of rotation with low bearing loads, resulting in a long life.
4) *Hub:* Aluminum forging connects the rotating shaft and rim.
5) *Shaft:* Steel forging rotates and maintains the rotor centerline.
6) *Brushless Permanent Magnetic Motor/Generator:* Efficiently converts the electrical energy into mechanical energy when the flywheel is charging, and back to electrical energy when discharging.

The basic energy recoupment topology for a flywheel energy storage system is configured as shown in Figure 16.2.

In this configuration, the storage system is designed to absorb or inject energy from the third rail as the train brakes or accelerates, respectively. The system can be controlled remotely via data links through SkyNet or it can be completely autonomous by simply sensing the conditions of the third rail itself and then locally computing reaction (or prediction) behaviors accordingly. Multiple units will be clustered together to operate independently or cooperate with communication through SmartBox/SkyNet, depending upon conditions.

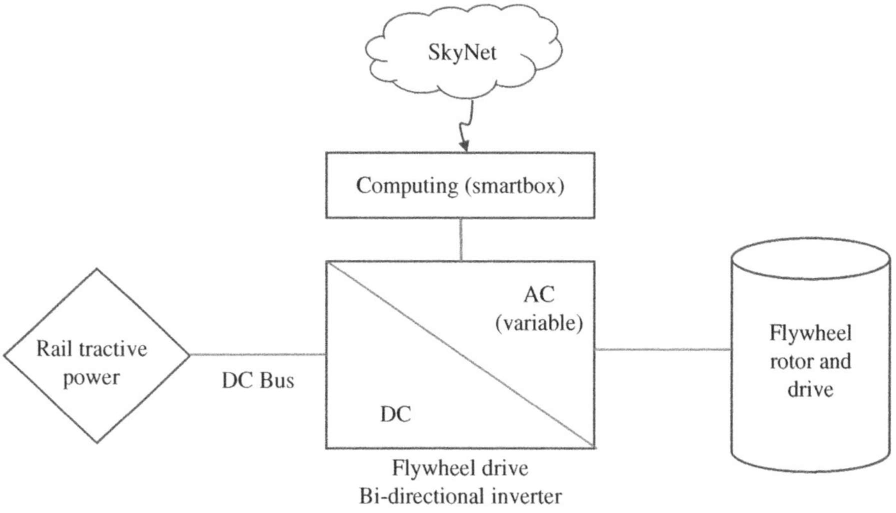

Figure 16.2 Topology of flywheel energy storage system.

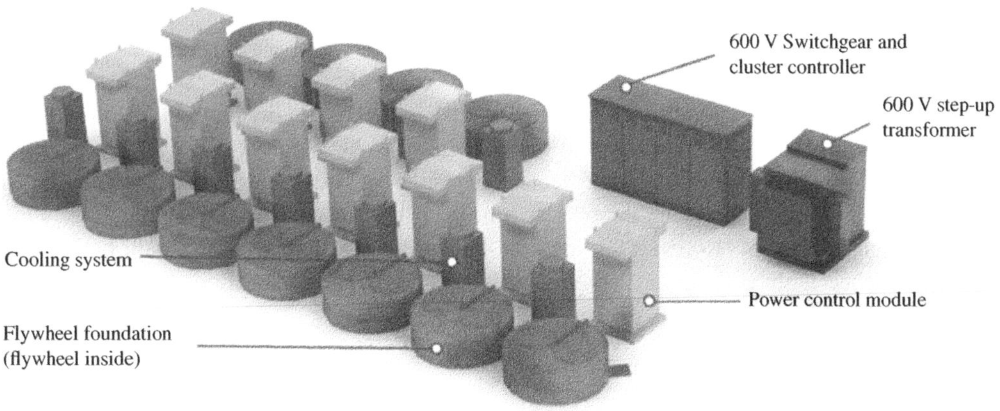

Figure 16.3 Typical grouping of KPS systems – can be arranged in a wide variety of configurations.

The Kinetic Power System can easily be customized for different applications. In the 1 MW wayside storage application discussed in this chapter, either six Beacon 400-167 flywheels or three Beacon 400-333 flywheels can be utilized. The principal difference is the electric motor/generator which in the former case is rated at 167 kW and, in the latter case, 333 kW. Hence, a 1 MW installation comprises six Beacon 400-167 flywheels or three Beacon 400-333 flywheels.

Installation is straightforward with only a standard construction backhoe/loader and forklift required. All equipment, including cable harnesses, come from Beacon as a complete kit.

The typical ground installation layout is as shown in Figures 16.3–16.8.

Figure 16.4 Flywheel during assembly.

16.3 Train Simulation Study

16.3.1 Synopsis

This study aims at simulating the flywheel-based wayside energy storage system (WESS) to be installed on the NYCT #7 line, under various deployment and operational scenarios. MATLAB/SIMULINK is used to carry out the proposed study [1, 2]. The main objective is to estimate the energy savings and peak demand reduction that can potentially be achieved as a result of flywheel WESS deployment.

16.3.2 Modeling Scope

The WESS is assumed to be located at 61st St-Woodside stop, about 580 ft away from the third rail. Tables 16.1 and 16.2 depict the substations and stops included in the model, respectively, as also illustrated in Figure 16.9 [3]. Table 16.2 also indicates the track distance between each stop and the last stop in Manhattan (34 St-Hudson Yards).

16.3.3 Modeling Scenarios

Table 16.3 lists the cases studied in this report. These cases include weekdays and weekends, both with and without WESS. We found through simulations that more savings can be achieved if the flywheel is only engaged when the train is nearby. Therefore, for the results presented in this report, it is assumed that the flywheel is activated when the train is running from the 52nd St passenger station to the 74th St passenger station. However, Section 16.3.4.4 compares the energy savings in both cases, when the flywheel is always engaged vs when the flywheel is only engaged when the train is nearby.

Figure 16.5 KPS systems being installed with the flywheel, PCM, cooling unit, and DCS.

Figure 16.6 Typical PCM electronics layout.

We also analyzed the impact of various chopper cutoff voltage settings. In addition, the impact of changing the flywheel converter location has been analyzed by comparing two cases, when the converter is collocated with the flywheel (580 ft away from the third rail) vs when it is located directly at the third rail. Finally, the impact of changing the flywheel size has been analyzed as well.

16.3.4 Results and Discussion

16.3.4.1 Transient Response

A train is assumed to run from 46th St-Bliss St passenger station toward Junction Blvd. During this trip, the train makes 9 acceleration/deceleration cycles. This case assumes no WESS is connected. The current profiles of Queens Blvd-39th Pl, 58th St, Roosevelt Ave-78th St, and Roosevelt Ave-Spruce St substations during those nine cycles are presented in Figure 16.10.

As depicted in Figure 16.10, when the train runs from 46th St toward 61st St, the substations located at Queens Blvd and 58th St contribute more than the other two substations, which are further away. In the beginning, the train's location is closer to Queens Blvd. Therefore, the Queens Blvd substation is contributing more than 58th St substation. As the train gets closer to 58th St, the current contribution from 58th St is increased while the current contribution from Queens Blvd is decreased.

Figure 16.7 Flywheel rotor package during final assembly.

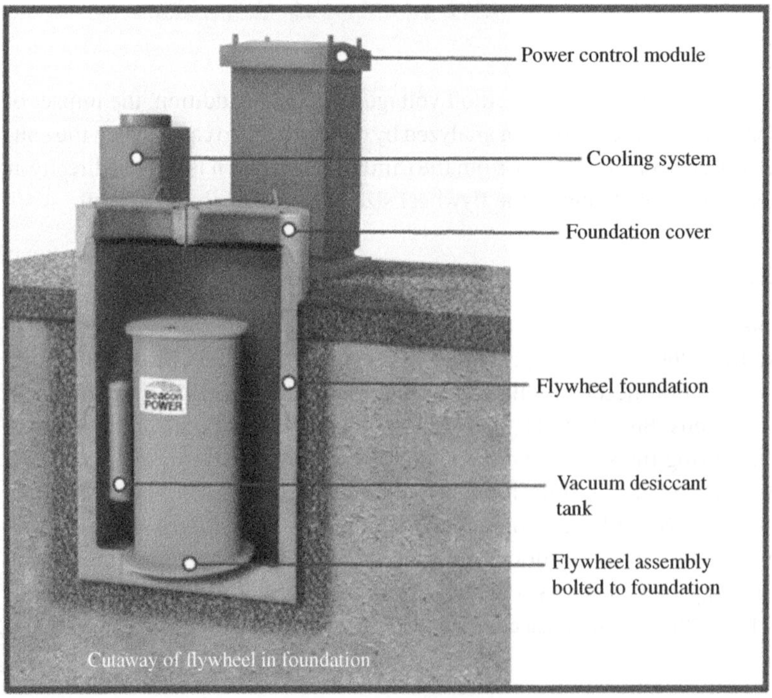

Figure 16.8 Cut-away view of foundation.

Table 16.1 Substations included in the model.

Substations	Location
Roosevelt Ave-Spruce St	Between Junction Blvd and 103rd St-Corona Plaza
Roosevelt Ave-78th St	Between 74th St-Broadway and 82nd St-Jackson Heights
58th St	Between 52nd St-Lincoln Ave and 61st St-Woodside
Queens Blvd-39th Pl	Between 33rd St-Rawson St and 40th St-Lowery St

Table 16.2 Passenger stations included in the model.

Stops	Track distance from 34 St-Hudson Yards Stop (ft)
Junction Blvd	42,640
90th St-Elmhurst Ave	40,650
82nd St-Jackson Heights	38,650
74th St-Broadway	36,560
69th St-Fisk Ave	35,150
61st St-Woodside	33,240
52nd St-Lincoln Ave	30,560
46th St-Bliss St	28,860
40th St-Lowery St	27,250

When the train gets closer to its final station (Junction Blvd in our simulated scenario), the current contribution from Queens Blvd is almost zero; the train is far away from this substation. The train current is supplied by the other two neighboring substations (Roosevelt Ave-78th St Substation and Roosevelt Ave-Spruce St substation).

The voltage profiles of the substations are presented in Figure 16.11. It can be observed that the voltage is generally within permissible limits. During train acceleration, the voltage of nearby substations drops to provide the energy required. During the deceleration phase, since the train releases its regenerative braking energy into the third rail, the voltage of the substations increase. These voltage changes are more prevalent when the train location is close to that of a substation.

To analyze the impact of energy storage on peak demand reduction and energy savings, a 1 MW flywheel WESS is integrated into the transit system model. The flywheel is located 580 ft away from the third rail at 61st St-Woodside passenger station. The train is again running from the 46th St passenger station to Junction Blvd passenger station.

The current profiles of Queens Blvd-39th Pl, 58th St, Roosevelt Ave-78th St, and Roosevelt Ave-Spruce St substations are presented in Figure 16.12.

Comparing Figures 16.10 and 16.12, it can be observed that by adding energy storage, the current contribution of 58th St substation during train acceleration is reduced. This impact will be reflected on the 24-hour interval metering profile for each substation and consequently the energy consumption for 24 hours. The WESS releases its energy during train acceleration and charges during train deceleration. The train acts as a source of energy for the WESS during deceleration. However, since

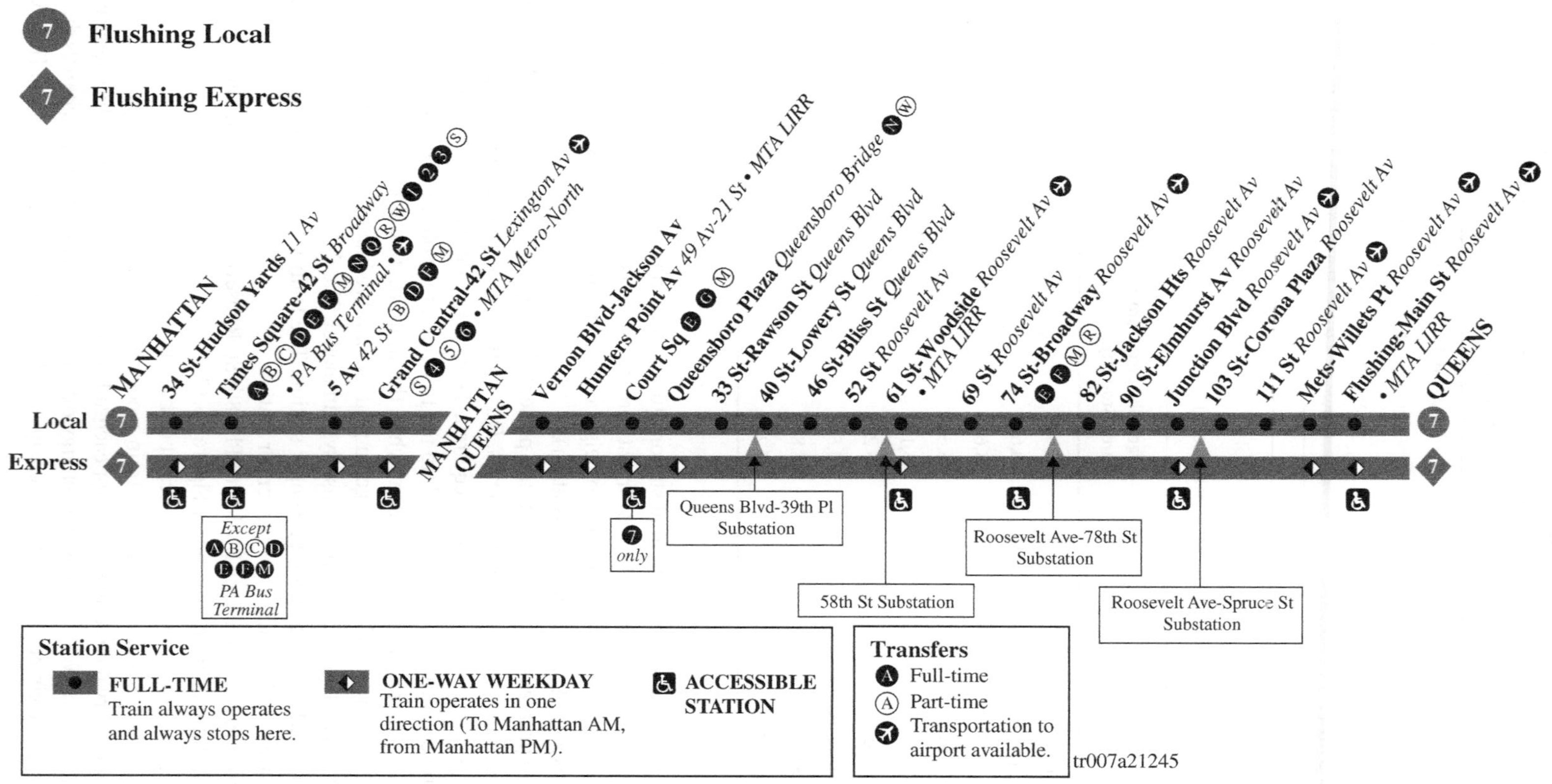

Figure 16.9 The NYCT #7 line under study.

Table 16.3 Modeling scenarios.

1, 2	Weekday 24 hours	With and without the WESS operating
3, 4	Saturday 24 hours	With and without the WESS operating
5, 6	Sunday 24 hours	With and without the WESS operating
7, 8	Weekday and weekend 24 hours with chopper cutoff set to 760 V	With and without the WESS operating
9, 10	Weekday and weekend 24 hours with chopper cutoff set to 800 V	With and without the WESS operating
11, 12	Weekday and weekend 24 hours with chopper cutoff set to 840 V	With and without the WESS operating
13, 14	Converter location	With WESS (580 ft away) vs at third rail

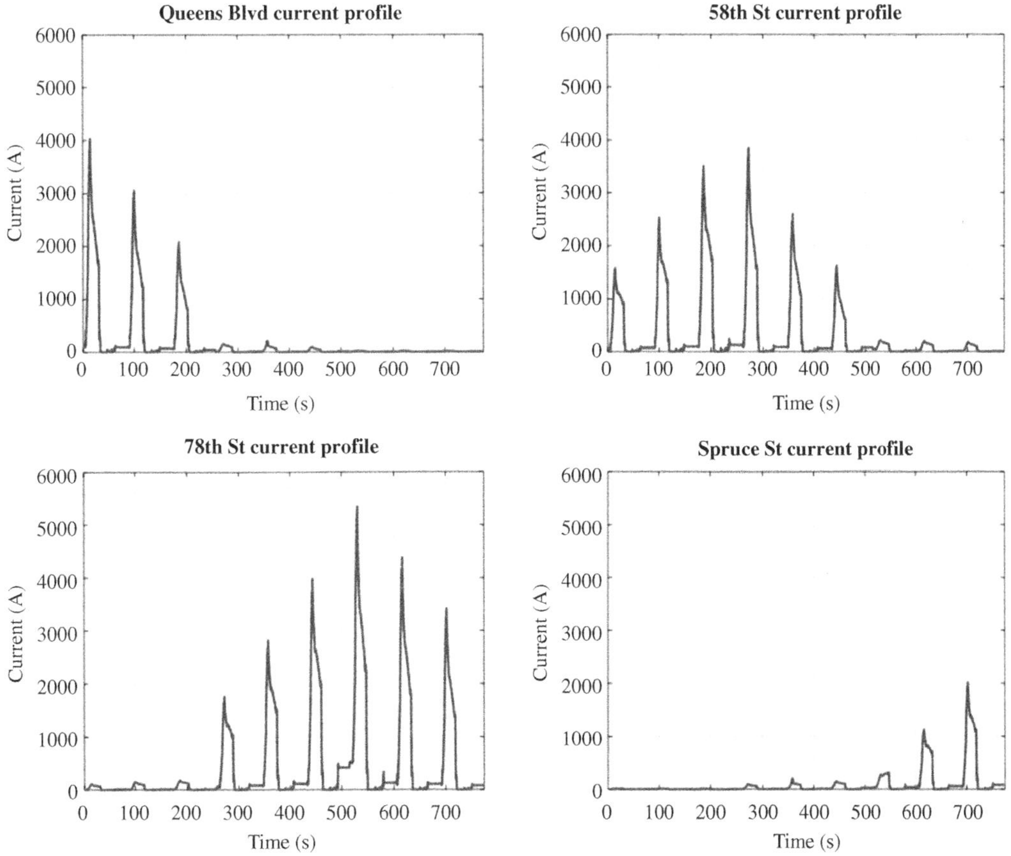

Figure 16.10 Current profiles of the substations without WESS.

the WESS is located near 58th St substation, it is observed that this substation sees the WESS as a load and contributes during the charging phase of the flywheels. This contribution can be further reduced by increasing the chopper voltage, as will be illustrated later in this report.

The voltage profiles of the substations during the same scenario are presented in Figure 16.13.

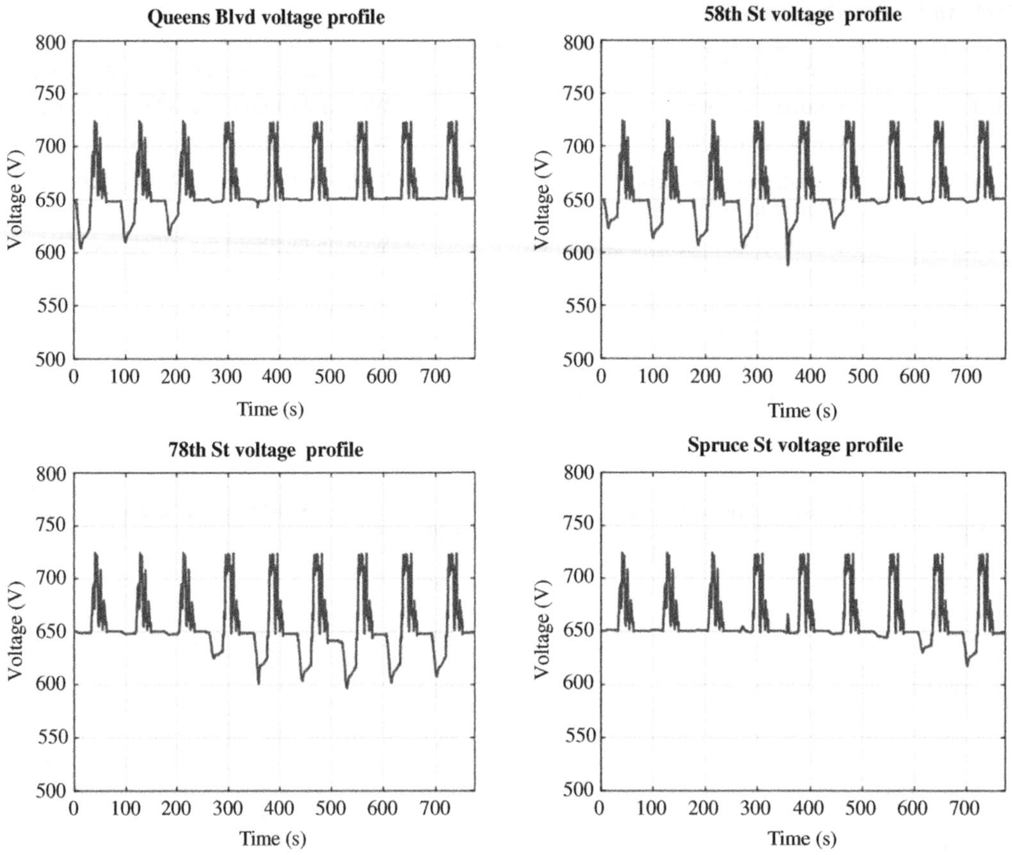

Figure 16.11 Voltage profiles of the substations' DC terminals without WESS.

The flywheel power and state of charge profiles are presented in Figures 16.14 and 16.15, respectively. During train acceleration and deceleration, the flywheel injects a constant 1 MW of power into the third rail. During the charging phase of energy storage, the flywheel's speed increases, and its state of charge rises. During the discharging phase, the flywheel's speed decreases, and its state of charge drops.

To analyze the impact of changing the converter location, two additional scenarios are considered. In the first scenario, the flywheel and the DC/DC converter are collocated 580 ft away from the third rail at 61st St-Woodside station. In the second scenario, the flywheel is located 580 ft away from the third rail, while the DC/DC converter is located at 61st St-Woodside station.

There is no significant impact on the substation peak power consumption. However, the power loss depends on the DC bus voltage level of the flywheel. In this analysis, the flywheel DC voltage is set to 800 V. Hence, the power loss is found to be less in the first scenario since there will be less wiring (the 580 ft) extended on the 800 V side instead of the lower voltage (higher current) third rail side. The following two equations present how power losses can be estimated, assuming 800 V, 650 V, and 43 μΩ/m, for the flywheel voltage, third rail voltage, and rail resistance, respectively. Note that this power loss ratio does not necessarily equal the energy loss ratio since calculating the energy losses must involve time. However, it indicates that the energy losses tend to be more if the DC–DC converter is located at the passenger station, not with the WESS.

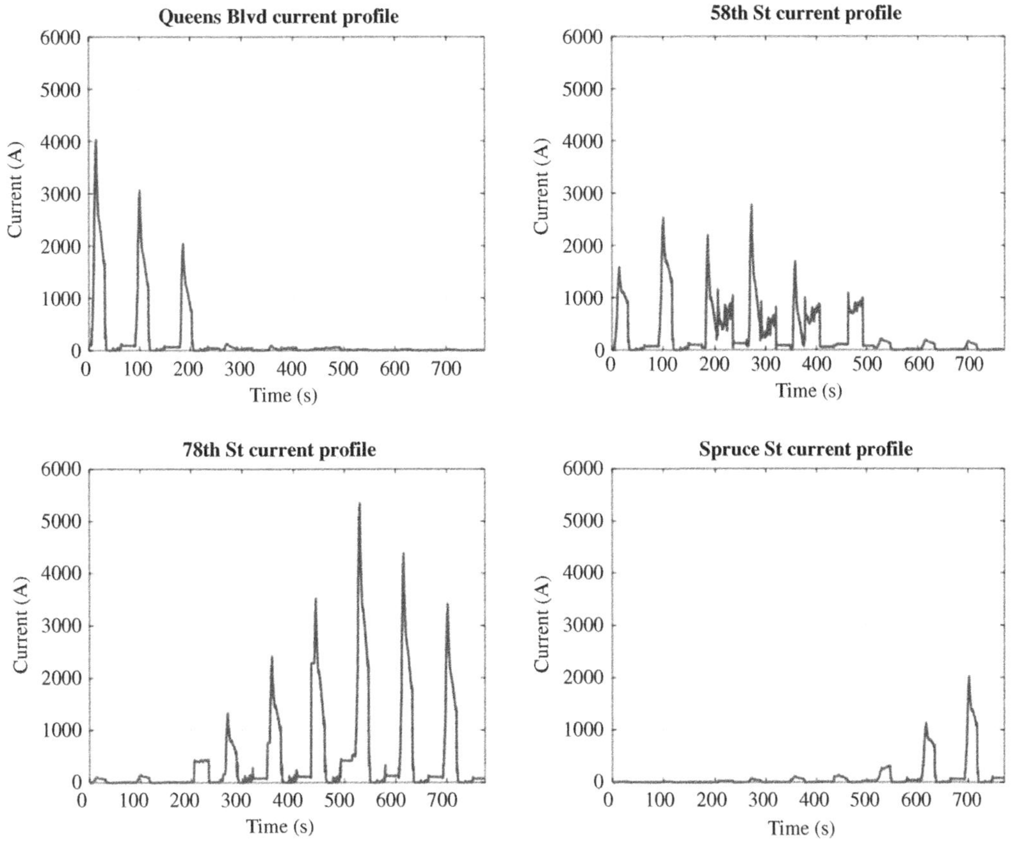

Figure 16.12 Current profiles of the substations with flywheel WESS.

$$P_{\text{Loss}_{\text{scenario1}}} = RI^2 = \left(43 * 10^{-6}\,\Omega/\text{m} * 580\,\text{ft} * 0.3\,\text{m/ft}\right)\left(\frac{1 * 10^6\,\text{W}}{800\,\text{V}}\right)^2 = 11.7\,\text{kW}$$

$$P_{\text{Loss}_{\text{scenario2}}} = RI^2 = \left(43 * 10^{-6}\,\Omega/\text{m} * 580\,\text{ft} * 0.3\,\text{m/ft}\right)\left(\frac{1 * 10^6\,\text{W}}{650\,\text{V}}\right)^2 = 17.7\,\text{kW}$$

16.3.4.2 24-hour Steady State Response

In order to analyze the impact on daily peak demand and energy consumption, the 24-hour operation of trains is also considered. The line 7 timetable/schedule is used to create the 24-hour power profile for the substations [4]. The energy savings, the power supplied by each substation, and the flywheel power are obtained by the transient analysis (e.g. Figures 16.10–16.15). Then, the train timetable is used to extrapolate the results to 24 hours, while taking into consideration the probability that trains exchange energy directly when they meet at passenger stations. Figure 16.16 illustrates part of this timetable during a weekday for Queens-bound trains (https://new.mta.info/document/9461). Similar timetables for Manhattan-bound trains and the operation of trains during Saturday and Sunday are also considered.

The 24-hour power profiles of 58th St and Roosevelt Ave-78th St substations, with and without WESS, are presented in Figures 16.17 and 16.18, respectively. The results show that by adding WESS, the peak power profile and 24-hour energy consumption of 58th St substation are both

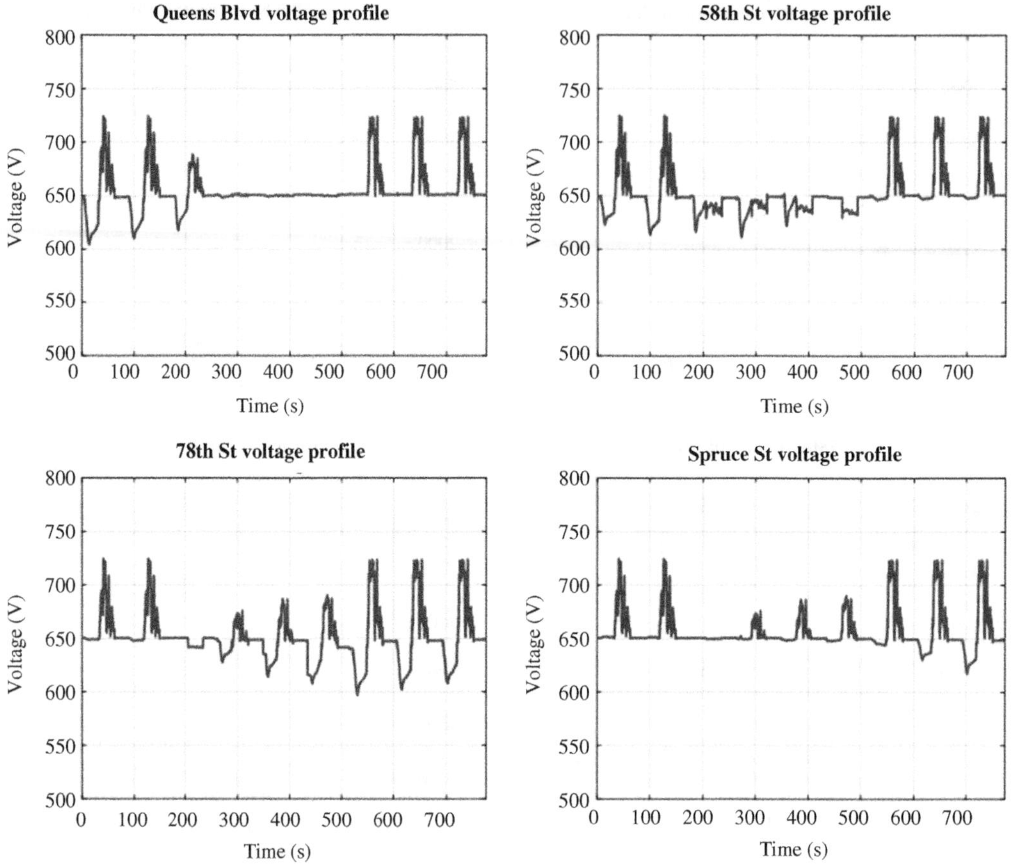

Figure 16.13 Voltage profiles of the substations' DC terminals with flywheel WESS.

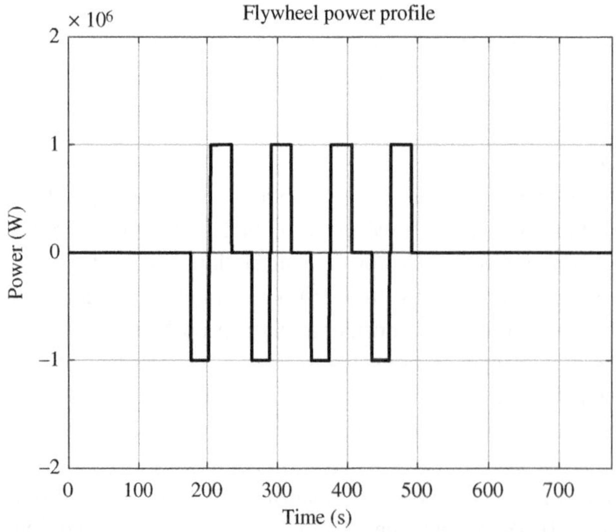

Figure 16.14 Power profile of flywheel energy storage.

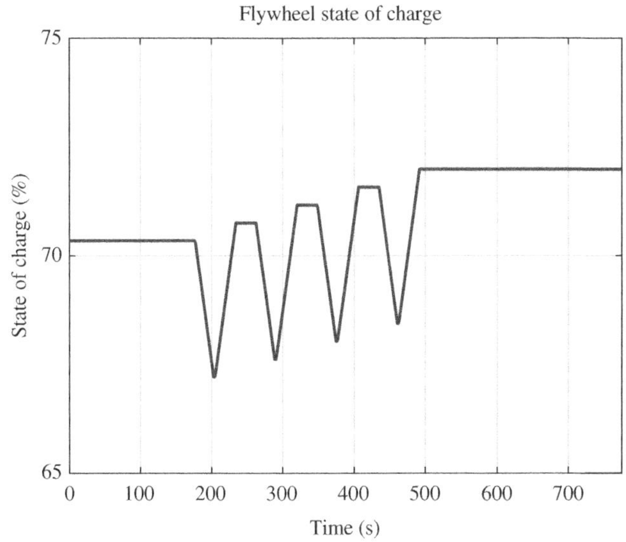

Figure 16.15 State of charge profile of flywheel energy storage.

Weekday Service

7 Queens-bound

From 34 St-Hudson Yards, Manhattan, to Flushing-Main St, Queens

34 St Hudsn Yds	Times Sq 42 St	5 Av	Grnd Central-42 St	Court Sq	Qnsboro Plz	Woodside 61 St	74 St Bdwy	111 St	Mets Willets Pt	Flushing Main St
1:02	1:05	1:06	1:07	1:15	1:16	1:23	1:27	1:32	1:34	1:37
1:17	1:20	1:21	1:22	1:30	1:31	1:38	1:42	1:47	1:49	1:52
1:37	1:40	1:41	1:42	1:50	1:51	1:58	2:02	2:07	2:09	2:12
1:57	2:00	2:01	2:02	2:10	2:11	2:18	2:22	2:27	2:29	2:32
2:17	2:20	2:21	2:22	2:30	2:31	2:38	2:42	2:47	2:49	2:52
2:37	2:40	2:41	2:42	2:50	2:51	2:58	3:02	3:07	3:09	3:12
2:57	3:00	3:01	3:02	3:10	3:11	3:18	3:22	3:27	3:29	3:32
3:17	3:20	3:21	3:22	3:30	3:31	3:38	3:42	3:47	3:49	3:52
3:37	3:40	3:41	3:42	3:50	3:51	3:58	4:02	4:07	4:09	4:12
3:57	4:00	4:01	4:02	4:10	4:11	4:18	4:22	4:27	4:29	4:32
4:17	4:20	4:21	4:22	4:30	4:31	4:38	4:42	4:47	4:49	4:52
4:37	4:40	4:41	4:42	4:50	4:51	4:58	5:02	5:07	5:09	5:12
4:57	5:00	5:01	5:02	5:10	5:11	5:18	5:22	5:27	5:29	5:32
5:12	5:15	5:17	5:18	5:25	5:27	5:34	5:37	5:42	5:44	5:48
5:27	5:30	5:32	5:33	5:40	5:42	5:49	5:52	5:57	5:59	6:03
5:37	5:40	5:42	5:43	5:50	5:52	5:59	6:02	6:07	6:09	6:13

Figure 16.16 Timetable for line 7 (Queens-bound) during weekday service.

reduced by 18.11%. The peak power demand and 24-hour energy consumption of Roosevelt Ave-78th St substation are reduced by 5.54%.

16.3.4.3 Effect of Changing Chopper Activation Voltage

For the cases just presented, the chopper activation voltage was set to 720 V. The same study is repeated for the cases where the chopper activation voltage is set to 800 V. The impact of this change on the current and voltage of the substations is presented in Figures 16.19 and 16.20, respectively.

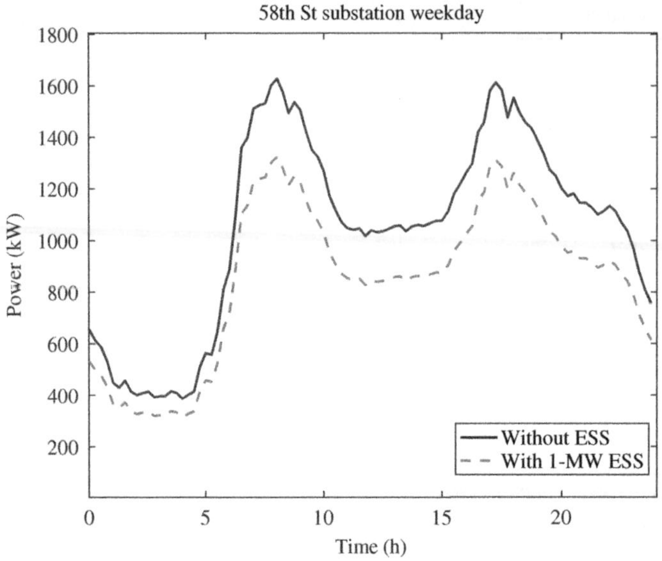

Figure 16.17 24-hour power profiles of 58th St substation.

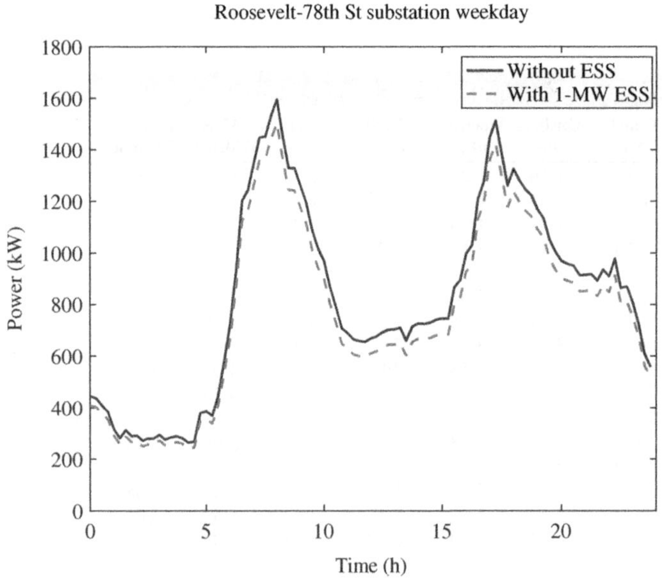

Figure 16.18 24-hour power profiles of Roosevelt Ave-78th substation.

The impact on the 24-hour power profiles of 58th St and Roosevelt Ave-78 St substations are presented in Figures 16.21 and 16.22, respectively.

As mentioned earlier, 58th St substation contributes during WESS charging. It can be observed from Figure 16.21 that this contribution can be reduced by increasing the train chopper voltage. When the chopper voltage is set to a higher value, the train can inject more current into the third rail, charging the WESS. This limits the energy contribution from 58th St substation during flywheel charging.

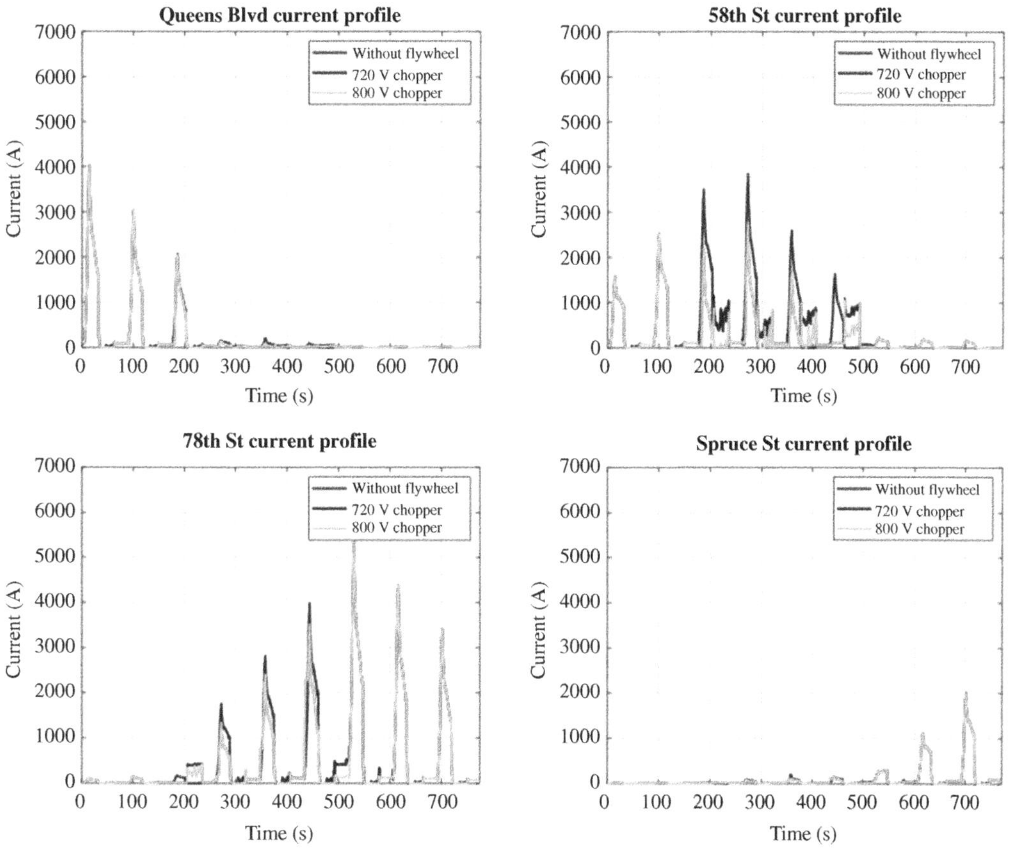

Figure 16.19 Current profiles of the substations.

The results show that by increasing the chopper activation voltage to 800 V, the peak power demand and 24-hour energy consumption of 58th St substation are both reduced by 29.491%, more than 150% of the savings achieved when the chopper activation voltage is set to 720 V.

Since the WESS is located near 58th St Substation, there is minimal contribution from Roosevelt Ave-78th St substation during WESS charging. Therefore, increasing the chopper activation voltage does not have a significant impact on the peak demand and energy consumption of that substation, as well as other substations. As can be seen, when the chopper voltage is set to 800 V, the peak demand reduction and 24-hour energy consumption of the Roosevelt-78th St substation is 8.053%. This represents approximately 2.5% more savings than what was achieved in the case with the 720 V chopper activation voltage.

The line 7 timetable has separate data for Saturday and Sunday. Therefore, separate simulation runs have been performed for each of them. The results for 58th St substation during Saturday and Sunday are presented in Figures 16.23–16.26.

The results show that by adding WESS, the peak power profile and 24-hour energy consumption of 58th St substation during Saturday and Sunday are again reduced by 18.11%. The peak power demand and 24-hour energy consumption of this substation, after increasing the chopper activation voltage to 800V, are reduced by 29.491%.

The 24-hour power profiles of Roosevelt Ave-78th substation during Saturday and Sunday are presented in Figures 16.27–16.30.

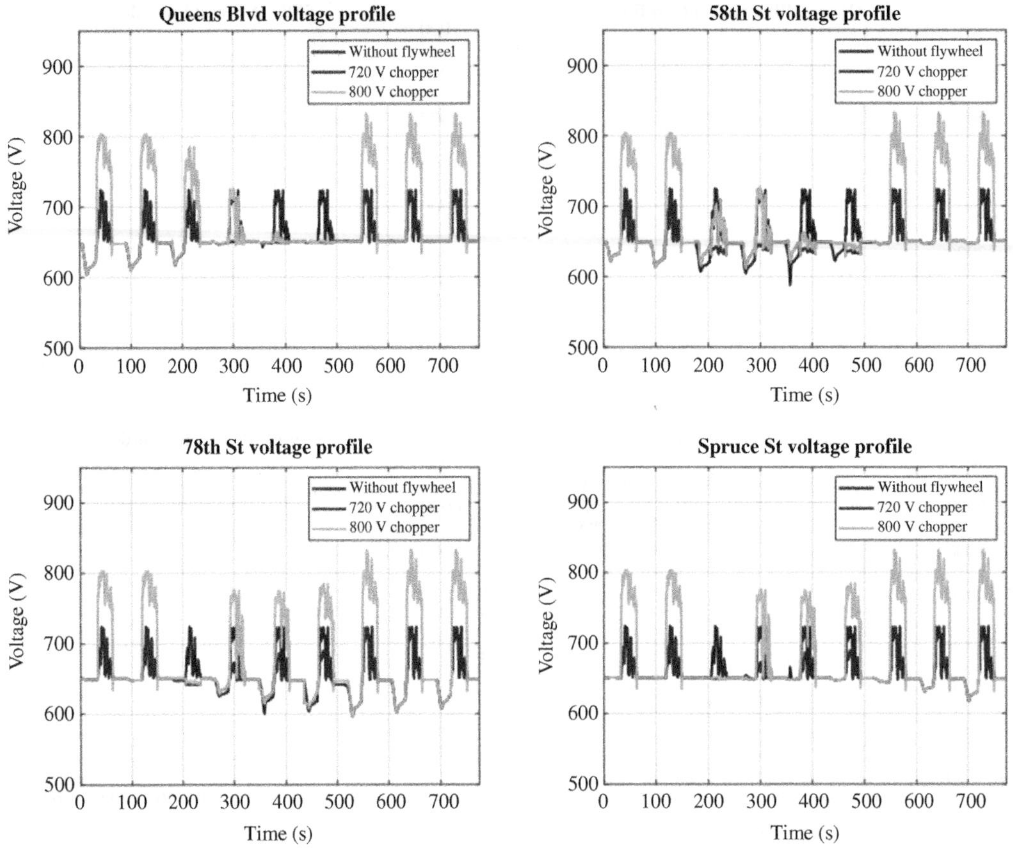

Figure 16.20 Voltage profiles of the substations.

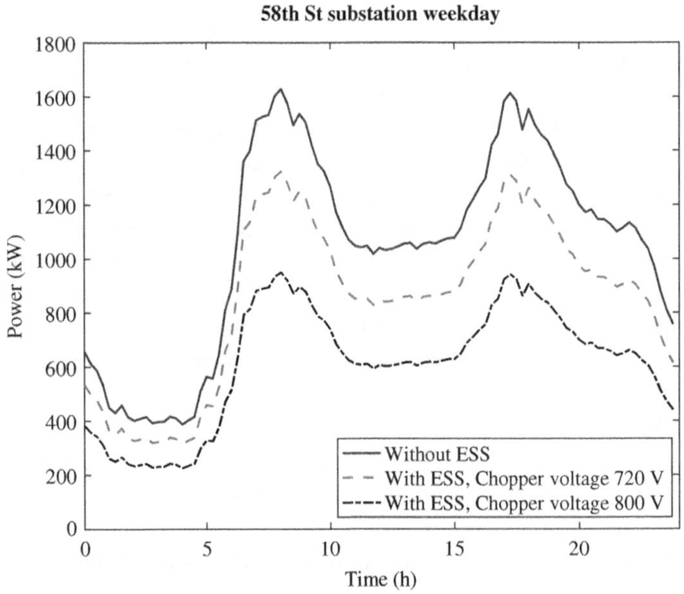

Figure 16.21 24-hour power profiles of 58th St substation, with chopper activation voltage set to 800 V.

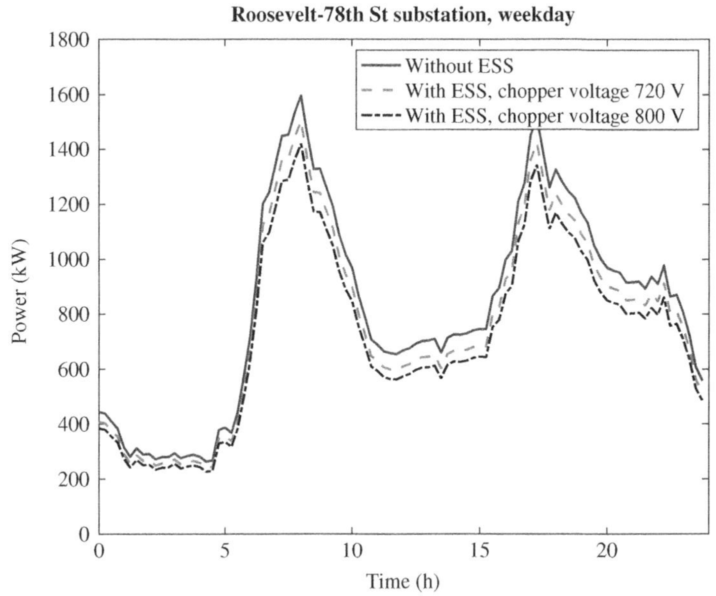

Figure 16.22 24-hour power profiles of Roosevelt Ave-78 St substation, with chopper activation voltage set to 800 V.

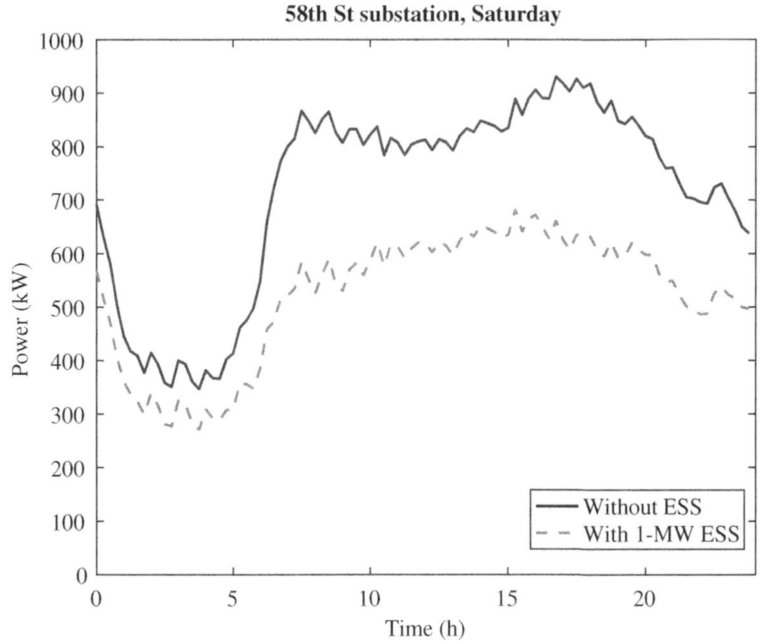

Figure 16.23 24-hour power profiles of 58th St substation (Saturday).

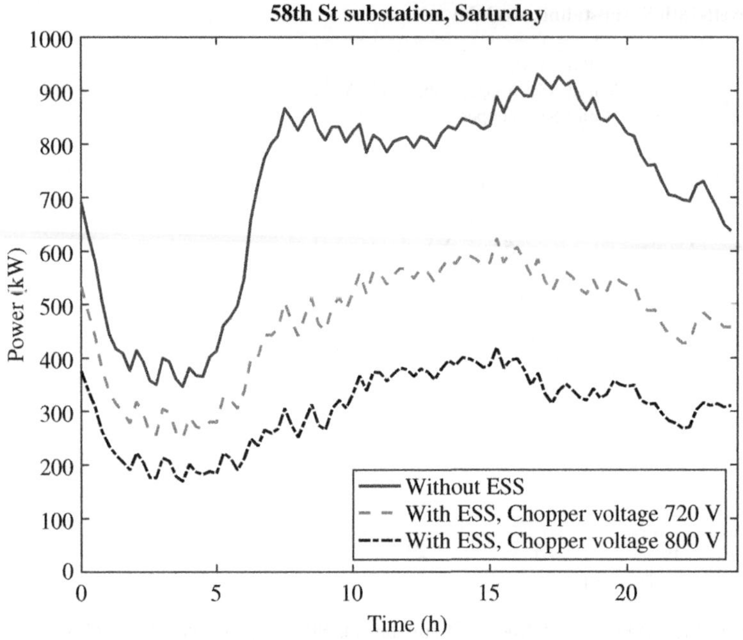

Figure 16.24 24-hour power profiles of 58th St substation (Saturday).

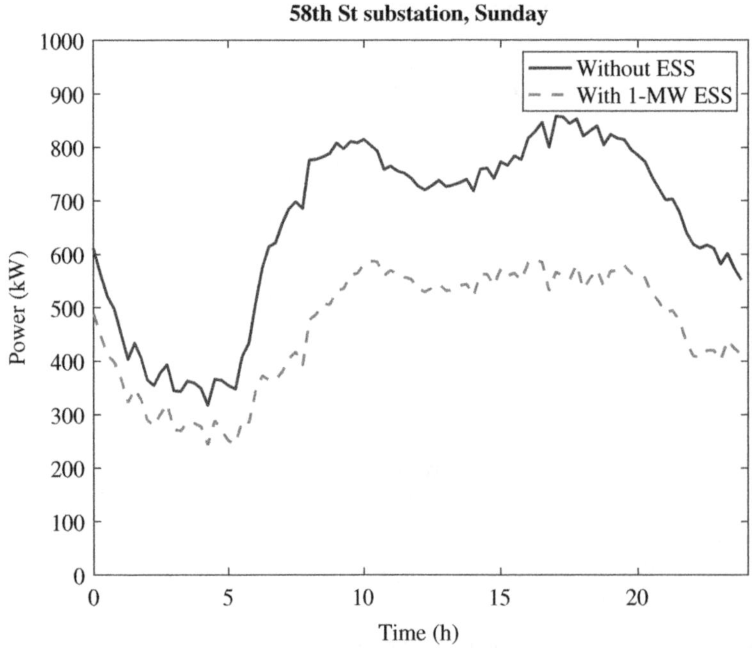

Figure 16.25 24-hour power profiles of 58th St substation (Sunday).

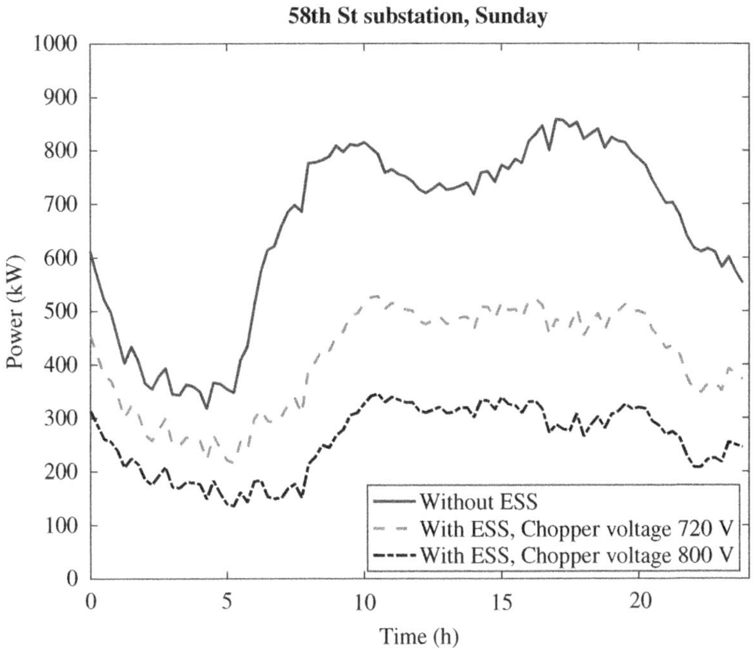

Figure 16.26 24-hour power profiles of 58th St substation (Sunday).

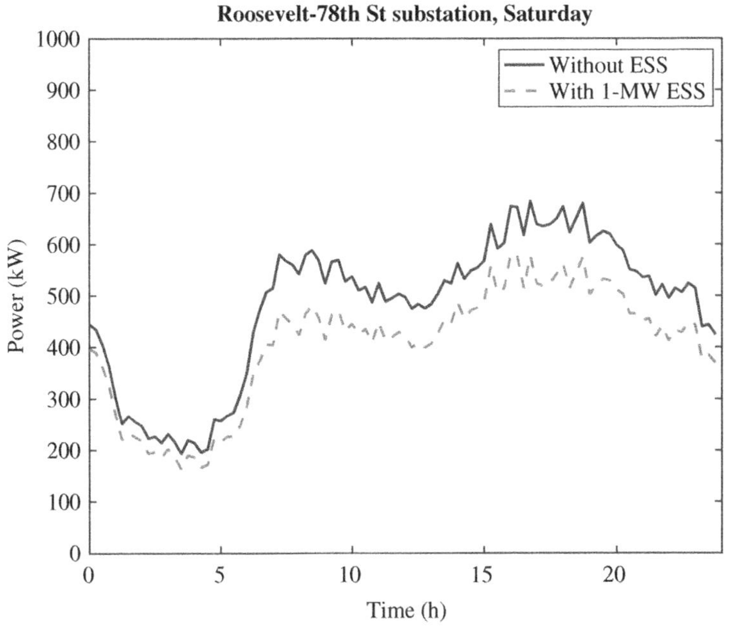

Figure 16.27 24-hour power profiles of Roosevelt Ave-78th substation (Saturday).

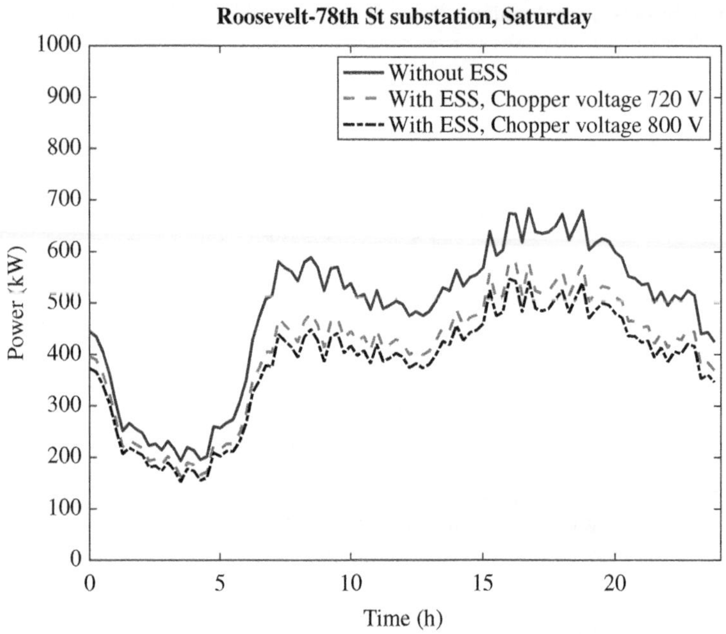

Figure 16.28 24-hour power profiles of Roosevelt Ave-78th substation (Saturday).

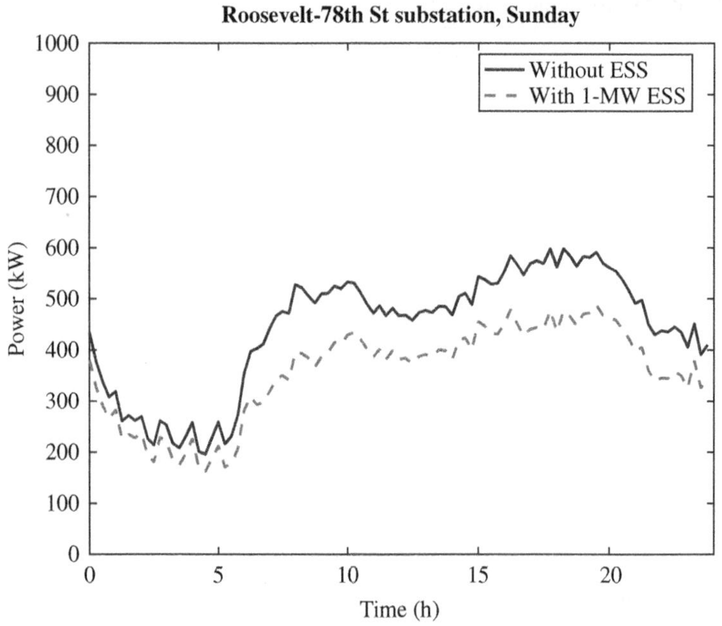

Figure 16.29 24-hour power profiles of Roosevelt Ave-78th substation (Sunday).

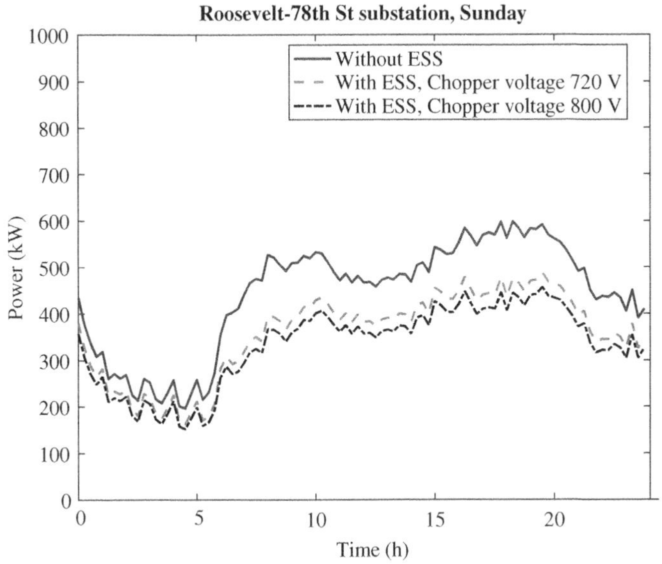

Figure 16.30 24-hour power profiles of Roosevelt Ave-78th substation (Sunday).

Table 16.4 Summary of results.

Substation	Scenarios	% Peak demand reduction	% Energy savings
58th St	Weekday with chopper voltage 720	18.11	18.11
	Weekday with chopper voltage 800	29.491	29.491
	Saturday with chopper voltage 720	18.11	18.11
	Saturday with chopper voltage 800	29.491	29.491
	Sunday with chopper voltage 720	18.11	18.11
	Sunday with chopper voltage 800	29.491	29.491
Roosevelt Ave-78th St	Weekday with chopper voltage 720	5.54	5.54
	Weekday with chopper voltage 800	8.053	8.053
	Saturday with chopper voltage 720	5.54	5.54
	Saturday with chopper voltage 800	8.053	8.053
	Sunday with chopper voltage 720	5.54	5.54
	Sunday with chopper voltage 800	8.053	8.053

The results show that by adding WESS, the peak power demand and 24-hour energy consumption of Roosevelt Ave-78th St substation during Saturday and Sunday are again reduced by 5.54%, and that increasing the chopper activation voltage does not have a significant impact on energy savings and peak demand reduction for this substation. The peak power demand and 24-hour energy consumption of this substation after increasing the chopper activation voltage to 800 V are reduced by 8.053% (approximately 2.5% increase in energy savings and peak demand reduction).

The percentage of peak power demand reduction and energy savings for all the studied scenarios is summarized in Table 16.4.

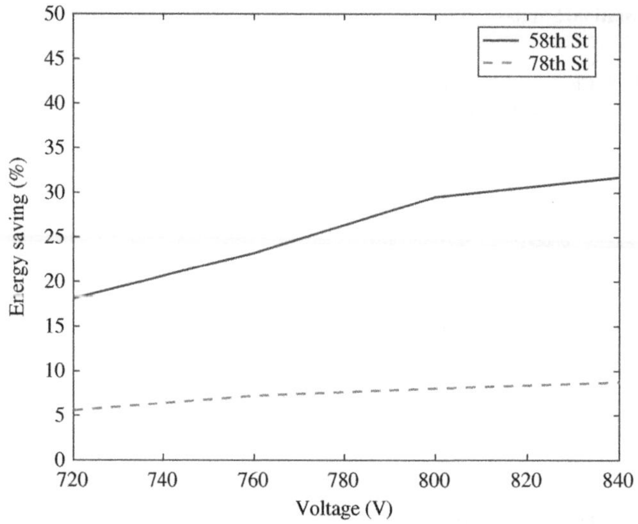

Figure 16.31 Energy savings versus chopper activation voltage.

To better understand the relationship between chopper cutoff voltage and energy savings, two more scenarios with the activation voltages set to 760 and 840 V have also been simulated. Figure 16.31 summarizes this comparison.

It can be observed that an increase in the activation voltage causes an increase in energy savings, as the train can inject more current into the third rail before the chopper kicks in. However, the relationship between cutoff voltage and energy savings is non-linear and cannot be related based only on voltage.

16.3.4.4 Engaging the flywheel all the time

We simulated different sizes of flywheel and chopper cut-off voltage setpoints under two operating scenarios: (i) when the flywheel is always activated; and (ii) when the flywheel is only activated when the train is nearby. A summary of the simulated cases is provided in Table 16.5. It can be seen that the energy savings increase from 9.91% to 18.11% at 58th St if the flywheel is only engaged when a train is nearby. This enables the flywheel to charge and discharge primarily from the train. When the flywheel is always connected, it tends to draw energy from the rectifier substations along with the train for charging, especially when the train is not close, offsetting the energy savings.

Table 16.5 also shows that increasing the chopper cut-off voltage increases the third rail receptivity to regenerative braking energy; therefore resulting in an increase in energy savings. A higher activation voltage gives more time for the train to inject regenerative energy into the third rail before the chopper kicks in. The impact of changing the flywheel size has also been investigated (Table 16.6). It was found that increasing the flywheel power capability from 0.9 to 1 MW slightly increases the energy savings. A further increase in the power to 1.8 MW causes the flywheel to draw energy for charging from the substations but this is not accompanied by a proportional increase in discharge to accelerating trains. As a result, the energy savings stay the same, or drop as in the case when the flywheel is only engaged when a train is nearby.

16.3.4.5 State of Charge Control

The results presented in previous figures were focused on transient behavior of the flywheel system within a granular timeframe. It is crucial to estimate the flywheel behavior over an extended period. In this section, we simulate the flywheel operation over 24 hours, along with its state of charge

Table 16.5 Activating the ESS only when a train is nearby vs. always activating it.

	Chopper voltage	Flywheel		QB_Energy savings (%)	58 St_Energy savings (%)	78 St_Energy savings (%)	Spruce St_Energy savings (%)
Case 0	720	No flywheel					
Case 1	720	1 MW (6 × 167 kw)	*Flywheel always connected*	0.469	9.91	2.14	0.322
Case 2	760	1 MW (6 × 167 kw)	*Flywheel always connected*	3.355	29.005	12.04	3.557
Case 3	800	1 MW (6 × 167 kw)	*Flywheel always connected*	4.066	35.357	14.511	4.25
Case 4	840	1 MW (6 × 167 kw)	*Flywheel always connected*	4.065	35.521	14.359	4.219
Case 5	720	0.9 MW (3 × 300 kw)	*Flywheel always connected*	0.485	7.62	1.367	0.331
Case 6	720	1.8 MW (6 × 300 kw)	*Flywheel always connected*	0.369	9.98	1.082	0.263
Case 7	800	0.9 MW (3 × 300 kw)	*Flywheel always connected*	3.888	33.465	13.813	4.096
Case 8	800	1.8 MW (6 × 300 kw)	*Flywheel always connected*	5.21	46.277	18.397	5.219
Case 9	720	1 MW (6 × 167 kw)	*Flywheel connected near train*	1.246	18.11	5.54	0.617
Case 10	760	1 MW (6 × 167 kw)	*Flywheel connected near train*	2.774	23.22	7.23	2.588
Case 11	800	1 MW (6 × 167 kw)	*Flywheel connected near train*	3.639	29.491	8.053	2.848
Case 12	840	1 MW (6 × 167 kw)	*Flywheel connected near train*	3.896	31.719	8.745	3.041
Case 13	720	0.9 MW (3 × 300 kw)	*Flywheel connected near train*	0.659	8.38	2.235	0.505
Case 14	720	1.8 MW (6 × 300 kw)	*Flywheel connected near train*	3.637	12.46	11.638	2.464

(Continued)

Table 16.5 (Continued)

	Chopper voltage	Flywheel		QB_Energy savings (%)	58 St_Energy savings (%)	78 St_Energy savings (%)	Spruce St_Energy savings (%)
Case 15	800	0.9 MW (3 × 300 kw)	*Flywheel connected near train*	3.067	28.602	7.389	2.74
Case 16	800	1.8 MW (6 × 300 kw)	*Flywheel connected near train*	10.12	36.33	29.21	8.23

Table 16.6 Impact of flywheel size on energy savings at 720 V chopper cutoff voltage.

	Energy saving (%)	
Size (MW)	**Flywheel always connected**	**Flywheel only connected when a train is nearby**
0.9	7.62	8.38
1	9.91	18.11
1.8	9.98	12.46

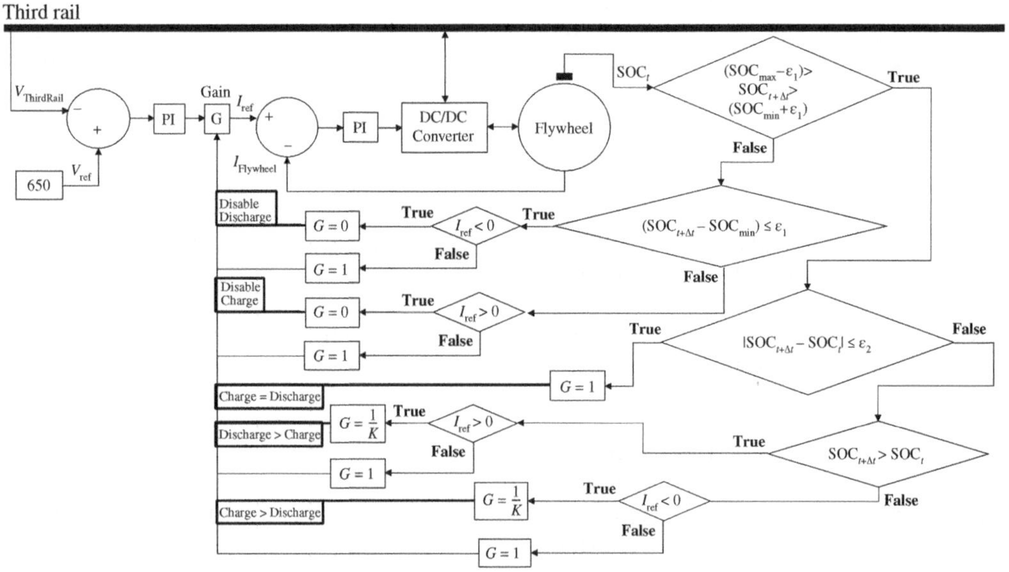

Figure 16.32 Flywheel state of charge control algorithm.

control algorithm depicted in Figure 16.32. This algorithm starts with a predefined desired SOC, lets the flywheel SOC deviate from the desired SOC until an allowable time (Δt) ends, and then it restores the SOC to its desired value. If the flywheel SOC at the end of Δt is greater than its desired value, the flywheel discharges more energy than what it charges with. On the other hand, if the SOC

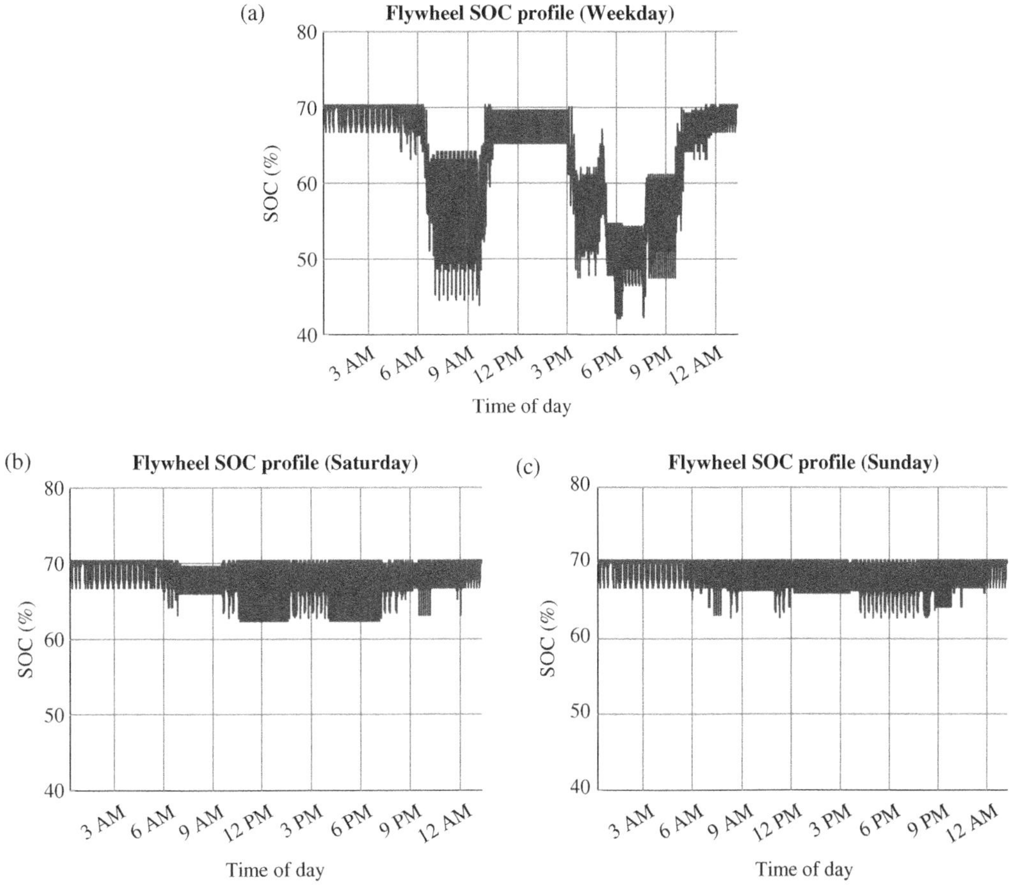

Figure 16.33 Flywheel state of charge over 24 hours corresponding to the case of 1 MW flywheel, 720 V chopper voltage, flywheel always connected (i.e. case 1 in Table 16.5): (a) weekday; (b) Saturday; and (c) Sunday.

at the end of Δt is less, the flywheel charges more. This process is controlled by a gain/factor (G) that either limits or magnifies the reference current. In addition, the algorithm continuously checks whether the SOC is within an acceptable band (between max and min, e.g. 65–75%) or not. The presented results in Figures 16.33 and 16.34 emulate a case with Δt of 5 minutes, ε_1 and ε_2 of about 5%, and K of 100.

Note that in Figure 16.32:

- $K > 1$
- I_{ref} is the flywheel's reference current
- $I_{Flywheel}$ is the flywheel's measured current ("+" if charging and "−" if discharging)
- V_{ref} is the flywheel reference voltage
- $V_{ThirdRail}$ is the third rail measured voltage
- ε_1 is the maximum allowable SOC deviation from SOC_{max} and SOC_{min}
- ε_2 is the maximum allowable deviation between SOC_t and $SOC_{t+\Delta t}$
- G is the gain value

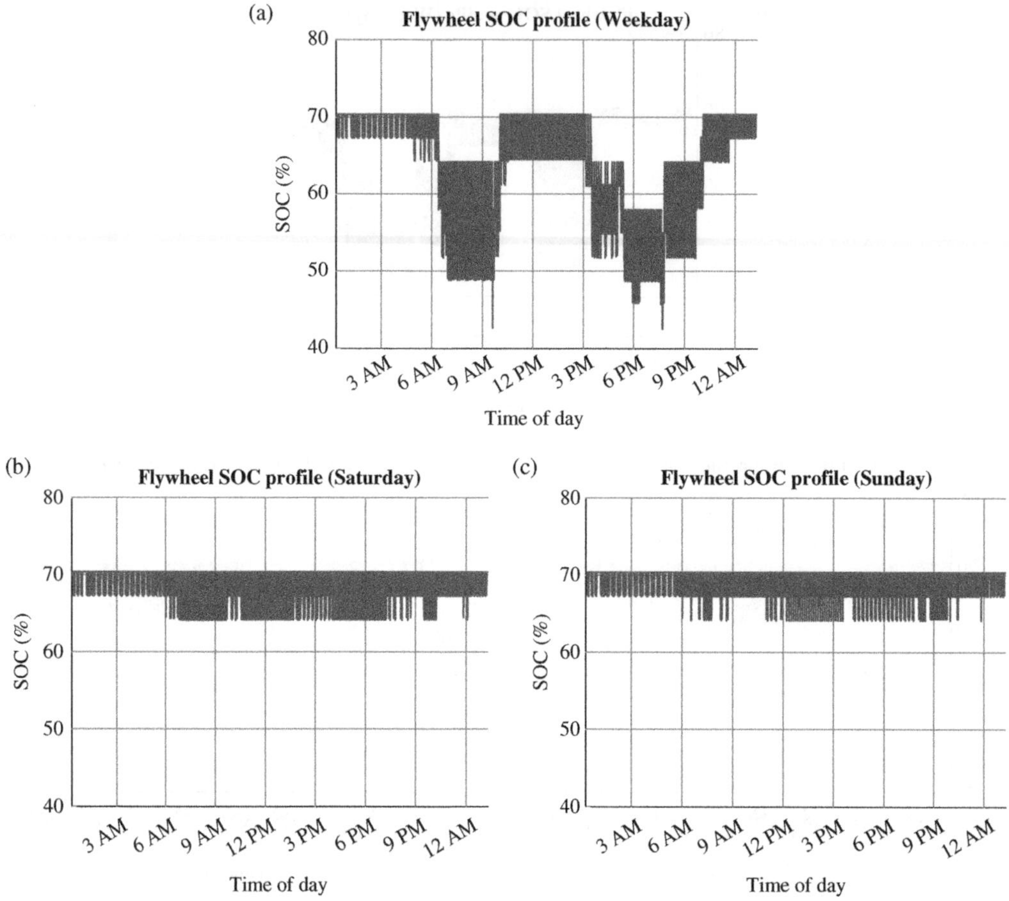

Figure 16.34 Flywheel state of charge over 24 hours corresponding to the case of 1 MW flywheel; 720 V chopper voltage; flywheel connected only when train is nearby (i.e. case 9 in Table 16.5): (a) weekday; (b) Saturday; and (c) Sunday.

16.4 1 MW Kinetic Energy Storage System Financial Results

16.4.1 Train Simulation Study

Amongst other things, the Train Simulation Study, just described, modeled the 24-hour power and energy profiles of #7 train for weekdays, Saturdays, and Sundays. It did this for a stretch of track bounded by the stops at 46th St-Bliss St and Junction Blvd. This stretch of track primarily draws power from four substations.

16.4.2 Cases Run

Cases were run not only for weekdays, Saturdays, and Sundays but also for different flywheel configurations, i.e. 1.0 MW – 3 × Beacon 400-333, 1 MW – 6 × Beacon 400-167, and 1.8 MW – 6 × Beacon 400-300 flywheels and for different chopper cutoff voltages. The second number in the flywheel designation refers to the motor power (kW). Financial savings, as shown next, are computed for the 1 MW case as that was the optimum case.

Outwardly, the Beacon 400-333 is identical to the Beacon 400-167. As stated earlier, the main difference is the motor's power rating. Recently, it has been determined that the Beacon 400-300 can be operated to produce 333 kW, such that three flywheels equal 1 MW. This (Beacon 400-333) is the designation used throughout this report.

A 6-flywheel installation requires about 4000 sq. ft. A 3-flywheel installation would require slightly more than half that area.

16.4.3 Capital Costs

Capital costs for the facility were developed from the NYSERDA project budget and the cost of the Beacon flywheels. The total cost estimated in the NYSERDA project budget is $3.43 million. This budget includes the cost of six Beacon 400-167 flywheels at $225,000 each. The cost of the Beacon 400-333 flywheels is expected to be 15–20% more expensive. It is still under development. A cost of $250,000 per flywheel was used in this analysis.

Three scenarios are evaluated: (i) the NYSERDA budget and the Beacon 400-167 flywheel; (ii) the NYSERDA budget less one-time costs ($1.067 million) and the Beacon 400-167 flywheel; and (iii) the NYSERDA budget less one-time costs ($1.067 million) and the Beacon 400-333 flywheel. The facility capital cost estimates are shown in Table 16.7. All are rated at 1 MW.

16.4.4 Estimation of Annual Energy and Demand

The 24-hour profiles are combined with the number of weekdays, Saturdays, Sundays, as well as holidays (these are shown in Table 16.8) to estimate annual values.

The power supplier to NYCT is the New York Power Authority (NYPA). NYCT has provided NYPA bills for 1/2022. From these bills, values for energy (kWh) and demand kW have been derived. The values are $0.0662/kWh and $33.1/kW.

The peak demand computation used by NYPA is as follows: the interval used to calculate the demand to be used for billing is the 30-minute demand using 5-minute rolling intervals. This means that you take six consecutive 5-minute intervals and sum them (for example, from 1/18/22 00:00 to 1/18/22 00:30) and then roll 5 minutes (e.g. 1/18/22 00:05 to 1/18/22 00:35) and keep rolling through the day for a month. The demand is the maximum of these individual summations, which is then divided by ½ (since the 30-minute sum is kWh over ½ hour).

Table 16.7 Facility capital cost estimates.

NYSERDA Project Cost Beacon 400-167 flywheel	$ 3,430,000
NYSERDA Project Cost less onetime costs Beacon 400-167 flywheel	$ 2,363,000
NYSERDA Project Cost less onetime costs Beacon 400-333 Flywheel	$ 1,511,000

Table 16.8 Annual number of days.

Annual Workdays	251
Annual Saturdays	52
Annual Sundays	52
Annual Holidays	10
Total	365

A simple numerical example is if the kWh is in each 5-minute interval for a 30-minute interval. Then, the total for each 1 kWh would be 6 kWh and the demand would be 12 kW (6 kWh/0.5).

The Train Simulation model performs the modeling in intervals of 15 minutes. Therefore, a running sum of two modeled intervals is used to simulate the NYPA calculation for 30-minute periods.

16.4.4.1 Results

Figure 16.21 from the Train Simulation Study shows the 24-hour power profile for a weekday with three operating modes (a) without the flywheel, (b) with the flywheel and the cutoff voltage set to 720 V, and (c) with the flywheel and the cutoff voltage increased to 800 V.

Table 16.9 summarizes the case results for tariff values as of 1/2022 and gives the simple payback time (project cost divided by savings) for the selected cases. Results are shown for both the Beacon 400-167 and Beacon 400-333 flywheels. As one can see from the values, there is considerable upside.

16.4.4.2 Emission Reduction

The reduction in energy usage also causes a reduction in carbon and other emissions, thereby reducing NYCT's carbon footprint. Table 16.10 shows the reduction in emissions of CO_2, SO_2, and NO_X. Both, the estimated reduction for the single 1 MW demonstration facility, as well as an estimate of the overall reduction if such facilities were implemented throughout the NYCT system are shown.

Table 16.9 Savings summary as on 1/2022.

		Savings summary 1/2022 Bills Capacity $33.1/kW Energy $0.0662/kWh				
		Annual energy savings ($)	**Annual demand savings ($)**	**Total annual savings ($)**	**Electricity cost**	**%**
58th Street substation	Without ESS				1,182,010.17	
	With ESS_Chopper activation voltage = 720 V	123,089	120,482	243,572		
	With ESS_Chopper activation voltage = 760 V	186,294	195,548	381,543		
	With ESS Chopper activation voltage = 800 V	246,873	267,496	514,369		
	With ESS_Chopper activation voltage = 840 V	259,902	282,970	542,873		
78th Street substation	Without ESS				1,034,195.24	
	With ESS_Chopper activation voltage = 720 V	38,266	36,483	74,749		

Table 16.9 (Continued)

		Savings summary 1/2022 Bills Capacity $33.1/kW Energy $0.0662/kWh				
		Annual energy savings ($)	**Annual demand savings ($)**	**Total annual savings ($)**	**Electricity cost**	**%**
	With ESS_Chopper activation voltage = 760 V	55,932	63,109	119,040		
	With ESS_Chopper activation voltage = 800 V	59,866	69,039	128,905		
	With ESS_Chopper activation voltage = 840 V	62,804	73,466	136,269		
Grand total					2,216,205.41	
	With ESS_Chopper activation voltage = 720 V	161,356	156,965	318,321		14.4%
	With ESS_Chopper activation voltage = 760 V	242,226	258,657	500,883		
	With ESS_Chopper activation voltage = 800 V	306,739	336,535	643,274		29.0%
	With ESS_Chopper activation voltage = 840 V	322,706	356,436	679,142		
				NYSERDA Project Cost Beacon 400-167 flywheel	NYSERDA Project Cost less one-time costs Beacon 400-167 flywheel	NYSERDA Project Cost less one-time costs Beacon 400-333 flywheel
Simple payback (years)	With ESS_Chopper activation voltage = 720 V			10.78	7.42	4.75
	With ESS_Chopper activation voltage = 800 V			5.33	3.67	235
NYSERDA Project Cost Beacon 400-167 flywheel	$ 3,430,000					

(*Continued*)

Table 16.9 (Continued)

			Savings summary 1/2022 Bills Capacity $33.1/kW Energy $0.0662/kWh				
			Annual energy savings ($)	**Annual demand savings ($)**	**Total annual savings ($)**	**Electricity cost**	**%**
NYSERDA Project Cost less one-time costs Beacon 400-167 flywheel	$	2,363,000					
NYSERDA Project cost less one-time costs Beacon 400-333 flywheel	$	1,511,000					

Table 16.10 Estimated reductions in emissions.

A single 1 MW installation			
With ESS_Chopper activation voltage = 720 V	CO_2 (MT)	459.4	
	SO_2 (MT)	0.1	
	NO_x (MT)	0.5	
With ESS_Chopper activation voltage = 800 V	CO_2 (MT)	873.4	
	SO_2 (MT)	0.3	
	NO_x (MT)	0.9	
If the same percentage savings are achieved throughout NYCT			
With ESS_Chopper activation voltage = 720 V	CO_2 (MT)	91,102.91	
	SO_2 (MT)	29.362	
	NO_x (MT)	96.16	
With ESS_Chopper activation voltage = 800 V	CO_2 (MT)	171,487.82	
	SO_2 (MT)	55.269	
	NO_x (MT)	181.00	
Source:			
https://www.eia.gov/electricity/data/state/		2019 data	New York

This data was derived from EIA data for New York for 2019. New York annual emissions of CO_2, SO_2, and NO_X from electric generation were divided by total New York electric generation to develop a coefficient in MT/MWh, which were then multiplied "by" the appropriate savings estimate.

References

1 Khodaparastan, M., Dutta, O., Saleh, M., and Mohamed, A. (2019). Modeling of DC electric rail transit systems with wayside energy storage. *IEEE Trans. Veh. Technol.* 68 (3): 2218–2228.

2 Khodaparastan, M., Brandauer, W., and Mohamed, A. (2019). Recuperation of regenerative braking energy in electric rail transit systems. *IEEE Trans. Int. Transport. Syst.* 20 (8): 2831–2847.

3 Available Online at: https://new.mta.info/map/5256 (accessed August 2022).

4 A. Mohamed, A. Reid, and T. Lamb. White Paper on Wayside Energy Storage for Regenerative Braking Energy Recuperation in the Electric Rail System. Consolidated Edison, Inc., 2018. Available online at: https://www.coned.com/-/media/files/coned/documents/our-energy-future/our-energy-projects/regenerative-braking-energy-recuperation.pdf (accessed August 2022).

17

Distributed Energy Resource Integration with Electrical Railway Systems: NYC Case Study

Rohama Ahmad, Jaskaran Singh, and Ahmed A. Mohamed

City University of New York, City College, Department of Electrical Engineering, New York, NY 10031, USA

17.1 Introduction

Power systems worldwide are striving to increase resiliency, as well as combat effects due to global warming. States/cities have set aggressive goals to drastically reduce greenhouse emissions within the upcoming years and this has led to new investments in renewable energy, as well as electric vehicles. As such, combatting global warming can be achieved through increased deployment of distributed energy resources (DERs), including energy storage systems (ESS) (batteries, flywheels, supercapacitors, etc.), as well as photovoltaics (PV), wind turbines, etc. By further decentralizing energy resources to a microgrid scale, the integration of DERs can be highly facilitated, at the same time enabling independent operation. However, microgrids are faced with economic and regulatory barriers. Hence, the concept of the "community microgrid" has recently arisen.

A community microgrid consists of multiple shared owners/participants, including residential/commercial facilities (e.g., university campuses, wastewater treatment plants, subway substations, fire stations, water-pumping stations, police departments, etc.) collectively sharing their resources. In a community microgrid setting, resources are virtually coordinated when connected to the main utility grid, however when disconnected from the main grid (islanded mode), demand is supplied locally, thereby still allowing independent operation. To efficiently achieve independent operation, optimal control strategies and energy management algorithms must be employed. The strategies and algorithms must take into account renewable energy resources, moving as well as stationary electric vehicles, controllable loads, and other major loads (e.g., electrified transportation and/or heating) [1].

Given the level of distribution, traditional hierarchy-based central control approaches would not be considered practical, since the bandwidth of data transfer would strain any communication technology. Grid operators would have to communicate with all the DERs in real-time, and given their uncertainty, the operators would have difficulty maintaining the load/generation balance. This would not only present local consumer-level challenges, but may also compromise the entire network's stability, resulting in more downtime (blackouts). Therefore, the proposed strategies would need to optimize the DERs locally, as well as offload the control and communication from the utility. If all of this is done successfully, DER integration can even be expanded to a larger scale, with the inclusion of more demanding networks, such as a subway system.

This chapter presents a case study, involving integration of various DERs with a subway system, all within a community microgrid setting. A coordinated control framework is proposed, and the environmental, as well as economic benefits highlighted.

Transportation Electrification: Breakthroughs in Electrified Vehicles, Aircraft, Rolling Stock, and Watercraft,
First Edition. Edited by Ahmed A. Mohamed, Ahmad Arshan Khan, Ahmed T. Elsayed, and Mohamed A. Elshaer.

17.2 DER Integration with Subway Systems

17.2.1 Regenerative Braking Energy Recuperation

Subway systems are already considered an efficient means of transportation due to their comparatively low per-capita energy consumption and greenhouse emissions. An advantage to their inclusion within a community microgrid setting is that there is a potential for further substantial savings if the regenerative braking energy produced by subway trains is more efficiently recuperated. Proper recycling of excess regenerative energy, through DER integration, can lead to reductions in overall power consumption (from the utility), as well as greenhouse gas emissions.

Most trains today are equipped with regenerative braking technology, which allows them to recover some of their kinetic energy when braking. Traction motors propelling the trains act as generators when braking, converting mechanical energy back to electrical energy. The electrical energy is used to power auxiliary loads, and any surplus of energy is sent back to the DC bus, i.e., the third rail circuit. If not properly recycled, this energy is simply dissipated as heat, through wayside "dumping" resistors. The wasted heat must be managed through proper ventilation in order to make nearby closed environments safer for users. Through DER integration, the amount of wasted heat can be reduced, and the associated energy used more efficiently elsewhere.

17.2.2 AC vs DC Integration

Depending on the electrical form of distribution (AC or DC), DER integration can lead to various benefits, as well as drawbacks. Power distribution networks around the world comprise multiple sub-levels of distribution, utilizing elements of AC, and, at times, even DC, circuitry. Figure 17.1 shows the electrical path from a substation in New York City (NYC) to its third rail.

As shown in Figure 17.1, NYC's power distribution network begins at the substation level, where an uncontrolled rectifier substation supplies the medium-level utility voltage (13.2-kVAC) to two three-phase transformers. The transformers further step down the voltage to about 465-VAC, as it makes its way to two full-bridge diode-based rectifiers (two for redundancy). A capacitive filter is connected between each of the transformers and rectifiers, and the output of the rectifiers is what forms the third rail voltage, as the positive (+) side is connected to the third rail, while the negative (−) side is connected to the running rails. The third rail voltage in NYC typically ranges between 620- to 650-VDC with no train load but fluctuates further as trains pass by.

For such a network, DERs can be integrated at the AC side (before the traction rectifiers), or the DC side (third rail side) of the network. Integration on the DC side has several advantages, such as enhanced overall efficiency due to the reduced number of conversion stages. PV cells generate DC electricity, and their inclusion within an AC network requires inverter technology, which does not produce pure sinusoidal waveforms. This introduces harmonics into the network, leading to further losses. Integration on the DC side also enhances the power-carrying capacity of cabling due to a lack of skin effect, thereby eliminating the need for var control.

Integration on the AC side has its own benefits, such as easier (and cheaper) voltage-conversion (transformers). Higher voltages allow for less currents, which makes a significant difference in long-distance transmission. However, for a subway system like in NYC, a substation is normally located between every 1–3 passenger stations with each passenger station being less than a mile away from the next. Therefore, it seems more beneficial to integrate DERs on the DC side when considering such a system. This also allows for easier integration of ESS, such as batteries, flywheels, and supercapacitors; which all store DC energy.

17.2.3 ESS Selection and Allocation

Supercapacitors and flywheels are primarily used for applications which require the release of a high amount of energy in a short duration of time. They work on a cycle-by-cycle basis and in such an

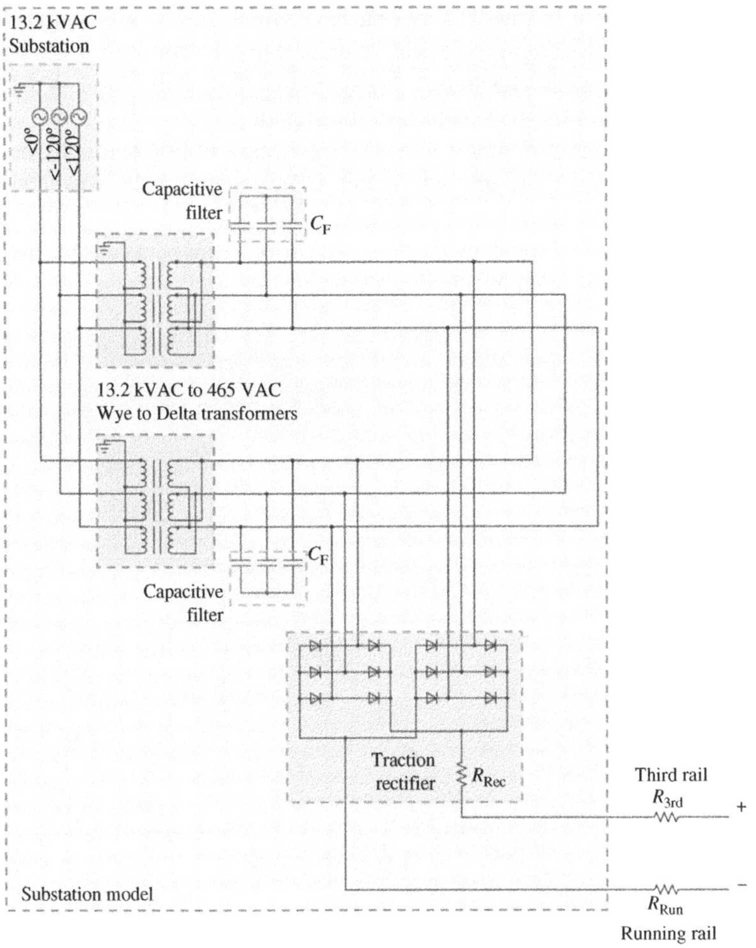

Figure 17.1 Electrical path from an NYC substation to the third rail.

application, would be sized just enough to recycle regenerative energy back and forth with each train's acceleration/deceleration. Placing either in tandem with the already existing rectifier substations would provide some level of load flexibility. Similarly, batteries would allow high energy storage but, due to their limited C-rate, would need to be largely oversized, and while this increases capital investment, it both elongates battery lifetime (low depth of discharge) and provides an opportunity for distribution grid services (i.e., demand response, etc.), in turn, providing even higher levels of load flexibility.

The ESS can be mounted on the trains themselves (onboard), within substations (offboard), or at strategic locations along the third rail (wayside). The disadvantage of onboard mounting is the need for space and the resulting increase in vehicle weight. Therefore, the latter two options are preferred and when applied within a community microgrid setting, more easily expand the shared network beyond just trains.

17.3 Case Study

A case study is presented using NYC's subway system, which is one of the busiest systems in the world and has the potential to integrate DERs, as well as benefit from their deployment.

17.3.1 NYC's Subway System

Public transportation in New York City is operated by the New York City Transit Authority (NYCTA), which falls under the Metropolitan Transit Authority (MTA). It is considered to be the busiest and largest transit system in North America, housing 200+ substations, as well as 400+ passenger stations, more than any other system in the world. Average daily ridership surpasses 5 million and average annual ridership surpasses 1.5 billion. The system is more than 100 years old and offers service on a 24/7 basis. Figure 17.2 shows a map of the system.

As depicted in the map, 36 different lines currently run in NYC, covering 4 of its 5 boroughs, as well as running above ground (bridges) and below ground (tunnels). The system is to be expanded as well within the upcoming years.

Figure 17.3 shows a diagram depicting the stops along just the 7-line, which runs between Queens and Manhattan, a part of which is also modeled in this case study.

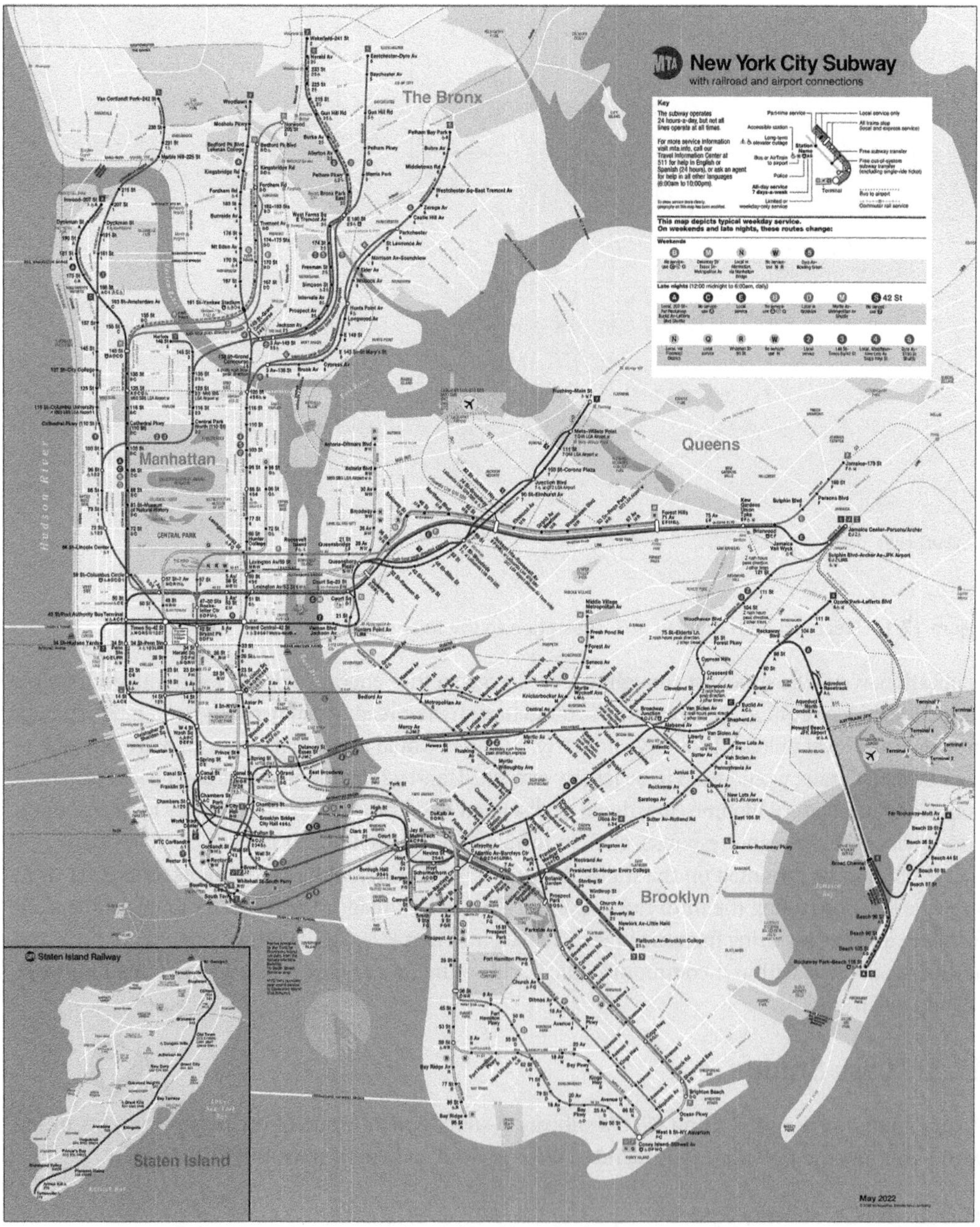

Figure 17.2 Map of NYC subway system. *Source:* Adapted with permission from https://new.mta.info/map/5256 (Metropolitan Transportation Authority).

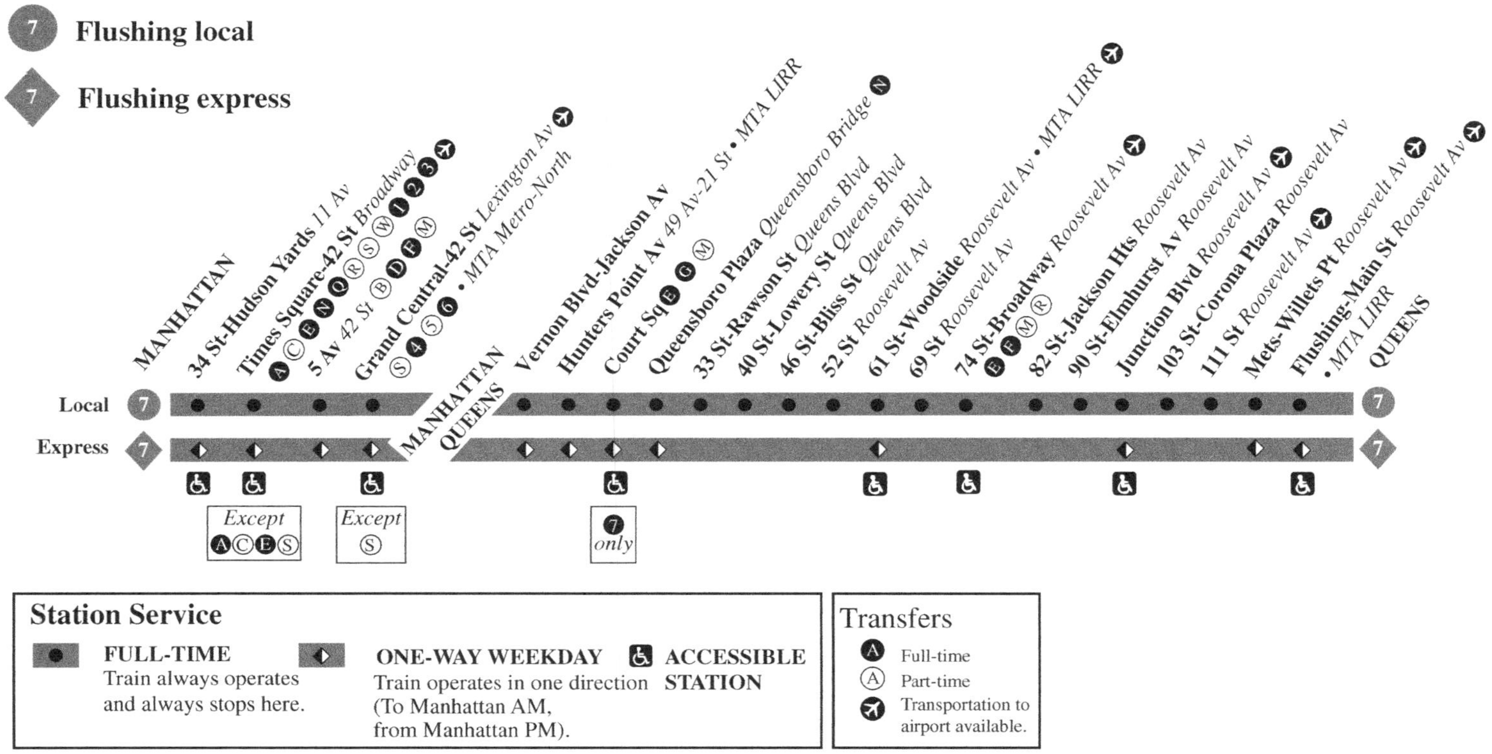

Figure 17.3 Diagram of 7-line (https://new-york.metro-map.net/nyc-subway-map-7-train-stops/).

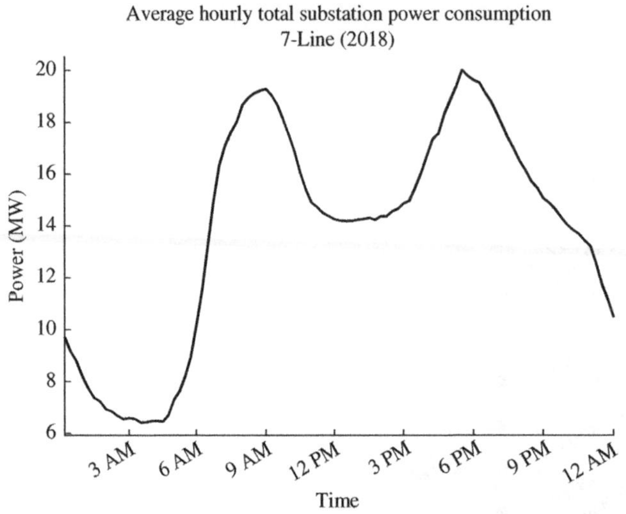

Figure 17.4 Average hourly total substation power consumption; 7-line (2018).

Figure 17.4 shows the average hourly total substation power consumption for the entire 7-line for the year 2018.

The NYC subway shows peaks in power during rush hour times when the most amount of passengers utilize it in order to go to and return from their work destinations. As visible in the figure, these rush hour times are typically between 6:30 AM and 9:30 AM for morning commute, and between 3:30 PM and 6:45 PM for evening commute, with the former primarily directed toward Manhattan, and the latter away from it. The borough of Manhattan houses most of the city's buildings and facilities, hence the most amount of people are employed within it. NYCT's 7-line consists of 22 passenger stations, powered by 13 substations, and along it, there is typically 1 substation located in between every 1–2 passenger stations, with each passenger station being approximately 2000-ft apart. 5 of the 13 substations are located in Manhattan, while the other 8 in Queens.

17.3.2 Model

In order to perform this case study, the subway network is modeled using MATLAB/Simulink and transient simulations are performed to test the system under varying operating conditions. Wayside DER integration is presented as a PV system located by the third rail nearby a station. Electric vehicle loads are added nearby as well to test the possibility of ondemand charging for customers within a community microgrid. Energy storage is added through the addition of a battery system. Figure 17.5 illustrates the system under study.

The train is modeled using the current profile shown in Figure 17.6, which consists of real-life current measurements taken onboard a 7-train moving from one station to the next.

The advantage of taking this approach is that it already takes into account train movements, onboard flow control, as well as variable resistance. The current profile shows a train accelerating (from about 5 to 33 seconds) as it draws power from the third rail. The train then decelerates (from about 33 to 63 seconds), injecting regenerative braking energy (negative current) back into the third rail, before coming to a stop.

The train also contains a braking chopper, i.e., an onboard protection system designed to disconnect the train due to an excessive rise in voltage.

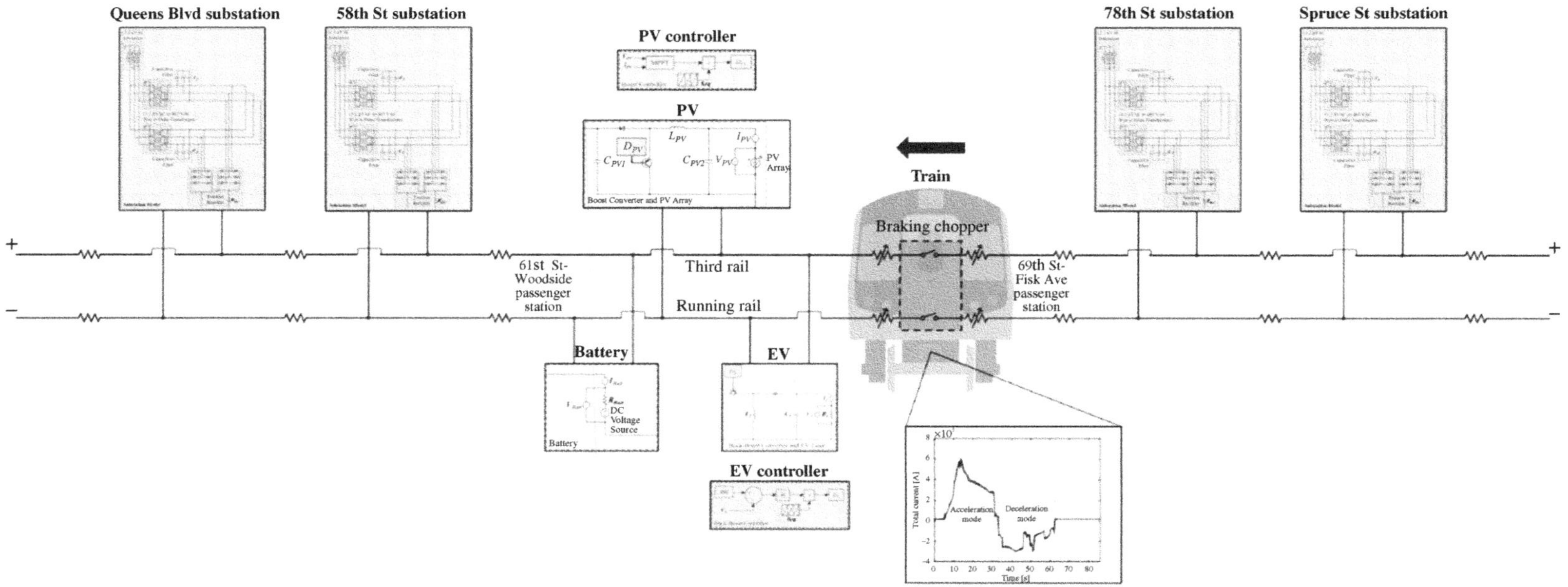

Figure 17.5 Subway system model.

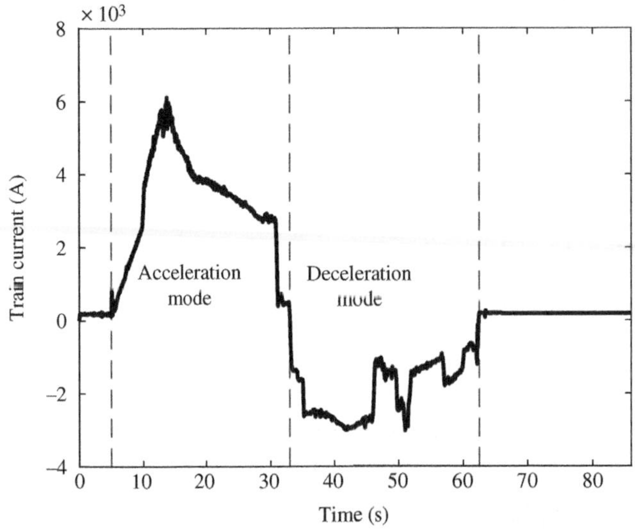

Figure 17.6 Train current profile input into model.

Figures 17.7–17.9 show kinematic profiles of the train's movement.

The 7-train is simulated to be traveling a distance of approximately 2000-ft, i.e., about 1-station distance, as it arrives at 61st St-Woodside Passenger Station, where the previously mentioned DERs are deployed.

To see the true benefits of DER deployment, first a case is run with all DERs disconnected. Figure 17.10 shows the resulting third rail voltage.

As visible in the figure, the third rail voltage is set to 650-VDC before any train movement occurs. Then as the train accelerates, it draws power, resulting in a voltage drop. The train then decelerates, sending regenerative braking energy back into the third rail, which causes voltage spikes as the voltage rises to about 720-VDC.

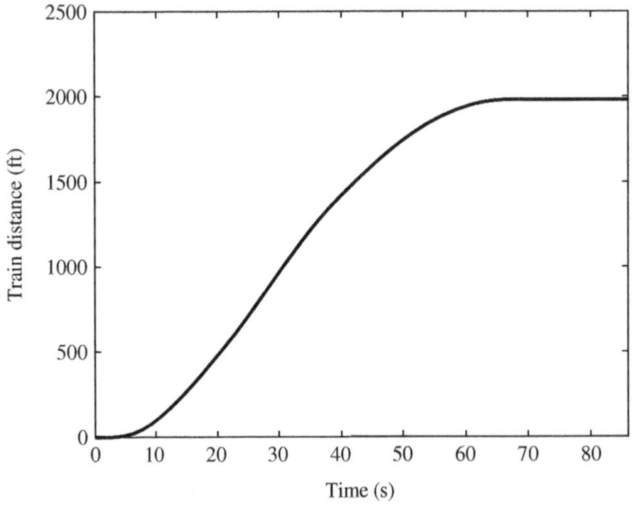

Figure 17.7 Train distance profile.

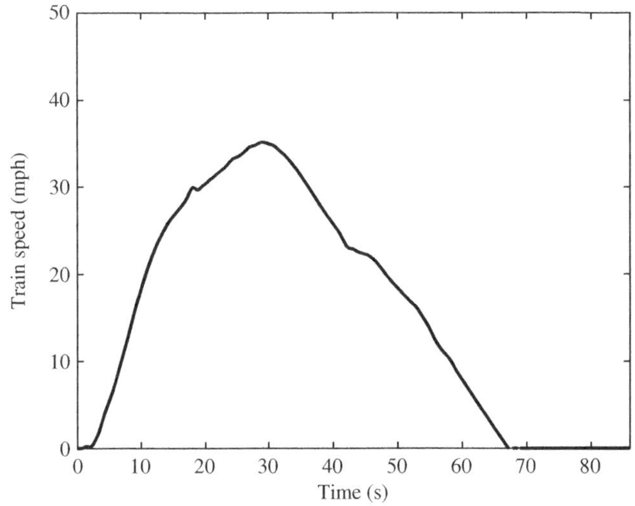

Figure 17.8 Train speed profile.

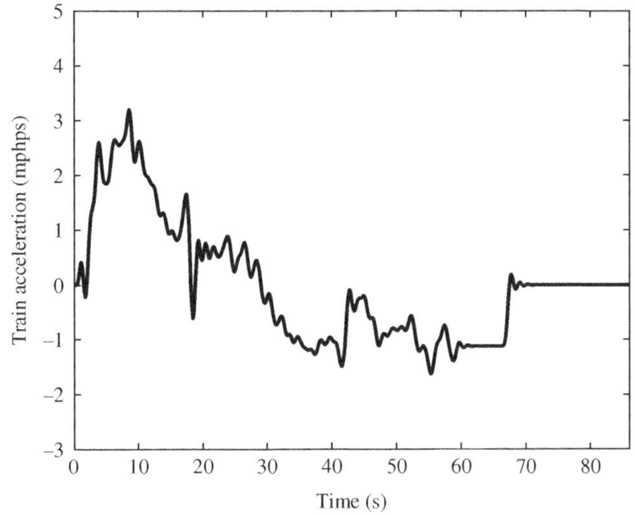

Figure 17.9 Train acceleration profile.

Figures 17.11 and 17.12 show the nearest four substations' current and power profiles, respectively.

As the train is simulated to be leaving from about 1-station distance away east of 61st St-Woodside passenger station (hence by 69th St-Fisk Ave passenger station), its nearest substation is 78th St Substation. Therefore, the train draws the most amount of power from that substation. The next closest substation, particularly when the train reaches 61st St-Woodside passenger station, is 58th St Substation, which provides the second most amount of power to the train. Further away, on the east and west sides respectively, are Spruce St and Queens Blvd substations,

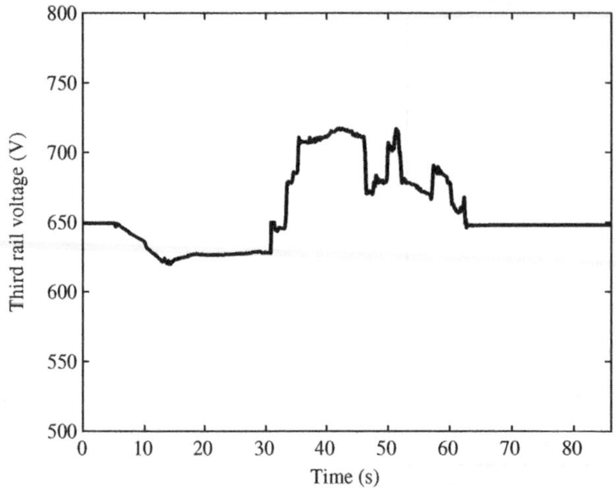

Figure 17.10 Train rail voltage (DERs disconnected).

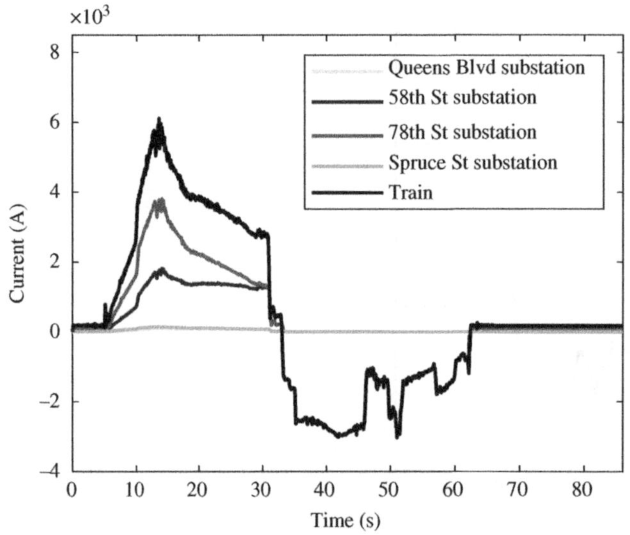

Figure 17.11 Substation current profiles (DERs disconnected)

which provide far less power to the train. The train's current and power profiles are also included in the figures to give an idea of the differences in magnitude. Note that each station's power and current measurements are taken directly where the substation is located, while the train's measurements are taken onboard, hence not stationary with respect to each station. Due to the rectifiers located at each substation, the regenerative energy sent back from the train doesn't make it back to its source to provide any benefits (no negative substation current or power values). It is through the integration of the DERs that the energy can be utilized rather than wasted elsewhere.

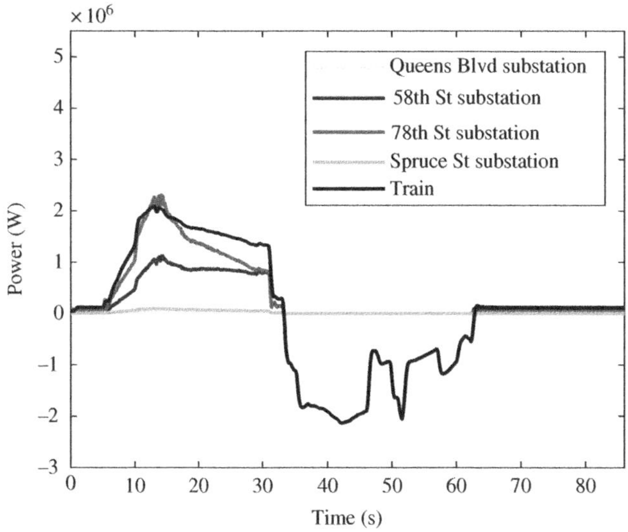

Figure 17.12 Substation power profiles (DERs disconnected).

17.3.3 DER Integration

The PV system consists of A10 Green Technology (A10J-S72-175) modules, forming 640 parallel strings, 10 series-connected modules per string, and 72 cells per module. The maximum power rating for each module itself is 175-W at standard test conditions (STC). The boost converter interfacing the PV system performs maximum power point tracking (MPPT), using the Perturb and Observe algorithm, with an initial duty cycle set to 0.87, initial voltage set to 80-V, and initial power set to 640-W. Figure 17.13 shows the elements of the PV system, as well as its controller.

The system is tested under varying operating conditions of temperature and solar radiation (Figures 17.14 and 17.15).

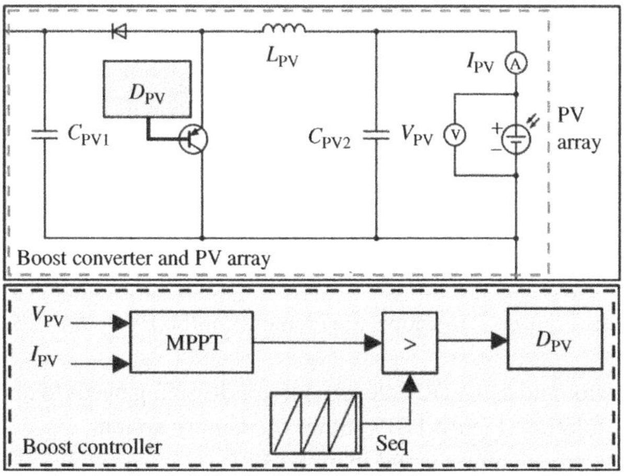

Figure 17.13 PV system.

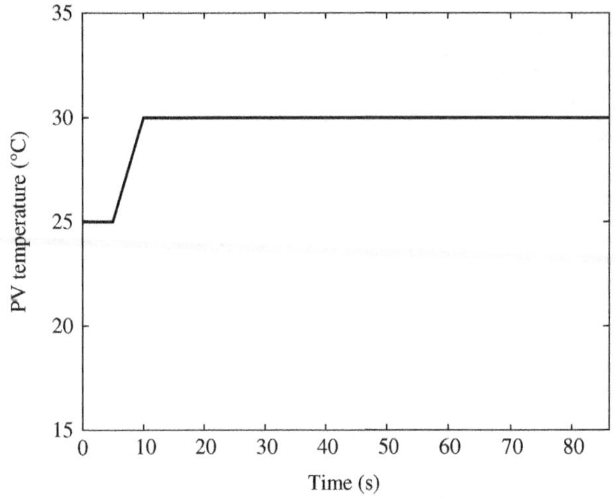

Figure 17.14 PV temperature.

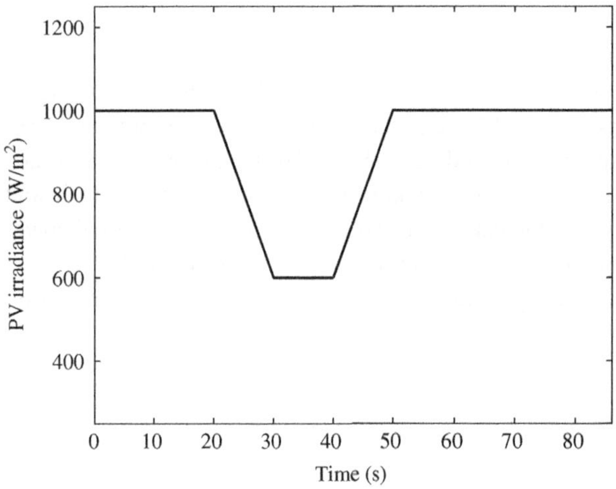

Figure 17.15 PV irradiance.

The EV system is modeled as a charging load regulated at 480-VDC. It consists of a buck-boost converter, controlled using a proportional-integral (PI) controller. The EV-load is varied between 0.5 and 1-MW to test different sizing options. Figure 17.16 shows the elements of the EV system, as well as its controller.

The battery system is modeled with a resistor and 650-VDC voltage source as its sizing needs to be determined based on the results of the simulation. Figure 17.17 shows the battery system.

The parameters of the different elements are all summarized in Table 17.1.

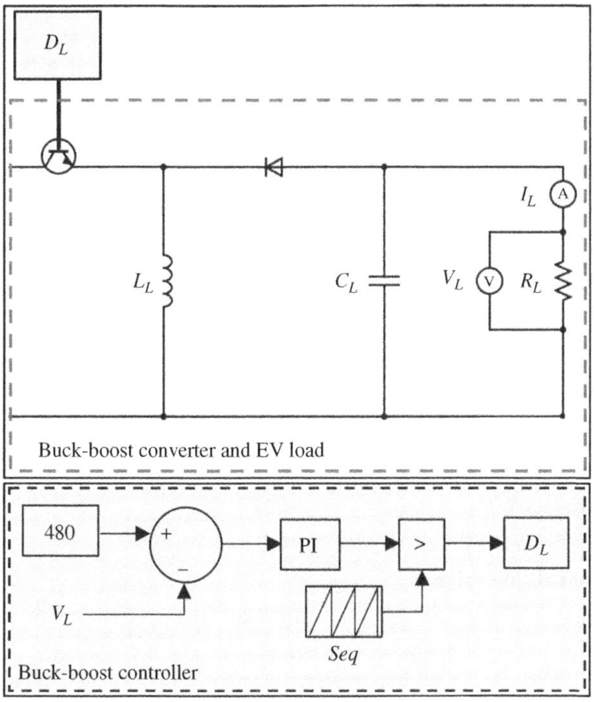

Figure 17.16 EV system.

Figure 17.17 Battery system.

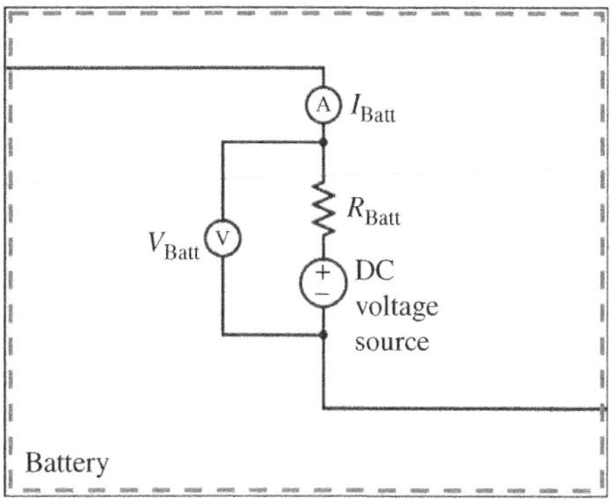

17.3.4 Results of DER Integration

Figure 17.18 shows the third rail voltage before and after integrating DERs into the subway network.

As visible in the figure, the third rail voltage no longer drops as much when the train accelerates and no longer rises as much when the train decelerates. This indicates that the train successfully draws its needed energy from the added elements (PV and battery), while the EV draws its needed energy from what is sent back by the train (along with the PV and battery).

Table 17.1 Parameters of system under study.

Symbol	Quantity	Value
C_F	Capacitive filter's capacitance	1 mF
R_{Rec}	Traction rectifier's internal resistance	1.08 mΩ
C_{PV1}	Boost converter capacitor 1's capacitance	43.2 mF
C_{PV2}	Boost converter capacitor 2's capacitance	43.2 mF
L_{PV}	Boost converter inductor's inductance	0.53 μH
R_{Batt}	Battery's internal resistance	5 mΩ
L_L	Buck-boost converter inductor's inductance	8 μH
C_L	Buck-boost converter capacitor's capacitance	1.2 F
R_L	EV load's equivalent resistance (1/2 MW)	0.46 Ω
R_L	EV load's equivalent resistance (3/4 MW)	0.31 Ω
R_L	EV load's equivalent resistance (1 MW)	0.23 Ω
K_p	Buck-boost PI controller's proportional gain value	0.001
K_i	Buck-boost PI controller's integral gain value	0.1
Seq	Repeating sequence's frequency	10 kHz
Seq	Repeating sequence's range	[0 1]

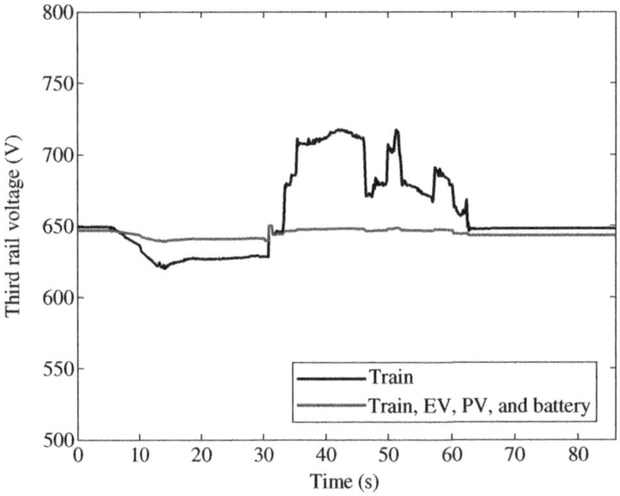

Figure 17.18 Third rail voltage comparison.

Figure 17.19 shows the EV voltage as it is successfully regulated at its desired value of 480-VDC.

Figure 17.20 shows the battery voltage, which drops as the train accelerates, and rises as the train decelerates. The voltage profile follows that of the train, and the battery even charges using the regenerative braking energy.

Figures 17.21 and 17.22 show the current and power profiles of the different elements, respectively, as well as of the neighboring substations.

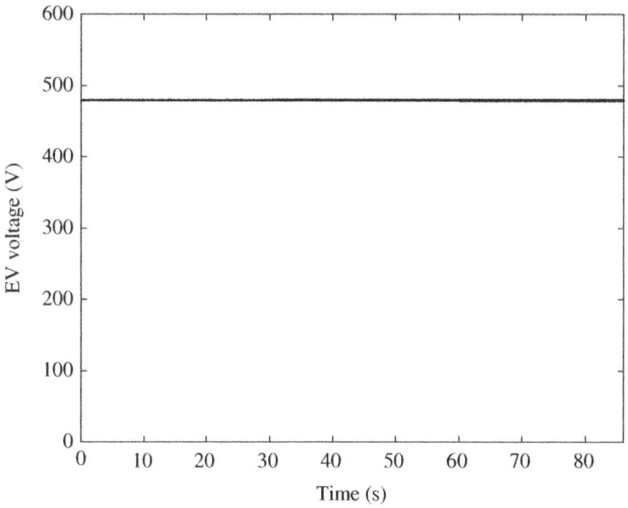

Figure 17.19 EV voltage.

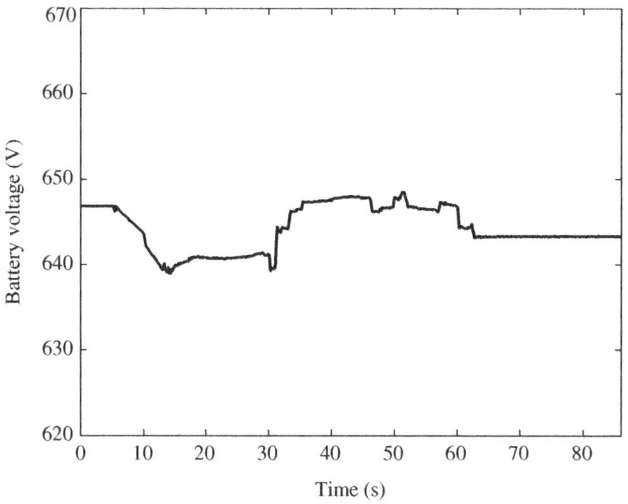

Figure 17.20 Battery voltage.

The PV's current and power profiles dip and rise as expected, with regard to the solar radiation and temperature changes, while the EV power remains constant at each stage. The train is successfully able to complete its trip as there is still a sufficient amount of power, as per charging demand. All elements work in sync and collectively stay at desired setpoints.

The biggest benefit of DER integration is seen at the substation level. Figure 17.23 shows a before-and-after power comparison between 58th St and 78th St substations, the two closest substations to the train and DERs.

58th St Substation is closest to where the DERs are located and a 50% drop is seen in its peak power output. The substation also successfully charges the EVs and battery toward the later

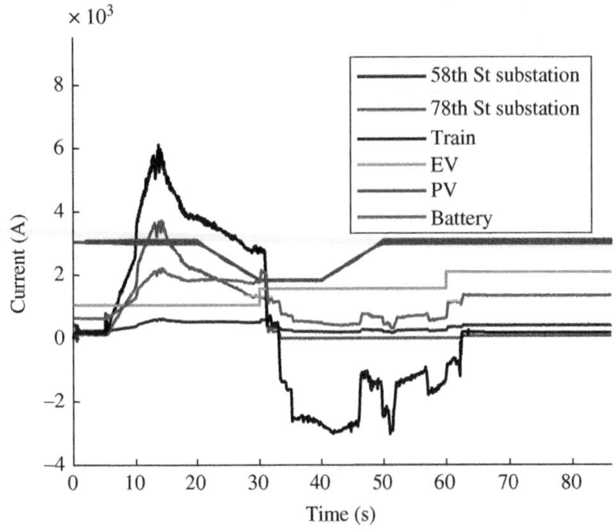

Figure 17.21 Substation current profiles (DERs connected).

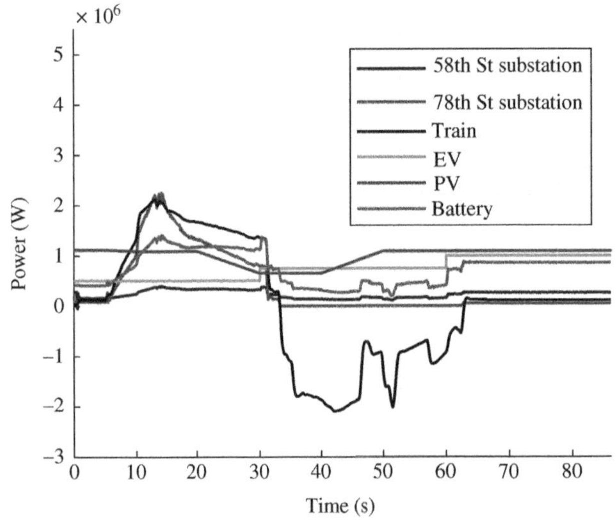

Figure 17.22 Substation power profiles (DERs connected).

part of the simulation when regenerative braking energy is no longer available (train has stopped). This presents a huge opportunity for savings, especially when expanded to a 24-hour scale.

Figure 17.24 shows the average hourly power consumption for 58th St Substation for the year 2018, and what a 50% reduction in peak power output could look like on a 24-hour scale.

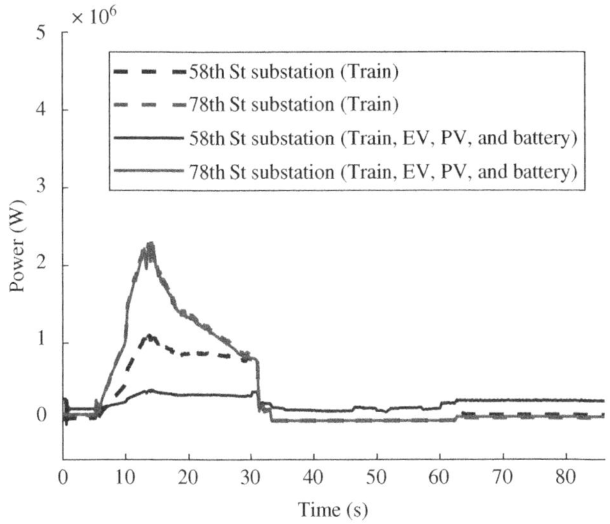

Figure 17.23 Substation power comparison.

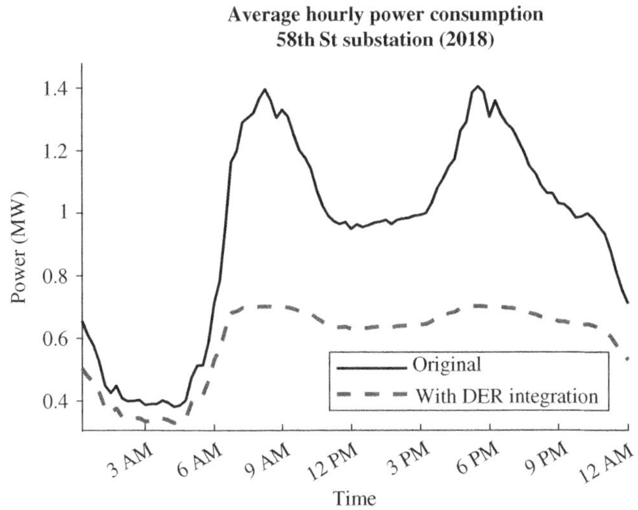

Figure 17.24 Average hourly power consumption; 58th St Substation (2018).

17.4 Conclusion

The results of the simulation prove the benefits of DER integration. Power consumption is significantly reduced from the substation level and on-demand EV charging is successfully available using pre-existing infrastructure. For a system, such as that of NYC, the costs of infrastructure upgrades can far exceed the possible savings brought forward through future plans for greenhouse gas emission reduction. Through wayside mounting, pre-existing equipment and structures can be used, with minimal changes done to the overall system. Third rail tracks expand through tunnels, above structures, under subway platforms, etc. Hence, strategic locations can be planned out when deciding where to place DERs. EV chargers can be located in nearby facilities, such as schools, bus depots, etc., and minimal cabling routed for ease of connection. For a majority of the subway lines, and especially for the 7-line, most substations are located outside. Hence, their rooftops, as well as those of nearby buildings, can be used to mount solar equipment. With the high-rise skyscrapers of NYC, minimal shade would cover any solar equipment, and within a community microgrid setting, multiple users can utilize the benefits of renewable energy.

Reference

1 Ahmad, R., Mohamed, A.A.A., Rezk, H., and Al-Dhaifallah, M. (2022). DC energy hubs for integration of community DERs, EVs, and subway systems. *Sustainability* 14: 1558. https://doi.org/10.3390/su14031558.

18

Challenges and State of the Art in the Agricultural Machinery Electrification

Luigi Alberti[1] *and Michele Mattetti*[2]

[1] *University of Padova, Department of Industrial Engineering, via Gradenigo 6/A, 35131 Padova, Italy*
[2] *University of Bologna, Department of Agricultural and Food Sciences, via Fanin 50, 40126 Bologna, Italy*

18.1 Introduction

The agriculture and forestry sectors are among the main responsible for global greenhouse gas emissions, with a contribution of more than 20% of carbon dioxide equivalent, 42% of methane, and 75% of nitrogen oxides (NOx).

Moreover, farm-gate emissions are expected to grow in the next decades [1]. Among the various agricultural activities, a considerable amount of emissions comes from internal combustion engines (ICEs), which are actually the most widespread power sources in the agriculture industry. Among them, diesel engines are the most common worldwide, both for moving self-propelled machinery, like tractors, harvesters and combines, and for stationary stand-alone power units. Diesel engine exhaust emissions are carcinogenic to humans [2], which makes them a hazard not only for the environment if not also for the operators of these machines. Besides, particulate matter (PM) produced by ICEs can deplete soil and damage some types of crops [3], causing a reduction in crop production and food quality in a long-term perspective [4].

Exhaust gas emissions are particularly critical when diesel engines are operated outside a certain range [5]. The maximum power region at high speed is one of the worst conditions in terms of exhaust emissions, but unfortunately, it also represents a typical working point in agriculture operations [6]. Up to 5% reduction in fuel consumption can be obtained by using an improved ICE control, with a benefit also on the total emissions [7]. Moreover, due to tractors' usual working cycles, idling condition has a relevant contribution in terms of environmental impact and engine life [8, 9]. So, idling reduction strategies and idling stop devices are very important in this application [10].

The number of agricultural tractors and machines have increased in the last years and continuous growth is expected in different regions around the world [11, 12]. Therefore, similar increases in terms of fuel consumption and emissions are expected. For these reasons, several treaties and agreements have been signed worldwide to impose new tighter emissions limits for ICEs of Non-Road Mobile Machinery (NRMM) category, to which agricultural vehicles belong. Among these new standards, there is the Stage V European regulation, which is particularly strict for machinery with diesel engines above 56 kW [13]. Other examples are Tiers 3–4 emission standards adopted by the Environmental Protection Agency (EPA) in the United States, which also introduced substantial

Transportation Electrification: Breakthroughs in Electrified Vehicles, Aircraft, Rolling Stock, and Watercraft,
First Edition. Edited by Ahmed A. Mohamed, Ahmad Arshan Khan, Ahmed T. Elsayed, and Mohamed A. Elshaer.

reductions in NOx and PM for diesel engines above 56 kW [14]. In order to meet these new limits on emissions, manufacturers are forced to equip new generation engines with selective catalytic reducers, diesel exhaust gas treatment, fluid tanks and particulate filters. Such components, in addition to an increased cost, make the diesel units bulkier, causing a reduction in the power density. So, whereas this is not a major concern for high-power row crop vehicles, the design of narrowly specialized tractors could become more challenging, due to strict size constraints on the vehicle chassis.

All these reasons encourage the manufacturers to introduce modifications in the current powertrains of agricultural machinery and to push forward industrial and academic research on this topic. Among various proposals, one feasible solution is the electrification of conventional drivetrains, following the trend in the automotive industry toward the development of hybrid electric and full-electric on-road vehicles. The introduction of electric drives in agricultural machinery can have many other relevant advantages in addition to fuel savings, emission reduction, and significant improvement in tank-to-wheel efficiency. The operating costs of the tractor could be substantially reduced not only because of lower fuel consumption but also thanks to reduced maintenance. Moreover, a further decrease in operating costs could be achieved with better economic exploitation of in-site renewable energy sources, like photovoltaic or bio-gas power plants, that could be used to recharge the on-board batteries of electrified tractors. The safety and driveability of the vehicle could be highly improved thanks to the possibility to use a lower number of gears or to develop drive-by-wire systems and automatic controls of the vehicle stability in a more effective way. Furthermore, electric drives could improve the performances of tractors and allow new functionalities, thus improving crop production. This is particularly true when they are implemented for autonomous and precision farming purposes [15]. Indeed, electric actuators are more compatible with automatic systems than their hydraulic counterparts, due to their ease of control and accuracy. Despite the earlier advantages, several challenges still need to be addressed in the electrification of agricultural machinery.

This chapter introduces the peculiarities of agricultural machinery and describes the main component of traditional machine configurations. After the description of the conventional powertrain of farming tractors and highlighting the most relevant challenges for electrification, a description of the electrically driven ICE auxiliaries is reported. A description of the various powertrain electrification architectures and the various components for tractor electrification is fully addressed in the next chapter. A portrait of state-of-the-art technology is given by looking both at various studies presented in literature and at prototypes and commercial models developed by manufacturers.

18.2 Conventional Powertrain and Electrification Challenges

The most widely adopted structure in conventional powertrain for mechanical front-wheel drive (MFWD) tractors is shown in Figure 18.1. The diesel ICE is the only source of power and it is adopted to supply all the loads through mechanical transmissions. Through the gearbox and differential, the power flows on the rear axle, i.e., to the wheels. Such a configuration allows a 4-wheel drive (4WD) when high traction efforts are required.

As in conventional cars, thanks to a clutch it is possible to disengage the ICE to allow gear change. Differently from on-road vehicles, however, besides the traction load a large amount of power is requested by external-connected or self-equipped implements that are required to operate particular tasks (e.g., power arrows, seeders, reapers, tillers, atomizers, etc.). At least a mechanical power

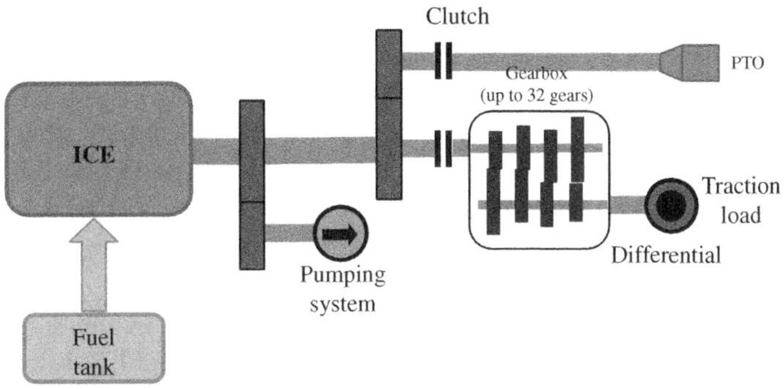

Figure 18.1 Traditional powertrain. Mechanical front-wheel drive tractor.

take-off (PTO) and a hydraulic power supply, both typically on the rear side of the vehicle, are available for external implements. The PTO is connected to the ICE shaft through a speed reducer and the hydraulic power is available in the form of pressurized oil for hydraulic actuators such as lifting mechanisms. An additional difference with respect to on-road vehicles regards also a significant diversity in the common power requirements. Farming vehicles usually operate in off-road dragging on non-compact soil trailers or particular ploughing tools. Therefore, they have very different traction effort and steering/braking power demands in comparison with cars or trucks.

With respect to other NRMM, agricultural machinery exhibit peculiarities that make their electrification very challenging. At first, they cover a really wide power range from a few tens of kW for specialized or family farming tractors up to 300 kW for open field tractors. Then, different arrangements characterize the conventional powertrain with different hydraulic systems and mechanical transmissions. Moreover, additional PTO in the front of the vehicle can be equipped. Furthermore, tractors are very versatile vehicles performing a great variety of duties and operations, which demand different power flows and large variability in load levels. For this reason, a rather complex gearbox with a huge number of gears is typically required to adapt the ICE characteristic to these highly varying operating conditions. In some cases, hydromechanic continuously variable transmissions (CVT) are used to reduce operator workload and improve vehicle functionality. This allows decoupling engine and vehicle speed [16].

Another important issue in agricultural machinery electrification is the identification of standard driving cycles. Due to the versatility of tractors, its definition is not as easy as for cars or other on-road vehicles.

Some regulations on emission limits for diesel engines for NRMM have been defined by EPA in cooperation with European Union. Such a non-road transient cycle (NRTC) standard is useful to test ICE and evaluate their conformity [17, 18]. Nevertheless, considering tractors, the operating cycle is defined only for heavy-duty row crop vehicles, and its torque and speed profiles are referred to as the ICE shaft. Therefore, information about the contribution of different types of loads is not given. For this reason, NRTC is not fully representative of a tractor working cycle. A step forward has been made by the German Agriculture Society with the definition of 14 working cycles for different operations, grouped in the DLG-PowerMix.

The lack of standard driving cycles is probably one of the major challenges in agricultural machinery electrification: it is not straightforward to define proper specifications for the design of electric components such as motors, drives, energy storage systems (ESSs), and so on. For these reasons, recent studies presented in literature take advantage of in-field measurements performed on specific tractors [19–22].

18.3 Electrification of Auxiliaries

Besides the main powertrain, also ICE auxiliaries can be electrified in order to improve their consumption reducing parasitic losses. When electrified, the auxiliaries can be controlled more effectively operating, for example, with a speed controlled independently from the ICE speed. Such an easier and less constrained control, when well exploited, can compensate loss arising from the additional power conversion. In addition, the speed decoupling of each auxiliary from the crankshaft speed enables new functionalities.

The two most demanding auxiliary loads are the HVAC (heating, ventilation and air-conditioning) compressor and the cooling fan: both are responsible for about 10% and 30% of ICE-rated power during onfield and idling operations, respectively [23]. An independently controlled fan speed enables higher airflow when the tractor operates with a high load at low speed improving cooling. Similarly, a lower airflow can be obtained when ICE operates in low-load high-speed conditions, reducing fuel consumption. As an additional feature, fan rotation can be easily reversed to clean the radiator.

A better control allows setting higher reference temperature, improving the efficiency. A further advantage provided by the adoption of an electric pump for the cooling system is the possibility to run it after engine power off, avoiding temperature hot-spots both in the turbocharger and in the engine block [24, 25].

To better clarify the benefits of auxiliaries' electrification, some examples from the literature are reported hereafter, although many are not in the agricultural sector. Nevertheless, they could be easily implemented on tractors.

In [26], an advanced cooling system on a heavy-duty truck has been considered. The vehicle is equipped with a 20 kW electric motor driving the fan, and an electric pump for the coolant circuit. Also the coolant bypass valve and the heat exchanger have been properly redesigned. Thanks to these modifications, a fuel consumption reduction of 2.7% during typical steady-state conditions has been found. Moreover, the ambient capability of the truck has been increased by more than 6 °C.

In [27], various cooling system configurations were analyzed for a heavy-duty diesel vehicle and a micro-hybrid truck:

1) A full-electric configuration with several electric fans, an electric pump, and a controllable bypass valve;
2) A configuration with electric fans but with a mechanical pump and a wax valve;
3) A system with a single mechanical fan and a mechanical pump, but with a controllable valve.

Thanks to electrification, in both cases it is possible to achieve a fuel saving of about 5%.

In [28], the performance of an electric fan on a 31 kW 2WD agricultural tractor was evaluated. In this specific example, the fuel saving varies between 2.2% and 8.1% depending on the operation. At the same time, an increase in the PTO torque of 4% has been measured.

Among the tractor manufacturers that proposed auxiliaries' electrification, John Deer launched in 2007 the 7430 and 7530 E-Premium tractors. Instead of a standard alternator, they were equipped with an enlarged 20 kW electric generator (EG) connected to the ICE flywheel, able to supply several auxiliaries driven by electric motors, such as fan, brake compressor, coolant pump, and air-conditioning compressor. The system was designed so as to allow for maximum cabin cooling even with ICE at idling conditions. In the 7530 E-Premium, a power socket was also included to supply portable working equipment, such as drilling, welding, and other tools. Such a power socket was the first example of electrical PTO (e-PTO) implemented on a tractor, at least in the authors'

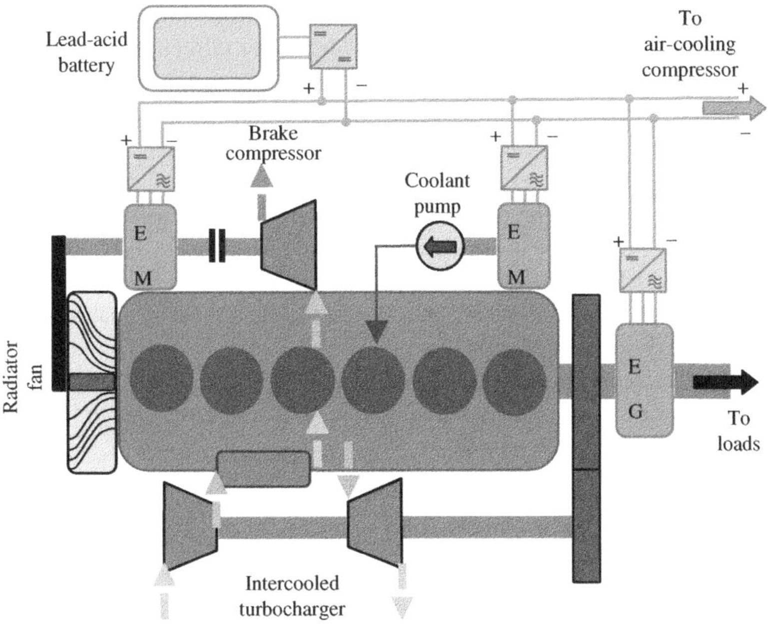

Figure 18.2 Outline of John Deere E-Premium diesel unit with electrified auxiliaries.

knowledge. Figure 18.2 shows the John Deere E-Premium platform with its electrified auxiliaries. According to a test conducted in [29] on the 7530 electrified and conventional variants, a 16% and 4% fuel consumption reduction was achieved, respectively in road transport and harrowing.

References

1 Tubiello, F. and Conchedda, G. (2021). Emissions due to agriculture global, regional and country trends. *FAO Food Nutr. Paper* 1: 03. https://www.fao.org/3/cb3808en/cb3808en.pdf.

2 IARC: Diesel engine exhaust carcinogenic (2012). https://www.iarc.who.int/wp-content/uploads/2018/07/pr213_E.pdf (accessed January 2021).

3 Rai, P.K. (2016). Impacts of particulate matter pollution on plants: Implications for environmental biomonitoring. *Ecotoxicol. Environ. Safety* 129: 120–136. https://doi.org/10.1016/j.ecoenv.2016.03.012.

4 Zhou, L., Chen, X., and Tian, X. (2018). The impact of fine particulate matter ($PM_{2.5}$) on China's agricultural production from 2001 to 2010. *J. Cleaner Prod.* 178: 133–141. https://doi.org/10.1016/j.jclepro.2017.12.204.

5 Imran, S., Emberson, D.R., Wen, D.S. et al. (2013). Performance and specific emissions contours of a diesel and RME fueled compression-ignition engine throughout its operating speed and power range. *Appl. Energy* 111: 771–777. https://doi.org/10.1016/j.apenergy.2013.04.040.

6 Janulevičius, A., Juostas, A., and Čiplienė, A. (2016). Estimation of carbon-oxide emissions of tractors during operation and correlation with the not-to-exceed zone. *Biosyst. Eng.* 147: 117–129. https://doi.org/10.1016/j.biosystemseng.2016.04.009.

7 Juostas, A. and Janulevičius, A. (2009). Evaluating working quality of tractors by their harmful impact on the environment. *J. Environ. Eng. Landsc. Manag.* 17 (2): 106–113. https://doi.org/10.3846/1648-6897.2009.17.106-113.

8 Molari, G., Mattetti, M., Lenzini, N., and Fiorati, S. (2019). An updated methodology to analyse the idling of agricultural tractors. *Biosyst. Eng.* 187: 160–170. https://doi.org/10.1016/j.biosystemseng.2019.09.001.

9 Perozzi, D., Mattetti, M., Molari, G., and Sereni, E. (2016). Methodology to analyse farm tractor idling time. *Biosyst. Eng.* 148: 81–89. https://doi.org/10.1016/j.biosystemseng.2016.05.007.

10 Rahman, S.M.A., Masjuki, H.H., Kalam, M.A. et al. (2013). Impact of idling on fuel consumption and exhaust emissions and available idle-reduction technologies for diesel vehicles–A review. *Energy Convers. Manag.* 74: 171–182. https://doi.org/10.1016/j.enconman.2013.05.019.

11 Agricultural tractor market size, share and trends analysis report by engine power, by region and Segment Forecasts, 2020–2027. (2020). https://www.grandviewresearch.com/industry-analysis/agricultural-tractors-market. (accessed January 2021).

12 Dyer, J.A. and Desjardins, R.L. (2006). Carbon dioxide emissions associated with the manufacturing of tractors and farm machinery in Canada. *Biosyst. Eng.* 93 (1): 107–118. https://doi.org/10.1016/j.biosystemseng.2005.09.011.

13 EU: Nonroad diesel engines. (2017). https://dieselnet.com/standards/eu/nonroad.php (accessed January 2021).

14 United States: Nonroad diesel engines. (2016). https://dieselnet.com/standards/us/nonroad.php (accessed January 2021).

15 Gonzalez-de-Soto, M., Emmi, L., Benavides, C. et al. (2016). Reducing air pollution with hybrid-powered robotic tractors for precision agriculture. *Biosyst. Eng.* 143: 79–94. ISSN 1537-5110. https://doi.org/10.1016/j.biosystemseng.2016. 01.008.

16 Renius, K.T. (2019). *Fundamentals of Tractor Design*. Springer https://doi.org/10.1007/978-3-030-32804-7.

17 EPA nonregulatory nonroad duty cycles. (2016). https://www.epa.gov/moves/epa-nonregulatory-nonroad-duty-cycles (accessed January 2021).

18 Nonroad transient cycle (NRTC). (2016). https://dieselnet.com/standards/cycles/nrtc.php (accessed January 2021).

19 Beligoj, M., Scolaro E., Alberti, L. et al. (2022). Feasibility evaluation of hybrid electric agricultural tractors based on life cycle cost analysis. *IEEE Access* 10: 28853–28867. doi: https://doi.org/10.1109/access.2022.3157635.

20 Scolaro, E., Beligoj, M., Estevez, M.P. et al. (2021). Electrification of agricultural machinery: a review. *IEEE Access* 9: 164520–164541. doi: https://doi.org/10.1109/ACCESS.2021.3135037.

21 Troncon, D., Alberti, L., and Matteti, M. (2019). A feasibility study for agriculture tractors electrification: duty cycles simulation and consumption comparison. *IEEE Transportation Electrification Conference and Expo (ITEC)*. https://doi.org/10.1109/itec.2019.8790502.

22 Troncon, D. and Alberti, L. (2020). Case of study of the electrification of a tractor: electric motor performance requirements and design. *Energies*. doi: https://doi.org/10.3390/en13092197.

23 Saetti, M., Matteti, M., Varani, M. et al. (2021). On the power demands of accessories on an agricultural tractor. *Biosyst. Eng.* 206: 109–122. ISSN 1537-5110. https://doi.org/10.1016/j.biosystemseng.2021.03.015.

24 Lin, H.-C., Chang, Y.-T., Tsai, G.-L. et al. (2014). Oil coking prevention using electric water pump for turbo-charge spark-ignition engines. *Math. Prob. Eng.* https://doi.org/10.1155/2014/498624.

25 Ribeiro, G.E., de Andrade Filho, A.P., and de Carvalho Meira, J.L. (2007). *Electric Water Pump for Engine Cooling*. SAE Technical Paper. SAE International. doi: https://doi.org/10.4271/2007-01-2785.

26 Slone, L. and Birkel, J. (2007). *Advanced Electric Systems and Aerodynamics for Efficiency Improvements in Heavy Duty Trucks*. Technical report. Caterpillar Inc.

27 Staunton, N., Pickert, V., and Maughan, R. (2008). Assessment of advanced thermal management systems for micro-hybrid trucks and heavy duty diesel vehicles. *2008 IEEE Vehicle Power and Propulsion Conference*. https://doi.org/10.1109/VPPC.2008.4677464.

28 Babu, R.M., Manikandan, S., and Nageshwara, R.P. (2019). *Electrical Operated Fan for Cooling System on Agricultural Tractors. SAE Technical Paper*. SAE International. https://doi.org/10.4271/2019-26-0079.

29 Pessina, D. and Facchinetti, D. (2009). Gemelli diversi. *Macchine Agricole Luglio*.

19

Electrification of Agricultural Machinery: Main Solutions and Components

Luigi Alberti[1] *and Diego Troncon*[2]

[1] *University of Padova, Department of Industrial Engineering, via Gradenigo 6/A, 35131 Padova, Italy*
[2] *CNH Industrial, Electrification System Integration, Viale delle Nazioni 55, 41122 Modena, Italy*

19.1 Powertrain Electrification

The adoption of electric drives in the drivetrains, which are functional for the main load of the vehicle is referred to as powertrain electrification. For agricultural tractors, a large amount of power demand comes from implements and so the main load is not only the traction effort as in other vehicles. As a result, also the drivetrains that deliver power to mechanical power take-off (PTO), hydraulic actuators, and hydraulic remotes are significant. On the contrary, the driving systems, such as steering and braking actuators, auxiliaries, the cabin air-conditioning and all the other systems that are functional to the driver comfort and the vehicle driveability are not considered as main loads, even though some of them could have a relevant power demand as described in the previous chapter.

Figure 19.1 shows the different electrification strategies that can be adopted for the electrification of a tractor. The main solutions result in a diesel-electric, hybrid-electric, or full-electric drivetrain. The various solutions are hereafter discussed.

19.1.1 Diesel-Electric and Hybrid-Electric Powertrains

When a diesel-electric drivetrain is adopted, all the power comes from an engine-driven generator. In hybrid-electric drivetrains, a second power source is present, which is able to store electrical energy (ESS: energy storage system) [1]. In agricultural tractors, the engine is usually fuelled with diesel, but some bio-fuel blends have been proposed to reduce emissions [2]. The main difference between a diesel-electric and a hybrid-electric powertrain consists in the presence of a bidirectional ESS that provides a significant amount of power, and is functional for the operation of the drivetrain. Hybrid electric and diesel-electric powertrains can be classified according to the power flow topology. Three fundamental architectures can be identified: series, parallel and their combination series-parallel. Series and series-parallel architectures can be diesel-electric or hybrid-electric thanks to the presence of an electric machine (EM) working mainly as a generator (EG) and mechanically connected to the diesel engine. Parallel architectures, instead, can be only hybrid electric because, in this architecture, an ESS is fundamental (typically a battery pack). When

Transportation Electrification: Breakthroughs in Electrified Vehicles, Aircraft, Rolling Stock, and Watercraft,
First Edition. Edited by Ahmed A. Mohamed, Ahmad Arshan Khan, Ahmed T. Elsayed, and Mohamed A. Elshaer.

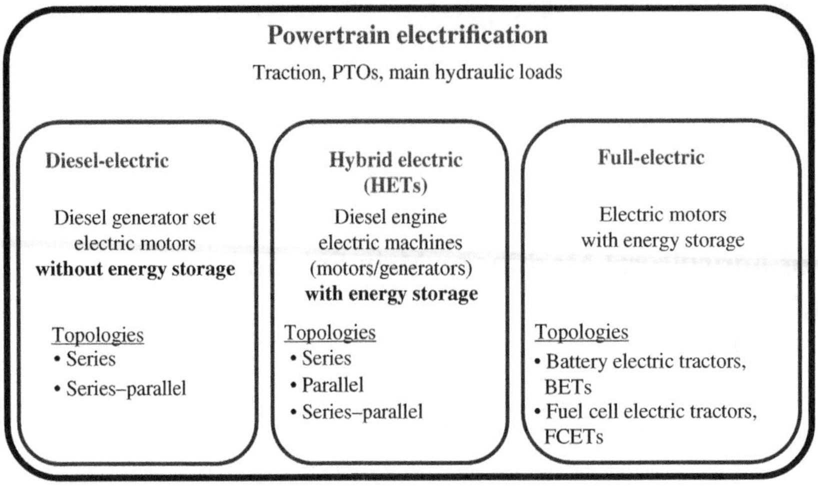

Figure 19.1 Classification of tractor drivetrain electrification.

the ESS can be recharged through an external power supply, the hybrid electric tractors (HETs) are defined as "plug-in" (PHETs: plug-in hybrid electric tractors).

Hybrid electric powertrains can be classified considering the so-called degree of hybridization (DoH), or hybridization factor H. The conventional DoH, widely approved for on-road vehicles, is defined as in (19.1) [3]:

$$H = \frac{P_{\mathrm{EM_T}}}{P_{\mathrm{EM_T}} + P_{\mathrm{ICE}}} \tag{19.1}$$

where $P_{\mathrm{EM_T}}$ is the rated power of the traction electric motor (EM) and P_{ICE} is the rated power of the combustion engine. The earlier presented definition, however, is not representative of powertrain electrification in the case of agricultural machinery where in addition to the traction drive, the drivetrains related to mechanical PTOs, hydraulic PTOs, and hydraulic actuators can be electrified as well. A different hybridization factor specifically defined for hybrid electric and diesel-electric non-road mobile machinery (NRMM) (H_{NRMM}) has been proposed in [4], which takes into account the electrification of those drivetrains that are functional to working tasks. In particular, the total hybridization factor H_{NRMM} is computed as the mean between two degrees of hybridization: the conventional one for the traction drive, H_{T}, and a second one for the working load drivetrains, H_{L}. This is expressed in (19.2):

$$\begin{cases} H_{\mathrm{T}} = \dfrac{P_{\mathrm{EM_T}}}{P_{\mathrm{EM_T}} + P_{\mathrm{ICE}}} \\ \\ H_{\mathrm{L}} = \dfrac{P_{\mathrm{EM_L}}}{P_{\mathrm{EM_L}} + P_{\mathrm{ICE}}} \end{cases} \Rightarrow \quad H_{\mathrm{NRMM}} = \frac{H_{\mathrm{T}} + H_{\mathrm{L}}}{2} \tag{19.2}$$

where $P_{\mathrm{EM_L}}$ is the total rated power of the electric drives on the working load drivetrains.

19.1.1.1 Series Architectures

The series configuration of diesel-electric and hybrid-electric drivetrains is shown in Figures 19.2 and 19.3, respectively. An electric generator is coupled to the ICEt, resulting in a diesel generator set

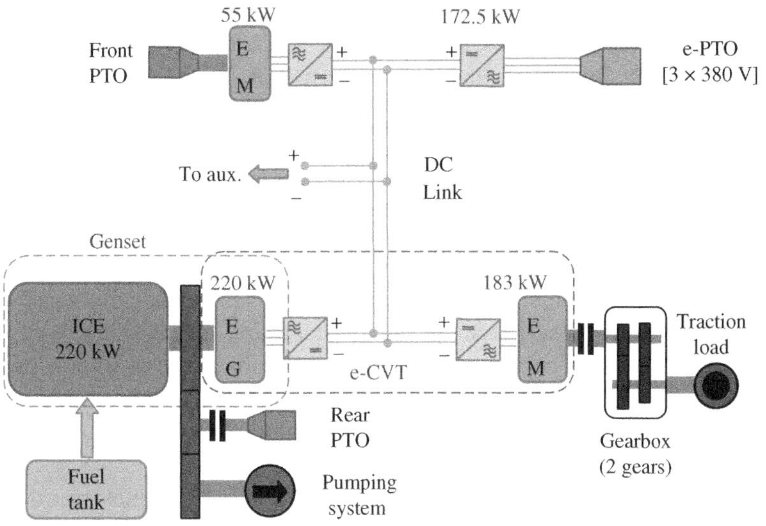

Figure 19.2 Series diesel-electric powertrain.

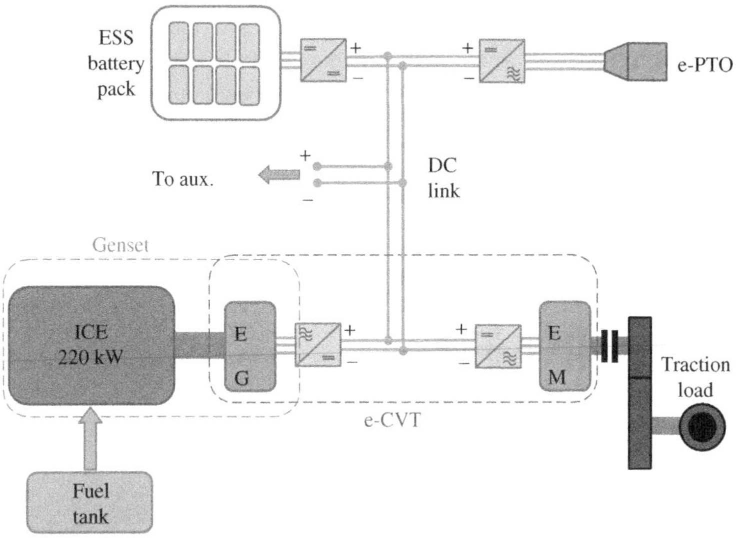

Figure 19.3 Series hybrid-electric powertrain.

(GENSET). The GENSET supplies all the power converters of the EMs through a DC bus. The coupling of a generator with an EM forms an electromechanical continuously variable transmission (e-CVT). In fact, the speed of the final drive can be adjusted continuously and its control is independent of the ICE speed.

With a series architecture, the ICE is completely decoupled from load. Therefore, the ICE can be operated always on its lowest brake-specific fuel consumption (BSFC) trajectory. As an additional advantage in fuel economy and for increasing performance, when adopted for tractors, such configuration allows to decouple all the loads from each other. As an example, the wheel speed could be independent of the PTO shaft speed, which affects the implementation performance.

The major disadvantage in the series architecture concerns the weight, the volume, and the cost of the main components, especially regarding EMs, which must be sized according to the entire power demand of each electrified drivetrain. In the series hybrid electric structure of Figure 19.3, an e-PTO is also included.

19.1.1.2 Parallel Architectures

Figure 19.4 shows a hybrid electric parallel architecture. In this case, an electric ESS is necessary. The EM is mechanically connected to the engine, and then to the final drive via the gearbox. With this architecture, the EM works mainly as a motor boosting the ICE during high-power duty operation. Operation as a generator can be adopted to recharge the ESS, for example, during low power operation in order to improve the overall energy management strategy. In the layout shown in Figure 19.4, a clutch is inserted between the ICE and the EM. In this way, it is also possible to decouple the two motors for a full-electric operation. In this case, however, limited power is available for the operation.

This is the easiest configuration to implement for the electrification of an existing vehicle because only one EM is needed and the ICE can be significantly downsized thanks to the possibility of torque boosting. Moreover, it can be realized without a complete redesign of the vehicle structure. On the other hand, battery recharging through ICE surplus power is very inefficient, and it should be avoided as much as possible. Then, the ESS should be able to store a large amount of energy to guarantee sufficient autonomy, i.e., Li-ion modules are usually chosen. The main drawback is due to the connection of the ICE with the load, which limits the possibility to operate the engine always at its lowest BSFC point. Nevertheless, it is possible to implement a load point shifting control strategy for the ICE, which tries to maintain as long as possible the highest efficient working point for each speed value. The electrification of a compact agricultural tractor for orchards and

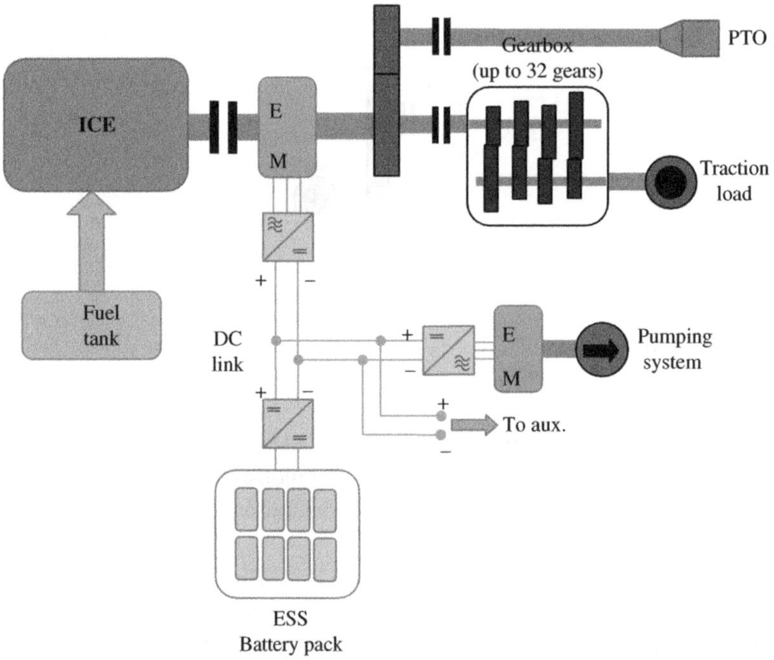

Figure 19.4 Parallel hybrid-electric powertrain.

vineyards has been presented in [5]. Figure 19.4 shows the parallel powertrain considered in the research. An existing traditional vehicle was converted to a parallel hybrid architecture. In particular, the original diesel engine of 77 kW (100 Hp) has been downsized to a 55 kW unit. In the saved space, a permanent magnet EM has been inserted so that the vehicle wheelbase was not increased. As far as the ESS system is concerned, a 25 kWh $LiFePo_4$ battery pack was considered suitable for the application. The feasibility of the hybrid tractor was assessed through simulated fuel consumption comparisons with its conventional counterpart, using different duty cycles identified from real working scenarios and in-field measurements [6]. From the simulations, the following remarks have been drawn:

- In hybrid mode, a consumption benefit is always achieved.
- A limited advantage is obtained for duty cycles with a large amount of power.
- The full-electric mode risks to be detrimental in heavy operating conditions due to the repeated energy switches and the high battery current.
- Significant benefits are obtained during cycles that need a low amount of energy (both in hybrid and full-electric mode).
- During low-power operations, a slow battery discharge allows to keep the engine off for a longer period.

These outcomes prove that through conversion of an existing platform, it is possible to develop an effective hybrid tractor with a relatively limited effort. Similar results are achieved also in other studies conducted on medium-sized tractors, which also consider alternative energy management strategies [7–9].

19.1.1.3 Series–Parallel Architectures

Series–parallel architectures try to mitigate the drawbacks of series and parallel structures. The basic idea is to avoid bulky EMs and double energy conversions, keeping at the same time the ICE speed decoupled from the load. This result can be achieved using planetary gear. Main differences between a planetary gear and conventional gear are that a planetary gear has coaxial arrangement of the shafts and load distribution to several planetary gears. The advantages of a planetary gear are higher volumetric power density and higher efficiency compared to conventional gear. Several configurations can be realized, depending on the position of the planetary gear in the transmission chain. Nevertheless, the resulting structure is more complex and two EMs, the motor and the generator are necessary.

In [10], the hybridization of a heavy-duty wheel loader has been considered. A similar topology could be implemented in tractors too. Figure 19.5 shows the proposed layout. The coupling shaft between the EG and the ICE is connected to the sun of the planetary gear, while the EM is linked to the ring shaft. The carrier shaft is connected to the wheels' axle differential. This powertrain exploits the benefits of a series architecture at low traction requirements and progressively takes advantage of the coupling of a parallel drivetrain as the power demand increases. The achieved results show that the ICE is always working close to the maximum efficiency point, while the battery SoC remains within the 40–60% range.

A different solution was proposed by [11] where the electrification of a 75 kW specialized tractor has been considered. Figure 19.6 shows the investigated e-CVT. The proposed layout is based on a coaxial and concentric arrangement of the electric machines connected to the planetary gear. In this way, a significant reduction of transmission displacement at the same power rate is achieved, with an increase in the e-CVT power density.

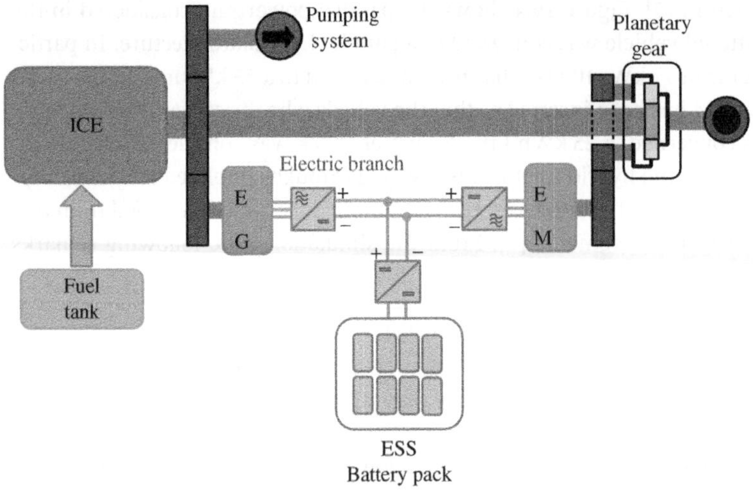

Figure 19.5 Input-coupled output-split series-parallel powertrain for a wheel loader.

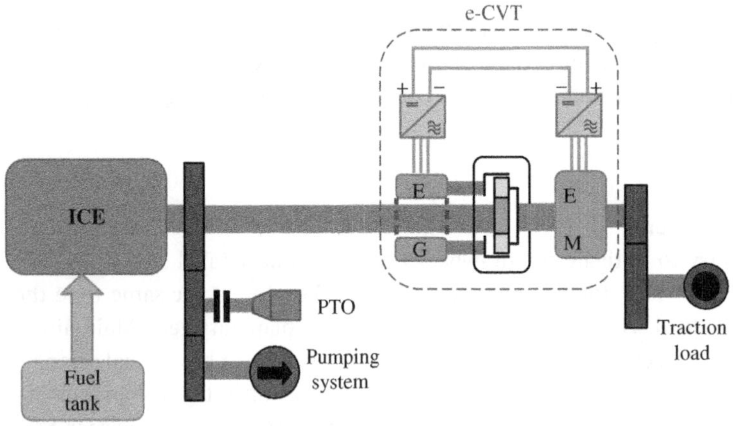

Figure 19.6 Power-split powertrain for a specialized tractor.

19.1.2 Full-Electric Powertrains

When the solely available power source is electric, a full-electric powertrain is achieved. In this case, the main possibilities are a rechargeable battery pack (BETs: Battery Electric Tractors), or a hydrogen-propelled powertrain (FCETs: Fuel Cell Electric Tractors). In the latter case, typically, a fuel cell stack is coupled with a battery or a supercapacitor bank to increase the performance.

19.1.3 Battery Electric Tractors (BETs)

The economical and technical feasibility of BETs has been presented in literature. In some cases, some prototypes have also been proposed on the market.

Figure 19.7 outlines the full-electric powertrain of a 9 kW prototype conceived and manufactured by [12]. The drivetrain is composed of two 4.5 kW induction motors (IMs) mounted on the rear wheels. Two independent inverters are used to supply motors so that the torque and speed of each wheel can be controlled independently. In this way, precise wheel-slip control can be achieved. As ESS, a lead-acid battery bank with four units was chosen.

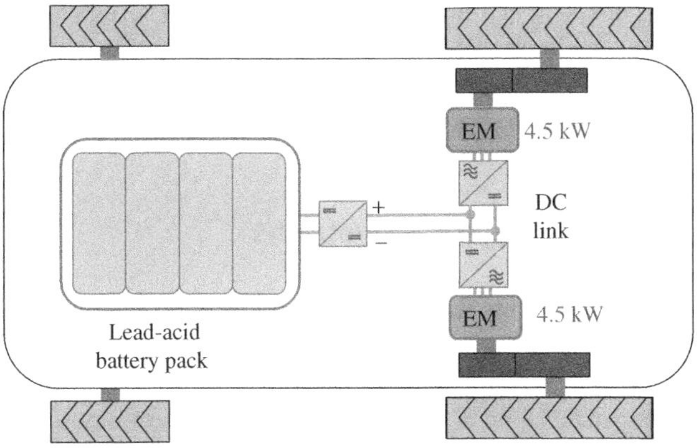

Figure 19.7 Full-electric powertrain for a small family farming tractor.

Two other examples are reported in [13, 14] where a single permanent magnet motor and a lead-acid battery were adopted. Authors report that the tractor is able to plough with a depth of about 5–6 cm at a constant speed of 5 km/h for more than 6 hours.

In [15], an innovative electrically driven tractor for family farming is proposed. It is a redesigned vehicle characterized by a four-wheel-drive chassis able to vary the track width while also changing the ground clearance. Many advantages are claimed by the proposers: a great adaptability thanks to the innovative mechanical layout; a fuel cost reduction of up to 81%; the possibility to make autonomous the machinery with a limited effort.

19.1.4 Fuel Cell Electric Tractors (FCETs)

The powertrain of New Holland NH2 fuel cell electric tractor 201 [16] presented at Agritechnica in 2011 is shown in Figure 19.8. The fuel cell stack is equipped with a proton exchange membrane

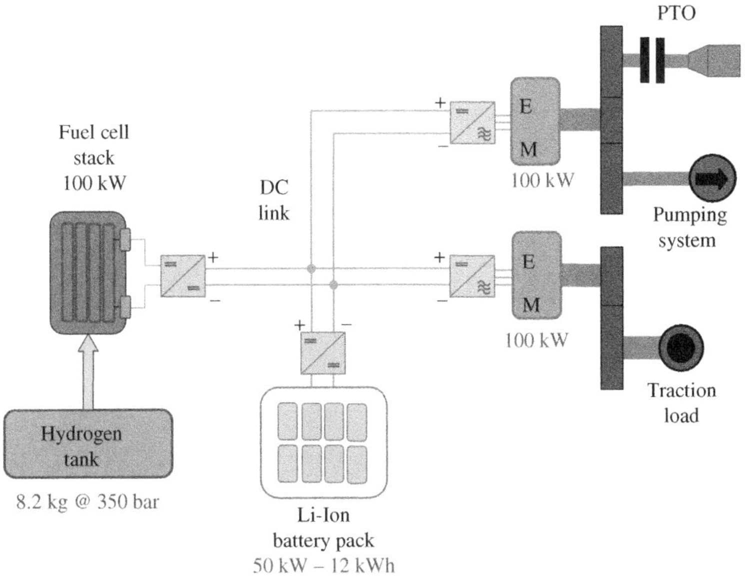

Figure 19.8 Full-electric powertrain of the hydrogen-propelled tractor, New Holland, NH2.

(PEM) rated 100 kW. The hydrogen tank is operated at a pressure of 350 bar holding 8.2 kg of gas. Two independent electric drives are adopted, one for the PTO and the hydraulic pump and another for the traction load. Besides the fuel cell, a 12 kWh Li-ion battery pack is adopted to improve the dynamic performances of the tractor, in particular, to preserve the fuel cell stack life.

19.2 Main Components for Tractors' Electric Drivetrains

The design and manufacturing of the main components for electric drivetrains of heavy-duty off-highway vehicles are under fast development. One of the main issues still open is the definition of proper design specifications, as well as a fair choice of device types and system settings. All this is also related to the complex system integration of this kind of vehicle. In the following sections, some proposals about the design and specifications of components for the electrification of agricultural tractors are summarized.

19.2.1 Electric Energy Storage Systems

Figure 19.9 shows the Ragone plot of the various possible ESS. Supercapacitors and capacitors have the highest specific power, while batteries are characterized by higher specific energy [17]. Among them, lead-acid modules are commonly implemented in low voltage systems, usually below 48 V, while Li-ion cells are commonly adopted in high voltage systems in automotive applications. Li-ion type is becoming the prominent battery technology in the next year thanks to its power-density and energy-density capability. Therefore, supercapacitors are typically used to balance high-varying transient power requirements and to recover more effectively kinetic energy. Battery packs, on the other hand, work as energy storage devices. To exploit the advantages of both technologies at the same time, batteries and supercapacitors can be combined in the same architecture.

As far as the charging technology is concerned, similar techniques adopted for on-road vehicles can be suitable for agricultural machinery too. As regards conductive wired charging, fast

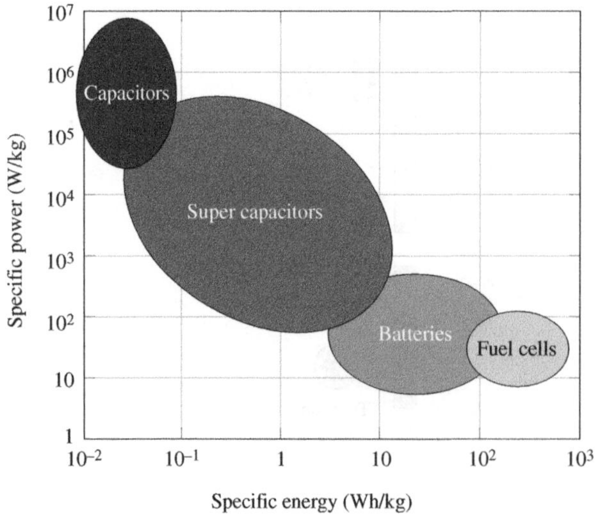

Figure 19.9 Ragone plot. Comparison between different power-generation devices and storage technologies widely used in electric and hybrid electric vehicles. Fuel cells are considered a typical storage device in vehicular applications.

high-voltage DC charging seems to be more convenient than low industrial voltage charging for agricultural machinery, not only because of the high charging power needed to supply large-capacity batteries in an acceptable working time but also thanks to some peculiarities of farmland. Indeed, high voltage levels could be available taking advantage of in-site renewable energy power plants, such as photovoltaic and biogas plants, and the DC/DC charger could be installed inside the farm instead of being on tractors.

19.2.2 Fuel Cells and Hydrogen Storage

Fuel cell powertrains seem to be more promising than batteries-powered drivetrains for off-road heavy vehicles' electrification thanks to their higher energy density, and to the faster refueling in comparison to battery charging. Nevertheless, in vehicle applications, supercapacitor banks or auxiliary batteries are mandatory to fulfill power peaks and fast transients.

19.2.3 Electric Machines

In HETs, the design of electric machines is particularly challenging due to the coexistence of high-power electric systems and diesel units in a limited space. This is particularly the case when the electrification of an existing platform is considered. This requirement is somewhat relaxed in a full-electric tractor where the redesign of the vehicle layout allows for a more flexible design.

In general, EMs for agricultural tractors must exhibit a high power density also to make electric drives more competitive with their hydraulic counterparts, which are widely used in tractors [18]. As a result, both the electric and magnetic loadings of the motor should be maximized to achieve high power densities. As a consequence, a higher thermal load must be dissipated and therefore liquid-cooled machines are typically adopted, for example, with an external water jacket.

A big challenge in the design of EM is the lack of precise duty cycles. This is due to the wide variety of tractor-operating conditions. The choice of proper torque requirements, for example, is not straightforward. Since, during various operating conditions, the machines are subjected to a changing torque demand, it could be a proper choice to fulfill them in overloading torque conditions. Nevertheless, in this case, the EM must be designed with a proper overloading capability in order to avoid any electromagnetic or thermal damage.

The choice of the electric machine type is not straightforward too. Various solutions have been proposed covering most of the state-of-the-art EM types for automotive applications. Generally, permanent magnet-synchronous motors exhibit the highest power densities. Nevertheless, vehicle manufacturers are greatly interested in the reduction of expensive rare-earth materials used in the magnets due to unreliable supply chains and price instability. Moreover, high-temperature class magnets should be used to realize the required overloading capability and particular attention must be given to avoiding their demagnetization. On the contrary, IMs do not suffer from demagnetization problems, and the thermal issues are limited only to slot liner insulation. So, they are cheaper, and they could exhibit better overloading capabilities even if their power density is lower. Moreover, IMs do not produce back-emf at the phase terminals when the rotor spins at no load, this brings some simplification in the machine integration. To mitigate the drawbacks of IMs, a synchronous motor with a wound rotor can be adopted. In this case, the performance is very close to that of a permanent magnet motor but without rare-earth materials. Nevertheless, a greater complexity due to the rotor winding which has to be supplied has to be considered in this case.

The choice of the machine type is strongly related to the system architecture and the load characteristics, especially on how the EM is mechanically connected to the load. When the EM is installed on the engine crankshaft, as the motor in parallel configuration of Figure 19.4 or the generator in series architecture of Figure 19.3, then high speed is not required (speed values not over

2600 rpm). Therefore, SPMs could be a proper choice. Also, when the EM is mechanically coupled through a variable speed transmission, especially in agriculture where the transmission covers a wide range of ratios, the constant power speed range is not required on the EM. Therefore, SPMs may be able to satisfy the load requirements. On the other hand, when the EM is mechanically coupled through speed reducers or gear trains with constant ratio or limited variation, as in power-split configurations of Figures 19.5 and 19.6 or in full-electric powertrains of Figures 19.7 and 19.8, a wide constant power speed range is needed. Thus, IPMs or IMs could be a better solution.

19.2.4 Power Converters

For the power electronics (PE) converters' choice, similar requirements already described for EMs apply. Nevertheless, some specific considerations can be made specifically for this drive component.

The PE must be sized considering the maximum current supplied at full load being its overloading capability typically very limited due to the shorter thermal time constant. Nevertheless, volume and weight must be minimized resulting in challenging requests in terms of efficiency and power density. Therefore, liquid cooling is typically adopted also for PE. Moreover, in hybrid configurations with a high level of integration, the same coolant for the engine and the electric drive can be adopted also for the PE.

References

1 Kebriaei, M., Niasar, A.H., and Asaei, B. (2015). Hybrid electric vehicles: an overview. *2015 International Conference on Connected Vehicles and Expo (ICCVE).* https://doi.org/10.1109/ICCVE.2015.84.

2 Lovarelli, D. and Bacenetti, J. (2019). Exhaust gases emissions from agricultural tractors: state of the art and future perspectives for machinery operators. *Biosyst. Eng.* 186: 204–213. https://doi.org/10.1016/j.biosystemseng.2019.07.011.

3 Govardhan, O.M. (2017). Fundamentals and classification of hybrid electric vehicles. *Int. J. Eng. Tech.* http://oaji.net/articles/2017/1992-1515159589.pdf.

4 Som, A. (2017). Trends and hybridization factor for heavy-duty working vehicles. In: *Hybrid Electric Vehicles*, Chapter 1 (ed. T. Donateo). IntechOpen. https://doi.org/10.5772/intechopen.68296.

5 Troncon, D., Alberti, L., Bolognani, S., et al. (2019). Electrification of agricultural machinery: a feasibility evaluation. *Fourteenth International Conference on Ecological Vehicles and Renewable Energies (EVER).* https://doi.org/10.1109/ever.2019.8813518.

6 Troncon, D., Alberti, L., and Matteti, M. (2019). A feasibility study for agriculture tractors electrification: duty cycles simulation and consumption comparison. *IEEE Transportation Electrification Conference and Expo (ITEC).* https://doi.org/10.1109/itec.2019.8790502.

7 Barthel, J., Gorges, D., Bell, M., and Munch, P. (2014). Energy management for hybrid electric tractors combining load point shifting, regeneration and boost. *2014 IEEE Vehicle Power and Propulsion Conference (VPPC).* https://doi.org/10.1109/VPPC.2014.7007061.

8 Bin, X., Hao, L., Zheng-He, S., and En-Rong, M. (2013). Powertrain system design of medium-sized hybrid electric tractor. *Inf. Technol. J.* https://doi.org/10.3923/itj.2013.7228.7233.

9 Mocera, F. and Som, A. (2020). Analysis of a parallel hybrid electric tractor for agricultural applications. *Energies.* https://doi.org/10.3390/en13123055.

10 Grammatico, S., Balluchi, A., and Cosoli, E. (2010). A series-parallel hybrid electric powertrain for industrial vehicles. *2010 IEEE Vehicle Power and Propulsion Conference.* https://doi.org/10.1109/VPPC.2010.5729045.

11 Rossi, C., Pontara, D., Falcomer, C. et al. (2021). A hybrid electric driveline for agricultural tractors based on an e-CVT power-split transmission. *Energies* 14 (21). https://doi.org/10.3390/en14216912https://www.mdpi.com/1996-1073/14/21/6912.

12 Melo, R.R., Antunes, F.L.M., Daher, S. et al. (2019). Conception of an electric propulsion system for a 9 kW electric tractor suitable for family farming. *IET Electr. Power Appl.* https://doi.org/10.1049/iet-epa.2019.0353.

13 Das, A., Jain, Y., Agrewale, M.R.B., et al. (2019). Design of a Concept Electric Mini Tractor. *2019 IEEE Transportation Electrification Conference (ITEC-India).* https://doi.org/10.1109/ITEC-India48457.2019.ITECINDIA2019-134.

14 Zhang, X. (2017). Design theory and performance analysis of electric tractor drive system. *Int. J. Eng. Res. Technol. (IJERT).* ISSN 2278-0181.

15 Gurusamy, S.K. and Devaradjane, G. (2015). Electrical tractive equipment design for small marginal farm mechanization. *2015 IEEE International Transportation Electrification Conference (ITEC).* https://doi.org/10.1109/ITEC-India.2015.7386863.

16 New Holland's NH2 fuel cell powered tractor to enter service. (2012). *Fuel Cells Bull.* 2012: 3–4. https://doi.org/10.1016/S1464-2859(12)70004-4.

17 Winter, M. and Brodd, R.J. (2004). What are batteries, fuel cells, and supercapacitors? *Chem. Rev.* https://doi.org/10.1021/cr020730k.

18 Scolaro, E., Alberti, L., and Barater, D. (2021). Electric drives for hybrid electric agricultural tractors. *2021 IEEE Workshop on Electrical Machines Design, Control and Diagnosis (WEMDCD)*, pp. 331–336. https://doi.org/10.1109/WEMDCD51469.2021.9425671.

20

Feasibility Evaluation of Hybrid Electric Agricultural Tractors Based on Life Cycle Cost Analysis

Luigi Alberti, Elia Scolaro, and Matteo Beligoj

University of Padova, Department of Industrial Engineering, via Gradenigo 6/A, 35131 Padova, Italy

This chapter presents a method to evaluate the economic feasibility of tractor powertrain electrification based on life cycle cost (LCC) analysis. For a parallel hybrid, the best combustion engine downsizing, among some discrete values, is evaluated. The methodology is applied to three case studies with different power levels and operating cycles: a 76 kW orchard tractor, a 175 kW row crop tractor with medium-duty use and a 210 kW row crop tractor with heavy-duty use. Fuel and electrical energy consumption are estimated through simulation. A range of powertrain component prices and fuel and electrical energy prices is taken into account in order to cover price uncertainty and show its effects. The results show that operating cost savings decrease when more power-intensive operations are performed. Considering a combination of system and energy prices deemed realistic by the authors, the operating cost savings, respectively for orchard, row crop medium-duty and row crop heavy-duty, are approximately 8%, 3%, and 0.5%, which result in 6%, 1%, and 0.1% LCC savings.

20.1 Introduction

In this chapter, a method to evaluate the economic feasibility of farming tractor powertrain electrification through a simplified life cycle cost (LCC) analysis is presented. This method, at the same time, is suitable for the determination of the best internal combustion engine (ICE) downsizing in the case of a parallel hybrid electric configuration. The proposed methodology is based on field measurements, due to the lack, at the time of writing, of standard load cycles that fully represent actual tractor operation. The LCC analysis is performed on three case studies, i.e. on three tractors with different power sizes and operating cycles, but it can be easily extended to other cases, for example, to different powertrain topologies.

A parallel hybrid architecture with a stepped transmission has been chosen, as it can be implemented without major changes to the overall tractor architecture, and as it offers a superior transmission efficiency compared to the series architecture. A Li-ion battery pack was selected due to its high energy density. More advanced hybrid battery-supercapacitor storage systems are also suitable for this application and their use could be beneficial as they exploit the advantages of both technologies. These hybrid storage systems, however, were not considered in this work as they would heavily increase the complexity of such a general analysis. The ICE was downsized in order to maintain the same capabilities of the base traditional tractor.

Transportation Electrification: Breakthroughs in Electrified Vehicles, Aircraft, Rolling Stock, and Watercraft, First Edition. Edited by Ahmed A. Mohamed, Ahmad Arshan Khan, Ahmed T. Elsayed, and Mohamed A. Elshaer.

20.2 Case Studies and Operating Cycles

In this chapter, three different case studies are considered. They are characterized by different power sizes and applications:

- Specialized orchard tractor with a rated power of 76 kW
- Row crop tractor, medium-duty use, with a rated power of 175 kW
- Row crop tractor, heavy-duty use, with a rated power of 210 kW

The simulations have been carried out by adopting specific working cycles developed on the basis of in-field measurements which have been specifically performed.

20.2.1 Orchard Tractor

For the specialized orchard tractor, the following operations were considered:

- Weeder, 14.3%, avg. pwr. = 35.6 kW
- Atomizer, 14.3%, avg. pwr. = 42.1 kW
- Grape harvester, 14.3%, avg. pwr. = 20.7 kW
- Plant-lifting plow, 28.6%, avg. pwr. = 11.0 kW
- Tying machine, 28.6%, avg. pwr. = 5.3 kW

Figure 20.1 shows an example of the ICE torque and speed profiles adopted for the LCC evaluation.

20.2.2 Row Crop Tractor's Medium-Duty Use

To properly represent the use of a row crop tractor under medium power duty, the following mix of operations is considered:

- Heavy plowing, 33.4%, avg. pwr. = 96.6 kW
- Medium plowing, 35.6%, avg. pwr. = 82.8 kW

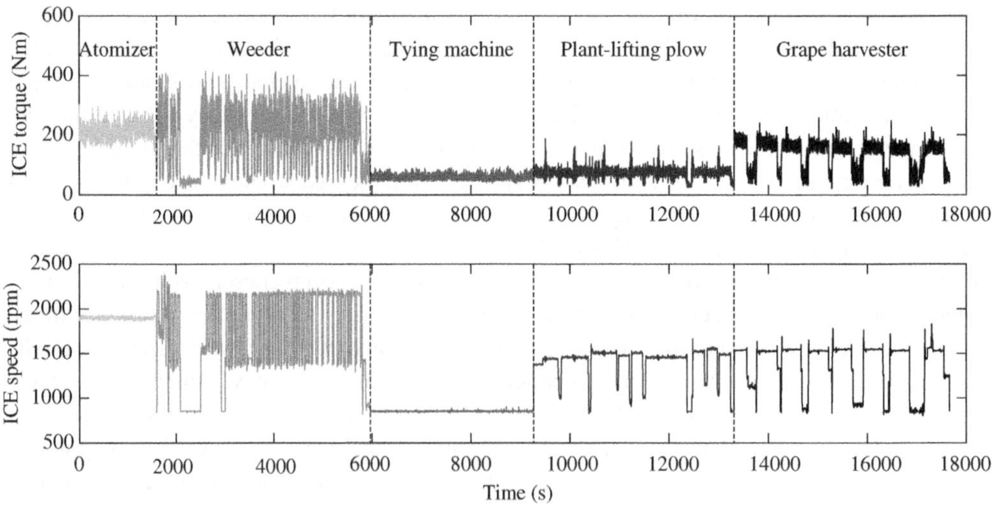

Figure 20.1 Orchard tractor load cycles.

- Rotary harrow, 17.8%, avg. pwr. = 114.7 kW
- Field transport + idle, 13.2%, avg. pwr. = 30.1 kW

20.2.3 Row Crop Tractor's Heavy-Duty Use

Finally, for the row crop tractor with heavy duty operation, the following mix of operations has been considered:

- Subsoiler, 10.3%, avg. pwr = 150.8 kW
- Cultivator, 12.3%, avg. pwr = 97.4 kW
- Heavy plowing, 18.7%, avg. pwr = 85.7 kW
- Tiller, 10.3%, avg. pwr = 145.5 kW
- Rotary harrow, 10.3%, avg. pwr = 122.7 kW
- Road transport, with and without trailer, 10.3%, avg. pwr = 63.6 kW
- Idle, 10.3%, avg. pwr = 12.7 kW

The load distribution of the three considered tractors is shown in Figure 20.2. The loads have been normalized by considering the maximum power of the original tractor P_{max}. In particular, Figure 20.2a shows the time fraction at each power level, whereas Figure 20.2b shows the cumulative distribution function (CDF). This quantity indicates the fraction of operating time that can be performed with a power level smaller than or equal to the one represented on the x-axis. From the CDF curves of Figure 20.2b, it is possible to note that the orchard tractor operates approximately 75–80% of the time at less than 40% of the maximum power. The two row crop tractors, on the other hand, exhibit a different operating power level. From these considerations, the orchard tractor category appears more promising for powertrain electrification.

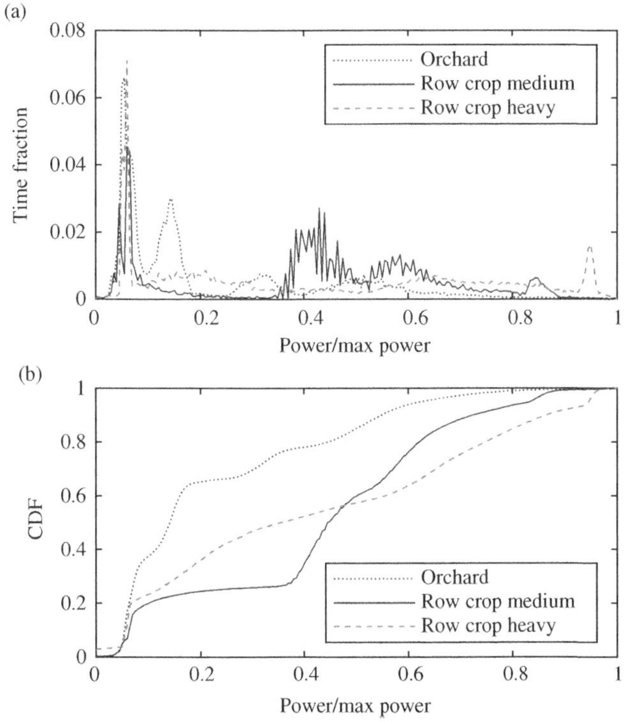

Figure 20.2 Load distribution of the three considered tractors.

20.3 System Modeling

A simulation model (Figure 20.3) is developed and used to compute fuel and electrical energy consumption, as well as the hybrid system specifications. In particular, a quasi-static model has been adopted to simulate both conventional and hybrid tractors. The load requirements of the tractors have been obtained from a specifically developed test during real operating conditions. In particular, measurements were referred to at the ICE shaft and considered with other data from the engine control unit (ECU).

In the following sections, the main details for each component of the model are given.

20.3.1 Internal Combustion Engine

The torque provided by the ICE, T_{ICE}, is determined by the power management algorithm and the engine speed is assumed to be equal to the field measurements. The fuel consumption estimation is based on ICE torque and speed, through a map of the brake-specific fuel consumption (BSFC). As an alternative, a polynomial-approximating function specifically tuned to have a good match with the consumption measured during field tests can be adopted.

20.3.2 Converter and Electric Machine

The torque of the electric machine (EM), T_{EM}, is considered the difference between the load torque and ICE torque. If necessary, a speed ratio (ICE speed/EM speed) can be considered to account for a coupling gearbox.

20.3.3 Battery

The adopted battery model is described in (20.1):

$$I_{cell} = \frac{U_{oc} - \sqrt{U_{oc}^2 - 4R_{cell}P_{cell}}}{2R_{cell}} \tag{20.1}$$

It is based on the steady-state model proposed in [1]. During the charging phase, both the cell current $I_{cell} > 0$ and cell power $P_{cell} > 0$. The other parameters are the battery internal resistance, R_{cell}, and the open-circuit voltage, U_{oc}. Both depend on the State of Charge (SoC), which is computed at time t as:

$$\text{SoC}(t) = \text{SoC}(t = 0) + \int_0^t \frac{I_{cell}(\tau)}{\text{Cap}_{cell}} d\tau \tag{20.2}$$

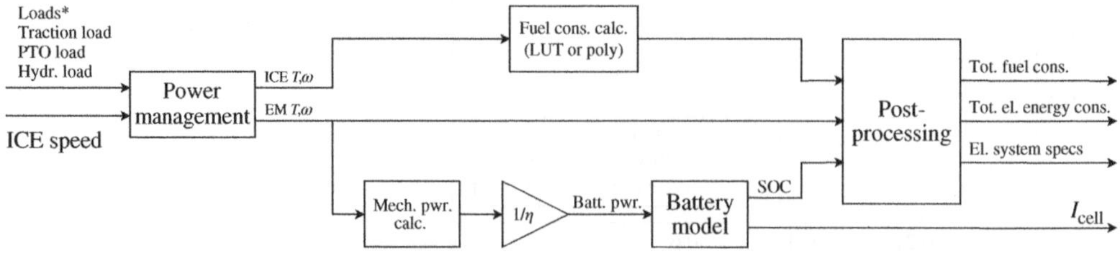

Figure 20.3 Model scheme. *Loads referred to ICE shaft.

being $\mathrm{Cap}_{\mathrm{cell}}$ the cell capacity in (Ah). The overall energy content of the battery, E, has been estimated as:

$$E = \mathrm{SoC} \cdot V_{\mathrm{nom}} \cdot \mathrm{Cap}_{\mathrm{cell}} \cdot N_{\mathrm{cell}} \tag{20.3}$$

with V_{nom} the nominal voltage of the cell and N_{cell} the total number of cells.

For the C-rate, the following limitations were assumed:

- continuous
  - charge 1C
  - discharge 3C
- peak ($\leq$10 s)
  - charge 3C
  - discharge 6C

20.3.4 Power Management

A simple rule-based power management strategy was chosen.

The powertrain can operate in three modes, depending on the required torque (and on the power management setting): when the required torque T_{req} is low, the ICE provides power for both tractor operation and battery charging, at medium torque the EM is not used, and at high torque, the EM is used to boost the ICE. The power management strategy is described by Eq. (20.4) and Figure 20.4. T_{Llim} and $T_{\mathrm{Hlim}}(\omega)$ are two thresholds. Boost mode is necessary to allow the hybrid tractor with a downsized ICE to perform the same tasks as the original tractor. Using the EM as a generator to charge the battery allows for frequent boost mode operation without the adoption of a large, bulky, and expensive battery. Additionally, a full-electric mode was considered only for the orchard tractor. $T_{\mathrm{Hlim}}(\omega)$ can correspond to the actual ICE torque limit or the ICE can be limited to a lower torque in order to use the energy stored in the battery to cover part of the load. Both torque thresholds are individual for each operating cycle.

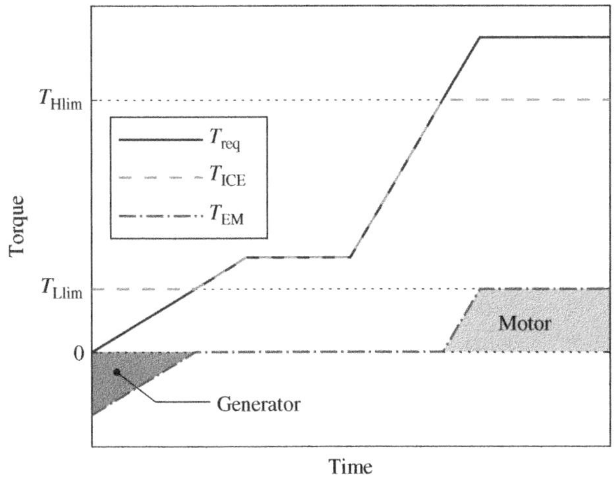

Figure 20.4 Power management strategy example with demo operating cycle. Constant T_{Hlim} is showed here for simplicity. Colored areas highlight where the EM acts as a generator or as a motor.

$$T_{\mathrm{ICE}} = \begin{cases} T_{\mathrm{Llim}}, & \text{if } T_{\mathrm{req}} \leq T_{\mathrm{Llim}} \\ T_{\mathrm{req}}, & \text{if } T_{\mathrm{Llim}} < T_{\mathrm{req}} \leq T_{\mathrm{Hlim}}(\omega) \\ T_{\mathrm{Hlim}}(\omega), & \text{if } T_{\mathrm{req}} > T_{\mathrm{Hlim}}(\omega) \end{cases}$$

$$T_{\mathrm{EM}} = \begin{cases} T_{\mathrm{req}} - T_{\mathrm{Llim}}, & \text{if } T_{\mathrm{req}} \leq T_{\mathrm{Llim}} \\ 0, & \text{if } T_{\mathrm{Llim}} < T_{\mathrm{req}} \leq T_{\mathrm{Hlim}}(\omega) \\ T_{\mathrm{req}} - T_{\mathrm{Hlim}}(\omega), & \text{if } T_{\mathrm{req}} > T_{\mathrm{Hlim}}(\omega) \end{cases} \tag{20.4}$$

20.3.5 CO_2 Emission Estimation

In the computations, only the CO_2 emissions due to fuel consumption, electrical energy consumption, and battery manufacturing are evaluated. In fact, they are considered the major causes of emission differences between hybrid and non-hybrid tractors.

For complete combustion, CO_2 emissions are assumed to be proportional to fuel consumption, electrical energy consumption, and battery capacity. The following emission factors have been considered:

- 3.92 kg $CO_2/kg_{\mathrm{Diesel}}$, derived from [2], considering the energy-based allocation method and lower heating value of diesel equal to 43.1 MJ/kg.
- 0.33 kg $CO_2/kWh_{\mathrm{El.energy}}$, from [3], considering the average between the EU-mix emission factors for low voltage supply of 2016 and 2030.
- 100 kg $CO_2/kWh_{\mathrm{Battery}}$, from [4], which reports a 61–106 kg CO_2/kWh.

These factors include extraction of raw materials, production, and transportation, delivery to the customer, and emissions from its combustion. Therefore, they can be considered representative of the whole process.

20.4 Design Specifications and Power Management Tuning

The simulation model has been used to identify the main design specifications of the hybrid powertrain. They are the battery capacity and power management parameters, the EM-rated torque, and power electronic specs, the power of the downsized ICE.

20.4.1 Battery Capacity Sizing and Power Management Tuning

The computation of the required battery capacity $\mathrm{Cap}_{\mathrm{req}}$ is summarized in (20.5):

$$\mathrm{Cap}_{\mathrm{req}} = \left[(E_{\mathrm{in}} - E_{\mathrm{fin}}) \cdot \mathrm{ceil}\left(\frac{t_{\mathrm{cycle}}}{t_{\mathrm{min}}}\right) + \right. \\ \left. + (E_{\max} - E_{\mathrm{in}}) + (E_{\mathrm{fin}} - E_{\min})\right] \cdot \frac{1}{\mathrm{DoD}} \tag{20.5}$$

where:

- E_{in} and E_{fin} are the battery energy levels respectively at the beginning and at the end of the operating cycle
- $E_{\max}$ and $E_{\min}$ are respectively the maximum and minimum battery energy levels throughout the operating cycle

- t_{cycle} is the duration of the considered operating cycle
- t_{min} is the minimum required operating time (8 hours).
- *DoD* is the depth of discharge, which is set to a conservative 60% to ensure long battery life and to cope with the strong reliability requirements of the agricultural industry [5].

The capacity value is influenced also by the power management tuning. The two torque thresholds, T_{Llim} and T_{Hlim}, have been chosen so as to minimize the required battery capacity. Moreover, if desired, they can also be set to maximize electrical energy usage if a certain cycle is not particularly intensive, complying at the same time with the battery C-rate limitations. Raising T_{Llim} reduces the required capacity while increasing fuel consumption and charging current. When the C-rate limitations are exceeded, a battery with a larger capacity should be adopted. After the determination for each load cycle of the T_{Llim} and battery capacity, the final battery capacity Cap_{act} can be selected as the maximum among the values required by each cycle. Then, the T_{Llim} threshold is individually adjusted for each cycle (except for the one that requires the maximum capacity) in order to properly exploit the battery minimizing the fuel consumption. Finally, the upper threshold $T_{Hlim}(\omega)$ can be adjusted. In particular, $T_{Hlim}(\omega)$ has been obtained by multiplying the actual ICE torque limit by a coefficient $k_H \leq 1$, which is determined by gradually reducing it from 1. Compared to the orchard tractor, the row crop tractor operating cycles are really power-demanding. Therefore, lowering the amount of torque that can be delivered by the combustion engine results in a rapid increase in battery capacity without a significant saving in operating costs. As a consequence, a factor $k_H = 1$ has been considered for row crop tractors, ensuring also a longer battery life thanks to the reduction of its charging cycles.

20.4.2 Electric Machine and Power Electronics Design Specs

To avoid an excessive oversizing of the EM, its design torque has been determined considering the thermal equivalent torque T_{eq} computed from the instantaneous EM torque T_{EM} as:

$$T_{eq} = \sqrt{\frac{1}{\tau}\int_0^\tau T_{EM}^2 \mathrm{d}t} \tag{20.6}$$

Reasonable values for the thermal time constant τ for water-cooled EMs has been considered, ranging from about 200 to 500 seconds. For the power converter, peak power has been used as a design requirement.

20.4.3 ICE Downsizing

Various levels of ICE downsizing have been considered for the two row crop tractors. A fine variation of ICE power level has been obtained using a coefficient, denoted as R, to downscale the torque curve of a base ICE.

For the orchard tractor instead, a single 55 kW engine has been considered for the hybrid variant. This value has been selected mainly to fulfill emission standard requirements. A similar approach has been adopted by several authors, for example, in [6–8]. Therefore, for the orchard tractor, only EM exploitation was varied keeping the same battery size, in order to keep it as small as possible. Three modes have been considered:

- Hybrid 1: use EM only in cycles in which ICE needs boosting; this mode is used to determine the battery capacity
- Hybrid 2: maximize EM usage; electric assistance in medium duty cycles (atomizer and harvester) and entire parts of low-power duty cycles (plant lifting plow and tying machine)

performed in purely electric mode (ICE off), until the battery reaches SoC = 20% (consistent with a 60% discharge from an 80% initial SoC)
- Hybrid Mix: intermediate case between *hybrid 1* and *hybrid 2*; purely electric operation is limited in order to reduce the initial tractor cost compared to *hybrid 1*. This means that operation with the plant-lifting plow is always performed with ICE on, as a higher power EM would be needed for purely electric operation.

20.5 Life Cycle Cost Analysis

An LCC analysis has been conducted to evaluate the economic feasibility of farming tractor powertrain electrification and to determine the best ICE downsizing considering various variables. LCC is considered the present value of all the costs that occur during the life cycle of a tractor. In the analysis presented here, only user costs have been considered. Maintenance and other costs have been assumed the same for both traditional and hybrid electric powertrains. Therefore, they have been ignored. As will be described in the following, the best downsizing coefficient *R* resulted quite close to 1. This justifies the choice of constant maintenance costs. The LCC is computed as follows:

$$\mathrm{LCC} = C_\mathrm{p} + \sum_{t=0}^{n} \frac{C_t}{(1+d)^t} \tag{20.7}$$

where:

- C_p is the initial cost (purchase)
- C_t are all the relevant costs that occur during the considered study period, fuel and electrical energy in this case
- t indicates the year when each cost occurs
- n is the study period, in this case, the service life. A general global engine life requirement is 6000 to 12,000 hours, increasing with the power level [9]. In this work, a 10,000-hours life was assumed, and life in years was computed by dividing the life in hours by the annual operating time.
- d is the discount rate used to compute the present value of future costs. For the present analysis, an 8% discount rate was chosen.

20.5.1 Tractor Components and Energy Pricing

In order to perform an effective LCC analysis, specific prices of the main tractor components, as well as energy costs (fuel and electricity), need to be determined. All the components shared by both hybrid and conventional tractors were ignored, and all prices were intended as consumer prices. For all the considered components, a price range was taken into account, as single values cannot be exactly determined and exceptional changes can happen, for example, due to the geopolitical situation. For the same reasons, and to cover for possible country-related variations, it is important to consider a cost range for fuel and electricity too. Moreover, the choice to consider a price range allows determining which elements have the higher impact on the LCC. Table 20.1 summarizes component prices and energy costs that are considered in this study. Reported ranges and standard prices are based on the cited references. Renius [9] reports the ICE cost as a fraction of the tractor cost. Thus, the ICE prices in Table 20.1 are based on a market analysis performed by the authors. Further details on all prices can be found in [10].

Table 20.1 Tractor components and energy pricing.

Component	Price range	Standard price[a]	Unit	References
ICE	50–300	200	€/kW	[9]
Battery	50–250	150	€/kWh	[11]
EM	15–60	35	€/kW	[12, 13]
PE	15–60	30	€/kW	[12, 14]
Fuel	0.70–1.30	0.90	€/l	[15]
Electricity	0.10–0.28	0.16	€/kWh	[16]

[a] Most realistic at the time of writing.

20.6 Results

The developed method described in the previous sections is suitable to cover a wide variation of engine downsizing degrees. Nevertheless, for the row crop tractors, some preliminary simulations have been carried out to evaluate the impact of R on the battery capacity. It has been found that ICE power reduction larger than 20% leads to a battery capacity exceeding 40 kWh. In turn, this leads to huge initial costs and it would not comply with packaging requirements (~200 Wh/l energy density [17]). On the contrary, if the downsizing degree R is very close to 1, tractor complexity is the same as with lower R values but the operating costs savings are negligible. Thus, for the row crop medium tractor, 0.8, 0.86, and 0.92 R values were considered, whereas for the heavy-duty one, the analysis was limited to two values: 0.88 and 0.92. These values are due to the specific engines that were used in simulation. The ICEs adopted for the non-hybrid variants were chosen to match the load cycle demands plus a very small margin. The ICE adopted for the medium non-hybrid variant had a 175 kW rated power. However, the hybrid variants were based on a 0.7, 0.75, and 0.8 downscaling of a 200 kW ICE, resulting in the reported R values. All the considered ICEs for the heavy tractor are based on the upscaling of a 175 kW engine and the R values are obtained similarly, by normalization to the power of the non-hybrid variant. As already said, for the hybrid orchard tractor, a single 55 kW ICE was considered. The capacity of the battery resulted similarly to that of the two larger tractors. This, however, should be considered acceptable as, thanks to the engine downsizing, less restrictive emission regulations apply with the possibility to remove some exhaust gas treatment devices clearing more space onboard and limiting costs.

20.6.1 Saving Each Cycle

The cost savings for each hybrid tractor are reported in Table 20.2 with reference to the traditional configuration. The larger savings are obtained for the orchard tractor. This is expected considering the operating cycles. As shown in Figure 20.2, in fact, the operations become more power-intensive moving from orchard to row crop heavy tractor. The savings are larger for low-power cycles, as the ICE operates in its low-efficiency region. Reducing the ICE size then results in better exploitation, shifting the operating point toward a region at higher efficiency. The ICE downsizing allows for about 15% savings in the very low-power orchard cycles (plant-lifting plow and tying machine). This is shown in the "Hyb. 1" column. In such a case, no EM utilization is considered when torque boost is not necessary.

Table 20.2 Tractor components and energy pricing.

Orchard				Row crop medium				Row crop heavy		
Opex % diff.	Hyb.1	Hyb. 2	Mix	Opex % diff.	$R = 0.8$	$R = 0.86$	$R = 0.92$	Opex % diff.	$R = 0.88$	$R = 0.92$
Weeder	−3.31%	−3.31%	−3.31%	Heavy plowing	−1.56%	−1.95%	−1.74%	Subsoiler	+1.46%	+0.54%
Atomizer	−0.96%	−2.17%	−2.17%	Medium plowing	−5.33%	−4.32%	−3.30%	Cultivator	−0.11%	−0.16%
Grape harvester	−7.72%	−8.43%	−8.43%	Rotary harrow	−4.67%	−3.98%	−3.17%	Heavy plowing	−0.99%	−0.71%
Plant-lifting plow	−14.85%	−17.99%	−14.85%	Field transport + idle	−1.42%	+0.35%	+2.57%	Tiller	−0.68%	−0.48%
Tying machine	−16.06%	−24.09%	−24.09%					Rotary harrow	−1.83%	−1.25%
								Road transport	−1.73%	−1.14%
								Idle	−3.00%	−1.92%
Average	−6.88%	−8.77%	−8.14%	Average	−3.60%	−3.14%	−2.41%	Average	−0.67%	−0.54%

Negative values represent savings compared to the traditional tractor.

Another interesting result is the higher costs of the hybrid variants for some load cycles. This is because numerous battery charging operations are required during the operation in order to ensure an adequate battery last to cover the minimum required working time. This lead to higher costs because the engine efficiency increase will not compensate for double energy conversion loss.

20.6.2 Varying Component and Energy Pricing – Convenience of the Hybrid Tractors

The two components that have the highest influence on the LCC and on the PayBack Time (PBT) are the ICE and the battery. This is because they are the components with the highest influence on the purchase price of a hybrid tractor, representing the major fraction of the purchase price (Capital expenditure, Capex).

PBT is defined as the time in which an investment becomes profitable. In particular, it has been computed as the minimum time after which a hybrid tractor becomes more convenient than the traditional and non-hybrid counterpart. It has been obtained by changing the reference period n in Eq. (20.7). PBT was used to highlight if the hybrid variant is convenient or not with different prices of the main components, as in Figures 20.5–20.7, which show the effect of ICE and battery price. The effect of the other components was not reported as it is much lower. Among the considered values, only three battery price levels, i.e. 100, 150, and 200 €/kWh, are shown. A low engine cost combined with a high battery cost results in a large purchase cost for the hybrid variant, which cannot be recovered during the tractor-operating life. This is particularly evident in the row crop heavy-duty tractor reported in Figure 20.7. This is due to the very low annual savings. In other words, the hybrid version is already more convenient from the beginning, or the price penalty will not be recovered during operational life.

Some small steps are visible in Figure 20.6. They are due to the change in the optimal R-value with the engine cost.

The influence of a change in the diesel and electrical energy prices on the LCC is shown in Figure 20.8. For the prices of ICE and electric components, "standard" values are assumed in this case.

When the diesel price is low and the electrical energy price high, a less intensive hybrid mode, with less electrical energy utilization, is preferable. Nevertheless, Hyb. 1 never becomes the most convenient mode, unless electrical energy price is very huge. As an important note, it results that the hybrid tractor always exhibits a smaller LCC than the non-hybrid counterpart.

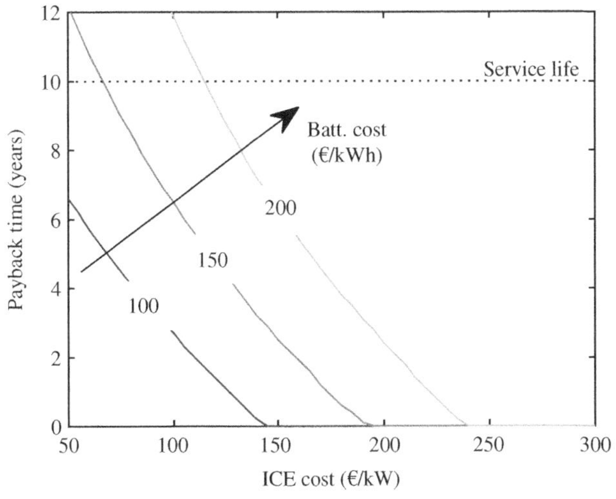

Figure 20.5 Orchard tractor PBT.

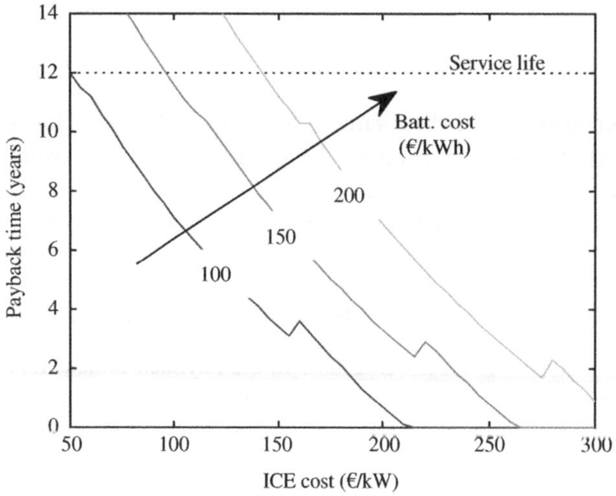

Figure 20.6 Row crop medium tractor PBT.

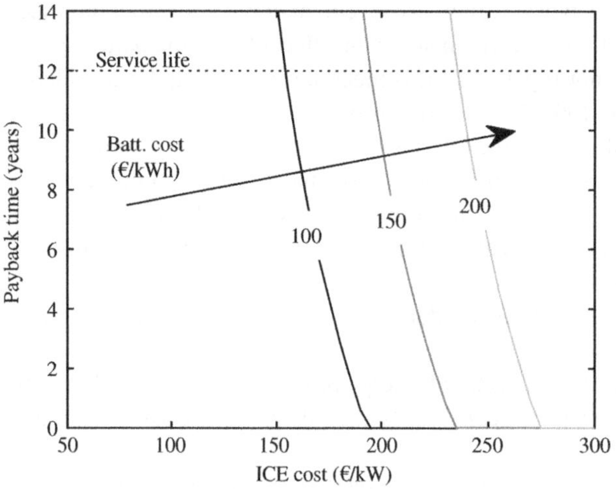

Figure 20.7 Row crop heavy tractor PBT.

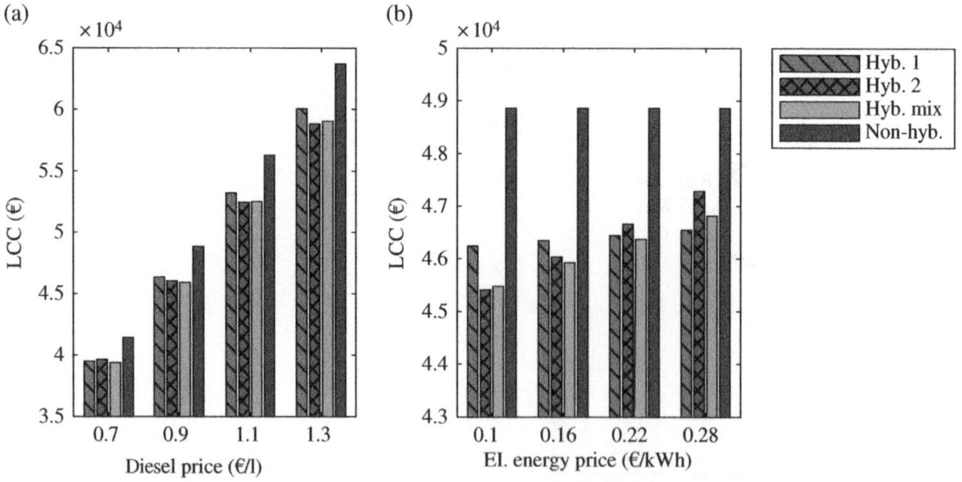

Figure 20.8 Orchard tractor LCC behavior with (a) varying diesel prices and fixed electrical energy price (0.16 €/kWh) and (b) varying electrical energy prices and fixed diesel price (0.9 €/l).

Table 20.3 Main specs and savings, "standard" price combination, best hybrid mode (orchard) or best downsizing *R* (row crop).

	Orchard	Row crop med.	Row crop heavy
	(Hyb. mix)	(*R* = 0.86)	(*R* = 0.92)
EM torque (Nm)			
Peak	147	382	471
Rated	63	157	118
EM power (kW)			
Peak	22	66	51
Rated	9	26	14
Batt. cap. (kWh)	21	24	14
Capex (€)	15,233; 15,442	36,433; 35,018	42,658; 42,022
Hyb.; non-hyb.	(−1.4%)	(+4.0%)	(+1.4%)
Opex (€/year)	4574; 4,980	15,747; 16,258	16,681; 16,771
Hyb.; non-hyb.	(−8.2%)	(−3.1%)	(−0.5%)
PBT (years)	0	3.3	9.6
Total CO_2	−9.7%	−3.1%	−0.5%

20.6.3 Specs and Savings Summary

The electric system specifications and savings are summarized in Table 20.3. The "standard" price combination has been adopted in the computations. Being the loads much lower in the orchard tractor, the specifications for the electric motor and related converter in this case are the least demanding. Nevertheless, the battery size is comparable in all the considered cases (probably due to the different load cycles). When the two row crop tractor specs are considered, it can be noted that in medium duty cases, higher ratings are required, except for peak torque. Even if this could initially appear not intuitive, it is consistent with the lower *R* index, i.e. a more downsized ICE.

The initial cost for the orchard tractor is lower when the electrified version is considered. An explanation of this is the fact that the cost for additional devices needed for the electrification of the powertrain, as well as the development cost, has been ignored. It is important to note that such aspects have a higher relative weight when a small tractor is considered. As a consequence, the obtained result should be considered too optimistic.

As far as the LCC is concerned, the savings for the orchard tractor are significant. On the other hand, they are more limited with the row crop medium tractor and quite negligible with the heavy-duty one. This is strictly related to the operational savings decrease moving to tractors with more power-intensive cycles.

20.7 Conclusion

This chapter presents a method to compute the economic feasibility of farming tractors' powertrain electrification. In particular, an LCC analysis was presented. Three different tractors with various power sizes and operating cycles have been considered. It has been shown that significant savings can be obtained from powertrain electrification when small specialized tractors, which perform a lot of low-power operations, are considered On larger size tractors, with more power-intensive duty cycle, operating cost savings are very small, resulting in limited LCC savings.

With the adopted method, it is then possible to evaluate the safety margin for which a hybrid tractor remains profitable, also taking into account variations in costs and prices.

References

1 Schaltz, E. (2011). Electrical vehicle design and modeling. In: *Electric Vehicles*, Chapter 1 (ed. S. Soylu). Rijeka: IntechOpen. https://doi.org/10.5772/20271.

2 Moretti, C., Moro, A., Edwards, R. et al. (2017). Analysis of standard and innovative methods for allocating upstream and refinery ghg emissions to oil products. *Appl. Energy* 206: 372–381. https://doi.org/10.1016/j.apenergy.2017.08.183.

3 Prussi, M., Yugo, M., De Prada, L. et al. (2020). JEC well-to-tank report v5. Publications Office of the European Union. ISBN: 978-92-76-19926-7. https://doi.org/10.2760/959137.

4 Emilsson, E. and Dahllf, L. (2019). Lithium-ion vehicle battery production status 2019 on energy use, CO_2 emissions, use of metals, products environmental footprint, and recycling. Technical report, IVL Swedish Environmental Research Institute.

5 Flint, J., Zhang, D., and Pei, X. (2014). Preliminary market analysis for a new hybrid electric farm tractor. *Proceedings of the 2014 International Conference on Global Economy, Commerce and Service Science*. https://doi.org/10.2991/gecss-14.2014.25.

6 Dalboni, M., Santarelli, P., Patroncini, P. et al. (2019). Electrification of a compact agricultural tractor: a successful case study. *2019 IEEE Transportation Electrification Conference and Expo (ITEC)*. https://doi.org/10.1109/ITEC.2019.8790496.

7 Troncon, D., Alberti, L., and Matteti, M. (2019). A feasibility study for agriculture tractors electrification: duty cycles simulation and consumption comparison. *IEEE Transportation Electrification Conference and Expo (ITEC)*. https://doi.org/10.1109/itec.2019.8790502.

8 Mocera, F. and Som, A. (2020). Analysis of a parallel hybrid electric tractor for agricultural applications. *Energies* 13 (12): 3055. doi: https://doi.org/10.3390/en13123055.

9 Renius, K.T. (2019). *Fundamentals of Tractor Design*. Springer. https://doi.org/10.1007/978-3-030-32804-7.

10 Beligoj, M., Scolaro, E., Alberti, L. et al. (2022). Feasibility evaluation of hybrid electric agricultural tractors based on life cycle cost analysis. *IEEE Access* 1. https://doi.org/10.1109/ACCESS.2022.3157635.

11 BloombergNEF. 2020 battery price survey. https://about.bnef.com/blog/battery-pack-prices-cited-below-100-kwh-for-the-first-time-in-2020-while-market-average-sits-at-137-kwh/. bnef.com (accessed October 2021).

12 Electrical and electronics technical team roadmap. Technical report, Vehicle Technologies Office, U.S. Department of Energy, October 2017. https://www.energy.gov/sites/prod/files/2017/11/f39/EETTRoadmap10-27-17.pdf.

13 Goss, J., Popescu, M., and Staton, D. (2013). A comparison of an interior permanent magnet and copper rotor induction motor in a hybrid electric vehicle application. *2013 International Electric Machines Drives Conference*, pp. 220–225. https://doi.org/10.1109/IEMDC.2013.6556256.

14 Murugesan, D.K. and Manickam, I. (2015). Reliability and cost analysis of different power inverter topologies in electric vehicles. *2015 IEEE Transportation Electrification Conference and Expo (ITEC)*, pp. 1–4. https://doi.org/10.1109/ITEC.2015.7165748.

15 Molari, G., Mattetti, M., Lenzini, N., and Fiorati, S. (2019). An updated methodology to analyse the idling of agricultural tractors. *Biosyst. Eng.* 187: 160–170. doi: https://doi.org/10.1016/j.biosystemseng.2019.09.001.

16 Prezzi e tariffe. https://www.arera.it/it/prezzi.htm#.arera.it (accessed August 2021).

17 Lbberding, H., Wessel, S., Offermanns, C. et al. (2020). From cell to battery system in bevs: analysis of system packing efficiency and cell types. *World Electr. Veh. J.* 11 (4): 77. https://doi.org/10.3390/wevj11040077.

21

Advances in Data-Driven Modeling and Control of Naval Power Systems

Javad Khazaei and Ali Hosseinipour

Lehigh University, Electrical and Computer Engineering, Bethlehem, PA 18015, USA

21.1 Introduction to DC Watercraft Systems

Medium-voltage DC (MVDC) distribution technology provides various advantages in applications such as railways, marine vessels, data centers, electric vehicle-charging stations, and wind collection grids [1]. MVDC shipboard microgrids particularly benefit from an easier connection of renewable energy sources and storage, which are inherently DC, elimination of reactive power control and phase synchronization, reduced bulk, and improved fuel economy and power quality compared with their AC counterparts [2].

Despite the aforementioned advantages, challenges arise in the design of MVDC naval microgrids, particularly with respect to specified power-sharing control and stability preservation of mixed sources, also known as hybrid microgrids [3]. There have been a multitude of efforts on steady-state power-sharing control in MVDC naval microgrids. Droop control is the most widely used method for power-sharing control of multiple sources [4]. Variations of droop control for the state of charge (SoC) balancing of battery energy storage systems (BESSs) in a shipboard microgrid and seaport microgrid have been reported in [5, 6]. Supervisory control levels are also developed for the improvement of primary level controllers and optimized power management of DC shipboard microgrids in the form of centralized [7, 8] and distributed [9, 10] controllers.

Several studies have focused on energy management of MVDC shipboard power system using hybrid energy storage devices and pulsed loads [11–14]. The proposed energy management techniques can be categorized into (i) centralized [13, 14], (ii) hierarchical [9, 15–17], and (iii) decentralized [18–26] methods. Generally, centralized and hierarchical designs ensure the energy management of resources in shipboard power systems, but the cost and complexity of the controller increase as more components are added to the system. This also defeats the purpose of having a modular design for advanced modular energy storage systems used in the marine corps. Furthermore, centralized and hierarchical designs require a communication network to function, which introduces a high risk of a single point of failure and cyberattacks. Therefore, decentralized approaches are the most suitable solutions for energy management of future MVDC shipboard power systems. Several recent studies focused on the application of decentralized control methods for MVDC shipboard power systems [18, 20–26]. These methods have the capability to successfully share the load between generation units and battery energy storage devices without the need for a communication network. A comparison between centralized and decentralized controllers in [22] showed that decentralized controllers are more robust when load variations occur. These studies,

Transportation Electrification: Breakthroughs in Electrified Vehicles, Aircraft, Rolling Stock, and Watercraft,
First Edition. Edited by Ahmed A. Mohamed, Ahmad Arshan Khan, Ahmed T. Elsayed, and Mohamed A. Elshaer.

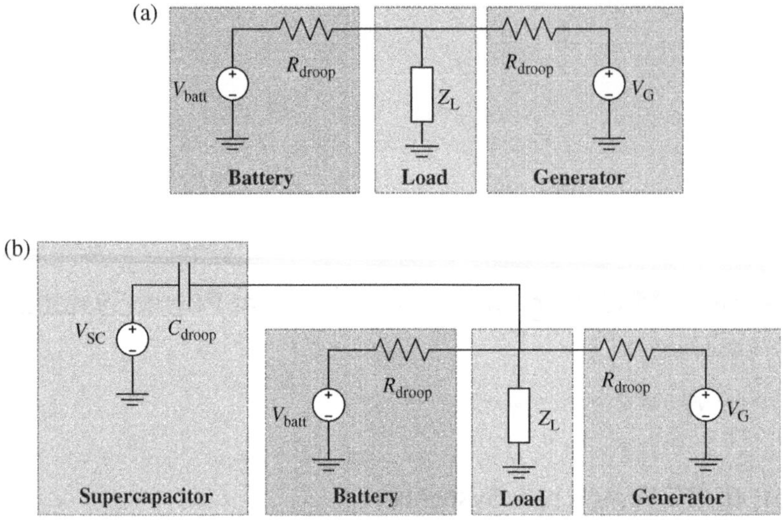

Figure 21.1 Equivalent circuits for conventional resistive droop control of batteries and generators in shipboard microgrids (a), and new decentralized control of hybrid energy storage units in shipboard microgrids using capacitive droop controller for SCs (b).

however, did not focus on hybrid storage units and mainly considered battery energy storage devices. In addition, the existing studies on decentralized power sharing of storage units normally rely on conventional resistive droop techniques (see the equivalent circuit in Figure 21.1a) that introduce high voltage dips in case of sudden load changes [12, 17, 27–33] (see the simulation results in Figure 21.2a).

The approach proposed in [34, 35] considered supercapacitors (SCs) for high-frequency transients and battery storage for steady-state operation. However, to enable the transition between SCs and battery storage devices, a state-logic model had to be used to delay the battery operation for 60 ms so that the SC manages a load transient within 60 ms. This method might not successfully manage the load during high-frequency load variations. It was recently shown in [36, 37] that capacitive droop controllers are more suitable for SCs in responding to fast changes in the loads such as pulsed power loads (PPLs) in MVDC power systems and do not have the complications of [34, 35] (see the equivalent diagram in Figure 21.1b). However, these controllers would introduce high-frequency components to the output response of the system (see the simulation results of output current of batteries as a result of capacitive droop controllers for SCs in Figure 21.2b).

The existing decentralized energy management techniques for hybrid storage units in MVDC shipboard power systems used a fixed droop characteristic that does not account for dynamic changes in the load or the SoC of storage devices. The conventional droop method can only achieve proportional power sharing in a steady state, but not during transients or in the case of energy storage systems with different characteristics or SoC. The existing decentralized adaptive droop-based energy management techniques are either designed for battery energy storage units only [38–40] or are very complex and hard to implement [34, 35]. These limitations call for a more robust droop control that resolve these challenges.

In addition, modeling of DC distribution systems and microgrids is typically conducted assuming full knowledge of detailed analytical models of switch mode power converters, also referred to as white-box models [41]. However, with the proliferation of commercial off-the-shelf converters, detailed information about converter parameters is usually not available [42]. This, in addition to the uncertainty in the parameters of a power converter, motivates utilizing system identification techniques to capture the dynamic behavior of converters by leveraging measurement data, which

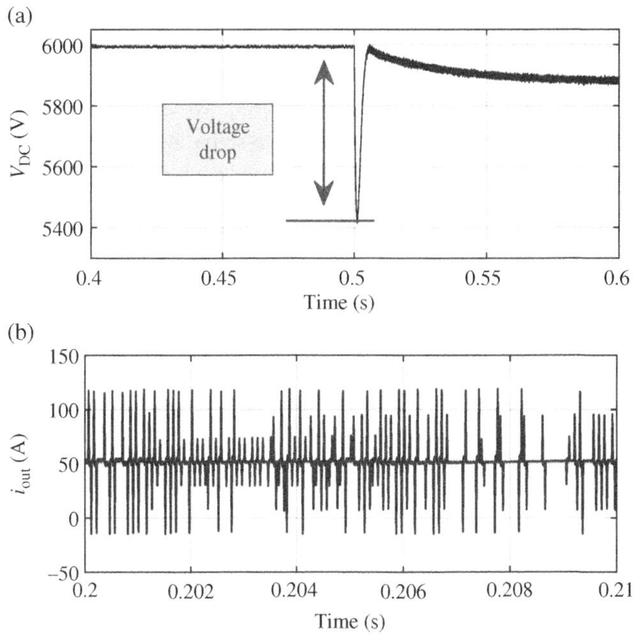

Figure 21.2 Limitations of existing studies on control of hybrid storage in MVDC shipboard power systems; voltage dip issue of conventional resistive droop control for battery energy storage devices (a), high-frequency variations resulted by capacitive droop control of supercapacitors (b).

at heart involves the injection of a frequency-rich signal to obtain the data representative of the system dynamics [43]. Applying system identification is motivated by the requirement of a model for control design, implementing adaptive control, monitoring of the system stability, and also as a step toward developing the digital twin of the converter and its interconnected system such as a microgrid [44].

Adopting system identification techniques, black-box and gray-box models of power converters can be obtained. Black-box modeling assumes no prior information about the topology or analytical equations of the system, relying purely on measured data for identification purposes [45, 46]. Due to the wide deployment of DC/DC converters in shipboard microgrids, they have been studied for nonlinear system identification using various black-box methods such as polytopic or Hammerstein methods [47–49]. However, these methods heavily rely on linear models as intermediate steps to estimate nonlinear dynamics, which renders them operating point-dependent and thus unable to represent full non-linear dynamics. Other non-linear black-box modeling methods include wavelet or dynamic neural networks [41, 50], but the main drawbacks of these methods are their computational cost, requirement for large training data points, and the lack of physical interpretability.

Gray-box identification methods rely on partial knowledge of the system model and dynamics, which then serves as a foundation to estimate the complete model of the system [51]. Gray-box identification of DC/DC buck converters is discussed in [52, 53] by using *a priori* knowledge of the static behavior of the converter. These methods improve on the locally accurate black-box models to arrive at a globally valid model at the cost of reducing the overall prediction accuracy. In [54], the prediction capability of the gray-box methods proposed in [52, 53] is improved by incorporating the *a priori* knowledge of the converter static behavior in the structure selection of the model. However, only the identification of voltage dynamics is addressed in [52–54] rather than a complete dynamic model. Moreover, the identified models are not physically interpretable. In [44], gray-box modeling using an iterative least-squares technique is conducted for a half-bridge converter to discover its state-space dynamic model. Despite being physically interpretable, prior knowledge of state-space equations and nominal parameters of the converter is required, which may not always be available, particularly in the case of commercial off-the-shelf converters. A hybrid Wiener-Hammerstein gray-box modeling method is also proposed in [55], which depends on the information provided in the datasheet of converters to predict the large-signal behavior, power consumption, and efficiency of DC/DC converters. In [56], a physics-informed deep neural network is proposed for estimating the parameters of a DC/DC buck converter. The method assumes prior knowledge of the parametric dynamic equations of the converter and also suffers from high computational costs and long training times.

To fill the gaps in the studies reviewed earlier, this chapter aims at:

- Reviewing the advanced system architecture for DC microgrids used in naval power systems.
- Designing a complex droop control that takes full advantage of BESSs' capabilities to contribute to both steady-state and medium-frequency power variations of MVDC shipboard microgrids, and enables sharing of the high-frequency load between SCs and steady-state load between auxiliary generators.
- Implementing a novel identification framework based on the sparse identification of nonlinear dynamics for DC/DC converters used in shipboard microgrids.

The rest of the chapter is organized as follows. Section 21.2 introduces state-of-the-art system architectures for DC watercrafts. In Section 21.3, the proposed complex droop control for hybrid storage systems in DC watercrafts is introduced. In Section 21.4, system identification of DC watercrafts is covered. Section 21.5 concludes the chapter.

21.2 Architectures for DC Shipboard Power Systems

21.2.1 Radial Topology

The typical configuration of a radial hybrid MVDC shipboard microgrid is illustrated in Figure 21.3. The synchronous generator (SG) is interfaced with a rectifier unit to provide the majority of power to balance the generation and demand. The BESS and SC are connected to the main DC bus through bidirectional DC/DC converters that enable charging and discharging power flow. The loads connected to the DC bus are of two types, namely resistive and converter-based electronic loads with constant power load (CPL) behavior [57]. PPLs with a sufficient pulse width can also be represented by CPLs [58].

21.2.2 Multi-Zone Topology

A multi-zone MVDC shipboard power system is shown in Figure 21.4. The ship service loads are distributed into n zones (only two zones are shown in the graph) and will be fed power from main and auxiliary generators as well as hybrid storage units. The system load will also include PPLs, and radar loads that will be distributed in various zones. The disconnect switches between the zones allow split-plant configuration for various testing and validation of contingencies during the operation of the system. The stability conditions during contingency should be designed and tested following the IEEE standard requirements to ensure a transient recovery time within 2 seconds [59]. The hybrid storage system will be designed to ensure continuous supply of power even when failures occur on storage units or other generators on the MVDC shipboard power system.

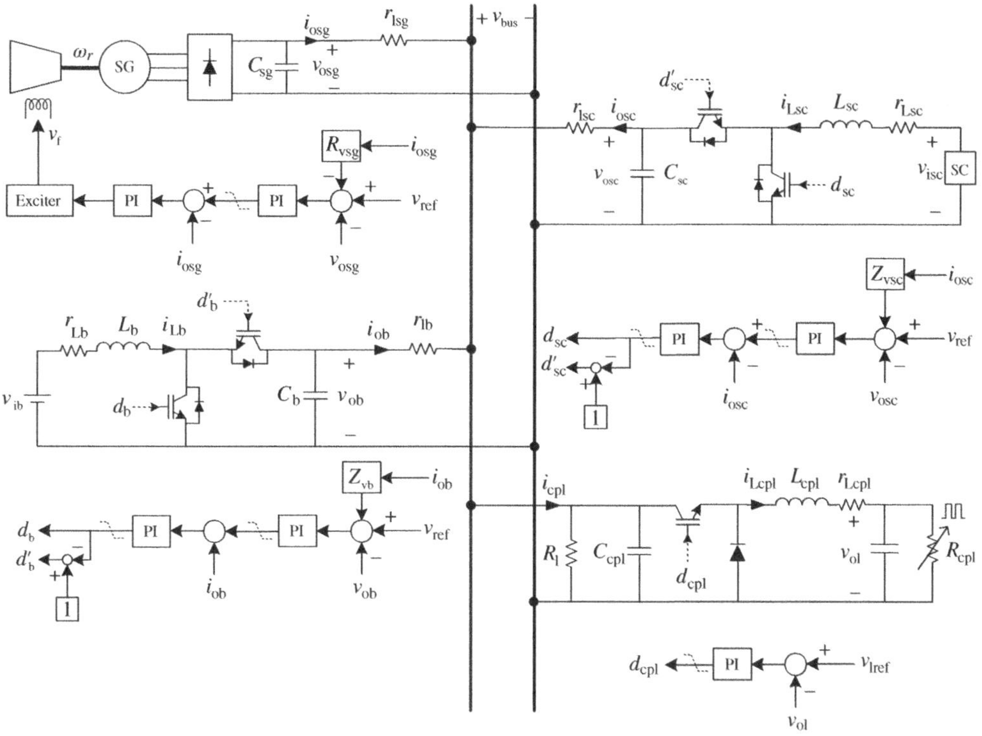

Figure 21.3 Radial MVDC shipboard microgrid.

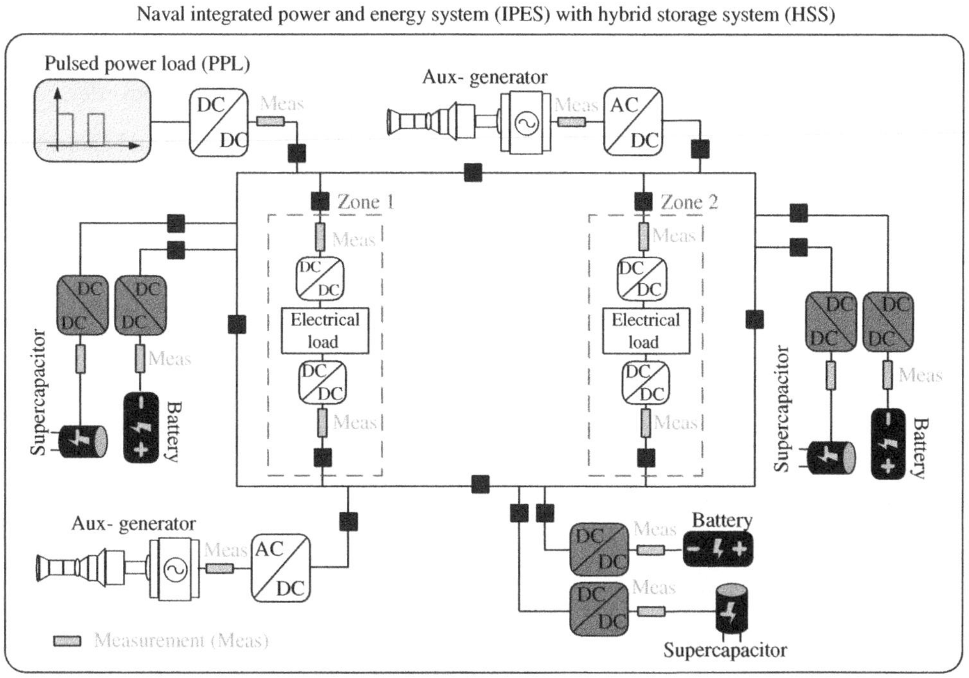

Figure 21.4 A multi-zone naval microgrid with hybrid energy storage.

21.3 Application of Hybrid Energy Storage in DC Watercrafts

For both SCs and battery storage units, bi-directional DC/DC converters with multi-loop controllers are considered. Primary controllers include current and voltage controllers for storage units and excitation system control for conventional generators. A generic structure of a bi-directional DC/DC converter with its multi-loop control for both SCs and battery storage units is illustrated in Figure 21.5. This control design adjusts the storage current using a current control loop while regulating the output converter voltage through a proportional integral (PI)-based voltage regulator. In the case of parallel converters sharing a common load, an outer controller will adjust the voltage reference provided to the voltage control loop (V_o^*) using the droop technique. The droop controller adjusts the reference voltage of the voltage control loop through a feedback loop that measures the converter current and multiplies it by a virtual impedance, which will be discussed in the next section.

21.3.1 Inner Control Loops

The current control loop is the most inner control loop and should be designed to exhibit a fast response. First, using the state-space dynamics of the converter with an output impedance (e.g. an RL impedance), the transfer functions between different signals in the network can be calculated for each storage unit. As an example, dynamics of the input current (I_1) in Figure 21.5 is represented by:

$$L_1 \frac{dI_1}{dt} = V_{\text{in}} - R_1 I_1 - d_1 V_{\text{out}} \tag{21.1}$$

which in a Laplace form is written as

$$I_1 = \frac{1}{L_1 s + R_1} [V_{\text{in}} - d_1 V_{\text{out}}] \tag{21.2}$$

Knowing that the inner current loop generates the duty cycle of the converter, e.g. $\left[k_{\text{pi}} + \frac{k_{\text{ii}}}{s}\right](I_1^* - I_1) = d_1$, the transfer function of current control loop is developed and presented in a block diagram as shown in Figure 21.6.

The open-loop and closed-loop transfer functions of the current controller loop can then be calculated from Figure 21.6. Next, using a defined control bandwidth for the current loop, the corner frequency of the PI compensator $\left(\text{i.e. } \frac{k_{\text{ii}}}{k_{\text{ip}}}\right)$ will be chosen to be less than the control bandwidth of the current loop, ω_i, e.g. 5–10 times lower. Next, knowing that the gain of the open-loop current transfer function (Figure 21.6) should be 0 dB at ω_i, the controller parameters are tuned.

The same approach is used to regulate the parameters of the voltage controller. A control diagram for the voltage control loop of the DC/DC converters for the storage units is depicted in Figure 21.7. The open-loop and closed-loop transfer functions of the voltage control loop are calculated from

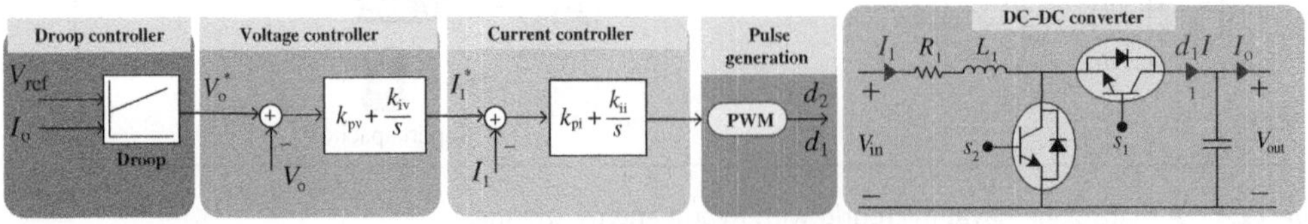

Figure 21.5 Multi-loop control diagram of DC/DC converters in shipboard microgrids.

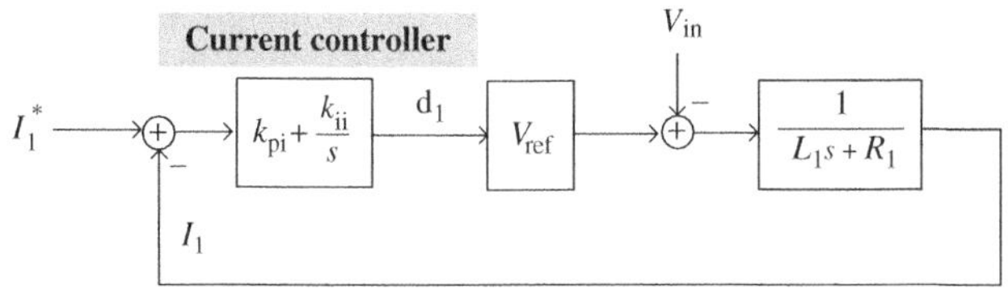

Figure 21.6 Current control loop for DC/DC converters.

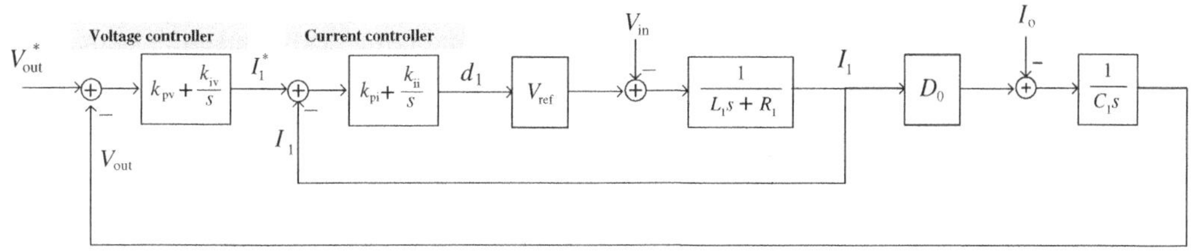

Figure 21.7 Control diagram of voltage control loop for storage units.

Figure 21.7. The parameters of the PI compensator for voltage regulator, i.e. k_{pv} and k_{iv}, are tuned using the control bandwidth of the voltage controller, ω_v. The voltage controller should be slower than the current controller. Normally, the bandwidth of the current controller should be selected 10 times higher than the voltage loop, i.e. $\omega_i = 10\omega_v$. Consequently, the corner frequency of the voltage controller, $\frac{k_{iv}}{k_{iv}}$, should be selected to be 5–10 times less than ω_v, i.e. $\frac{k_{ii}}{k_{ip}} = \eta\omega_v$, where η is selected from 5 to 10. In addition, the gain of the open-loop transfer function from Figure 21.7 should be 0 at ω_v. Solving these two equations will provide the regulator gains. The damping will also be regulated by choosing a larger η value.

21.3.2 Generator Control

For parallel operation of generators, automatic voltage regulators (AVRs) are used to regulate the DC-link voltage at the DC bus. In case of multiple generators running in parallel, the DC voltage is shared between the generators' excitation systems using a resistive droop control methodology [60].

This concept is illustrated in Figure 21.8, where V_{Go} is the DC voltage at the common node and I_{Go} is the generator's DC current. The conventional droop control design for the generators includes a virtual resistive droop control that can be used for steady-state power sharing between parallel SGs [36, 37]. The droop control for the generators is formulated as:

$$V_{Go,i} = V_{ref} - R_{G,i} I_{Go,i} \tag{21.3}$$

where V_{ref} is the reference DC bus voltage, $V_{Go,i}$, $R_{G,i}$, and $I_{Go,i}$ are the ith generator output voltage, resistive droop coefficient, and output current, respectively. The droop gains for the generators will be designed by:

$$R_{G,i} = \frac{\Delta V_{max}}{I_{i,max}} \tag{21.4}$$

where ΔV_{max} is the maximum allowable voltage dip on the DC link and $I_{i,max}$ is the rated current for generators. The performance of primary loops for the storage units and generators will be validated through rigorous time-domain simulations and frequency responses of open-loop and closed-loop transfer functions before designing the outer loops in Task 2. The PI's preliminary work in inner

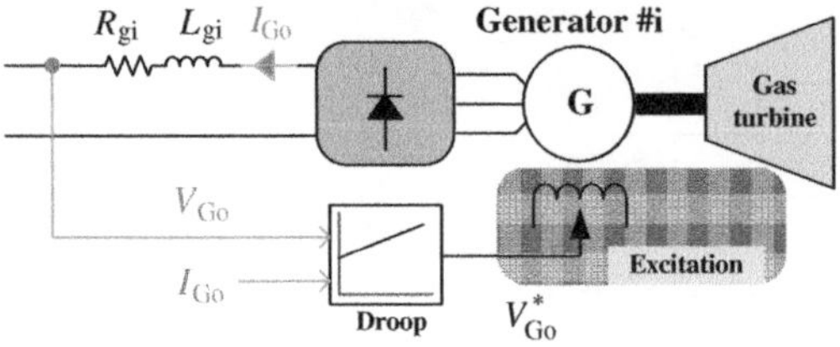

Figure 21.8 Generator control with droop capability.

current and voltage control design for power converters in smart grids [61–71] will facilitate the successful delivery of the objectives defined in this task.

21.3.3 Resistive-Capacitive Droop Control

Before explaining the proposed control design for parallel and series RC droop controllers for storage units, a background on capacitive droop controllers and numerical simulations on performance of resistive and capacitive droop controllers are provided. Considering a conventional resistive droop controller for generators (following Eq. (21.3)) and battery storage units, dynamics of the droop controllers for batteries can be formulated as:

$$V_{\mathrm{Bo},i} = V_{\mathrm{ref}} - R_{\mathrm{B},i} I_{\mathrm{Bo},i} \tag{21.5}$$

where $V_{\mathrm{Bo},i}$, $R_{\mathrm{B},i}$, and $I_{\mathrm{Bo},i}$ are battery storage ith voltage, virtual resistive droop coefficient, and current, respectively. For the SCs, a capacitive droop control was recently proposed that uses a capacitive virtual impedance that only responds during transients and acts as an open circuit in a steady state. Therefore, the SC only responds to the load change during transients [36]. The capacitive droop control for SCs can be formulated as:

$$V_{\mathrm{SCo},i} = V_{\mathrm{ref}} - \frac{1}{C_{\mathrm{SC},i} s} I_{\mathrm{SCo},i} \tag{21.6}$$

where $V_{\mathrm{SCo},i}$, $C_{\mathrm{SC},i}$, and $I_{\mathrm{SCo},i}$ are the ith SC output voltage, capacitive droop coefficient, and output current, respectively. Consider a simplified single-line diagram of an MVDC shipboard system with a generator, a battery storage system, an SC, and a load as illustrated in Figure 21.9.

As can be seen in Figure 21.9a, the components are linked to the DC bus through resistive cables (R_{g2}, R_{b2}, and R_{sc2}), V_{ref} is the voltage setpoint of each power source, and R_{g1}, R_{b1}, and C_{sc1} are resistive and capacitive droop gains of generator, battery, and SC, respectively. The droop gains are normally larger than the line impedances in MVDC shipboard system. Therefore, to simplify the derivations, the line impedances can be ignored. With simple Kirchhoff's voltage and current laws (KVL and KCL) applied to Figure 21.9a, current sharing between sources is derived as

$$I_{\mathrm{G1}} = \frac{R_{\mathrm{B1}}}{R_{\mathrm{B1}} + R_{\mathrm{G1}}} \frac{I_{\mathrm{o}}}{R_{\mathrm{eq}} C_{\mathrm{SC1}} s + 1} \tag{21.7}$$

$$I_{\mathrm{B1}} = \frac{R_{\mathrm{G1}}}{R_{\mathrm{B1}} + R_{\mathrm{G1}}} \frac{I_{\mathrm{o}}}{R_{\mathrm{eq}} C_{\mathrm{SC1}} s + 1} \tag{21.8}$$

$$I_{\mathrm{SC1}} = \frac{R_{\mathrm{eq}} C_{\mathrm{SC1}} s}{R_{\mathrm{eq}} C_{\mathrm{SC1}} s + 1} I_{\mathrm{o}}, \quad R_{\mathrm{eq}} = R_{\mathrm{B1}} \| R_{\mathrm{G1}} \tag{21.9}$$

where I_{o} is the common load current. A simulation study was conducted to show the load current sharing between the sources as shown in Figure 21.9b. By taking a look at the time-domain simulations of a load change event in Figure 21.9b, it is observed that the SC immediately responds to the load change event and gradually reduces its current to zero, while the battery storage and generator share the load current based on their droop gains. The droop gains for the battery storage can be designed by:

$$R_i = \frac{\Delta V_{\mathrm{max}}}{I_{i,\mathrm{max}}} \tag{21.10}$$

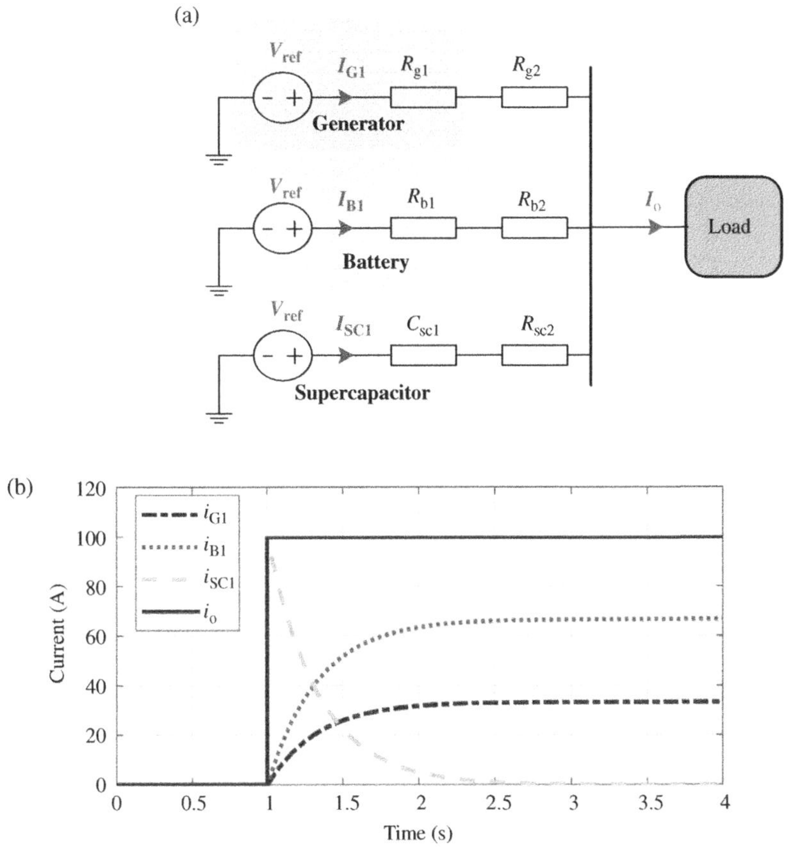

Figure 21.9 Simulation results on power sharing with resistive and capacitive droop controllers (R_{G1} = 200 Ω, R_{B1} = 100 Ω, τ = 0.33 seconds, C_{SC1} = 5 mF).

where ΔV_{max} is the maximum allowable voltage dip on the DC link and $I_{i,\ max}$ is the rated current for batteries. The power-sharing speed can be regulated by modifying the time constant ($\tau = R_{eq}C_{SC1}$) of the control design. By calculating the resistive droop gain of the battery and selecting the time constant, τ for desirable response, the virtual capacitance for the SC can then be tuned to achieve a desirable dynamic response. Detailed stability analysis can be conducted to find the stable region for various droop gain designs.

21.3.4 Proposed Complex Droop Control

The proposed control scheme is designed to ensure that SGs only respond to low-frequency fluctuations of power, BESSs respond to medium-frequency power variations as well as perform as backups to SGs, and SCs only pick up high-frequency power transients in the system. To this end, a complex droop control scheme is proposed, which can be represented by:

$$Z_g = R_g \tag{21.11}$$

$$Z_{vb} = \left(r_{db} + \frac{1}{C_{vb}s}\right) \Big\| R_{vb} \tag{21.12}$$

$$Z_{vsc} = r_{dsc} + \frac{1}{C_{vsc}s} \tag{21.13}$$

where Z_g, Z_{vb}, and Z_{vsc} are the droop gains for the SG, BESS, and SC, respectively.

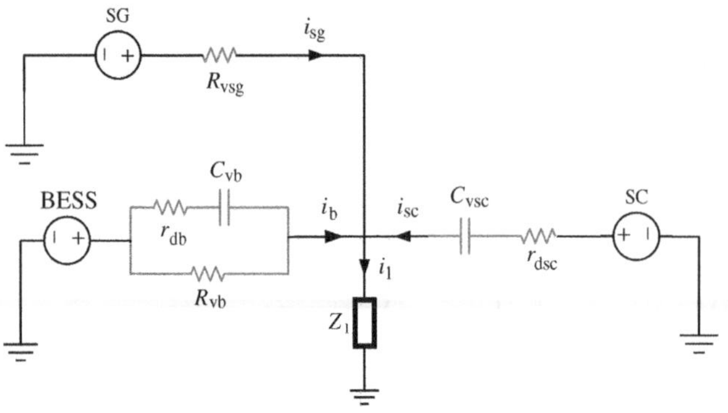

Figure 21.10 Simplified equivalent circuit of the MVDC microgrid with the proposed complex droop control scheme.

R_g and R_{vb} are virtual resistances for the SG and BESS, respectively. C_{vb} and C_{vsc} are virtual capacitances for the BESS and SC, respectively. Damping resistances for the BESS and SC are also represented by r_{db} and r_{dsc}, respectively. As the time constant of voltage and current controllers are much faster than the droop control, the equivalent circuit of the system can be illustrated as in Figure 21.10. It will be shown later that the damping resistances (r_{db} and r_{dsc}) possess small values and thus will have minimum effect on the designed dynamic power sharing. Neglecting the effect of these damping resistances, the output currents of the source-side converters can be obtained by solving the circuit in Figure 21.10 as

$$i_{sg} = \frac{1}{R_{vsg}(1 + (C_{vb} + C_{vsc}s))} i_l \tag{21.14}$$

$$i_b = \frac{1 + R_{vb}C_{vb}s}{R_{vb}(1 + (C_{vb} + C_{vsc}s))} i_l \tag{21.15}$$

$$i_{sc} = \frac{C_{vsc}s}{1 + (C_{vb} + C_{vsc}s)} i_l \tag{21.16}$$

where i_{sg}, i_b, and i_{sc} are the output currents of the SG, BESS, and SC converters, respectively. i_l is the aggregated load current. Line impedances are also neglected in Figure 21.10 due to short cable lengths in an MVDC shipboard microgrid [72].

It can be seen from (21.14)–(21.16) that the SG current responds to low-pass-filtered components of the load, while the SC picks up the high-pass-filtered load current. On the other hand, the BESS current consists of both low- and high-pass-filtered components of the load. This is the main advantage of the proposed control that enables the BESS to contribute to both steady-state and dynamic voltage stabilization. A direct comparison between the proposed method and the method of [36] is conducted in Figure 21.11. In both methods, dynamic power sharing is achieved between the system sources as the SG responds to the low-frequency component of the power disturbance, while the BESS and SC tackle the medium- and high-frequency components of the power, respectively. However, it can be seen in Figure 21.11c that the BESS does not contribute to the steady-state voltage regulation in the method of [36], while in the proposed method, the BESS provides backup power in steady-state in addition to addressing the medium-frequency disturbance according to Figure 21.11b. This results in improved voltage regulation as shown in Figure 21.11a and reduction in the SG required power, comparing Figure 21.11b,c.

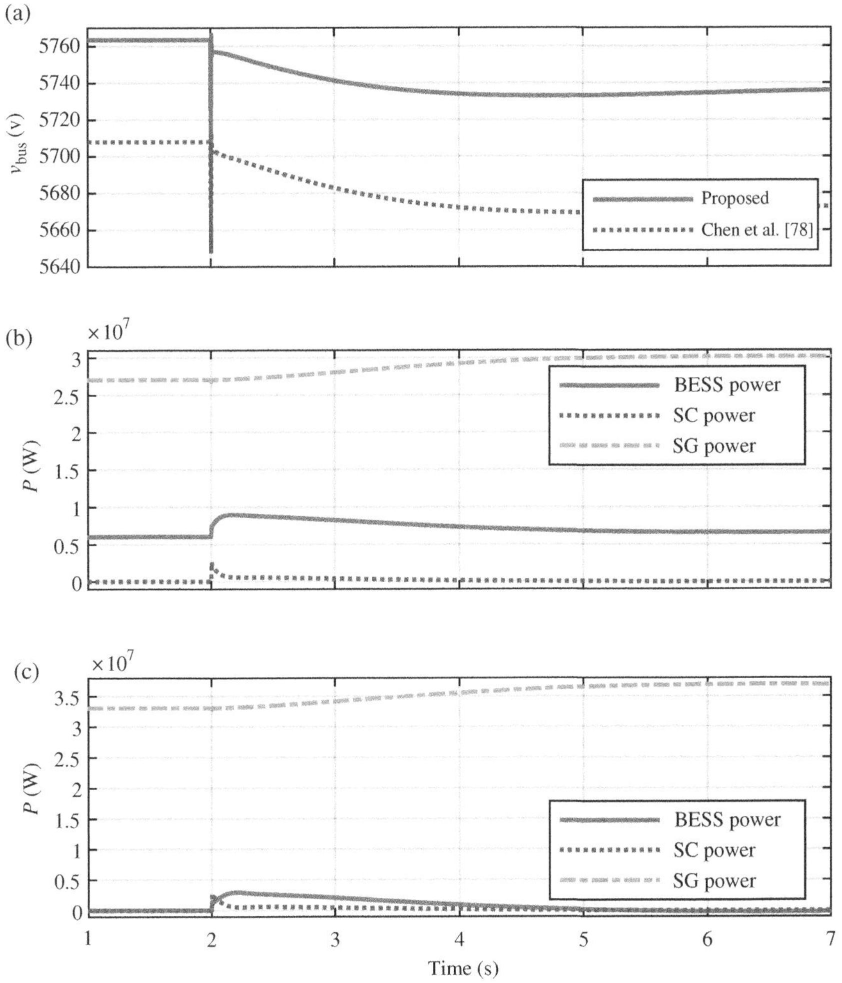

Figure 21.11 Transient response of (a) DC bus voltage, (b) sources power in the proposed method, and (c) sources power in the method of [36], subjected to a pulsed load change (C_{vsc} = 5 F, C_{vb} = 20 F, r_{db} = 0.01 Ω, r_{dsc} = 0.01 Ω). *Source:* Adapted from Chen et al. [36].

21.4 Sparse Identification of Nonlinear Dynamics of DC/DC Converters in Watercrafts

In this section, a novel approach is proposed to identify the nonlinear dynamics of DC/DC converters in shipboard microgrids using sparse regression techniques and by only relying on available state measurements. The underlying principle of the sparse identification method is that non-homogeneous dynamical systems of the form $x = f(x, u)$, given in (21.17), typically have only a few terms on the right-hand side of their state-space model [73].

$$\frac{\mathrm{d}}{\mathrm{d}t}\mathbf{x}(t) = \dot{\mathbf{x}}(t) = \mathbf{f}(\mathbf{x}(t), \mathbf{u}(t)) \tag{21.17}$$

where $\mathbf{x}(t) \in \mathbb{R}^n$ is the state vector, $\mathbf{u}(t) \in \mathbb{R}^q$ is the input or control vector, and $\mathbf{f}(\mathbf{x}(t), \mathbf{u}(t)) : \mathbb{R}^n \times \mathbb{R}^q \rightarrow \mathbb{R}^n$.

The function $\mathbf{f}$ only consists of a few active terms from the space of possible right-hand side functions. The possible right-hand side terms can be represented by a library of candidate functions including polynomials or sinusoids, $\mathbf{\Theta} \in \mathbb{R}^{m \times p}$, required for approximation of the dynamics of **f**. This library comprises a total of p candidate terms including linear and nonlinear functions of **x** and **u**. In order to obtain the proper coefficients of candidate functions in $\mathbf{\Theta}$, m time-series snapshots of the state vector **x** and the input vector **u** are collected either through simulations or experiments. Then, they are arranged into matrices of the forms

$$\mathbf{X} = \begin{bmatrix} \mathbf{x_1} & \mathbf{x_2} & \dots & \mathbf{x_m} \end{bmatrix}^{\mathrm{T}} \text{and } \mathbf{U} = \begin{bmatrix} \mathbf{u_1} & \mathbf{u_2} & \dots & \mathbf{u_m} \end{bmatrix}^{\mathrm{T}} \tag{21.18}$$

Therefore, for a system with n states, q inputs, and m time-series snapshots of data, **X** and **U** can be written as

$$\mathbf{X} = \begin{bmatrix} \mathbf{x}^{\mathrm{T}}(t_1) \\ \mathbf{x}^{\mathrm{T}}(t_2) \\ \vdots \\ \mathbf{x}^{\mathrm{T}}(t_m) \end{bmatrix} = \begin{bmatrix} x_1(t_1) & x_2(t_1) & \cdots & x_n(t_1) \\ x_1(t_2) & x_2(t_2) & \cdots & x_n(t_2) \\ \vdots & \vdots & \ddots & \vdots \\ x_1(t_m) & x_2(t_m) & \cdots & x_n(t_m) \end{bmatrix} \tag{21.19}$$

$$\mathbf{U} = \begin{bmatrix} \mathbf{u}^{\mathrm{T}}(t_1) \\ \mathbf{u}^{\mathrm{T}}(t_2) \\ \vdots \\ \mathbf{u}^{\mathrm{T}}(t_m) \end{bmatrix} = \begin{bmatrix} u_1(t_1) & u_2(t_1) & \cdots & u_q(t_1) \\ u_1(t_2) & u_2(t_2) & \cdots & u_q(t_2) \\ \vdots & \vdots & \ddots & \vdots \\ u_1(t_m) & u_2(t_m) & \cdots & u_q(t_m) \end{bmatrix} \tag{21.20}$$

The collected state and input measurements are then compiled into the library $\mathbf{\Theta}$ as given in (21.21), where higher-order polynomials are denoted by functions **P2**, **P3**, etc. For example, **P2** (**X**, **U**) includes candidate functions comprising second-order quadratic functions of states and inputs as (21.22).

$$\mathbf{\Theta}(\mathbf{X}, \mathbf{U}) = \begin{bmatrix} | & | & | & | & | & | & | & | & | & | & | \\ \mathbf{1} & \mathbf{X} & \mathbf{U} & \mathbf{P2}(\mathbf{X}) & \mathbf{P2}(\mathbf{X{,}U}) & \mathbf{P2}(\mathbf{U}) & \dots & \sin(\mathbf{X}) & \sin(\mathbf{U}) & \sin(\mathbf{P2}(\mathbf{X{,}U})) & \dots \\ | & | & | & | & | & | & | & | & | & | & | \end{bmatrix} \tag{21.21}$$

As can be seen in (21.21), the library of candidate functions can include constant, polynomial, and trigonometric terms. A general guideline is to first include simple functions such as polynomials, and then add more complex terms such as trigonometric functions to the library. However, due to the gray-box nature of the identification method, partial knowledge of the dynamical system can be leveraged to decide what functions to include in the library [74]. A sample term from the library of (21.21) representing can be written as

$$\mathbf{P}_2(\mathbf{X}, \mathbf{U}) = \begin{bmatrix} x_1(t_1)u_1(t_1) & x_1(t_1)u_2(t_1) & \cdots & x_n(t_1)u_q(t_1) \\ x_1(t_2)u_1(t_2) & x_1(t_2)u_2(t_2) & \cdots & x_n(t_2)u_q(t_2) \\ \vdots & \vdots & \ddots & \vdots \\ x_1(t_m)u_1(t_m) & x_1(t_m)u_2(t_m) & \cdots & x_n(t_m)u_q(t_m) \end{bmatrix} \tag{21.22}$$

The obtained data matrices then can be used to represent the dynamical system of (21.17) as

$$\dot{\mathbf{X}} = \mathbf{\Theta}(\mathbf{X}, \mathbf{U})\mathbf{\Xi} \tag{21.23}$$

where $\mathbf{\Xi}$ is the matrix of coefficients for the candidate functions in $\mathbf{\Theta}$ to be obtained using sparse regression.

21.4.1 Smoothing Data for Derivative Estimation

There is usually some noise in the data collected for identification. As a result of using numerical differentiation methods, identification performance may be compromised due to increased noise levels. Filtering the noisy data is therefore an important step before estimating the time derivative of states. To smooth data, various approaches such as the Savitzky–Golay filter, the total variation regularized derivative, and the integral formulation of the governing equations are proposed [75]. However, in [75], it was mentioned that for relatively clean data, no smoothing is required since it will result in lost information in the measurement signals. In this study, since the measurement data used for the identification of the converter are already filtered averages of the switching waveforms and preprocessed for anomalies, there is no need to apply extra filtering steps.

21.4.2 Estimating the Time Derivative Matrix $\dot{\mathbf{X}}$

The time derivative of the state matrix, i.e. $\dot{\mathbf{X}} = [\dot{\mathbf{x}}_1 \dot{\mathbf{x}}_2 \dots \dot{\mathbf{x}}_m]^{\mathrm{T}}$, is required to complete the data collection for the identification process. $\dot{\mathbf{X}}$ is typically approximated by numerical differentiation methods such as finite difference schemes. Considering a smooth function in the neighborhood of point x, the derivatives can be approximated using Taylor series expansion at specified mesh points. Due to the transient nature of capacitor current and inductor voltages in the DC/DC converters in watercrafts, a central finite difference of fourth order is used to approximate the derivatives more accurately, which is given by

$$\dot{\mathbf{X}} \approx \frac{8\mathbf{X}(t_{i+1}) - 8\mathbf{X}(t_{i-1}) + \mathbf{X}(t_{i-2}) - \mathbf{X}(t_{i+2})}{12h} \tag{21.24}$$

where $\mathbf{X}(t_{i+1})$ is the time-series sample $i+1$ and h is the mesh spacing, which is considered the same as the sampling time of the simulation in this study, i.e. 10 μs.

21.4.3 Identification by Sparse Regression

The last step in identifying the dynamics of DC/DC converters using available measurements is solving for the coefficients of sparse gain matrix in Ξ that decide what terms are active in the $\dot{\mathbf{X}}$ dynamics. The goal of sparse regression is to obtain the fewest terms in Ξ that result in a close fit for the studied dynamical system using the gathered data. To obtain the sparse gains, the following optimization model needs to be solved

$$\xi_{\mathbf{k}} = \arg\min_{\hat{\xi}_{\mathbf{k}}} \|\dot{\mathbf{X}}_k - \Theta(\mathbf{X},\ \mathbf{U})\hat{\xi}_{\mathbf{k}}\|_2 + \lambda\|\hat{\xi}_{\mathbf{k}}\|_0 \tag{21.25}$$

where $\xi_{\mathbf{k}}$ and $\dot{\mathbf{X}}_k$ are the k-th columns of the $\dot{\mathbf{X}}$ and Ξ, respectively. The objective function in (21.25) comprises two norm functions. The L2 norm denoted by $\|.\|_2$, solves the least-squares problem to minimize the error between numerically calculated derivatives and predicted derivatives of the DC/DC converter, while the L0 norm, $\|.\|_0$, tries to minimize the number of nonzero elements in $\xi_{\mathbf{k}}$, which, in turn, promotes the sparsity in the coefficients matrix. Furthermore, λ is the sparsity-promoting hyperparameter that can be tuned to result in the best estimation.

The optimization problem of (21.25) is solved using sequential thresholded least squares [73] and the algorithm for this approach is given in Algorithm 21.1. The algorithm first performs the least-squares regression to find a nonsparse initial guess of coefficients $\hat{\xi}_{\mathbf{k}}$. The nonsparse $\hat{\xi}_{\mathbf{k}}$ will include some very small terms, which are then hard thresholded and zeroed out. This procedure is iteratively repeated by the algorithm until convergence. The choice of λ is crucial for convergence of the

Algorithm 21.1 Sequential Thresholded Least Squares

Input: Measurements $\mathbf{X}$, $\mathbf{U}$
Input: Estimated derivatives $\dot{\mathbf{X}}$
Procedure LEAST-SQUARE
$\Xi = \boldsymbol{\Theta}^{-1}\dot{\mathbf{X}}$ (initial guess using least-squares)
  for $k = 1 : \mathrm{p}$ (number of iterations)
    Set λ (sparsity hyperparameter)
    $\mathrm{ind_{small}} \leftarrow |\Xi| < \lambda$
    $0 \leftarrow \Theta(\mathrm{ind_{small}})$
    for $k = 1 : n$ **do** (n dimension of state $\mathbf{X}$)
      $\mathrm{ind_{big}} \neq \mathrm{ind_{small}}(:, \mathrm{k})$
      $\Xi(ind_{big}, k) = \Theta(:, ind_{big})^{-1}\dot{\mathbf{X}}(:, k)$
    end for
  end for
Output: sparse matrix Ξ

algorithm. While increasing λ results in sparser ξ_k as the algorithm tries to minimize the objective function in (21.25), it can lead to an underfit model with a high error. On the other hand, small values of λ increase the complexity of the model by increasing the number of nonzero coefficient while achieving a smaller error. Therefore, λ is tuned such that a good balance between sparsity and the least-squares error is achieved.

21.4.4 Dynamic Model of the DC/DC Converters

To identify the dynamics of DC/DC converters in watercrafts using measurements, a bidirectional half-bridge converter is considered for dynamics identification. The power and control stages of the converter are demonstrated in Figure 21.5. The converter operates under voltage mode control with a cascaded structure. The voltage reference is dictated by the conventional droop control, enabling decentralized power sharing among parallel converters [76]. The average dynamics of the converter to be discovered can be written as

$$\frac{d\bar{i}_L}{dt} = \frac{1}{L}\left(\bar{v}_{\mathrm{in}} - r\bar{i}_{\mathrm{L}} - \bar{v}_{\mathrm{o}} + \bar{d}\bar{v}_{\mathrm{o}}\right) \tag{21.26}$$

$$\frac{d\bar{v}_{\mathrm{o}}}{dt} = \frac{1}{C}\left(\bar{i}_{\mathrm{L}} - \bar{d}\bar{i}_{\mathrm{L}} - \bar{i}_{\mathrm{o}}\right) \tag{21.27}$$

where $\bar{i}_{\mathrm{L}}$ and $\bar{v}_{\mathrm{o}}$ are the state variables, and $\bar{d}$, $\bar{i}_{\mathrm{o}}$, and $\bar{v}_{\mathrm{in}}$ are the system inputs. The bar sign indicates an averaged value over a switching period of the converter. According to Figure 21.5, the duty cycle d is decided by the output of the control system as

$$d = \left[\left(v_{\mathrm{o}}^{*} - v_{\mathrm{o}}\right)G_{\mathrm{v}}(s) - i_{\mathrm{L}}\right]G_{\mathrm{i}}(s) \tag{21.28}$$

$$G_{\mathrm{v}}(s) = k_{\mathrm{pv}} + k_{\mathrm{iv}}/s \tag{21.29}$$

$$G_{i}(s) = k_{\mathrm{pi}} + k_{\mathrm{ii}}/s \tag{21.30}$$

where k_{pv} and k_{iv} are the proportional and integral gains for the outer voltage controller, and k_{pi} and k_{ii} are the gains for the inner current controller. These gains are designed using frequency response methods to ensure stable operation of the cascaded control structure as discussed in the previous section. In addition, v_o^* is decided by the droop control as

$$v_o^* = v_{ref} - Z_v i_o \tag{21.31}$$

where Z_v is the droop gain.

21.4.5 Case Studies

The first step in the identification process is to collect measurement data. To this end, an external current source that perturbs the system with the injected current i_{inj} is applied to the output terminals of the DC/DC converter. The external source injects a swept-sine signal at different steady-state operating points of the converter. In this work, the amplitude of the excitation signal is maintained at $0.05|i_o|$, so that the system is not significantly perturbed while the amplitude is large enough to generate reasonable excitation samples for identification purposes. To decide the library of candidate terms $\boldsymbol{\Theta}$, a basic knowledge of the system dynamics should be available. The candidate terms considered in this study are polynomials including (21.26) and (21.27), i.e. $\bar{v}_{in}$, $\bar{i}_L$, $\bar{v}_o$, $\overline{d\bar{v}_o}$, $\overline{d\bar{i}_L}$, and $\bar{i}_o$. Before applying the sparse identification technique, the collected data should be grouped into training and testing datasets. Then by applying the sparse regression algorithm in (21.25), the identified dynamic equations of the DC/DC converter result in (21.32) and (21.33).

$$\frac{d\bar{i}_L}{dt} = M\bar{v}_{in} + N\bar{i}_L + P\bar{v}_o + q\overline{d\bar{v}_o} \tag{21.32}$$

$$\frac{d\bar{v}_o}{dt} = f\bar{i}_L + g\overline{d\bar{i}_L} + h\bar{i}_o \tag{21.33}$$

Table 21.1 compares the coefficients of the nonlinear terms in (21.32) and (21.33) for the analytical and estimated model of the CUT obtained with 75% of data as the training dataset. The resulting coefficients can be seen to be very close to their actual counterparts.

21.4.6 Time-Domain Verification

In this case, the nonlinear dynamic model of the converter is developed in MATLAB and the dynamic response of the system with the estimated parameters in Table 21.1 is compared with its analytical model. The results are illustrated in Figure 21.12. Two step changes are applied to

Table 21.1 Coefficients of the CUT dynamic model.

Coefficients	Analytical value	Estimated value
M	370.3704	372.6335
N	−37.037	−18.5521
P	−370.3704	−372.7679
q	370.3704	372.8829
f	5.2762×10^4	5.2768×10^4
g	-5.2762×10^4	-5.2751×10^4
h	-5.2762×10^4	-5.2762×10^4

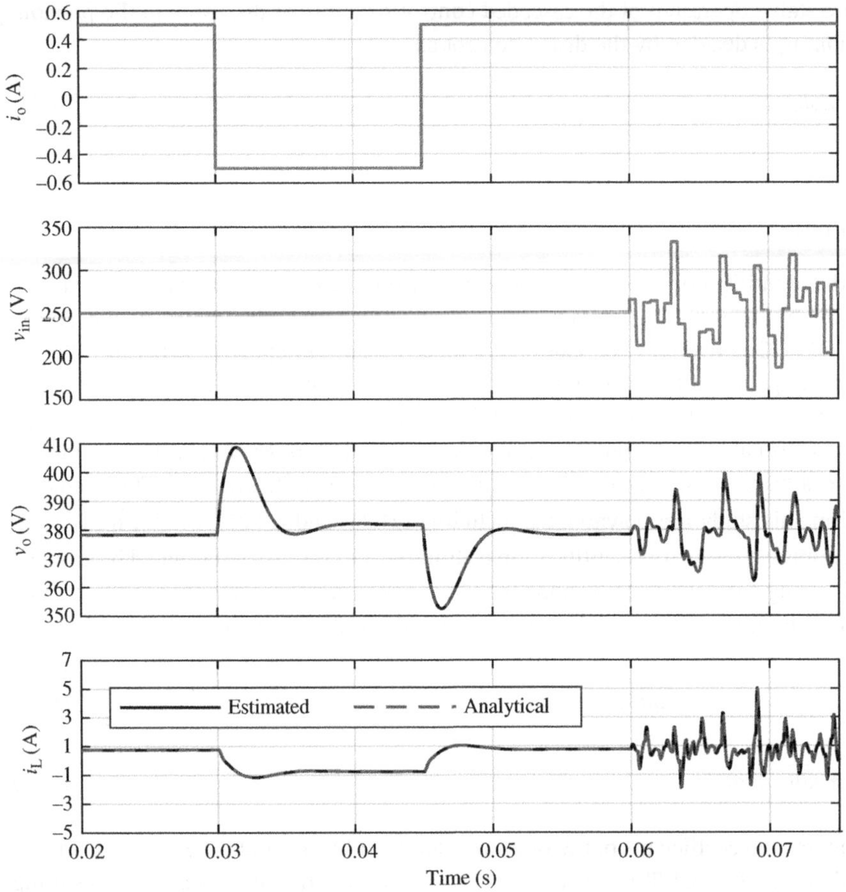

Figure 21.12 Analytical dynamic response of the DC/DC converters compared with the estimated model under disturbances.

the output current at $t = 0.03$ s and $t = 0.045$ s such that a transition between the charging and discharging modes of the DC/DC converter is realized. Furthermore, at $t = 0.06$ s, a white noise perturbation is applied to the input voltage to emulate the volatility of an input source. The estimated model by the proposed method can be seen to closely follow the analytical model, which confirms the accurate identification of the DC/DC converter dynamic model using the proposed sparse regression technique.

21.5 Conclusion and Future Work

In this chapter, the problem of control and system identification of DC watercrafts has been studied. To resolve the existing power-sharing challenges in DC shipboard microgrids with conventional generation and hybrid energy storage systems, complex droop control is proposed that provides multiple degrees of freedom to adjust steady-state and dynamic power sharing between hybrid storage composed of batteries and SCs as well as conventional generation units over the entire frequency range while taking into account the operating modes of the system. The main advantage of the proposed control over the existing methods is that it enables the BESS to address steady-state and medium-frequency power variations at the same time, which reduces the need for backup power from SGs, takes full advantage of BESS dynamic response capabilities, and improves voltage regulation. It also enables the SC to enhance its contribution to high-frequency oscillations damping. The results of the study provide guidelines for the design of filter-less power management in DC watercrafts. In addition, to address the system identification problem of DC watercrafts, a novel system identification framework for direct discovery of the non-linear dynamic model of droop-controlled DC/DC converters is proposed using a sparse regression technique. By collecting the measurement data online and without the need to interrupt the system, the training phase can successfully identify nonlinear dynamics of the DC watercrafts with a small approximation error. Time-domain simulations demonstrate a close fit of the identified model to its analytical counterpart.

Future works may include hardware validation and testing of the proposed methods on the converters with different control schemes such as current mode control operating in different watercraft topologies.

References

1 Steinke, J.K., Maibach, P., Ortiz, G. et al. (2019). Mvdc applications and technology. *PCIM Europe 2019; International Exhibition and Conference for Power Electronics, Intelligent Motion, Renewable Energy and Energy Management*, pp. 1–8.

2 Jin, Z., Sulligoi, G., Cuzner, R. et al. (2016). Next-generation shipboard dc power system: introduction smart grid and dc microgrid technologies into maritime electrical networks. *IEEE Electrification Mag.* 4 (2): 45–57.

3 Ghimire, P., Park, D., Zadeh, M.K. et al. (2019). Shipboard electric power conversion: Ssystem architecture, applications, control, and challenges [technology leaders]. *IEEE Electrification Mag.* 7 (4): 6–20.

4 Othman, M.B., Reddy, N.P., Ghimire, P. et al. (2019). A hybrid power system laboratory: testing electric and hybrid propulsion. *IEEE Electrification Mag.* 7 (4): 89–97.

5 Xu, L., Wei, B., Yu, Y. et al. (2021). Coordinated control of diesel generators and batteries in dc hybrid electric shipboard power system. *Energies* 14 (19): [Online]. Available:. https://www.mdpi.com/1996-1073/14/19/6246.

6 Mutarraf, M.U., Terriche, Y., Nasir, M. et al. (2021). A communication-less multimode control approach for adaptive power sharing in ship-based seaport microgrid. *IEEE Trans. Transport. Electrification* 7 (4): 3070–3082.

7 Ghimire, P., Zadeh, M., Pedersen, E., and Thorstensen, J. (2021). Dynamic modeling, simulation, and testing of a marine dc hybrid power system. *IEEE Trans. Transport. Electrification* 7 (2): 905–919.

8 Chua, L.W.Y., Tjahjowidodo, T., Seet, G.G.L., and Chan, R. (2018). Implementation of optimization-based power management for all-electric hybrid vessels. *IEEE Access* 6: 74339–74354.

9 Edrington, C.S., Ozkan, G., Papari, B. et al. (2020). Distributed energy management for ship power systems with distributed energy storage. *J. Mar. Eng. Technol.* 19 (suppl 1): 31–44.

10 Edrington, C.S., Ozkan, G., Papari, B., and Perkins, D. (2021). Distributed adaptive power management for medium voltage ship power systems. *J. Mar. Eng. Technol.* 0 (0): 1–16. [Online]. Available: https://doi.org/10.1080/20464177.2021.1894783.

11 Lashway, C.R., Elsayed, A.T., and Mohammed, O.A. (2016). Hybrid energy storage management in ship power systems with multiple pulsed loads. *Electr. Power Syst. Res.* 141: 50–62.

12 Khan, M.M.S., Faruque, M.O., and Newaz, A. (2017). Fuzzy logic based energy storage management system for mvdc power system of all electric ship. *IEEE Trans. Energy Convers.* 32 (2): 798–809.

13 Mo, R. and Li, H. (2016). Hybrid energy storage system with active filter function for shipboard mvdc system applications based on isolated modular multilevel dc/dc converter. *IEEE J. Emerg. Sel. Top. Power Electron.* 5 (1): 79–87.

14 Khan, M.M.S. and Faruque, M. (2016). Management of hybrid energy storage systems for mvdc power system of all electric ship. *2016 North American Power Symposium (NAPS)*. IEEE, pp. 1–6.

15 Jin, Z., Meng, L., Vasquez, J.C., and Guerrero, J.M. (2017). Frequency-division power sharing and hierarchical control design for dc shipboard microgrids with hybrid energy storage systems. *2017 IEEE Applied Power Electronics Conference and Exposition (APEC)*. IEEE, pp. 3661–3668.

16 Jin, Z., Meng, L., Guerrero, J.M., and Han, R. (2017). Hierarchical control design for a shipboard power system with dc distribution and energy storage aboard future more-electric ships. *IEEE Trans. Ind. Inf.* 14 (2): 703–714.

17 Lai, K. and Illindala, M.S. (2018). A distributed energy management strategy for resilient shipboard power system. *Appl. Energy* 228: 821–832.

18 Zhu, L., Liu, J., Cupelli, M., and Monti, A. (2013). Decentralized linear quadratic gaussian control of multi-generator mvdc shipboard power system with constant power loads. in *2013 IEEE Electric Ship Technologies Symposium (ESTS)*. IEEE, pp. 308–313.

19 Cupelli, M., Monti, A., De Din, E., and Sulligoi, G. (2016). Case study of voltage control for mvdc microgrids with constant power loads - comparison between centralized and decentralized control strategies. *2016 18th Mediterranean Electrotechnical Conference (MELECON)*, pp. 1–6.

20 Faddel, S., Youssef, T.A., and Mohammed, O. (2018). Decentralized controller for energy storage management on mvdc ship power system with pulsed loads. *2018 IEEE Transportation Electrification Conference and Expo (ITEC)*. IEEE, pp. 254–259.

21 Faddel, S., El Hariri, M., and Mohammed, O. (2018). Intelligent control framework for energy storage management on mvdc ship power system. *2018 IEEE International Conference on Environment and Electrical Engineering and 2018 IEEE Industrial and Commercial Power Systems Europe (EEEIC/I&CPS Europe)*. IEEE, pp. 1–6.

22 Cupelli, M., Monti, A., De Din, E., and Sulligoi, G. (2016). Case study of voltage control for mvdc microgrids with constant power loads-comparison between centralized and decentralized control strategies. *2016 18th Mediterranean Electrotechnical Conference (MELECON)*. IEEE, pp. 1–6.

23 Saad, A.A., Faddel, S., Youssef, T., and Mohammed, O. (2019). Small-signal model predictive control based resilient energy storage management strategy for all electric ship mvdc voltage stabilization. *J. Energy Storage* 21: 370–382.

24 Mildt, D. and Kubo, R. (2017). Decentralized hybrid switching control of multiconverter mvdc shipboard power systems. *IECON 2017-43rd Annual Conference of the IEEE Industrial Electronics Society*. IEEE, pp. 6795–6800.

25 Mills, A.J. and Ashton, R.W. (2017). Adaptive, sparse, and multi-rate lqr control of an mvdc shipboard power system with constant power loads. *2017 IEEE International Conference on Industrial Technology (ICIT)*. IEEE, pp. 498–503.

26 Cupelli, M., Gurumurthy, S.K., and Monti, A. (2017). Modelling and control of single phase dab based mvdc shipboard power system. *IECON 2017-43rd Annual Conference of the IEEE Industrial Electronics Society*. IEEE, pp. 6813–6819.

27 Cooper, S. and Nehrir, H. (2017). Ensuring stability in a multi-zone mvdc shipboard power system. *2017 IEEE Electric Ship Technologies Symposium (ESTS)*. IEEE, pp. 380–387.

28 Shen, Q. (2012). Distributed control approach for power and energy management in a notional shipboard power system. PhD thesis. Florida State University.

29 Vu, T.V., Perkins, D., Gonsoulin, D., et al. (2018). Large-scale distributed control for mvdc ship power systems. *IECON 2018-44th Annual Conference of the IEEE Industrial Electronics Society*. IEEE, pp. 3431–3436.

30 Ni, K., Hu, Y., and Li, X. (2017). An overview of design, control, power management, system stability and reliability in electric ships. *Power Electron. Drives* 2 (2): 5–29.

31 Mutarraf, M.U., Terriche, Y., Niazi, K.A.K. et al. (2018). Energy storage systems for shipboard microgridsa review. *Energies* 11 (12): 3492.

32 Mutarraf, M.U., Terriche, Y., Niazi, K.A.K. et al. (2019). Control of hybrid diesel/pv/battery/ultra-capacitor systems for future shipboard microgrids. *Energies* 12 (18): 3460.

33 Elsayed, A.T., Elsayad, N., and Mohammed, O.A. (2016). Pareto based optimal sizing and energy storage mix in ship power systems. *2016 IEEE Industry Applications Society Annual Meeting*. IEEE, pp. 1–6.

34 Faddel, S., Saad, A.A., Youssef, T., and Mohammed, O. (2019). Decentralized control algorithm for the hybrid energy storage of shipboard power system. *IEEE J. Emerg. Sel. Top. Power Electron.* 8 (1): 720–731.

35 Faddel, S., Saad, A.A., and Mohammed, O. (2018). Decentralized energy management of hybrid energy storage on mvdc shipboard power system. *2018 IEEE Industry Applications Society Annual Meeting (IAS)*. IEEE, pp. 1–7.

36 Chen, X., Zhou, J., Shi, M. et al. (2019). A novel virtual resistor and capacitor droop control for hess in medium-voltage dc system. *IEEE Trans. Power Syst.* 34 (4): 2518–2527.

37 Xu, Q., Hu, X., Wang, P. et al. (2016). A decentralized dynamic power sharing strategy for hybrid energy storage system in autonomous dc microgrid. *IEEE Trans. Ind. Electron.* 64 (7): 5930–5941.

38 Lu, X., Sun, K., Guerrero, J.M. et al. (2013). State-of-charge balance using adaptive droop control for distributed energy storage systems in dc microgrid applications. *IEEE Trans. Ind. Electron.* 61 (6): 2804–2815.

39 Hoang, K.D. and Lee, H.-H. (2018). Accurate power sharing with balanced battery state of charge in distributed dc microgrid. *IEEE Trans. Ind. Electron.* 66 (3): 1883–1893.

40 Kim, H.-J., Chun, C.Y., Lee, K.-J. et al. (2015). Control strategy of multiple energy storages system for dc microgrid. *2015 9th International Conference on Power Electronics and ECCE Asia (ICPE-ECCE Asia)*. IEEE, pp. 1750–1755.

41 Rojas-Dueas, G., Riba, J.-R., and Moreno-Eguilaz, M. (2021). Black-box modeling of dcdc converters based on wavelet convolutional neural networks. *IEEE Trans. Instrum. Meas.* 70: 1–9.

42 Francs-Roger, A., Anvari-Moghaddam, A., Rodrguez-Daz, E. et al. (2018). Dynamic assessment of cots converters-based dc integrated power systems in electric ships. *IEEE Trans. Ind. Informat.* 14 (12): 5518–5529.

43 Al-Greer, M., Armstrong, M., Ahmeid, M., and Giaouris, D. (2019). Advances on system identification techniques for dcdc switch mode power converter applications. *IEEE Trans. Power Electron.* 34 (7): 6973–6990.

44 Wunderlich, A., Booth, K., and Santi, E. (2021). Hybrid analytical and data-driven modeling techniques for digital twin applications. *2021 IEEE Electric Ship Technologies Symposium (ESTS)*, pp. 1–7.

45 Arnedo, L., Boroyevich, D., Burgos, R., and Wang, F. (2008). Un-terminated frequency response measurements and model order reduction for black-box terminal characterization models. *2008 Twenty-Third Annual IEEE Applied Power Electronics Conference and Exposition*, pp. 1054–1060.

46 Cvetkovic, I., Boroyevich, D., Mattavelli, P. et al. (2013). Unterminated small-signal behavioral model of dcdc converters. *IEEE Trans. Power Electron.* 28 (4): 1870–1879.

47 Arnedo, L., Boroyevich, D., Burgos, R., and Wang, F. (2008). Polytopic black-box modeling of dc-dc converters. *2008 IEEE Power Electronics Specialists Conference*, pp. 1015–1021.

48 Francs, A., Asensi, R., and Uceda, J. (2019). Blackbox polytopic model with dynamic weighting functions for dc-dc converters. *IEEE Access* 7: 160263–160273.

49 Valdivia, V., Barrado, A., Lzaro, A. et al. (2014). Black-box behavioral modeling and identification of dcdc converters with input current control for fuel cell power conditioning. *IEEE Trans. Ind. Electron.* 61 (4): 1891–1903.

50 Wunderlich, A. and Santi, E. (2021). Digital twin models of power electronic converters using dynamic neural networks. *2021 IEEE Applied Power Electronics Conference and Exposition (APEC)*, pp. 2369–2376.

51 Francs, A., Asensi, R., Garca, S. et al. (2018). Modeling electronic power converters in smart dc microgridsan overview. *IEEE Trans. Smart Grid* 9 (6): 6274–6287.

52 Aguirre, L., Donoso-Garcia, P., and Santos-Filho, R. (2000). Use of a priori information in the identification of global nonlinear models-a case study using a buck converter. *IEEE Trans. Circuits Syst. I* 47 (7): 1081–1085.

53 Correa, M., Aguirre, L., and Saldanha, R. (2002). Using steady-state prior knowledge to constrain parameter estimates in nonlinear system identification. *IEEE Trans. Circuits Syst. I* 49 (9): 1376–1381.

54 Hafiz, F., Swain, A., Mendes, E.M.A.M., and Aguirre, L.A. (2020). Multiobjective evolutionary approach to grey-box identification of buck converter. *IEEE Trans. Circuits Syst. I* 67 (6): 2016–2028.

55 Oliver, J.A., Prieto, R., Cobos, A. et al. (2009). Hybrid wiener-hammerstein structure for grey-box modeling of dc-dc converters. *2009 Twenty-Fourth Annual IEEE Applied Power Electronics Conference and Exposition*, pp. 280–285.

56 Zhao, S., Peng, Y., Zhang, Y., and Wang, H. (2022). Parameter estimation of power electronic converters with physics-informed machine learning. *IEEE Trans. Power Electron.* 37 (10): 1–1.

57 Javaid, U., Freijedo, F.D., Dujic, D., and van der Merwe, W. (2017). Dynamic assessment of sourceload interactions in marine mvdc distribution. *IEEE Trans. Ind. Electron.* 64 (6): 4372–4381.

58 Weaver, W.W., Robinett, R.D., Wilson, D.G., and Matthews, R.C. (2017). Metastability of pulse power loads using the hamiltonian surface shaping method. *IEEE Trans. Energy Convers.* 32 (2): 820–828.

59 I. Standard, “60092-101, 2002” (2002). Electrical installations in ships. Definitions and general requirement.

60 Zahedi, B. and Norum, L.E. (2013). Voltage regulation and power sharing control in ship lvdc power distribution systems. *2013 15th European Conference on Power Electronics and Applications (EPE)*. IEEE, pp. 1–8.

61 Khazaei, J. (2020). Optimal flow of mvdc shipboard microgrids with hybrid storage enhanced with capacitive and resistive droop controllers. *IEEE Trans. Power Syst.* 99: 1.

62 Khazaei, J. and Nguyen, D.H. (2017). Multi-agent consensus design for heterogeneous energy storage devices with droop control in smart grids. *IEEE Trans. Smart Grid* 10 (2): 1395–1404.

63 Nguyen, D.H. and Khazaei, J. (2017). Multi-agent time-delayed fast consensus design for distributed battery energy storage systems. *IEEE Trans. Sustain. Energy* 9 (3): 1397–1406.

64 Khazaei, J., Nguyen, D.H., and Thao, N.G.M. (2017). Primary and secondary voltage/frequency controller design for energy storage devices using consensus theory. *2017 IEEE 6th International Conference on Renewable Energy Research and Applications (ICRERA)*, pp. 447–453.

65 Khazaei, J. and Miao, Z. (2018). Consensus control for energy storage systems. *IEEE Trans. Smart Grid* 9 (4): 3009–3017.

66 Khazaei, J. and Nguyen, D.H. (2019). Multi-agent consensus design for heterogeneous energy storage devices with droop control in smart grids. *IEEE Trans. Smart Grid* 10 (2): 1395–1404.

67 Nguyen, D.H. and Khazaei, J. (2018). Multiagent time-delayed fast consensus design for distributed battery energy storage systems. *IEEE Trans. Sustainable Energy* 9 (3): 1397–1406.

68 Khazaei, J. and Nguyen, D.H. (2019). Distributed consensus for output power regulation of dfigs with on-site energy storage. *IEEE Trans. Energy Convers.* 34 (2): 1043–1051.

69 Nguyen, D.H., Khazaei, J., Stewart, S.W., and Annoni, J. (2019). Distributed cooperative control of wind farms with on-site battery energy storage systems. *Advanced Control and Optimization Paradigms for Wind Energy Systems*. Springer, pp. 41–66.

70 Adeyemo, O., Idowu, P., Asrari, A., and Khazaei, J. (2018). Reactive power control for multiple batteries connected in parallel using modified power factor method. *2018 North American Power Symposium (NAPS)*. IEEE, pp. 1–6.

71 Khazaei, J., Nguyen, D.H., and Asrari, A. (2019). Consensus-based demand response of pmsg wind turbines with distributed energy storage considering capability curves. *IEEE Trans. Sustainable Energy* 11 (4): 2315–2325.

72 Babaei, M., Shi, J., and Abdelwahed, S. (2018). A survey on fault detection, isolation, and reconfiguration methods in electric ship power systems. *IEEE Access* 6: 9430–9441.

73 Brunton, S.L., Proctor, J.L., and Kutz, J.N. (2016). Discovering governing equations from data by sparse identification of nonlinear dynamical systems. *Proc. Natl. Acad. Sci.* 113 (15): 3932–3937.

74 Fasel, U., Kaiser, E., Kutz, J.N. et al. (2021). Sindy with control: A tutorial. *arXiv preprint arXiv:2108.13404*.

75 de Silva, B.M., Higdon, D.M., Brunton, S.L., and Kutz, J.N. (2020). Discovery of physics from data: Universal laws and discrepancies. *Front. Artif. Int.* 3: 25.

76 Guerrero, J.M., Chandorkar, M., Lee, T.-L., and Loh, P.C. (2013). Advanced control architectures for intelligent microgridspart i: decentralized and hierarchical control. *IEEE Trans. Ind. Electron.* 60 (4): 1254–1262.

22

Shipboard DC Hybrid Power Systems: Pathway to Electrification and Decarbonization

Mehdi Zadeh[1] *and Pramod Ghimire*[1,2]

[1] *Norwegian University of Science and Technology (NTNU), Department of Marine Technology, Trondheim, Trøndelag, 7050 Norway*
[2] *Kongsberg Digital (KDI), Maritime Simulation, Horten, Vestfold, 3190 Norway*

22.1 Introduction

The significance of ocean space is increasing as it provides various opportunities and resources, be it food and minerals or the potential of renewable energy harvesting [1]. Further, it also provides a significant contribution to global trade as 80–90% of global trade takes place through maritime transportation [2]. Therefore, it is also considered the *backbone of global trade and economy* [3]. Further, the trend of global trade through maritime transportation is increasing. This upward trend increases not only the opportunities but also the challenges of *sustainability of the blue economy*. The major challenges come with energy efficiency improvement and emission reduction of the watercraft systems.

For the efficient utilization of resources and opportunities in the ocean space, various watercraft types are in operation based on the nature of the resources or the opportunities. In other words, various types of watercraft are in existence to accomplish dedicated tasks or purposes in the ocean space. These watercrafts propel through the power produced by the prime movers. The emission-free sails and oars were initially used for the ship's propulsion [4]. However, the controlled propulsion and navigation in many circumstances were challenging. Further, with the development of prime movers like steam and diesel engines, the propulsion units were driven by the power produced by these prime movers [4]. The propulsion system is further divided into different types based on coupling between the prime movers and the propulsion unit, such as mechanical, electrical, and hybrid propulsion systems.

The propulsion unit has been conventionally driven through a gearbox or directly by the prime mover in the mechanical propulsion. In contrast, in the electric propulsion system, the propulsion unit is driven through an electric motor, which is fed by an electric distribution system, the so-called switchboard, and the switchboard itself is fed by the source of electrical power, which is usually a synchronous generator coupled with the prime mover, which is usually a combustion engine. The coupled engine and generator system is also called the engine-genset. In a hybrid mechanical-electrical propulsion, the propulsion unit is coupled mechanically to the prime mover and electrically to the electrical power generator and captures the positive aspects of both propulsion systems.

The electrical power is used to accomplish various objectives onboard a vessel. The commercial use of electricity in the form of light bulbs started on the vessel *SS Columbia* [5]. It has been followed by a series of innovative developments of the electrical equipment, machinery, and systems down

Transportation Electrification: Breakthroughs in Electrified Vehicles, Aircraft, Rolling Stock, and Watercraft,
First Edition. Edited by Ahmed A. Mohamed, Ahmad Arshan Khan, Ahmed T. Elsayed, and Mohamed A. Elshaer.

the timeline, which is ever-increasing the electrical power demand onboard a vessel [6]. Some of the latest examples are autonomous vessels, fully electric vessels based on batteries and fuel cells (FCs), and hybrid power systems [7].

22.2 Shipboard Power System Architectures

As in the land-based power system, AC power system architecture has been dominating onboard a vessel. The dominance comes with the well-developed control and safety systems together with the major AC components such as electric generators and motors. On the other hand, thanks to the emerging use of energy carriers such as batteries, supercapacitors, and FCs, the relevance of the DC power system is increasing as all these carriers operate in DC systems.

The AC and DC power systems primarily differ with the physical positioning of the power converters. The development of efficient power converters has enabled the power conversion between AC and DC. Both AC and DC power systems onboard a vessel have merits and demerits. A general comparison of AC and DC systems is summarized in Table 22.1 [8] based on [6, 7, 9–14].

22.2.1 AC Switchboards

In an AC-type onboard power system, the main switchboard is energized with an AC voltage of 50 or 60 Hz. As the system frequency must be maintained, the prime movers in this system usually need to operate at a nearly constant speed. In addition to the system frequency, system voltage is also controlled in this system. The presence of reactive power increases complexity; however, it

Table 22.1 Comparative analysis of AC and DC power systems.

Characteristics	AC system	DC system
Reactive power	✓	✗
Voltage control	✓	✓
Frequency control	✓	✗
Synchronization	✓	✗
Harmonic distortion	✓	✗
Ripples	✗	✓
Variable-speed engine	✗	✓
Bulky components	✓	✗
Easy voltage transformation	✓	✗
Easy ESD connections	✗	✓
Fuel-saving potential	✗	✓
Well-developed control systems	✓	✓
Well-developed safety systems	✓	✗
Easy fault detection	✓	✗
Low maintenance	✓	✗
Better component availability	✓	✗
Proper standardization	✓	✗

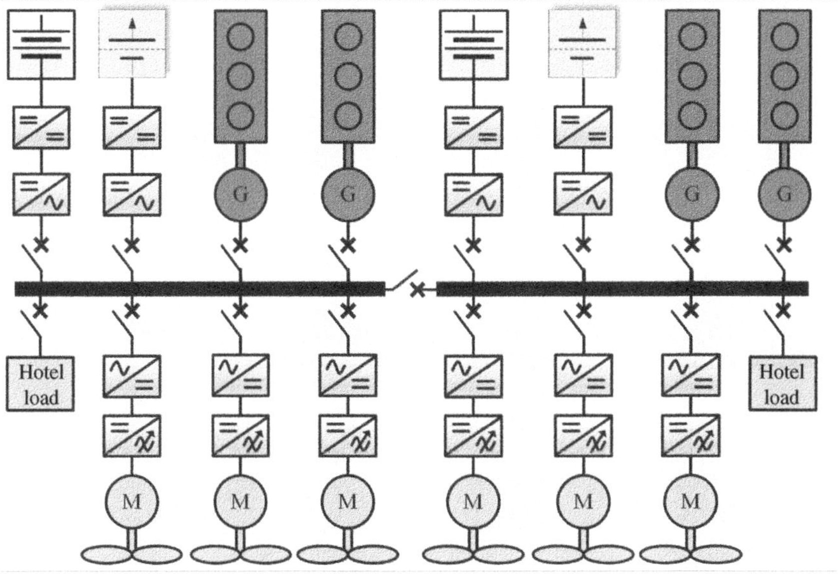

Figure 22.1 Typical marine AC hybrid power system architecture.

facilitates voltage regulation. Further, voltage, frequency, and phase synchronization are required in this system. A single line diagram (SLD) of a typical AC hybrid power system including the energy storage system (ESS) is shown in Figure 22.1. As batteries, supercapacitors, and FCs operate in a DC system, the connection of these energy carriers to the main switchboard requires power conversions, thereby reducing some flexibility. On the other hand, the voltage transformation is relatively more straightforward in an AC system, thereby increasing flexibility. However, the bulky AC components (voltage transformers and harmonic filters) increase the dead weight for the voltage transformation and reduction of harmonic distortions. Further, well-developed safety and control systems increase the robustness of the AC system. However, differential protection is increasingly used in AC systems, especially in complex systems with varying generator power ratings.

22.2.2 DC Power System

In the onboard DC power systems, the main switchboard is energized with the DC voltage. As the system is free of frequency, the prime movers are usually operated at a variable speed to attain optimal fuel consumption. This system is relatively simple as phase and frequency synchronization, and harmonic filters are not required. However, the rectification of AC to DC may generate ripples that usually require capacitive filtering. SLD of a typical DC hybrid power system with the emerging energy carriers and ESS is shown in Figure 22.2.

Renewable energy sources and emerging energy storages such as batteries, supercapacitors, hybrid ESSs, and FCs can be integrated into the DC system with less energy conversion as these sources are inherently DC. However, DC-to-DC converters may be required to adapt the voltage levels [15]. In commercial vessels, the major electrical power producers are still synchronous generators; the DC system requires the rectification of the AC voltages produced by the gensets. However, the rectification part of the variable speed drives of the propulsion units and other electric machinery are not required. The hotel loads are usually connected to an AC system that requires voltage inversion in the DC system. Though well-developed control systems for DC systems are

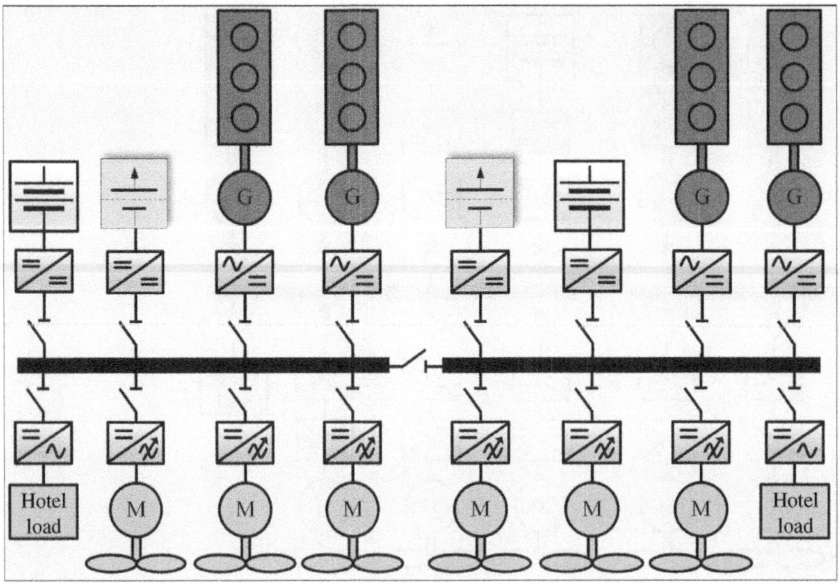

Figure 22.2 Typical marine DC hybrid power system architecture.

available, the safety system (fault detection, location, and isolation) is still a major challenge for DC systems with higher voltage levels.

22.2.3 Hybrid AC–DC Power System

The major difference between AC and DC power systems comes down to the positioning of the power converters. Extra power conversion is required in the variable speed drives in the AC system, whereas it is required for the electrical generator output in the DC system. In the hybrid AC–DC power system, two main switchboards are maintained, one in the AC system and the other in the DC system, as shown in Figure 22.3.

These AC and DC switchboards or grids are then connected through a bidirectional power converter. The advantage of this architecture comes with minimal power conversions. The electrical generators and hotel loads can be combined in the AC system, whereas the propulsion loads, batteries, supercapacitors, and FCs can be connected to the DC main switchboard. However, the major benefit of the DC system due to the variable speed operation of the combustion engines is not realizable.

22.3 Shipboard DC Power System Topologies

As in marine AC power systems, different topologies can also be adopted in DC power systems. The selection of the bus topology depends on the high-level control and fault management design, requirement from the class society, and the functional requirement of the type of vessel itself based on the installed power. Some relevant DC bus topologies for the shipboard power system are introduced here with simplified illustrations based on [6, 16–22].

Figure 22.4 depicts a simplified DC bus topology with a single main bus supplied by the generators, battery, and an FC. The propulsion and hotel loads are also connected to the same bus. This

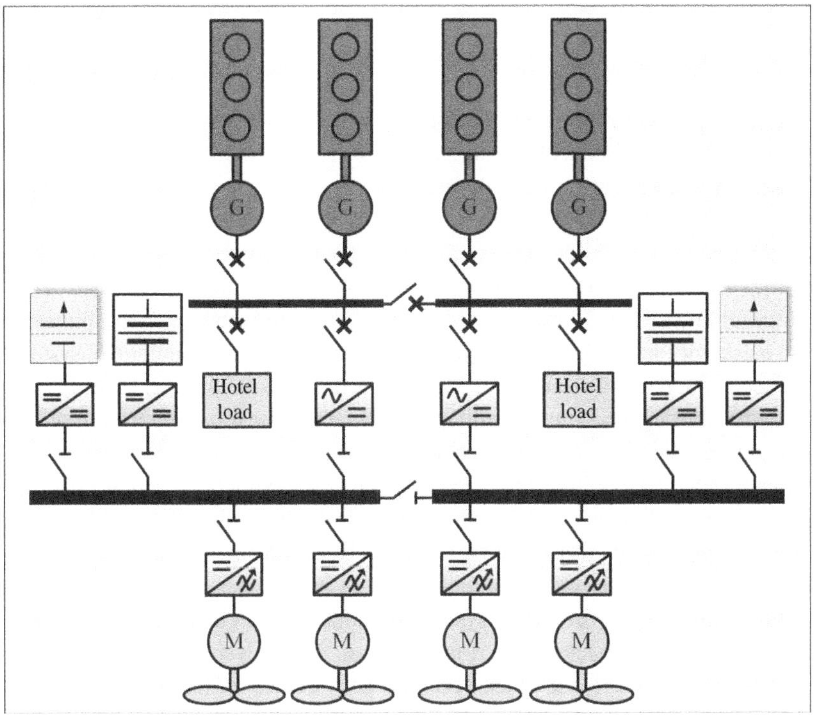

Figure 22.3 Typical marine AC–DC hybrid power system architecture.

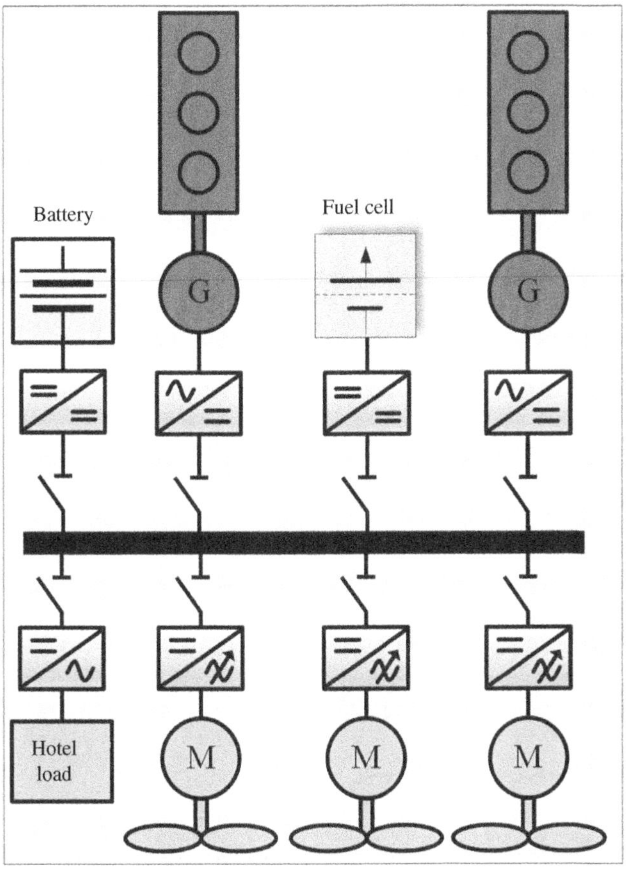

Figure 22.4 Simple DC bus topology.

topology is simple in nature and the high-level control can be relatively simple. However, the redundancy in the system is compromised as fault diagnosis and isolation become relatively complex. This type of topology can be suitable for smaller vessels.

Zonal bus topology can be implemented to avoid the drawbacks of the simple DC bus topology. In the zonal bus topology, the main bus is divided into two or more zones using a bus-tie breaker. The power generators, energy storage devices, and loads are distributed to different zones. This enhances the redundancy and resilience of the onboard power system and enables the implementation of the protection selectivity in the DC system [23]. However, the high-level control system may be relatively complex as the control strategy for a common bus (bus-tie breaker closed), and sectionalized bus (bus-tie breaker opened) may be different. The common zonal separation of the bus can be based on the physical location onboard a vessel such as port, starboard, bow, and stern. Figure 22.5 shows a typical zonal bus topology with each generator supplying the bus in each zone. The battery and FC are connected to zones 1 and 2, respectively. Similarly, loads are also distributed to the bus in each zone. The number of zones can be designed based on the installed power and required level of redundancy.

To further increase the redundancy in the power system, the topology can be extended to a ring bus topology, where the buses are connected in a ring structure. The power generators, storage devices, and loads are then distributed to each bus. The high-level control of this topology is relatively complex; however, the degree of reliability is higher or failure identification and isolation

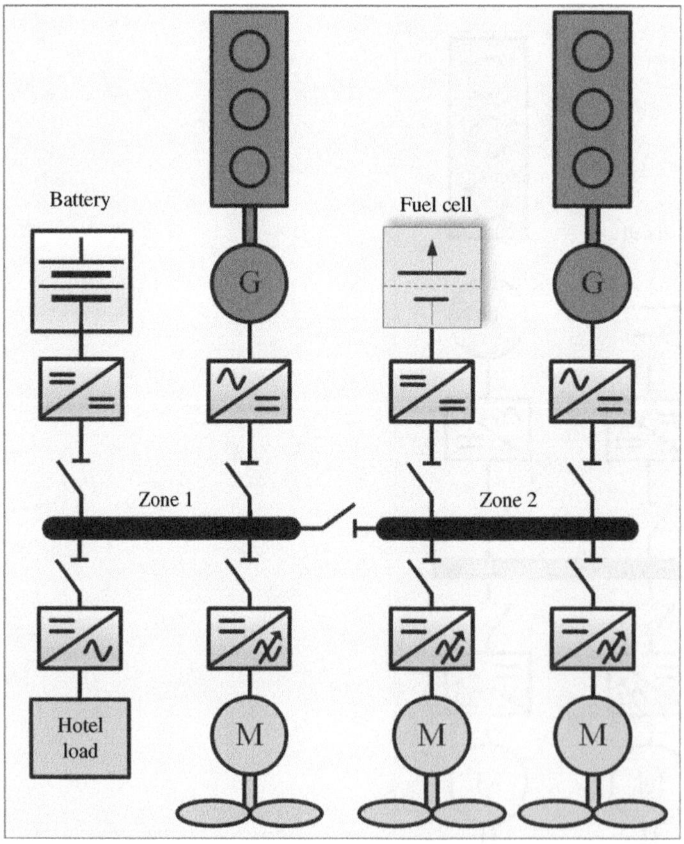

Figure 22.5 Zonal bus topology.

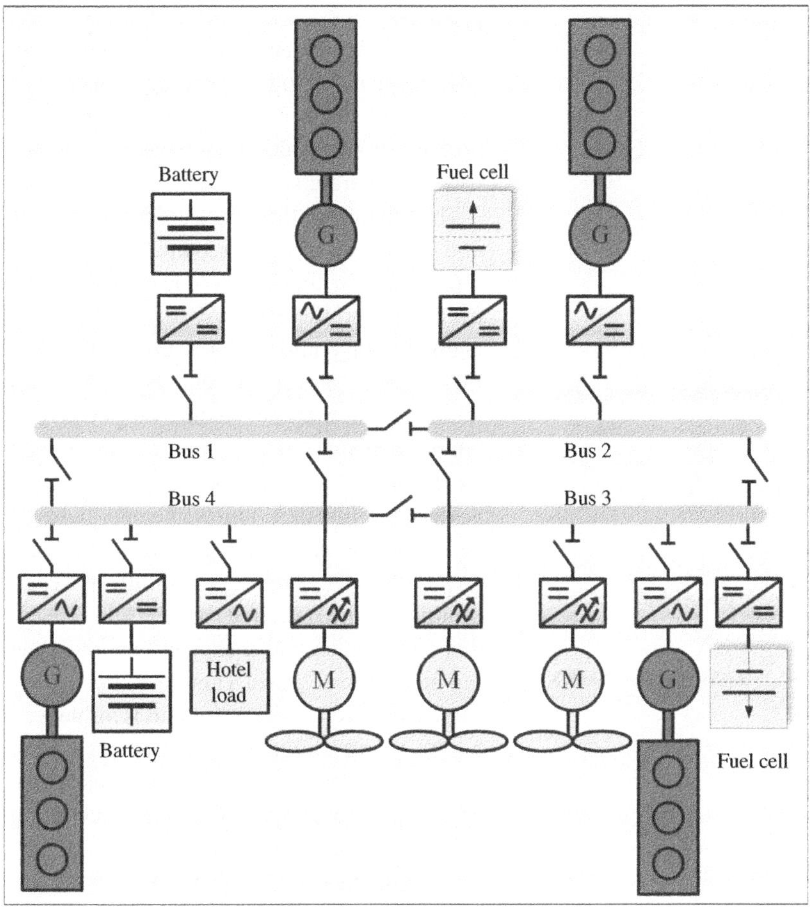

Figure 22.6 Ring bus topology.

are relatively easier in this system. Figure 22.6 presents the schematic of a ring bus topology. This type of topology can be implemented in deep-sea vessels, which also need to be operating in harsh weather and sea conditions.

Figure 22.7 illustrates the multi-voltage DC bus topology where zonal buses with two different voltage levels are connected through a common grid converter. The power generators can be operating in medium-voltage DC (MVDC), whereas battery and FCs can be working in low-voltage DC (LVDC). The load in the power system can also be segregated into different voltage levels based on their rated power capacity so that the current through the equipment can be in the range where safety and stability of the system are ensured. The segregated loads can then be connected to the buses with appropriate voltage levels.

22.4 Energy Storage and Alternative Energy Sources in Shipboard Power System

Though green and renewable energy sources were initially used in historical watercraft [4], fossil fuels are the dominating source of energy in watercraft systems. The green and renewable energy sources such as solar, wind, and wave energy include a high degree of uncertainty for the

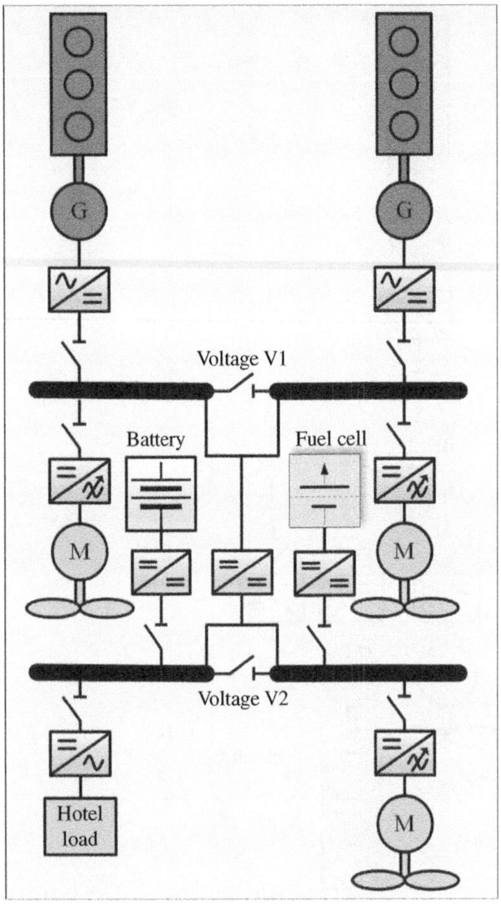

Figure 22.7 Multi-voltage bus topology.

prediction, due to which controlled propulsion had been a challenge. On the other hand, advancements in internal combustion engines turned out to be reliable energy conversion devices, due to which they have been widely used in watercraft.

The mission of the vessel defines its energy and power requirements, which are usually represented by time series of load power, thus called load profile. However, the ability of the energy storage and power sources to deliver the required power and energy depends on their gravimetric and volumetric power and energy density. These properties are defined by the weight and volume constraints of the vessels. For instance, renewable energy-based zero-emission propulsion with solar and wind power is attractive in the green shift. However, producing the required power and energy demand by such power sources is not feasible for many vessel types. Therefore, the energy carrier selection in the watercraft is influenced by the energy and power requirement for the mission, the weight and volume constraints, as well as rules and regulations concerning ship design and operation.

The gravimetric energy and power density of the common energy carriers in the watercraft systems are presented in a Ragone plot in Figure 22.8 [24, 25]. From the perspective of energy and power density, combustion engines can be considered a promising alternative. However, alternatives to combustion engines are considered due to the harmful emissions from fossil fuels and relatively low efficiency of the engines. Therefore, an FC is considered an attractive alternative [26] where reliable control and safety systems are under development, though the FC propulsion can still be considered premature both in terms of power or energy capacity and fuel transportation. Batteries are considered a good option; however, they lack sufficient energy content for all vessel types. On the other hand, supercapacitors can only provide high power for short time intervals. Therefore, to achieve emission reduction and efficiency improvement in watercraft, the combination of these power sources is considered the most viable option with the available technologies in the field [6, 8, 27].

22.4.1 Energy Storages

Energy storage devices such as batteries and supercapacitors are increasingly used in watercraft, mostly hybridizing the power system. Depending on the energy density properties, their use may vary a bit. However, batteries are mostly in use. Therefore, a high-level control system such as power and energy management system (PEMS) should include the defined control algorithm or strategies to

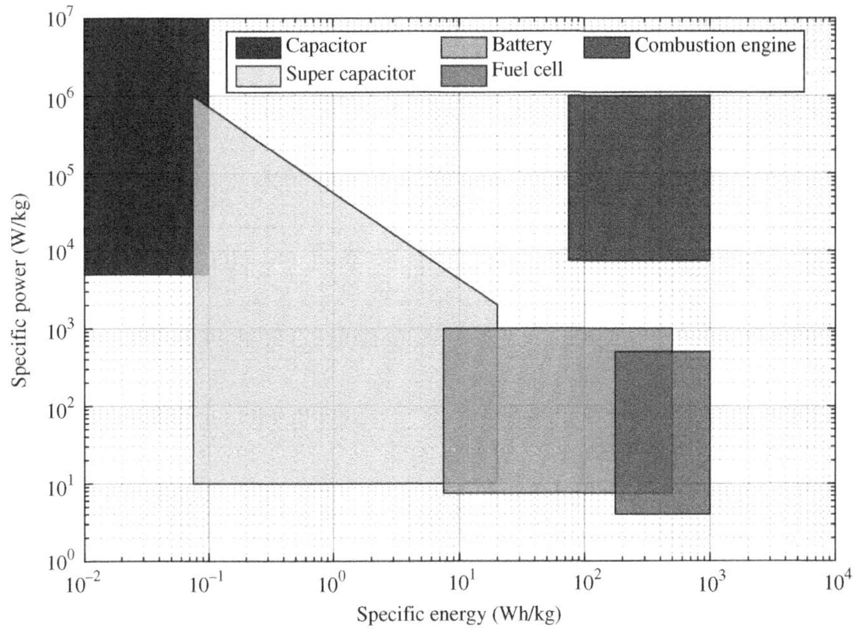

Figure 22.8 Ragone plot showing typical gravimetric energy and power density ranges for energy carriers.

achieve these functional applications of energy storage devices. The major functional applications of the battery in watercraft systems are further described in Section 22.5.

22.4.2 Fuel Cell

FCs, as electrochemical power sources, convert the chemical energy of the fuel directly into electric power without a prime mover. FCs are attractive for electric propulsion due to several main reasons such as compatibility with H_2 and gaseous fuels, smooth efficiency profile and potentially higher energy efficiency compared to gas engines, low vibrations, mechanical impacts, and low maintenance requirements compared to rotatory machinery. On the other hand, FCs have often a much lower lifetime compared to engines and even batteries.

FC systems are recently under development for marine propulsion based on both low-temperature FCs, e.g. proton-exchange membrane FCs (PEMFCs), and high-temperature FCs, e.g. solid oxide FCs (SOFCs) and molten carbonate FC (MCFC). The low-temperature FCs are preferred in terms of safety, load tracking, and weight and space saving thanks to smaller auxiliary systems, while high-temperature FCs can provide higher electrical efficiency thanks to their high-temperature exhaust gas that enables combined heating and power (CHP) in a so-called co-generation system.

A typical FC powertrain with DC power architecture is shown in Figure 22.9. In this scheme, a DC–DC power interface is always required for isolating the FC from the main DC grid, regulating

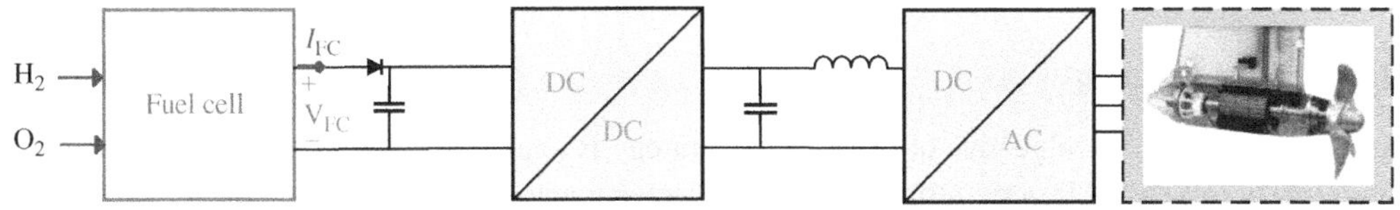

Figure 22.9 Schematic of a fuel cell powertrain. *Source:* Marine Engineering Education.

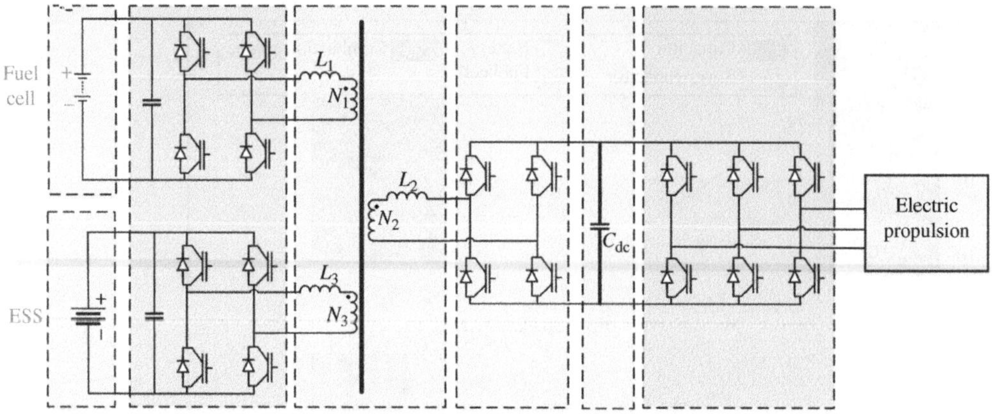

Figure 22.10 Integrated FC and ESS topology.

the voltage on the DC bus independently from the FC, and boosting the output voltage of the FC, which is usually very lower than the main DC bus due to the inherent low voltage of FC stacks [26].

FCs can also be integrated with the ESS (batteries and supercapacitors) via a three-winding isolation transformer as shown in Figure 22.10.

22.5 High-Level Control of Energy Storage Systems

The ESS can act both as an energy supplier and consumer. It enables flexibility in the power system. The extra power generated can be stored in energy storage devices. On the other hand, ESSs can supply power when available power is deficient. However, they can neither store nor supply infinite power because of the size and capacity constraints. Therefore, implementing the proper strategies to manage energy utilization in storage devices is crucial. Some of those energy management strategies or high-level control approaches for the energy storage devices in the DC shipboard power system are discussed next based on [24, 28, 29].

22.5.1 Peak Shaving

The marine loads are usually fluctuating in nature due to different circumstances; the environmental condition is one. These fluctuating loads can make the power system parameters such as bus voltage and engine load fluctuate. The fluctuating generator loads may lead to unnecessary start and stop of engine-generator sets when the load-dependent start-stop is activated. On the other hand, energy storage devices like batteries can shave the load peaks of the generator, as shown in Figure 22.11. Depending on the loading conditions, the battery may charge or discharge. It ensures that the engine-generator sets operate between the predefined load limits.

22.5.2 Load Leveling

As in peak shaving, the objective of this control strategy is to avoid the fluctuating loading of the generator. In this case, the generator supplies the average load power. At the same time, the battery takes care of the load variations, i.e. battery charges if the load demand is lower than average load power and discharges if the load is less than the average load power, as shown in Figure 22.12.

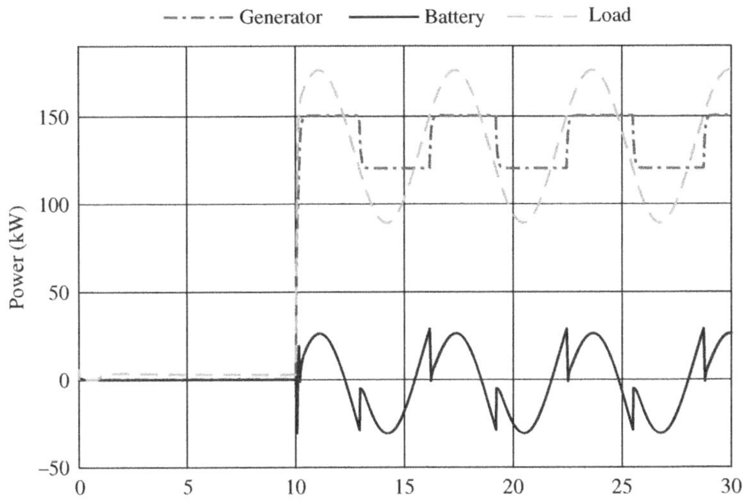

Figure 22.11 Load sharing through peak-shaving strategy.

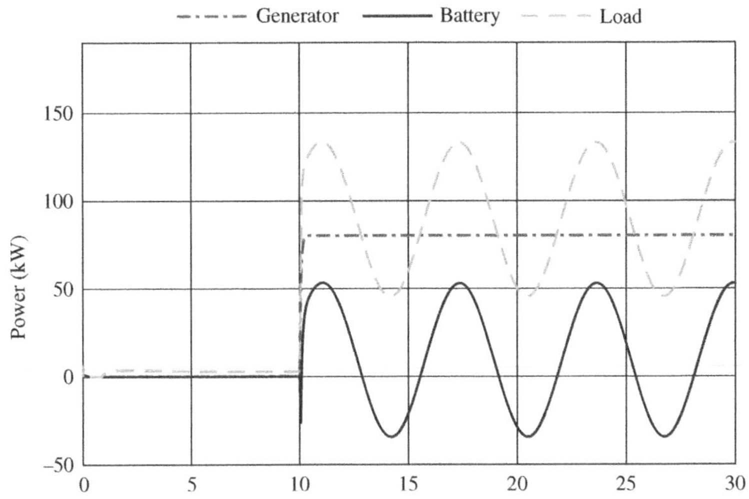

Figure 22.12 Load sharing through load-leveling strategy.

It limits the fluctuations in the engine-generator sets; however, the battery capacity required to level the generator load is usually higher than in peak shaving.

22.5.3 Zero Emission

In the restricted areas where it is not allowed to operate fossil-fueled conventional engines, the batteries can be used to supply the total load demand in the vessel, zero-emission operating mode, as shown in Figure 22.13. Usually, the battery has limited capacity compared to conventional engines. Therefore, it is crucial to design the battery and other component capacities such that they can provide power to the required propulsion and critical auxiliaries. In addition, proper planning for zero-emission operating mode is necessary to ensure that the battery contains sufficient charge.

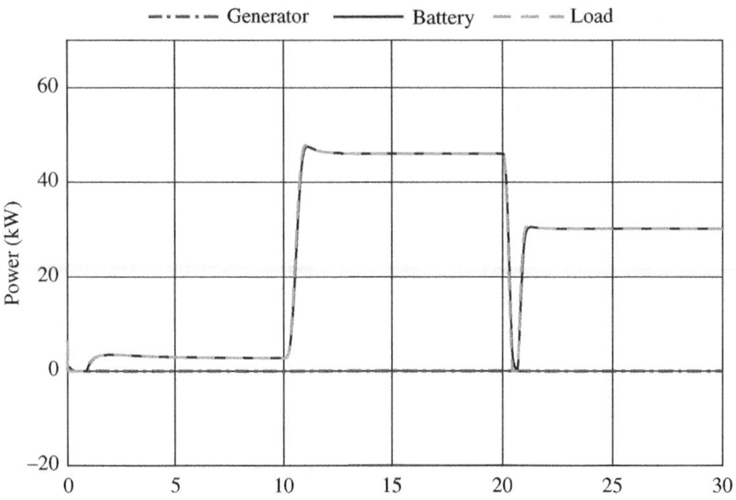

Figure 22.13 Zero-emission operation.

22.5.4 Battery Charging

As the charge contained in a battery decreases with its discharging, it is essential to charge the battery, especially before zero-emission operation. Figure 22.14 shows that the generator is supplying power not only to the load but also to recharge the battery.

22.5.5 Strategic Loading

The batteries contribute to load sharing with gensets. This battery function enables a reduction of fuel consumption by allowing engines to operate in fuel-optimal regions. This is by the fact that the engine fuel consumption and efficiency are functions of the engine loading as maximum continuous rating (MCR). Figure 22.15 shows the curve of the fuel consumption for a typical marine combustion engine, namely specific fuel oil consumption (SFOC); the SFOC represents the

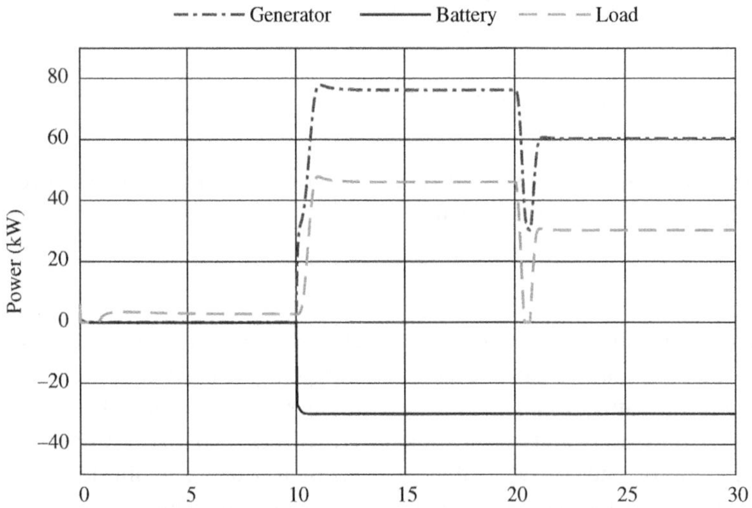

Figure 22.14 Battery-charging operation.

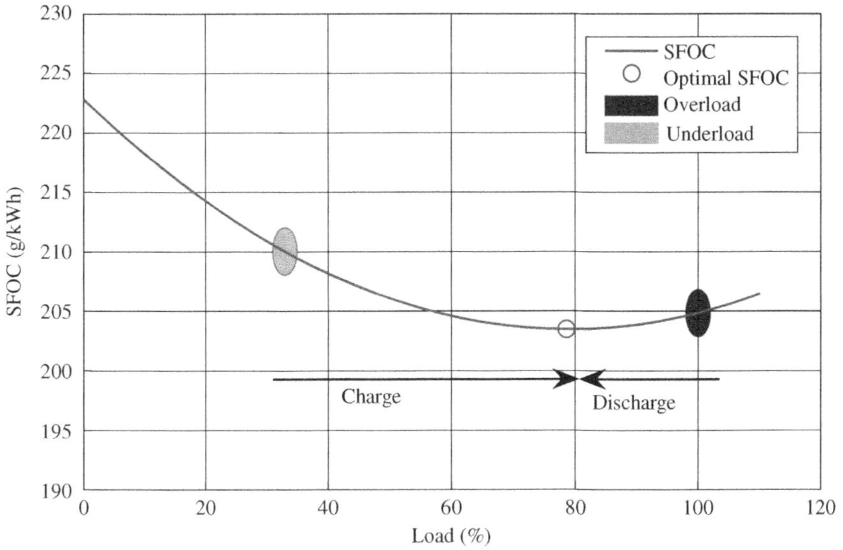

Figure 22.15 Strategic loading operation.

consumption as a gram of fuel per kWh energy generated and transferred through the engine shaft. The given curve indicates the SFOC as a function of engine loading in MCR percentage.

As shown in the figure, the consumption has a nonlinear relation with loading and the high- or low-load condition may result in higher consumption and lower efficiency. The optimal efficiency may be associated with a specific loading, which is, in this case, around 80% of MCR; this specific operating point results in minimum fuel consumption and maximum efficiency. Here, the battery can help keep the engine loading in the optimal operating point or optimal region of operation by supplying the surplus load. The battery can then be charged in low-load conditions and discharged under high-load operation. In the figure, the overloading and underloading conditions are indicated by low and high MCR percentages, e.g. 35% and 100%, to elaborate on the effect of loading on the SFOC and efficiency. Compared to the light grey and dark grey ovals, on the grey circle which is around 80%, the fuel consumption is obviously reduced. The so-called optimal operation point can be achieved thanks to battery operation and strategic loading. This control strategy enables to reduce the fuel consumption and emissions [29], whereas the battery gets repetitive charge–discharge cycles [30].

22.5.6 Enhanced Dynamic Performance

Usually, the large engine has a slower response due to higher inertia. Similarly, gas engines and FCs also have a slow response. In contrast, marine loads fluctuate and can increase or decrease abruptly based on the mission profile or due to environmental conditions. This high load change can make the system unstable, leading to a blackout [29]. Therefore, the battery can compensate for the slower response of engines and FCs as it has a faster response and increases safety and robustness.

22.5.7 Spinning Reserve

Depending on the different classifications of the class societies, the vessels need an extra level of redundancy. Therefore, additional generators are connected to the bus in a conventional power system to maintain redundancy levels. However, operating an extra generator reduces the generator load, which means the loading in all engine-generator sets becomes less than the fuel-optimal region. In contrast, the use of batteries as a spinning reserve increases robustness and decreases fuel consumption and emissions [29].

22.6 Load Sharing in DC Power System

Droop control is an established control technique in the shipboard power system. In an AC power system, active and reactive power are managed through frequency and voltage droop controls. However, the DC power system is frequency-free and does not require reactive power-sharing. But, active power-sharing has to be performed and is usually done through the voltage droop control. The droop control is in principle a proportional controller that finds a voltage setpoint for the generator's automatic voltage regulator (AVR). This voltage setpoint is calculated based on the slope of the droop line and the actual load in the generators. The droop control can be implemented in the distributed control for each generator. For equal load-sharing among the generators, the slope of the droop line or droop-sharing coefficient needs to be similar, as shown in Figure 22.16 [24].

On the other hand, different droop coefficients for the generators make unequal power-sharing, as shown in Figure 22.17. Further, it can also be extended for sharing with the battery where the battery is charging when the power is negative and discharging when the power is positive.

22.7 Efficiency Improvement and Emission Reduction Potentials

Energy efficiency quantifies the conversion of input energy to the intended output. For example, the chemical energy contained in the fuel is the input energy to the system. In contrast, the electric power consumption and mechanical energy output from the propulsion units are the output energy of the power system [8, 31]. Therefore, the energy efficiency is dependent on fuel consumption, and the lower the fuel consumption for intended output, the higher the energy efficiency. Therefore, the vessel's energy efficiency has been one of the crucial factors during the selection of new technology or the comparison with other technologies. At the same time, regulatory bodies are making stringent rules and regulations to reduce emissions from the maritime sector and improve energy efficiency [24]. Therefore, energy efficiency improvement also makes the industry competitive with other transportation sectors and reduces emission, a function of fuel consumption.

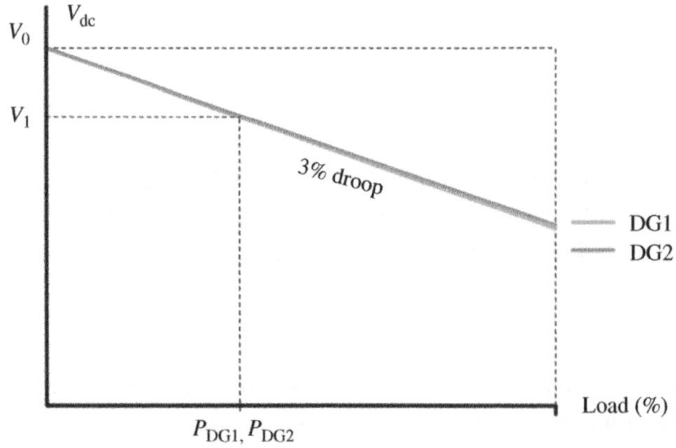

Figure 22.16 Equal sharing through droop control. *Source:* Othman et al. [24]/IEEE.

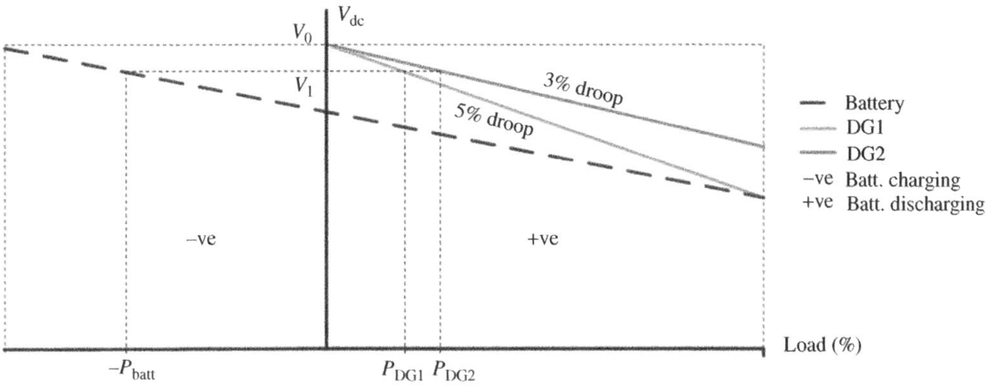

Figure 22.17 Unequal sharing among generators and battery using droop control. *Source:* Othman et al. [24]/IEEE.

The system efficiency can be determined based on the power or energy in the system. The power efficiency is an instantaneous efficiency of the system, whereas energy efficiency is the time average of the power efficiency as presented in Eqs. (22.1) and (22.2), where P_0 is output power, P_i is the input power, η_p is power efficiency, and η_e is energy efficiency.

$$\eta_p(t) = \frac{P_0(t)}{P_i(t)} \tag{22.1}$$

$$\eta_e(t) = \frac{\int P_0(t)dt}{\int P_i(t)dt} \tag{22.2}$$

Conventionally, the rated efficiencies for the components have been used to determine the power system efficiency [32]. On the other hand, loading conditions for the components impact their efficiency [33]. Therefore, the losses or efficiency determination for the components and the complete power system needs to be dynamic with the loading conditions [8, 31]. In addition, to make a fairly realistic estimation, the availability of an actual load profile is essential.

Further, one of the benefits of the DC power system in the maritime industry is improved efficiency. Operating the engines in variable speed operation helps reduce fuel consumption, especially in low-load conditions [34]. The comparative study of energy efficiency in different power system architectures in a cruise ship shows that the DC power system with variable speed engines provides slightly improved efficiency compared to the AC system [8]. It also showed that the DC power system is slightly more efficient than AC in both calmer and rough sea conditions. The improvement in efficiency reflects less fuel consumption to do the intended work than AC, which means less emission and improved decarbonization in the maritime industry.

22.8 Case Studies

A case study is presented in this section to elaborate efficiency aspects of the onboard power system. A typical hybrid power system with AC and DC switchboards is shown in Figure 22.18. As shown in the SLD, the switchboard is split into two sections for the sake of redundancy and improved resilience. Based on the vessel classification, it can also be split into 3 or 4 sections. In the sectionized switchboard, the sections can be connected through switchgears, namely bus-tie breakers. A typical

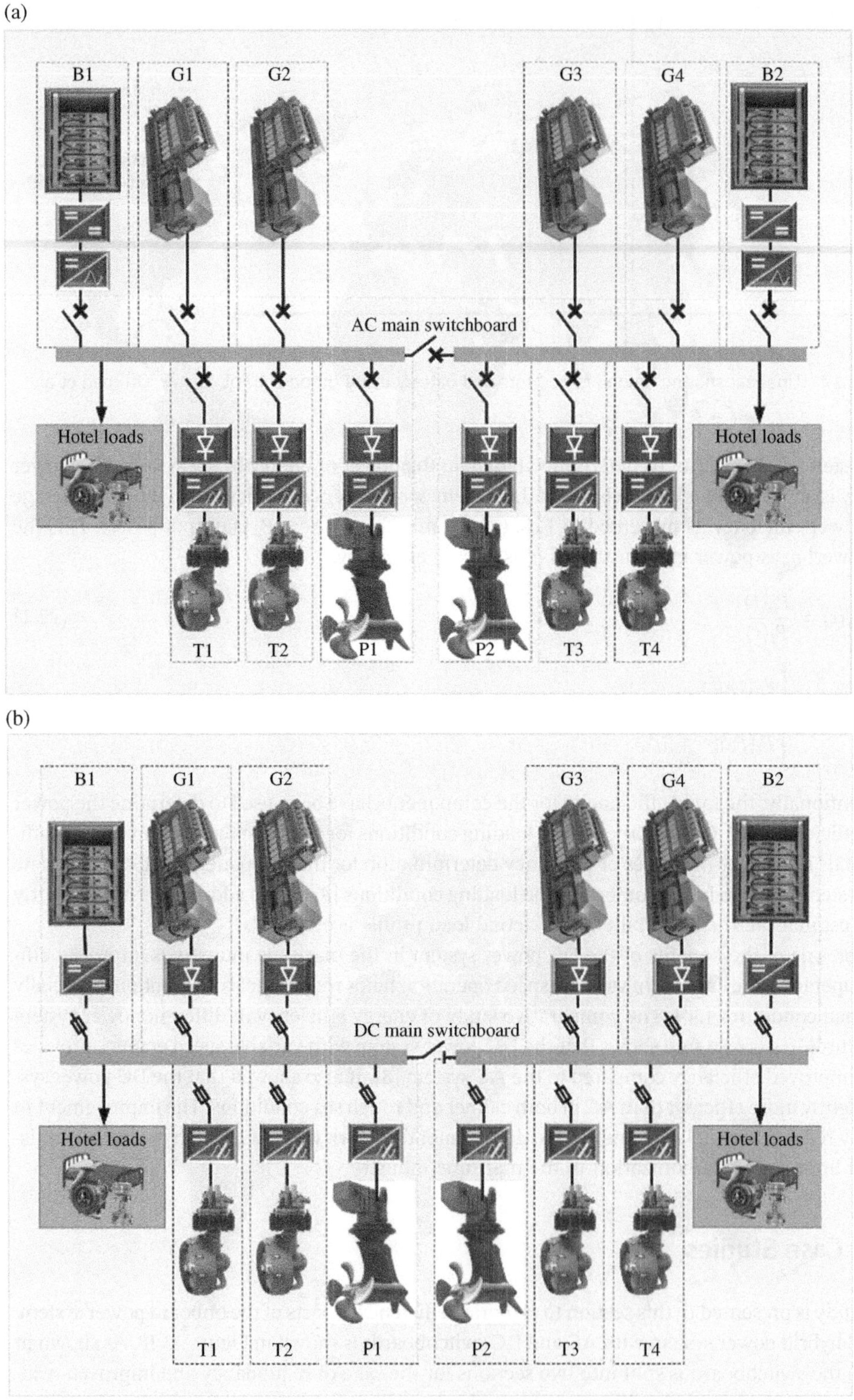

Figure 22.18 Typical onboard hybrid power system: (a) AC switchboard and (b) DC switchboard.

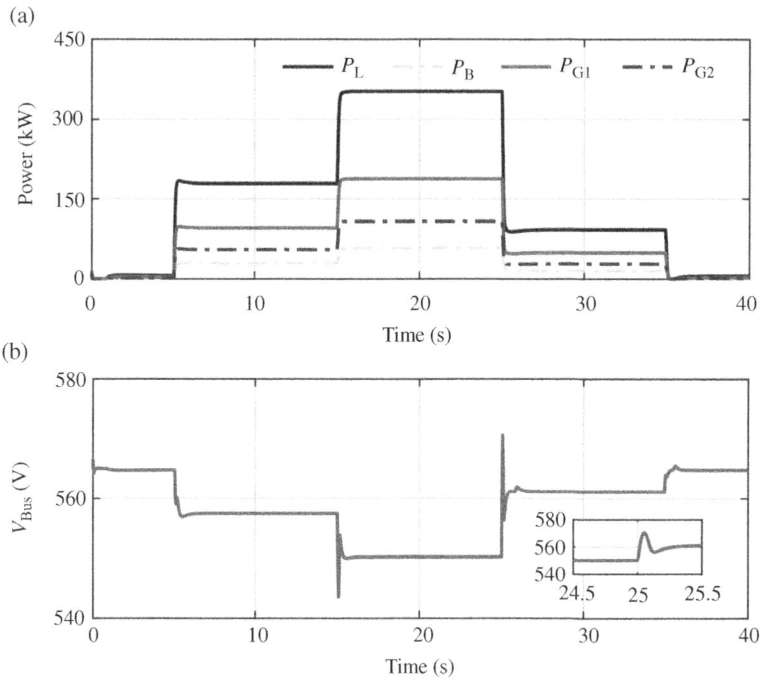

Figure 22.19 (a) Typical load profile with power distribution between gensets and battery. (b) DC bus voltage on the main switchboard and fluctuations corresponding to the load changes.

vessel load profile is shown in Figure 22.19a with load sharing between gensets and the battery on one DC bus section. Figure 22.19b also demonstrates the DC bus voltage variations on the main DC switchboard. In this case, the rated DC voltage is assumed to be equal to 565 V and the voltage at the battery end equal to 346 V [27].

To analyze the energy efficiency, the whole vessel powertrain shall be considered including the power sources (gensets and battery), power distribution system, and the propulsion load. The efficiency of the engine-genset can be estimated with the SFOC of the engine as a function of the engine loading and shaft speed. An example of the SFOC surface with the two variables deduced from experimental data is shown in Figure 22.20a; the corresponding surface of energy efficiency is also given in Figure 22.20b. As shown in these graphs, the fuel consumption and hence the efficiency, is a function of the output power and the shaft rotational speed. Therefore, the efficiency of the engine-genset can be optimized by keeping the engine loading and speed within the optimal range. Certain engine speed needs to be maintained to attain engine-operational parameters such as torque limit. An example of the engine speed setting based on the engine loading is given in Figure 22.20c; the experimental data are fitted to a continuous curve based on Eq. (22.3).

$$\Omega(L) = -0.0001L^3 + 0.02396L^2 - 0.9323L + 70 \tag{22.3}$$

where the angular velocity Ω is given in percentage of the rated speed, and the load L is given in percentage of the MCR.

Besides, the battery efficiency is also a function of the battery state of charge (SoC) and C-rate as demonstrated in Figure 22.21. Usually, the battery charging efficiency is different from discharge efficiency, as given in Figures 22.21a,b. From the system point of view, the ESS can provide higher efficiency by shifting the engine operating point of the engines toward the optimal region.

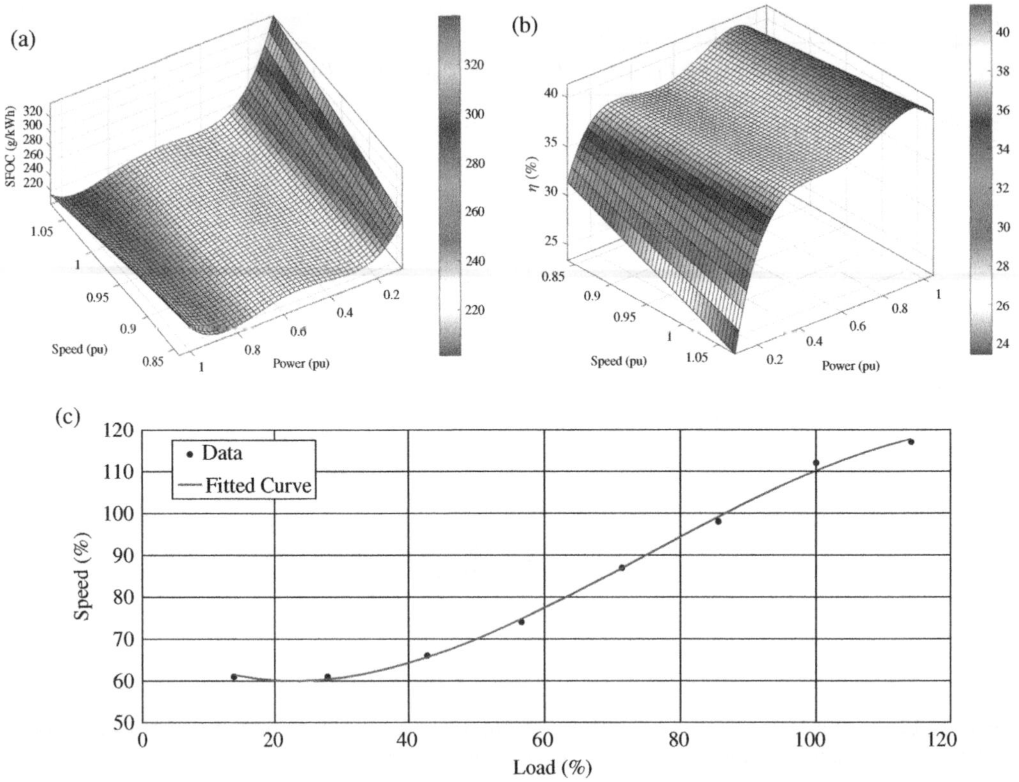

Figure 22.20 Engine-operational characteristics: (a) SOFC, (b) fuel efficiency, and (c) engine speed variation and loading.

22.8.1 Case Study 1 – Cruise Vessel

A case study of a cruise vessel with hybrid power systems and an extended load profile (170 hours) is given in Figure 22.22 [8]. Here, the dynamic energy efficiency of the total ship electric powertrain is given with the measured load profile and the power system architecture presented in Figure 22.18. Here three main cases are compared such as AC switchboard, DC switchboard with fixed speed engines (FSDC), and DC switchboard with variable speed engines (VSDC). An instantaneous power efficiency curve is given in Figure 22.22a while the energy efficiency for the whole profile and extended for various environmental conditions is given in Figure 22.22b. It is seen that a marginal efficiency advantage is achieved with the variable speed engine and DC power system. This is due to the load correction of the engine-gensets with the variable speed. Regarding the weather conditions, the energy efficiency trend for the power system architectures is similar in all three cases, which shows that all power system architectures are similarly affected by the weather conditions. Therefore, the weather condition itself cannot be the deciding factor in selecting power system architecture. Further, it is observed that the energy efficiency for all three architectures in the calm sea condition is slightly less than in the real case. In contrast, it is slightly higher in the rough sea condition than in the real case. It signifies that the varying loads due to weather conditions are well handled by the battery such that diesel engines are less affected by the load fluctuations, which is one of the significant benefits of hybridization.

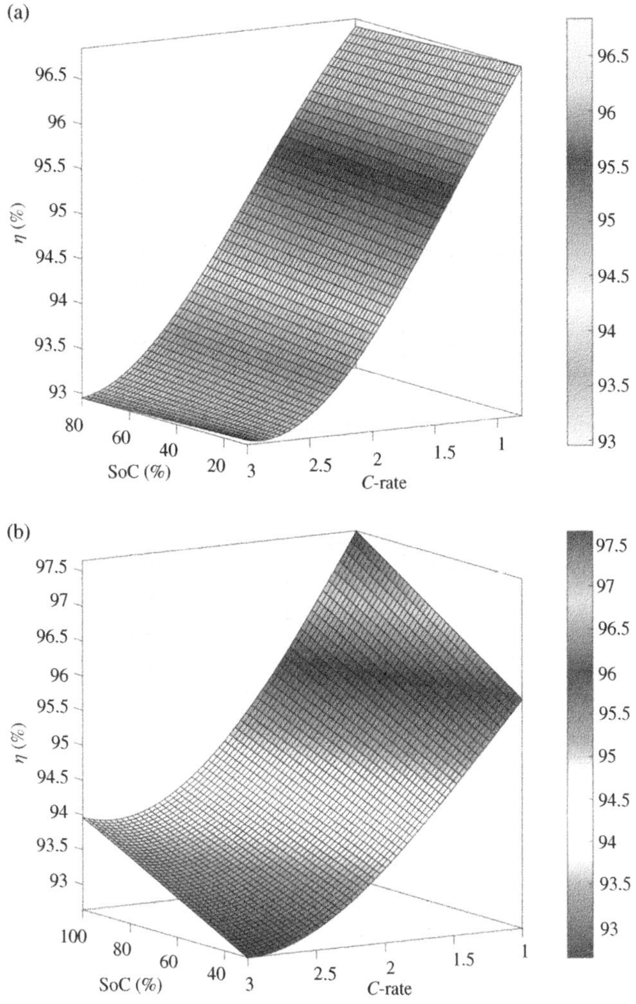

Figure 22.21 Efficiency of the battery system as a function of SOC and *C*-rate: (a) charging and (b) discharging.

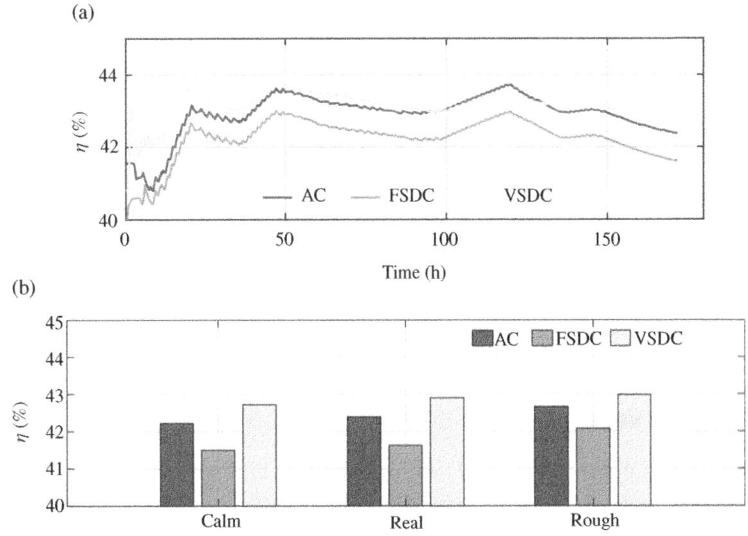

Figure 22.22 Efficiency analysis of the cruise ship with the hybrid power system and various power system architectures. (a) Instantaneous power efficiency. (b) Energy efficiency of the studied power system architectures in different sea conditions.

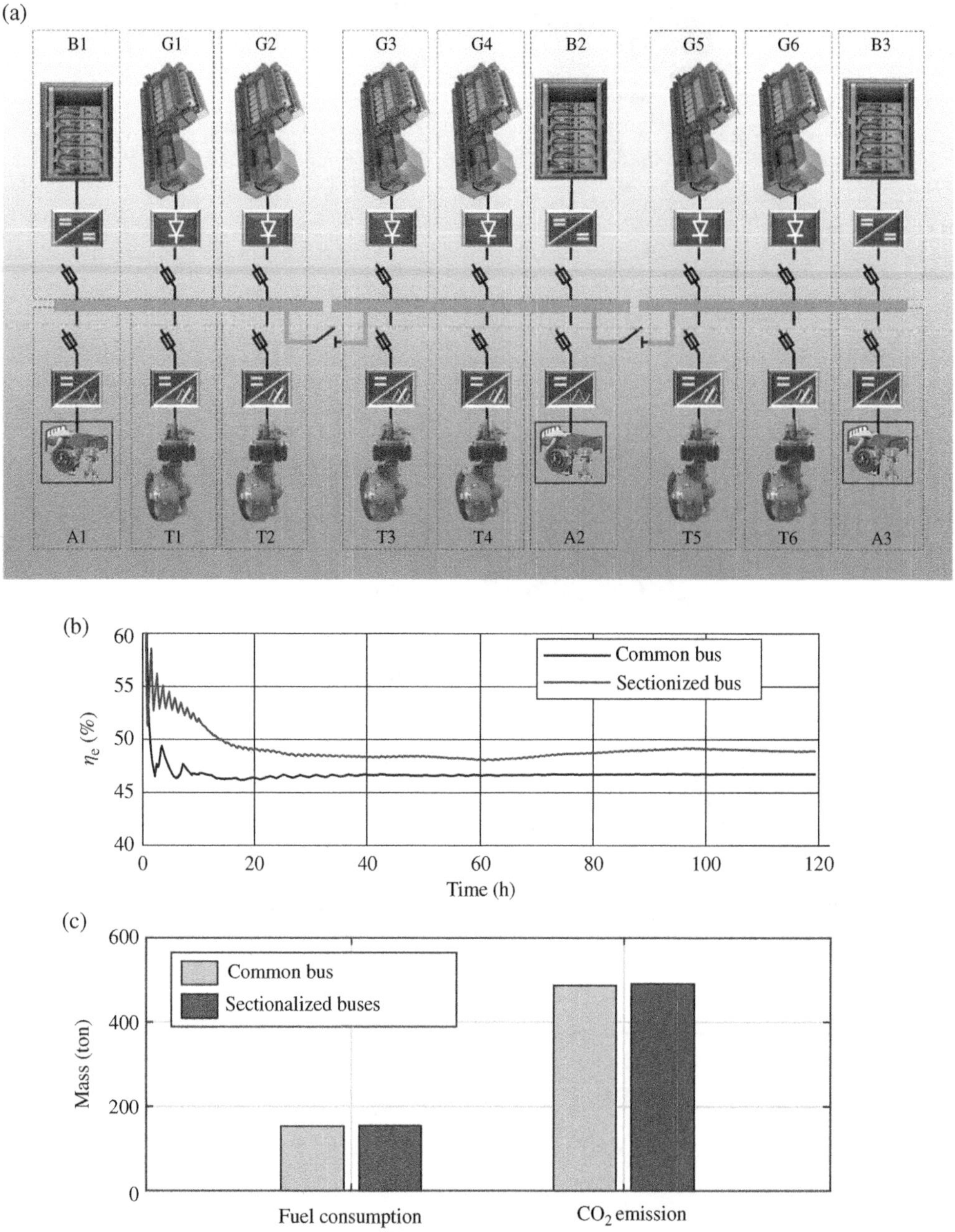

Figure 22.23 Load profile-based efficiency and emission analysis: (a) SLD of the DC hybrid power system for the offshore vessel, (b) energy efficiency variation against time, and (c) the total accumulated fuel consumption and CO_2 emissions.

22.8.2 Case Study 2 – Offshore Vessel

Another case study for an offshore support vessel (OSV) with a battery hybrid power system is performed with a measured load profile for almost 120 hours [35]; the corresponding SLD is shown in Figure 22.23a. In this case, the simulations are performed for two different cases, namely "common bus" referring to the case that the bus-tie breakers are closed, and "sectionized bus" referring to the case that both bus-tie breakers are open, resulting in separated DC bus sections. The efficiency analysis has been performed considering the entire electric powertrain from the source to the load; the average efficiency or energy efficiency during the total simulation period and with the common and sectionized bus is shown in Figure 22.23b. Then, the total fuel consumption and corresponding CO_2 emissions are presented in Figure 22.23c.

The average energy efficiencies calculated for the total load profile are approximately 46% and 48% for the common and sectionized bus, respectively. However, the fuel consumptions and CO_2 emissions for both cases are similar. It indicates that the battery in the sectionized bus supplies more energy compared to the one in the common bus. Therefore, the battery SoC is dropped during the operation with a sectionized bus. It is important to equalize the battery SoC during the operation in order to have a fair comparison of energy efficiency. In this case, equalization is applied to the fuel and emission calculation, and hence, a realistic picture is obtained. An essential aspect of hybridization is battery sizing, meaning that the capacity of the battery and its maximum *C*-rate shall be chosen based on the battery function and, consequently, energy and power demand. An oversized battery may result in higher losses and reduced efficiency in addition to the higher investment cost.

References

1 United Nations. (2013). Green Economy in a Blue World. www.unep.org/ (accessed 12 July 2021).

2 United Nations. (2014). The Oceans Economy: Opportunities and Challenges for Small Island Developing States. *United Nations Conference on Trade and Development UNCTAD*, p. 40.

3 United Nations. (2016). Maritime Transport Is 'Backbone of Global Trade and the Global Economy. https://www.un.org/press/en/2016/sgsm18129.doc.htm (accessed 12 July 2021).

4 Geertsma, R.D.D., Negenborn, R.R.R., Visser, K., and Hopman, J.J.J. (2017). Design and control of hybrid power and propulsion systems for smart ships: a review of developments. *Appl. Energy* 194: 30–54. https://doi.org/10.1016/j.apenergy.2017.02.060.

5 Rodskar, E., Johansen, T.A., Skjong, E. et al. (2015). The marine vessel's electrical power system: from its birth to present day. *Proc. IEEE* 103 (12): 2410–2424. https://doi.org/10.1109/jproc.2015.2496722.

6 Ghimire, P., Park, D., Zadeh, M. et al. (2019). Shipboard electric power conversion: system architecture, applications, control, and challenges [technology leaders]. *IEEE Electrification Mag.* 7 (4): 6–20. https://doi.org/10.1109/MELE.2019.2943948.

7 Skjong, E., Volden, R., Rodskar, E. et al. (2016). Past, present, and future challenges of the marine vessel' s electrical power system. *IEEE Trans. Transport. Electrification* 2 (4): 522–537. https://doi.org/10.1109/TTE.2016.2552720.

8 Ghimire, P., Zadeh, M., Thorstensen, J., and Pedersen, E. (2021). Data-driven efficiency modeling and analysis of all-electric ship powertrain; a comparison of power system architectures. *IEEE Trans. Transport. Electrification* 1–1. https://doi.org/10.1109/TTE.2021.3123886.

9 Zahedi, B., Norum, L.E., and Ludvigsen, K.B. (2014). Optimized efficiency of all-electric ships by dc hybrid power systems. *J. Power Sources* 255: 341–354. https://doi.org/10.1016/j.jpowsour.2014.01.031.

10 Simmonds, O.J. (2014). DC: is it the alternative choice for naval power distribution? *J. Mar. Eng. Technol.* 13 (3): 37–43. https://doi.org/10.1080/20464177.2014.11658120.

11 Kanellos, F.D., Tsekouras, G.J., and Prousalidis, J. (2015). Onboard DC grid employing smart grid technology: Challenges, state of the art and future prospects. *IET Electr. Syst. Transport.* 5 (1): 1–11. https://doi.org/10.1049/iet-est.2013.0056.

12 Justo, J.J., Mwasilu, F., Lee, J., and Jung, J.W. (2013). AC-microgrids versus DC-microgrids with distributed energy resources: a review. *Renewable and Sustainable Energy Reviews* 24. Pergamon: 387–405. https://doi.org/10.1016/j.rser.2013.03.067.

13 Planas, E., Andreu, J., Gárate, J.I. et al. (2015). AC and DC technology in microgrids: a review. *Renew. Sustain. Energy Rev.* 43: 726–749. https://doi.org/10.1016/j.rser.2014.11.067.

14 Kim, K., Park, K., Roh, G., and Chun, K. (2018). DC-grid system for ships: a study of benefits and technical considerations. *J. Int. Marit. Saf. Environ. Affairs Shipping* 2 (1): 1–12. https://doi.org/10.1080/25725084.2018.1490239.

15 Zadeh, M., Gavagsaz-Ghoachani, R., Pierfederici, S. et al. (2016). Energy management and stabilization of a hybrid DC microgrid for transportation applications. *IEEE Applied Power Electronics Conference and Exposition (APEC'16).*

16 Jin, Z., Sulligoi, G., Cuzner, R. et al. (2016). Next-generation shipboard DC power system: introduction smart grid and dc microgrid technologies into maritime electrical netowrks. *IEEE Electrification Mag.* 4 (2): 45–57. https://doi.org/10.1109/MELE.2016.2544203.

17 DNV GL. (2015). Recommended practice DNV GL AS dynamic positioning vessel design philosophy guidelines. http://www.dnvgl.com (accessed 6 February 2022).

18 Coffey, S., Timmers, V., Li, R. et al. (2021). Review of MVDC applications, technologies, and future prospects. *Energies* 14 (24): 8294. https://doi.org/10.3390/EN14248294.

19 Vu, T.v., Gonsoulin, D., Perkins, D. et al. (2017). Distributed control implementation for zonal MVDC ship power systems. *2017 IEEE Electric Ship Technologies Symposium, ESTS 2017*, pp. 539–543. https://doi.org/10.1109/ESTS.2017.8069334.

20 Javaid, U., Dujic, D., and van der Merwe, W. (2015). MVDC marine electrical distribution: are we ready?. *IECON 2015 - 41st Annual Conference of the IEEE Industrial Electronics Society*, pp. 823–828. https://doi.org/10.1109/IECON.2015.7392201.

21 Tessarolo, A., Castellan, S., Menis, R., and Sulligoi, G. (2013). Electric generation technologies for all-electric ships with medium-voltage DC power distribution systems. *2013 IEEE Electric Ship Technologies Symposium, ESTS 2013*, pp. 275–281. https://doi.org/10.1109/ESTS.2013.6523746.

22 McCoy, T.J. (2002). Trends in ship electric propulsion. *Proc. IEEE Power Eng. Soc. Trans. Distrib. Conf.* 1 (Summer): 343–346. https://doi.org/10.1109/PESS.2002.1043247.

23 Kim, S., Kim, S.-N., and Dujic, D. (2019). Extending protection selectivity in DC shipboard power systems by means of additional bus capacitance. *IEEE Trans. Ind. Electron.* 67 (5): 3673–3683.

24 Othman, M.B., Reddy, N.P., Ghimire, P. et al. (2019). A hybrid power system laboratory: testing electric and hybrid propulsion. *IEEE Electrification Mag.* 7 (4): 89–97. https://doi.org/10.1109/MELE.2019.2943982.

25 Moura, S.J., Siegel, J.B., Siegel, D.J. et al. (2010). Education on vehicle electrification: battery systems, fuel cells, and hydrogen. *2010 IEEE Vehicle Power and Propulsion Conference, VPPC 2010*. https://doi.org/10.1109/VPPC.2010.5729150.

26 Shakeri, N., Zadeh, M., and Bremnes Nielsen, J. (2020). Hydrogen fuel cells for ship electric propulsion: moving toward greener ships. *IEEE Electrification Mag.* 8 (2): 27–43. https://doi.org/10.1109/MELE.2020.2985484.

27 Ghimire, P., Zadeh, M., Pedersen, E., and Thorstensen, J. (2021). Dynamic modeling, simulation, and testing of a marine DC hybrid power system. *IEEE Trans. Transport. Electrification* 7 (2): 905–919. https://doi.org/10.1109/TTE.2020.3023896.

28 Ghimire, P., Reddy,N.P., Zadeh, M.K. et al. (2020). Dynamic modeling and real-time simulation of a ship hybrid power system using a mixed-modeling approach. *2020 IEEE Transportation Electrification Conference & Expo (ITEC)*, pp. 1–6. https://doi.org/10.1109/itec48692.2020.9161520.

29 Sorensen, A.J., Skjetne, R., Bo, T. et al. (2017). Toward safer, smarter, and greener ships: using hybrid marine power plants. *IEEE Electrification Mag.* 5 (3): 68–73. https://doi.org/10.1109/mele.2017.2718861.

30 Miyazaki, M.R., Sorensen, A.J., Lefebvre, N. et al. (2016). Hybrid modeling of strategic loading of a marine hybrid power plant with experimental validation. *IEEE Access* 4: 8793–8804. https://doi.org/10.1109/ACCESS.2016.2629000.

31 Ghimire, P., Zadeh, M., Pedersen, E., and Thorstensen, J. (2021). Dynamic efficiency modeling of a marine DC hybrid power system. *2021 IEEE Applied Power Electronics Conference and Exposition (APEC)*, pp. 855–862. https://doi.org/10.1109/APEC42165.2021.9487343.

32 Ådnanes, A.K. (2003). Maritime electrical installations and diesel electric propulsion. *Oslo*. http://www.trpa.org/wp-content/uploads/ABB-AS_2003.pdf (accessed 2 September 2019).

33 Rasmussen, N. (2007). Electrical efficiency modeling for data centers. http://www.eei.org/magazine/editorial_content/nonav_stories/2004-01-01-NT.htm (accessed 18 June 2020).

34 Skjong, E., Johansen, T.A., Molinas, M., and Sorensen, A.J. (2017). Approaches to economic energy management in diesel-electric marine vessels. *IEEE Trans. Transport. Electrification* 3 (1): 22–35. https://doi.org/10.1109/TTE.2017.2648178.

35 Ghimire, P., Karimi, S., Zadeh, M. et al. (2022). Model-based efficiency and emissions evaluation of a marine hybrid power system with load profile. *Electric Power Syst. Res.* 212: 108530. ISSN 0378-7796. https://doi.org/10.1016/j.epsr.2022.108530.

Index

Note: Page numbers in *italics* denote figures and page numbers in **bold** denote tables.

Transportation Electrification: Breakthroughs in Electrified Vehicles, Aircraft, Rolling Stock, and Watercraft, First Edition. Edited by Ahmed A. Mohamed, Ahmad Arshan Khan, Ahmed T. Elsayed, and Mohamed A. Elshaer.

d

f

q

r

IEEE Press Series on Power and Energy Systems

Series Editor: Ganesh Kumar Venayagamoorthy, Clemson University, Clemson, South Carolina, USA.

The mission of the IEEE Press Series on Power and Energy Systems is to publish leading-edge books that cover a broad spectrum of current and forward-looking technologies in the fast-moving area of power and energy systems including smart grid, renewable energy systems, electric vehicles and related areas. Our target audience includes power and energy systems professionals from academia, industry and government who are interested in enhancing their knowledge and perspectives in their areas of interest.

1. *Electric Power Systems: Design and Analysis, Revised Printing*
   Mohamed E. El-Hawary

2. *Power System Stability*
   Edward W. Kimbark

3. *Analysis of Faulted Power Systems*
   Paul M. Anderson

4. *Inspection of Large Synchronous Machines: Checklists, Failure Identification, and Troubleshooting*
   Isidor Kerszenbaum

5. *Electric Power Applications of Fuzzy Systems*
   Mohamed E. El-Hawary

6. *Power System Protection*
   Paul M. Anderson

7. *Subsynchronous Resonance in Power Systems*
   Paul M. Anderson, B.L. Agrawal, J.E. Van Ness

8. *Understanding Power Quality Problems: Voltage Sags and Interruptions*
   Math H. Bollen

9. *Analysis of Electric Machinery*
   Paul C. Krause, Oleg Wasynczuk, and S.D. Sudhoff

10. *Power System Control and Stability, Revised Printing*
    Paul M. Anderson, A.A. Fouad

11. *Principles of Electric Machines with Power Electronic Applications, Second Edition*
    Mohamed E. El-Hawary

12. *Pulse Width Modulation for Power Converters: Principles and Practice*
    D. Grahame Holmes and Thomas Lipo

13. *Analysis of Electric Machinery and Drive Systems, Second Edition*
    Paul C. Krause, Oleg Wasynczuk, and S.D. Sudhoff

14. *Risk Assessment for Power Systems: Models, Methods, and Applications*
    Wenyuan Li

15. *Optimization Principles: Practical Applications to the Operations of Markets of the Electric Power Industry*
Narayan S. Rau

16. *Electric Economics: Regulation and Deregulation*
Geoffrey Rothwell and Tomas Gomez

17. *Electric Power Systems: Analysis and Control*
Fabio Saccomanno

18. *Electrical Insulation for Rotating Machines: Design, Evaluation, Aging, Testing, and Repair*
Greg C. Stone, Edward A. Boulter, Ian Culbert, and Hussein Dhirani

19. *Signal Processing of Power Quality Disturbances*
Math H. J. Bollen and Irene Y. H. Gu

20. *Instantaneous Power Theory and Applications to Power Conditioning*
Hirofumi Akagi, Edson H.Watanabe and Mauricio Aredes

21. *Maintaining Mission Critical Systems in a 24/7 Environment*
Peter M. Curtis

22. *Elements of Tidal-Electric Engineering*
Robert H. Clark

23. *Handbook of Large Turbo-Generator Operation and Maintenance, Second Edition*
Geoff Klempner and Isidor Kerszenbaum

24. *Introduction to Electrical Power Systems*
Mohamed E. El-Hawary

25. *Modeling and Control of Fuel Cells: Distributed Generation Applications*
M. Hashem Nehrir and Caisheng Wang

26. *Power Distribution System Reliability: Practical Methods and Applications*
Ali A. Chowdhury and Don O. Koval

27. *Introduction to FACTS Controllers: Theory, Modeling, and Applications*
Kalyan K. Sen, Mey Ling Sen

28. *Economic Market Design and Planning for Electric Power Systems*
James Momoh and Lamine Mili

29. *Operation and Control of Electric Energy Processing Systems*
James Momoh and Lamine Mili

30. *Restructured Electric Power Systems: Analysis of Electricity Markets with Equilibrium Models*
Xiao-Ping Zhang

31. *An Introduction to Wavelet Modulated Inverters*
S.A. Saleh and M.A. Rahman

32. *Control of Electric Machine Drive Systems*
Seung-Ki Sul

33. *Probabilistic Transmission System Planning*
Wenyuan Li

34. *Electricity Power Generation: The Changing Dimensions*
Digambar M. Tagare

35. *Electric Distribution Systems*
Abdelhay A. Sallam and Om P. Malik

36. *Practical Lighting Design with LEDs*
Ron Lenk, Carol Lenk

37. *High Voltage and Electrical Insulation Engineering*
Ravindra Arora and Wolfgang Mosch

38. *Maintaining Mission Critical Systems in a 24/7 Environment, Second Edition*
Peter Curtis

39. *Power Conversion and Control of Wind Energy Systems*
Bin Wu, Yongqiang Lang, Navid Zargari, Samir Kouro

40. *Integration of Distributed Generation in the Power System*
Math H. Bollen, Fainan Hassan

41. *Doubly Fed Induction Machine: Modeling and Control for Wind Energy Generation Applications*
Gonzalo Abad, Jesús López, Miguel Rodrigues, Luis Marroyo, and Grzegorz Iwanski

42. *High Voltage Protection for Telecommunications*
Steven W. Blume

43. *Smart Grid: Fundamentals of Design and Analysis*
James Momoh

44. *Electromechanical Motion Devices, Second Edition*
Paul Krause Oleg, Wasynczuk, Steven Pekarek

45. *Electrical Energy Conversion and Transport: An Interactive Computer-Based Approach, Second Edition*
George G. Karady, Keith E. Holbert

46. *ARC Flash Hazard and Analysis and Mitigation*
J.C. Das

47. *Handbook of Electrical Power System Dynamics: Modeling, Stability, and Control*
Mircea Eremia, Mohammad Shahidehpour

48. *Analysis of Electric Machinery and Drive Systems, Third Edition*
Paul C. Krause, Oleg Wasynczuk, S.D. Sudhoff, Steven D. Pekarek

49. *Extruded Cables for High-Voltage Direct-Current Transmission: Advances in Research and Development*
Giovanni Mazzanti, Massimo Marzinotto

50. *Power Magnetic Devices: A Multi-Objective Design Approach*
S.D. Sudhoff

51. *Risk Assessment of Power Systems: Models, Methods, and Applications, Second Edition*
Wenyuan Li

52. *Practical Power System Operation*
Ebrahim Vaahedi

53. *The Selection Process of Biomass Materials for the Production of Bio-Fuels and Co-Firing*
Najib Altawell

54. *Electrical Insulation for Rotating Machines: Design, Evaluation, Aging, Testing, and Repair, Second Edition*
Greg C. Stone, Ian Culbert, Edward A. Boulter, and Hussein Dhirani

55. *Principles of Electrical Safety*
Peter E. Sutherland

56. *Advanced Power Electronics Converters: PWM Converters Processing AC Voltages*
Euzeli Cipriano dos Santos Jr., Edison Roberto Cabral da Silva

57. *Optimization of Power System Operation, Second Edition*
Jizhong Zhu

58. *Power System Harmonics and Passive Filter Designs*
J.C. Das

59. *Digital Control of High-Frequency Switched-Mode Power Converters*
Luca Corradini, Dragan Maksimoviæ, Paolo Mattavelli, Regan Zane

60. *Industrial Power Distribution, Second Edition*
Ralph E. Fehr, III

61. *HVDC Grids: For Offshore and Supergrid of the Future*
Dirk Van Hertem, Oriol Gomis-Bellmunt, Jun Liang

62. *Advanced Solutions in Power Systems: HVDC, FACTS, and Artificial Intelligence*
Mircea Eremia, Chen-Ching Liu, Abdel-Aty Edris

63. *Operation and Maintenance of Large Turbo-Generators*
Geoff Klempner, Isidor Kerszenbaum

64. *Electrical Energy Conversion and Transport: An Interactive Computer-Based Approach*
George G. Karady, Keith E. Holbert

65. *Modeling and High-Performance Control of Electric Machines*
John Chiasson

66. *Rating of Electric Power Cables in Unfavorable Thermal Environment*
George J. Anders

67. *Electric Power System Basics for the Nonelectrical Professional*
Steven W. Blume

68. *Modern Heuristic Optimization Techniques: Theory and Applications to Power Systems*
Kwang Y. Lee, Mohamed A. El-Sharkawi

69. *Real-Time Stability Assessment in Modern Power System Control Centers*
Savu C. Savulescu

70. *Optimization of Power System Operation*
Jizhong Zhu

71. *Insulators for Icing and Polluted Environments*
Masoud Farzaneh, William A. Chisholm

72. *PID and Predictive Control of Electric Devices and Power Converters Using MATLAB®/Simulink®*
Liuping Wang, Shan Chai, Dae Yoo, Lu Gan, Ki Ng

73. *Power Grid Operation in a Market Environment: Economic Efficiency and Risk Mitigation*
Hong Chen

74. *Electric Power System Basics for Nonelectrical Professional, Second Edition*
Steven W. Blume

75. *Energy Production Systems Engineering Thomas*
Howard Blair

76. *Model Predictive Control of Wind Energy Conversion Systems*
Venkata Yaramasu, Bin Wu

77. *Understanding Symmetrical Components for Power System Modeling*
J.C. Das

78. *High-Power Converters and AC Drives, Second Edition*
Bin Wu, Mehdi Narimani

79. *Current Signature Analysis for Condition Monitoring of Cage Induction Motors: Industrial Application and Case Histories*
William T. Thomson, Ian Culbert

80. *Introduction to Electric Power and Drive Systems*
Paul Krause, Oleg Wasynczuk, Timothy O'Connell, Maher Hasan

81. *Instantaneous Power Theory and Applications to Power Conditioning, Second Edition*
Hirofumi, Edson Hirokazu Watanabe, Mauricio Aredes

82. *Practical Lighting Design with LEDs, Second Edition*
Ron Lenk, Carol Lenk

83. *Introduction to AC Machine Design*
Thomas A. Lipo

84. *Advances in Electric Power and Energy Systems: Load and Price Forecasting*
Mohamed E. El-Hawary

85. *Electricity Markets: Theories and Applications*
Jeremy Lin, Jernando H. Magnago

86. *Multiphysics Simulation by Design for Electrical Machines, Power Electronics and Drives*
Marius Rosu, Ping Zhou, Dingsheng Lin, Dan M. Ionel, Mircea Popescu, Frede Blaabjerg, Vandana Rallabandi, David Staton

87. *Modular Multilevel Converters: Analysis, Control, and Applications*
Sixing Du, Apparao Dekka, Bin Wu, Navid Zargari

88. *Electrical Railway Transportation Systems*
Morris Brenna, Federica Foiadelli, Dario Zaninelli

89. *Energy Processing and Smart Grid*
James A. Momoh

90. *Handbook of Large Turbo-Generator Operation and Maintenance, 3rd Edition*
Geoff Klempner, Isidor Kerszenbaum

91. *Advanced Control of Doubly Fed Induction Generator for Wind Power Systems*
Dehong Xu, Dr. Frede Blaabjerg, Wenjie Chen, Nan Zhu

92. *Electric Distribution Systems, 2nd Edition*
Abdelhay A. Sallam, Om P. Malik

93. *Power Electronics in Renewable Energy Systems and Smart Grid: Technology and Applications*
Bimal K. Bose

94. *Distributed Fiber Optic Sensing and Dynamic Rating of Power Cables*
Sudhakar Cherukupalli, and George J Anders

95. *Power System and Control and Stability, Third Edition*
Vijay Vittal, James D. McCalley, Paul M. Anderson, and A. A. Fouad

96. *Electromechanical Motion Devices: Rotating Magnetic Field-Based Analysis and Online Animations, Third Edition*
Paul Krause, Oleg Wasynczuk, Steven D. Pekarek, and Timothy O'Connell

97. *Applications of Modern Heuristic Optimization Methods in Power and Energy Systems*
Kwang Y. Lee and Zita A. Vale

98. *Handbook of Large Hydro Generators: Operation and Maintenance*
Glenn Mottershead, Stefano Bomben, Isidor Kerszenbaum, and Geoff Klempner

99. *Advances in Electric Power and Energy: Static State Estimation*
Mohamed E. El-hawary

100. *Arc Flash Hazard Analysis and Mitigation, Second Edition*
J.C. Das

101. *Maintaining Mission Critical Systems in a 24/7 Environment, Third Edition*
Peter M. Curtis

102. *Real-Time Electromagnetic Transient Simulation of AC-DC Networks*
Venkata Dinavahi and Ning Lin

103. *Probabilistic Power System Expansion Planning with Renewable Energy Resources and Energy Storage Systems*
Jaeseok Choi and Kwang Y. Lee

104. *Power Magnetic Devices: A Multi-Objective Design Approach, Second Edition*
Scott D. Sudhoff

105. *Optimal Coordination of Power Protective Devices with Illustrative Examples*
Ali R. Al-Roomi

106. *Resilient Control Architectures and Power Systems*
Craig Rieger, Ronald Boring, Brian Johnson, and Timothy McJunkin

107. *AlternativeLiquid Dielectrics for High Voltage Transformer Insulation Systems: Performance Analysis and Applications*
Edited by U. Mohan Rao, I. Fofana, and R. Sarathi

108. *Introduction to the Analysis of Electromechanical Systems*
Paul C. Krause, Oleg Wasynczuk, and Timothy O'Connell

109. *Power Flow Control Solutions for a Modern Grid using SMART Power Flow Controllers*
Kalyan K. Sen and Mey Ling Sen

110. *Power System Protection: Fundamentals and Applications*
John Ciufo and Aaron Cooperberg

111. *Soft-Switching Technology for Three-phase Power Electronics Converters*
Dehong Xu, Rui Li, Ning He, Jinyi Deng, and Yuying Wu

112. *Power System Protection, Second Edition*
Paul M. Anderson, Charles Henville, Rasheek Rifaat, Brian Johnson, and Sakis Meliopoulos

113. *High Voltage and Electrical Insulation Engineering, Second Edition*
Ravindra Arora and Wolfgang Mosch.

114. *Modeling and Control of Modern Electrical Energy Systems*
Masoud Karimi-Ghartemani

115. *Control of Power Electronic Converters with Microgrid Applications*
Armdam Ghosh and Firuz Zare

116. *Coordinated Operation and Planning of Modern Heat and Electricity Incorporated Networks*
Mohammadreza Daneshvar, Behnam Mohammadi-Ivatloo, and Kazem Zare

117. *Smart Energy for Transportation and Health in a Smart City*
Chun Sing Lai, Loi Lei Lai, and Qi Hong Lai

118. *Wireless Power Transfer: Principles and Applications*
Zhen Zhang and Hongliang Pang

119. *Intelligent Data Mining and Analysis in Power and Energy Systems: Models and Applications for Smarter Efficient Power Systems*
Zita Vale, Tiago Pinto, Michael Negnevitsky, and Ganesh Kumar Venayagamoorthy

120. *Introduction to Modern Analysis of Electric Machines and Drives*
Paul C. Krause and Thomas C. Krause

121. *Electromagnetic Analysis and Condition Monitoring of Synchronous Generators*
Hossein Ehya and Jawad Faiz.

122. *Transportation Electrification: Breakthroughs in Electrified Vehicles, Aircraft, Rolling Stock, and Watercraft*
Ahmed A. Mohamed, Ahmad Arshan Khan, Ahmed T. Elsayed, Mohamed A. Elshaer

Printed and bound by CPI Group (UK) Ltd, Croydon, CR0 4YY
16/04/2025
14658590-0005